Definition of Absolute Value

$$|x| = \begin{cases} x \text{ when } x \geq 0 \\ -x \text{ when } x < 0 \end{cases}$$

Absolute Value Equations

- If $k \geq 0$, then $|x| = k$ is equivalent to $x = k$ or $x = -k$.
- If a and b are algebraic expressions, $|a| = |b|$ is equivalent to $a = b$ or $a = -b$.

Absolute Value Inequalities

- If $k > 0$, then $|x| < k$ is equivalent to $-k < x < k$.
- If $k > 0$, then $|x| > k$ is equivalent to $x > k$ or $x < -k$.

These two properties hold for $\leq$ and $\geq$ also.

Circles

The standard form of an equation of a circle with center (h, k) and radius r:

$$(x - h)^2 + (y - k)^2 = r^2$$

The standard form of an equation of a circle with center $(0, 0)$ and radius r:

$$x^2 + y^2 = r^2$$

Quadratic Functions

A **quadratic function** is a second-degree polynomial function in one variable of the form

$$f(x) = ax^2 + bx + c \text{ or } y = ax^2 + bx + c,$$

where a, b, and c are real numbers and $a \neq 0$.

The graph of a quadratic function of the form

$$f(x) = ax^2 + bx + c \quad (a \neq 0)$$

is a **parabola** with vertex at

$$\left(-\frac{b}{2a}, c - \frac{b^2}{4a}\right)$$

- If $a > 0$, the parabola **opens upward.**
- If $a < 0$, the parabola **opens downward.**

The **standard form of an equation of a quadratic function** is

$$y = f(x) = a(x - h)^2 + k \quad (a \neq 0)$$

The vertex is at (h, k).
- The parabola opens upward when $a > 0$ and downward when $a < 0$.
- The axis of symmetry of the parabola is the vertical line graph of the equation $x = h$.

Tests for Symmetry

Test for x-axis symmetry:
To test for x-axis symmetry, replace y with $-y$. If the resulting equation is equivalent to the original one, the graph is symmetric about the x-axis.

Test for y-axis symmetry:
To test for y-axis symmetry, replace x with $-x$. If the resulting equation is equivalent to the original one, the graph is symmetric about the y-axis.

Test for origin symmetry:
To test for symmetry about the origin, replace x with $-x$ and y with $-y$. If the resulting equation is equivalent to the original one, the graph is symmetric about the origin.

Graphs of Common Functions

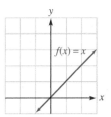
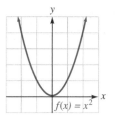

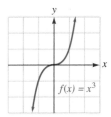
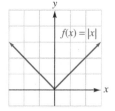

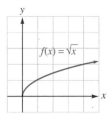
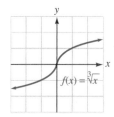

Greatest-Integer Function

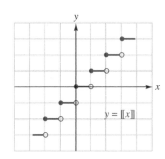

Interest Formulas

Compound interest formula:
If P dollars are deposited in an account earning interest at an annual rate r, compounded n times each year, the amount A in the account after t years is given by

$$A = P\left(1 + \frac{r}{n}\right)^{nt}$$

Continuous compound interest formula:
If P dollars are deposited in an account earning interest at an annual rate r, compounded continuously, the amount A after t years is given by the formula

$$A = Pe^{rt}$$

Special Version of Gustafson College Algebra

for MTH 103 College Algebra

Contributing Authors: Sue Allen & Pavel Sikorskii

Gustafson | Hughes

CENGAGE
Learning·

Australia • Brazil • Japan • Korea • Mexico • Singapore • Spain • United Kingdom • United States

CENGAGE
Learning·

Special Version of Gustafson College Algebra for MTH 103 College Algebra

College Algebra, Eleventh Edition
Gustafson | Hughes

© 2013, 2010 Cengage Learning. All rights reserved.

Senior Manager, Student Engagement:

Linda deStefano

Janey Moeller

Manager, Student Engagement:

Julie Dierig

Marketing Manager:

Rachael Kloos

Manager, Production Editorial:

Kim Fry

Manager, Intellectual Property Project Manager:

Brian Methe

Senior Manager, Production and Manufacturing:

Donna M. Brown

Manager, Production:

Terri Daley

For product information and technology assistance, contact us at
Cengage Learning Customer & Sales Support, 1-800-354-9706

For permission to use material from this text or product,
submit all requests online at **cengage.com/permissions**
Further permissions questions can be emailed to
permissionrequest@cengage.com

This book contains select works from existing Cengage Learning resources and was produced by Cengage Learning Custom Solutions for collegiate use. As such, those adopting and/or contributing to this work are responsible for editorial content accuracy, continuity and completeness.

Compilation © 2014 Cengage Learning

ISBN-13: 978-1-305-30087-3

ISBN-10: 1-305-30087-4

WCN: 01-100-101

Cengage Learning

5191 Natorp Boulevard
Mason, Ohio 45040
USA

Cengage Learning is a leading provider of customized learning solutions with office locations around the globe, including Singapore, the United Kingdom, Australia, Mexico, Brazil, and Japan. Locate your local office at:
international.cengage.com/region.

Cengage Learning products are represented in Canada by Nelson Education, Ltd. For your lifelong learning solutions, visit **www.cengage.com/custom.**
Visit our corporate website at **www.cengage.com.**

Printed in the United States of America

Brief Contents

Preface

To the Instructor

It is with great delight that we present the eleventh edition of *College Algebra.* This edition maintains the same philosophy of the highly successful previous editions but is enhanced to meet the current expectations of today's students and instructors. Our goal is to increase students' problem-solving skills while at the same time preparing them for success in trigonometry, calculus, statistics, or other disciplines of study. The textbook has been revised for greater clarity of design and instruction.

Our goal is to write a textbook that

- presents solid mathematics written in a way that is easy to understand for students with a broad range of abilities and backgrounds;
- emphasizes the important concept of a function;
- uses real-life applications to motivate learning and problem solving;
- improves critical-thinking abilities in all students;
- develops algebra skills needed for future success in mathematics courses.

We believe that we have accomplished this goal through a successful blending of content and pedagogy. We present a thorough coverage of classic college algebra topics, incorporated into a contemporary framework of tested teaching strategies. Features are written to appeal to both students and instructors. In keeping with the spirit of the NCTM and AMATYC standards, this book emphasizes conceptual understanding, problem solving, and the appropriate use of technology.

New Features

- **Strategy Boxes** *To enable students to build on their mathematical reasoning and approach problems with confidence,* Strategy boxes offer problem-solving techniques and steps at appropriate points in the material.

Strategy for Solving Quadratic Inequalities	**Method 1: Constructing a Table and Testing Numbers**
	• Solve the quadratic equation and use the roots of the equation to establish intervals on a number line.
	• Construct a table. To do so, write down each interval, select a number to test from each interval, test the selected value to determine if it satisfies the inequality, and then write the result.
	• Use the results from the table and write the solution of the quadratic inequality.
	Method 2: Constructing a Sign Graph
	• Solve the quadratic equation and use the roots of the equation to establish intervals on the number line.

- **Caution Boxes** *To alert students to common errors and misunderstandings, and reinforce correct mathematics,* Caution boxes appear throughout the text.

Caution	Be sure to write the Quadratic Formula correctly. Do **not** write the Quadratic Formula as $$x = -b \pm \frac{\sqrt{b^2 - 4ac}}{2a}$$

- **Now Try Exercises**   *To provide students an additional opportunity to assess their understanding of the concept related to each worked example*, a reference to an exercise follows all Examples and Self Check problems. These references also show students a correspondence between the examples in the book and the exercises sets.

EXAMPLE 3   **Finding the Domain of a Function**

Find the domain of the function defined by the equation $y = \dfrac{1}{x^2 - 5x - 6}$.

SOLUTION   We can factor the denominator to see what values of x will give 0's in the denominator. These values are not in the domain.

$$x^2 - 5x - 6 = 0$$
$$(x - 6)(x + 1) = 0$$
$$x - 6 = 0 \quad \text{or} \quad x + 1 = 0$$
$$x = 6 \qquad\qquad x = -1$$

The domain is $(-\infty, -1) \cup (-1, 6) \cup (6, \infty)$. ∎

Self Check 3   Find the domain of the function defined by the equation $y = \dfrac{2}{x^2 - 16}$.

Now Try Exercise 41.

- **Titled Examples**   *To clearly identify the topic and purpose of each example*, descriptive titles have been added to example identifiers.

Continued and Updated Features

We have kept and updated the pedagogical features that made the previous editions of the book so successful.

- **Student-Friendly Writing Style**   *To alleviate student anxiety about reading a mathematics textbook*, the exposition is clear, concise, and reader-friendly. The writing level is informal yet accurate. Students and instructors alike should find the reading both interesting and inviting.

- **Careers and Mathematics Chapter Openers**   *To encourage students to explore careers that use mathematics and make a connection between math and real life*, each chapter opens with Careers and Mathematics. New, exciting careers are featured in this edition. These snapshots include information on how the professionals use math in their work and who employs them. Most information is taken from the *Occupational Outlook Handbook.* A web address is provided for students to learn more about the career.

CAREERS AND MATHEMATICS: **Epidemiologists**

Epidemiologists investigate and describe the determinants and distribution of disease, disability, and other health outcomes. They also develop means for prevention and control. *Applied epidemiologists* typically work for state health agencies and are responsible for responding to disease outbreaks and determining the cause and method of containment. *Research epidemiologists* work in laboratories studying ways to prevent future outbreaks. This career can be quite rewarding, both mentally and financially. Epidemiologists spend a lot of time saving lives and finding solutions for better health.

- **Section Openers** *To pique interest and motivate students to read the material,* each section begins with a contemporary photo and a real-life application that will appeal to students of varied interests.

- **Numbered Objectives** *To keep students focused,* numbered learning objectives are given at the beginning of each section and appear as subheadings in the section.

4.3 Logarithmic Functions and Their Graphs

In this section, we will learn to

1. Evaluate logarithms.
2. Evaluate common logarithms.
3. Evaluate natural logarithms.
4. Graph logarithmic functions.
5. Use transformations to graph logarithmic functions.

Guests aboard the Royal Caribbean's cruise ship *Freedom of the Seas* can now "hang ten" while out to sea. The flowrider surf simulator allows riders to body board surf against a wave-like water flow of 34,000 gallons per minute.

- **Example Structure** *To help students gain a deeper understanding of how to solve each problem,* solutions begin with a stated approach. The examples are engaging, and step-by-step solutions with annotations are provided.

- **Application Examples** *To answer the student question, When will I ever use this math?* applications from a wide range of disciplines demonstrate how mathematics is used to solve real problems. These applications motivate the student and help students become better problem solvers. An eye-catching photo accompanies many of these modern examples.

- **Self Checks** *To actively reinforce student understanding of concepts and example solutions,* each example is followed immediately by a Self Check exercise. The answers for students are offered at the end of each section. To assist instructors, the answers to Self Checks appear next to the problem in the AIE, printed in blue.

- **Comments** *To provide additional insights into specific content,* Comment boxes appear throughout the textbook. These noteworthy statements provide clarification on a specific step or concept in an example. Some offer a tip for studying the material.

- **Accents on Technology and Calculators** *To encourage students to become intelligent users of technology and grasp concepts graphically,* Accents on Technology appear throughout the textbook. These illustrate and guide the use of a TI-84 graphing calculator for specific problems, and many are new. Although graphing calculators are incorporated into the book, their use is not required. All graphing topics are fully discussed in traditional ways.

ACCENT ON TECHNOLOGY   **Exponential Regression**

The table below shows the cooling temperatures of a hot cup of coffee after it is made.

Time in minutes	Temperature °F
0	179.5
5	168.7
8	158.1
11	149.2
15	141.7
18	134.6
22	125.4
25	123.5
30	116.3
34	113.2

- **Getting Ready Exercises** *To test student understanding of concepts and proper use of mathematical vocabulary*, each problem set begins with Getting Ready exercises. Students should be able to answer these fill-in-the blank questions before moving on to the Practice exercises.

- **Comprehensive Exercise Sets** *To improve mathematical skills and cement understanding,* the exercise sets progress from routine to more challenging. The mathematics in each exercise set is sound, but not so rigorous that it will confuse students. All exercise sets include Getting Ready, Practice, Application, Discovery and Writing, and Review problems. The book contains more than 4,000 exercises, many of which are new.

- **Interesting and Contemporary Application Exercises** *To continue the emphasis on solving problems through realistic applications,* new Application exercises have been added and others updated. All Application problems have titles.

- **Chapter Reviews and Chapter Tests** *To give students the best opportunity for study and exam preparation,* each chapter closes with a Chapter Review and Chapter Test. Chapter Reviews are comprehensive and consist of three parts: definitions and concepts, examples, and review exercises. Chapter Tests cover all the important topics and yet are brief enough to emulate a real-time test, so students can practice not only the math but also their test-taking aptitude. As an additional quick reference, endpapers offer the important formulas and graphs developed in the book.

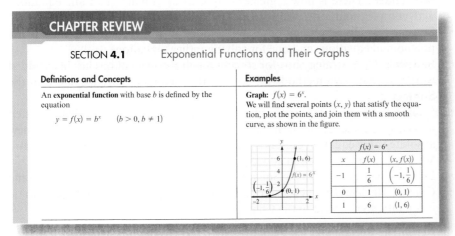

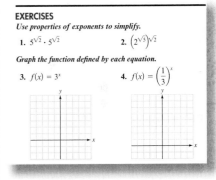

- **Cumulative Reviews** *To reinforce student learning and improve student retention of concepts,* Cumulative Review exercises appear after every two chapters. These comprehensive reviews revisit all the essential topics covered in prior chapters.

Content Changes

Each section in the text has been edited to fine-tune the presentation of topics for better flow of concepts and for clarity. There are many new exercises and applications. Some changes made to specific chapters include:

Chapter 0: A Review of Basic Algebra
- Section 0.1—Four new exercises have been added. Students are asked to determine whether the decimal form of a given fraction terminates or repeats.
- Section 0.2—The Accents on Technology in this section now show TI-84 Plus graphing calculator screen shots. Two contemporary applications were added.
- Section 0.3—Four new exercises were added to provide students additional practice rationalizing the denominator.
- Section 0.4—Four new exercises were added. Two exercises ask students to multiply binomials containing radicals. Two exercises are long division problems and help improve proficiency of this skill.

- Section 0.5—Two new exercises, factoring the difference of two squares, were added. Examples 11 and 12 were reversed for a better flow of the content of the section.
- Section 0.6—A new Strategy box has been added to clearly exhibit two methods for simplifying complex fractions.

Chapter 1: Equations and Inequalities

- Section 1.1—The title of the section has been changed to "Linear Equations and Rational Equations." The title better reflects the content in the section. Students are now asked to identify restrictions on the values of variables for only linear and rational equations; radical equations have been deleted.
- Section 1.2—Seven Self Checks have been added to this section. These are contemporary applications. Two new Application exercises have been added.
- Section 1.3—The discriminant material has been rewritten for greater clarity. Eight additional Square Root Property exercises and two new Quadratic Formula exercises have been added. These help strengthen students' ability to solve equations and simplify radicals.
- Section 1.4—Six Self Checks have been added to this section. These are contemporary applications. Two new Application exercises were also added.
- Section 1.5—Example 11, "Factoring the Sum of Two Squares," is new. An Accent on Technology showing operations of complex numbers has been added. The exercise set has been extensively revised and 42 new exercises have been added. There is now a greater emphasis on solving quadratic equations with complex solutions, using the methods previously taught.
- Section 1.6—Two Strategy boxes were added in this section, one for solving polynomial equations and one for solving radical equations.
- Section 1.7—A Strategy box for solving quadratic inequalities has been added to the section. Two methods for solving the inequalities are displayed in the Strategy box.
- Section 1.8—The forms of the absolute value inequalities have been rewritten to clearly include less than or equal to and greater than or equal to. Two new absolute value equation exercises have been added and two new Discovery and Writing exercises have been added.
- End of Chapter—Four new Square Root Property exercises were added to the Chapter Review. Two new exercises, a compound inequality and a quadratic inequality, were added to the Chapter Test.

Chapter 2: The Rectangular Coordinate System and Graphs of Equations

- Section 2.1—The Accent on Technology "Finding Intercepts Using Zoom and Trace" has been streamlined for efficiency. The distance and midpoint exercises have been revised.
- Section 2.2—Four new exercises, corresponding to Example 5, have been added.
- Section 2.3—A new Accent on Technology, "Linear Regression," has been added. Two new exercises have also been added.
- Section 2.4—Four contemporary applications involving circles have been added. Two Discovery and Writing exercises have been added.
- Section 2.5—Two new inverse variation exercises have been added.

Chapter 3: Functions

- Section 3.1—In this section, a new Accent on Technology, "Evaluating a Function," was added. Four new domain exercises, involving square roots or cube roots, were added. Two new function exercises were added. The functions have two x terms. Also, two cube-root graphing exercises were added. Two contemporary applications were also added.
- Section 3.2—A new Accent on Technology, "Finding the Maximum Point or Minimum Point (Vertex) of a Parabola," was added.
- Section 3.3—The definitions of even and odd functions are formally defined. The definitions of increasing and decreasing are formally defined. Six exercises

were added to identify the graph of a function as being even, odd, or neither. A new Accent on Technology, "Piecewise-Defined Functions," now appears in the section.

- Section 3.4—The title of the section has been changed to "Transformations of the Graphs of Functions." An Accent on Technology box has been added to illustrate translations.
- Section 3.5—There is one new Accent on Technology in this section. Self Checks 5–10 are new. The applications have been revised in this section.
- Section 3.6—There are now Accents on Technology demonstrating combinations of functions and the domain of composite functions. Sixteen new exercises were added. These include eight exercises asking students to identify function values given a graph and two contemporary applications.

Chapter 4: Exponential and Logarithmic Functions

- Section 4.1—Example 1, "Approximating Exponential Expressions," is new. The Compound Interest Formula now uses n for the number of years instead of k. A new Accent on Technology, "Graphing Exponential Functions," has been added.
- Section 4.2—A new Accent on Technology, "Exponential Regression," has been added.
- Section 4.3—Two new examples have been added to cover converting from exponential form to logarithmic form and converting from logarithmic form to exponential form. TI-84 Plus graphing calculator screens are now shown extensively throughout this section. Six new graphing exercises have been added.
- Section 4.4—The section opener has been revised and includes Haiti and Japan.
- Section 4.5—Three new Accent on Technology boxes were added. These illustrate the Product, Quotient, and Power Properties of Logarithms and using the Change-of-Base Formula to graph a logarithmic function.
- Section 4.6—The Accents on Technology are new in this section. They are "Verifying Solutions of an Exponential Equation," "Finding Approximate Solutions to an Exponential Equation," and "Finding Approximate Solutions to a Logarithmic Equation." Example 9, "Solving Logarithmic Equations," is new. A summary of strategies used to solve exponential and logarithmic equations is now given. Eight new exercises have been added to help students master solving exponential equations using like bases.
- End of Chapter—One new exercise has been added to the Chapter Review.

Chapter 5: Solving Polynomial Equations

- Section 5.1—Four new exercises were added. These require synthetic division using complex numbers.
- Section 5.2—A Strategy box for finding a polynomial equation when given a partial list of roots has been added.
- Section 5.3—A new Accent on Technology, "Using the Table Feature on a Graphing Calculator to Find Roots of a Polynomial Equation," has been added. The exercises have been reordered based on the degree of the polynomial equation. Two new polynomial equation exercises have been added. These are fifth-degree equations and the leading coefficient is not one. Two contemporary applications have also been added.
- Section 5.4—Two new exercises have been added.
- End of Chapter—Three new exercises have been added to the Chapter Review. These exercises require the Rational Roots Theorem. In the Chapter Test, two exercises were changed from logarithms to natural logarithms.

Chapter 6: Linear Systems

- Section 6.1—A new Accent on Technology, "Solving a System of Linear Equations," has been added. Self Check 9 was added. There are four new exercises that have been added. These require graphing and have either no solution or an infinite number of solutions. Two contemporary applications have been added.

- Section 6.2—A new Accent on Technology, "Reduce a Matrix to Row Reduced Echelon Form," has been added.
- Section 6.3—Self Check 6 has been added.
- Section 6.4—Self Check 6 has been added.
- Section 6.5—Self Check 7 has been added.
- Section 6.6—Property boxes have been added to display the various types of partial-fraction decomposition problems that surface and the techniques used to solve the problem
- Section 6.8—Self Checks 3, 4, and 5 have been added.

Chapter 7: Conic Sections and Quadratic Systems
- Section 7.1—Six new exercises have been added. Four ask the student to identify the conic as a circle or a parabola. There are two new exercises asking students to use a graphing calculator to graph a parabola.
- Section 7.2—There are nine new exercises. Six ask students to identify the conic as a circle, parabola, or ellipse. Two ask students to use a graphing calculator and graph an ellipse. One contemporary application has been added.
- Section 7.3—Ten new exercises have been added. Eight exercises ask students to identify the conic as a circle, parabola, ellipse, or hyperbola. Two ask students to graph a hyperbola using a graphing calculator.

Chapter 8: Sequences, Series, and Probability
- Section 8.1—A new Accent on Technology, "Factorials," has been added.
- Section 8.2—A new Accent on Technology, "Sequences, Series, and Summation," has been added.
- Section 8.3—The section title has been revised. It is now titled "Arithmetic Sequences and Series."
- Section 8.4—Self Checks 6, 7, and 8 have been added. These are contemporary applications.
- Section 8.6—Two Accents on Technology have been added. One involves permutations and the other involves combinations. Three contemporary applications and two new exercises have been added.

Chapter 9: The Mathematics of Finance
*This chapter is now available online at cengagebrain.com. Students will also have access to Answers to Selected Exercises for this chapter, and instructors will have access to the Annotated Instructor's Edition of this chapter.

Organization and Coverage

This text can be used in a variety of ways. To maintain optimum flexibility, many chapters are sufficiently independent to allow you to pick and choose topics that are relevant to your students. After teaching Chapters 0–3 in order, you can teach Chapters 4–9 in any order.

Ancillaries for the Instructor

Enhanced WebAssign® (ISBN-10: 0-538-73810-3; ISBN-13: 978-0-538-73810-1)
Exclusively from Cengage Learning, Enhanced WebAssign offers an extensive online program for College Algebra to encourage the practice that's so critical for concept mastery. The meticulously crafted pedagogy and exercises in this text become even more effective in Enhanced WebAssign, supplemented by multimedia tutorial support and immediate feedback as students complete their assignments. Algorithmic problems allow you to assign unique versions to each student. The Practice Another Version feature (activated at your discretion) allows students to attempt the questions with new sets of values until they feel confident enough to work the original problem.

Students benefit from a new YouBook with highlighting and search features; Personal Study Plans (based on diagnostic quizzing) that identify chapter topics they still need to master; and links to video solutions, interactive tutorials, and even live online help.

PowerLecture with ExamView® (ISBN-10: 1-133-10345-6; ISBN-13: 978-1-133-10345-5)

This CD-ROM provides the instructor with dynamic media tools for teaching. Create, deliver, and customize tests (both print and online) in minutes with ExamView Computerized Testing Featuring Algorithmic Equations. Easily build solution sets for homework or exams using Solution Builder's online solutions manual. Microsoft® PowerPoint® lecture slides and figures from the book are also included on this CD-ROM.

Solution Builder (www.cengage.com/solutionbuilder)

This online instructor database offers complete worked solutions to all exercises in the text, allowing you to create customized, secure solutions printouts (in PDF format) matched exactly to the problems you assign in class.

Complete Solutions Manual (ISBN-10: 1-133-10342-1; ISBN-13: 978-1-133-10342-4)

This manual contains solutions to all exercises from the text, including Chapter Review Exercises, Chapter Tests, and Cumulative Review Exercises.

Test Bank (ISBN-10: 1-133-10343-X; ISBN-13: 978-1-133-10343-1)

The test bank includes six tests per chapter as well as three final exams. The tests are made up of a combination of multiple-choice, free-response, true/false, and fill-in-the-blank questions.

Ancillaries for the Student

Enhanced WebAssign® (ISBN-10: 0-538-73810-3; ISBN-13: 978-0-538-73810-1)

Exclusively from Cengage Learning, Enhanced WebAssign offers an extensive online program for College Algebra to encourage the practice that's so critical for concept mastery. You'll receive multimedia tutorial support as you complete your assignments. You'll also benefit from a new Premium eBook with highlighting and search features; Personal Study Plans (based on diagnostic quizzing) that identify chapter topics you still need to master; and links to video solutions, interactive tutorials, and even live online help.

Student Solutions Manual (ISBN-10: 1-133-10347-2; ISBN-13: 978-1-133-10347-9)

Go beyond the answers—see what it takes to get there and improve your grade! This manual provides worked-out, step-by-step solutions to the odd-numbered problems in the text. This gives you the information you need to truly understand how these problems are solved.

Text-Specific DVDs (ISBN 10: 1-133-10344-8; ISBN-13: 978-1-133-10344-8)

These text-specific instructional videos provide students with visual reinforcement of concepts and explanations given in easy-to-understand terms with detailed examples and sample problems. A flexible format offers versatility for quickly accessing topics or catering lectures to self-paced, online, or hybrid courses. Closed captioning is provided for the hearing impaired.

To get access, visit CengageBrain.com

CengageBrain.com

Visit www.cengagebrain.com to access additional course materials and companion resources. At the CengageBrain.com home page, search for the ISBN of your title (from the back cover of your book) using the search box at the top of the page. This will take you to the product page where free companion resources can be found.

To the Student

Congratulations! You now own a state-of-art textbook that has been written especially for you. We have written a book that you can easily read and understand and that you will find interesting as well. The book includes carefully written sections that will appeal to you because of our use of popular culture and contemporary examples. The book also contains real-life applications that help you see how mathematics is currently used in our world today.

Our goal is for you to use this textbook as we designed it—as a study tool—so that you will be successful in learning mathematics and reaching your career goals. So what are you waiting for? Grab your favorite pencil and some paper and dive in! You can begin reading, working exercises, and mastering skills today. We wish you the very best on this exciting mathematics journey you are about to take.

Acknowledgments

We are grateful to the following people who reviewed previous editions of the text or the current manuscript in its various stages. All of them had valuable suggestions that have been incorporated into this book.

Catherine Aguilar-Morgan, New Mexico State University–Alamogordo

Ebrahim Ahmadizadeh, Northampton Community College

Ricardo Alfaro, University of Michigan–Flint

Richard Andrews, University of Wisconsin

James Arnold, University of Wisconsin

Ronald Atkinson, Tennessee State University

Wilson Banks, Illinois State University

Chad Bemis, Riverside Community College

Anjan Biswas, Tennessee State College

Jerry Bloomberg, Essex Community College

Elaine Bouldin, Middle Tennessee State University

Dale Boye, Schoolcraft College

Eddy Joe Brackin, University of North Alabama

Susan Williams Brown, Gadsden State Community College

Jana Bryant, Manatee Community College

Lee R. Clancy, Golden West College

Krista Blevins Cohlmia, Odessa College

Dayna Coker, Southwestern Oklahoma State University–Sayre campus

Jan Collins, Embry Riddle College

Cecilia Cooper, William & Harper College

John S. Cross, University of Northern Iowa

Charles D. Cunningham, Jr., James Madison University

M. Hilary Davies, University of Alaska–Anchorage

Elias Deeba, University of Houston–Downtown

Grace DeVelbiss, Sinclair Community College

Lena Dexter, Faulkner State Junior College

Emily Dickinson, University of Arkansas

Mickey P. Dunlap, University of Tennessee, Martin

Gerard G. East, Southwestern Oklahoma State University

Eric Ellis, Essex Community College

Eunice F. Everett, Seminole Community College

Dale Ewen, Parkland College

Harold Farmer, Wallace Community College–Hanceville

Ronald J. Fischer, Evergreen Valley College

Mary Jane Gates, University of Arkansas at Little Rock

Lee R. Gibson, University of Louisville

Marvin Goodman, Monmouth College

Edna Greenwood, Tarrant County College

Jerry Gustafson, Beloit College

Jerome Hahn, Bradley University

Douglas Hall, Michigan State University

Robert Hall, University of Wisconsin

David Hansen, Monterey Peninsula College

Shari Harris, John Wood Community College

Kevin Hastings, University of Delaware

William Hinrichs, Rock Valley College

Arthur M. Hobbs, Texas A&M University

Jack E. Hofer, California Polytechnic State University

Ingrid Holzner, University of Wisconsin

Wayne Humphrey, Cisco College

Warren Jaech, Tacoma Community College

Joy St. John Johnson, Alabama A&M University

Nancy Johnson, Broward Community College

Patricia H. Jones, Methodist College

William B. Jones, University of Colorado

Barbara Juister, Elgin Community College

David Kinsey, University of Southern Indiana

Helen Kriegsman, Pittsburg State University

Marjorie O. Labhart, University of Southern Indiana

Betty J. Larson, South Dakota State University

Paul Lauritsen, Brown College

Jaclyn LeFebvre, Illinois Central College

Susan Loveland, University of Alaska–Anchorage

James Mark, Eastern Arizona College

Marcel Maupin, Oklahoma State University, Oklahoma City

Robert O. McCoy, University of Alaska–Anchorage

Judy McKinney, California Polytechnic Institute at Pomona

Sandra McLaurin, University of North Carolina

Marcus McWaters, University of Southern Florida

Donna Menard, University of Massachusetts, Dartmouth

James W. Mettler, Pennsylvania State University

Eldon L. Miller, University of Mississippi

Stuart E. Mills, Louisiana State University, Shreveport

Mila Mogilevskaya, Wichita State University

Gilbert W. Nelson, North Dakota State

Marie Neuberth, Catonsville City College

Charles Odion, Houston Community College

C. Altay Özgener, State College of Florida

Anthony Peressini, University of Illinois

David L. Phillips, University of Southern Colorado

William H. Price, Middle Tennessee State University

Ronald Putthoff, University of Southern Mississippi

Brooke P. Quinlan, Hillsborough Community College

Leela Rakesh, Carnegie Mellon University

Janet P. Ray, Seattle Central Community College

Robert K. Rhea, J. Sargeant Reynolds Community College

Barbara Riggs, Tennessee Technological University

Minnie Riley, Hinds Community College

Renee Roames, Purdue University

Paul Schaefer, SUNY, Geneseo

Vincent P. Schielack, Jr., Texas A&M University

Robert Sharpton, Miami Dade Community College

L. Thomas Shiflett, Southwest Missouri State University

Richard Slinkman, Bemidji State University

Merreline Smith, California Polytechnic Institute at Pomona

John Snyder, Sinclair Community College

Sandra L. Spain, Thomas Nelson Community College

Warren Strickland, Del Mar College

Paul K. Swets, Angelo State College

Ray Tebbetts, San Antonio College

Faye Thames, Lamar State University

Douglas Tharp, University of Houston–Downtown

Carolyn A. Wailes, University of Alabama, Birmingham

Carol M. Walker, Hinds Community College

William Waller, University of Houston–Downtown

Richard H. Weil, Brown College

Carroll G. Wells, Western Kentucky University

William H. White, University of South Carolina at Spartanburg

Clifton Whyburn, University of Houston

Charles R. Williams, Midwestern State University

Harry Wolff, University of Wisconsin

Roger Zarnowski, Angelo State University

Albert Zechmann, University of Nebraska

We wish to thank the staff at Cengage Learning, especially Gary Whalen and Jennifer Risden for their support in the production process. Cynthia Ashton, Sabrina Black, and Lynh Pham have worked tirelessly to pull together all the pieces behind the scenes. Leslie Lahr, our development editor, provided constant encouragement and valuable input. We appreciate you so much for being by our side during this entire journey. We also thank Lori Heckelman for her fine artwork, and special thanks goes to both Rhoda Bontrager and Craig Beffa at Graphic World for providing excellent publishing services. We appreciate the important work of accuracy reviewing done by John Samons.

R. David Gustafson

Jeffrey D. Hughes

About the Authors

R. David Gustafson is professor emeritus of mathematics at Rock Valley College in Illinois and coauthor of several best-selling math texts, including Gustafson/Frisk's *Beginning Algebra, Intermediate Algebra, Beginning and Intermediate Algebra: A Combined Approach, College Algebra,* and the Tussy/Gustafson developmental mathematics series. His numerous professional honors include Rock Valley Teacher of the Year and Rockford's Outstanding Educator of the Year. He earned a Master of Arts from Rockford College in Illinois, as well as a Master of Science from Northern Illinois University.

Jeff Hughes is a mathematics instructor at Hinds Community College in Mississippi and has degrees from Mississippi College and the University of Mississippi. He has worked at Hinds CC for 23 years and his favorite courses to teach are College Algebra, the Calculus Sequence, and Differential Equations. He sees teaching as a privilege and enjoys the challenge of finding new ways to present topics while engaging students. He has been the recipient of Hinds CC's Outstanding Academic Instructor Award twice and is active in AMATYC. He currently serves as president of LaMsMATYC, the Mississippi-Louisiana affiliate of AMATYC. He has worked part-time as Minister of Students at his church for over 15 years and enjoys traveling. He taught English in China for seven summers with ELIC, the English Language Institute/China.

About the Cover

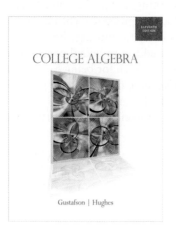

The wonderful cover art is by Van Evan Fuller. Van is not only a brilliant artist but also a gifted writer. He lives in Baton Rouge, Louisiana, and is a graduate of LSU. The cover art was initially created as a traditional, though digital, abstract. The artist then processed it through deforming software into which he programmed a series of variables. After creating about a dozen such compositions, all based on the same abstract painting, the final design was created using four of these. In the art, numerous variations of mathematical curves such as circles, parabolas, ellipses, and hyperbolas can be seen intertwined in an amazing, colorful array. In the words of the artist, "Every beautiful thing is somehow rooted in mathematics, and the expression of mathematics is always beautiful." The authors were struck by this beauty and the way Van was able to capture it. You can see more of Van's creative designs at his online gallery at www.vanevanfuller.com.

A Review of Basic Algebra

0

CAREERS AND MATHEMATICS:

© Istockphoto.com/Lucas Rucchin

Pharmacist

Pharmacists distribute prescription drugs to individuals. They also advise patients, physicians, and other healthcare workers on the selection, dosages, interactions, and side effects of medications. They also monitor patients to ensure that they are using their medications safely and effectively. Some pharmacists specialize in oncology, nuclear pharmacy, geriatric pharmacy, or psychiatric pharmacy.

Education and Mathematics Required

- Pharmacists are required to possess a Pharm.D. degree from an accredited college or school of pharmacy. This degree generally takes four years to complete. To be admitted to a Pharm.D. program, at least two years of college must be completed, which includes courses in the natural sciences, mathematics, humanities, and the social sciences. A series of examinations must also be passed to obtain a license to practice pharmacy.
- College Algebra, Trigonometry, Statistics, and Calculus I are courses required for admission to a Pharm.D. program.

How Pharmacists Use Math and Who Employs Them

- Pharmacists use math throughout their work to calculate dosages of various drugs. These dosages are based on weight and whether the medication is given in pill form, by infusion, or intravenously.
- Most pharmacists work in a community setting, such as a retail drugstore, or in a healthcare facility, such as a hospital.

Career Outlook and Salary

- Employment of pharmacists is expected to grow by 17 percent between 2008 and 2018, which is faster than the average for all occupations.
- Median annual wages of wage and salary pharmacists is approximately $106,410.

For more information see: www.bls.gov/oco

In this chapter, we review many concepts and skills learned in previous algebra courses. Be sure to master this material now, because it is the basis for the rest of this course.

0.1 Sets of Real Numbers

In this section, we will learn to

1. Identify sets of real numbers.

2. Identify properties of real numbers.

3. Graph subsets of real numbers on the number line.

4. Graph intervals on the number line.

5. Define absolute value.

6. Find distances on the number line.

5	3			7				
6			1	9	5			
	9	8				6		
8				6				3
4			8		3			1
7				2				6
	6					2	8	
			4	1	9			5
				8			7	9

SUDOKU

Sudoku, a game that involves number placement, is very popular. The objective is to fill a 9 by 9 grid so that each column, each row and each of the 3 by 3 blocks contains the numbers from 1 to 9. A partially completed Sudoku grid is shown in the margin.

To solve Sudoku puzzles, logic and the set of numbers, $\{1, 2, 3, 4, 5, 6, 7, 8, 9\}$ are used. Sets of numbers are important in mathematics, and we begin our study of algebra with this topic.

A **set** is a collection of objects, such as a set of dishes or a set of golf clubs. The set of vowels in the English language can be denoted as $\{a, e, i, o, u\}$, where the braces $\{\ \}$ are read as "the set of."

If every member of one set B is also a member of another set A, we say that B is a **subset** of A. We can denote this by writing $B \subset A$, where the symbol $\subset$ is read as "is a subset of." (See Figure 0.1 below.) If set B equals set A, we can write $B \subseteq A$.

If A and B are two sets, we can form a new set consisting of all members that are in set A or set B or both. This set is called the **union** of A and B. We can denote this set by writing, $A \cup B$ where the symbol $\cup$ is read as "union." (See Figure 0.1 below.)

We can also form the set consisting of all members that are in both set A and set B. This set is called the **intersection** of A and B. We can denote this set by writing $A \cap B$, where the symbol $\cap$ is read as "intersection." (See Figure 0.1 below.)

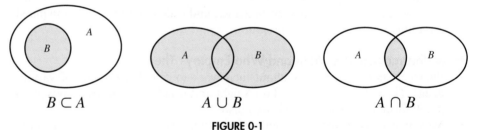

$$B \subset A \qquad A \cup B \qquad A \cap B$$

FIGURE 0-1

EXAMPLE 1 **Understanding Subsets and Finding the Union and Intersection of Two Sets**

Let $A = \{a, e, i\}$, $B = \{c, d, e\}$, and $V = \{a, e, i, o, u\}$.

a. Is $A \subset V$?   **b.** Find $A \cup B$.   **c.** Find $A \cap B$.

SOLUTION **a.** Since each member of set A is also a member of set V, $A \subset V$.

b. The union of set A and set B contains the members of set A, set B, or both. Thus, $A \cup B = \{a, c, d, e, i\}$.

c. The intersection of set A and set B contains the members that are in both set A and set B. Thus, $A \cap B = \{e\}$.

Self Check 1 **a.** Is $B \subset V$?   **b.** Find $B \cup V$.
c. Find $A \cap V$

Now Try Exercise 33.

1. Identify Sets of Real Numbers

There are several sets of numbers that we use in everyday life.

Basic Sets of Numbers

Natural numbers
The numbers that we use for counting: $\{1, 2, 3, 4, 5, 6, \dots\}$

Whole numbers
The set of natural numbers including 0: $\{0, 1, 2, 3, 4, 5, 6, \dots\}$

Integers
The set of whole numbers and their negatives:
$\{\dots, -5, -4, -3, -2, -1, 0, 1, 2, 3, 4, 5, \dots\}$

In the definitions above, each group of three dots (called an *ellipsis*) indicates that the numbers continue forever in the indicated direction.

Two important subsets of the natural numbers are the *prime* and *composite* numbers. A **prime number** is a natural number greater than 1 that is divisible only by itself and 1. A **composite number** is a natural number greater than 1 that is not prime.

- **The set of prime numbers**: $\{2, 3, 5, 7, 11, 13, 17, 19, 23, 29, 31, \dots\}$
- **The set of composite numbers**: $\{4, 6, 8, 9, 10, 12, 14, 15, 16, 18, 20, 21, \dots\}$

Two important subsets of the set of integers are the *even* and *odd integers*. The **even integers** are the integers that are exactly divisible by 2. The **odd integers** are the integers that are not exactly divisible by 2.

- **The set of even integers**: $\{\dots, -10, -8, -6, -4, -2, 0, 2, 4, 6, 8, 10, \dots\}$
- **The set of odd integers**: $\{\dots, -9, -7, -5, -3, -1, 1, 3, 5, 7, 9, \dots\}$

So far, we have listed numbers inside braces to specify sets. This method is called the **roster method**. When we give a rule to determine which numbers are in a set, we are using **set-builder notation**. To use set-builder notation to denote the set of prime numbers, we write

$$\{x \mid x \text{ is a prime number}\}$$

variable   such   rule that determines
that   membership in the set

Read as "the set of all numbers x such that x is a prime number." Recall that when a letter stands for a number, it is called a variable.

Caution

Remember that the denominator of a fraction can **never** be 0.

The fractions of arithmetic are called *rational numbers*.

Rational Numbers

Rational numbers are fractions that have an integer numerator and a nonzero integer denominator. Using set-builder notation, the rational numbers are

$$\left\{ \frac{a}{b} \mid a \text{ is an integer and } b \text{ is a nonzero integer} \right\}$$

Rational numbers can be written as fractions or decimals. Some examples of rational numbers are

$$5 = \frac{5}{1}, \qquad \frac{3}{4} = 0.75, \qquad -\frac{1}{3} = -0.333\dots, \qquad -\frac{5}{11} = -0.454545\dots$$

The = sign indicates that two quantities are equal.

These examples suggest that the decimal forms of all rational numbers are either *terminating decimals* or *repeating decimals*.

EXAMPLE 2   **Determining whether the Decimal Form of a Fraction Terminates or Repeats**

Determine whether the decimal form of each fraction terminates or repeats:

a. $\dfrac{7}{16}$ **b.** $\dfrac{65}{99}$

SOLUTION   In each case, we perform a long division and write the quotient as a decimal.

a. To change $\frac{7}{16}$ to a decimal, we perform a long division to get $\frac{7}{16} = 0.4375$. Since 0.4375 terminates, we can write $\frac{7}{16}$ as a terminating decimal.

b. To change $\frac{65}{99}$ to a decimal, we perform a long division to get $\frac{65}{99} = 0.656565\ldots$. Since $0.656565\ldots$ repeats, we can write $\frac{65}{99}$ as a repeating decimal.

We can write repeating decimals in compact form by using an overbar. For example, $0.656565\ldots = 0.\overline{65}$. ∎

Self Check 2   Determine whether the decimal form of each fraction terminates or repeats:

a. $\dfrac{38}{99}$ **b.** $\dfrac{7}{8}$

Now Try Exercise 35.

Some numbers have decimal forms that neither terminate nor repeat. These nonterminating, nonrepeating decimals are called **irrational numbers**. Three examples of irrational numbers are

$$1.010010001000010\ldots \qquad \sqrt{2} = 1.414213562\ldots \qquad \text{and} \qquad \pi = 3.141592654\ldots$$

The union of the set of rational numbers (the terminating and repeating decimals) and the set of irrational numbers (the nonterminating, nonrepeating decimals) is the set of *real numbers* (the set of all decimals).

Real Numbers   A **real number** is any number that is rational or irrational. Using set-builder notation, the set of real numbers is

$$\{x \mid x \text{ is a rational or an irrational number}\}$$

EXAMPLE 3   **Classifying Real Numbers**

In the set $\left\{-3, -2, 0, \frac{1}{2}, 1, \sqrt{5}, 2, 4, 5, 6\right\}$, list all

a. even integers   **b.** prime numbers   **c.** rational numbers

SOLUTION   We will check to see whether each number is a member of the set of even integers, the set of prime numbers, and the set of rational numbers.

a. even integers: $-2, 0, 2, 4, 6$

b. prime numbers: $2, 5$

c. rational numbers: $-3, -2, 0, \frac{1}{2}, 1, 2, 4, 5, 6$ ∎

Self Check 3   In the set in Example 3, list all   **a.** odd integers   **b.** composite numbers
c. irrational numbers.

Now Try Exercise 43.

Figure 0-2 shows how the previous sets of numbers are related.

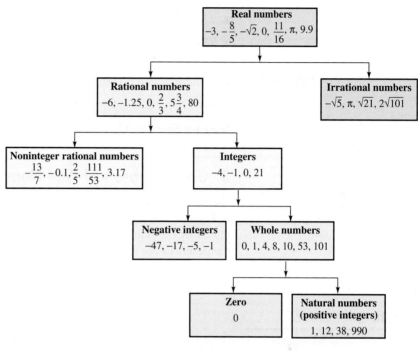

FIGURE 0-2

2. Identify Properties of Real Numbers

When we work with real numbers, we will use the following properties.

Properties of Real Numbers   If a, b, and c are real numbers,

The Commutative Properties for Addition and Multiplication

$$a + b = b + a \qquad\qquad ab = ba$$

The Associative Properties for Addition and Multiplication

$$(a + b) + c = a + (b + c) \qquad (ab)c = a(bc)$$

The Distributive Property of Multiplication over Addition or Subtraction

$$a(b + c) = ab + ac \qquad \text{or} \qquad a(b - c) = ab - ac$$

The Double Negative Rule

$$-(-a) = a$$

Caution   When the Associative Property is used, the order of the real numbers does **not** change. The real numbers that occur within the parentheses change.

The Distributive Property also applies when more than two terms are within parentheses.

EXAMPLE 4   **Identifying Properties of Real Numbers**

Determine which property of real numbers justifies each statement.

a. $(9 + 2) + 3 = 9 + (2 + 3)$ **b.** $3(x + y + 2) = 3x + 3y + 3 \cdot 2$

SOLUTION   We will compare the form of each statement to the forms listed in the properties of real numbers box.

a. This form matches the Associative Property of Addition.

b. This form matches the Distributive Property.   ∎

Self Check 4   Determine which property of real numbers justifies each statement:

a. $mn = nm$ **b.** $(xy)z = x(yz)$ **c.** $p + q = q + p$

Now Try Exercise 17.

3. Graph Subsets of Real Numbers on the Number Line

We can graph subsets of real numbers on the **number line**. The number line shown in Figure 0-3 continues forever in both directions. The **positive numbers** are represented by the points to the right of 0, and the **negative numbers** are represented by the points to the left of 0.

Comment

Zero is neither positive nor negative.

FIGURE 0-3

Figure 0-4(a) shows the graph of the natural numbers from 1 to 5. The point associated with each number is called the *graph* of the number, and the number is called the *coordinate* of its point.

Figure 0-4(b) shows the graph of the prime numbers that are less than 10.

Figure 0-4(c) shows the graph of the integers from -4 to 3.

Figure 0-4(d) shows the graph of the real numbers $-\frac{7}{3}, -\frac{3}{4}, 0.\overline{3}$, and $\sqrt{2}$.

Comment

$\sqrt{2}$ can be shown as the diagonal of a square with sides of length 1.

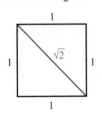

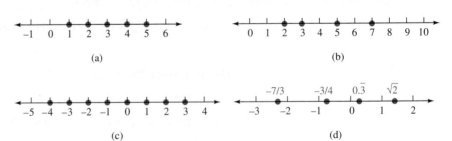

(a)   (b)

(c)   (d)

FIGURE 0-4

The graphs in Figure 0-4 suggest that there is a **one-to-one correspondence** between the set of real numbers and the points on a number line. This means that to each real number there corresponds exactly one point on the number line, and to each point on the number line there corresponds exactly one real-number coordinate.

EXAMPLE 5   **Graphing a Set of Numbers on a Number Line**

Graph the set $\left\{-3, -\frac{4}{3}, 0, \sqrt{5}\right\}$.

SOLUTION   We will mark (plot) each number on the number line. To the nearest tenth, $\sqrt{5} = 2.2$.

Self Check 5   Graph the set $\left\{-2, \frac{3}{4}, \sqrt{3}\right\}$. (*Hint*: To the nearest tenth, $\sqrt{3} = 1.7$.)
Now Try Exercise 51.

4. Graph Intervals on the Number Line

To show that two quantities are not equal, we can use an **inequality symbol.**

Symbol	Read as	Examples		
$\neq$	"is not equal to"	$5 \neq 8$	and	$0.25 \neq \frac{1}{3}$
$<$	"is less than"	$12 < 20$	and	$0.17 < 1.1$
$>$	"is greater than"	$15 > 9$	and	$\frac{1}{2} > 0.2$
$\leq$	"is less than or equal to"	$25 \leq 25$	and	$1.7 \leq 2.3$
$\geq$	"is greater than or equal to"	$19 \geq 19$	and	$15.2 \geq 13.7$
$\approx$	"is approximately equal to"	$\sqrt{2} \approx 1.414$	and	$\sqrt{3} \approx 1.732$

It is possible to write an inequality with the inequality symbol pointing in the opposite direction. For example,

- $12 < 20$ is equivalent to $20 > 12$
- $2.3 \geq -1.7$ is equivalent to $-1.7 \leq 2.3$

In Figure 0-3, the coordinates of points get larger as we move from left to right on a number line. Thus, if a and b are the coordinates of two points, the one to the right is the greater. This suggests the following facts:

- If $a > b$, point a lies to the right of point b on a number line.
- If $a < b$, point a lies to the left of point b on a number line.

Figure 0-5(a) shows the graph of the *inequality* $x > -2$ (or $-2 < x$). This graph includes all real numbers x that are greater than -2. The parenthesis at -2 indicates that -2 is not included in the graph. Figure 0-5(b) shows the graph of $x \leq 5$ (or $5 \geq x$). The bracket at 5 indicates that 5 is included in the graph.

(a)   (b)

FIGURE 0-5

Sometimes two inequalities can be written as a single expression called a **compound inequality.** For example, the compound inequality

$$5 < x < 12$$

is a combination of the inequalities $5 < x$ and $x < 12$. It is read as "5 is less than x, and x is less than 12," and it means that x is between 5 and 12. Its graph is shown in Figure 0-6.

FIGURE 0-6

The graphs shown in Figures 0-5 and 0-6 are portions of a number line called **intervals**. The interval shown in Figure 0-7(a) is denoted by the inequality $-2 < x < 4$, or in **interval notation** as $(-2, 4)$. The parentheses indicate that the endpoints are not included. The interval shown in Figure 0-7(b) is denoted by the inequality $x > 1$, or as $(1, \infty)$ in interval notation.

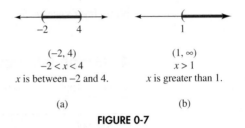

$(-2, 4)$
$-2 < x < 4$
x is between -2 and 4.

$(1, \infty)$
$x > 1$
x is greater than 1.

(a)   (b)

FIGURE 0-7

Caution

> The symbol ∞ (infinity) is **not** a real number. It is used to indicate that the graph in Figure 0-7(b) extends infinitely far to the right.

A compound inequality such as $-2 < x < 4$ can be written as two separate inequalities:

$$x > -2 \quad \text{and} \quad x < 4$$

This expression represents the **intersection** of two intervals. In interval notation, this expression can be written as

$$(-2, \infty) \cap (-\infty, 4) \qquad \text{Read the symbol } \cap \text{ as "intersection."}$$

Since the graph of $-2 < x < 4$ will include all points whose coordinates satisfy both $x > -2$ and $x < 4$ at the same time, its graph will include all points that are larger than -2 but less than 4. This is the interval $(-2, 4)$, whose graph is shown in Figure 0-7(a).

EXAMPLE 6 **Writing an Inequality in Interval Notation and Graphing the Inequality**

Write the inequality $-3 < x < 5$ in interval notation and graph it.

SOLUTION This is the interval $(-3, 5)$. Its graph includes all real numbers between -3 and 5, as shown in Figure 0-8.

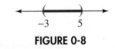

FIGURE 0-8

Self Check 6 Write the inequality $-2 < x \le 5$ in interval notation and graph it.

Now Try Exercise 63.

If an interval extends forever in one direction, it is called an **unbounded interval**.

Unbounded Intervals		
Interval	**Inequality**	**Graph**
(a, ∞)	$x > a$	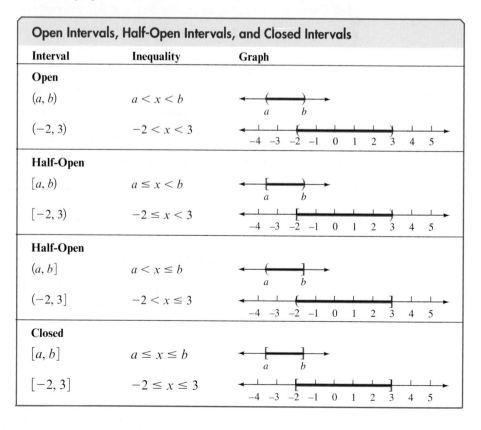
$(2, \infty)$	$x > 2$	
$[a, \infty)$	$x \geq a$	
$[2, \infty)$	$x \geq 2$	
$(-\infty, a)$	$x < a$	
$(-\infty, 2)$	$x < 2$	
$(-\infty, a]$	$x \leq a$	
$(-\infty, 2]$	$x \leq 2$	
$(-\infty, \infty)$	$-\infty < x < \infty$	

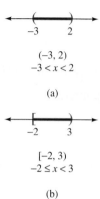

$(-3, 2)$
$-3 < x < 2$

(a)

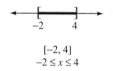

$[-2, 3)$
$-2 \leq x < 3$

(b)

FIGURE 0-9

A bounded interval with no endpoints is called an **open interval**. Figure 0-9(a) shows the open interval between -3 and 2. A bounded interval with one endpoint is called a **half-open interval**. Figure 0-9(b) shows the half-open interval between -2 and 3, including -2.

Intervals that include two endpoints are called **closed intervals**. Figure 0-10 shows the graph of a closed interval from -2 to 4.

Open Intervals, Half-Open Intervals, and Closed Intervals		
Interval	**Inequality**	**Graph**
Open		
(a, b)	$a < x < b$	
$(-2, 3)$	$-2 < x < 3$	
Half-Open		
$[a, b)$	$a \leq x < b$	
$[-2, 3)$	$-2 \leq x < 3$	
Half-Open		
$(a, b]$	$a < x \leq b$	
$(-2, 3]$	$-2 < x \leq 3$	
Closed		
$[a, b]$	$a \leq x \leq b$	
$[-2, 3]$	$-2 \leq x \leq 3$	

$[-2, 4]$
$-2 \leq x \leq 4$

FIGURE 0-10

EXAMPLE 7   **Writing an Inequality in Interval Notation and Graphing the Inequality**

Write the inequality $3 \leq x$ in interval notation and graph it.

SOLUTION   The inequality $3 \leq x$ can be written in the form $x \geq 3$. This is the interval $[3, \infty)$. Its graph includes all real numbers greater than or equal to 3, as shown in Figure 0-11.

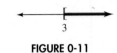

FIGURE 0-11

Self Check 7   Write the inequality $5 > x$ in interval notation and graph it.

Now Try Exercise 65.

EXAMPLE 8   **Writing an Inequality in Interval Notation and Graphing the Inequality**

Write the inequality $5 \geq x \geq -1$ in interval notation and graph it.

SOLUTION   The inequality $5 \geq x \geq -1$ can be written in the form

$$-1 \leq x \leq 5$$

This is the interval $[-1, 5]$. Its graph includes all real numbers from -1 to 5. The graph is shown in Figure 0-12.

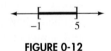

FIGURE 0-12

Self Check 8   Write the inequality $0 \leq x \leq 3$ in interval notation and graph it.

Now Try Exercise 69.

The expression

$$x < -2 \text{ or } x \geq 3 \qquad \text{Read as ``}x\text{ is less than } -2 \text{ or } x \text{ is greater than or equal to 3.''}$$

represents the *union* of two intervals. In interval notation, it is written as

$$(-\infty, -2) \cup [3, \infty) \qquad \text{Read the symbol } \cup \text{ as ``union.''}$$

Its graph is shown in Figure 0-13.

FIGURE 0-13

5. Define Absolute Value

The **absolute value** of a real number x (denoted as $|x|$) is the distance on a number line between 0 and the point with a coordinate of x. For example, points with coordinates of 4 and -4 both lie four units from 0, as shown in Figure 0-14. Therefore, it follows that

$$|-4| = |4| = 4$$

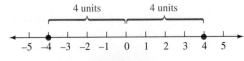

FIGURE 0-14

In general, for any real number x,

$$|-x| = |x|$$

We can define absolute value algebraically as follows.

Absolute Value   If x is a real number, then

$$|x| = x \qquad \text{when } x \geq 0$$

$$|x| = -x \qquad \text{when } x < 0$$

Caution

> Remember that x is **not** always positive and $-x$ is **not** always negative.

This definition indicates that when x is positive or 0, then x is its own absolute value. However, when x is negative, then $-x$ (which is positive) is its absolute value. Thus, $|x|$ is always nonnegative.

$$|x| \geq 0 \qquad \text{for all real numbers } x$$

EXAMPLE 9   **Using the Definition of Absolute Value**

Write each number without using absolute value symbols:

a. $|3|$ **b.** $|-4|$ **c.** $|0|$ **d.** $-|-8|$

SOLUTION   In each case, we will use the definition of absolute value.

a. $|3| = 3$ **b.** $|-4| = 4$ **c.** $|0| = 0$ **d.** $-|-8| = -(8) = -8$

Self Check 9   Write each number without using absolute value symbols:
a. $|-10|$ **b.** $|12|$ **c.** $-|6|$

Now Try Exercise 85.

In Example 10, we must determine whether the number inside the absolute value is positive or negative.

EXAMPLE 10   **Simplifying an Expression with Absolute Value Symbols**

Write each number without using absolute value symbols:

a. $|\pi - 1|$ **b.** $|2 - \pi|$ **c.** $|2 - x|$ if $x \geq 5$

SOLUTION   **a.** Since $\pi \approx 3.1416$, $\pi - 1$ is positive, and $\pi - 1$ is its own absolute value.

$$|\pi - 1| = \pi - 1$$

b. Since $2 - \pi$ is negative, its absolute value is $-(2 - \pi)$.

$$|2 - \pi| = -(2 - \pi) = -2 - (-\pi) = -2 + \pi = \pi - 2$$

c. Since $x \geq 5$, the expression $2 - x$ is negative, and its absolute value is $-(2 - x)$.

$$|2 - x| = -(2 - x) = -2 + x = x - 2 \text{ provided } x \geq 5$$

Self Check 10   Write each number without using absolute value symbols. (*Hint*: $\sqrt{5} \approx 2.236$.)
a. $\left|2 - \sqrt{5}\right|$ **b.** $|2 - x|$ if $x \leq 1$

Now Try Exercise 89.

6. Find Distances on the Number Line

On the number line shown in Figure 0-15, the distance between the points with coordinates of 1 and 4 is $4 - 1$, or 3 units. However, if the subtraction were done in the other order, the result would be $1 - 4$, or -3 units. To guarantee that the distance between two points is always positive, we can use absolute value symbols. Thus, the distance d between two points with coordinates of 1 and 4 is

$$d = |4 - 1| = |1 - 4| = 3$$

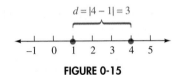

$d = |4 - 1| = 3$

FIGURE 0-15

In general, we have the following definition for the distance between two points on the number line.

Distance between Two Points   If a and b are the coordinates of two points on the number line, the distance between the points is given by the formula

$$d = |b - a|$$

EXAMPLE 11   **Finding the Distance between Two Points on a Number Line**

Find the distance on a number line between points with coordinates of **a.** 3 and 5 **b.** -2 and 3   **c.** -5 and -1

SOLUTION   We will use the formula for finding the distance between two points.

a. $d = |5 - 3| = |2| = 2$

b. $d = |3 - (-2)| = |3 + 2| = |5| = 5$

c. $d = |-1 - (-5)| = |-1 + 5| = |4| = 4$ ■

Self Check 11   Find the distance on a number line between points with coordinates of **a.** 4 and 10 **b.** -2 and -7

Now Try Exercise 99.

Self Check Answers   **1. a.** no   **b.** {a, c, d, e, i, o, u}   **c.** {a, e, i}   **2. a.** repeats
b. terminates   **3. a.** $-3, 1, 5$ **b.** 4, 6   **c.** $\sqrt{5}$ **4. a.** Commutative
Property of Multiplication   **b.** Associative Property of Multiplication
c. Commutative Property of Addition   **5.**

6. $(-2, 5]$ **7.** $(-\infty, 5)$

8. $[0, 3]$ **9. a.** 10   **b.** 12   **c.** -6

10. a. $\sqrt{5} - 2$ **b.** $2 - x$ **11. a.** 6   **b.** 5

Exercises **0.1**

Getting Ready
You should be able to complete these vocabulary and concept statements before you proceed to the practice exercises.

Fill in the blanks.

1. A _____ is a collection of objects.
2. If every member of one set B is also a member of a second set A, then B is called a _____ of A.
3. If A and B are two sets, the set that contains all members that are in sets A and B or both is called the _____ of A and B.
4. If A and B are two sets, the set that contains all members that are in both sets is called the _____ of A and B.
5. A real number is any number that can be expressed as a _____ .
6. A _____ is a letter that is used to represent a number.
7. The smallest prime number is ____ .
8. All integers that are exactly divisible by 2 are called _____ integers.
9. Natural numbers greater than 1 that are not prime are called _____ numbers.
10. Fractions such as $\frac{2}{3}$, $\frac{8}{2}$, and $-\frac{7}{9}$ are called _____ numbers.
11. Irrational numbers are _____ that don't terminate and don't repeat.
12. The symbol _____ is read as "is less than or equal to."
13. On a number line, the _____ numbers are to the left of 0.
14. The only integer that is neither positive nor negative is ____ .
15. The Associative Property of Addition states that $(x + y) + z =$ _____ .
16. The Commutative Property of Multiplication states that $xy =$ _____ .
17. Use the Distributive Property to complete the statement: $5(m + 2) =$ _____ .
18. The statement $(m + n)p = p(m + n)$ illustrates the _____ Property of _____ .
19. The graph of an _____ is a portion of a number line.
20. The graph of an open interval has _____ endpoints.
21. The graph of a closed interval has _____ endpoints.
22. The graph of a _____ interval has one endpoint.
23. Except for 0, the absolute value of every number is _____ .

24. The _____ between two distinct points on a number line is always positive.

Let

$\mathbf{N} =$ *the set of natural numbers*
$\mathbf{W} =$ *the set of whole numbers*
$\mathbf{Z} =$ *the set of integers*
$\mathbf{Q} =$ *the set of rational numbers*
$\mathbf{R} =$ *the set of real numbers*

Determine whether each statement is true or false. Read the symbol $\subset$ as "is a subset of."

25. $\mathbf{N} \subset \mathbf{W}$ 26. $\mathbf{Q} \subset \mathbf{R}$
27. $\mathbf{Q} \subset \mathbf{N}$ 28. $\mathbf{Z} \subset \mathbf{Q}$
29. $\mathbf{W} \subset \mathbf{Z}$ 30. $\mathbf{R} \subset \mathbf{Z}$

Practice
Let $A = \{a, b, c, d, e\}$, $B = (d, e, f, g\}$, and $C = \{a, c, e, f\}$. Find each set.

31. $A \cup B$ 32. $A \cap B$
33. $A \cap C$ 34. $B \cup C$

Determine whether the decimal form of each fraction terminates or repeats.

35. $\dfrac{9}{16}$ 36. $\dfrac{3}{8}$

37. $\dfrac{3}{11}$ 38. $\dfrac{5}{12}$

Consider the following set:
$\{-5, -4, -\frac{2}{3}, 0, 1, \sqrt{2}, 2, 2.75, 6, 7\}.$

39. Which numbers are natural numbers?
40. Which numbers are whole numbers?
41. Which numbers are integers?
42. Which numbers are rational numbers?

43. Which numbers are irrational numbers?
44. Which numbers are prime numbers?
45. Which numbers are composite numbers?
46. Which numbers are even integers?
47. Which numbers are odd integers?
48. Which numbers are negative numbers?

Graph each subset of the real numbers on a number line.

49. The natural numbers between 1 and 5

50. The composite numbers less than 10

51. The prime numbers between 10 and 20

52. The integers from –2 to 4

53. The integers between –5 and 0

54. The even integers between –9 and –1

55. The odd integers between –6 and 4

56. -0.7, 1.75, and $3\frac{7}{8}$

Write each inequality in interval notation and graph the interval.

57. $x > 2$ **58.** $x < 4$

59. $0 < x < 5$ **60.** $-2 < x < 3$

61. $x > -4$ **62.** $x < 3$

63. $-2 \le x < 2$ **64.** $-4 < x \le 1$

65. $x \le 5$ **66.** $x \ge -1$

67. $-5 < x \le 0$ **68.** $-3 \le x < 4$

69. $-2 \le x \le 3$ **70.** $-4 \le x \le 4$

71. $6 \ge x \ge 2$ **72.** $3 \ge x \ge -2$

Write each pair of inequalities as the intersection of two intervals and graph the result.

73. $x > -5$ and $x < 4$

74. $x \ge -3$ and $x < 6$

75. $x \ge -8$ and $x \le -3$

76. $x > 1$ and $x \le 7$

Write each inequality as the union of two intervals and graph the result.

77. $x < -2$ or $x > 2$

78. $x \le -5$ or $x > 0$

79. $x \le -1$ or $x \ge 3$

80. $x < -3$ or $x \ge 2$

Write each expression without using absolute value symbols.

81. $|13|$ **82.** $|-17|$
83. $|0|$ **84.** $-|63|$
85. $-|-8|$ **86.** $|-25|$
87. $-|32|$ **88.** $-|-6|$
89. $|\pi - 5|$ **90.** $|8 - \pi|$
91. $|\pi - \pi|$ **92.** $|2\pi|$
93. $|x + 1|$ and $x \ge 2$ **94.** $|x + 1|$ and $x \le -2$

95. $|x - 4|$ and $x < 0$ **96.** $|x - 7|$ and $x > 10$

Find the distance between each pair of points on the number line.

97. 3 and 8   **98.** -5 and 12
99. -8 and -3 **100.** 6 and -20

Applications

101. What subset of the real numbers would you use to describe the populations of several cities?

102. What subset of the real numbers would you use to describe the subdivisions of an inch on a ruler?

103. What subset of the real numbers would you use to report temperatures in several cities?

104. What subset of the real numbers would you use to describe the financial condition of a business?

Discovery and Writing

105. Explain why $-x$ could be positive.

106. Explain why every integer is a rational number.

107. Is the statement $|ab| = |a| \cdot |b|$ always true? Explain.

108. Is the statement $\left|\dfrac{a}{b}\right| = \dfrac{|a|}{|b|}$ $(b \ne 0)$ always true? Explain.

109. Is the statement $|a + b| = |a| + |b|$ always true? Explain.

110. Under what conditions will the statement given in Exercise 109 be true?

111. Explain why it is incorrect to write $a < b > c$ if $a < b$ and $b > c$.

112. Explain why $|b - a| = |a - b|$.

0.2 Integer Exponents and Scientific Notation

In this section, we will learn to

1. Define natural-number exponents.

2. Apply the rules of exponents.

3. Apply the rules for order of operations to evaluate expressions.

4. Express numbers in scientific notation.

5. Use scientific notation to simplify computations.

The number of cells in the human body is approximated to be one hundred trillion or 100,000,000,000,000. One hundred trillion is $(10)(10)(10) \cdots (10)$, where ten occurs fourteen times. Fourteen factors of ten can be written as 10^{14}.

In this section, we will use integer exponents to represent repeated multiplication of numbers.

1. Define Natural-Number Exponents

When two or more quantities are multiplied together, each quantity is called a *factor* of the product. The exponential expression x^4 indicates that x is to be used as a factor four times.

$$x^4 = \overbrace{x \cdot x \cdot x \cdot x}^{4 \text{ factors of } x}$$

In general, the following is true.

Natural-Number Exponents   For any natural number n,

$$x^n = \overbrace{x \cdot x \cdot x \cdot \cdots \cdot x}^{n \text{ factors of } x}$$

In the **exponential expression** x^n, x is called the **base**, and n is called the **exponent** or the **power** to which the base is raised. The expression x^n is called a **power of x**. From the definition, we see that a natural-number exponent indicates how many times the base of an exponential expression is to be used as a factor in a product. If an exponent is 1, the 1 is usually not written:

$$x^1 = x$$

EXAMPLE 1   **Using the Definition of Natural-Number Exponents**

Write each expression without using exponents:

a. 4^2 **b.** $(-4)^2$ **c.** -5^3 **d.** $(-5)^3$ **e.** $3x^4$ **f.** $(3x)^4$

SOLUTION   In each case, we apply the definition of natural-number exponents.

a. $4^2 = 4 \cdot 4 = 16$ Read 4^2 as "four squared."

b. $(-4)^2 = (-4)(-4) = 16$ Read $(-4)^2$ as "negative four squared."

c. $-5^3 = -5(5)(5) = -125$ Read -5^3 as "the negative of five cubed."

d. $(-5)^3 = (-5)(-5)(-5) = -125$ Read $(-5)^3$ as "negative five cubed."

e. $3x^4 = 3 \cdot x \cdot x \cdot x \cdot x$ Read $3x^4$ as "3 times x to the fourth power."

f. $(3x)^4 = (3x)(3x)(3x)(3x) = 81 \cdot x \cdot x \cdot x \cdot x$ Read $(3x)^4$ as "3x to the fourth power." ∎

Self Check 1   Write each expression without using exponents:
a. 7^3 **b.** $(-3)^2$ **c.** $5a^3$ **d.** $(5a)^4$

Now Try Exercise 19.

Comment   Note the distinction between ax^n and $(ax)^n$:

$$\overset{n \text{ factors of } x}{\overbrace{ax^n = a \cdot x \cdot x \cdot x \cdot \;\cdots\; \cdot x}}$$ $$\qquad (ax)^n = \overset{n \text{ factors of } ax}{\overbrace{(ax)(ax)(ax) \cdot \;\cdots\; \cdot (ax)}}$$

Also note the distinction between $-x^n$ and $(-x)^n$:

$$-x^n = -(\overset{n \text{ factors of } x}{\overbrace{x \cdot x \cdot x \cdot \;\cdots\; \cdot x}})$$ $$\qquad (-x)^n = \overset{n \text{ factors of } -x}{\overbrace{(-x)(-x)(-x) \cdot \;\cdots\; \cdot (-x)}}$$

ACCENT ON TECHNOLOGY   **Using a Calculator to Find Powers**

We can use a graphing calculator to find powers of numbers. For example, consider 2.35^3.

- Input 2.35 and press the [^] key.
- Input 3 and press [ENTER].

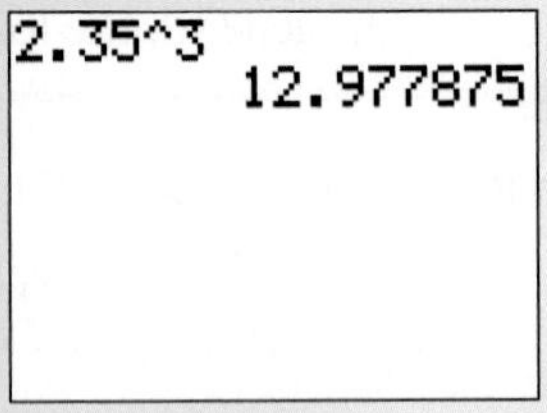

```
2.35^3
        12.977875
```

Comment

To find powers on a scientific calculator use the $\boxed{y^x}$ key.

FIGURE 0-16

We see that $2.35^3 = 12.977875$ as the figure shows above.

2. Apply the Rules of Exponents

We begin the review of the rules of exponents by considering the product $x^m x^n$. Since x^m indicates that x is to be used as a factor m times, and since x^n indicates that x is to be used as a factor n times, there are $m + n$ factors of x in the product $x^m x^n$.

$$x^m x^n = \overset{m+n \text{ factors of } x}{\overbrace{\overset{m \text{ factors of } x}{\overbrace{x \cdot x \cdot x \cdot \;\cdots\; \cdot x}} \; \overset{n \text{ factors of } x}{\overbrace{x \cdot x \cdot x \cdot \;\cdots\; \cdot x}}}} = x^{m+n}$$

This suggests that to multiply exponential expressions with the same base, we *keep the base and add the exponents.*

Product Rule for Exponents	If m and n are natural numbers, then

$$x^m x^n = x^{m+n}$$

Comment

The Product Rule applies to exponential expressions with the same base. A product of two powers with different bases, such as $x^4 y^3$, cannot be simplified.

To find another property of exponents, we consider the exponential expression $(x^m)^n$. In this expression, the exponent n indicates that x^m is to be used as a factor n times. This implies that x is to be used as a factor mn times.

$$(x^m)^n = \overbrace{\underbrace{(x^m)(x^m)(x^m) \cdot \cdots \cdot (x^m)}_{n \text{ factors of } x^m}}^{mn \text{ factors of } x} = x^{mn}$$

This suggests that to raise an exponential expression to a power, we *keep the base and multiply the exponents.*

To raise a product to a power, we raise each factor to that power.

$$(xy)^n = \overbrace{(xy)(xy)(xy) \cdot \cdots \cdot (xy)}^{n \text{ factors of } xy} = \overbrace{(x \cdot x \cdot x \cdot \cdots \cdot x)}^{n \text{ factors of } x}\overbrace{(y \cdot y \cdot y \cdot \cdots \cdot y)}^{n \text{ factors of } y} = x^n y^n$$

To raise a fraction to a power, we raise both the numerator and the denominator to that power. If $y \neq 0$, then

$$\left(\frac{x}{y}\right)^n = \overbrace{\left(\frac{x}{y}\right)\left(\frac{x}{y}\right)\left(\frac{x}{y}\right) \cdot \cdots \cdot \left(\frac{x}{y}\right)}^{n \text{ factors of } \frac{x}{y}}$$

$$= \frac{\overbrace{xxx \cdot \cdots \cdot x}^{n \text{ factors of } x}}{\underbrace{yyy \cdot \cdots \cdot y}_{n \text{ factors of } y}}$$

$$= \frac{x^n}{y^n}$$

The previous three results are called the *Power Rules of Exponents.*

Power Rules of Exponents	If m and n are natural numbers, then

$$(x^m)^n = x^{mn} \qquad (xy)^n = x^n y^n \qquad \left(\frac{x}{y}\right)^n = \frac{x^n}{y^n} \qquad (y \neq 0)$$

EXAMPLE 2 **Using Exponent Rules to Simplify Expressions with Natural-Number Exponents**

Simplify: **a.** $x^5 x^7$ **b.** $x^2 y^3 x^5 y$ **c.** $(x^4)^9$ **d.** $(x^2 x^5)^3$

e. $\left(\dfrac{x}{y^2}\right)^5$ **f.** $\left(\dfrac{5x^2 y}{z^3}\right)^2$

SOLUTION In each case, we will apply the appropriate rule of exponents.

a. $x^5 x^7 = x^{5+7} = x^{12}$ **b.** $x^2 y^3 x^5 y = x^{2+5} y^{3+1} = x^7 y^4$

c. $(x^4)^9 = x^{4 \cdot 9} = x^{36}$ **d.** $(x^2 x^5)^3 = (x^7)^3 = x^{21}$

e. $\left(\dfrac{x}{y^2}\right)^5 = \dfrac{x^5}{(y^2)^5} = \dfrac{x^5}{y^{10}}$ $(y \neq 0)$

f. $\left(\dfrac{5x^2 y}{z^3}\right)^2 = \dfrac{5^2 (x^2)^2 y^2}{(z^3)^2} = \dfrac{25 x^4 y^2}{z^6}$ $(z \neq 0)$

Self Check 2 Simplify:

a. $(y^3)^2$ **b.** $(a^2 a^4)^3$ **c.** $(x^2)^3 (x^3)^2$ **d.** $\left(\dfrac{3 a^3 b^2}{c^3}\right)^3$ $(c \neq 0)$

Now Try Exercise 49.

If we assume that the rules for natural-number exponents hold for exponents of 0, we can write

$$x^0 x^n = x^{0+n} = x^n = 1 x^n$$

Since $x^0 x^n = 1 x^n$, it follows that if $x \neq 0$, then $x^0 = 1$.

Zero Exponent $x^0 = 1$ $(x \neq 0)$

If we assume that the rules for natural-number exponents hold for exponents that are negative integers, we can write

$$x^{-n} x^n = x^{-n+n} = x^0 = 1 \quad (x \neq 0)$$

However, we know that

$$\frac{1}{x^n} \cdot x^n = 1 \quad (x \neq 0) \qquad \frac{1}{x^n} \cdot x^n = \frac{x^n}{x^n}, \text{ and any nonzero number divided by itself is 1.}$$

Since $x^{-n} x^n = \frac{1}{x^n} \cdot x^n$, it follows that $x^{-n} = \frac{1}{x^n}$ $(x \neq 0)$.

Negative Exponents If n is an integer and $x \neq 0$, then

$$x^{-n} = \frac{1}{x^n} \qquad \text{and} \qquad \frac{1}{x^{-n}} = x^n$$

Because of the previous definitions, all of the rules for natural-number exponents will hold for integer exponents.

EXAMPLE 3 **Simplifying Expressions with Integer Exponents**

Simplify and write all answers without using negative exponents:

a. $(3x)^0$ **b.** $3(x^0)$ **c.** x^{-4} **d.** $\dfrac{1}{x^{-6}}$ **e.** $x^{-3} x$ **f.** $(x^{-4} x^8)^{-5}$

SOLUTION We will use the definitions of zero exponent and negative exponents to simplify each expression.

a. $(3x)^0 = 1$ **b.** $3(x^0) = 3(1) = 3$ **c.** $x^{-4} = \dfrac{1}{x^4}$ **d.** $\dfrac{1}{x^{-6}} = x^6$

e. $x^{-3}x = x^{-3+1}$

$\qquad = x^{-2}$

$\qquad = \dfrac{1}{x^2}$

f. $(x^{-4}x^8)^{-5} = (x^4)^{-5}$

$\qquad = x^{-20}$

$\qquad = \dfrac{1}{x^{20}}$

Self Check 3   Simplify and write all answers without using negative exponents:

a. $7a^0$ **b.** $3a^{-2}$ **c.** $a^{-4}a^2$ **d.** $(a^3a^{-7})^3$

Now Try Exercise 59.

To develop the Quotient Rule for Exponents, we proceed as follows:

$$\frac{x^m}{x^n} = x^m\left(\frac{1}{x^n}\right) = x^m x^{-n} = x^{m+(-n)} = x^{m-n} \quad (x \neq 0)$$

This suggests that to divide two exponential expressions with the same nonzero base, we *keep the base and subtract the exponent in the denominator from the exponent in the numerator*.

Quotient Rule for Exponents   If m and n are integers, then

$$\frac{x^m}{x^n} = x^{m-n} \quad (x \neq 0)$$

EXAMPLE 4   **Simplifying Expressions with Integer Exponents**

Simplify and write all answers without using negative exponents:

a. $\dfrac{x^8}{x^5}$ **b.** $\dfrac{x^2 x^4}{x^{-5}}$

SOLUTION   We will apply the Product and Quotient Rules of Exponents.

a. $\dfrac{x^8}{x^5} = x^{8-5}$

$\qquad = x^3$

b. $\dfrac{x^2 x^4}{x^{-5}} = \dfrac{x^6}{x^{-5}}$

$\qquad = x^{6-(-5)}$

$\qquad = x^{11}$

Self Check 4   Simplify and write all answers without using negative exponents:

a. $\dfrac{x^{-6}}{x^2}$ **b.** $\dfrac{x^4 x^{-3}}{x^2}$

Now Try Exercise 69.

EXAMPLE 5   **Simplifying Expressions with Integer Exponents**

Simplify and write all answers without using negative exponents:

a. $\left(\dfrac{x^3 y^{-2}}{x^{-2} y^3}\right)^{-2}$ **b.** $\left(\dfrac{x}{y}\right)^{-n}$

SOLUTION   We will apply the appropriate rules of exponents.

$$\textbf{a.} \quad \left(\frac{x^3y^{-2}}{x^{-2}y^3}\right)^{-2} = (x^{3-(-2)}y^{-2-3})^{-2}$$

$$= (x^5y^{-5})^{-2}$$

$$= x^{-10}y^{10}$$

$$= \frac{y^{10}}{x^{10}}$$

$$\textbf{b.} \quad \left(\frac{x}{y}\right)^{-n} = \frac{x^{-n}}{y^{-n}}$$

$$= \frac{x^{-n}x^ny^n}{y^{-n}x^ny^n} \qquad \text{Multiply numerator and denominator by 1 in the form } \frac{x^ny^n}{x^ny^n}.$$

$$= \frac{x^0y^n}{y^0x^n} \qquad x^{-n}x^n = x^0 \text{ and } y^{-n}y^n = y^0.$$

$$= \frac{y^n}{x^n} \qquad x^0 = 1 \text{ and } y^0 = 1.$$

$$= \left(\frac{y}{x}\right)^n$$

Self Check 5   Simplify and write all answers without using negative exponents:

$$\textbf{a.} \quad \left(\frac{x^4y^{-3}}{x^{-3}y^2}\right)^2 \qquad \textbf{b.} \quad \left(\frac{2a}{3b}\right)^{-3}$$

Now Try Exercise 75.

Part b of Example 5 establishes the following rule.

A Fraction to a Negative Power   If n is a natural number, then

$$\left(\frac{x}{y}\right)^{-n} = \left(\frac{y}{x}\right)^n \qquad (x \neq 0 \quad \text{and} \quad y \neq 0)$$

---

3. Apply the Rules for Order of Operations to Evaluate Expressions

When several operations occur in an expression, we must perform the operations in the following order to get the correct result.

Strategy for Evaluating Expressions Using Order of Operations   If an expression does not contain grouping symbols such as parentheses or brackets, follow these steps:

1. Find the values of any exponential expressions.
2. Perform all multiplications and/or divisions, working from left to right.
3. Perform all additions and/or subtractions, working from left to right.

- If an expression contains grouping symbols such as parentheses, brackets, or braces, use the rules above to perform the calculations within each pair of grouping symbols, working from the innermost pair to the outermost pair.
- In a fraction, simplify the numerator and the denominator of the fraction separately. Then simplify the fraction, if possible.

Comment   Many students remember the Order of Operations Rule with the acronym **PEMDAS:**

- **P**arentheses
- **E**xponents
- **M**ultiplication
- **D**ivision
- **A**ddition
- **S**ubtraction

For example, to simplify $\dfrac{3[4 - (6 + 10)]}{2^2 - (6 + 7)}$, we proceed as follows:

$$\frac{3[4 - (6 + 10)]}{2^2 - (6 + 7)} = \frac{3(4 - 16)}{2^2 - (6 + 7)} \qquad \text{Simplify within the inner parentheses: } 6 + 10 = 16.$$

$$= \frac{3(-12)}{2^2 - 13} \qquad \text{Simplify within the parentheses: } 4 - 16 = -12, \text{ and } 6 + 7 = 13.$$

$$= \frac{3(-12)}{4 - 13} \qquad \text{Evaluate the power: } 2^2 = 4.$$

$$= \frac{-36}{-9} \qquad 3(-12) = -36.\ 4 - 13 = -9.$$

$$= 4$$

EXAMPLE 6   **Evaluating Algebraic Expressions**

If $x = -2$, $y = 3$, and $z = -4$, evaluate

a. $-x^2 + y^2 z$ **b.** $\dfrac{2z^3 - 3y^2}{5x^2}$

SOLUTION   In each part, we will substitute the numbers for the variables, apply the rules of order of operations, and simplify.

a. $-x^2 + y^2 z = -(-2)^2 + 3^2(-4)$

$$= -(4) + 9(-4) \qquad \text{Evaluate the powers.}$$

$$= -4 + (-36) \qquad \text{Do the multiplication.}$$

$$= -40 \qquad \text{Do the addition.}$$

b. $\dfrac{2z^3 - 3y^2}{5x^2} = \dfrac{2(-4)^3 - 3(3)^2}{5(-2)^2}$

$$= \frac{2(-64) - 3(9)}{5(4)} \qquad \text{Evaluate the powers.}$$

$$= \frac{-128 - 27}{20} \qquad \text{Do the multiplications.}$$

$$= \frac{-155}{20} \qquad \text{Do the subtraction.}$$

$$= -\frac{31}{4} \qquad \text{Simplify the fraction.}$$

Self Check 6   If $x = 3$ and $y = -2$, evaluate $\dfrac{2x^2 - 3y^2}{x - y}$.

Now Try Exercise 91.

4. Express Numbers in Scientific Notation

Scientists often work with numbers that are very large or very small. These numbers can be written compactly by expressing them in *scientific notation*.

Scientific Notation   A number is written in *scientific notation* when it is written in the form

$$N \times 10^n$$

where $1 \leq |N| < 10$ and n is an integer.

Light travels 29,980,000,000 centimeters per second.

To express this number in scientific notation, we must write it as the product of a number between 1 and 10 and some integer power of 10. The number 2.998 lies between 1 and 10. To get 29,980,000,000, the decimal point in 2.998 must be moved ten places to the right. This is accomplished by multiplying 2.998 by 10^{10}.

Standard notation $\longrightarrow$ $29{,}980{,}000{,}000 = 2.998 \times 10^{10}$ $\longleftarrow$ Scientific notation

One meter is approximately 0.0006214 mile. To express this number in scientific notation, we must write it as the product of a number between 1 and 10 and some integer power of 10. The number 6.214 lies between 1 and 10. To get 0.0006214, the decimal point in 6.214 must be moved four places to the left. This is accomplished by multiplying 6.214 by $\frac{1}{10^4}$ or by multiplying 6.214 by 10^{-4}.

Standard notation $\longrightarrow$ $0.0006214 = 6214 \times 10^{-4}$ $\longleftarrow$ Scientific notation

To write each of the following numbers in scientific notation, we start to the right of the first nonzero digit and count to the decimal point. The exponent gives the number of places the decimal point moves, and the sign of the exponent indicates the direction in which it moves.

a. $3\,7\,2\,0\,0\,0 = 3.72 \times 10^5$ 5 places to the right.

b. $0\,.\,0\,0\,0\,5\,3\,7 = 5.37 \times 10^{-4}$ 4 places to the left.

c. $7.36 = 7.36 \times 10^0$ No movement of the decimal point.

EXAMPLE 7   **Writing Numbers in Scientific Notation**

Write each number in scientific notation: **a.** 62,000   **b.** −0.0027

SOLUTION   **a.** We must express 62,000 as a product of a number between 1 and 10 and some integer power of 10. This is accomplished by multiplying 6.2 by 10^4.

$$62{,}000 = 6.2 \times 10^4$$

b. We must express −0.0027 as a product of a number whose absolute value is between 1 and 10 and some integer power of 10. This is accomplished by multiplying −2.7 by 10^{-3}.

$$-0.0027 = -2.7 \times 10^{-3}$$

Self Check 7   Write each number in scientific notation:
a. −93,000,000   **b.** 0.0000087

Now Try Exercise 103.

EXAMPLE 8   **Writing Numbers in Standard Notation**

Write each number in standard notation:

a. 7.35×10^2 **b.** 3.27×10^{-5}

SOLUTION   **a.** The factor of 10^2 indicates that 7.35 must be multiplied by 2 factors of 10. Because each multiplication by 10 moves the decimal point one place to the right, we have

$$7.35 \times 10^2 = 735$$

b. The factor of 10^{-5} indicates that 3.27 must be divided by 5 factors of 10. Because each division by 10 moves the decimal point one place to the left, we have

$$3.27 \times 10^{-5} = 0.0000327$$

■

Self Check 8   Write each number in standard notation: **a.** 6.3×10^3 **b.** 9.1×10^{-4}

Now Try Exercise 111.

5. Use Scientific Notation to Simplify Computations

Another advantage of scientific notation becomes evident when we multiply and divide very large and very small numbers.

EXAMPLE 9   **Using Scientific Notation to Simplify Computations**

Use scientific notation to calculate $\dfrac{(3,400,000)(0.00002)}{170,000,000}$.

SOLUTION   After changing each number to scientific notation, we can do the arithmetic on the numbers and the exponential expressions separately.

$$\frac{(3,400,000)(0.00002)}{170,000,000} = \frac{(3.4 \times 10^6)(2.0 \times 10^{-5})}{1.7 \times 10^8}$$
$$= \frac{6.8}{1.7} \times 10^{6+(-5)-8}$$
$$= 4.0 \times 10^{-7}$$
$$= 0.0000004$$

■

Self Check 9   Use scientific notation to simplify $\dfrac{(192,000)(0.0015)}{(0.0032)(4,500)}$.

Now Try Exercise 119.

ACCENT ON TECHNOLOGY   **Scientific Notation**

Graphing calculators will often give an answer in scientific notation. For example, consider 21^8.

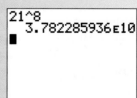

FIGURE 0-17

We see in the figure that the answer given is 3.782285936E10, which means $3.782285936 \times 10^{10}$.

Comment   To calculate an expression like $\dfrac{21^8}{0.000000000061}$ on a scientific calculator, it is necessary to convert the denominator to scientific notation because the number has too many digits to fit on the screen.

- For scientific notation, we enter these numbers and press these keys: 6.1 $\boxed{\text{EXP}}$ 11 $\boxed{+/-}$.

- To evaluate the expression above, we enter these numbers and press these keys:

    21 $\boxed{y^x}$ 8 $\boxed{=}$ $\boxed{\div}$ 6.1 $\boxed{\text{EXP}}$ 11 $\boxed{+/-}$ $\boxed{=}$

The display will read $\boxed{6.200468748^{20}}$. In standard notation, the answer is approximately 620,046,874,800,000,000,000.

Self Check Answers
1. a. $7 \cdot 7 \cdot 7 = 343$ b. $(-3)(-3) = 9$ c. $5 \cdot a \cdot a \cdot a$
d. $(5a)(5a)(5a)(5a) = 625 \cdot a \cdot a \cdot a \cdot a$ 2. a. y^6 b. a^{18} c. x^{12}
d. $\dfrac{27a^9 b^6}{c^9}$ 3. a. 7 b. $\dfrac{3}{a^2}$ c. $\dfrac{1}{a^2}$ d. $\dfrac{1}{a^{12}}$ 4. a. $\dfrac{1}{x^8}$ b. $\dfrac{1}{x}$
5. a. $\dfrac{x^{14}}{y^{10}}$ b. $\dfrac{27b^3}{8a^3}$ 6. $\dfrac{6}{5}$ 7. a. -9.3×10^7 b. 8.7×10^{-6}
8. a. $6,300$ b. 0.00091 9. 20

Exercises 0.2

Getting Ready
You should be able to complete these vocabulary and concept statements before you proceed to the practice exercises.

Fill in the blanks.

1. Each quantity in a product is called a _____ of the product.

2. A _____ number exponent tells how many times a base is used as a factor.

3. In the expression $(2x)^3$, ____ is the exponent and ____ is the base.

4. The expression x^n is called an _____ expression.

5. A number is in _____ notation when it is written in the form $N \times 10^n$, where $1 \le |N| < 10$ and n is an _____.

6. Unless _____ indicate otherwise, _____ are performed before additions.

Complete each exponent rule. Assume $x \ne 0$.

7. $x^m x^n =$ _____

8. $(x^m)^n =$ _____

9. $(xy)^n =$ _____

10. $\dfrac{x^m}{x^n} =$ _____

11. $x^0 =$ ___

12. $x^{-n} =$ _____

Practice
Write each number or expression without using exponents.

13. 13^2

14. 10^3

15. -5^2

16. $(-5)^2$

17. $4x^3$

18. $(4x)^3$

19. $(-5x)^4$

20. $-6x^2$

21. $-8x^4$

22. $(-8x)^4$

Write each expression using exponents.

23. $7xxx$

24. $-8yyyy$

25. $(-x)(-x)$

26. $(2a)(2a)(2a)$

27. $(3t)(3t)(-3t)$

28. $-(2b)(2b)(2b)(2b)$

29. $xxxyy$

30. $aaabbbb$

Use a calculator to simplify each expression.

31. 2.2^3

32. 7.1^4

33. -0.5^4

34. $(-0.2)^4$

Simplify each expression. Write all answers without using negative exponents. Assume that all variables are restricted to those numbers for which the expression is defined.

35. $x^2 x^3$

36. $y^3 y^4$

37. $(z^2)^3$

38. $(t^6)^7$

39. $(y^5 y^2)^3$

40. $(a^3 a^6)a^4$

41. $(z^2)^3(z^4)^5$

42. $(t^3)^4(t^5)^2$

43. $(a^2)^3(a^4)^2$

44. $(a^2)^4(a^3)^3$

45. $(3x)^3$

46. $(-2y)^4$

47. $(x^2 y)^3$

48. $(x^3 z^4)^6$

49. $\left(\dfrac{a^2}{b}\right)^3$

50. $\left(\dfrac{x}{y^3}\right)^4$

51. $(-x)^0$

52. $4x^0$

53. $(4x)^0$

54. $-2x^0$

55. z^{-4}

56. $\dfrac{1}{t^{-2}}$

57. $y^{-2}y^{-3}$

58. $-m^{-2}m^3$

59. $(x^3x^{-4})^{-2}$

60. $(y^{-2}y^3)^{-4}$

61. $\dfrac{x^7}{x^3}$

62. $\dfrac{r^5}{r^2}$

63. $\dfrac{a^{21}}{a^{17}}$

64. $\dfrac{t^{13}}{t^4}$

65. $\dfrac{(x^2)^2}{x^2x}$

66. $\dfrac{s^9s^3}{(s^2)^2}$

67. $\left(\dfrac{m^3}{n^2}\right)^3$

68. $\left(\dfrac{t^4}{t^3}\right)^3$

69. $\dfrac{(a^3)^{-2}}{aa^2}$

70. $\dfrac{r^9r^{-3}}{(r^{-2})^3}$

71. $\left(\dfrac{a^{-3}}{b^{-1}}\right)^{-4}$

72. $\left(\dfrac{t^{-4}}{t^{-3}}\right)^{-2}$

73. $\left(\dfrac{r^4r^{-6}}{r^3r^{-3}}\right)^2$

74. $\dfrac{(x^{-3}x^2)^2}{(x^2x^{-5})^{-3}}$

75. $\left(\dfrac{x^5y^{-2}}{x^{-3}y^2}\right)^4$

76. $\left(\dfrac{x^{-7}y^5}{x^7y^{-4}}\right)^3$

77. $\left(\dfrac{5x^{-3}y^{-2}}{3x^2y^{-3}}\right)^{-2}$

78. $\left(\dfrac{3x^2y^{-5}}{2x^{-2}y^{-6}}\right)^{-3}$

79. $\left(\dfrac{3x^5y^{-3}}{6x^{-5}y^3}\right)^{-2}$

80. $\left(\dfrac{12x^{-4}y^3z^{-5}}{4x^4y^{-3}z^5}\right)^3$

81. $\dfrac{(8^{-2}z^{-3}y)^{-1}}{(5y^2z^{-2})^3(5yz^{-2})^{-1}}$

82. $\dfrac{(m^{-2}n^3p^4)^{-2}(mn^{-2}p^3)^4}{(mn^{-2}p^3)^{-4}(mn^2p)^{-1}}$

Simplify each expression.

83. $-\dfrac{5[6^2+(9-5)]}{4(2-3)^2}$

84. $\dfrac{6[3-(4-7)^2]}{-5(2-4^2)}$

Let $x = -2$, $y = 0$, and $z = 3$ and evaluate each expression.

85. x^2

86. $-x^2$

87. x^3

88. $-x^3$

89. $(-xz)^3$

90. $-xz^3$

91. $\dfrac{-(x^2z^3)}{z^2-y^2}$

92. $\dfrac{z^2(x^2-y^2)}{x^3z}$

93. $5x^2-3y^3z$

94. $3(x-z)^2+2(y-z)^3$

95. $\dfrac{-3x^{-3}z^{-2}}{6x^2z^{-3}}$

96. $\dfrac{(-5x^2z^{-3})^2}{5xz^{-2}}$

Express each number in scientific notation.

97. 372,000

98. 89,500

99. $-177,000,000$

100. $-23,470,000,000$

101. 0.007

102. 0.00052

103. -0.000000693

104. -0.000000089

105. one trillion

106. one millionth

Express each number in standard notation.

107. 9.37×10^5

108. 4.26×10^9

109. 2.21×10^{-5}

110. 2.774×10^{-2}

111. 0.00032×10^4

112. $9,300.0 \times 10^{-4}$

113. -3.2×10^{-3}

114. -7.25×10^3

Use the method of Example 9 to do each calculation. Write all answers in scientific notation.

115. $\dfrac{(65,000)(45,000)}{250,000}$

116. $\dfrac{(0.000000045)(0.00000012)}{45,000,000}$

117. $\dfrac{(0.00000035)(170,000)}{0.00000085}$

118. $\dfrac{(0.0000000144)(12,000)}{600,000}$

119. $\dfrac{(45,000,000,000)(212,000)}{0.00018}$

120. $\dfrac{(0.00000000275)(4,750)}{500,000,000,000}$

Applications

Use scientific notation to compute each answer. Write all answers in scientific notation.

121. **Speed of sound** The speed of sound in air is 3.31×10^4 centimeters per second. Compute the speed of sound in meters per minute.

122. **Volume of a box** Calculate the volume of a box that has dimensions of 6,000 by 9,700 by 4,700 millimeters.

123. **Mass of a proton** The mass of one proton is 0.0000000000000000000000000167248 gram. Find the mass of one billion protons.

124. Speed of light The speed of light in a vacuum is approximately 30,000,000,000 centimeters per second. Find the speed of light in miles per hour. (160,934.4 cm = 1 mile.)

125. Astronomy The distance d, in miles, of the nth planet from the sun is given by the formula

$$d = 9{,}275{,}200\left[3(2^{n-2}) + 4\right]$$

To the nearest million miles, find the distance of Earth and the distance of Mars from the sun. Give each answer in scientific notation.

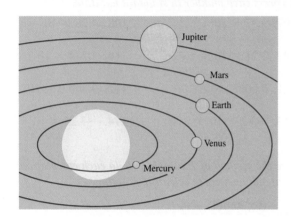

126. License Plates The number of different license plates of the form three digits followed by three letters, as in the illustration, is $10 \cdot 10 \cdot 10 \cdot 26 \cdot 26 \cdot 26$. Write this expression using exponents. Then evaluate it and express the result in scientific notation.

Discovery and Writing

127. New way to the center of the earth The spectacular "blue marble' image is the most detailed true-color image of the entire Earth to date. A new NASA-developed technique estimates Earth's center of

mass within 1 millimeter (0.04 inch) a year by using a combination of four space-based techniques.

The distance from the Earth's center to the North Pole (the **polar radius**) measures approximately 6,356.750 km, and the distance from the center to the equator (the **equatorial radius**) measures approximately 6,378.135 km. Express each distance using scientific notation.

128. Refer to Exercise 127. Given that 1 km is approximately equal to 0.62 miles, use scientific notation to express each distance in miles.

Write each expression with a single base.

129. $x^n x^2$

130. $\dfrac{x^m}{x^3}$

131. $\dfrac{x^m x^2}{x^3}$

132. $\dfrac{x^{3m+5}}{x^2}$

133. $x^{m+1} x^3$

134. $a^{n-3} a^3$

135. Explain why $-x^4$ and $(-x)^4$ represent different numbers.

136. Explain why 32×10^2 is not in scientific notation.

Review

137. Graph the interval $(-2, 4)$.

138. Graph the interval $(-\infty, -3] \cup [3, \infty)$.

139. Evaluate $|\pi - 5|$.

140. Find the distance between -7 and -5 on the number line.

0.3 Rational Exponents and Radicals

In this section, we will learn to

1. Define rational exponents whose numerators are 1.
2. Define rational exponents whose numerators are not 1.
3. Define radical expressions.
4. Simplify and combine radicals.
5. Rationalize denominators and numerators.

Medioimages/Photodisc/Jupiter Images

"Dead Man's Curve" is a 1964 hit song by the rock and roll duo Jan Berry and Dean Torrence. The song details a teenage drag race that ends in an accident. Today, dead man's curve is a commonly used expression given to dangerous curves on our roads. Every curve has a "critical speed." If we exceed this speed, regardless of how skilled a driver we are, we will lose control of the vehicle.

The radical expression $3.9\sqrt{r}$ gives the critical speed in miles per hour when we travel a curved road with a radius of r feet. A knowledge of square roots and radicals is important and used in the construction of safe highways and roads. We will study the topic of radicals in this section.

1. Define Rational Exponents Whose Numerators Are 1

If we apply the rule $(x^m)^n = x^{mn}$ to $(25^{1/2})^2$, we obtain

$$(25^{1/2})^2 = 25^{(1/2)2} \qquad \text{Keep the base and multiply the exponents.}$$

$$= 25^1 \qquad \tfrac{1}{2} \cdot 2 = 1$$

$$= 25$$

> **Caution**
>
> In the expression $a^{1/n}$, there is **no** real-number nth root of a when n is even and $a < 0$. For example, $(-64)^{1/2}$ is **not** a real number, because the square of **no** real number is -64.

Thus, $25^{1/2}$ is a real number whose square is 25. Although both $5^2 = 25$ and $(-5)^2 = 25$, we define $25^{1/2}$ to be the positive real number whose square is 25:

$$25^{1/2} = 5 \qquad \text{Read } 25^{1/2} \text{ as "the square root of 25."}$$

In general, we have the following definition.

Rational Exponents   If $a \geq 0$ and n is a natural number, then $a^{1/n}$ (read as "the nth root of a") is the nonnegative real number b such that

$$b^n = a$$

Since $b = a^{1/n}$, we have $b^n = (a^{1/n})^n = a$.

EXAMPLE 1   Simplifying Expressions with Rational Exponents

In each case, we will apply the definition of rational exponents.

a. $16^{1/2} = 4$ Because $4^2 = 16$. Read $16^{1/2}$ as "the square root of 16."

b. $27^{1/3} = 3$ Because $3^3 = 27$. Read $27^{1/3}$ as "the cube root of 27."

c. $\left(\dfrac{1}{81}\right)^{1/4} = \dfrac{1}{3}$ Because $\left(\dfrac{1}{3}\right)^4 = \dfrac{1}{81}$. Read $\left(\dfrac{1}{81}\right)^{1/4}$ as "the fourth root of $\dfrac{1}{81}$."

d. $-32^{1/5} = -(32^{1/5})$ Read $32^{1/5}$ as "the fifth root of 32."

$\phantom{-32^{1/5}} = -(2)$ Because $2^5 = 32$.

$\phantom{-32^{1/5}} = -2$

Self Check 1   Simplify: **a.** $100^{1/2}$ **b.** $243^{1/5}$

Now Try Exercise 15.

If n is even in the expression $a^{1/n}$ and the base contains variables, we often use absolute value symbols to guarantee that an even root is nonnegative.

$(49x^2)^{1/2} = 7|x|$ Because $(7|x|)^2 = 49x^2$. Since x could be negative, absolute value symbols are necessary to guarantee that the square root is nonnegative.

$(16x^4)^{1/4} = 2|x|$ Because $(2|x|)^4 = 16x^4$. Since x could be negative, absolute value symbols are necessary to guarantee that the fourth root is nonnegative.

$(729x^{12})^{1/6} = 3x^2$ Because $(3x^2)^6 = 729x^{12}$. Since x^2 is always nonnegative, no absolute value symbols are necessary. Read $(729x^{12})^{1/6}$ as "the sixth root of $729x^{12}$."

If n is an odd number in the expression $a^{1/n}$, the base a can be negative.

EXAMPLE 2   **Simplifying Expressions with Rational Exponents**

Simplify by using the definition of rational exponent.

a. $(-8)^{1/3} = -2$ Because $(-2)^3 = -8$.

b. $(-3,125)^{1/5} = -5$ Because $(-5)^5 = -3,125$.

c. $\left(-\dfrac{1}{1,000}\right)^{1/3} = -\dfrac{1}{10}$ Because $\left(-\dfrac{1}{10}\right)^3 = -\dfrac{1}{1,000}$.

Self Check 2   Simplify: **a.** $(-125)^{1/3}$ **b.** $(-100,000)^{1/5}$

Now Try Exercise 19.

Caution

If n is odd in the expression $a^{1/n}$, we **don't** need to use absolute value symbols, because odd roots can be negative.

$(-27x^3)^{1/3} = -3x$ Because $(-3x)^3 = -27x^3$.

$(-128a^7)^{1/7} = -2a$ Because $(-2a)^7 = -128a^7$.

We summarize the definitions concerning $a^{1/n}$ as follows.

Summary of Definitions of $a^{1/n}$

If n is a natural number and a is a real number in the expression $a^{1/n}$, then

If $a \geq 0$, then $a^{1/n}$ is the nonnegative real number b such that $b^n = a$.

If $a < 0$ $\begin{cases} \text{and } n \text{ is odd, then } a^{1/n} \text{ is the real number } b \text{ such that } b^n = a. \\ \text{and } n \text{ is even, then } a^{1/n} \text{ is not a real number.} \end{cases}$

The following chart also shows the possibilities that can occur when simplifying $a^{1/n}$.

Strategy for Simplifying Expressions of the Form $a^{1/n}$

a	n	$a^{1/n}$	Examples
$a = 0$	n is a natural number.	$0^{1/n}$ is the real number 0 because $0^n = 0$.	$0^{1/2} = 0$ because $0^2 = 0$. $0^{1/5} = 0$ because $0^5 = 0$.
$a > 0$	n is a natural number.	$a^{1/n}$ is the non-negative real number such that $(a^{1/n})^n = a$.	$16^{1/2} = 4$ because $4^2 = 16$. $27^{1/3} = 3$ because $3^3 = 27$.
$a < 0$	n is an odd natural number.	$a^{1/n}$ is the real number such that $(a^{1/n})^n = a$.	$(-32)^{1/5} = -2$ because $(-2)^5 = -32$. $(-125)^{1/3} = -5$ because $(-5)^3 = -125$.
$a < 0$	n is an even natural number.	$a^{1/n}$ is not a real number.	$(-9)^{1/2}$ is not a real number. $(-81)^{1/4}$ is not a real number.

2. Define Rational Exponents Whose Rational Exponents Are Not 1

The definition of $a^{1/n}$ can be extended to include rational exponents whose numerators are not 1. For example, $4^{3/2}$ can be written as either

$(4^{1/2})^3$ or $(4^3)^{1/2}$ Because of the Power Rule, $(x^m)^n = x^{mn}$.

This suggests the following rule.

Rule for Rational Exponents

If m and n are positive integers, the fraction $\frac{m}{n}$ is in lowest terms, and $a^{1/n}$ is a real number, then

$$a^{m/n} = (a^{1/n})^m = (a^m)^{1/n}$$

In the previous rule, we can view the expression $a^{m/n}$ in two ways:

1. $(a^{1/n})^m$: the mth power of the nth root of a
2. $(a^m)^{1/n}$: the nth root of the mth power of a

For example, $(-27)^{2/3}$ can be simplified in two ways:

$$(-27)^{2/3} = [(-27)^{1/3}]^2 \qquad \text{or} \qquad (-27)^{2/3} = [(-27)^2]^{1/3}$$

$$= (-3)^2 \qquad\qquad\qquad\qquad = (729)^{1/3}$$

$$= 9 \qquad\qquad\qquad\qquad\qquad = 9$$

As this example suggests, it is usually easier to take the root of the base first to avoid large numbers.

Comment

It is helpful to think of the phrase *power over root* when we see a rational exponent. The numerator of the fraction represents the *power* and the denominator represents the *root*. Begin with the *root* when simplifying to avoid large numbers.

Negative Rational Exponents

If m and n are positive integers, the fraction $\frac{m}{n}$ is in lowest terms and $a^{1/n}$ is a real number, then

$$a^{-m/n} = \frac{1}{a^{m/n}} \quad \text{and} \quad \frac{1}{a^{-m/n}} = a^{m/n} \qquad (a \neq 0)$$

EXAMPLE 3   **Simplifying Expressions with Rational Exponents**

We will apply the rules for rational exponents.

a. $25^{3/2} = (25^{1/2})^3$

$\qquad = 5^3$

$\qquad = 125$

b. $\left(-\dfrac{x^6}{1{,}000}\right)^{2/3} = \left[\left(-\dfrac{x^6}{1{,}000}\right)^{1/3}\right]^2$

$\qquad\qquad\qquad = \left(-\dfrac{x^2}{10}\right)^2$

$\qquad\qquad\qquad = \dfrac{x^4}{100}$

c. $32^{-2/5} = \dfrac{1}{32^{2/5}}$

$\qquad\quad = \dfrac{1}{(32^{1/5})^2}$

$\qquad\quad = \dfrac{1}{2^2}$

$\qquad\quad = \dfrac{1}{4}$

d. $\dfrac{1}{81^{-3/4}} = 81^{3/4}$

$\qquad\quad = (81^{1/4})^3$

$\qquad\quad = 3^3$

$\qquad\quad = 27$

Self Check 3   Simplify: **a.** $49^{3/2}$ **b.** $16^{-3/4}$ **c.** $\dfrac{1}{(27x^3)^{-2/3}}$

Now Try Exercise 43.

Because of the definition, rational exponents follow the same rules as integer exponents.

EXAMPLE 4   **Using Exponent Rules to Simplify Expressions with Rational Exponents**

Simplify each expression. Assume that all variables represent positive numbers, and write answers without using negative exponents.

a. $(36x)^{1/2} = 36^{1/2}x^{1/2}$

$\qquad\qquad = 6x^{1/2}$

b. $\dfrac{(a^{1/3}b^{2/3})^6}{(y^3)^2} = \dfrac{a^{6/3}b^{12/3}}{y^6}$

$\qquad\qquad\qquad = \dfrac{a^2b^4}{y^6}$

c. $\dfrac{a^{x/2}a^{x/4}}{a^{x/6}} = a^{x/2 + x/4 - x/6}$

$\qquad\qquad = a^{6x/12 + 3x/12 - 2x/12}$

$\qquad\qquad = a^{7x/12}$

d. $\left[\dfrac{-c^{-2/5}}{c^{4/5}}\right]^{5/3} = (-c^{-2/5 - 4/5})^{5/3}$

$\qquad\qquad = [(-1)(c^{-6/5})]^{5/3}$

$\qquad\qquad = (-1)^{5/3}(c^{-6/5})^{5/3}$

$\qquad\qquad = -1c^{-30/15}$

$\qquad\qquad = -c^{-2}$

$\qquad\qquad = -\dfrac{1}{c^2}$

Self Check 4 Use the directions for Example 4:

a. $\left(\dfrac{y^2}{49}\right)^{1/2}$ b. $\dfrac{b^{3/7}b^{2/7}}{b^{4/7}}$ c. $\dfrac{(9r^2s)^{1/2}}{rs^{-3/2}}$

Now Try Exercise 59.

3. Define Radical Expressions

Radical signs can also be used to express roots of numbers.

Definition of $\sqrt[n]{a}$ If n is a natural number greater than 1 and if $a^{1/n}$ is a real number, then

$$\sqrt[n]{a} = a^{1/n}$$

In the **radical expression** $\sqrt[n]{a}$, the symbol $\sqrt{}$ is the **radical sign**, a is the **radicand**, and n is the **index** (or the **order**) of the radical expression. If the order is 2, the expression is a **square root**, and we do not write the index.

$$\sqrt{a} = \sqrt[2]{a}$$

If the index of a radical is 3, we call the radical a **cube root**.

nth Root of a Nonnegative Number If n is a natural number greater than 1 and $a \geq 0$, then $\sqrt[n]{a}$ is the nonnegative number whose nth power is a.

$$\left(\sqrt[n]{a}\right)^n = a$$

Caution

In the expression $\sqrt[n]{a}$, there is **no** real-number nth root of a when n is even and $a < 0$. For example, $\sqrt{-64}$ is **not** a real number, because the square of **no** real number is -64.

If 2 is substituted for n in the equation $\left(\sqrt[n]{a}\right)^n = a$, we have

$$\left(\sqrt[2]{a}\right)^2 = \left(\sqrt{a}\right)^2 = \sqrt{a}\sqrt{a} = a \text{ for } a \geq 0$$

This shows that if a number a can be factored into two equal factors, either of those factors is a square root of a. Furthermore, if a can be factored into n equal factors, any one of those factors is an nth root of a.

If n is an odd number greater than 1 in the expression $\sqrt[n]{a}$, the radicand can be negative.

EXAMPLE 5 **Finding *n*th Roots of Real Numbers**

We apply the definitions of cube root and fifth root.

a. $\sqrt[3]{-27} = -3$ — Because $(-3)^3 = -27$.

b. $\sqrt[3]{-8} = -2$ — Because $(-2)^3 = -8$.

c. $\sqrt[3]{-\dfrac{27}{1,000}} = -\dfrac{3}{10}$ — Because $\left(-\dfrac{3}{10}\right)^3 = -\dfrac{27}{1,000}$.

d. $-\sqrt[5]{-243} = -\left(\sqrt[5]{-243}\right)$
$$= -(-3)$$
$$= 3$$

Self Check 5 Find each root: **a.** $\sqrt[3]{216}$ **b.** $\sqrt[5]{-\dfrac{1}{32}}$

Now Try Exercise 69.

We summarize the definitions concerning $\sqrt[n]{a}$ as follows.

Summary of Definitions of $\sqrt[n]{a}$ If *n* is a natural number greater than 1 and *a* is a real number, then

If $a \geq 0$, then $\sqrt[n]{a}$ is the nonnegative real number such that $\left(\sqrt[n]{a}\right)^n = a$.

If $a < 0$ $\begin{cases} \text{and } n \text{ is odd, then } \sqrt[n]{a} \text{ is the real number such that } \left(\sqrt[n]{a}\right)^n = a. \\ \text{and } n \text{ is even, then } \sqrt[n]{a} \text{ is not a real number.} \end{cases}$

The following chart also shows the possibilities that can occur when simplifying $\sqrt[n]{a}$.

Strategy for Simplifying Expressions of the Form $\sqrt[n]{a}$

a	n	$\sqrt[n]{a}$	Examples
$a = 0$	*n* is a natural number greater than 1.	$\sqrt[n]{0}$ is the real number 0 because $0^n = 0$.	$\sqrt[3]{0} = 0$ because $0^3 = 0$. $\sqrt[5]{0} = 0$ because $0^5 = 0$.
$a > 0$	*n* is a natural number greater than 1.	$\sqrt[n]{a}$ is the non-negative real number such that $\left(\sqrt[n]{a}\right)^n = a$.	$\sqrt{16} = 4$ because $4^2 = 16$. $\sqrt[3]{27} = 3$ because $3^3 = 27$.
$a < 0$	*n* is an odd natural number greater than 1.	$\sqrt[n]{a}$ is the real number such that $\left(\sqrt[n]{a}\right)^n = a$.	$\sqrt[5]{-32} = -2$ because $(-2)^5 = -32$. $\sqrt[3]{-125} = -5$ because $(-5)^3 = -125$.
$a < 0$	*n* is an even natural number.	$\sqrt[n]{a}$ is not a real number.	$\sqrt{-9}$ is not a real number. $\sqrt[4]{-81}$ is not a real number.

We have seen that if $a^{1/n}$ is real, then $a^{m/n} = (a^{1/n})^m = (a^m)^{1/n}$.
This same fact can be stated in radical notation.

$$a^{m/n} = (\sqrt[n]{a})^m = \sqrt[n]{a^m}$$

Thus, *the mth power of the nth root of a* is the same as *the nth root of the mth power of a*. For example, to find $\sqrt[3]{27^2}$, we can proceed in either of two ways:

$$\sqrt[3]{27^2} = \left(\sqrt[3]{27}\right)^2 = 3^2 = 9 \text{ or } \sqrt[3]{27^2} = \sqrt[3]{729} = 9$$

By definition, $\sqrt{a^2}$ represents a nonnegative number. If a could be negative, we must use absolute value symbols to guarantee that $\sqrt{a^2}$ will be nonnegative. Thus, if a is unrestricted,

$$\sqrt{a^2} = |a|$$

A similar argument holds when the index is any even natural number. The symbol $\sqrt[4]{a^4}$, for example, means the *positive* fourth root of a^4. Thus, if a is unrestricted,

$$\sqrt[4]{a^4} = |a|$$

EXAMPLE 6   **Simplifying Radical Expressions**

If x is unrestricted, simplify **a.** $\sqrt[6]{64x^6}$ **b.** $\sqrt[3]{x^3}$ **c.** $\sqrt{9x^8}$

SOLUTION   We apply the definitions of sixth roots, cube roots, and square roots.

a. $\sqrt[6]{64x^6} = 2|x|$ Use absolute value symbols to guarantee that the result will be nonnegative.

b. $\sqrt[3]{x^3} = x$ Because the index is odd, no absolute value symbols are needed.

c. $\sqrt{9x^8} = 3x^4$ Because $3x^4$ is always nonnegative, no absolute value symbols are needed. ■

Self Check 6   Use the directions for Example 6:

a. $\sqrt[4]{16x^4}$ **b.** $\sqrt[3]{27y^3}$ **c.** $\sqrt[4]{x^8}$

Now Try Exercise 73.

---

4. Simplify and Combine Radicals

Many properties of exponents have counterparts in radical notation. For example, since $a^{1/n}b^{1/n} = (ab)^{1/n}$ and $\frac{a^{1/n}}{b^{1/n}} = \left(\frac{a}{b}\right)^{1/n}$ and $(b \neq 0)$, we have the following.

Multiplication and Division Properties of Radicals   If all expressions represent real numbers,

$$\sqrt[n]{a}\sqrt[n]{b} = \sqrt[n]{ab} \qquad\qquad \frac{\sqrt[n]{a}}{\sqrt[n]{b}} = \sqrt[n]{\frac{a}{b}} \qquad (b \neq 0)$$

In words, we say

The product of two nth roots is equal to the nth root of their product.

The quotient of two nth roots is equal to the nth root of their quotient.

Caution

These properties involve the nth root of the product of two numbers or the nth root of the quotient of two numbers. There is **no** such property for sums or differences. For example,

$$\sqrt{9+4} \neq \sqrt{9} + \sqrt{4}, \text{ because}$$

$$\sqrt{9+4} = \sqrt{13} \quad \text{but} \quad \sqrt{9} + \sqrt{4} = 3 + 2 = 5$$

and $\sqrt{13} \neq 5$. In general,

$$\cancel{\sqrt{a+b} = \sqrt{a} + \sqrt{b}} \quad \text{and} \quad \cancel{\sqrt{a-b} = \sqrt{a} - \sqrt{b}}$$

Numbers that are squares of positive integers, such as

1, 4, 9, 16, 25, and 36

are called **perfect squares**. Expressions such as $4x^2$ and $\frac{1}{9}x^6$ are also perfect squares, because each one is the square of another expression with integer exponents and rational coefficients.

$$4x^2 = (2x)^2 \text{ and } \frac{1}{9}x^6 = \left(\frac{1}{3}x^3\right)^2$$

Numbers that are cubes of positive integers, such as

1, 8, 27, 64, 125, and 216

are called **perfect cubes**. Expressions such as $64x^3$ and $\frac{1}{27}x^9$ are also perfect cubes, because each one is the cube of another expression with integer exponents and rational coefficients.

$$64x^3 = (4x)^3 \text{ and } \frac{1}{27}x^9 = \left(\frac{1}{3}x^3\right)^3$$

There are also perfect fourth powers, perfect fifth powers, and so on.

We can use perfect powers and the Multiplication Property of Radicals to simplify many radical expressions. For example, to simplify $\sqrt{12x^5}$, we factor $12x^5$ so that one factor is the largest perfect square that divides $12x^5$. In this case, it is $4x^4$. We then rewrite $12x^5$ as $4x^4 \cdot 3x$ and simplify.

$$\sqrt{12x^5} = \sqrt{4x^4 \cdot 3x} \quad \text{Factor } 12x^5 \text{ as } 4x^4 \cdot 3x.$$

$$= \sqrt{4x^4}\sqrt{3x} \quad \text{Use the Multiplication Property of Radicals: } \sqrt{ab} = \sqrt{a}\sqrt{b}.$$

$$= 2x^2\sqrt{3x} \quad \sqrt{4x^4} = 2x^2$$

To simplify $\sqrt[3]{432x^9y}$, we find the largest perfect-cube factor of $432x^9y$ (which is $216x^9$) and proceed as follows:

$$\sqrt[3]{432x^9y} = \sqrt[3]{216x^9 \cdot 2y} \quad \text{Factor } 432x^9y \text{ as } 216x^9 \cdot 2y.$$

$$= \sqrt[3]{216x^9}\sqrt[3]{2y} \quad \text{Use the Multiplication Property of Radicals: } \sqrt[3]{ab} = \sqrt[3]{a}\sqrt[3]{b}.$$

$$= 6x^3\sqrt[3]{2y} \quad \sqrt[3]{216x^9} = 6x^3$$

Radical expressions with the same index and the same radicand are called **like** or **similar radicals**. We can combine the like radicals in $3\sqrt{2} + 2\sqrt{2}$ by using the Distributive Property.

$$3\sqrt{2} + 2\sqrt{2} = (3+2)\sqrt{2}$$

$$= 5\sqrt{2}$$

This example suggests that to combine like radicals, we *add their numerical coefficients and keep the same radical.*

When radicals have the same index but different radicands, we can often change them to equivalent forms having the same radicand. We can then combine them. For example, to simplify $\sqrt{27} - \sqrt{12}$, we simplify both radicals and combine like radicals.

$$\sqrt{27} - \sqrt{12} = \sqrt{9 \cdot 3} - \sqrt{4 \cdot 3} \qquad \text{Factor 27 and 12.}$$
$$= \sqrt{9}\sqrt{3} - \sqrt{4}\sqrt{3} \qquad \sqrt{ab} = \sqrt{a}\sqrt{b}$$
$$= 3\sqrt{3} - 2\sqrt{3} \qquad \sqrt{9} = 3 \text{ and } \sqrt{4} = 2.$$
$$= \sqrt{3} \qquad \text{Combine like radicals.}$$

EXAMPLE 7   **Adding and Subtracting Radical Expressions**

Simplify: **a.** $\sqrt{50} + \sqrt{200}$ **b.** $3z\sqrt[5]{64z} - 2\sqrt[5]{2z^6}$

SOLUTION   We will simplify each radical expression and then combine like radicals.

a. $\sqrt{50} + \sqrt{200} = \sqrt{25 \cdot 2} + \sqrt{100 \cdot 2}$
$$= \sqrt{25}\sqrt{2} + \sqrt{100}\sqrt{2}$$
$$= 5\sqrt{2} + 10\sqrt{2}$$
$$= 15\sqrt{2}$$

b. $3z\sqrt[5]{64z} - 2\sqrt[5]{2z^6} = 3z\sqrt[5]{32 \cdot 2z} - 2\sqrt[5]{z^5 \cdot 2z}$
$$= 3z\sqrt[5]{32}\,\sqrt[5]{2z} - 2\sqrt[5]{z^5}\sqrt[5]{2z}$$
$$= 3z(2)\sqrt[5]{2z} - 2z\sqrt[5]{2z}$$
$$= 6z\sqrt[5]{2z} - 2z\sqrt[5]{2z}$$
$$= 4z\sqrt[5]{2z}$$

Self Check 7   Simplify: **a.** $\sqrt{18} - \sqrt{8}$ **b.** $2\sqrt[3]{81a^4} + a\sqrt[3]{24a}$

Now Try Exercise 85.

5. Rationalize Denominators and Numerators

By **rationalizing the denominator**, we can write a fraction such as

$$\frac{\sqrt{5}}{\sqrt{3}}$$

as a fraction with a rational number in the denominator. All that we must do is multiply both the numerator and the denominator by $\sqrt{3}$. (Note that $\sqrt{3}\sqrt{3}$ is the rational number 3.)

$$\frac{\sqrt{5}}{\sqrt{3}} = \frac{\sqrt{5}\sqrt{3}}{\sqrt{3}\sqrt{3}} = \frac{\sqrt{15}}{3}$$

To rationalize the numerator, we multiply both the numerator and the denominator by $\sqrt{5}$. (Note that $\sqrt{5}\sqrt{5}$ is the rational number 5.)

$$\frac{\sqrt{5}}{\sqrt{3}} = \frac{\sqrt{5}\sqrt{5}}{\sqrt{3}\sqrt{5}} = \frac{5}{\sqrt{15}}$$

EXAMPLE 8 **Rationalizing the Denominator of a Radical Expression**

Rationalize each denominator and simplify. Assume that all variables represent positive numbers.

a. $\dfrac{1}{\sqrt{7}}$ **b.** $\sqrt[3]{\dfrac{3}{4}}$ **c.** $\sqrt{\dfrac{3}{x}}$ **d.** $\sqrt{\dfrac{3a^3}{5x^5}}$

SOLUTION We will multiply both the numerator and the denominator by a radical that will make the denominator a rational number.

a. $\dfrac{1}{\sqrt{7}} = \dfrac{1\sqrt{7}}{\sqrt{7}\sqrt{7}}$

$= \dfrac{\sqrt{7}}{7}$

b. $\sqrt[3]{\dfrac{3}{4}} = \dfrac{\sqrt[3]{3}}{\sqrt[3]{4}}$

$= \dfrac{\sqrt[3]{3}\sqrt[3]{2}}{\sqrt[3]{4}\sqrt[3]{2}}$

Multiply numerator and denominator by $\sqrt[3]{2}$, because $\sqrt[3]{4}\sqrt[3]{2} = \sqrt[3]{8} = 2$.

$= \dfrac{\sqrt[3]{6}}{\sqrt[3]{8}}$

$= \dfrac{\sqrt[3]{6}}{2}$

c. $\sqrt{\dfrac{3}{x}} = \dfrac{\sqrt{3}}{\sqrt{x}}$

$= \dfrac{\sqrt{3}\sqrt{x}}{\sqrt{x}\sqrt{x}}$

$= \dfrac{\sqrt{3x}}{x}$

d. $\sqrt{\dfrac{3a^3}{5x^5}} = \dfrac{\sqrt{3a^3}}{\sqrt{5x^5}}$

$= \dfrac{\sqrt{3a^3}\sqrt{5x}}{\sqrt{5x^5}\sqrt{5x}}$

Multiply numerator and denominator by $\sqrt{5x}$, because $\sqrt{5x^5}\sqrt{5x} = \sqrt{25x^6} = 5x^3$.

$= \dfrac{\sqrt{15a^3x}}{\sqrt{25x^6}}$

$= \dfrac{\sqrt{a^2}\sqrt{15ax}}{5x^3}$

$= \dfrac{a\sqrt{15ax}}{5x^3}$

■

Self Check 8 Use the directions for Example 8:

a. $\dfrac{6}{\sqrt{6}}$ **b.** $\sqrt[3]{\dfrac{2}{5x}}$

Now Try Exercise 101.

EXAMPLE 9 **Rationalizing the Numerator of a Radical Expression**

Rationalize each numerator and simplify. Assume that all variables represent positive numbers: **a.** $\dfrac{\sqrt{x}}{7}$ **b.** $\dfrac{2\sqrt[3]{9x}}{3}$

SOLUTION We will multiply both the numerator and the denominator by a radical that will make the numerator a rational number.

a. $\dfrac{\sqrt{x}}{7} = \dfrac{\sqrt{x} \cdot \sqrt{x}}{7\sqrt{x}}$ **b.** $\dfrac{2\sqrt[3]{9x}}{3} = \dfrac{2\sqrt[3]{9x} \cdot \sqrt[3]{3x^2}}{3\sqrt[3]{3x^2}}$

$\qquad\quad = \dfrac{x}{7\sqrt{x}}$ $\qquad\qquad\quad = \dfrac{2\sqrt[3]{27x^3}}{3\sqrt[3]{3x^2}}$

$\qquad\qquad\qquad\qquad\qquad\quad = \dfrac{2(3x)}{3\sqrt[3]{3x^2}}$

$\qquad\qquad\qquad\qquad\qquad\quad = \dfrac{2x}{\sqrt[3]{3x^2}}$ Divide out the 3's.

Self Check 9   Use the directions for Example 9:

a. $\dfrac{\sqrt{2x}}{5}$ **b.** $\dfrac{3\sqrt[3]{2y^2}}{6}$

Now Try Exercise 111.

After rationalizing denominators, we often can simplify an expression.

EXAMPLE 10   **Rationalizing Denominators and Simplifying**

Simplify: $\sqrt{\dfrac{1}{2}} + \sqrt{\dfrac{1}{8}}$.

SOLUTION   We will rationalize the denominators of each radical and then combine like radicals.

$$\sqrt{\dfrac{1}{2}} + \sqrt{\dfrac{1}{8}} = \dfrac{1}{\sqrt{2}} + \dfrac{1}{\sqrt{8}} \qquad\qquad \sqrt{\dfrac{1}{2}} = \dfrac{\sqrt{1}}{\sqrt{2}} = \dfrac{1}{\sqrt{2}}; \sqrt{\dfrac{1}{8}} = \dfrac{\sqrt{1}}{\sqrt{8}} = \dfrac{1}{\sqrt{8}}$$

$$= \dfrac{1\sqrt{2}}{\sqrt{2}\sqrt{2}} + \dfrac{1\sqrt{2}}{\sqrt{8}\sqrt{2}}$$

$$= \dfrac{\sqrt{2}}{2} + \dfrac{\sqrt{2}}{\sqrt{16}}$$

$$= \dfrac{\sqrt{2}}{2} + \dfrac{\sqrt{2}}{4}$$

$$= \dfrac{3\sqrt{2}}{4}$$

Self Check 10   Simplify: $\sqrt[3]{\dfrac{x}{2}} - \sqrt[3]{\dfrac{x}{16}}$.

Now Try Exercise 115.

Another property of radicals can be derived from the properties of exponents. If all of the expressions represent real numbers,

$$\sqrt[n]{\sqrt[m]{x}} = \sqrt[n]{x^{1/m}} = (x^{1/m})^{1/n} = x^{1/(mn)} = \sqrt[mn]{x}$$

$$\sqrt[m]{\sqrt[n]{x}} = \sqrt[m]{x^{1/n}} = (x^{1/n})^{1/m} = x^{1/(nm)} = \sqrt[mn]{x}$$

These results are summarized in the following *theorem* (a fact that can be proved).

Theorem   If all of the expressions involved represent real numbers, then

$$\sqrt[m]{\sqrt[n]{x}} = \sqrt[n]{\sqrt[m]{x}} = \sqrt[mn]{x}$$

We can use the previous theorem to simplify many radicals. For example,

$$\sqrt[3]{\sqrt{8}} = \sqrt{\sqrt[3]{8}} = \sqrt{2}$$

Rational exponents can be used to simplify many radical expressions, as shown in the following example.

EXAMPLE 11 **Simplifying Radicals Using the Previously Stated Theorem**

Simplify. Assume that x and y are positive numbers.

a. $\sqrt[6]{4}$ **b.** $\sqrt[12]{x^3}$ **c.** $\sqrt[9]{8y^3}$

SOLUTION   In each case, we will write the radical as an exponential expression, simplify the resulting expression, and write the final result as a radical.

a. $\sqrt[6]{4} = 4^{1/6} = (2^2)^{1/6} = 2^{2/6} = 2^{1/3} = \sqrt[3]{2}$

b. $\sqrt[12]{x^3} = x^{3/12} = x^{1/4} = \sqrt[4]{x}$

c. $\sqrt[9]{8y^3} = (2^3 y^3)^{1/9} = (2y)^{3/9} = (2y)^{1/3} = \sqrt[3]{2y}$

Self Check 11   Simplify: **a.** $\sqrt[4]{4}$ **b.** $\sqrt[9]{27x^3}$

Now Try Exercise 117.

Self Check Answers   **1. a.** 10   **b.** 3   **2. a.** -5 **b.** -10 **3. a.** 343   **b.** $\dfrac{1}{8}$

c. $9x^2$ **4. a.** $\dfrac{y}{7}$ **b.** $b^{1/7}$ **c.** $3s^2$ **5. a.** 6   **b.** $-\dfrac{1}{2}$ **6. a.** $2|x|$

b. $3y$ **c.** x^2 **7. a.** $\sqrt{2}$ **b.** $8a\sqrt[3]{3a}$ **8. a.** $\sqrt{6}$ **b.** $\dfrac{\sqrt[3]{50x^2}}{5x}$

9. a. $\dfrac{2x}{5\sqrt{2x}}$ **b.** $\dfrac{y}{\sqrt[3]{4y}}$ **10.** $\dfrac{\sqrt[3]{4x}}{4}$ **11. a.** $\sqrt{2}$ **b.** $\sqrt[3]{3x}$

Exercises **0.3**

Getting Ready
You should be able to complete these vocabulary and concept statements before you proceed to the practice exercises.

Fill in the blanks.

1. If $a = 0$ and n is a natural number, then $a^{1/n} =$ ___ .

2. If $a > 0$ and n is a natural number, then $a^{1/n}$ is a _____ number.

3. If $a < 0$ and n is an even number, then $a^{1/n}$ is _____ a real number.

4. $6^{2/3}$ can be written as _____ or _____ .

5. $\sqrt[n]{a} =$ _____

6. $\sqrt[n]{a^2} =$ _____

7. $\sqrt[n]{a}\sqrt[n]{b} =$ _____

8. $\sqrt[n]{\dfrac{a}{b}} =$ _____

9. $\sqrt{x + y}$ _____ $\sqrt{x} + \sqrt{y}$

10. $\sqrt[m]{\sqrt[n]{x}}$ or $\sqrt[n]{\sqrt[m]{x}}$ can be written as _____ .

Practice
Simplify each expression.

11. $9^{1/2}$

12. $8^{1/3}$

13. $\left(\dfrac{1}{25}\right)^{1/2}$

14. $\left(\dfrac{16}{625}\right)^{1/4}$

15. $-81^{1/4}$

16. $-\left(\dfrac{8}{27}\right)^{1/3}$

17. $(10,000)^{1/4}$

18. $1,024^{1/5}$

19. $\left(-\dfrac{27}{8}\right)^{1/3}$

20. $-64^{1/3}$

21. $(-64)^{1/2}$

22. $(-125)^{1/3}$

Simplify each expression. Use absolute value symbols when necessary.

23. $(16a^2)^{1/2}$

24. $(25a^4)^{1/2}$

25. $(16a^4)^{1/4}$

26. $(-64a^3)^{1/3}$

27. $(-32a^5)^{1/5}$

28. $(64a^6)^{1/6}$

29. $(-216b^6)^{1/3}$

30. $(256t^8)^{1/4}$

31. $\left(\dfrac{16a^4}{25b^2}\right)^{1/2}$

32. $\left(-\dfrac{a^5}{32b^{10}}\right)^{1/5}$

33. $\left(-\dfrac{1,000x^6}{27y^3}\right)^{1/3}$

34. $\left(\dfrac{49t^2}{100z^4}\right)^{1/2}$

Simplify each expression. Write all answers without using negative exponents.

35. $4^{3/2}$

36. $8^{2/3}$

37. $-16^{3/2}$

38. $(-8)^{2/3}$

39. $-1,000^{2/3}$

40. $100^{3/2}$

41. $64^{-1/2}$

42. $25^{-1/2}$

43. $64^{-3/2}$

44. $49^{-3/2}$

45. $-9^{-3/2}$

46. $(-27)^{-2/3}$

47. $\left(\dfrac{4}{9}\right)^{5/2}$

48. $\left(\dfrac{25}{81}\right)^{3/2}$

49. $\left(-\dfrac{27}{64}\right)^{-2/3}$

50. $\left(\dfrac{125}{8}\right)^{-4/3}$

Simplify each expression. Assume that all variables represent positive numbers. Write all answers without using negative exponents.

51. $(100s^4)^{1/2}$

52. $(64u^6v^3)^{1/3}$

53. $(32y^{10}z^5)^{-1/5}$

54. $(625a^4b^8)^{-1/4}$

55. $(x^{10}y^5)^{3/5}$

56. $(64a^6b^{12})^{5/6}$

57. $(r^8s^{16})^{-3/4}$

58. $(-8x^9y^{12})^{-2/3}$

59. $\left(-\dfrac{8a^6}{125b^9}\right)^{2/3}$

60. $\left(\dfrac{16x^4}{625y^8}\right)^{3/4}$

61. $\left(\dfrac{27r^6}{1,000s^{12}}\right)^{-2/3}$

62. $\left(-\dfrac{32m^{10}}{243n^{15}}\right)^{-2/5}$

63. $\dfrac{a^{2/5}a^{4/5}}{a^{1/5}}$

64. $\dfrac{x^{6/7}x^{3/7}}{x^{2/7}x^{5/7}}$

Simplify each radical expression.

65. $\sqrt{49}$

66. $\sqrt{81}$

67. $\sqrt[3]{125}$

68. $\sqrt[3]{-64}$

69. $\sqrt[3]{-125}$

70. $\sqrt[5]{-243}$

71. $\sqrt[5]{-\dfrac{32}{100,000}}$

72. $\sqrt[4]{\dfrac{256}{625}}$

Simplify each expression, using absolute value symbols when necessary. Write answers without using negative exponents.

73. $\sqrt{36x^2}$

74. $-\sqrt{25y^2}$

75. $\sqrt{9y^4}$

76. $\sqrt{a^4b^8}$

77. $\sqrt[3]{8y^3}$

78. $\sqrt[3]{-27z^9}$

79. $\sqrt[4]{\dfrac{x^4y^8}{z^{12}}}$

80. $\sqrt[5]{\dfrac{a^{10}b^5}{c^{15}}}$

Simplify each expression. Assume that all variables represent positive numbers so that no absolute value symbols are needed.

81. $\sqrt{8} - \sqrt{2}$

82. $\sqrt{75} - 2\sqrt{27}$

83. $\sqrt{200x^2} + \sqrt{98x^2}$

84. $\sqrt{128a^3} - a\sqrt{162a}$

85. $2\sqrt{48y^5} - 3y\sqrt{12y^3}$

86. $y\sqrt{112y} + 4\sqrt{175y^3}$

87. $2\sqrt[3]{81} + 3\sqrt[3]{24}$

88. $3\sqrt[4]{32} - 2\sqrt[4]{162}$

89. $\sqrt[4]{768z^5} + \sqrt[4]{48z^5}$

90. $-2\sqrt[5]{64y^2} + 3\sqrt[5]{486y^2}$

91. $\sqrt{8x^2y} - x\sqrt{2y} + \sqrt{50x^2y}$

92. $3x\sqrt{18x} + 2\sqrt{2x^3} - \sqrt{72x^3}$

93. $\sqrt[3]{16xy^4} + y\sqrt[3]{2xy} - \sqrt[3]{54xy^4}$

94. $\sqrt[4]{512x^5} - \sqrt[4]{32x^5} + \sqrt[4]{1,250x^5}$

Rationalize each denominator and simplify. Assume that all variables represent positive numbers.

95. $\dfrac{3}{\sqrt{3}}$

96. $\dfrac{6}{\sqrt{5}}$

97. $\dfrac{2}{\sqrt{x}}$

98. $\dfrac{8}{\sqrt{y}}$

99. $\dfrac{2}{\sqrt[3]{2}}$

100. $\dfrac{4d}{\sqrt[3]{9}}$

101. $\dfrac{5a}{\sqrt[3]{25a}}$

102. $\dfrac{7}{\sqrt[3]{36c}}$

103. $\dfrac{2b}{\sqrt[4]{3a^2}}$

104. $\sqrt{\dfrac{x}{2y}}$

105. $\sqrt[3]{\dfrac{2u^4}{9v}}$

106. $\sqrt[3]{-\dfrac{3s^5}{4r^2}}$

Rationalize each numerator and simplify. Assume that all variables are positive numbers.

107. $\dfrac{\sqrt{5}}{10}$

108. $\dfrac{\sqrt{y}}{3}$

109. $\dfrac{\sqrt[3]{9}}{3}$

110. $\dfrac{\sqrt[3]{16b^2}}{16}$

111. $\dfrac{\sqrt[5]{16b^3}}{64a}$

112. $\sqrt{\dfrac{3x}{57}}$

Rationalize each denominator and simplify.

113. $\sqrt{\dfrac{1}{3}} - \sqrt{\dfrac{1}{27}}$

114. $\sqrt[3]{\dfrac{1}{2}} + \sqrt[3]{\dfrac{1}{16}}$

115. $\sqrt{\dfrac{x}{8}} - \sqrt{\dfrac{x}{2}} + \sqrt{\dfrac{x}{32}}$

116. $\sqrt[3]{\dfrac{y}{4}} + \sqrt[3]{\dfrac{y}{32}} - \sqrt[3]{\dfrac{y}{500}}$

Simplify each radical expression.

117. $\sqrt[4]{9}$

118. $\sqrt[6]{27}$

119. $\sqrt[10]{16x^6}$

120. $\sqrt[6]{27x^9}$

Discovery and Writing

We often can multiply and divide radicals with different indices. For example, to multiply $\sqrt{3}$ by $\sqrt[3]{5}$, we first write each radical as a sixth root

$$\sqrt{3} = 3^{1/2} = 3^{3/6} = \sqrt[6]{3^3} = \sqrt[6]{27}$$

$$\sqrt[3]{5} = 5^{1/3} = 5^{2/6} = \sqrt[6]{5^2} = \sqrt[6]{25}$$

and then multiply the sixth roots.

$$\sqrt{3}\sqrt[3]{5} = \sqrt[6]{27}\sqrt[6]{25} = \sqrt[6]{(27)(25)} = \sqrt[6]{675}$$

Division is similar.

Use this idea to write each of the following expressions as a single radical.

121. $\sqrt{2}\sqrt[3]{2}$

122. $\sqrt{3}\sqrt[3]{5}$

123. $\dfrac{\sqrt[4]{3}}{\sqrt{2}}$

124. $\dfrac{\sqrt[3]{2}}{\sqrt{5}}$

125. For what values of x does $\sqrt[4]{x^4} = x$? Explain.

126. If all of the radicals involved represent real numbers and $y \neq 0$, explain why

$$\sqrt[n]{\dfrac{x}{y}} = \dfrac{\sqrt[n]{x}}{\sqrt[n]{y}}$$

127. If all of the radicals involved represent real numbers and there is no division by 0, explain why

$$\left(\dfrac{x}{y}\right)^{-m/n} = \sqrt[n]{\dfrac{y^m}{x^m}}$$

128. The definition of $x^{m/n}$ requires that $\sqrt[n]{x}$ be a real number. Explain why this is important. (Hint: Consider what happens when n is even, m is odd, and x is negative.)

Review

129. Write $-2 < x \leq 5$ using interval notation.

130. Write the expression $|3 - x|$ without using absolute value symbols. Assume that $x > 4$.

Evaluate each expression when $x = -2$ and $y = 3$.

131. $x^2 - y^2$

132. $\dfrac{xy + 4y}{x}$

133. Write 617,000,000 in scientific notation.

134. Write 0.00235×10^4 in standard notation.

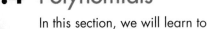

0.4 Polynomials

In this section, we will learn to

1. Define polynomials.
2. Add and subtract polynomials.
3. Multiply polynomials.
4. Rationalize denominators.
5. Divide polynomials.

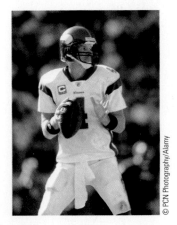

© PCN Photography/Alamy

Football is one of the most popular sports in the United States. Brett Favre, one of the most talented quarterbacks ever to play the game, holds the record for the most career NFL touchdown passes, the most NFL pass completions, and the most passing yards. He led the Green Bay Packers to the Super Bowl and most recently played for the Minnesota Vikings. His nickname is Gunslinger.

An algebraic expression can be used to model the trajectory or path of a football when passed by Brett Favre. Suppose the height in feet, t seconds after the football leaves Brett's hand, is given by the algebraic expression

$$-0.1t^2 + t + 5.5$$

At a time of $t = 3$ seconds, we see that the height of the football is

$$-0.1(3)^2 + 3 + 5.5 = 7.6 \text{ ft.}$$

Algebraic expressions like $-0.1t^2 + t + 5.5$ are called **polynomials,** and we will study them in this section.

1. Define Polynomials

A **monomial** is a real number or the product of a real number and one or more variables with whole-number exponents. The number is called the **coefficient** of the variables. Some examples of monomials are

$$3x, \qquad 7ab^2, \qquad -5ab^2c^4, \qquad x^3, \qquad \text{and} \qquad -12$$

with coefficients of 3, 7, -5, 1, and -12, respectively.

The **degree** of a monomial is the sum of the exponents of its variables. All nonzero constants (except 0) have a degree of 0.

The degree of $3x$ is 1.   The degree of $7ab^2$ is 3.

The degree of $-5ab^2c^4$ is 7.   The degree of x^3 is 3.

The degree of -12 is 0 (since $-12 = -12x^0$).   0 has no defined degree.

A monomial or a sum of monomials is called a **polynomial**. Each monomial in that sum is called a **term** of the polynomial. A polynomial with two terms is called a **binomial**, and a polynomial with three terms is called a **trinomial**.

Monomials	Binomials	Trinomials
$3x^2$	$2a + 3b$	$x^2 + 7x - 4$
$-25xy$	$4x^3 - 3x^2$	$4y^4 - 2y + 12$
a^2b^3c	$-2x^3 - 4y^2$	$12x^3y^2 - 8xy - 24$

The **degree of a polynomial** is the degree of the term in the polynomial with highest degree. The only polynomial with no defined degree is 0, which is called the **zero polynomial**. Here are some examples.

- $3x^2y^3 + 5xy^2 + 7$ is a trinomial of 5th degree, because its term with highest degree (the first term) is 5.
- $3ab + 5a^2b$ is a binomial of degree 3.
- $5x + 3y^2 + \sqrt[4]{3}z^4 - \sqrt{7}$ is a polynomial, because its variables have whole-number exponents. It is of degree 4.
- $-7y^{1/2} + 3y^2 + \sqrt[5]{3}z$ is not a polynomial, because one of its variables (y in the first term) does not have a whole-number exponent.

If two terms of a polynomial have the same variables with the same exponents, they are **like** or **similar** terms. To combine the like terms in the sum $3x^2y + 5x^2y$ or the difference $7xy^2 - 2xy^2$, we use the Distributive Property:

$$3x^2y + 5x^2y = (3 + 5)x^2y \qquad\qquad 7xy^2 - 2xy^2 = (7 - 2)xy^2$$
$$= 8x^2y \qquad\qquad\qquad\qquad = 5xy^2$$

This illustrates that *to combine like terms, we add (or subtract) their coefficients and keep the same variables and the same exponents.*

2. Add and Subtract Polynomials

Recall that we can use the Distributive Property to remove parentheses enclosing the terms of a polynomial. When the sign preceding the parentheses is +, we simply drop the parentheses:

$$+(a + b - c) = +1(a + b - c)$$
$$= 1a + 1b - 1c$$
$$= a + b - c$$

When the sign preceding the parentheses is −, we drop the parentheses and the − sign and change the sign of each term within the parentheses.

$$-(a + b - c) = -1(a + b - c)$$
$$= -1a + (-1)b - (-1)c$$
$$= -a - b + c$$

We can use these facts to add and subtract polynomials. *To add (or subtract) polynomials, we remove parentheses (if necessary) and combine like terms.*

EXAMPLE 1   **Adding Polynomials**

Add: $(3x^3y + 5x^2 - 2y) + (2x^3y - 5x^2 + 3x)$.

SOLUTION   To add the polynomials, we remove parentheses and combine like terms.

$$(3x^3y + 5x^2 - 2y) + (2x^3y - 5x^2 + 3x)$$
$$= 3x^3y + 5x^2 - 2y + 2x^3y - 5x^2 + 3x$$
$$= 3x^3y + 2x^3y + 5x^2 - 5x^2 - 2y + 3x \qquad \text{Use the Commutative Property to rearrange terms.}$$
$$= 5x^3y - 2y + 3x \qquad \text{Combine like terms.}$$

We can add the polynomials in a vertical format by writing like terms in a column and adding the like terms, column by column.

$$
\begin{array}{l}
3x^3y + 5x^2 - 2y \\
\underline{2x^3y - 5x^2 \quad\;\; + 3x} \\
5x^3y \quad\quad\; - 2y + 3x
\end{array}
$$

■

Self Check 1   Add: $(4x^2 + 3x - 5) + (3x^2 - 5x + 7)$.

Now Try Exercise 21.

EXAMPLE 2   **Subtracting Polynomials**

Subtract: $(2x^2 + 3y^2) - (x^2 - 2y^2 + 7)$.

SOLUTION   To subtract the polynomials, we remove parentheses and combine like terms.

$$(2x^2 + 3y^2) - (x^2 - 2y^2 + 7)$$
$$= 2x^2 + 3y^2 - x^2 + 2y^2 - 7$$
$$= 2x^2 - x^2 + 3y^2 + 2y^2 - 7 \qquad \text{Use the Commutative Property to rearrange terms.}$$
$$= x^2 + 5y^2 - 7 \qquad \text{Combine like terms.}$$

We can subtract the polynomials in a vertical format by writing like terms in a column and subtracting the like terms, column by column.

$$2x^2 + 3y^2 \qquad\qquad 2x^2 - x^2 = (2 - 1)x^2$$
$$\underline{-(x^2 - 2y^2 + 7)} \qquad 3y^2 - (-2)y^2 = 3y^2 + 2y^2 = (3 + 2)y^2$$
$$x^2 + 5y^2 - 7 \qquad\qquad 0 - 7 = -7$$

■

Self Check 2   Subtract: $(4x^2 + 3x - 5) - (3x^2 - 5x + 7)$.

Now Try Exercise 23.

We can also use the Distributive Property to remove parentheses enclosing several terms that are multiplied by a constant. For example,

$$4(3x^2 - 2x + 6) = 4(3x^2) - 4(2x) + 4(6)$$
$$= 12x^2 - 8x + 24$$

This example suggests that *to add multiples of one polynomial to another, or to subtract multiples of one polynomial from another, we remove parentheses and combine like terms.*

EXAMPLE 3   **Using the Distributive Property and Combining Like Terms**

Simplify: $7x(2y^2 + 13x^2) - 5(xy^2 - 13x^3)$.

SOLUTION $7x(2y^2 + 13x^2) - 5(xy^2 - 13x^3)$

$$= 14xy^2 + 91x^3 - 5xy^2 + 65x^3 \qquad \text{Use the Distributive Property to remove parentheses.}$$
$$= 14xy^2 - 5xy^2 + 91x^3 + 65x^3 \qquad \text{Use the Commutative Property to rearrange terms.}$$
$$= 9xy^2 + 156x^3 \qquad\qquad\qquad\qquad \text{Combine like terms.}$$

■

Self Check 3   Simplify: $3(2b^2 - 3a^2b) + 2b(b + a^2)$.

Now Try Exercise 30.

3. Multiply Polynomials

To find the product of $3x^2y^3z$ and $5xyz^2$, we proceed as follows:

$$(3x^2y^3z)(5xyz^2) = 3 \cdot x^2 \cdot y^3 \cdot z \cdot 5 \cdot x \cdot y \cdot z^2$$
$$= 3 \cdot 5 \cdot x^2 \cdot x \cdot y^3 \cdot y \cdot z \cdot z^2 \qquad \text{Use the Commutative Property to rearrange terms.}$$
$$= 15x^3y^4z^3$$

This illustrates that *to multiply two monomials, we multiply the coefficients and then multiply the variables.*

To find the product of a monomial and a polynomial, we use the Distributive Property.

$$3xy^2(2xy + x^2 - 7yz) = 3xy^2(2xy) + (3xy^2)(x^2) - (3xy^2)(7yz)$$
$$= 6x^2y^3 + 3x^3y^2 - 21xy^3z$$

This illustrates that *to multiply a polynomial by a monomial, we multiply each term of the polynomial by the monomial.*

To multiply one binomial by another, we use the Distributive Property twice.

EXAMPLE 4   **Multiplying Binomials**

Multiply: **a.** $(x + y)(x + y)$ **b.** $(x - y)(x - y)$ **c.** $(x + y)(x - y)$

SOLUTION   **a.** $(x + y)(x + y) = (x + y)x + (x + y)y$

$$= x^2 + xy + xy + y^2$$

$$= x^2 + 2xy + y^2$$

b. $(x - y)(x - y) = (x - y)x - (x - y)y$

$$= x^2 - xy - xy + y^2$$

$$= x^2 - 2xy + y^2$$

c. $(x + y)(x - y) = (x + y)x - (x + y)y$

$$= x^2 + xy - xy - y^2$$

$$= x^2 - y^2$$

■

Self Check 4   Multiply: **a.** $(x + 2)(x + 2)$ **b.** $(x - 3)(x - 3)$
c. $(x + 4)(x - 4)$

Now Try Exercise 45.

The products in Example 4 are called **special products**. Because they occur so often, it is worthwhile to learn their forms.

Special Product Formulas $(x + y)^2 = (x + y)(x + y) = x^2 + 2xy + y^2$

$(x - y)^2 = (x - y)(x - y) = x^2 - 2xy + y^2$

$(x + y)(x - y) = x^2 - y^2$

Caution

Remember that $(x + y)^2$ and $(x - y)^2$ have trinomials for their products and that

$$(x + y)^2 \neq x^2 + y^2 \qquad \text{and} \qquad (x - y)^2 \neq x^2 - y^2$$

For example,

$$(3 + 5)^2 \neq 3^2 + 5^2 \qquad \text{and} \qquad (3 - 5)^2 \neq 3^2 - 5^2$$

$$8^2 \neq 9 + 25 \qquad \qquad (-2)^2 \neq 9 - 25$$

$$64 \neq 34 \qquad \qquad 4 \neq -16$$

We can use the **FOIL method** to multiply one binomial by another. FOIL is an acronym for **F**irst terms, **O**uter terms, **I**nner terms, and **L**ast terms. To use this method to multiply $3x - 4$ by $2x + 5$, we write

First terms   Last terms

$$(3x - 4)(2x + 5) = 3x(2x) + 3x(5) - 4(2x) - 4(5)$$

Inner terms

$$= 6x^2 + 15x - 8x - 20$$

Outer terms

$$= 6x^2 + 7x - 20$$

In this example,

- the product of the first terms is $6x^2$,
- the product of the outer terms is $15x$,
- the product of the inner terms is $-8x$, and
- the product of the last terms is -20.

The resulting like terms of the product are then combined.

EXAMPLE 5   **Using the FOIL Method to Multiply Polynomials**

Use the FOIL method to multiply: $\left(\sqrt{3} + x\right)\left(2 - \sqrt{3}x\right)$.

SOLUTION $(\sqrt{3} + x)(2 - \sqrt{3}x) = 2\sqrt{3} - \sqrt{3}\sqrt{3}x + 2x - x\sqrt{3}x$

$$= 2\sqrt{3} - 3x + 2x - \sqrt{3}x^2$$

$$= 2\sqrt{3} - x - \sqrt{3}x^2$$

Self Check 5   Multiply: $\left(2x + \sqrt{3}\right)\left(x - \sqrt{3}\right)$.

Now Try Exercise 63.

To multiply a polynomial with more than two terms by another polynomial, we multiply each term of one polynomial by each term of the other polynomial and combine like terms whenever possible.

EXAMPLE 6   **Multiplying Polynomials**

Multiply: **a.** $(x + y)(x^2 - xy + y^2)$ **b.** $(x + 3)^3$

SOLUTION   **a.** $(x + y)(x^2 - xy + y^2) = x^3 - x^2y + xy^2 + yx^2 - xy^2 + y^3$

$$= x^3 + y^3$$

b. $(x + 3)^3 = (x + 3)(x + 3)^2$

$$= (x + 3)(x^2 + 6x + 9)$$

$$= x^3 + 6x^2 + 9x + 3x^2 + 18x + 27$$

$$= x^3 + 9x^2 + 27x + 27$$

Self Check 6   Multiply: $(x + 2)(2x^2 + 3x - 1)$.

Now Try Exercise 67.

We can use a vertical format to multiply two polynomials, such as the polynomials given in Self Check 6. We first write the polynomials as follows and draw a line beneath them. We then multiply each term of the upper polynomial by each term of the lower polynomial and write the results so that like terms appear in each column. Finally, we combine like terms column by column.

$$2x^2 + 3x - 1$$
$$\underline{\qquad\qquad x + 2}$$
$$4x^2 + 6x - 2 \qquad \text{Multiply } 2x^2 + 3x - 1 \text{ by 2.}$$
$$\underline{2x^3 + 3x^2 - \ x \qquad\quad \text{Multiply } 2x^2 + 3x - 1 \text{ by } x.}$$
$$2x^3 + 7x^2 + 5x - 2 \qquad \text{In each column, combine like terms.}$$

If n is a whole number, the expressions $a^n + 1$ and $2a^n - 3$ are polynomials and we can multiply them as follows:

$$(a^n + 1)(2a^n - 3) = 2a^{2n} - 3a^n + 2a^n - 3$$
$$= 2a^{2n} - a^n - 3 \qquad \text{Combine like terms.}$$

We can also use the methods previously discussed to multiply expressions that are not polynomials, such as $x^{-2} + y$ and $x^2 - y^{-1}$.

$$(x^{-2} + y)(x^2 - y^{-1}) = x^{-2+2} - x^{-2}y^{-1} + x^2y - y^{1-1}$$
$$= x^0 - \frac{1}{x^2y} + x^2y - y^0$$
$$= 1 - \frac{1}{x^2y} + x^2y - 1 \qquad x^0 = 1 \text{ and } y^0 = 1.$$
$$= x^2y - \frac{1}{x^2y}$$

4. Rationalize Denominators

If the denominator of a fraction is a binomial containing square roots, we can use the product formula $(x + y)(x - y)$ to rationalize the denominator. For example, to rationalize the denominator of

$$\frac{6}{\sqrt{7} + 2} \qquad \begin{array}{l} \text{To rationalize a denominator means to change} \\ \text{the denominator into a rational number.} \end{array}$$

we multiply the numerator and denominator by $\sqrt{7} - 2$ and simplify.

$$\frac{6}{\sqrt{7} + 2} = \frac{6(\sqrt{7} - 2)}{(\sqrt{7} + 2)(\sqrt{7} - 2)} \qquad \frac{\sqrt{7} - 2}{\sqrt{7} - 2} = 1$$
$$= \frac{6(\sqrt{7} - 2)}{7 - 4}$$
$$= \frac{6(\sqrt{7} - 2)}{3} \qquad \text{Here the denominator is a rational number.}$$
$$= 2(\sqrt{7} - 2)$$

In this example, we multiplied both the numerator and the denominator of the given fraction by $\sqrt{7} - 2$. This binomial is the same as the denominator of the given fraction $\sqrt{7} + 2$, except for the sign between the terms. Such binomials are called **conjugate binomials** or **radical conjugates**.

| Conjugate Binomials | **Conjugate binomials** are binomials that are the same except for the sign between their terms. The conjugate of $a + b$ is $a - b$, and the conjugate of $a - b$ is $a + b$. |

EXAMPLE 7   **Rationalizing the Denominator of a Radical Expression**

Rationalize the denominator: $\dfrac{\sqrt{3x} - \sqrt{2}}{\sqrt{3x} + \sqrt{2}}$ $(x > 0)$.

SOLUTION   We multiply the numerator and the denominator by $\sqrt{3x} - \sqrt{2}$ (the conjugate of $\sqrt{3x} + \sqrt{2}$) and simplify.

$$\frac{\sqrt{3x} - \sqrt{2}}{\sqrt{3x} + \sqrt{2}} = \frac{\left(\sqrt{3x} - \sqrt{2}\right)\left(\sqrt{3x} - \sqrt{2}\right)}{\left(\sqrt{3x} + \sqrt{2}\right)\left(\sqrt{3x} - \sqrt{2}\right)} \qquad \frac{\sqrt{3x} - \sqrt{2}}{\sqrt{3x} - \sqrt{2}} = 1$$

$$= \frac{\sqrt{3x}\sqrt{3x} - \sqrt{3x}\sqrt{2} - \sqrt{2}\sqrt{3x} + \sqrt{2}\sqrt{2}}{\left(\sqrt{3x}\right)^2 - \left(\sqrt{2}\right)^2}$$

$$= \frac{3x - \sqrt{6x} - \sqrt{6x} + 2}{3x - 2}$$

$$= \frac{3x - 2\sqrt{6x} + 2}{3x - 2}$$

Self Check 7   Rationalize the denominator: $\dfrac{\sqrt{x} + 2}{\sqrt{x} - 2}$.

Now Try Exercise 89.

In calculus, we often rationalize a numerator.

EXAMPLE 8   **Rationalizing the Numerator of a Radical Expression**

Rationalize the numerator: $\dfrac{\sqrt{x + h} - \sqrt{x}}{h}$.

SOLUTION   To rid the numerator of radicals, we multiply the numerator and the denominator by the conjugate of the numerator and simplify.

$$\frac{\sqrt{x + h} - \sqrt{x}}{h} = \frac{\left(\sqrt{x + h} - \sqrt{x}\right)\left(\sqrt{x + h} + \sqrt{x}\right)}{h\left(\sqrt{x + h} + \sqrt{x}\right)} \qquad \frac{\sqrt{x + h} + \sqrt{x}}{\sqrt{x + h} + \sqrt{x}} = 1$$

$$= \frac{x + h - x}{h\left(\sqrt{x + h} + \sqrt{x}\right)} \qquad \text{Here the numerator has no radicals.}$$

$$= \frac{h}{h\left(\sqrt{x + h} + \sqrt{x}\right)}$$

$$= \frac{1}{\sqrt{x + h} + \sqrt{x}} \qquad \text{Divide out the common factor of } h.$$

Self Check 8   Rationalize the numerator: $\dfrac{\sqrt{4 + h} - 2}{h}$.

Now Try Exercise 99.

5. Divide Polynomials

To divide monomials, we write the quotient as a fraction and simplify by using the rules of exponents. For example,

$$\frac{6x^2y^3}{-2x^3y} = -3x^{2-3}y^{3-1}$$

$$= -3x^{-1}y^2$$

$$= -\frac{3y^2}{x}$$

To divide a polynomial by a monomial, we write the quotient as a fraction, write the fraction as a sum of separate fractions, and simplify each one. For example, to divide $8x^5y^4 + 12x^2y^5 - 16x^2y^3$ by $4x^3y^4$, we proceed as follows:

$$\frac{8x^5y^4 + 12x^2y^5 - 16x^2y^3}{4x^3y^4} = \frac{8x^5y^4}{4x^3y^4} + \frac{12x^2y^5}{4x^3y^4} + \frac{-16x^2y^3}{4x^3y^4}$$

$$= 2x^2 + \frac{3y}{x} - \frac{4}{xy}$$

To divide two polynomials, we can use long division. To illustrate, we consider the division

$$\frac{2x^2 + 11x - 30}{x + 7}$$

which can be written in long division form as

$$x + 7\overline{)2x^2 + 11x - 30}$$

The binomial $x + 7$ is called the **divisor**, and the trinomial $2x^2 + 11x - 30$ is called the **dividend**. The final answer, called the **quotient**, will appear above the long division symbol.

We begin the division by asking "What monomial, when multiplied by x, gives $2x^2$?" Because $x \cdot 2x = 2x^2$, the answer is $2x$. We place $2x$ in the quotient, multiply each term of the divisor by $2x$, subtract, and bring down the -30.

$$\begin{array}{r} 2x \\ x + 7\overline{)2x^2 + 11x - 30} \\ \underline{2x^2 + 14x} \\ -\ 3x - 30 \end{array}$$

We continue the division by asking "What monomial, when multiplied by x, gives $-3x$?" We place the answer, -3, in the quotient, multiply each term of the divisor by -3, and subtract. This time, there is no number to bring down.

$$\begin{array}{r} 2x -\ \ 3 \\ x + 7\overline{)2x^2 + 11x - 30} \\ \underline{2x^2 + 14x} \\ -\ 3x - 30 \\ \underline{-\ 3x - 21} \\ -9 \end{array}$$

Because the degree of the remainder, -9, is less than the degree of the divisor, the division process stops, and we can express the result in the form

$$\text{quotient} + \frac{\text{remainder}}{\text{divisor}}$$

Thus,

$$\frac{2x^2 + 11x - 30}{x + 7} = 2x - 3 + \frac{-9}{x + 7}$$

EXAMPLE 9   **Using Long Division to Divide Polynomials**

Divide $6x^3 - 11$ by $2x + 2$.

SOLUTION   We set up the division, leaving spaces for the missing powers of x in the dividend.

Comment

In Example 9, we could write the missing powers of x using coefficients of 0.

$$2x + 2 \overline{)6x^3 + 0x^2 + 0x - 11}$$

$$2x + 2 \overline{)6x^3 \qquad\quad - 11}$$

The division process continues as usual, with the following results:

$$
\begin{array}{r}
3x^2 - 3x + 3 \\
2x + 2 \overline{)6x^3 \qquad\qquad\quad - 11} \\
\underline{6x^3 + 6x^2} \\
-6x^2 \\
\underline{-6x^2 - 6x} \\
+6x - 11 \\
\underline{+6x + 6} \\
-17
\end{array}
$$

Thus, $\dfrac{6x^3 - 11}{2x + 2} = 3x^2 - 3x + 3 + \dfrac{-17}{2x + 2}$.

Self Check 9   Divide: $3x + 1 \overline{)9x^2 - 1}$.

Now Try Exercise 113.

EXAMPLE 10   **Using Long Division to Divide Polynomials**

Divide $-3x^3 - 3 + x^5 + 4x^2 - x^4$ by $x^2 - 3$.

SOLUTION   The division process works best when the terms in the divisor and dividend are written with their exponents in descending order.

$$
\begin{array}{r}
x^3 - x^2 + 1 \\
x^2 - 3 \overline{)x^5 - x^4 - 3x^3 + 4x^2 \qquad - 3} \\
\underline{x^5 \qquad\quad - 3x^3} \\
-x^4 \qquad\qquad + 4x^2 \\
\underline{-x^4 \qquad\qquad + 3x^2} \\
x^2 - 3 \\
\underline{x^2 - 3} \\
0
\end{array}
$$

Thus, $\dfrac{-3x^3 - 3 + x^5 + 4x^2 - x^4}{x^2 - 3} = x^3 - x^2 + 1$.

Self Check 10   Divide: $x^2 + 1 \overline{)3x^2 - x + 1 - 2x^3 + 3x^4}$.

Now Try Exercise 115.

Self Check Answers
1. $7x^2 - 2x + 2$ 2. $x^2 + 8x - 12$ 3. $8b^2 - 7a^2b$
4. a. $x^2 + 4x + 4$ b. $x^2 - 6x + 9$ c. $x^2 - 16$ 5. $2x^2 - x\sqrt{3} - 3$
6. $2x^3 + 7x^2 + 5x - 2$ 7. $\dfrac{x + 4\sqrt{x} + 4}{x - 4}$ 8. $\dfrac{1}{\sqrt{4 + h} + 2}$
9. $3x - 1$ 10. $3x^2 - 2x + \dfrac{x + 1}{x^2 + 1}$

Exercises **0.4**

Getting Ready

You should be able to complete these vocabulary and concept statements before you proceed to the practice exercises.

Fill in the blanks.

1. A _____ is a real number or the product of a real number and one or more _____.
2. The _____ of a monomial is the sum of the exponents of its _____.
3. A _____ is a polynomial with three terms.
4. A _____ is a polynomial with two terms.
5. A monomial is a polynomial with _____ term.
6. The constant 0 is called the _____ polynomial.
7. Terms with the same variables with the same exponents are called _____ terms.
8. The _____ of a polynomial is the same as the degree of its term of highest degree.
9. To combine like terms, we add their _____ and keep the same _____ and the same exponents.
10. The conjugate of $3\sqrt{x} + 2$ is _____.

Determine whether the given expression is a polynomial. If so, tell whether it is a monomial, a binomial, or a trinomial, and give its degree.

11. $x^2 + 3x + 4$
12. $5xy - x^3$
13. $x^3 + y^{1/2}$
14. $x^{-3} - 5y^{-2}$
15. $4x^2 - \sqrt{5x^3}$
16. x^2y^3
17. $\sqrt{15}$
18. $\dfrac{5}{x} + \dfrac{x}{5} + 5$
19. 0
20. $3y^3 - 4y^2 + 2y + 2$

Practice

Perform the operations and simplify.

21. $(x^3 - 3x^2) + (5x^3 - 8x)$
22. $(2x^4 - 5x^3) + (7x^3 - x^4 + 2x)$
23. $(y^5 + 2y^3 + 7) - (y^5 - 2y^3 - 7)$
24. $(3t^7 - 7t^3 + 3) - (7t^7 - 3t^3 + 7)$
25. $2(x^2 + 3x - 1) - 3(x^2 + 2x - 4) + 4$
26. $5(x^3 - 8x + 3) + 2(3x^2 + 5x) - 7$

27. $8(t^2 - 2t + 5) + 4(t^2 - 3t + 2) - 6(2t^2 - 8)$

28. $-3(x^3 - x) + 2(x^2 + x) + 3(x^3 - 2x)$
29. $y(y^2 - 1) - y^2(y + 2) - y(2y - 2)$
30. $-4a^2(a + 1) + 3a(a^2 - 4) - a^2(a + 2)$

31. $xy(x - 4y) - y(x^2 + 3xy) + xy(2x + 3y)$

32. $3mn(m + 2n) - 6m(3mn + 1) - 2n(4mn - 1)$

33. $2x^2y^3(4xy^4)$
34. $-15a^3b(-2a^2b^3)$

35. $-3m^2n(2mn^2)\left(-\dfrac{mn}{12}\right)$

36. $-\dfrac{3r^2s^3}{5}\left(\dfrac{2r^2s}{3}\right)\left(\dfrac{15rs^2}{2}\right)$

37. $-4rs(r^2 + s^2)$
38. $6u^2v(2uv^2 - y)$
39. $6ab^2c(2ac + 3bc^2 - 4ab^2c)$

40. $-\dfrac{mn^2}{2}(4mn - 6m^2 - 8)$

41. $(a + 2)(a + 2)$ 42. $(y - 5)(y - 5)$

43. $(a - 6)^2$ 44. $(t + 9)^2$

45. $(x + 4)(x - 4)$ 46. $(z + 7)(z - 7)$

47. $(x - 3)(x + 5)$

48. $(z + 4)(z - 6)$

49. $(u + 2)(3u - 2)$

50. $(4x + 1)(2x - 3)$

51. $(5x - 1)(2x + 3)$

52. $(4x - 1)(2x - 7)$

53. $(3a - 2b)^2$

54. $(4a + 5b)(4a - 5b)$

55. $(3m + 4n)(3m - 4n)$

56. $(4r + 3s)^2$

57. $(2y - 4x)(3y - 2x)$

58. $(-2x + 3y)(3x + y)$

59. $(9x - y)(x^2 - 3y)$

60. $(8a^2 + b)(a + 2b)$

61. $(5z + 2t)(z^2 - t)$

62. $(y - 2x^2)(x^2 + 3y)$

63. $\left(\sqrt{5} + 3x\right)\left(2 - \sqrt{5x}\right)$

64. $\left(\sqrt{2} + x\right)\left(3 + \sqrt{2x}\right)$

65. $(3x - 1)^3$

66. $(2x - 3)^3$

67. $(3x + 1)(2x^2 + 4x - 3)$

68. $(2x - 5)(x^2 - 3x + 2)$

69. $(3x + 2y)(2x^2 - 3xy + 4y^2)$

70. $(4r - 3s)(2r^2 + 4rs - 2s^2)$

Multiply the expressions as you would multiply polynomials.

71. $2y^n(3y^n + y^{-n})$

72. $3a^{-n}(2a^n + 3a^{n-1})$

73. $-5x^{2n}y^n(2x^{2n}y^{-n} + 3x^{-2n}y^n)$

74. $-2a^{3n}b^{2n}(5a^{-3n}b - ab^{-2n})$

75. $(x^n + 3)(x^n - 4)$

76. $(a^n - 5)(a^n - 3)$

77. $(2r^n - 7)(3r^n - 2)$

78. $(4z^n + 3)(3z^n + 1)$

79. $x^{1/2}(x^{1/2}y + xy^{1/2})$

80. $ab^{1/2}(a^{1/2}b^{1/2} + b^{1/2})$

81. $(a^{1/2} + b^{1/2})(a^{1/2} - b^{1/2})$

82. $(x^{3/2} + y^{1/2})^2$

Rationalize each denominator.

83. $\dfrac{2}{\sqrt{3} - 1}$

84. $\dfrac{1}{\sqrt{5} + 2}$

85. $\dfrac{3x}{\sqrt{7} + 2}$

86. $\dfrac{14y}{\sqrt{2} - 3}$

87. $\dfrac{x}{x - \sqrt{3}}$

88. $\dfrac{y}{2y + \sqrt{7}}$

89. $\dfrac{y + \sqrt{2}}{y - \sqrt{2}}$

90. $\dfrac{x - \sqrt{3}}{x + \sqrt{3}}$

91. $\dfrac{\sqrt{2} - \sqrt{3}}{1 - \sqrt{3}}$

92. $\dfrac{\sqrt{3} - \sqrt{2}}{1 + \sqrt{2}}$

93. $\dfrac{\sqrt{x} - \sqrt{y}}{\sqrt{x} + \sqrt{y}}$

94. $\dfrac{\sqrt{2x} + y}{\sqrt{2x} - y}$

Rationalize each numerator.

95. $\dfrac{\sqrt{2} + 1}{2}$

96. $\dfrac{\sqrt{x} - 3}{3}$

97. $\dfrac{y - \sqrt{3}}{y + \sqrt{3}}$

98. $\dfrac{\sqrt{a} - \sqrt{b}}{\sqrt{a} + \sqrt{b}}$

99. $\dfrac{\sqrt{x + 3} - \sqrt{x}}{3}$

100. $\dfrac{\sqrt{2 + h} - \sqrt{2}}{h}$

Perform each division and write all answers without using negative exponents.

101. $\dfrac{36a^2b^3}{18ab^6}$

102. $\dfrac{-45r^2s^5t^3}{27r^6s^2t^8}$

103. $\dfrac{16x^6y^4z^9}{-24x^9y^6z^0}$

104. $\dfrac{32m^6n^4p^2}{26m^6n^7p^2}$

105. $\dfrac{5x^3y^2 + 15x^3y^4}{10x^2y^3}$

106. $\dfrac{9m^4n^9 - 6m^3n^4}{12m^3n^3}$

107. $\dfrac{24x^5y^7 - 36x^2y^5 + 12xy}{60x^5y^4}$

108. $\dfrac{9a^3b^4 + 27a^2b^4 - 18a^2b^3}{18a^2b^7}$

Perform each division. If there is a nonzero remainder, write the answer in quotient $+ \frac{\text{remainder}}{\text{divisor}}$ form.

109. $x + 3\overline{)3x^2 + 11x + 6}$

110. $3x + 2\overline{)3x^2 + 11x + 6}$

111. $2x - 5\overline{)2x^2 - 19x + 37}$

112. $x - 7\overline{)2x^2 - 19x + 35}$

113. $\dfrac{2x^3 + 1}{x - 1}$

114. $\dfrac{2x^3 - 9x^2 + 13x - 20}{2x - 7}$

115. $x^2 + x - 1\overline{)x^3 - 2x^2 - 4x + 3}$

116. $x^2 - 3\overline{)x^3 - 2x^2 - 4x + 5}$

117. $\dfrac{x^5 - 2x^3 - 3x^2 + 9}{x^3 - 2}$

118. $\dfrac{x^5 - 2x^3 - 3x^2 + 9}{x^3 - 3}$

119. $\dfrac{x^5 - 32}{x - 2}$

120. $\dfrac{x^4 - 1}{x + 1}$

121. $11x - 10 + 6x^2 \overline{)36x^4 - 121x^2 + 120 + 72x^3 - 142x}$

122. $x + 6x^2 - 12 \overline{)-121x^2 + 72x^3 - 142x + 120 + 36x^4}$

Applications

123. Geometry Find an expression that represents the area of the brick wall.

$(x - 2)$ ft

$(x + 5)$ ft

124. Geometry The area of the triangle shown in the illustration is represented as $(x^2 + 3x - 40)$ square feet. Find an expression that represents its height.

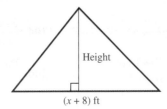

Height

$(x + 8)$ ft

125. Gift Boxes The corners of a 12 in.-by-12 in. piece of cardboard are folded inward and glued to make a box. Write a polynomial that represents the volume of the resulting box.

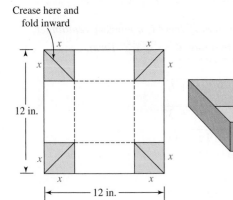

Crease here and fold inward

x

12 in.

12 in.

126. Travel Complete the following table, which shows the rate (mph), time traveled (hr), and distance traveled (mi) by a family on vacation.

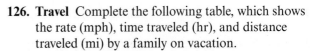

r	$\cdot$	t	$=$	d
$3x + 4$				$3x^2 + 19x + 20$

Discovery and Writing

127. Show that a trinomial can be squared by using the formula
$(a + b + c)^2 = a^2 + b^2 + c^2 + 2ab + 2bc + 2ac.$

128. Show that $(a + b + c + d)^2 = a^2 + b^2 + c^2 + d^2 + 2ab + 2ac + 2ad + 2bc + 2bd + 2cd.$

129. Explain the FOIL method.

130. Explain how to rationalize the numerator of $\dfrac{\sqrt{x} + 2}{x}$.

131. Explain why $(a + b)^2 \neq a^2 + b^2$.

132. Explain why $\sqrt{a^2 + b^2} \neq \sqrt{a^2} + \sqrt{b^2}$.

Review

Simplify each expression. Assume that all variables represent positive numbers.

133. $9^{3/2}$

134. $\left(\dfrac{8}{125}\right)^{-2/3}$

135. $\left(\dfrac{625x^4}{16y^8}\right)^{3/4}$

136. $\sqrt{80x^4}$

137. $\sqrt[3]{16ab^4} - b\sqrt[3]{54ab}$

138. $x\sqrt[4]{1,280x} + \sqrt[4]{80x^5}$

SOLUTION   Since the first term is $2x^2$, the first terms of the binomial factors must be $2x$ and x:

$$2x^2 - x - 6 = (2x \qquad)(x \qquad)$$

The product of the last terms must be -6, and the sum of the products of the outer terms and the inner terms must be $-x$. Since the only factorization of -6 that will cause this to happen is $3(-2)$, we have

$$2x^2 - x - 6 = (2x + 3)(x - 2) \qquad \text{We can check by multiplying.}$$

Self Check 8   Factor: $6x^2 - x - 2$.

Now Try Exercise 33.

It is not easy to give specific rules for factoring trinomials, because some guess-work is often necessary. However, the following hints are helpful.

Strategy for Factoring a General Trinomial with Integer Coefficients

1. Write the trinomial in descending powers of one variable.
2. Factor out any greatest common factor, including -1 if that is necessary to make the coefficient of the first term positive.
3. When the sign of the first term of a trinomial is $+$ and the sign of the third term is $+$, the sign between the terms of each binomial factor is the same as the sign of the middle term of the trinomial.

   When the sign of the first term is $+$ and the sign of the third term is $-$, one of the signs between the terms of the binomial factors is $+$ and the other is $-$.
4. Try various combinations of first terms and last terms until you find one that works. If no possibilities work, the trinomial is prime.
5. Check the factorization by multiplication.

EXAMPLE 9   **Factoring a Trinomial Completely**

Factor: $10xy + 24y^2 - 6x^2$.

SOLUTION   We write the trinomial in descending powers of x and then factor out the common factor of -2.

$$10xy + 24y^2 - 6x^2 = -6x^2 + 10xy + 24y^2$$
$$= -2(3x^2 - 5xy - 12y^2)$$

Since the sign of the third term of $3x^2 - 5xy - 12y^2$ is $-$, the signs between the binomial factors will be opposite. Since the first term is $3x^2$, the first terms of the binomial factors must be $3x$ and x:

$$-2(3x^2 - 5xy - 12y^2) = -2(3x \qquad)(x \qquad)$$

The product of the last terms must be $-12y^2$, and the sum of the outer terms and the inner terms must be $-5xy$. Of the many factorizations of $-12y^2$, only $4y(-3y)$ leads to a middle term of $-5xy$. So we have

$$10xy + 24y^2 - 6x^2 = -6x^2 + 10xy + 24y^2$$
$$= -2(\mathbf{3x^2 - 5xy - 12y^2})$$
$$= -2(\mathbf{3x + 4y})(\mathbf{x - 3y}) \qquad \text{We can check by multiplying.}$$

Self Check 9 Factor: $-6x^2 - 15xy - 6y^2$.

Now Try Exercise 69.

5. Factor Trinomials by Grouping

Another way of factoring trinomials involves factoring by grouping. This method can be used to factor trinomials of the form $ax^2 + bx + c$. For example, to factor $6x^2 + 5x - 6$, we note that $a = 6$, $b = 5$, and $c = -6$, and proceed as follows:

1. Find the product ac: $6(-6) = -36$. This number is called the **key number**.
2. Find two factors of the key number (-36) whose sum is $b = 5$. Two such numbers are 9 and -4.

$$9(-4) = -36 \quad \text{and} \quad 9 + (-4) = 5$$

3. Use the factors 9 and -4 as coefficients of two terms to be placed between $6x^2$ and -6.

$$6x^2 + 5x - 6 = 6x^2 + 9x - 4x - 6$$

4. Factor by grouping:

$$6x^2 + 9x - 4x - 6 = 3x(2x + 3) - 2(2x + 3)$$
$$= (2x + 3)(3x - 2) \qquad \text{Factor out } 2x + 3.$$

EXAMPLE 10 **Factoring a Trinomial by Grouping**

Factor: $15x^2 + x - 2$.

SOLUTION Since $a = 15$ and $c = -2$ in the trinomial, $ac = -30$. We now find factors of -30 whose sum is $b = 1$. Such factors are 6 and -5. We use these factors as coefficients of two terms to be placed between $15x^2$ and -2.

$$15x^2 + 6x - 5x - 2$$

Finally, we factor by grouping.

$$3x(5x + 2) - (5x + 2) = (5x + 2)(3x - 1)$$

■

Self Check 10 Factor: $15a^2 + 17a - 4$.

Now Try Exercise 39.

We can often factor polynomials with variable exponents. For example, if n is a natural number,

$$a^{2n} - 5a^n - 6 = (a^n + 1)(a^n - 6)$$

because

$$(a^n + 1)(a^n - 6) = a^{2n} - 6a^n + a^n - 6$$
$$= a^{2n} - 5a^n - 6 \qquad \text{Combine like terms.}$$

6. Factor the Sum and Difference of Two Cubes

Two other types of factoring involve binomials that are the sum or the difference of two cubes. Like the difference of two squares, they can be factored by using a formula.

Factoring the Sum and Difference of Two Cubes

$$x^3 + y^3 = (x + y)(x^2 - xy + y^2)$$

$$x^3 - y^3 = (x - y)(x^2 + xy + y^2)$$

EXAMPLE 11   **Factoring the Sum of Two Cubes**

Factor: $27x^6 + 64y^3$.

SOLUTION   We can write this expression as the sum of two cubes and factor it as follows:

$$27x^6 + 64y^3 = (3x^2)^3 + (4y)^3$$

$$= (3x^2 + 4y)[(3x^2)^2 - (3x^2)(4y) + (4y)^2]$$

$$= (3x^2 + 4y)(9x^4 - 12x^2y + 16y^2) \qquad \text{We can check by multiplying.}$$

Self Check 11   Factor: $8a^3 + 1{,}000b^6$.

Now Try Exercise 41.

EXAMPLE 12   **Factoring the Difference of Two Cubes**

Factor: $x^3 - 8$.

SOLUTION   This binomial can be written as $x^3 - 2^3$, which is the difference of two cubes. Substituting into the formula for the difference of two cubes gives

$$x^3 - 2^3 = (x - 2)(x^2 + 2x + 2^2)$$

$$= (x - 2)(x^2 + 2x + 4) \qquad \text{We can check by multiplying.}$$

Self Check 12   Factor: $p^3 - 64$.

Now Try Exercise 43.

7. Factor Miscellaneous Polynomials

EXAMPLE 13   **Factoring a Miscellaneous Polynomial**

Factor: $x^2 - y^2 + 6x + 9$.

SOLUTION   Here we will factor a trinomial and a difference of two squares.

$$x^2 - y^2 + 6x + 9 = x^2 + 6x + 9 - y^2 \qquad \text{Use the Commutative Property to rearrange the terms.}$$

$$= (x + 3)^2 - y^2 \qquad \text{Factor } x^2 + 6x + 9.$$

$$= (x + 3 + y)(x + 3 - y) \qquad \text{Factor the difference of two squares.}$$

We could try to factor this expression in another way.

$$x^2 - y^2 + 6x + 9 = (x + y)(x - y) + 3(2x + 3) \qquad \text{Factor } x^2 - y^2 \text{ and } 6x + 9.$$

However, we are unable to finish the factorization. If grouping in one way doesn't work, try various other ways.

Self Check 13   Factor: $a^2 + 8a - b^2 + 16$.

Now Try Exercise 97.

EXAMPLE 14   **Factoring a Miscellaneous Trinomial**

Factor: $z^4 - 3z^2 + 1$.

SOLUTION   This trinomial cannot be factored as the product of two binomials, because no combination will give a middle term of $-3z^2$. However, if the middle term were $-2z^2$, the trinomial would be a perfect square, and the factorization would be easy:

$$z^4 - 2z^2 + 1 = (z^2 - 1)(z^2 - 1)$$
$$= (z^2 - 1)^2$$

We can change the middle term in $z^4 - 3z^2 + 1$ to $-2z^2$ by adding z^2 to it. However, to make sure that adding z^2 does not change the value of the trinomial, we must also subtract z^2. We can then proceed as follows.

$$\begin{aligned} z^4 - 3z^2 + 1 &= z^4 - 3z^2 + z^2 + 1 - z^2 &&\text{Add and subtract } z^2. \\ &= z^4 - 2z^2 + 1 - z^2 &&\text{Combine } -3z^2 \text{ and } z^2. \\ &= (z^2 - 1)^2 - z^2 &&\text{Factor } z^4 - 2z^2 + 1. \\ &= (z^2 - 1 + z)(z^2 - 1 - z) &&\text{Factor the difference of two squares.} \end{aligned}$$

In this type of problem, we will always try to add and subtract a perfect square in hopes of making a perfect-square trinomial that will lead to factoring a difference of two squares. ∎

Self Check 14   Factor: $x^4 + 3x^2 + 4$.

Now Try Exercise 103.

It is helpful to identify the problem type when we must factor polynomials that are given in random order.

Factoring Strategy   1. Factor out all common monomial factors.
2. If an expression has two terms, check whether the problem type is
   a. The difference of two squares:
   $$x^2 - y^2 = (x + y)(x - y)$$
   b. The sum of two cubes:
   $$x^3 + y^3 = (x + y)(x^2 - xy + y^2)$$
   c. The difference of two cubes:
   $$x^3 - y^3 = (x - y)(x^2 + xy + y^2)$$
3. If an expression has three terms, try to factor it as a *trinomial.*
4. If an expression has four or more terms, try factoring by *grouping.*
5. Continue until each individual factor is prime.
6. Check the results by multiplying.

Self Check Answers   **1.** $4a(a - 2b)$ **2.** $a^2b^2c^2(1 + abc)$ **3.** $(x + y)(x + 2)$
4. $(3a + 4b)(3a - 4b)$ **5.** $(a^2 + 9b^2)(a + 3b)(a - 3b)$
6. $-3(x + 2)(x - 2)$ **7.** $(p - 6)(p + 1)$ **8.** $(3x - 2)(2x + 1)$
9. $-3(x + 2y)(2x + y)$ **10.** $(3a + 4)(5a - 1)$
11. $8(a + 5b^2)(a^2 - 5ab^2 + 25b^4)$ **12.** $(p - 4)(p^2 + 4p + 16)$
13. $(a + 4 + b)(a + 4 - b)$ **14.** $(x^2 + 2 + x)(x^2 + 2 - x)$

Exercises **0.5**

Getting Ready
You should be able to complete these vocabulary and concept statements before you proceed to the practice exercises.

Fill in the blanks.

1. When polynomials are multiplied together, each polynomial is a _____ of the product.
2. If a polynomial cannot be factored using _____ coefficients, it is called a _____ polynomial.

Complete each factoring formula.

3. $ax + bx =$ _____
4. $x^2 - y^2 =$ _____
5. $x^2 + 2xy + y^2 =$ _____
6. $x^2 - 2xy + y^2 =$ _____
7. $x^3 + y^3 =$ _____
8. $x^3 - y^3 =$ _____

Practice
In each expression, factor out the greatest common monomial.

9. $3x - 6$
10. $5y - 15$
11. $8x^2 + 4x^3$
12. $9y^3 + 6y^2$
13. $7x^2y^2 + 14x^3y^2$
14. $25y^2z - 15yz^2$

In each expression, factor by grouping.

15. $a(x + y) + b(x + y)$
16. $b(x - y) + a(x - y)$
17. $4a + b - 12a^2 - 3ab$
18. $x^2 + 4x + xy + 4y$

In each expression, factor the difference of two squares.

19. $4x^2 - 9$
20. $36z^2 - 49$
21. $4 - 9r^2$
22. $16 - 49x^2$
23. $81x^4 - 1$
24. $81 - x^4$
25. $(x + z)^2 - 25$
26. $(x - y)^2 - 9$

In each expression, factor the trinomial.

27. $x^2 + 8x + 16$
28. $a^2 - 12a + 36$
29. $b^2 - 10b + 25$
30. $y^2 + 14y + 49$
31. $m^2 + 4mn + 4n^2$
32. $r^2 - 8rs + 16s^2$
33. $12x^2 - xy - 6y^2$
34. $8x^2 - 10xy - 3y^2$

In each expression, factor the trinomial by grouping.

35. $x^2 + 10x + 21$
36. $x^2 + 7x + 10$
37. $x^2 - 4x - 12$
38. $x^2 - 2x - 63$
39. $6p^2 + 7p - 3$
40. $4q^2 - 19q + 12$

In each expression, factor the sum of two cubes.

41. $t^3 + 343$
42. $r^3 + 8s^3$

In each expression, factor the difference of two cubes.

43. $8z^3 - 27$
44. $125a^3 - 64$

Factor each expression completely. If an expression is prime, so indicate.

45. $3a^2bc + 6ab^2c + 9abc^2$
46. $5x^3y^3z^3 + 25x^2y^2z^2 - 125xyz$
47. $3x^3 + 3x^2 - x - 1$
48. $4x + 6xy - 9y - 6$
49. $2txy + 2ctx - 3ty - 3ct$
50. $2ax + 4ay - bx - 2by$
51. $ax + bx + ay + by + az + bz$
52. $6x^2y^3 + 18xy + 3x^2y^2 + 9x$
53. $x^2 - (y - z)^2$
54. $z^2 - (y + 3)^2$
55. $(x - y)^2 - (x + y)^2$
56. $(2a + 3)^2 - (2a - 3)^2$
57. $x^4 - y^4$
58. $z^4 - 81$
59. $3x^2 - 12$
60. $3x^3y - 3xy$
61. $18xy^2 - 8x$
62. $27x^2 - 12$
63. $x^2 - 2x + 15$
64. $x^2 + x + 2$
65. $-15 + 2a + 24a^2$
66. $-32 - 68x + 9x^2$
67. $6x^2 + 29xy + 35y^2$
68. $10x^2 - 17xy + 6y^2$
69. $12p^2 - 58pq - 70q^2$
70. $3x^2 - 6xy - 9y^2$
71. $-6m^2 + 47mn - 35n^2$
72. $-14r^2 - 11rs + 15s^2$
73. $-6x^3 + 23x^2 + 35x$
74. $-y^3 - y^2 + 90y$

75. $6x^4 - 11x^3 - 35x^2$

76. $12x + 17x^2 - 7x^3$

77. $x^4 + 2x^2 - 15$

78. $x^4 - x^2 - 6$

79. $a^{2n} - 2a^n - 3$

80. $a^{2n} + 6a^n + 8$

81. $6x^{2n} - 7x^n + 2$

82. $9x^{2n} + 9x^n + 2$

83. $4x^{2n} - 9y^{2n}$

84. $8x^{2n} - 2x^n - 3$

85. $10y^{2n} - 11y^n - 6$

86. $16y^{4n} - 25y^{2n}$

87. $2x^3 + 2{,}000$

88. $3y^3 + 648$

89. $(x + y)^3 - 64$

90. $(x - y)^3 + 27$

91. $64a^6 - y^6$

92. $a^6 + b^6$

93. $a^3 - b^3 + a - b$

94. $(a^2 - y^2) - 5(a + y)$

95. $64x^6 + y^6$

96. $z^2 + 6z + 9 - 225y^2$

97. $x^2 - 6x + 9 - 144y^2$

98. $x^2 + 2x - 9y^2 + 1$

99. $(a + b)^2 - 3(a + b) - 10$

100. $2(a + b)^2 - 5(a + b) - 3$

101. $x^6 + 7x^3 - 8$

102. $x^6 - 13x^4 + 36x^2$

103. $x^4 + x^2 + 1$

104. $x^4 + 3x^2 + 4$

105. $x^4 + 7x^2 + 16$

106. $y^4 + 2y^2 + 9$

107. $4a^4 + 1 + 3a^2$

108. $x^4 + 25 + 6x^2$

109. Candy To find the amount of chocolate used in the outer coating of one of the malted-milk balls shown, we can find the volume V of the chocolate shell using the formula $V = \frac{4}{3}\pi r_1^3 - \frac{4}{3}\pi r_2^3$. Factor the expression on the right side of the formula.

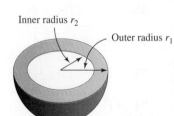

Inner radius r_2

Outer radius r_1

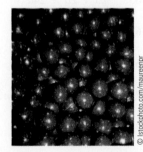

110. Movie Stunts The formula that gives the distance a stuntwoman is above the ground t seconds after she falls over the side of a 144-foot tall building is $f = 144 - 16t^2$. Factor the right side.

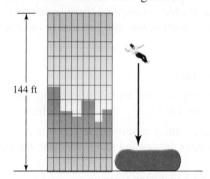

144 ft

Discovery and Writing

111. Explain how to factor the difference of two squares.

112. Explain how to factor the difference of two cubes.

113. Explain how to factor $a^2 - b^2 + a + b$.

114. Explain how to factor $x^2 + 2x + 1$.

Factor the indicated monomial from the given expression.

115. $3x + 2; 2$

116. $5x - 3; 5$

117. $x^2 + 2x + 4; 2$

118. $3x^2 - 2x - 5; 3$

119. $a + b; a$

120. $a - b; b$

121. $x + x^{1/2}; x^{1/2}$

122. $x^{3/2} - x^{1/2}; x^{1/2}$

123. $2x + \sqrt{2}y; \sqrt{2}$

124. $\sqrt{3}a - 3b; \sqrt{3}$

125. $ab^{3/2} - a^{3/2}b; ab$

126. $ab^2 + b; b^{-1}$

Factor each expression by grouping three terms and two terms.

127. $x^2 + x - 6 + xy - 2y$

128. $2x^2 + 5x + 2 - xy - 2y$

129. $a^4 + 2a^3 + a^2 + a + 1$

130. $a^4 + a^3 - 2a^2 + a - 1$

Review

131. Which natural number is neither prime nor composite?

132. Graph the interval $[-2,3)$.

133. Simplify: $(x^3x^2)^4$.

134. Simplify: $\dfrac{(a^3)^3(a^2)^4}{(a^2a^3)^3}$.

135. $\left(\dfrac{3x^4x^3}{6x^{-2}x^4}\right)^0$

136. Simplify: $\sqrt{20x^5}$.

137. Simplify: $\sqrt{20x} - \sqrt{125x}$.

138. Rationalize the denominator: $\dfrac{3}{\sqrt[3]{3}}$.

0.6 Rational Expressions

In this section, we will learn to

1. Define rational expressions.

2. Simplify rational expressions.

3. Multiple and divide rational expressions.

4. Add and subtract rational expressions.

5. Simplify complex fractions.

Abercrombie and Fitch (A&F) is a very successful American clothing company founded in 1892 by David Abercrombie and Ezra Fitch. Today the stores are popular shopping destinations for university students wanting to keep up with the latest styles and trends.

Photo by Tim Boyle/Getty Images

Suppose that a clothing manufacturer finds that the cost in dollars of producing x fleece vintage shirts is given by the algebraic expression $13x + 1,000$. The average cost of producing each shirt could be obtained by dividing the production cost, $13x + 1,000$, by the number of shirts produced, x. The algebraic fraction

$$\frac{13x + 1,000}{x}$$

represents the average cost per shirt. We see that the average cost of producing 200 shirts would be $18.

$$\frac{13(\mathbf{200}) + 1,000}{\mathbf{200}} = 18$$

An understanding of algebraic fractions is important in solving many real-life problems.

1. Define Rational Expressions

If x and y are real numbers, the quotient $\frac{x}{y}$ ($y \neq 0$) is called a **fraction**. The number x is called the **numerator**, and the number y is called the **denominator**.

Algebraic fractions are quotients of algebraic expressions. If the expressions are polynomials, the fraction is called a **rational expression**. The first two of the following algebraic fractions are rational expressions. The third is not, because the numerator and denominator are not polynomials.

Caution

Remember that the denominator of a fraction **cannot** be zero.

$$\frac{5y^2 + 2y}{y^2 - 3y - 7} \qquad \frac{8ab^2 - 16c^3}{2x + 3} \qquad \frac{x^{1/2} + 4x}{x^{3/2} - x^{1/2}}$$

We summarize some of the properties of fractions as follows:

Properties of Fractions   If a, b, c, and d are real numbers and no denominators are 0, then

Equality of Fractions

$$\frac{a}{b} = \frac{c}{d} \quad \text{if and only if} \quad ad = bc$$

Fundamental Property of Fractions

$$\frac{ax}{bx} = \frac{a}{b}$$

Multiplication and Division of Fractions

$$\frac{a}{b} \cdot \frac{c}{d} = \frac{ac}{bd} \quad \text{and} \quad \frac{a}{b} \div \frac{c}{d} = \frac{a}{b} \cdot \frac{d}{c} = \frac{ad}{bc}$$

Addition and Subtraction of Fractions

$$\frac{a}{b} + \frac{c}{b} = \frac{a+c}{b} \quad \text{and} \quad \frac{a}{b} - \frac{c}{b} = \frac{a-c}{b}$$

The first two examples illustrate each of the previous properties of fractions.

EXAMPLE 1   **Illustrating the Properties of Fractions**

Assume that no denominators are 0.

a. $\dfrac{2a}{3} = \dfrac{4a}{6}$ Because $2a(6) = 3(4a)$

b. $\dfrac{6xy}{10xy} = \dfrac{3(2xy)}{5(2xy)}$ Factor the numerator and denominator and divide out the common factors.

$= \dfrac{3}{5}$

Self Check 1   **a.** Is $\dfrac{3y}{5} = \dfrac{15z}{25}$?   **b.** Simplify: $\dfrac{15a^2b}{25ab^2}$.

Now Try Exercise 9.

EXAMPLE 2   **Illustrating the Properties of Fractions**

Assume that no denominators are 0.

a. $\dfrac{2r}{7s} \cdot \dfrac{3r}{5s} = \dfrac{2r \cdot 3r}{7s \cdot 5s}$

$= \dfrac{6r^2}{35s^2}$

b. $\dfrac{3mn}{4pq} \div \dfrac{2pq}{7mn} = \dfrac{3mn}{4pq} \cdot \dfrac{7mn}{2pq}$

$= \dfrac{21m^2n^2}{8p^2q^2}$

c. $\dfrac{2ab}{5xy} + \dfrac{ab}{5xy} = \dfrac{2ab + ab}{5xy}$

$= \dfrac{3ab}{5xy}$

d. $\dfrac{6uv^2}{7w^2} - \dfrac{3uv^2}{7w^2} = \dfrac{6uv^2 - 3uv^2}{7w^2}$

$= \dfrac{3uv^2}{7w^2}$

Self Check 2   Perform each operation:

a. $\dfrac{3a}{5b} \cdot \dfrac{2a}{7b}$ b. $\dfrac{2ab}{3rs} \div \dfrac{2rs}{4ab}$ c. $\dfrac{5pq}{3t} + \dfrac{3pq}{3t}$ d. $\dfrac{5mn^2}{3w} - \dfrac{mn^2}{3w}$

Now Try Exercise 19.

To add or subtract rational expressions with unlike denominators, we write each expression as an equivalent expression with a common denominator. We can then add or subtract the expressions. For example,

$$\frac{3x}{5} + \frac{2x}{7} = \frac{3x(7)}{5(7)} + \frac{2x(5)}{7(5)} \qquad\qquad \frac{4a^2}{15} - \frac{3a^2}{10} = \frac{4a^2(2)}{15(2)} - \frac{3a^2(3)}{10(3)}$$

$$= \frac{21x}{35} + \frac{10x}{35} \qquad\qquad\qquad\quad = \frac{8a^2}{30} - \frac{9a^2}{30}$$

$$= \frac{21x + 10x}{35} \qquad\qquad\qquad\quad = \frac{8a^2 - 9a^2}{30}$$

$$= \frac{31x}{35} \qquad\qquad\qquad\qquad\quad = \frac{-a^2}{30}$$

$$\qquad\qquad\qquad\qquad\qquad\qquad\qquad = -\frac{a^2}{30}$$

A rational expression is **in lowest terms** if all factors common to the numerator and the denominator have been removed. To **simplify a rational expression** means to write it in lowest terms.

2. Simplify Rational Expressions

To simplify rational expressions, we use the Fundamental Property of Fractions. This enables us to divide out all factors that are common to the numerator and the denominator.

EXAMPLE 3   **Simplifying a Rational Expression**

Simplify: $\dfrac{x^2 - 9}{x^2 - 3x} \ (x \neq 0, 3).$

SOLUTION   We factor the difference of two squares in the numerator, factor out x in the denominator, and divide out the common factor of $x - 3$.

$$\frac{x^2 - 9}{x^2 - 3x} = \frac{(x + 3)\cancel{(x - 3)}}{x\cancel{(x - 3)}} \qquad \frac{x-3}{x-3} = 1$$

$$= \frac{x + 3}{x}$$

■

Self Check 3   Simplify: $\dfrac{a^2 - 4a}{a^2 - a - 12} \ (a \neq 4, -3).$

Now Try Exercise 23.

We will encounter the following properties of fractions in the next examples.

Properties of Fractions   If a and b represent real numbers and there are no divisions by 0, then

- $\dfrac{a}{1} = a$

- $\dfrac{a}{a} = 1$

- $\dfrac{a}{b} = \dfrac{-a}{-b} = -\dfrac{a}{-b} = -\dfrac{-a}{b}$

- $-\dfrac{a}{b} = \dfrac{a}{-b} = \dfrac{-a}{b} = -\dfrac{-a}{-b}$

EXAMPLE 4   **Simplifying a Rational Expression**

Simplify: $\dfrac{x^2 - 2xy + y^2}{y - x}$ $(x \neq y)$.

SOLUTION   We factor the trinomial in the numerator, factor -1 from the denominator, and divide out the common factor of $x - y$.

$$\frac{x^2 - 2xy + y^2}{y - x} = \frac{(x - y)\cancel{(x - y)}}{-1\cancel{(x - y)}} \qquad \frac{x - y}{x - y} = 1$$

$$= \frac{x - y}{-1}$$

$$= -\frac{x - y}{1}$$

$$= -(x - y)$$

■

Self Check 4   Simplify: $\dfrac{a^2 - ab - 2b^2}{2b - a}$ $(2b - a \neq 0)$.

Now Try Exercise 25.

EXAMPLE 5   **Simplifying a Rational Expression**

Simplify: $\dfrac{x^2 - 3x + 2}{x^2 - x - 2}$ $(x \neq 2, -1)$.

SOLUTION   We factor the numerator and denominator and divide out the common factor of $x - 2$.

$$\frac{x^2 - 3x + 2}{x^2 - x - 2} = \frac{(x - 1)\cancel{(x - 2)}}{(x + 1)\cancel{(x - 2)}} \qquad \frac{x - 2}{x - 2} = 1$$

$$= \frac{x - 1}{x + 1}$$

■

Self Check 5   Simplify: $\dfrac{a^2 + 3a - 4}{a^2 + 2a - 3}$ $(a \neq 1, -3)$.

Now Try Exercise 27.

3. Multiply and Divide Rational Expressions

EXAMPLE 6   **Multiplying Rational Expressions**

Multiply: $\dfrac{x^2 - x - 2}{x^2 - 1} \cdot \dfrac{x^2 + 2x - 3}{x - 2}$ $(x \neq 1, -1, 2)$.

SOLUTION   To multiply the rational expressions, we multiply the numerators, multiply the denominators, and divide out the common factors.

$$\frac{x^2 - x - 2}{x^2 - 1} \cdot \frac{x^2 + 2x - 3}{x - 2} = \frac{(x^2 - x - 2)(x^2 + 2x - 3)}{(x^2 - 1)(x - 2)}$$

$$= \frac{\cancel{(x - 2)}\cancel{(x + 1)}\cancel{(x - 1)}(x + 3)}{\cancel{(x + 1)}\cancel{(x - 1)}\cancel{(x - 2)}} \qquad \frac{x - 2}{x - 2} = 1, \frac{x + 1}{x + 1} = 1, \frac{x - 1}{x - 1} = 1$$

$$= x + 3$$

Self Check 6   Simplify: $\dfrac{x^2 - 9}{x^2 - x} \cdot \dfrac{x - 1}{x^2 - 3x}$ $(x \neq 0, 1, 3)$.

Now Try Exercise 33.

EXAMPLE 7   **Dividing Rational Expressions**

Divide: $\dfrac{x^2 - 2x - 3}{x^2 - 4} \div \dfrac{x^2 + 2x - 15}{x^2 + 3x - 10}$ $(x \neq 2, -2, 3, -5)$.

SOLUTION   To divide the rational expressions, we multiply by the reciprocal of the second rational expression. We then simplify by factoring the numerator and denominator and dividing out the common factors.

$$\frac{x^2 - 2x - 3}{x^2 - 4} \div \frac{x^2 + 2x - 15}{x^2 + 3x - 10} = \frac{x^2 - 2x - 3}{x^2 - 4} \cdot \frac{x^2 + 3x - 10}{x^2 + 2x - 15}$$

$$= \frac{(x^2 - 2x - 3)(x^2 + 3x - 10)}{(x^2 - 4)(x^2 + 2x - 15)}$$

$$= \frac{\cancel{(x - 3)}(x + 1)\cancel{(x - 2)}\cancel{(x + 5)}}{(x + 2)\cancel{(x - 2)}\cancel{(x + 5)}\cancel{(x - 3)}} \qquad \frac{x - 3}{x - 3} = 1, \frac{x - 2}{x - 2} = 1, \frac{x + 5}{x + 5} = 1$$

$$= \frac{x + 1}{x + 2}$$

Self Check 7   Simplify: $\dfrac{a^2 - a}{a + 2} \div \dfrac{a^2 - 2a}{a^2 - 4}$ $(a \neq 0, 2, -2)$.

Now Try Exercise 39.

EXAMPLE 8   **Using Multiplication and Division to Simplify a Rational Expression**

Simplify: $\dfrac{2x^2 - 5x - 3}{3x - 1} \cdot \dfrac{3x^2 + 2x - 1}{x^2 - 2x - 3} \div \dfrac{2x^2 + x}{3x}$ $\left(x \neq \dfrac{1}{3}, -1, 3, 0, -\dfrac{1}{2}\right)$.

SOLUTION   We can change the division to a multiplication, factor, and simplify.

$$\frac{2x^2 - 5x - 3}{3x - 1} \cdot \frac{3x^2 + 2x - 1}{x^2 - 2x - 3} \div \frac{2x^2 + x}{3x}$$

$$= \frac{2x^2 - 5x - 3}{3x - 1} \cdot \frac{3x^2 + 2x - 1}{x^2 - 2x - 3} \cdot \frac{3x}{2x^2 + x}$$

$$= \frac{(2x^2 - 5x - 3)(3x^2 + 2x - 1)(3x)}{(3x - 1)(x^2 - 2x - 3)(2x^2 + x)}$$

$$= \frac{\cancel{(x - 3)}\cancel{(2x + 1)}\cancel{(3x - 1)}\cancel{(x + 1)}3x}{\cancel{(3x - 1)}\cancel{(x + 1)}\cancel{(x - 3)}x\cancel{(2x + 1)}} \qquad \frac{x - 3}{x - 3} = 1, \frac{2x + 1}{2x + 1} = 1, \frac{3x - 1}{3x - 1} = 1, \frac{x + 1}{x + 1} = 1, \frac{x}{x} = 1$$

$$= 3$$

Self Check 8   Simplify: $\dfrac{x^2 - 25}{x - 2} \div \dfrac{x^2 - 5x}{x^2 - 2x} \cdot \dfrac{x^2 + 2x}{x^2 + 5x}$ $(x \neq 0, 2, 5, -5)$.

Now Try Exercise 45.

4. Add and Subtract Rational Expressions

To add (or subtract) rational expressions with like denominators, we add (or subtract) the numerators and keep the common denominator.

EXAMPLE 9   **Adding Rational Expression with Like Deonominators**

Add: $\dfrac{2x + 5}{x + 5} + \dfrac{3x + 20}{x + 5}$ $(x \neq -5)$.

SOLUTION $\dfrac{2x + 5}{x + 5} + \dfrac{3x + 20}{x + 5} = \dfrac{5x + 25}{x + 5}$ Add the numerators and keep the common denominator.

$$= \dfrac{5\cancel{(x + 5)}}{1\cancel{(x + 5)}}$$ Factor out 5 and divide out the common factor of $x + 5$.

$$= 5$$

Self Check 9   Add: $\dfrac{3x - 2}{x - 2} + \dfrac{x - 6}{x - 2}$ $(x \neq 2)$.

Now Try Exercise 49.

To add (or subtract) rational expressions with unlike denominators, we must find a common denominator, called the **least** (or lowest) **common denominator (LCD)**. Suppose the unlike denominators of three rational expressions are 12, 20, and 35. To find the LCD, we first find the prime factorization of each number.

$$12 = 4 \cdot 3 \qquad\qquad 20 = 4 \cdot 5 \qquad\qquad 35 = 5 \cdot 7$$
$$= 2^2 \cdot 3 \qquad\qquad\quad = 2^2 \cdot 5$$

Because the LCD is the smallest number that can be divided by 12, 20, and 35, it must contain factors of 2^2, 3, 5, and 7. Thus, the

$$\text{LCD} = 2^2 \cdot 3 \cdot 5 \cdot 7 = 420$$

That is, 420 is the smallest number that can be divided without remainder by 12, 20, and 35.

When finding an LCD, we always factor each denominator and then create the LCD by using each factor the greatest number of times that it appears in any one denominator. The product of these factors is the LCD.

Comment

Remember to find the *least* common denominator, use each factor the *greatest* number of times that it occurs.

This rule also applies if the unlike denominators of the rational expressions contain variables. Suppose the unlike denominators are $x^2(x - 5)$ and $x(x - 5)^3$. To find the LCD, we use x^2 and $(x - 5)^3$. Thus, the LCD is the product $x^2(x - 5)^3$.

EXAMPLE 10   **Adding Rational Expressions with Unlike Denominators**

Add: $\dfrac{1}{x^2 - 4} + \dfrac{2}{x^2 - 4x + 4}$ $(x \neq 2, -2)$.

SOLUTION　We factor each denominator and find the LCD.

$$x^2 - 4 = (x + 2)(x - 2)$$

$$x^2 - 4x + 4 = (x - 2)(x - 2) = (x - 2)^2$$

The LCD is $(x + 2)(x - 2)^2$. We then write each rational expression with its denominator in factored form, convert each rational expression into an equivalent expression with a denominator of $(x + 2)(x - 2)^2$, add the expressions, and simplify.

$$\frac{1}{x^2 - 4} + \frac{2}{x^2 - 4x + 4} = \frac{1}{(x + 2)(x - 2)} + \frac{2}{(x - 2)(x - 2)}$$

$$= \frac{1(x - 2)}{(x + 2)(x - 2)(x - 2)} + \frac{2(x + 2)}{(x - 2)(x - 2)(x + 2)} \qquad \frac{x - 2}{x - 2} = 1, \frac{x + 2}{x + 2} = 1$$

$$= \frac{1(x - 2) + 2(x + 2)}{(x + 2)(x - 2)(x - 2)}$$

$$= \frac{x - 2 + 2x + 4}{(x + 2)(x - 2)(x - 2)}$$

$$= \frac{3x + 2}{(x + 2)(x - 2)^2}$$

Comment

Always attempt to simplify the final result. In this case, the final fraction is already in lowest terms.

Self Check 10　Add: $\dfrac{3}{x^2 - 6x + 9} + \dfrac{1}{x^2 - 9}$ $(x \neq 3, -3)$.

Now Try Exercise 59.

EXAMPLE 11　**Combining and Simplifying Rational Expressions with Unlike Denominators**

Simplify: $\dfrac{x - 2}{x^2 - 1} - \dfrac{x + 3}{x^2 + 3x + 2} + \dfrac{3}{x^2 + x - 2}$ $(x \neq 1, -1, -2)$.

SOLUTION　We factor the denominators to find the LCD.

$$x^2 - 1 = (x + 1)(x - 1)$$

$$x^2 + 3x + 2 = (x + 2)(x + 1)$$

$$x^2 + x - 2 = (x + 2)(x - 1)$$

The LCD is $(x + 1)(x + 2)(x - 1)$. We now write each rational expression as an equivalent expression with this LCD, and proceed as follows:

$$\frac{x - 2}{x^2 - 1} - \frac{x + 3}{x^2 + 3x + 2} + \frac{3}{x^2 + x - 2}$$

$$= \frac{x - 2}{(x + 1)(x - 1)} - \frac{x + 3}{(x + 1)(x + 2)} + \frac{3}{(x - 1)(x + 2)}$$

$$= \frac{(x - 2)(x + 2)}{(x + 1)(x - 1)(x + 2)} - \frac{(x + 3)(x - 1)}{(x + 1)(x + 2)(x - 1)} + \frac{3(x + 1)}{(x - 1)(x + 2)(x + 1)} \qquad \frac{x + 2}{x + 2} = 1, \frac{x - 1}{x - 1} = 1, \frac{x + 1}{x + 1} = 1$$

$$= \frac{(x^2 - 4) - (x^2 + 2x - 3) + (3x + 3)}{(x + 1)(x + 2)(x - 1)}$$

$$= \frac{x^2 - 4 - x^2 - 2x + 3 + 3x + 3}{(x + 1)(x + 2)(x - 1)}$$

$$= \frac{x + 2}{(x + 1)(x + 2)(x - 1)}$$

$$= \frac{1}{(x + 1)(x - 1)} \qquad \text{Divide out the common factor of } x + 2. \ \frac{x + 2}{x + 2} = 1$$

Self Check 11   Simplify: $\dfrac{4y}{y^2 - 1} - \dfrac{2}{y + 1} + 2 \ (y \neq 1, -1)$.

Now Try Exercise 61.

5. Simplify Complex Fractions

A **complex fraction** is a fraction that has a fraction in its numerator or a fraction in its denominator. There are two methods generally used to simplify complex fractions. These are stated here for you.

Strategies for Simplifying Complex Fractions

Method 1: Multiply the Complex Fraction by 1

- Determine the LCD of all fractions in the complex fraction.
- Multiply both numerator and denominator of the complex fraction by the LCD. Note that when we multiply by $\frac{LCD}{LCD}$ we are multiplying by 1.

Method 2: Simplify the Numerator and Denominator and then Divide

- Simplify the numerator and denominator so that both are single fractions.
- Perform the division by multiplying the numerator by the reciprocal of the denominator.

EXAMPLE 12   **Simplifying Complex Fractions**

Simplify: $\dfrac{\dfrac{1}{x} + \dfrac{1}{y}}{\dfrac{x}{y}} \ (x, y \neq 0)$.

Method 1: We note that the LCD of the three fractions in the complex fraction is xy. So we multiply the numerator and denominator of the complex fraction by xy and simplify:

$$\frac{\dfrac{1}{x} + \dfrac{1}{y}}{\dfrac{x}{y}} = \frac{xy\left(\dfrac{1}{x} + \dfrac{1}{y}\right)}{xy\left(\dfrac{x}{y}\right)} = \frac{\dfrac{xy}{x} + \dfrac{xy}{y}}{\dfrac{xyx}{y}} = \frac{y + x}{x^2}$$

Method 2: We combine the fractions in the numerator of the complex fraction to obtain a single fraction over a single fraction.

$$\frac{\dfrac{1}{x} + \dfrac{1}{y}}{\dfrac{x}{y}} = \frac{\dfrac{1(y)}{x(y)} + \dfrac{1(x)}{y(x)}}{\dfrac{x}{y}} = \frac{\dfrac{y + x}{xy}}{\dfrac{x}{y}}$$

Then we use the fact that any fraction indicates a division:

$$\frac{\dfrac{y + x}{xy}}{\dfrac{x}{y}} = \frac{y + x}{xy} \div \frac{x}{y} = \frac{y + x}{xy} \cdot \frac{y}{x} = \frac{(y + x)y}{xyx} = \frac{y + x}{x^2}$$

Self Check 12   Simplify: $\dfrac{\dfrac{1}{x} - \dfrac{1}{y}}{\dfrac{1}{x} + \dfrac{1}{y}}$ $(x, y \neq 0)$.

Now Try Exercise 81.

Self Check Answers   **1. a.** no   **b.** $\dfrac{3a}{5b}$ **2. a.** $\dfrac{6a^2}{35b^2}$ **b.** $\dfrac{4a^2b^2}{3r^2s^2}$ **c.** $\dfrac{8pq}{3t}$ **d.** $\dfrac{4mn^2}{3w}$

3. $\dfrac{a}{a+3}$ **4.** $-(a+b)$ **5.** $\dfrac{a+4}{a+3}$ **6.** $\dfrac{x+3}{x^2}$ **7.** $a-1$

8. $x+2$ **9.** 4   **10.** $\dfrac{4x+6}{(x+3)(x-3)^2}$ **11.** $\dfrac{2y}{y-1}$ **12.** $\dfrac{y-x}{y+x}$

Exercises **0.6**

Getting Ready
You should be able to complete these vocabulary and concept statements before you proceed to the practice exercises.

Fill in the blanks.
1. In the fraction $\frac{a}{b}$, a is called the _____.
2. In the fraction $\frac{a}{b}$, b is called the _____.
3. $\frac{a}{b} = \frac{c}{d}$ if and only if _____.
4. The denominator of a fraction can never be _____.

Complete each formula.
5. $\dfrac{a}{b} \cdot \dfrac{c}{d} =$ _____
6. $\dfrac{a}{b} \div \dfrac{c}{d} =$ _____
7. $\dfrac{a}{b} + \dfrac{c}{b} =$ _____
8. $\dfrac{a}{b} - \dfrac{c}{b} =$ _____

Determine whether the fractions are equal. Assume that no denominators are 0.
9. $\dfrac{8x}{3y}$, $\dfrac{16x}{6y}$
10. $\dfrac{3x^2}{4y^2}$, $\dfrac{12y^2}{16x^2}$
11. $\dfrac{25xyz}{12ab^2c}$, $\dfrac{50a^2bc}{24xyz}$
12. $\dfrac{15rs^2}{4rs^2}$, $\dfrac{37.5a^3}{10a^3}$

Practice
Simplify each rational expression. Assume that no denominators are 0.

13. $\dfrac{7a^2b}{21ab^2}$
14. $\dfrac{35p^3q^2}{49p^4q}$

Perform the operations and simplify, whenever possible. Assume that no denominators are 0.

15. $\dfrac{4x}{7} \cdot \dfrac{2}{5a}$
16. $\dfrac{-5y}{2z} \cdot \dfrac{4}{y^2}$
17. $\dfrac{8m}{5n} \div \dfrac{3m}{10n}$
18. $\dfrac{15p}{8q} \div \dfrac{-5p}{16q^2}$
19. $\dfrac{3z}{5c} + \dfrac{2z}{5c}$
20. $\dfrac{7a}{4b} - \dfrac{3a}{4b}$
21. $\dfrac{15x^2y}{7a^2b^3} - \dfrac{x^2y}{7a^2b^3}$
22. $\dfrac{8rst^2}{15m^4t^2} + \dfrac{7rst^2}{15m^4t^2}$

Simplify each fraction. Assume that no denominators are 0.

23. $\dfrac{2x-4}{x^2-4}$
24. $\dfrac{x^2-16}{x^2-8x+16}$
25. $\dfrac{4-x^2}{x^2-5x+6}$
26. $\dfrac{25-x^2}{x^2+10x+25}$
27. $\dfrac{6x^3+x^2-12x}{4x^3+4x^2-3x}$
28. $\dfrac{6x^4-5x^3-6x^2}{2x^3-7x^2-15x}$
29. $\dfrac{x^3-8}{x^2+ax-2x-2a}$
30. $\dfrac{xy+2x+3y+6}{x^3+27}$

Perform the operations and simplify, whenever possible. Assume that no denominators are 0.

31. $\dfrac{x^2-1}{x} \cdot \dfrac{x^2}{x^2+2x+1}$

32. $\dfrac{y^2 - 2y + 1}{y} \cdot \dfrac{y + 2}{y^2 + y - 2}$

33. $\dfrac{3x^2 + 7x + 2}{x^2 + 2x} \cdot \dfrac{x^2 - x}{3x^2 + x}$

34. $\dfrac{x^2 + x}{2x^2 + 3x} \cdot \dfrac{2x^2 + x - 3}{x^2 - 1}$

35. $\dfrac{x^2 + x}{x - 1} \cdot \dfrac{x^2 - 1}{x + 2}$

36. $\dfrac{x^2 + 5x + 6}{x^2 + 6x + 9} \cdot \dfrac{x + 2}{x^2 - 4}$

37. $\dfrac{2x^2 + 32}{8} \div \dfrac{x^2 + 16}{2}$

38. $\dfrac{x^2 + x - 6}{x^2 - 6x + 9} \div \dfrac{x^2 - 4}{x^2 - 9}$

39. $\dfrac{z^2 + z - 20}{z^2 - 4} \div \dfrac{z^2 - 25}{z - 5}$

40. $\dfrac{ax + bx + a + b}{a^2 + 2ab + b^2} \div \dfrac{x^2 - 1}{x^2 - 2x + 1}$

41. $\dfrac{3x^2 + 5x - 2}{x^3 + 2x^2} \div \dfrac{6x^2 + 13x - 5}{2x^3 + 5x^2}$

42. $\dfrac{x^2 + 13x + 12}{8x^2 - 6x - 5} \div \dfrac{2x^2 - x - 3}{8x^2 - 14x + 5}$

43. $\dfrac{x^2 + 7x + 12}{x^3 - x^2 - 6x} \cdot \dfrac{x^2 - 3x - 10}{x^2 + 2x - 3} \cdot \dfrac{x^3 - 4x^2 + 3x}{x^2 - x - 20}$

44. $\dfrac{x(x - 2) - 3}{x(x + 7) - 3(x - 1)} \cdot \dfrac{x(x + 1) - 2}{x(x - 7) + 3(x + 1)}$

45. $\dfrac{x^2 - 2x - 3}{21x^2 - 50x - 16} \cdot \dfrac{3x - 8}{x - 3} \div \dfrac{x^2 + 6x + 5}{7x^2 - 33x - 10}$

46. $\dfrac{x^3 + 27}{x^2 - 4} \div \left(\dfrac{x^2 + 4x + 3}{x^2 + 2x} \div \dfrac{x^2 + x - 6}{x^2 - 3x + 9} \right)$

47. $\dfrac{3}{x + 3} + \dfrac{x + 2}{x + 3}$

48. $\dfrac{3}{x + 1} + \dfrac{x + 2}{x + 1}$

49. $\dfrac{4x}{x - 1} - \dfrac{4}{x - 1}$

50. $\dfrac{6x}{x - 2} - \dfrac{3}{x - 2}$

51. $\dfrac{2}{5 - x} + \dfrac{1}{x - 5}$

52. $\dfrac{3}{x - 6} - \dfrac{2}{6 - x}$

53. $\dfrac{3}{x + 1} + \dfrac{2}{x - 1}$

54. $\dfrac{3}{x + 4} + \dfrac{x}{x - 4}$

55. $\dfrac{a + 3}{a^2 + 7a + 12} + \dfrac{a}{a^2 - 16}$

56. $\dfrac{a}{a^2 + a - 2} + \dfrac{2}{a^2 - 5a + 4}$

57. $\dfrac{x}{x^2 - 4} - \dfrac{1}{x + 2}$

58. $\dfrac{b^2}{b^2 - 4} - \dfrac{4}{b^2 + 2b}$

59. $\dfrac{3x - 2}{x^2 + 2x + 1} - \dfrac{x}{x^2 - 1}$

60. $\dfrac{2t}{t^2 - 25} - \dfrac{t + 1}{t^2 + 5t}$

61. $\dfrac{2}{y^2 - 1} + 3 + \dfrac{1}{y + 1}$

62. $2 + \dfrac{4}{t^2 - 4} - \dfrac{1}{t - 2}$

63. $\dfrac{1}{x - 2} + \dfrac{3}{x + 2} - \dfrac{3x - 2}{x^2 - 4}$

64. $\dfrac{x}{x - 3} - \dfrac{5}{x + 3} + \dfrac{3(3x - 1)}{x^2 - 9}$

65. $\left(\dfrac{1}{x - 2} + \dfrac{1}{x - 3} \right) \cdot \dfrac{x - 3}{2x}$

66. $\left(\dfrac{1}{x + 1} - \dfrac{1}{x - 2} \right) \div \dfrac{1}{x - 2}$

67. $\dfrac{3x}{x - 4} - \dfrac{x}{x + 4} - \dfrac{3x + 1}{16 - x^2}$

68. $\dfrac{7x}{x - 5} + \dfrac{3x}{5 - x} + \dfrac{3x - 1}{x^2 - 25}$

69. $\dfrac{1}{x^2 + 3x + 2} - \dfrac{2}{x^2 + 4x + 3} + \dfrac{1}{x^2 + 5x + 6}$

70. $\dfrac{-2}{x - y} + \dfrac{2}{x - z} - \dfrac{2z - 2y}{(y - x)(z - x)}$

71. $\dfrac{3x - 2}{x^2 + x - 20} - \dfrac{4x^2 + 2}{x^2 - 25} + \dfrac{3x^2 - 25}{x^2 - 16}$

72. $\dfrac{3x + 2}{8x^2 - 10x - 3} + \dfrac{x + 4}{6x^2 - 11x + 3} - \dfrac{1}{4x + 1}$

Simplify each complex fraction. Assume that no denominators are 0.

73. $\dfrac{\dfrac{3a}{b}}{\dfrac{6ac}{b^2}}$

74. $\dfrac{\dfrac{3t^2}{9x}}{\dfrac{t}{18x}}$

75. $\dfrac{\dfrac{3a^2b}{ab}}{27}$

76. $\dfrac{\dfrac{3u^2v}{4t}}{3uv}$

77. $\dfrac{\dfrac{x-y}{ab}}{\dfrac{y-x}{ab}}$

78. $\dfrac{\dfrac{x^2-5x+6}{2x^2y}}{\dfrac{x^2-9}{2x^2y}}$

79. $\dfrac{\dfrac{1}{x}+\dfrac{1}{y}}{xy}$

80. $\dfrac{xy}{\dfrac{11}{x}+\dfrac{11}{y}}$

81. $\dfrac{\dfrac{1}{x}+\dfrac{1}{y}}{\dfrac{1}{x}-\dfrac{1}{y}}$

82. $\dfrac{\dfrac{1}{x}-\dfrac{1}{y}}{\dfrac{1}{x}+\dfrac{1}{y}}$

83. $\dfrac{\dfrac{3a}{b}-\dfrac{4a^2}{x}}{\dfrac{1}{b}+\dfrac{1}{ax}}$

84. $\dfrac{1-\dfrac{x}{y}}{\dfrac{x^2}{y^2}-1}$

85. $\dfrac{x+1-\dfrac{6}{x}}{x+5+\dfrac{6}{x}}$

86. $\dfrac{2z}{1-\dfrac{3}{z}}$

87. $\dfrac{3xy}{1-\dfrac{1}{xy}}$

88. $\dfrac{x-3+\dfrac{1}{x}}{-\dfrac{1}{x}-x+3}$

89. $\dfrac{3x}{x+\dfrac{1}{x}}$

90. $\dfrac{2x^2+4}{2+\dfrac{4x}{5}}$

91. $\dfrac{\dfrac{x}{x+2}-\dfrac{2}{x-1}}{\dfrac{3}{x+2}+\dfrac{x}{x-1}}$

92. $\dfrac{\dfrac{2x}{x-3}+\dfrac{1}{x-2}}{\dfrac{3}{x-3}-\dfrac{x}{x-2}}$

Write each expression without using negative exponents, and simplify the resulting complex fraction. Assume that no denominators are 0.

93. $\dfrac{1}{1+x^{-1}}$

94. $\dfrac{y^{-1}}{x^{-1}+y^{-1}}$

95. $\dfrac{3(x+2)^{-1}+2(x-1)^{-1}}{(x+2)^{-1}}$

96. $\dfrac{2x(x-3)^{-1}-3(x+2)^{-1}}{(x-3)^{-1}(x+2)^{-1}}$

Applications

97. Engineering The stiffness k of the shaft shown in the illustration is given by the following formula where k_1 and k_2 are the individual stiffnesses of each section. Simplify the complex fraction on the right side of the formula.

$$k=\dfrac{1}{\dfrac{1}{k_1}+\dfrac{1}{k_2}}$$

Section 1   Section 2

98. Electronics The combined resistance R of three resistors with resistances of R_1, R_2, and R_3 is given by the following formula. Simplify the complex fraction on the right side of the formula.

$$R=\dfrac{1}{\dfrac{1}{R_1}+\dfrac{1}{R_2}+\dfrac{1}{R_3}}$$

Discovery and Writing

Simplify each complex fraction. Assume that no denominators are 0.

99. $\dfrac{x}{1+\dfrac{1}{3x^{-1}}}$

100. $\dfrac{ab}{2+\dfrac{3}{2a^{-1}}}$

101. $\dfrac{1}{1+\dfrac{1}{1+\dfrac{1}{x}}}$

102. $\dfrac{y}{2+\dfrac{2}{2+\dfrac{2}{y}}}$

103. Explain why the formula $\dfrac{a}{b}+\dfrac{c}{d}=\dfrac{ad+bc}{bd}$ is valid.

104. Explain why the formula $\dfrac{a}{b}\div\dfrac{c}{d}=\dfrac{a}{b}\cdot\dfrac{d}{c}$ is valid.

105. Explain the Commutative Property of Addition and explain why it is useful.

106. Explain the Distributive Property and explain why it is useful.

Review

Write each expression without using absolute value symbols.

107. $|-6|$

108. $|5-x|$, given that $x<0$

Simplify each expression.

109. $\left(\dfrac{x^3y^{-2}}{x^{-1}y}\right)^{-3}$

110. $(27x^6)^{2/3}$

111. $\sqrt{20}-\sqrt{45}$

112. $2(x^2+4)-3(2x^2+5)$

CHAPTER REVIEW

SECTION **0.1**   Sets of Real Numbers

Definitions and Concepts	Examples
Natural numbers: The numbers that we count with.	1, 2, 3, 4, 5, 6, 7, 8, 9, 10, ...
Whole numbers: The natural numbers and 0.	0, 1, 2, 3, 4, 5, 6, 7, 8, 9, 10, ...
Integers: The whole numbers and the negatives of the natural numbers.	... , $-5, -4, -3, -2, -1, 0, 1, 2, 3, 4, 5,$ ...
Rational numbers: $\{x\|x$ can be written in the form $\frac{a}{b}$ $(b \neq 0)$, where a and b are integers.$\}$	$2 = \frac{2}{1}, -5 = -\frac{5}{1}, \frac{2}{3}, -\frac{7}{3}, 0.25 = \frac{1}{4}$
All decimals that either terminate or repeat.	$\frac{3}{4} = 0.25, \frac{13}{5} = 2.6, \frac{2}{3} = 0.666... = 0.\overline{6}, \frac{7}{11} = 0.\overline{63}$
Irrational numbers: Nonrational real numbers. All decimals that neither terminate nor repeat.	$\sqrt{2}, -\sqrt{26}, \pi, 0.232232223$ ...
Real numbers: Any number that can be expressed as a decimal.	$-6, -\frac{9}{13}, 0, \pi, \sqrt{31}, 10.73$
Prime numbers: A natural number greater than 1 that is divisible only by itself and 1.	2, 3, 5, 7, 11, 13, 17, ...
Composite numbers: A natural number greater than 1 that is not prime.	4, 6, 8, 9, 10, 12, 14, 15, ...
Even integers: The integers that are exactly divisible by 2.	..., $-6, -4, -2, 0, 2, 4, 6,$ ...
Odd integers: The integers that are not exactly divisible by 2.	..., $-5, -3, -1, 1, 3, 5,$ ...
Associative Properties: of addition $(a + b) + c = a + (b + c)$	$5 + (4 + 7) = (5 + 4) + 7$
of multiplication $(ab)c = a(bc)$	$(5 \cdot 4) \cdot 7 = 5(4 \cdot 7)$
Commutative Properties: of addition $a + b = b + a$	$4 + 7 = 7 + 4$ $1.7 + 2.5 = 2.5 + 1.7$
of multiplication $ab = ba$	$4 \cdot 7 = 7 \cdot 4$ $1.7(2.5) = 2.5(1.7)$
Distributive Property: $a(b + c) = ab + ac$	$3(x + 6) = 3x + 3 \cdot 6$ $0.2(y - 10) = 0.2y - 0.2(10)$
Double Negative Rule: $-(-a) = a$	$-(-5) = 5$ $-(-a) = a$
Open intervals have no endpoints.	$(-3, 2)$
Closed intervals have two endpoints.	$[-3, 2]$
Half-open intervals have one endpoint.	$(-3, 2]$

Definitions and Concepts	Examples
Absolute value:	
If $x \geq 0$, then $\lvert x \rvert = x$.	$\lvert 7 \rvert = 7 \quad \lvert 3.5 \rvert = 3.5 \quad \left\lvert \dfrac{7}{2} \right\rvert = \dfrac{7}{2} \quad \lvert 0 \rvert = 0$
If $x < 0$, then $\lvert x \rvert = -x$.	$\lvert -7 \rvert = 7 \quad \lvert -3.5 \rvert = 3.5 \quad \left\lvert -\dfrac{7}{2} \right\rvert = \dfrac{7}{2}$
Distance: The **distance** between points a and b on a number line is $d = \lvert b - a \rvert$.	The distance on the number line between points with coordinates of -3 and 2 is $d = \lvert 2 - (-3) \rvert = \lvert 5 \rvert = 5$.

EXERCISES

Consider the set $\left\{ -6, -3, 0, \frac{1}{2}, 3, \pi, \sqrt{5}, 6, 8 \right\}$. List the numbers in this set that are

1. natural numbers.

2. whole numbers.

3. integers.

4. rational numbers.

5. irrational numbers.

6. real numbers.

Consider the set $\left\{ -6, -3, 0, \frac{1}{2}, 3, \pi, \sqrt{5}, 6, 8 \right\}$. List the numbers in this set that are

7. prime numbers.

8. composite numbers.

9. even integers.

10. odd integers.

Determine which property of real numbers justifies each statement.

11. $(a + b) + 2 = a + (b + 2)$

12. $a + 7 = 7 + a$

13. $4(2x) = (4 \cdot 2)x$

14. $3(a + b) = 3a + 3b$

15. $(5a)7 = 7(5a)$

16. $(2x + y) + z = (y + 2x) + z$

17. $-(-6) = 6$

Graph each subset of the real numbers:

18. the prime numbers between 10 and 20

19. the even integers from 6 to 14

Graph each interval on the number line.

20. $-3 < x \leq 5$

21. $x \geq 0$ or $x < -1$

22. $(-2, 4]$

23. $(-\infty, 2) \cap (-5, \infty)$

24. $(-\infty, -4) \cup [6, \infty)$

Write each expression without absolute value symbols.

25. $\lvert 6 \rvert$ 26. $\lvert -25 \rvert$

27. $\lvert 1 - \sqrt{2} \rvert$ 28. $\lvert \sqrt{3} - 1 \rvert$

29. On a number line, find the distance between points with coordinates of -5 and 7.

SECTION **0.2**   Integer Exponents and Scientific Notation

Definitions and Concepts	Examples
Natural-number exponents: $$x^n = \overbrace{x \cdot x \cdot x \cdots x}^{n \text{ factors of } x}$$	$$x^5 = x \cdot x \cdot x \cdot x \cdot x$$

Rules of exponents:
If there are no divisions by 0,

- $x^m x^n = x^{m+n}$
- $(x^m)^n = x^{mn}$
- $(xy)^n = x^n y^n$
- $\left(\dfrac{x}{y}\right)^n = \dfrac{x^n}{y^n}$
- $x^0 = 1 \ (x \neq 0)$
- $x^{-n} = \dfrac{1}{x^n}$
- $\dfrac{x^m}{x^n} = x^{m-n}$
- $\left(\dfrac{x}{y}\right)^{-n} = \left(\dfrac{y}{x}\right)^n$

Examples:

$$x^2 x^5 = x^{2+5} = x^7 \qquad (x^2)^7 = x^{2 \cdot 7} = x^{14}$$

$$(xy)^5 = x^5 y^5 \qquad \left(\frac{x}{y}\right)^5 = \frac{x^5}{y^5}$$

$$6^0 = 1 \qquad 6^{-2} = \frac{1}{6^2} = \frac{1}{36}$$

$$\frac{x^6}{x^4} = x^{6-4} = x^2 \qquad \left(\frac{x}{6}\right)^{-3} = \left(\frac{6}{x}\right)^3 = \frac{6^3}{x^3} = \frac{216}{x^3}$$

Scientific notation:
A number is written in **scientific notation** when it is written in the form $N \times 10^n$, where $1 \leq |N| < 10$.

Write each number in scientific notation.
$$386{,}000 = 3.86 \times 10^5 \qquad 0.0025 = 2.5 \times 10^{-3}$$

Write each number in standard form.
$$7.3 \times 10^3 = 7{,}300 \qquad 5.25 \times 10^{-4} = 0.000525$$

EXERCISES

Write each expression without using exponents.

30. $-5a^3$

31. $(-5a)^2$

Write each expression using exponents.

32. $3ttt$

33. $(-2b)(3b)$

Simplify each expression.

34. $n^2 n^4$

35. $(p^3)^2$

36. $(x^3 y^2)^4$

37. $\left(\dfrac{a^4}{b^2}\right)^3$

38. $(m^{-3} n^0)^2$

39. $\left(\dfrac{p^{-2} q^2}{2}\right)^3$

40. $\dfrac{a^5}{a^8}$

41. $\left(\dfrac{a^2}{b^3}\right)^{-2}$

42. $\left(\dfrac{3x^2 y^{-2}}{x^2 y^2}\right)^{-2}$

43. $\left(\dfrac{a^{-3} b^2}{ab^{-3}}\right)^{-2}$

44. $\left(\dfrac{-3x^3 y}{xy^3}\right)^{-2}$

45. $\left(-\dfrac{2m^{-2} n^0}{4m^2 n^{-1}}\right)^{-3}$

46. If $x = -3$ and $y = 3$, evaluate $-x^2 - xy^2$

Write each number in scientific notation.

47. 6,750

48. 0.00023

Write each number in standard notation.

49. 4.8×10^2

50. 0.25×10^{-3}

51. Use scientific notation to simplify $\dfrac{(45{,}000)(350{,}000)}{0.000105}$.

SECTION 0.3   Rational Exponents and Radicals

Definitions and Concepts	Examples
Summary of $a^{1/n}$ definitions: • If $a \geq 0$, then $a^{1/n}$ is the nonnegative number b such that $b^n = a$. • If $a < 0$ and n is odd, then $a^{1/n}$ is the real number b such that $b^n = a$. • If $a < 0$ and n is even, then $a^{1/n}$ is not a real number.	$16^{1/2} = 4$ because $4^2 = 16$ $(-27)^{1/3} = -3$ because $(-3)^3 = -27$ $(-16)^{1/2}$ is not a real number because no real number squared is -16.
Rule for rational exponents: If m and n are positive integers, $\frac{m}{n}$ is in lowest terms, and $a^{1/n}$ is a real number, then $a^{m/n} = (a^{1/n})^m = (a^m)^{1/n}$	$8^{2/3} = (8^{1/3})^2 = 2^2 = 4$ or $(8^2)^{1/3} = 64^{1/3} = 4$
Definition of $\sqrt[n]{a}$: $\sqrt[n]{a} = a^{1/n}$	$\sqrt[3]{125} = 125^{1/3} = 5$
Properties of radicals: If all radicals are real numbers and there are no divisions by 0, then • $\sqrt[n]{ab} = \sqrt[n]{a}\sqrt[n]{b}$ • $\sqrt[n]{\dfrac{a}{b}} = \dfrac{\sqrt[n]{a}}{\sqrt[n]{b}}$ • $\sqrt[m]{\sqrt[n]{a}} = \sqrt[n]{\sqrt[m]{a}} = \sqrt[mn]{a}$	$\sqrt[5]{32x^{10}} = \sqrt[5]{32}\sqrt[5]{x^{10}} = 2x^2$ $\sqrt[4]{\dfrac{x^{12}}{625}} = \dfrac{\sqrt[4]{x^{12}}}{\sqrt[4]{625}} = \dfrac{x^{12/4}}{5} = \dfrac{x^3}{5}$ $\sqrt[3]{\sqrt{64}} = \sqrt[3]{8} = 2 \quad \sqrt{\sqrt[3]{64}} = \sqrt{4} = 2 \quad \sqrt[2 \cdot 3]{64} = \sqrt[6]{64} = 2$

EXERCISES

Simplify each expression, if possible.

52. $121^{1/2}$

53. $\left(\dfrac{27}{125}\right)^{1/3}$

54. $(32x^5)^{1/5}$

55. $(81a^4)^{1/4}$

56. $(-1,000x^6)^{1/3}$

57. $(-25x^2)^{1/2}$

58. $(x^{12}y^2)^{1/2}$

59. $\left(\dfrac{x^{12}}{y^4}\right)^{-1/2}$

60. $\left(\dfrac{-c^{2/3}c^{5/3}}{c^{-2/3}}\right)^{1/3}$

61. $\left(\dfrac{a^{-1/4}a^{3/4}}{a^{9/2}}\right)^{-1/2}$

Simplify each expression.

62. $64^{2/3}$

63. $32^{-3/5}$

64. $\left(\dfrac{16}{81}\right)^{3/4}$

65. $\left(\dfrac{32}{243}\right)^{2/5}$

66. $\left(\dfrac{8}{27}\right)^{-2/3}$

67. $\left(\dfrac{16}{625}\right)^{-3/4}$

68. $(-216x^3)^{2/3}$

69. $\dfrac{p^{a/2}p^{a/3}}{p^{a/6}}$

Simplify each expression.

70. $\sqrt{36}$

71. $-\sqrt{49}$

72. $\sqrt{\dfrac{9}{25}}$

73. $\sqrt[3]{\dfrac{27}{125}}$

74. $\sqrt{x^2y^4}$

75. $\sqrt[3]{x^3}$

76. $\sqrt[4]{\dfrac{m^8n^4}{p^{16}}}$

77. $\sqrt[5]{\dfrac{a^{15}b^{10}}{c^5}}$

Simplify and combine terms.

78. $\sqrt{50} + \sqrt{8}$

79. $\sqrt{12} + \sqrt{3} - \sqrt{27}$

80. $\sqrt[3]{24x^4} - \sqrt[3]{3x^4}$

Rationalize each denominator.

81. $\dfrac{\sqrt{7}}{\sqrt{5}}$

82. $\dfrac{8}{\sqrt{8}}$

83. $\dfrac{1}{\sqrt[3]{2}}$

84. $\dfrac{2}{\sqrt[3]{25}}$

Rationalize each numerator.

85. $\dfrac{\sqrt{2}}{5}$

86. $\dfrac{\sqrt{5}}{5}$

87. $\dfrac{\sqrt{2x}}{3}$

88. $\dfrac{3\sqrt[3]{7x}}{2}$

SECTION 0.4   Polynomials

Definitions and Concepts	Examples
Monomial: A polynomial with one term.	$2, 3x, 4x^2y, -x^3y^2z$
Binomial: A polynomial with two terms.	$2t + 3, 3r^2 - 6r, 4m + 5n$
Trinomial: A polynomial with three terms.	$3p^2 - 7p + 8, 3m^2 + 2n - p$
The **degree of a monomial** is the sum of the exponents on its variables.	The degree of $4x^2y^3$ is $2 + 3 = 5$
The **degree of a polynomial** is the degree of the term in the polynomial with highest degree.	The degree of the first term of $3p^2q^3 - 6p^3q^4 + 9pq^2$ is $2 + 3 = 5$, the degree of the second term is $3 + 4 = 7$, and the degree of the third term is $1 + 2 = 3$. The degree of the polynomial is the largest of these. It is 7.
Multiplying a monomial times a polynomial: $a(b + c + d + \cdots) = ab + ac + ad + \cdots$	$3(x + 2) = 3x + 6$ $2x(3x^2 - 2y + 3) = 6x^3 - 4xy + 6x$
Addition and subtraction of polynomials: To add or subtract polynomials, remove parentheses and combine like terms.	Add: $(3x^2 + 5x) + (2x^2 - 2x)$. $\begin{aligned}(3x^2 + 5x) + (2x^2 - 2x) &= 3x^2 + 5x + 2x^2 - 2x\\ &= 3x^2 + 2x^2 + 5x - 2x\\ &= 5x^2 + 3x\end{aligned}$ Subtract: $(4a^2 - 5b) - (3a^2 - 7b)$. $\begin{aligned}(4a^2 - 5b) - (3a^2 - 7b) &= 4a^2 - 5b - 3a^2 + 7b\\ &= 4a^2 - 3a^2 - 5b + 7b\\ &= a^2 + 2b\end{aligned}$
Special products: $(x + y)^2 = x^2 + 2xy + y^2$ $(x - y)^2 = x^2 - 2xy + y^2$ $(x + y)(x - y) = x^2 - y^2$	$(2m + 3)^2 = 4m^2 + 12m + 9$ $(4t - 3s)^2 = 16t^2 - 24ts + 9s^2$ $(2m + n)(2m - n) = 4m^2 - 2mn + 2mn - n^2 = 4m^2 - n^2$
Multiplying a binomial times a binomial: Use the FOIL method.	$\begin{aligned}(2a + b)(a - b) &= 2a(a) + 2a(-b) + ba + b(-b)\\ &= 2a^2 - 2ab + ab - b^2\\ &= 2a^2 - ab - b^2\end{aligned}$

Definitions and Concepts	Examples
The **conjugate** of $a + b$ is $a - b$.	
To rationalize the denominator of a radical expression, multiply both the numerator and denominator of the rational expression by the conjugate of the denominator.	Rationalize the denominator: $\dfrac{x}{\sqrt{x} + 2}$. $$\frac{x}{\sqrt{x} + 2} = \frac{x(\sqrt{x} - 2)}{(\sqrt{x} + 2)(\sqrt{x} - 2)}$$ Multiply the numerator and denominator by the conjugate of $\sqrt{x} + 2$. $$= \frac{x\sqrt{x} - 2x}{x - 4}$$ Simplify.
Division of polynomials: To divide polynomials, use long division.	Divide: $2x + 3 \overline{)6x^3 + 7x^2 - x + 3}$. $$\begin{array}{r} 3x^2 - x + 1 \\ 2x + 3 \overline{)6x^3 + 7x^2 - x + 3} \\ \underline{6x^3 + 9x^2} \\ -2x^2 - x \\ \underline{-2x^2 - 3x} \\ +2x + 3 \\ \underline{+2x + 3} \\ 0 \end{array}$$

EXERCISES

Give the degree of each polynomial and tell whether the polynomial is a monomial, a binomial, or a trinomial.

89. $x^3 - 8$

90. $8x - 8x^2 - 8$

91. $\sqrt{3}x^2$

92. $4x^4 - 12x^2 + 1$

Perform the operations and simplify.

93. $2(x + 3) + 3(x - 4)$

94. $3x^2(x - 1) - 2x(x + 3) - x^2(x + 2)$

95. $(3x + 2)(3x + 2)$

96. $(3x + y)(2x - 3y)$

97. $(4a + 2b)(2a - 3b)$

98. $(z + 3)(3z^2 + z - 1)$

99. $(a^n + 2)(a^n - 1)$

100. $\left(\sqrt{2} + x\right)^2$

101. $\left(\sqrt{2} + 1\right)\left(\sqrt{3} + 1\right)$

102. $\left(\sqrt[3]{3} - 2\right)\left(\sqrt[3]{9} + 2\sqrt[3]{3} + 4\right)$

Rationalize each denominator.

103. $\dfrac{2}{\sqrt{3} - 1}$

104. $\dfrac{-2}{\sqrt{3} - \sqrt{2}}$

105. $\dfrac{2x}{\sqrt{x} - 2}$

106. $\dfrac{\sqrt{x} - \sqrt{y}}{\sqrt{x} + \sqrt{y}}$

Rationalize each numerator.

107. $\dfrac{\sqrt{x} + 2}{5}$

108. $\dfrac{1 - \sqrt{a}}{a}$

Perform each division.

109. $\dfrac{3x^2y^2}{6x^3y}$

110. $\dfrac{4a^2b^3 + 6ab^4}{2b^2}$

111. $2x + 3 \overline{)2x^3 + 7x^2 + 8x + 3}$

112. $x^2 - 1 \overline{)x^5 + x^3 - 2x - 3x^2 - 3}$

SECTION 0.5   Factoring Polynomials

Definitions and Concepts	Examples
Factoring out a common monomial:	
$ab + ac = a(b + c)$	$3p^3 - 6p^2q + 9p = 3p(p^2 - 2pq + 3)$
Factoring the difference of two squares:	
$x^2 - y^2 = (x + y)(x - y)$	$4x^2 - 9 = (2x + 3)(2x - 3)$
Factoring trinomials:	
• Trinomial squares	
$\quad x^2 + 2xy + y^2 = (x + y)^2$	$9a^2 + 12ab + 4b^2 = (3a + 2b)(3a + 2b) = (3a + 2b)^2$
$\quad x^2 - 2xy + y^2 = (x - y)^2$	$r^2 - 4rs + 4s^2 = (r - 2s)(r - 2s) = (r - 2s)^2$
• To factor general trinomials use trial and error or grouping.	$6x^2 - 5x - 6 = (2x - 3)(3x + 2)$
Factoring the sum and difference of two cubes:	
$x^3 + y^3 = (x + y)(x^2 - xy + y^2)$	$r^3 + 8 = (r + 2)(r^2 - 2r + 4)$
$x^3 - y^3 = (x - y)(x^2 + xy + y^2)$	$27a^3 - 8b^3 = (3a - 2b)(9a^2 + 6ab + 4b^2)$

EXERCISES

Factor each expression completely, if possible.

113. $3t^3 - 3t$

114. $5r^3 - 5$

115. $6x^2 + 7x - 24$

116. $3a^2 + ax - 3a - x$

117. $8x^3 - 125$

118. $6x^2 - 20x - 16$

119. $x^2 + 6x + 9 - t^2$

120. $3x^2 - 1 + 5x$

121. $8z^3 + 343$

122. $1 + 14b + 49b^2$

123. $121z^2 + 4 - 44z$

124. $64y^3 - 1,000$

125. $2xy - 4zx - wy + 2zw$ **126.** $x^8 + x^4 + 1$

SECTION 0.6   Algebraic Fractions

Definitions and Concepts	Examples
Properties of fractions: If there are no divisions by 0, then	
• $\dfrac{a}{b} = \dfrac{c}{d}$ if and only if $ad = bc$.	$\dfrac{3x}{4} = \dfrac{6x}{8}$ because $(3x)8$ and $4(6x)$ both equal $24x$.
• $\dfrac{a}{b} = \dfrac{ax}{bx}$	$\dfrac{6x^2}{8x^3} = \dfrac{3 \cdot 2 \cdot \cancel{x} \cdot \cancel{x}}{2 \cdot 4 \cdot \cancel{x} \cdot \cancel{x} \cdot x} = \dfrac{3}{4x}$
• $\dfrac{a}{b} \cdot \dfrac{c}{d} = \dfrac{ac}{bd}$	$\dfrac{3p}{2q} \cdot \dfrac{2p}{6q} = \dfrac{3p \cdot 2p}{2q \cdot 6q} = \dfrac{3p \cdot 2p}{2q \cdot 3 \cdot 2q} = \dfrac{p^2}{2q^2}$
• $\dfrac{a}{b} \div \dfrac{c}{d} = \dfrac{ad}{bc}$	$\dfrac{2t}{3s} \div \dfrac{2t}{6s} = \dfrac{2t}{3s} \cdot \dfrac{6s}{2t} = \dfrac{2t \cdot 6s}{3s \cdot 2t} = \dfrac{2t \cdot 2 \cdot 3\cancel{s}}{3\cancel{s}2\cancel{t}} = 2$

Definitions and Concepts	Examples
• $\dfrac{a}{b} + \dfrac{c}{b} = \dfrac{a+c}{b}$ • $\dfrac{a}{b} - \dfrac{c}{b} = \dfrac{a-c}{b}$	$\dfrac{x}{4} + \dfrac{y}{4} = \dfrac{x+y}{4}$ $\dfrac{3p}{2q} - \dfrac{p}{2q} = \dfrac{3p-p}{2q} = \dfrac{2p}{2q} = \dfrac{p}{q}$
• $a \cdot 1 = a$ • $\dfrac{a}{1} = a$ • $\dfrac{a}{a} = 1$	$7 \cdot 1 = 1$ $\dfrac{7}{1} = 7$ $\dfrac{7}{7} = 1$
• $\dfrac{a}{b} = \dfrac{-a}{-b} = -\dfrac{a}{-b} = -\dfrac{-a}{b}$	$\dfrac{7}{2} = \dfrac{-7}{-2} = -\dfrac{7}{-2} = -\dfrac{-7}{2}$
• $-\dfrac{a}{b} = \dfrac{a}{-b} = \dfrac{-a}{b} = -\dfrac{-a}{-b}$	$-\dfrac{7}{2} = \dfrac{7}{-2} = \dfrac{-7}{2} = -\dfrac{-7}{-2}$
Simplifying rational expressions: To simplify a rational expression, factor the numerator and denominator, if possible, and divide out factors that are common to the numerator and denominator.	To simplify $\dfrac{2-x}{2x-4}$, factor -1 from the numerator and 2 from the denominator to get $$\dfrac{2-x}{2x-4} = \dfrac{-1(-2+x)}{2(x-2)} = \dfrac{-1(x-2)}{2(x-2)} = \dfrac{-1}{2} = -\dfrac{1}{2}$$
Adding or subtracting rational expressions: To add or subtract rational expressions with unlike denominators, find the LCD of the expressions, write each expression with a denominator that is the LCD, add or subtract the expressions, and simplify the result, if possible.	$\dfrac{2x}{x+2} - \dfrac{2x}{x-3} = \dfrac{2x(x-3)}{(x+2)(x-3)} - \dfrac{2x(x+2)}{(x-3)(x+2)}$ $= \dfrac{2x(x-3) - 2x(x+2)}{(x+2)(x-3)}$ $= \dfrac{2x^2 - 6x - 2x^2 - 4x}{(x+2)(x-3)}$ $= \dfrac{-10x}{(x+2)(x-3)}$
Complex fractions: To simplify complex fractions, multiply the numerator and denominator of the complex fraction by the LCD of all the fractions.	$\dfrac{1 - \dfrac{y}{2}}{\dfrac{1}{y} + \dfrac{1}{2}} = \dfrac{2y\left(1 - \dfrac{y}{2}\right)}{2y\left(\dfrac{1}{y} + \dfrac{1}{2}\right)} = \dfrac{2y - y^2}{2 + y} = \dfrac{y(2-y)}{2+y}$

EXERCISES

Simplify each rational expression.

127. $\dfrac{2-x}{x^2 - 4x + 4}$

128. $\dfrac{a^2 - 9}{a^2 - 6a + 9}$

Perform each operation and simplify. Assume that no denominators are 0.

129. $\dfrac{x^2 - 4x + 4}{x + 2} \cdot \dfrac{x^2 + 5x + 6}{x - 2}$

130. $\dfrac{2y^2 - 11y + 15}{y^2 - 6y + 8} \cdot \dfrac{y^2 - 2y - 8}{y^2 - y - 6}$

131. $\dfrac{2t^2 + t - 3}{3t^2 - 7t + 4} \div \dfrac{10t + 15}{3t^2 - t - 4}$

132. $\dfrac{p^2 + 7p + 12}{p^3 + 8p^2 + 4p} \div \dfrac{p^2 - 9}{p^2}$

133. $\dfrac{x^2 + x - 6}{x^2 - x - 6} \cdot \dfrac{x^2 - x - 6}{x^2 + x - 2} \div \dfrac{x^2 - 4}{x^2 - 5x + 6}$

134. $\left(\dfrac{2x + 6}{x + 5} \div \dfrac{2x^2 - 2x - 4}{x^2 - 25}\right) \dfrac{x^2 - x - 2}{x^2 - 2x - 15}$

135. $\dfrac{2}{x - 4} + \dfrac{3x}{x + 5}$

136. $\dfrac{5x}{x - 2} - \dfrac{3x + 7}{x + 2} + \dfrac{2x + 1}{x + 2}$

137. $\dfrac{x}{x - 1} + \dfrac{x}{x - 2} + \dfrac{x}{x - 3}$

138. $\dfrac{x}{x + 1} - \dfrac{3x + 7}{x + 2} + \dfrac{2x + 1}{x + 2}$

139. $\dfrac{3(x + 1)}{x} - \dfrac{5(x^2 + 3)}{x^2} + \dfrac{x}{x + 1}$

140. $\dfrac{3x}{x + 1} + \dfrac{x^2 + 4x + 3}{x^2 + 3x + 2} - \dfrac{x^2 + x - 6}{x^2 - 4}$

Simplify each complex fraction. Assume that no denominators are 0.

141. $\dfrac{\dfrac{5x}{2}}{\dfrac{3x^2}{8}}$

142. $\dfrac{\dfrac{3x}{y}}{\dfrac{6x}{y^2}}$

143. $\dfrac{\dfrac{1}{x} + \dfrac{1}{y}}{x - y}$

144. $\dfrac{x^{-1} + y^{-1}}{y^{-1} - x^{-1}}$

CHAPTER TEST

Consider the set $\{-7, -\tfrac{2}{3}, 0, 1, 3, \sqrt{10}, 4\}.$

1. List the numbers in the set that are odd integers.

2. List the numbers in the set that are prime numbers.

Determine which property justifies each statement.

3. $(a + b) + c = (b + a) + c$

4. $a(b + c) = ab + ac$

Graph each interval on a number line.

5. $-4 < x \le 2$

6. $(-\infty, -3) \cup [6, \infty)$

Write each expression without using absolute value symbols.

7. $|-17|$

8. $|x - 7|$, when $x < 0$

Find the distance on a number line between points with the following coordinates.

9. -4 and 12

10. -20 and -12

Simplify each expression. Assume that all variables represent positive numbers, and write all answers without using negative exponents.

11. $x^4 x^5 x^2$

12. $\dfrac{r^2 r^3 s}{r^4 s^2}$

13. $\dfrac{(a^{-1}a^2)^{-2}}{a^{-3}}$

14. $\left(\dfrac{x^0 x^2}{x^{-2}}\right)^6$

Write each number in scientific notation.

15. $450,000$

16. 0.000345

Write each number in standard notation.

17. 3.7×10^3

18. 1.2×10^{-3}

Simplify each expression. Assume that all variables represent positive numbers, and write all answers without using negative exponents.

19. $(25a^4)^{1/2}$

20. $\left(\dfrac{36}{81}\right)^{3/2}$

21. $\left(\dfrac{8t^6}{27s^9}\right)^{-2/3}$

22. $\sqrt[3]{27a^6}$

23. $\sqrt{12} + \sqrt{27}$

24. $2\sqrt[3]{3x^4} - 3x\sqrt[3]{24x}$

25. Rationalize the denominator: $\dfrac{x}{\sqrt{x} - 2}$.

26. Rationalize the numerator: $\dfrac{\sqrt{x} - \sqrt{y}}{\sqrt{x} + \sqrt{y}}$.

Perform each operation.

27. $(a^2 + 3) - (2a^2 - 4)$

28. $(3a^3b^2)(-2a^3b^4)$

29. $(3x - 4)(2x + 7)$

30. $(a^n + 2)(a^n - 3)$

31. $(x^2 + 4)(x^2 - 4)$

32. $(x^2 - x + 2)(2x - 3)$

33. $x - 3 \overline{)6x^2 + x - 23}$

34. $2x - 1 \overline{)2x^3 + 3x^2 - 1}$

Factor each polynomial.

35. $3x + 6y$

36. $x^2 - 100$

37. $10t^2 - 19tw + 6w^2$

38. $3a^3 - 648$

39. $x^4 - x^2 - 12$

40. $6x^4 + 11x^2 - 10$

Perform each operation and simplify if possible. Assume that no denominators are 0.

41. $\dfrac{x}{x + 2} + \dfrac{2}{x + 2}$

42. $\dfrac{x}{x + 1} - \dfrac{x}{x - 1}$

43. $\dfrac{x^2 + x - 20}{x^2 - 16} \cdot \dfrac{x^2 - 25}{x - 5}$

44. $\dfrac{x + 2}{x^2 + 2x + 1} \div \dfrac{x^2 - 4}{x + 1}$

Simplify each complex fraction. Assume that no denominators are 0.

45. $\dfrac{\dfrac{1}{a} + \dfrac{1}{b}}{\dfrac{1}{b}}$

46. $\dfrac{x^{-1}}{x^{-1} + y^{-1}}$

Equations and Inequalities

1

CAREERS AND MATHEMATICS:　Marketing

© Istockphoto.com/Arthur Kwiatkowski

People who work in the field of marketing coordinate their companies' market research, marketing strategy, sales, advertising, promotion, pricing, product development, and public relations activities. Advertising managers direct a firm's promotional campaign while marketing managers promote products and services. Promotions managers work on incentives to increase sales while public relations managers plan and direct the public image for the employer. Sales managers plan and direct the distribution of the product to the customer.

Education and Mathematics Required
*   For marketing, sales, and promotions management positions, employers often prefer a bachelor's or master's degree in business administration with an emphasis on marketing. For advertising management positions, some employers prefer a bachelor's degree in advertising or journalism. For public relations management positions, some employers prefer a bachelor's or master's degree in public relations or journalism.
*   College Algebra, Business Calculus I and II, and Economic Statistics are required.

How Marketing Managers Use Math and Who Employs Them
*   Statistics is used for predicting sales and effectiveness of advertising campaigns.
*   These managers are found in virtually every industry. They are employed in wholesale trade, retail trade, manufacturing, the finance and insurance industries, as well as professional, scientific, and technical services, public and private educational services, and healthcare.

Career Outlook and Salary
*   Overall employment of advertising, marketing, promotions, public relations, and sales managers is expected to increase by 13 percent through 2018.
*   The median annual wages in May 2008 were $80,220 for advertising and promotions managers, $108,580 for marketing managers, $97,260 for sales managers, and $89,430 for public relations managers.

For more information see: www.bls.gov/oco

The main topic of this chapter is equations—one of the most important concepts in algebra. Equations are used in almost every academic discipline and especially in chemistry, physics, medicine, computer science, and business.

1.1 Linear Equations and Rational Equations

In this section, we will learn to

1. Use the properties of equality.
2. Solve linear equations.
3. Solve rational equations.
4. Solve formulas for a specific variable.

It's been said that "weddings today are as beautiful as they are expensive." Suppose a couple budgets $3,000 for their wedding reception at a historic home. If $600 is charged for renting the home and there is a $24-per-person fee for food and beverages, how many guests can the couple accommodate at their reception?

If we let the **variable** x represent the number of guests, the expression $600 + 24x$ represents the cost for the reception. That is, $600 for the home rental plus $24 times the number of guests, x. We want to know the value of x that makes the expression equal $3,000.

We can write the statement $600 + 24x = 3000$ to indicate that the two quantities are equal. A statement indicating that two quantities are equal is called an **equation.** In this section, we will review how to solve equations of this type.

If $x = 100$, the equation is true because when we substitute 100 for x, we obtain a true statement.

$$600 + 24(\mathbf{100}) = 3000$$

$$600 + 2400 = 3000$$

$$3000 = 3000$$

The couple can accommodate 100 guests at their wedding reception.

An equation can be either true or false. For example, the equation $2 + 2 = 4$ is true, and the equation $2 + 3 = 6$ is false. An equation such as $3x - 2 = 10$ can be true or false depending on the value of x.

If $x = 4$, the equation is true, because 4 satisfies the equation.

$$3x - 2 = 10$$

$$3(\mathbf{4}) - 2 \stackrel{?}{=} 10 \qquad \text{Substitute 4 for } x. \text{ Read } \stackrel{?}{=} \text{ as "is possibly equal to."}$$

$$12 - 2 \stackrel{?}{=} 10$$

$$10 = 10$$

This equation is false for all other values of x.

Any number that satisfies an equation is called a **solution** or **root** of the equation. The set of all solutions of an equation is called its **solution set.** We have seen that the solution set of $3x - 2 = 10$ is $\{4\}$. To **solve** an equation means to find its solution set.

There can be restrictions on the values of a variable. For example, in the fraction

$$\frac{x^2 + 4}{x - 2}$$

we cannot replace x with 2, because that would make the denominator equal to 0.

EXAMPLE 1   **Finding the Restrictions on the Values of a Variable**

Find the restrictions on the values of b in the equation: $\dfrac{7b}{b + 6} = \dfrac{2}{b - 1}$.

SOLUTION For $\frac{7b}{b+6}$ to be a real number, b cannot be -6 because that would make the denominator 0. For $\frac{2}{b-1}$ to be a real number, b cannot be 1 because that would make the denominator 0. Thus, the values of b are restricted to the set of all real numbers except 1 or -6.

∎

Self Check 1 Find the restrictions on a: $\dfrac{5}{a+5} = \dfrac{3a}{a-2}$.

Now Try Exercise 13.

There are three types of equations: identities, contradictions, and conditional equations. These are defined and illustrated in the following table.

Type of Equation	Definition	Example
Identity	Every acceptable real number replacement for the variable is a solution.	$x^2 - 9 = (x+3)(x-3)$ Every real number x is a solution.
Contradiction	No real number is a solution.	$x = x + 1$ The equation has no solution. No real number can be 1 greater than itself.
Conditional Equation	Solution set contains some but not all real numbers.	$3x - 2 = 10$ The equation has one solution, the number 4.

Two equations with the same solution set are called **equivalent equations.**

1. Use the Properties of Equality

There are certain properties of equality that we can use to transform equations into equivalent but less complicated equations. If we use these properties, the resulting equations will be equivalent and will have the same solution set.

Properties of Equality

The Addition and Subtraction Properties
If a, b, and c are real numbers and $a = b$, then

$$a + c = b + c \qquad \text{and} \qquad a - c = b - c$$

The Multiplication and Division Properties
If a, b, and c are real numbers and $a = b$, then

$$ac = bc \qquad \text{and} \qquad \frac{a}{c} = \frac{b}{c} \qquad (c \neq 0)$$

The Substitution Property
In an equation, a quantity may be substituted for its equal without changing the truth of the equation.

2. Solve Linear Equations

The easiest equations to solve are the **first-degree** or **linear equations.** Since these equations involve first-degree polynomials, they are also called **first-degree polynomial equations.**

Linear Equations   A **linear equation in one variable** (say, x) is any equation that can be written in the form

$$ax + b = 0 \qquad (a \text{ and } b \text{ are real numbers and } a \neq 0)$$

To solve the linear equation $2x + 3 = 0$, we subtract 3 from both sides of the equation and divide both sides by 2.

$$2x + 3 = 0$$

$$2x + 3 - 3 = 0 - 3 \qquad \text{To undo the addition of 3, subtract 3 from both sides.}$$

$$2x = -3$$

$$\frac{2x}{2} = -\frac{3}{2} \qquad \text{To undo the multiplication by 2, divide both sides by 2.}$$

$$x = -\frac{3}{2}$$

To verify that $-\frac{3}{2}$ satisfies the equation, we substitute $-\frac{3}{2}$ for x and simplify:

$$2x + 3 = 0$$

$$2\left(-\frac{3}{2}\right) + 3 \overset{?}{=} 0 \qquad \text{Substitute } -\frac{3}{2} \text{ for } x.$$

$$-3 + 3 \overset{?}{=} 0 \qquad 2\left(-\frac{3}{2}\right) = -3$$

$$0 = 0$$

Since both sides of the equation are equal, the solution checks.

In Exercise 74, you will be asked to solve the general linear equation $ax + b = 0$ for x, thereby showing that every conditional linear equation has exactly one solution.

EXAMPLE 2   **Solving a Linear Equation**

Find the solution set: $3(x + 2) = 5x + 2$.

SOLUTION   We proceed as follows:

$$3(x + 2) = 5x + 2$$

$$3x + 6 = 5x + 2 \qquad \text{Use the Distributive Property and remove parentheses.}$$

$$3x + 6 - 3x = 5x - 3x + 2 \qquad \text{Subtract } 3x \text{ from both sides.}$$

$$6 = 2x + 2 \qquad \text{Combine like terms.}$$

$$6 - 2 = 2x + 2 - 2 \qquad \text{Subtract 2 from both sides.}$$

$$4 = 2x \qquad \text{Simplify.}$$

$$\frac{4}{2} = \frac{2x}{2} \qquad \text{Divide both sides by 2.}$$

$$2 = x \qquad \text{Simplify.}$$

Since all of the above equations are equivalent, the solution set of the original equation is $\{2\}$. Verify that 2 satisfies the equation.   ∎

Self Check 2   Find the solution set: $4(x - 3) = 7x - 3$.

Now Try Exercise 35.

ACCENT ON TECHNOLOGY   **Checking Solutions to Linear Equations**

Using the table feature on a graphing calculator, we can easily check the solution of a linear equation. We enter each side of the equation into the graph editor; go to the table, and then enter the value of x that we found as the solution. The value of both entries in the table should be the same. These steps are shown in Figure 1-1 below for Example 2. We see that both Y_1 and Y_2 equal 12 when $x = 2$. This verifies the solution.

Enter each side of the equation.   **Go to the table and enter the value $x = 2$.**

   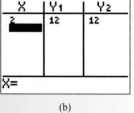

(a)   (b)

FIGURE 1-1

EXAMPLE 3   **Solving a Linear Equation with Fractions**

Find the solution set: $\dfrac{3}{2}y - \dfrac{2}{3} = \dfrac{1}{5}y$.

SOLUTION   To clear the equation of fractions, we multiply both sides by the least common denominator (LCD) of the three fractions and proceed as follows:

$$\frac{3}{2}y - \frac{2}{3} = \frac{1}{5}y$$

$$30\left(\frac{3}{2}y - \frac{2}{3}\right) = 30\left(\frac{1}{5}y\right) \qquad \text{Multiply both sides by 30, the LCD of } \frac{3}{2}, \frac{2}{3}, \text{ and } \frac{1}{5}.$$

$$45y - 20 = 6y \qquad \text{Remove parentheses and simplify.}$$

$$45y - 20 + 20 = 6y + 20 \qquad \text{Add 20 to both sides.}$$

$$45y = 6y + 20 \qquad \text{Simplify.}$$

$$45y - 6y = 6y - 6y + 20 \qquad \text{Subtract } 6y \text{ from both sides.}$$

$$39y = 20 \qquad \text{Combine like terms.}$$

$$\frac{39y}{39} = \frac{20}{39} \qquad \text{Divide both sides by 39.}$$

$$y = \frac{20}{39} \qquad \text{Simplify.}$$

The solution set is $\left\{\dfrac{20}{39}\right\}$. Verify that $\dfrac{20}{39}$ satisfies the equation.   ∎

Self Check 3   Find the solution set: $\dfrac{2}{3}p - 3 = \dfrac{p}{6}$.

Now Try Exercise 43.

EXAMPLE 4 **Solving Linear Equations**

Solve: **a.** $3(x + 5) = 3(1 + x)$ **b.** $5 + 5(x + 2) - 2x = 3x + 15$

SOLUTION **a.**

$$3(x + 5) = 3(1 + x)$$
$$3x + 15 = 3 + 3x \qquad \text{Remove parentheses.}$$
$$3x - 3x + 15 = 3 + 3x - 3x \qquad \text{Subtract } 3x \text{ from both sides.}$$
$$15 = 3 \qquad \text{Combine like terms.}$$

Since $15 = 3$ is false, the equation has no roots. Its solution set is the empty set, which is denoted as $\varnothing$. This equation is a contradiction.

b. $5 + 5(x + 2) - 2x = 3x + 15$

$$5 + 5x + 10 - 2x = 3x + 15 \qquad \text{Remove parentheses.}$$
$$3x + 15 = 3x + 15 \qquad \text{Simplify.}$$

Because both sides of the final equation are identical, every value of x will make the equation true. The solution set is the set of all real numbers. This equation is an identity.

∎

Self Check 4 Solve: **a.** $-2(x - 4) + 6x = 4(x + 1)$
b. $2(x + 1) + 4 = 2(x + 3)$

Now Try Exercise 51.

3. Solve Rational Equations

Rational equations are equations that contain rational expressions. Some examples of rational equations are

$$\frac{2}{x - 3} = 7, \qquad \frac{x + 1}{x - 2} = \frac{3}{x - 2}, \qquad \text{and} \qquad \frac{x + 2}{x + 3} + \frac{1}{x^2 + 2x - 3} = 1$$

When solving these equations, we will multiply both sides by a quantity containing a variable. When we do this, we could inadvertently multiply both sides of an equation by 0 and obtain a solution that makes the denominator of a fraction 0. In this case, we have found a false solution, called an **extraneous solution.** These solutions do not satisfy the equation and must be discarded.

The following equation has an extraneous solution.

$$\frac{x + 1}{x - 2} = \frac{3}{x - 2}$$
$$(x - 2)\left(\frac{x + 1}{x - 2}\right) = (x - 2)\left(\frac{3}{x - 2}\right) \qquad \text{Multiply both sides by } x - 2.$$
$$x + 1 = 3 \qquad \frac{x - 2}{x - 2} = 1$$
$$x = 2 \qquad \text{Subtract 1 from both sides.}$$

If we check by substituting 2 for x, we obtain 0's in the denominator. Thus, 2 is not a root. The solution set is $\varnothing$.

EXAMPLE 5 **Solving a Rational Equation**

Solve: $\dfrac{x + 2}{x + 3} + \dfrac{1}{x^2 + 2x - 3} = 1$.

SOLUTION   Note that x cannot be -3, because that would cause the denominator of the first fraction to be 0. To find other restrictions, we factor the trinomial in the denominator of the second fraction.

$$x^2 + 2x - 3 = (x + 3)(x - 1)$$

Since the denominator will be 0 when $x = -3$ or $x = 1$, x cannot be -3 or 1.

$$\frac{x + 2}{x + 3} + \frac{1}{x^2 + 2x - 3} = 1$$

$$\frac{x + 2}{x + 3} + \frac{1}{(x + 3)(x - 1)} = 1 \qquad \text{Factor } x^2 + 2x - 3.$$

$$(x + 3)(x - 1)\left[\frac{x + 2}{x + 3} + \frac{1}{(x + 3)(x - 1)}\right] = (x + 3)(x - 1)1 \qquad \text{Multiply both sides by } (x + 3)(x - 1).$$

$$\cancel{(x + 3)}(x - 1)\left(\frac{x + 2}{\cancel{x + 3}}\right) + \cancel{(x + 3)(x - 1)}\frac{1}{\cancel{(x + 3)(x - 1)}} = (x + 3)(x - 1)1 \qquad \text{Remove brackets.}$$

$$(x - 1)(x + 2) + 1 = (x + 3)(x - 1) \qquad \text{Simplify.}$$

$$x^2 + x - 2 + 1 = x^2 + 2x - 3 \qquad \text{Multiply the binomials.}$$

$$x - 1 = 2x - 3 \qquad \begin{array}{l}\text{Subtract } x^2 \text{ from both sides}\\ \text{and combine like terms.}\end{array}$$

$$2 = x \qquad \text{Add 3 and subtract } x \text{ from both sides.}$$

Because 2 is a meaningful replacement for x, it is a root. However, it is a good idea to check it.

$$\frac{x + 2}{x + 3} + \frac{1}{x^2 + 2x - 3} = 1$$

$$\frac{2 + 2}{2 + 3} + \frac{1}{2^2 + 2(2) - 3} \stackrel{?}{=} 1 \qquad \text{Substitute 2 for } x.$$

$$\frac{4}{5} + \frac{1}{5} \stackrel{?}{=} 1$$

$$1 = 1$$

Since 2 satisfies the equation, it is a root.

Self Check 5   Solve: $\dfrac{3}{5} + \dfrac{7}{x + 2} = 2$.

Now Try Exercise 61.

4. Solve Formulas for a Specific Variable

Many equations, called **formulas,** contain several variables. For example, the formula that converts degrees Celsius to degrees Fahrenheit is $F = \frac{9}{5}C + 32$. If we want to change a large number of Fahrenheit readings to degrees Celsius, it would be tedious to substitute each value of F into the formula and then repeatedly solve it for C. It is better to solve the formula for C, substitute the values for F, and evaluate C directly.

EXAMPLE 6   **Solving a Formula for a Specific Variable**

Solve $F = \dfrac{9}{5}C + 32$ for C.

SOLUTION   We use the same methods as for solving linear equations.

$$F = \frac{9}{5}C + 32$$

$$F - 32 = \frac{9}{5}C \qquad \text{Subtract 32 from both sides.}$$

$$\frac{5}{9}(F - 32) = \frac{5}{9}\left(\frac{9}{5}C\right) \qquad \text{Multiply both sides by } \frac{5}{9}.$$

$$\frac{5}{9}(F - 32) = C \qquad \text{Simplify.}$$

This result can also be written in the form $C = \dfrac{5F - 160}{9}$.

Self Check 6   Solve $C = \dfrac{5}{9}(F - 32)$ for F.

Now Try Exercise 79.

EXAMPLE 7   **Solving a Formula for a Specific Variable**

The formula $A = P + Prt$ is used to find the amount of money in a savings account at the end of a specified time. A represents the amount, P represents the principal (the original deposit), r represents the rate of simple interest per unit of time, and t represents the number of units of time. Solve this formula for P.

SOLUTION   We factor P from both terms on the right side of the equation and proceed as follows:

$$A = P + Prt$$

$$A = P(1 + rt) \qquad \text{Factor out } P.$$

$$\frac{A}{1 + rt} = P \qquad \text{Divide both sides by } 1 + rt.$$

$$P = \frac{A}{1 + rt}$$

Self Check 7   Solve $pq = fq + fp$ for f.

Now Try Exercise 85.

Self Check Answers   **1.** all real numbers except 2 or -5 **2.** $\{-3\}$ **3.** $\{6\}$
4. a. no solution   **b.** all real numbers   **5.** 3   **6.** $F = \dfrac{9}{5}C + 32$
7. $f = \dfrac{pq}{q + p}$

Exercises **1.1**

Getting Ready

You should be able to complete these vocabulary and concept statements before you proceed to the practice exercises.

Fill in the blanks.

1. If a number satisfies an equation, it is called a _____ or a _____ of the equation.
2. If an equation is true for all values of its variable, it is called an _____.
3. A contradiction is an equation that is true for _____ values of its variable.
4. A _____ equation is true for some values of its variable and is not true for others.
5. An equation of the form $ax + b = 0$ is called a _____ equation.
6. If an equation contains rational expressions, it is called a _____ equation.
7. A conditional linear equation has _____ root.
8. The _____ of a fraction can never be 0.

Practice

Each quantity represents a real number. Find any restrictions on x.

9. $x + 3 = 1$

10. $\dfrac{1}{2}x - 7 = 14$

11. $\dfrac{1}{x} = 12$

12. $\dfrac{3}{x - 2} = 9x$

13. $\dfrac{8}{x - 3} = \dfrac{5}{x + 2}$

14. $\dfrac{x}{x - 3} = -\dfrac{4}{x + 4}$

15. $\dfrac{1}{x - 3} = \dfrac{5x}{x^2 - 16}$

16. $\dfrac{1}{x^2 - 3x - 4} = \dfrac{5}{x} + 2$

Solve each equation, if possible. Classify each one as an identity, a conditional equation, or a contradiction.

17. $2x + 5 = 15$

18. $3x + 2 = x + 8$

19. $2(n + 2) - 5 = 2n$

20. $3(m + 2) = 2(m + 3) + m$

21. $\dfrac{x + 7}{2} = 7$

22. $\dfrac{x}{2} - 7 = 14$

23. $2(a + 1) = 3(a - 2) - a$

24. $x^2 = (x + 4)(x - 4) + 16$

25. $3(x - 3) = \dfrac{6x - 18}{2}$

26. $x(x + 2) = (x + 1)^2$

27. $\dfrac{3}{b - 3} = 1$

28. $x^2 - 8x + 15 = (x - 3)(x + 5)$

29. $2x^2 + 5x - 3 = (2x - 1)(x + 3)$

30. $2x^2 + 5x - 3 = 2x\left(x + \dfrac{19}{2}\right)$

Solve each equation. If an equation has no solution, so indicate.

31. $2x + 7 = 10 - x$

32. $9a - 3 = 15 + 3a$

33. $5(x - 2) = 2(x + 4)$

34. $5(r - 4) = -5(r - 4)$

35. $7(2x + 5) - 6(x + 8) = 7$

36. $6(x - 5) - 4(x + 2) = -1$

37. $\dfrac{5}{3}z - 8 = 7$

38. $\dfrac{4}{3}y + 12 = -4$

39. $\dfrac{z}{5} + 2 = 4$

40. $\dfrac{3p}{7} - p = -4$

41. $\dfrac{3x - 2}{3} = 2x + \dfrac{7}{3}$

42. $\dfrac{7}{2}x + 5 = x + \dfrac{15}{2}$

43. $\dfrac{3x + 1}{20} = \dfrac{1}{2}$

44. $2x - \dfrac{7}{6} + \dfrac{x}{6} = \dfrac{4x + 3}{6}$

45. $\dfrac{3 + x}{3} + \dfrac{x + 7}{2} = 4x + 1$

46. $2(2x + 1) - \dfrac{3x}{2} = \dfrac{-3(4 + x)}{2}$

47. $\dfrac{3}{2}(3x - 2) - 10x - 4 = 0$

48. $\dfrac{a(a - 3) + 5}{7} = \dfrac{(a - 1)^2}{7}$

49. $\dfrac{(y + 2)^2}{3} = y + 2 + \dfrac{y^2}{3}$

50. $(t + 1)(t - 1) = (t + 2)(t - 3) + 4$

51. $x(x + 2) = (x + 1)^2 - 1$

52. $(x - 2)(x - 3) = (x + 3)(x + 4)$

53. $2(s + 2) + (s + 3)^2 = s(s + 5) + 2\left(\dfrac{17}{2} + s\right)$

54. $\dfrac{3}{x} + \dfrac{1}{2} = \dfrac{4}{x}$

55. $\dfrac{2}{x + 1} + \dfrac{1}{3} = \dfrac{1}{x + 1}$

56. $\dfrac{3}{x - 2} + \dfrac{1}{x} = \dfrac{3}{x - 2}$

57. $\dfrac{9t + 6}{t(t + 3)} = \dfrac{7}{t + 3}$

58. $x + \dfrac{2(-2x + 1)}{3x + 5} = \dfrac{3x^2}{3x + 5}$

59. $\dfrac{2}{(a - 7)(a + 2)} = \dfrac{4}{(a + 3)(a + 2)}$

60. $\dfrac{2}{n - 2} + \dfrac{1}{n + 1} = \dfrac{1}{n^2 - n - 2}$

61. $\dfrac{2x + 3}{x^2 + 5x + 6} + \dfrac{3x - 2}{x^2 + x - 6} = \dfrac{5x - 2}{x^2 - 4}$

62. $\dfrac{3x}{x^2 + x} - \dfrac{2x}{x^2 + 5x} = \dfrac{x + 2}{x^2 + 6x + 5}$

63. $\dfrac{3x + 5}{x^3 + 8} + \dfrac{3}{x^2 - 4} = \dfrac{2(3x - 2)}{(x - 2)(x^2 - 2x + 4)}$

64. $\dfrac{1}{n + 8} - \dfrac{3n - 4}{5n^2 + 42n + 16} = \dfrac{1}{5n + 2}$

65. $\dfrac{1}{11 - n} - \dfrac{2(3n - 1)}{-7n^2 + 74n + 33} = \dfrac{1}{7n + 3}$

66. $\dfrac{4}{a^2 - 13a - 48} - \dfrac{2}{a^2 - 18a + 32} = \dfrac{1}{a^2 + a - 6}$

67. $\dfrac{5}{y + 4} + \dfrac{2}{y + 2} = \dfrac{6}{y + 2} - \dfrac{1}{y^2 + 6y + 8}$

68. $\dfrac{6}{2a - 6} - \dfrac{3}{3 - 3a} = \dfrac{1}{a^2 - 4a + 3}$

69. $\dfrac{3y}{6 - 3y} + \dfrac{2y}{2y + 4} = \dfrac{8}{4 - y^2}$

70. $\dfrac{3 + 2a}{a^2 + 6 + 5a} - \dfrac{2 - 3a}{a^2 - 6 + a} = \dfrac{5a - 2}{a^2 - 4}$

71. $\dfrac{a}{a + 2} - 1 = -\dfrac{3a + 2}{a^2 + 4a + 4}$

72. $\dfrac{x - 1}{x + 3} + \dfrac{x - 2}{x - 3} = \dfrac{1 - 2x}{3 - x}$

Solve each formula for the specified variable.

73. $k = 2.2p$; p

74. $ax + b = 0$; x

75. $P = 2l + 2w$; w

76. $V = \dfrac{1}{3}\pi r^2 h$; h

77. $V = \dfrac{1}{3}\pi r^2 h$; r^2

78. $z = \dfrac{x - \mu}{\sigma}$; μ

79. $P_n = L + \dfrac{si}{f}$; s

80. $P_n = L + \dfrac{si}{f}$; f

81. $F = \dfrac{mMg}{r^2}$; m

82. $\dfrac{1}{f} = \dfrac{1}{p} + \dfrac{1}{q}$; f

83. $\dfrac{x}{a} + \dfrac{y}{b} = 1$; y

84. $\dfrac{x}{a} - \dfrac{y}{b} = 1$; a

85. $\dfrac{1}{r} = \dfrac{1}{r_1} + \dfrac{1}{r_2}$; r

86. $\dfrac{1}{r} = \dfrac{1}{r_1} + \dfrac{1}{r_2}$; r_1

87. $l = a + (n - 1)d$; n

88. $l = a + (n - 1)d$; d

89. $a = (n - 2)\dfrac{180}{n}$; n

90. $S = \dfrac{a - lr}{1 - r}$; a

91. $R = \dfrac{1}{\dfrac{1}{r_1} + \dfrac{1}{r_2} + \dfrac{1}{r_3}}$; r_1

92. $R = \dfrac{1}{\dfrac{1}{r_1} + \dfrac{1}{r_2} + \dfrac{1}{r_3}}$; r_3

Discovery and Writing

93. Explain why a conditional linear equation always has exactly one root.

94. Define an extraneous solution and explain how such a solution occurs.

Review

Simplify each expression. Use absolute value symbols when necessary.

95. $(25x^2)^{1/2}$

96. $\left(\dfrac{25p^2}{16q^4}\right)^{1/2}$

97. $\left(\dfrac{125x^3}{8y^6}\right)^{-2/3}$

98. $\left(-\dfrac{27y^3}{1{,}000x^6}\right)^{1/3}$

99. $\sqrt{25y^2}$

100. $\sqrt[3]{-125y^9}$

101. $\sqrt[4]{\dfrac{a^4 b^{12}}{z^8}}$

102. $\sqrt[5]{\dfrac{x^{10}y^5}{z^{15}}}$

1.2 Applications of Linear Equations

In this section, we will learn to

1. Solve number problems.
2. Solve geometric problems.
3. Solve investment problems.
4. Solve break-point analysis problems.
5. Solve shared-work problems.
6. Solve mixture problems.
7. Solve uniform motion problems.

John Shearer/WireImage

Tim McGraw is one of the most successful country music singers. He is married to country singer Faith Hill and is the son of former baseball player Tug McGraw. His albums sales have exceeded 40 million copies and his concert tours have been attended by thousands of fans.

Suppose you hear that for one day only Tim McGraw concert tickets can be purchased for 30% off the original price. Knowing this is a bargain, you immediately purchase a ticket for $28, the selling price. Later that day, this question comes to mind: What was the original price of the concert ticket?

A linear equation can be used to model this problem. We can let x represent the original price of the concert ticket and subtract 30% of x (the discount) and we will get the selling price $28. The linear equation is $x - 0.3x = 28$. We can solve this equation to determine the original price of the concert ticket.

$$x - 0.3x = 28$$
$$0.7x = 28 \qquad \text{Combine like terms.}$$
$$x = \frac{28}{0.7} \qquad \text{Divide both sides by 0.7.}$$
$$x = 40$$

The original price of the concert ticket is $40.

In this section, we will use the equation-solving techniques discussed in the previous section to solve applied problems (often called *word problems*). To solve these problems, we must translate the verbal description of the problem into an equation. The process of finding the equation that describes the words of the problem is called **mathematical modeling.** The equation itself is often called a **mathematical model** of the situation described in the word problem.

The following list of steps provides a strategy to follow when we try to find the equation that models an applied problem.

Strategy for Modeling Equations

1. Analyze the problem to see what you are to find. Often, drawing a diagram or making a table will help you visualize the facts.

2. Pick a variable to represent the quantity that is to be found, and write a sentence telling what that variable represents. Express all other quantities mentioned in the problem as expressions involving this single variable.

3. Find a way to express a quantity in two different ways. This might involve a formula from geometry, finance, or physics.

4. Form an equation indicating that the two quantities found in Step 3 are equal.

5. Solve the equation.

6. Answer the questions asked in the problem.

7. Check the answers in the words of the problem.

This list does not apply to all situations, but it can be used for a wide range of problems with only slight modifications.

1. Solve Number Problems

EXAMPLE 1   Solving a Number Problem

A student has scores of 74%, 78%, and 70% on three exams. What score is needed on a fourth exam for the student to earn an average grade of 80%?

SOLUTION   To find an equation that models the problem, we can let x represent the required grade on the fourth exam. The average grade will be one-fourth of the sum of the four grades. We know this average is to be 80.

The average of the four grades	equals	the required average grade.
$\dfrac{74 + 78 + 70 + x}{4}$	$=$	80

We can solve this equation for x.

$$\frac{222 + x}{4} = 80 \qquad \text{74 + 78 + 70 = 222}$$

$$222 + x = 320 \qquad \text{Multiply both sides by 4.}$$

$$x = 98 \qquad \text{Subtract 222 from both sides.}$$

To earn an average of 80%, the student must score 98% on the fourth exam.   ∎

Self Check 1   A student scores 82%, 96%, 91%, and 92% on four college algebra exams. What score is needed on a fifth exam for the student to earn an average grade of 90%?

Now Try Exercise 9.

2. Solve Geometric Problems

EXAMPLE 2   Solving a Geometric Problem

A city ordinance requires a man to install a fence around the swimming pool shown in Figure 1-2. He wants the border around the pool to be of uniform width. If he has 154 feet of fencing, find the width of the border.

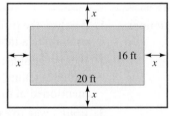

FIGURE 1-2

SOLUTION   We can let x represent the width of the border. The distance around the large rectangle, called its **perimeter,** is given by the formula $P = 2l + 2w$, where l is the length, $20 + 2x$, and w is the width, $16 + 2x$. Since the man has 154 feet of fencing, the perimeter will be 154 feet. To find an equation that models the problem, we substitute these values into the formula for perimeter.

$$P = 2l + 2w$$ The formula for the perimeter of a rectangle.

$$154 = 2(20 + 2x) + 2(16 + 2x)$$ Substitute 154 for P, $20 + 2x$ for l, and $16 + 2x$ for w.

$$154 = 40 + 4x + 32 + 4x$$ Use the Distributive Property to remove parentheses.

$$154 = 72 + 8x$$ Combine like terms.

$$82 = 8x$$ Subtract 72 from both sides.

$$10\frac{1}{4} = x$$ Divide both sides by 8.

This border will be $10\frac{1}{4}$ feet wide.   ∎

Self Check 2   In Example 2, if 168 feet of fencing is available, find the border's width.

Now Try Exercise 13.

3. Solve Investment Problems

EXAMPLE 3   **Solving an Investment Problem**

A woman invested $10,000, part at 9% and the rest at 14%. If the annual income from these investments is $1,275, how much did she invest at each rate?

SOLUTION   We can let x represent the amount invested at 9%. Then $10,000 - x$ represents the amount invested at 14%. Since the annual income from any investment is the product of the interest rate and the amount invested, we have the following information.

Type of investment	Rate	Amount invested	Interest earned
9% investment	0.09	x	$0.09x$
14% investment	0.14	$10,000 - x$	$0.14(10,000 - x)$

The total income from these two investments can be expressed in two ways: as $1,275 and as the sum of the incomes from the two investments.

The income from the 9% investment	plus	the income from the 14% investment	equals	the total income.
$0.09x$	$+$	$0.14(10,000 - x)$	$=$	1,275

We can solve this equation for x.

$$0.09x + 0.14(10,000 - x) = 1,275$$

$$9x + 14(10,000 - x) = 127,500$$ To eliminate the decimal points, multiply both sides by 100.

$$9x + 140,000 - 14x = 127,500$$ Use the Distributive Property to remove parentheses.

$$-5x + 140,000 = 127,500$$ Combine like terms.

$$-5x = -12,500$$ Subtract 140,000 from both sides.

$$x = 2,500$$ Divide both sides by -5.

The amount invested at 9% was $2,500, and the amount invested at 14% was $7,500 ($10,000 - $2,500). These amounts are correct, because 9% of $2,500 is $225, 14% of $7,500 is $1,050, and the sum of these amounts is $1,275.   ∎

Self Check 3   A man invests $12,000, part at 7% and the rest at 9%. If the annual income from these investments is $965, how much was invested at each rate?

Now Try Exercise 21.

4. Solve Break-Point Analysis Problems

Running a machine involves two costs—**setup costs** and **unit costs.** Setup costs include the cost of installing a machine and preparing it to do a job. Unit cost is the cost to manufacture one item, which includes the costs of material and labor.

EXAMPLE 4   **Solving a Break-Point Analysis Problem**

Suppose that one machine has a setup cost of $400 and a unit cost of $1.50, and a second machine has a setup cost of $500 and a unit cost of $1.25. Find the **break point** (the number of units manufactured at which the cost on each machine is the same).

SOLUTION   We can let x represent the number of items to be manufactured. The cost C_1 of using machine 1 is

$$C_1 = 400 + 1.5x$$

and the cost C_2 of using machine 2 is

$$C_2 = 500 + 1.25x$$

The break point occurs when these two costs are equal.

The cost of using machine 1	equals	the cost of using machine 2.
$400 + 1.5x$	$=$	$500 + 1.25x$

We can solve this equation for x.

$$400 + 1.5x = 500 + 1.25x$$

$1.5x = 100 + 1.25x$ Subtract 400 from both sides.

$0.25x = 100$ Subtract $1.25x$ from both sides.

$x = 400$ Divide both sides by 0.25.

The break point is 400 units. This result is correct, because it will cost the same amount to manufacture 400 units with either machine.

$$C_1 = \$400 + \$1.5(400) = \$1,000 \quad \text{and} \quad C_2 = \$500 + \$1.25(400) = \$1,000$$

Self Check 4   An ATM has a setup cost of $3,000 and operating costs averaging $1 per transaction. Another ATM machine has a setup cost of $3,500 and an operating cost of $0.50 per transaction. Find the number of transactions at which the costs for each ATM is the same.

Now Try Exercise 29.

5. Solve Shared-Work Problems

EXAMPLE 5   **Solving a Shared-Work Problem**

The Toll Way Authority needs to pave 100 miles of interstate highway before freezing temperatures come in about 60 days. Sjostrom and Sons has estimated that it can do the job in 110 days. Scandroli and Sons has estimated that it can do the job in 140 days. If the authority hires both contractors, will the job get done in time?

SOLUTION   Since Sjostrom can do the job in 110 days, they can do $\frac{1}{110}$ of the job in one day. Since Scandroli can do the job in 140 days, they can do $\frac{1}{140}$ of the job in one day. If we let n represent the number of days it will take to pave the highway if both contractors work together, they can do $\frac{1}{n}$ of the job in one day. The work that they can do together in one day is the sum of what each can do in one day.

The part Sjostrom can pave in one day	plus	the part Scandroli can pave in one day	equals	the part they can pave together in one day.
$\dfrac{1}{110}$	$+$	$\dfrac{1}{140}$	$=$	$\dfrac{1}{n}$

We can solve this equation for n.

$$\frac{1}{110} + \frac{1}{140} = \frac{1}{n}$$

$$(110)(140)n\left(\frac{1}{110} + \frac{1}{140}\right) = (110)(140)n\left(\frac{1}{n}\right) \qquad \text{Multiply both sides by } (110)(140)n \text{ to eliminate the fractions.}$$

$$\frac{(110)(140)n}{110} + \frac{(110)(140)n}{140} = \frac{(110)(140)n}{n} \qquad \text{Use the Distributive Property to remove parentheses.}$$

$$140n + 110n = 15{,}400 \qquad \frac{110}{110} = 1, \frac{140}{140} = 1, \text{ and } \frac{n}{n} = 1.$$

$$250n = 15{,}400 \qquad \text{Combine like terms.}$$

$$n = 61.6 \qquad \text{Divide both sides by 250.}$$

It will take the contractors about 62 days to pave the highway. With any luck, the job will be done in time.   ∎

Self Check 5   John and Eric both work at Firestone Auto Care. John can install a new set of tires in 45 minutes. Eric is faster and can install a set in 30 minutes. If John and Eric work together, how long will it take them to install one set of tires?

Now Try Exercise 33.

6. Solve Mixture Problems

EXAMPLE 6   **Solving a Mixture Problem**

A container is partially filled with 20 liters of whole milk containing 4% butterfat. How much 1% milk must be added to obtain a mixture that is 2% butterfat?

SOLUTION   Since the first container shown in Figure 1-3 contains 20 liters of 4% milk, it contains $0.04(20)$ liters of butterfat. To this amount, we will add the contents of the second container, which holds $0.01(l)$ liters of butterfat.

The sum of these two amounts will equal the number of liters of butterfat in the third container, which is $0.02(20 + l)$ liters of butterfat. This information is presented in table form in Figure 1-4.

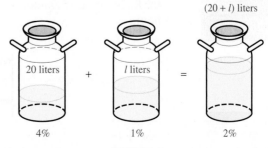

FIGURE 1-3

The butterfat in the 4% milk	plus	the butterfat in the 1% milk	equals	the butterfat in the 2% milk.
4% of 20 liters	+	1% of l liters	=	2% of $(20 + l)$ liters

	Percentage of butterfat	·	Amount of milk	=	Amount of butterfat
4 % milk	0.04		20		0.04(20)
1% milk	0.01		l		0.01(l)
2% milk	0.02		$20 + l$		0.02(20 + l)

FIGURE 1-4

We can solve this equation for l.

$$0.04(20) + 0.01(l) = 0.02(20 + l)$$
$$4(20) + l = 2(20 + l) \qquad \text{Multiply both sides by 100.}$$
$$80 + l = 40 + 2l \qquad \text{Remove parentheses.}$$
$$40 = l \qquad \text{Subtract 40 and } l \text{ from both sides.}$$

To dilute the 20 liters of 4% milk to a 2% mixture, 40 liters of 1% milk must be added. To check, we note that the final mixture contains $0.02(60) = 1.2$ liters of pure butterfat, and that this is equal to the amount of pure butterfat in the 4% milk and the 1% milk; $0.04(20) + 0.01(40) = 1.2$ liters. ∎

Self Check 6   Milk containing 4% butterfat is mixed with 8 gallons of milk containing 1% butterfat to make a low-fat cottage cheese mixture containing 2% butterfat. How many gallons of the richer milk is used?

Now Try Exercise 39.

7. Solve Uniform Motion Problems

EXAMPLE 7   Solving a Uniform Motion Problem

A man leaves home driving his Ford F-150 truck at the rate of 50 mph. When his daughter discovers than he has forgotten his wallet, she drives her Ford Mustang after him at a rate of 65 mph. How long will it take her to catch her dad if he had a 15-minute head start?

SOLUTION   Uniform motion problems are based on the formula $d = rt$, where d is the distance, r is the rate, and t is the time. We can draw a diagram and organize the information given in the problem in a chart as shown in Figures 1-5 and 1-6. In the chart, t represents the number of hours the daughter must drive to overtake her father. Because the father has a 15-minute, or $\frac{1}{4}$ hour, head start, he has been on the road for $\left(t + \frac{1}{4}\right)$ hours.

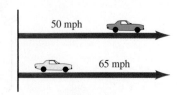

50 mph

65 mph

FIGURE 1-5

Patrick Mezirka/Shutterstock.com

	d	$=$	r	$\cdot$	t
Man	$50\left(t + \frac{1}{4}\right)$		50		$t + \frac{1}{4}$
Daughter	$65t$		65		t

FIGURE 1-6

We can set up the following equation and solve it for t.

The distance the man drives	equals	the distance the daughter drives.
$50\left(t + \dfrac{1}{4}\right)$	$=$	$65t$

We can solve this equation for t.

$$50\left(t + \frac{1}{4}\right) = 65t$$

$$50t + \frac{25}{2} = 65t \qquad \text{Remove parentheses.}$$

$$\frac{25}{2} = 15t \qquad \text{Subtract } 50t \text{ from both sides.}$$

$$\frac{5}{6} = t \qquad \text{Divide both sides by 15 and simplify.}$$

It will take the daughter $\frac{5}{6}$ hour, or 50 minutes, to overtake her dad. ∎

Self Check 7   On a reality television show, an officer traveling in a police cruiser at 90 mph pursues Jennifer who has a 3-minute head start. If the officer overtakes Jennifer in 12 minutes, how fast is Jennifer traveling?

Now Try Exercise 53.

Self Check Answers   **1.** 89   **2.** 12 feet   **3.** $5,750 at 7%; $6,250 at 9%   **4.** 1,000
5. 18 minutes   **6.** 4   **7.** 72 mph

Exercises **1.2**

Getting Ready
You should be able to complete these vocabulary and concept statements before you proceed to the practice exercises.

Fill in the blanks.

1. To average n scores, _____ the scores and divide by n.

2. The formula for the _____ of a rectangle is $P = 2l + 2w$.

3. The simple annual interest earned on an investment is the product of the interest rate and the _____ invested.

4. The number of units manufactured at which the cost on two machines is equal is called the _____.

5. Distance traveled is the product of the _____ and the _____.

6. 5% of 30 liters is _____ liters.

Practice
Solve each problem.

7. **Test scores** Tate scored 5 points higher on his midterm and 13 points higher on his final than he did on his first exam. If his mean (average) score was 90, what was his score on the first exam?

8. **Test scores** Courtney took four tests in science class. On each successive test, her score improved by 3 points. If her mean score was 69.5%, what did she score on the first test?

9. **Teacher certification** On the Illinois certification test for teachers specializing in learning disabilities, a teacher earned the scores shown in the accompanying table. What was the teacher's score in program development?

Human development with special needs	82
Assessment	90
Program development and instruction	?
Professional knowledge and legal issues	78
AVERAGE SCORE	86

10. **Golfing** Par on a golf course is 72. If a golfer shot rounds of 76, 68, and 70 in a tournament, what will she need to shoot on the final round to average par?

11. **Replacing locks** A locksmith at Pop-A-Lock charges $40 plus $28 for each lock installed. How many locks can be replaced for $236?

12. **Delivering ads** A University of Florida student earns $20 per day delivering advertising brochures door-to-door, plus 75¢ for each person he interviews. How many people did he interview on a day when he earned $56?

13. **Electronic LED billboard** An electronic LED billboard in Times Square is 26 feet taller than it is wide. If its perimeter is 92 feet, find the dimensions of the billboard.

14. **Hockey rink** A National Hockey League rink is 115 feet longer than it is wide. If the perimeter of the rink is 570 feet, find the dimensions of the rink?

15. **Width of a picture frame** The picture frame with the dimensions shown in the illustration was built with 14 feet of framing material. Find x its width.

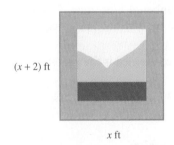

$(x + 2)$ ft

x ft

16. **Fencing a garden** If a gardener fences in the total rectangular area shown in the illustration instead of just the square area, he will need twice as much fencing to enclose the garden. How much fencing will he need?

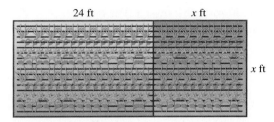

24 ft x ft x ft

17. **Wading pool dimensions** The area of the triangular swimming pool shown in the illustration is doubled by adding a rectangular wading pool. Find the dimensions of the wading pool. (*Hint:* The area of a triangle $= \frac{1}{2}bh$, and the area of a rectangle $= lw$.)

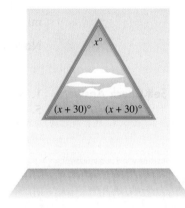

Wading pool   20 ft   Swimming pool

x ft   16 ft

18. **House construction** A builder wants to install a triangular window with the angles shown in the illustration. What angles will he have to cut to make the window fit? (*Hint:* The sum of the angles in a triangle equals 180°.)

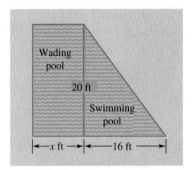

$x°$

$(x + 30)°$ $(x + 30)°$

19. **Length of a living room** If a carpenter adds a porch with dimensions shown in the illustration to the living room, the living area will be increased by 50%. Find the length of the living room.

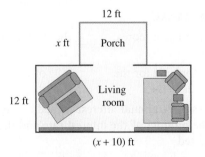

12 ft x ft   Porch

12 ft   Living room

$(x + 10)$ ft

20. Depth of water in a trough The trough in the illustration has a cross-sectional area of 54 square inches. Find the depth, d, of the trough. (*Hint:* Area of a trapezoid $= \frac{1}{2}h(b_1 + b_2)$.)

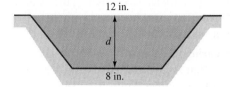

21. Investment problem An executive invests $22,000, some at 7% and some at 6% annual interest. If he receives an annual return of $1,420, how much is invested at each rate?

22. Financial planning After inheriting some money, a woman wants to invest enough to have an annual income of $5,000. If she can invest $20,000 at 9% annual interest, how much more will she have to invest at 7% to achieve her goal? (See the table.)

Type	Rate	Amount	Income
9% investment	0.09	20,000	.09(20,000)
7% investment	0.07	x	.07x

23. Investment problem Equal amounts are invested at 6%, 7%, and 8% annual interest. If the three investments yield a total of $2,037 annual interest, find the total investment.

24. Investment problem A woman invests $37,000, part at 8% and the rest at $9\frac{1}{2}$% annual interest. If the $9\frac{1}{2}$% investment provides $452.50 more income than the 8% investment, how much is invested at each rate?

25. Ticket sales A full-price ticket for a college basketball game costs $2.50, and a student ticket costs $1.75. If 585 tickets were sold, and the total receipts were $1,217.25, how many tickets were student tickets?

26. Ticket sales Of the 800 tickets sold to a movie, 480 were full-price tickets costing $7 each. If the gate receipts were $4,960, what did a student ticket cost?

27. Discounts After being discounted 20%, a weather radio sells for $63.96. Find the original price.

28. Markups A merchant increases the wholesale cost of a Maytag washing machine by 30% to determine the selling price. If the washer sells for $588.90, find the wholesale cost.

29. Break-point analysis A machine to mill a brass plate has a setup cost of $600 and a unit cost of $3 for each plate manufactured. A bigger machine has a setup cost of $800 but a unit cost of only $2 for each plate manufactured. Find the break point.

30. Break-point analysis A machine to manufacture fasteners has a setup cost of $1,200 and a unit cost of $0.005 for each fastener manufactured. A newer machine has a setup cost of $1,500 but a unit cost of only $0.0015 for each fastener manufactured. Find the break point.

31. Computer sales A computer store has fixed costs of $8,925 per month and a unit cost of $850 for every computer it sells. If the store can sell all the computers it can get for $1,275 each, how many must be sold for the store to break even? (*Hint:* The break-even point occurs when costs equal income.)

32. Restaurant management A restaurant has fixed costs of $137.50 per day and an average unit cost of $4.75 for each meal served. If a typical meal costs $6, how many customers must eat at the restaurant each day for the owner to break even?

33. Roofing houses Kyle estimates that it will take him 7 days to roof his house. A professional roofer estimates that it will take him 4 days to roof the same house. How long will it take if they work together?

34. Sealing asphalt One crew can seal a parking lot in 8 hours and another in 10 hours. How long will it take to seal the parking lot if the two crews work together?

35. Mowing lawns Julie can mow a lawn with a lawn tractor in 2 hours, and her husband can mow the same lawn with a push mower in 4 hours. How long will it take to mow the lawn if they work together?

36. Filling swimming pools A garden hose can fill a swimming pool in 3 days, and a larger hose can fill the pool in 2 days. How long will it take to fill the pool if both hoses are used?

37. Filling swimming pools An empty swimming pool can be filled in 10 hours. When full, the pool can be drained in 19 hours. How long will it take to fill the empty pool if the drain is left open?

38. Preparing seafood Kevin stuffs shrimp in his job as a seafood chef. He can stuff 1,000 shrimp in 6 hours. When his sister helps him, they can stuff 1,000 shrimp in 4 hours. If Kevin gets sick, how long will it take his sister to stuff 500 shrimp?

39. Diluting solutions How much water should be added to 20 ounces of a 15% solution of alcohol to dilute it to a 10% solution?

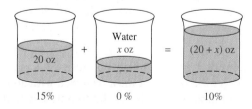

40. Increasing concentrations The beaker shown below contains a 2% saltwater solution.
   a. How much water must be boiled away to increase the concentration of the salt solution from 2% to 3%?
   b. Where on the beaker would the new water level be?

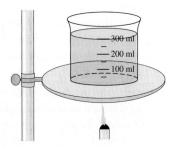

41. Winterizing cars A car radiator has a 6-liter capacity. If the liquid in the radiator is 40% antifreeze, how much liquid must be replaced with pure antifreeze to bring the mixture up to a 50% solution?

42. Mixing milk If a bottle holding 3 liters of milk contains $3\frac{1}{2}\%$ butterfat, how much skimmed milk must be added to dilute the milk to 2% butterfat?

43. Preparing solutions A nurse has 1 liter of a solution that is 20% alcohol. How much pure alcohol must she add to bring the solution up to a 25% concentration?

44. Diluting solutions If there are 400 cubic centimeters of a chemical in 1 liter of solution, how many cubic centimeters of water must be added to dilute it to a 25% solution? (*Hint:* 1,000 cc = 1 liter.)

45. Cleaning swimming pools A swimming pool contains 15,000 gallons of water. How many gallons of chlorine must be added to "shock the pool" and bring the water to a $\frac{3}{100}\%$ solution?

46. Mixing fuels An automobile engine can run on a mixture of gasoline and a substitute fuel. If gas costs $3.50 per gallon and the substitute fuel costs $2 per gallon, what percent of a mixture must be substitute fuel to bring the cost down to $2.75 per gallon?

47. Evaporation How many liters of water must evaporate to turn 12 liters of a 24% salt solution into a 36% solution?

48. Increasing concentrations A beaker contains 320 ml of a 5% saltwater solution. How much water should be boiled away to increase the concentration to 6%?

49. Lowering fat How many pounds of extra-lean hamburger that is 7% fat must be mixed with 30 pounds of hamburger that is 15% fat to obtain a mixture that is 10% fat?

50. Dairy foods How many gallons of cream that is 22% butterfat must be mixed with milk that is 2% butterfat to get 20 gallons of milk containing 4% butterfat?

51. Mixing solutions How many gallons of a 5% alcohol solution must be mixed with 90 gallons of 1% solution to obtain a 2% solution?

52. Preparing medicines A doctor prescribes an ointment that is 2% hydrocortisone. A pharmacist has 1% and 5% concentrations in stock. How much of each should the pharmacist use to make a 1-ounce tube?

53. Driving rates John drove to Daytona Beach, Florida, in 5 hours. When he returned, there was less traffic, and the trip took only 3 hours. If John averaged 26 mph faster on the return trip, how fast did he drive each way?

54. Distance problem Allison drove home at 60 mph, but her brother Austin, who left at the same time, could drive at only 48 mph. When Allison arrived, Austin still had 60 miles to go. How far did Allison drive?

55. Distance problem Two cars leave Hinds Community College traveling in opposite directions. One car travels at 60 mph and the other at 64 mph. In how many hours will they be 310 miles apart?

56. Bank robbery Some bank robbers leave town, speeding at 70 mph. Ten minutes later, the police give chase, traveling at 78 mph. How long, after the robbery, will it take the police to overtake the robbers?

57. Jogging problem Two Michigan State University cross-country runners are 440 yards apart and are running toward each other, one at 8 mph and the other at 10 mph. In how many seconds will they meet?

58. Driving rates One morning, Justin drove 5 hours before stopping to eat lunch at Pizza Hut. After lunch, he increased his speed by 10 mph. If he completed a 430-mile trip in 8 hours of driving time, how fast did he drive in the morning?

59. Boating problem A Johnson motorboat goes 5 miles upstream in the same time it requires to go 7 miles downstream. If the river flows at 2 mph, find the speed of the boat in still water.

60. Wind velocity A plane can fly 340 mph in still air. If it can fly 200 miles downwind in the same amount of time it can fly 140 miles upwind, find the velocity of the wind.

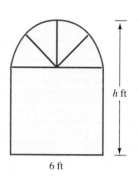

61. **Feeding cattle** A cattleman wants to mix 2,400 pounds of cattle feed that is to be 14% protein. Barley (11.7% protein) will make up 25% of the mixture. The remaining 75% will be made up of oats (11.8% protein) and soybean meal (44.5% protein). How many pounds of each will he use?

62. **Feeding cattle** If the cattleman in Exercise 61 wants only 20% of the mixture to be barley, how many pounds of each should he use?

Use a calculator to help solve each problem.

63. **Machine tool design** 712.51 cubic millimeters of material was removed by drilling the blind hole as shown in the illustration. Find the depth of the hole. (*Hint:* The volume of a cylinder is given by $V = \pi r^2 h$.)

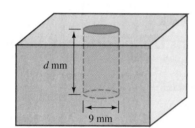

64. **Architecture** The Norman window with dimensions as shown is a rectangle topped by a semicircle. If the area of the window is 68.2 square feet, find its height h.

Discovery and Writing

65. Consider the strategy you use to solve investment and uniform motion problems. Describe any similarities you observe in these problem types.

66. Which type of application was hardest for you to solve? Why? What strategy or approach works best for you when approaching solving this problem?

Review
Factor each expression.

67. $x^2 - 2x - 63$

68. $2x^2 + 11x - 21$

69. $9x^2 - 12x - 5$

70. $9x^2 - 2x - 7$

71. $x^2 + 6x + 9$

72. $x^2 - 10x + 25$

73. $x^3 + 8$

74. $27a^3 - 64$

1.3 Quadratic Equations

In this section, we will learn to

1. Solve quadratic equations using factoring and the Square Root Property.
2. Solve quadratic equations using completing the square.
3. Solve quadratic equations using the Quadratic Formula.
4. Determine the easiest strategy to use to solve a quadratic equation.
5. Solve formulas for a variable that is squared.
6. Define and use the discriminant.
7. Write rational equations in quadratic form and solve the equations.

Fenway Park, America's most beloved ballpark, is home to the Boston Red Sox baseball club. The park opened in 1912 and is the oldest major league baseball stadium.

In baseball, the distance between home plate and first base is 90 feet and the distance between first base and second base is 90 feet. To find the distance between home plate and second base, we can use the *Pythagorean Theorem*, which states that

The sum of the squares of the two legs of a right triangle is equal to the square of its hypotenuse.

Because home plate, first base, and second base form a right triangle, we can let x represent the distance between home plate and second base (the **hypotenuse** of the right triangle) and 90 feet represent the length of each leg. We can then apply the Pythagorean Theorem and write the equation $90^2 + 90^2 = x^2$. To find the distance between home plate and second base, we must solve this equation.

$$90^2 + 90^2 = x^2$$

$$8{,}100 + 8{,}100 = x^2 \qquad \text{Square 90 two times.}$$

$$16{,}200 = x^2 \qquad \text{Simplify.}$$

To find x, we must determine what positive number squared gives 16,200. From Chapter 0, we know that this number is the square root of 16,200.

$$\sqrt{16{,}200} \approx 127.3 \qquad \text{Use a calculator and round to the nearest tenth.}$$

To the nearest tenth, the distance between home plate and second base is 127.3 feet.

Since this equation contains the term x^2, it is an example of a new type of equation, called a *quadratic* equation. In this section, we will learn several strategies for solving these equations.

1. Solve Quadratic Equations Using Factoring and the Square Root Property

Polynomial equations such as $2x^2 + 11x - 21 = 0$ and $3x^2 - x + 2 = 0$ are called *quadratic* or *second-degree* equations.

Quadratic Equation   A **quadratic equation** is an equation that can be written in the form $ax^2 + bx + c = 0$, where a, b, and c are real numbers and $a \neq 0$.

To solve quadratic equations by factoring, we can use the following theorem.

Zero-Factor Theorem   If a and b are real numbers, and if $ab = 0$, then

$$a = 0 \qquad \text{or} \qquad b = 0$$

PROOF   Suppose that $ab = 0$. If $a = 0$, we are finished, because at least one of a or b is 0.

If $a \neq 0$, then a has a reciprocal $\frac{1}{a}$, and we can multiply both sides of the equation $ab = 0$ by $\frac{1}{a}$ to obtain

$$ab = 0$$

$$\frac{1}{a}(ab) = \frac{1}{a}(0) \qquad \text{Multiply both sides by } \frac{1}{a}.$$

$$\left(\frac{1}{a} \cdot a\right)b = 0 \qquad \text{Use the Associative Property to group } \frac{1}{a} \text{ and } a \text{ together.}$$

$$1b = 0 \qquad \frac{1}{a} \cdot a = 1$$

$$b = 0$$

Thus, if $a \neq 0$, then b must be 0, and the theorem is proved.

EXAMPLE 1   **Solving a Quadratic Equation by Factoring**

Solve: $2x^2 - 9x - 35 = 0$.

SOLUTION   The left side can be factored and written as

$$(2x + 5)(x - 7) = 0$$

This product can be 0 if and only if one of the factors is 0. So we can use the Zero-Factor Theorem and set each factor equal to 0. We can then solve each equation for x.

$$2x + 5 = 0 \qquad \text{or} \qquad x - 7 = 0$$
$$2x = -5 \qquad\qquad\qquad x = 7$$
$$x = -\frac{5}{2}$$

Because $(2x + 5)(x - 7) = 0$ only if one of its factors is zero, $-\frac{5}{2}$ and 7 are the only solutions of the equation.

Verify that each one satisfies the equation. ∎

Self Check 1   Solve: $6x^2 + 7x - 3 = 0$.

Now Try Exercise 17.

Comment

The Zero-Factor Theorem can be used only when there is a constant term of 0 on the right side of the equation.

In many quadratic equations, the quadratic expression does not factor over the set of integers. For example, the left side of $x^2 - 5x + 3 = 0$ is a prime polynomial and cannot be factored over the set of integers.

To develop a method to solve these equations, we consider the equation $x^2 = c$. If c is positive, it has two real roots that can be found by adding $-c$ to both sides, factoring $x^2 - c$ over the set of real numbers, setting each factor equal to 0, and solving for x.

$$x^2 = c$$
$$x^2 - c = 0 \qquad\qquad \text{Subtract } c \text{ from both sides.}$$
$$x^2 - \left(\sqrt{c}\right)^2 = 0 \qquad\qquad \left(\sqrt{c}\right)^2 = c$$
$$\left(x - \sqrt{c}\right)\left(x + \sqrt{c}\right) = 0 \qquad\qquad \text{Factor the difference of two squares.}$$
$$x - \sqrt{c} = 0 \quad \text{or} \quad x + \sqrt{c} = 0 \qquad\qquad \text{Set each factor equal to 0.}$$
$$x = \sqrt{c} \quad \Big| \quad x = -\sqrt{c}$$

The roots of $x^2 = c$ are $x = \sqrt{c}$ and $x = -\sqrt{c}$. This fact is summarized in the **Square Root Property.**

Square Root Property   If $c > 0$, the equation $x^2 = c$ has two real roots:

$$x = \sqrt{c} \quad \text{or} \quad x = -\sqrt{c}$$

EXAMPLE 2   **Solving a Quadratic Equation by Using the Square Root Property**

Solve: $x^2 - 8 = 0$.

SOLUTION   We solve for x^2 and use the Square Root Property.

$$x^2 - 8 = 0$$

$$x^2 = 8$$

$$x = \sqrt{8} \quad \text{or} \quad x = -\sqrt{8}$$

$$x = 2\sqrt{2} \quad \Big| \quad x = -2\sqrt{2} \qquad \sqrt{8} = \sqrt{4}\sqrt{2} = 2\sqrt{2}$$

Verify that each root satisfies the equation.   ∎

Self Check 2   Solve: $x^2 - 12 = 0$.

Now Try Exercise 21.

EXAMPLE 3   Solve: $(x + 4)^2 = 27$.

SOLUTION   Again, we will use the Square Root Property.

$$(x + 4)^2 = 27$$

$$x + 4 = \sqrt{27} \qquad \text{or} \quad x + 4 = -\sqrt{27}$$

$$x + 4 = 3\sqrt{3} \qquad \qquad x + 4 = -3\sqrt{3} \qquad \sqrt{27} = \sqrt{9}\sqrt{3} = 3\sqrt{3}$$

$$x = -4 + 3\sqrt{3} \quad \Big| \quad x = -4 - 3\sqrt{3}$$

Verify that each root satisfies the equation.   ∎

Self Check 3   Solve: $(2x + 5)^2 = 45$.

Now Try Exercise 33.

2. Solve Quadratic Equations Using Completing the Square

Another way to solve quadratic equations is called **completing the square.** This method is based on the following products:

$$x^2 + 2ax + a^2 = (x + a)^2 \qquad \text{and} \qquad x^2 - 2ax + a^2 = (x - a)^2$$

The trinomials $x^2 + 2ax + a^2$ and $x^2 - 2ax + a^2$ are perfect-square trinomials, because each one factors as the square of a binomial. In each case, the coefficient of the first term is 1. If we take one-half of the coefficient of x in the middle term and square it, we obtain the third term.

$$\left[\frac{1}{2}(2a)\right]^2 = a^2 \qquad \text{and} \qquad \left[\frac{1}{2}(-2a)\right]^2 = (-a)^2 = a^2$$

This suggests that to make $x^2 + bx$ a perfect-square trinomial, we find one-half of b, square it, and add the result to the binomial. For example, to make $x^2 + 10x$ a perfect-square trinomial, we find one-half of 10 to get 5, square 5 to get 25, and add 25 to $x^2 + 10x$.

$$x^2 + 10x + \left[\frac{1}{2}(10)\right]^2 = x^2 + 10x + (5)^2$$

$$= x^2 + 10x + 25 \qquad \text{Note that } x^2 + 10x + 25 = (x + 5)^2.$$

To make $x^2 - 11x$ a perfect-square trinomial, we find one-half of -11 to get $-\frac{11}{2}$, square $-\frac{11}{2}$ to get $\frac{121}{4}$, and add $\frac{121}{4}$ to $x^2 - 11x$.

$$x^2 - 11x + \left[\frac{1}{2}(-11)\right]^2 = x^2 - 11x + \left(-\frac{11}{2}\right)^2$$

$$= x^2 - 11x + \frac{121}{4} \qquad \text{Note that } x^2 - 11x + \frac{121}{4} = \left(x - \frac{11}{2}\right)^2.$$

To solve a quadratic equation in x by completing the square, we follow these steps.

Strategy for Completing the Square	1. If the coefficient of x^2 is not 1, make it 1 by dividing both sides of the equation by the coefficient of x^2.
	2. If necessary, add a number to both sides of the equation to get the constant on the right side of the equation.
	3. Complete the square on x:
	a. Identify the coefficient of x, take one-half of it, and square the result.
	b. Add the number found in part a to both sides of the equation.
	4. Factor the perfect-square trinomial and combine like terms.
	5. Solve the resulting quadratic equation by using the Square Root Property.

To use completing the square to solve $x^2 - 10x + 24 = 0$, we note that the coefficient of x^2 is 1. We move on to Step 2 and subtract 24 from both sides to get the constant term on the right side of the equal sign.

$$x^2 - 10x = -24$$

We can then complete the square by adding $\left[\frac{1}{2}(-10)\right]^2 = 25$ to both sides.

$$x^2 - 10x + \mathbf{25} = -24 + \mathbf{25}$$

$$x^2 - 10x + 25 = 1 \qquad \text{Simplify on the right side.}$$

We then factor the perfect square trinomial on the left side.

$$(x - 5)^2 = 1$$

Finally, we use the Square Root Property to solve this equation.

$$x - 5 = 1 \qquad \text{or} \qquad x - 5 = -1$$
$$x = 6 \qquad\qquad\qquad x = 4$$

EXAMPLE 4   Solving a Quadratic Equation by Completing the Square

Use completing the square to solve $x^2 + 4x - 6 = 0$.

SOLUTION   Here the coefficient of x^2 is already 1. We move to Step 2 and add 6 to both sides to isolate the binomial $x^2 + 4x$.

$$x^2 + 4x = 6$$

We then find the number to add to both sides by completing the square. Since one-half of 4 (the coefficient of x) is 2 and $2^2 = 4$, we add 4 to both sides.

$$x^2 + 4x + 4 = 6 + 4 \qquad \text{Add 4 to both sides.}$$

$$x^2 + 4x + 4 = 10$$

$$(x + 2)^2 = 10 \qquad \text{Factor } x^2 + 4x + 4.$$

$$x + 2 = \sqrt{10} \quad \text{or} \quad x + 2 = -\sqrt{10} \qquad \text{Use the Square Root Property.}$$

$$x = -2 + \sqrt{10} \quad \Big| \quad x = -2 - \sqrt{10}$$

Verify that each root satisfies the original equation.

■

Self Check 4   Solve: $x^2 - 2x - 9 = 0$.

Now Try Exercise 49.

EXAMPLE 5   **Solving a Quadratic Equation by Completing the Square**

Use completing the square to solve $x(x + 3) = 9$.

SOLUTION   We remove parentheses to get

$$x^2 + 3x = 9$$

Since the coefficient of x^2 is 1 and the constant is on the right side, we move to Step 3 and find the number to be added to both sides to complete the square. Since one-half of 3 (the coefficient of x) is $\frac{3}{2}$ and the square of $\frac{3}{2}$ is $\frac{9}{4}$, we add $\frac{9}{4}$ to both sides.

$$x^2 + 3x + \frac{9}{4} = 9 + \frac{9}{4} \qquad \text{Add } \frac{9}{4} \text{ to both sides.}$$

$$\left(x + \frac{3}{2}\right)^2 = \frac{45}{4} \qquad \text{Factor } x^2 + 3x + \frac{9}{4} \text{ and combine terms on right.}$$

$$x + \frac{3}{2} = \frac{\sqrt{45}}{2} \quad \text{or} \quad x + \frac{3}{2} = -\frac{\sqrt{45}}{2} \qquad \text{Use the Square Root Property.}$$

$$x + \frac{3}{2} = \frac{3\sqrt{5}}{2} \quad \text{or} \quad x + \frac{3}{2} = -\frac{3\sqrt{5}}{2} \qquad \sqrt{45} = \sqrt{9}\sqrt{5} = 3\sqrt{5}$$

$$x = \frac{-3 + 3\sqrt{5}}{2} \quad \Big| \quad x = \frac{-3 - 3\sqrt{5}}{2} \qquad \text{Subtract } \frac{3}{2} \text{ from both sides. and combine.}$$

Verify that each root satisfies the original equation.

■

Self Check 5   Solve: $x(x + 5) = 1$.

Now Try Exercise 51.

EXAMPLE 6   **Solving a Quadratic Equation by Completing the Square**

Use completing the square to solve $6x^2 + 5x - 6 = 0$.

SOLUTION   We begin by dividing both sides of the equation by 6 to make the coefficient of x^2 equal to 1. Then we proceed as follows:

$$6x^2 + 5x - 6 = 0$$

$$x^2 + \frac{5}{6}x - 1 = 0 \qquad \text{Divide both sides by 6.}$$

$$x^2 + \frac{5}{6}x = 1 \qquad \text{Add 1 to both sides.}$$

$$x^2 + \frac{5}{6}x + \frac{25}{144} = 1 + \frac{25}{144} \qquad \text{Add } \left(\frac{1}{2} \cdot \frac{5}{6}\right)^2, \text{ or } \frac{25}{144}, \text{ to both sides.}$$

$$\left(x + \frac{5}{12}\right)^2 = \frac{169}{144} \qquad \text{Factor } x^2 + \frac{5}{6}x + \frac{25}{144}.$$

We now apply the Square Root Property.

$$x + \frac{5}{12} = \sqrt{\frac{169}{144}} \quad \text{or} \quad x + \frac{5}{12} = -\sqrt{\frac{169}{144}}$$

$$x + \frac{5}{12} = \frac{13}{12} \qquad\qquad x + \frac{5}{12} = -\frac{13}{12}$$

$$x = \frac{8}{12} \qquad\qquad\qquad x = -\frac{18}{12}$$

$$x = \frac{2}{3} \qquad\qquad\qquad x = -\frac{3}{2}$$

Verify that each root satisfies the original equation.

Self Check 6   Solve: $2x^2 - 5x - 3 = 0$.

Now Try Exercise 57.

3. Solve Quadratic Equations Using the Quadratic Formula

We can solve the equation $ax^2 + bx + c = 0$ $(a \neq 0)$ by completing the square. The result will be a formula that we can use to solve quadratic equations.

$$ax^2 + bx + c = 0$$

$$\frac{ax^2}{a} + \frac{b}{a}x + \frac{c}{a} = \frac{0}{a} \qquad\qquad \text{Divide both sides by } a.$$

$$x^2 + \frac{b}{a}x = -\frac{c}{a} \qquad\qquad \text{Simplify and subtract } \frac{c}{a} \text{ from both sides.}$$

$$x^2 + \frac{b}{a}x + \frac{b^2}{4a^2} = \frac{b^2}{4a^2} - \frac{4ac}{4aa} \qquad \begin{array}{l} \text{Add } \frac{b^2}{4a^2} \text{ to both sides and multiply the numerator and} \\ \text{denominator of } \frac{c}{a} \text{ by } 4a. \end{array}$$

$$\left(x + \frac{b}{2a}\right)^2 = \frac{b^2 - 4ac}{4a^2} \qquad \begin{array}{l} \text{Factor the left side and add the fractions on the} \\ \text{right side.} \end{array}$$

We can now use the Square Root Property.

$$x + \frac{b}{2a} = \sqrt{\frac{b^2 - 4ac}{4a^2}} \qquad \text{or} \quad x + \frac{b}{2a} = -\sqrt{\frac{b^2 - 4ac}{4a^2}}$$

$$x = -\frac{b}{2a} + \frac{\sqrt{b^2 - 4ac}}{2a} \qquad\qquad x = -\frac{b}{2a} - \frac{\sqrt{b^2 - 4ac}}{2a}$$

$$x = \frac{-b + \sqrt{b^2 - 4ac}}{2a} \qquad\qquad x = \frac{-b - \sqrt{b^2 - 4ac}}{2a}$$

These values of x are the two roots of the equation $ax^2 + bx + c = 0$. They are usually combined into a single expression, called the **Quadratic Formula.**

Quadratic Formula   The solutions of the general quadratic equation, $ax^2 + bx + c = 0$, are

$$x = \frac{-b \pm \sqrt{b^2 - 4ac}}{2a} \quad (a \neq 0)$$

The Quadratic Formula should be read twice, once using the $+$ sign and once using the $-$ sign. The Quadratic Formula implies that

$$x = \frac{-b + \sqrt{b^2 - 4ac}}{2a} \quad \text{or} \quad x = \frac{-b - \sqrt{b^2 - 4ac}}{2a}$$

Caution   Be sure to write the Quadratic Formula correctly. Do **not** write the Quadratic Formula as

$$x = -b \pm \frac{\sqrt{b^2 - 4ac}}{2a}$$

EXAMPLE 7   **Solving a Quadratic Equation Using the Quadratic Formula**

Use the Quadratic Formula to solve $x^2 - 5x + 3 = 0$.

SOLUTION   In this equation $a = 1$, $b = -5$, and $c = 3$. We will substitute these values into the Quadratic Formula.

$$x = \frac{-b \pm \sqrt{b^2 - 4ac}}{2a} \qquad \text{This is the Quadratic Formula.}$$

$$x = \frac{-(-5) \pm \sqrt{(-5)^2 - 4(1)(3)}}{2(1)} \qquad \text{Substitute 1 for } a, -5 \text{ for } b, \text{ and 3 for } c.$$

$$x = \frac{5 \pm \sqrt{13}}{2} \qquad (-5)^2 - 4(1)(3) = 25 - 12 = 13$$

Both values satisfy the original equation.

Self Check 7   Solve: $3x^2 - 5x + 1 = 0$.

Now Try Exercise 65.

EXAMPLE 8   **Solving a Quadratic Equation Using the Quadratic Formula**

Use the Quadratic Formula to solve $2x^2 + 8x - 7 = 0$.

SOLUTION   In this equation, $a = 2$, $b = 8$, and $c = -7$. We will substitute these values into the Quadratic Formula.

$$x = \frac{-b \pm \sqrt{b^2 - 4ac}}{2a} \qquad \text{This is the Quadratic Formula.}$$

$$x = \frac{-8 \pm \sqrt{8^2 - 4(2)(-7)}}{2(2)} \qquad \text{Substitute 2 for } a, 8 \text{ for } b, \text{ and } -7 \text{ for } c.$$

$$x = \frac{-8 \pm \sqrt{120}}{4} \qquad 8^2 - 4(2)(-7) = 64 + 56 = 120$$

$$x = \frac{-8 \pm 2\sqrt{30}}{4} \qquad \sqrt{120} = \sqrt{4 \cdot 30} = 2\sqrt{30}$$

$$x = \frac{2\left(-4 \pm \sqrt{30}\right)}{4} \qquad \text{Factor out 2 in the numerator.}$$

$$x = \frac{-4 \pm \sqrt{30}}{2} \qquad \text{Simplify.}$$

Both values satisfy the original equation.   ■

Self Check 8   Solve: $4x^2 + 16x - 13 = 0$.

Now Try Exercise 69.

4. Determine the Easiest Strategy to Use To Solve a Quadratic Equation

So far, we have solved quadratic equations by *factoring*, by the *square root method*, by *completing the square*, and by the *Quadratic Formula*. With so many methods available, it is useful to think about which one will be the easiest way to solve a specific quadratic equation. Although we have used completing the square to develop the Quadratic Formula, it is usually the most complicated way to solve a quadratic equation. Therefore, unless specified, we will usually not use this method. However, we will complete the square again later in the book to write certain equations in specific forms.

The following chart summarizes the different types of quadratic equations that can occur, a suggested method for solving them, and an example.

Type of Quadratic Equation	Easiest Strategy to Solve It	Example
Equations of the form $ax^2 + bx + c = 0$ where the left side factors easily.	Use factoring and the Zero-Factor Theorem.	Solve: $6x^2 - 11x + 3 = 0$ $(2x - 3)(3x - 1) = 0$ $2x - 3 = 0$ or $3x - 1 = 0$ $x = \dfrac{3}{2}$ $\quad\Big\vert\quad$ $x = \dfrac{1}{3}$
Equations of the form $ax^2 + bx = 0$ where the constant term is missing and the left side factors easily.	Use factoring and the Zero-Factor Theorem.	Solve: $9x^2 + 6x = 0$ $3x(3x + 2) = 0$ $3x = 0$ or $3x + 2 = 0$ $x = 0$ $\quad\Big\vert\quad$ $x = -\dfrac{2}{3}$
Equations of the form $ax^2 - c = 0$ where the term involving x is missing and the left side factors easily.	Use factoring and the Zero-Factor Theorem.	Solve: $4x^2 - 9 = 0$ $(2x + 3)(2x - 3) = 0$ $2x + 3 = 0$ or $2x - 3 = 0$ $x = -\dfrac{3}{2}$ $\quad\Big\vert\quad$ $x = \dfrac{3}{2}$
Equations of the form $ax^2 - c = 0$ or $x^2 = k$ where k is a contant.	Use the Square Root Property.	Solve: $2x^2 - 5 = 0$ $x^2 = \dfrac{5}{2}$ $x = \pm\sqrt{\dfrac{5}{2}}$ $x = \pm\dfrac{\sqrt{10}}{2}$

(continued)

Type of Quadratic Equation	Easiest Strategy to Solve It	Example
Equations of the form $ax^2 + bx + c = 0$ where the left side cannot be factored easily or cannot be factored at all.	Use the Quadratic Formula.	Solve: $3x^2 - x - 5 = 0$. $x = \dfrac{-b \pm \sqrt{b^2 - 4ac}}{2a}$ $x = \dfrac{-(-1) \pm \sqrt{(-1)^2 - 4(3)(-5)}}{2(3)}$ $\begin{aligned} a &= 3 \\ b &= -1 \\ c &= -5 \end{aligned}$ $x = \dfrac{1 \pm \sqrt{1 + 60}}{6}$ $x = \dfrac{1 \pm \sqrt{61}}{6}$

5. Solve Formulas for a Variable That Is Squared

Many formulas involve quadratic equations. For example, if an object is fired straight up into the air with an initial velocity of 88 feet per second, its height is given by the formula $h = 88t - 16t^2$, where h represents its height (in feet) and t represents the elapsed time (in seconds) since it was fired.

To solve this formula for t, we use the Quadratic Formula.

$$h = 88t - 16t^2$$

$$16t^2 - 88t + h = 0 \qquad \text{Add } 16t^2 \text{ and } -88t \text{ to both sides.}$$

$$t = \frac{-(-88) \pm \sqrt{(-88)^2 - 4(16)(h)}}{2(16)} \qquad \text{Substitute into the Quadratic Formula.}$$

$$t = \frac{88 \pm \sqrt{7{,}744 - 64h}}{32} \qquad \text{Simplify.}$$

6. Define and Use the Discriminant

We can predict what type of roots a quadratic equation will have before we solve it. Suppose that the coefficients a, b, and c in the equation $ax^2 + bx + c = 0$ $(a \neq 0)$ are real numbers. Then the two roots of the equation are given by the Quadratic Formula

$$x = \frac{-b \pm \sqrt{b^2 - 4ac}}{2a} \qquad (a \neq 0)$$

The value of $b^2 - 4ac$, called the **discriminant,** determines the nature of the roots. The possibilities are summarized in the table as follows.

Discriminant	Number of Roots and Type of Roots
0	One repeated rational root
Positive and a perfect square	Two different rational roots
Positive and not a perfect square	Two different irrational roots
Negative	No real number roots

EXAMPLE 9   **Using the Discriminant to Determine the Number and Type of Roots of a Quadratic Equation**

Determine the number and type of roots of $3x^2 + 4x + 1 = 0$.

SOLUTION   We calculate the discriminant $b^2 - 4ac$.

$$b^2 - 4ac = 4^2 - 4(3)(1) \qquad \text{Substitute 4 for } b, 3 \text{ for } a, \text{ and } 1 \text{ for } c.$$

$$= 16 - 12$$

$$= 4$$

Since a, b, and c are real numbers and the discriminant is positive and a perfect square, there will be two different rational roots.   ∎

Self Check 9   Determine the number and type of the roots of $4x^2 - 3x - 2 = 0$.

Now Try Exercise 87.

EXAMPLE 10   ### Using the Discriminant to Find the Constant k

If k is a constant, many quadratic equations are represented by the equation

$$(k - 2)x^2 + (k + 1)x + 4 = 0$$

Find the values of k that will give an equation with roots that are equal rational numbers.

SOLUTION   We calculate the discriminant $b^2 - 4ac$ and set it equal to 0.

$$b^2 - 4ac = (k + 1)^2 - 4(k - 2)(4)$$

$$0 = k^2 + 2k + 1 - 16k + 32$$

$$0 = k^2 - 14k + 33$$

$$0 = (k - 3)(k - 11)$$

$$k - 3 = 0 \quad \text{or} \quad k - 11 = 0$$

$$k = 3 \qquad \qquad k = 11$$

When $k = 3$ or $k = 11$, the equation will have equal roots. As a check, we let $k = 3$ and note that the equation $(k - 2)x^2 + (k + 1)x + 4 = 0$ becomes

$$(3 - 2)x^2 + (3 + 1)x + 4 = 0$$

$$x^2 + 4x + 4 = 0$$

The roots of this equation are equal rational numbers, as expected:

$$x^2 + 4x + 4 = 0$$

$$(x + 2)(x + 2) = 0$$

$$x + 2 = 0 \quad \text{or} \quad x + 2 = 0$$

$$x = -2 \qquad \qquad x = -2$$

Similarly, $k = 11$ will give an equation with equal rational roots.   ∎

Self Check 10   Find k such that $(k - 2)x^2 - (k + 3)x + 9 = 0$ will have equal roots.

Now Try Exercise 93.

7. Write Rational Equations in Quadratic Form and Solve the Equations

If an equation can be written in quadratic form, it can be solved with the techniques used for solving quadratic equations.

EXAMPLE 11   Solving a Rational Equation

Solve: $\dfrac{1}{x-1} + \dfrac{3}{x+1} = 2$.

SOLUTION   Since neither denominator can be zero, $x \neq 1$ and $x \neq -1$. If either number appears as a root, it must be discarded.

$$\dfrac{1}{x-1} + \dfrac{3}{x+1} = 2$$

$$(x-1)(x+1)\left[\dfrac{1}{x-1} + \dfrac{3}{x+1}\right] = (x-1)(x+1)2 \qquad \text{Multiply both sides by } (x-1)(x+1).$$

$$(x+1) + 3(x-1) = 2(x^2 - 1) \qquad \text{Remove brackets and simplify.}$$

$$4x - 2 = 2x^2 - 2 \qquad \text{Remove parentheses and simplify.}$$

$$0 = 2x^2 - 4x \qquad \text{Add } 2 - 4x \text{ to both sides.}$$

The resulting equation is a quadratic equation that we can solve by factoring.

$$2x^2 - 4x = 0$$

$$2x(x-2) = 0 \qquad \text{Factor } 2x^2 - 4x.$$

$$2x = 0 \quad \text{or} \quad x - 2 = 0$$

$$x = 0 \qquad\qquad x = 2$$

Verify these results by checking each root in the original equation. ∎

Self Check 11   Solve: $\dfrac{1}{x-1} + \dfrac{2}{x+1} = 1$.

Now Try Exercise 109.

Self Check Answers   1. $\dfrac{1}{3}, -\dfrac{3}{2}$ 2. $2\sqrt{3}, -2\sqrt{3}$ 3. $\dfrac{-5 + 3\sqrt{5}}{2}, \dfrac{-5 - 3\sqrt{5}}{2}$

4. $1 + \sqrt{10}, 1 - \sqrt{10}$ 5. $\dfrac{-5 + \sqrt{29}}{2}, \dfrac{-5 - \sqrt{29}}{2}$ 6. $3, -\dfrac{1}{2}$

7. $\dfrac{5 \pm \sqrt{13}}{6}$ 8. $\dfrac{-4 \pm \sqrt{29}}{2}$ 9. two different irrational roots

10. $3, 27$ 11. $0, 3$

Exercises **1.3**

Getting Ready
You should be able to complete these vocabulary and concept statements before you proceed to the practice exercises.

Fill in the blanks.

1. A quadratic equation is an equation that can be written in the form _____, where $a \neq 0$.

2. If a and b are real numbers and _____, then $a = 0$ or $b = 0$.

3. If $c > 0$, the equation $x^2 = c$ has two roots. They are $x =$ _____ and $x =$ _____.

4. The Quadratic Formula is $(a \neq 0)$.   _____

5. If a, b, and c are real numbers and if $b^2 - 4ac = 0$, the two roots of the quadratic equation are repeated _____.

6. If a, b, and c are real numbers and $b^2 - 4ac < 0$, the two roots of the quadratic equation are _____.

Practice

Solve each equation by factoring.

7. $x^2 - x - 6 = 0$

8. $x^2 + 8x + 15 = 0$

9. $x^2 - 144 = 0$

10. $x^2 + 4x = 0$

11. $2x^2 + x - 10 = 0$

12. $3x^2 + 4x - 4 = 0$

13. $5x^2 - 13x + 6 = 0$

14. $2x^2 + 5x - 12 = 0$

15. $15x^2 + 16x = 15$

16. $6x^2 - 25x = -25$

17. $12x^2 + 9 = 24x$

18. $24x^2 + 6 = 24x$

Use the Square Root Property to solve each equation.

19. $x^2 = 9$

20. $x^2 = 64$

21. $y^2 - 50 = 0$

22. $x^2 - 75 = 0$

23. $2x^2 = 40$

24. $5x^2 = 400$

25. $4x^2 = 7$

26. $16x^2 = 11$

27. $2x^2 - 13 = 0$

28. $-3x^2 = -11$

29. $(x - 1)^2 = 4$

30. $(y + 2)^2 - 49 = 0$

31. $(x + 1)^2 - 8 = 0$

32. $(y + 2)^2 - 98 = 0$

33. $(2x + 1)^2 = 27$

34. $(5y + 2)^2 - 48 = 0$

Complete the square to make each a perfect-square trinomial.

35. $x^2 + 6x$

36. $x^2 + 8x$

37. $x^2 - 4x$

38. $x^2 - 12x$

39. $a^2 + 5a$

40. $t^2 + 9t$

41. $r^2 - 11r$

42. $s^2 - 7s$

43. $y^2 + \dfrac{3}{4}y$

44. $p^2 + \dfrac{3}{2}p$

45. $q^2 - \dfrac{1}{5}q$

46. $m^2 - \dfrac{2}{3}m$

Solve each equation by completing the square.

47. $x^2 - 8x + 15 = 0$

48. $x^2 + 10x + 21 = 0$

49. $x^2 + 12x = -8$

50. $x^2 - 6x = -1$

51. $x^2 + 5 = -5x$

52. $x^2 + 1 = -4x$

53. $2x^2 - 20x = -49$

54. $4x^2 + 8x = 7$

55. $3x^2 = 1 - 4x$

56. $3x^2 + 4x = 5$

57. $2x^2 = 3x + 1$

58. $2x^2 + 5x = 14$

Use the Quadratic Formula to solve each equation.

59. $x^2 - 12 = 0$

60. $x^2 - 60 = 0$

61. $x^2 - 25x = 0$

62. $x^2 + x = 0$

63. $2x^2 - x - 15 = 0$

64. $6x^2 + x - 2 = 0$

65. $3x^2 = -5x - 1$

66. $2x^2 = 5x + 11$

67. $x^2 + 1 = -7x$

68. $13x^2 + 1 = -10x$

69. $3x^2 + 6x = -1$

70. $2x(x + 3) = -1$

71. $5x\left(x + \dfrac{1}{5}\right) = 3$

72. $7x^2 = 2x + 2$

Solve each formula for the indicated variable.

73. $h = \dfrac{1}{2}gt^2$; t

74. $x^2 + y^2 = r^2$; x

75. $h = 64t - 16t^2$; t

76. $y = 16x^2 - 4$; x

77. $\dfrac{x^2}{a^2} + \dfrac{y^2}{b^2} = 1$; y

78. $\dfrac{x^2}{a^2} - \dfrac{y^2}{b^2} = 1$; x

79. $\dfrac{x^2}{a^2} - \dfrac{y^2}{b^2} = 1$; a

80. $\dfrac{x^2}{a^2} - \dfrac{y^2}{b^2} = 1$; b

81. $x^2 + xy - y^2 = 0$; x

82. $x^2 - 3xy + y^2 = 0$; y

Use the discriminant to determine the number and type of roots. Do not solve the equation.

83. $x^2 + 6x + 9 = 0$

84. $-3x^2 + 2x = 21$

85. $3x^2 - 2x + 5 = 0$

86. $9x^2 + 42x + 49 = 0$

87. $10x^2 + 29x = 21$

88. $10x^2 + x = 21$

89. $x^2 - 5x + 2 = 0$

90. $-8x^2 - 2x = 13$

91. Does $1{,}492x^2 + 1{,}984x - 1{,}776 = 0$ have any roots that are real numbers?

92. Does $2{,}004x^2 + 10x + 1{,}994 = 0$ have any roots that are real numbers?

93. Find two values of k such that $x^2 + kx + 3k - 5 = 0$ will have two roots that are equal.

94. For what value(s) of b will the solutions of $x^2 - 2bx + b^2 = 0$ be equal?

Change each rational equation to quadratic form and solve it by the most efficient method.

95. $x + 1 = \dfrac{12}{x}$

96. $x - 2 = \dfrac{15}{x}$

97. $8x - \dfrac{3}{x} = 10$

98. $15x - \dfrac{4}{x} = 4$

99. $\dfrac{5}{x} = \dfrac{4}{x^2} - 6$

100. $\dfrac{6}{x^2} + \dfrac{1}{x} = 12$

101. $x\left(30 - \dfrac{13}{x}\right) = \dfrac{10}{x}$

102. $x\left(20 - \dfrac{17}{x}\right) = \dfrac{10}{x}$

103. $\dfrac{1}{x} + \dfrac{3}{x + 2} = 2$

104. $\dfrac{1}{x - 1} + \dfrac{1}{x - 4} = \dfrac{5}{4}$

105. $\dfrac{1}{x + 1} + \dfrac{5}{2x - 4} = 1$

106. $\dfrac{x(2x + 1)}{x - 2} = \dfrac{10}{x - 2}$

107. $x + 1 + \dfrac{x + 2}{x - 1} = \dfrac{3}{x - 1}$

108. $\dfrac{1}{4 - y} = \dfrac{1}{4} + \dfrac{1}{y + 2}$

109. $\dfrac{24}{a} - 11 = \dfrac{-12}{a + 1}$

110. $\dfrac{(a - 2)(a + 4)}{10} = \dfrac{a(a - 3)}{5}$

111. $\dfrac{4 + a}{2a} = \dfrac{a - 2}{3}$

112. $\dfrac{36}{b} - 17 = \dfrac{-24}{b + 1}$

Discovery and Writing

113. If r_1 and r_2 are the roots of $ax^2 + bx + c = 0$, show that $r_1 + r_2 = -\dfrac{b}{a}$.

114. If r_1 and r_2 are the roots of $ax^2 + bx + c = 0$, show that $r_1 r_2 = \dfrac{c}{a}$.

In Exercises 115 and 116, a stone is thrown straight upward, higher than the top of a tree. The stone is even with the top of the tree at time t_1 on the way up and at time t_2 on the way down. If the height of the tree is h feet, both t_1 and t_2 are solutions of $h = v_0 t - 16t^2$.

115. Show that the tree is $16t_1 t_2$ feet tall.

116. Show that v_0 is $16(t_1 + t_2)$ feet per second.

117. Explain why the Zero-Factor Theorem is true.

118. Explain how to complete the square on $x^2 - 17x$.

Quadratic equations can be solved automatically by using a computer program, such as Excel.

119. Solve: $2x^2 - 3x - 4 = 0$.
  a. Open an Excel spreadsheet. In cell B1, enter the left side of the equation as
  $$= 2{*}A1{\wedge}2 - 3{*}A1 - 4$$
  (Cell A1 is reserved for the value of x.)
  b. After pressing ENTER, you will see the number -4 in cell B1. Since cell A1 is empty, its value is considered to be 0 and the value of the quadratic is -4 when $x = 0$.
  c. To solve the equation, enter a guess for the solution in cell A1. To find a positive solution, enter 1 as a guess in cell A1.
  d. Click inside cell B1. On the Menu bar, look under the Tools menu for SOLVER. (If there is no SOLVER, choose Add-Ins and then, inside the dialog box, select Solver Add-In and click OK.) After choosing SOLVER, a parameters window will appear. Inside the window, in front of Equal To, select Value Of, and be sure it is followed by the number 0. In the GUESS box, enter $\$A\$1$. Then click Solve. When the Solver Results dialog box opens, be sure that Keep Solver Solution is selected, and click OK. The solution of the equation appears in cell A1.
  e. Find the negative solution of the equation.

120. Use Excel to solve $6x^2 + 13x - 5 = 0$.

Review

Simplify each expression.

121. $5x(x - 2) - x(3x - 2)$

122. $(x + 3)(x - 9) - x(x - 5)$

123. $(m + 3)^2 - (m - 3)^2$

124. $[(y + z)(y - z)]^2$

125. $\sqrt{50x^3} - x\sqrt{8x}$

126. $\dfrac{2x}{\sqrt{5} - 2}$

1.4 Applications of Quadratic Equations

In this section, we will learn to

1. Solve geometric problems.
2. Solve uniform motion problems.
3. Solve falling body problems.
4. Solve business problems.
5. Solve shared-work problems.

The Grand Canyon Skywalk is a tourist attraction located along the Colorado River in the state of Arizona. The glass walkway is shaped like a horseshoe and is 4,000 feet above the floor of the canyon. The Grand Canyon, known for its overwhelming size and beautiful landscape, is awe-inspiring and one of our nation's most astounding natural wonders.

If a Clif energy bar is accidentally dropped over the side of the skywalk, how long will it take it to hit the canyon floor?

If t represents the time in seconds, the quadratic equation $-16t^2 + 4,000 = 0$ models the time it takes the energy bar to fall to the canyon floor. We can solve this equation by using the Square Root Property

$$-16t^2 + 4,000 = 0$$

$$-16t^2 = -4,000 \qquad \text{Subtract 4,000 from both sides.}$$

$$t^2 = 250 \qquad \text{Divide both sides by } -16.$$

$$t = \pm\sqrt{250} \qquad \text{Use the Square Root Property.}$$

$$t \approx \pm 15.8 \qquad \text{Round to the nearest tenth.}$$

Because time cannot be negative, we disregard the negative answer. The time it will take the energy bar to reach the canyon floor is about 15.8 seconds.

As this example illustrates, the solutions of many problems involve quadratic equations.

1. Solve Geometric Problems

EXAMPLE 1   **Solving an Area Problem**

The length of a rectangle exceeds its width by 3 feet. If its area is 40 square feet, find its dimensions.

SOLUTION   To find an equation that models the problem, we can let w represent the width of the rectangle. Then, $w + 3$ will represent its length (see Figure 1-7). Since the formula for the area of a rectangle is $A = lw$ (area = length × width), the area of the rectangle is $(w + 3)w$, which is equal to 40.

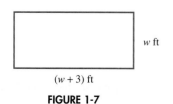

w ft

$(w + 3)$ ft

FIGURE 1-7

The length of the rectangle	times	the width of the rectangle	equals	the area of the rectangle.
$(w + 3)$	·	w	=	40

We can solve this equation for w.

$$(w + 3)w = 40$$

$$w^2 + 3w = 40$$

$$w^2 + 3w - 40 = 0 \qquad \text{Subtract 40 from both sides.}$$

$$(w - 5)(w + 8) = 0 \qquad \text{Factor.}$$

$$w - 5 = 0 \quad \text{or} \quad w + 8 = 0$$

$$w = 5 \qquad \qquad w = -8$$

When $w = 5$, the length is $w + 3 = 8$. The solution -8 must be discarded, because a rectangle cannot have a negative width.

We can verify that this solution is correct by observing that a rectangle with dimensions of 5 feet by 8 feet has an area of 40 square feet. ∎

Self Check 1   The length of a rectangle exceeds its width by 10 feet. If its area is 375 square feet, find its dimensions.

Now Try Exercise 5.

EXAMPLE 2   Solve a Right Triangle Problem

On a college campus, a sidewalk 85 meters long (represented by the red lines in Figure 1-8) joins a dormitory building D with the student center C. However, the students prefer to walk directly from D to C. If segment DC is 65 meters long, how long is each piece of the existing sidewalk?

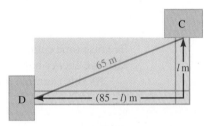

FIGURE 1-8

SOLUTION   We note that the triangle shown in the figure is a right triangle, with a **hypotenuse** that is 65 meters long. If we let the shorter leg of the triangle be l meters long, the length of the longer leg will be $(85 - l)$ meters. By the **Pythagorean Theorem,** we know that the sum of the squares of the two legs of a right triangle is equal to the square of the hypotenuse. Thus, we can form the equation

$$l^2 + (85 - l)^2 = 65^2 \qquad \text{In a right triangle, } a^2 + b^2 = c^2.$$

which we can solve as follows.

$$l^2 + 7{,}225 - 170l + l^2 = 4{,}225 \qquad \text{Expand } (85 - l)^2.$$

$$2l^2 - 170l + 3{,}000 = 0 \qquad \text{Combine like terms and subtract 4,225 from both sides.}$$

$$l^2 - 85l + 1{,}500 = 0 \qquad \text{Divide both sides by 2.}$$

Since the left side is difficult to factor, we will solve this equation using the Quadratic Formula.

$$l = \frac{-b \pm \sqrt{b^2 - 4ac}}{2a}$$

$$l = \frac{-(-85) \pm \sqrt{(-85)^2 - 4(1)(1{,}500)}}{2(1)}$$

$$l = \frac{85 \pm \sqrt{1{,}225}}{2}$$

$$l = \frac{85 \pm 35}{2}$$

$$l = \frac{85 + 35}{2} \quad \text{or} \quad l = \frac{85 - 35}{2}$$

$$= 60 \qquad\qquad\qquad = 25$$

The length of the shorter leg is 25 meters. The length of the longer leg is $(85 - 25)$ meters, or 60 meters. ∎

Self Check 2   The length of a video screen is 21 feet shorter than its height. If the diagonal of the screen is 39 feet, find the height of the screen.

Now Try Exercise 15.

2. Solve Uniform Motion Problems

EXAMPLE 3   **Solving a Uniform Motion Problem**

A man drives 600 miles to a convention. On the return trip, he is able to increase his speed by 10 mph and save 2 hours of driving time. How fast did he drive in each direction?

SOLUTION   We can let s represent the car's speed (in mph) driving to the convention. On the return trip, his speed was $s + 10$ mph. Recall that the distance traveled by an object moving at a constant rate for a certain time is given by the formula $d = rt$. If we divide both sides of this formula by r, we will have a formula for time.

$$t = \frac{d}{r}$$

We can organize the information given in this problem as shown in the following table.

	d	$=$	r	$\cdot$	t
Outbound trip	600		s		$\dfrac{600}{s}$
Return trip	600		$s + 10$		$\dfrac{600}{s + 10}$

Although neither the outbound nor the return travel time is given, we know the difference of those times.

The longer time of the outbound trip	minus	the shorter time of the return trip	equals	the difference in travel times.
$\dfrac{600}{s}$	$-$	$\dfrac{600}{s + 10}$	$=$	2

We can solve this equation for s.

$$\frac{600}{s} - \frac{600}{s+10} = 2$$

$$s(s+10)\left(\frac{600}{s} - \frac{600}{s+10}\right) = s(s+10)(2)$$ Multiply both sides by $s(s+10)$ to clear the equation of fractions.

$$600(s+10) - 600s = 2s(s+10)$$ Simplify.

$$600s + 6{,}000 - 600s = 2s^2 + 20s$$ Remove parentheses.

$$6{,}000 = 2s^2 + 20s$$ Combine like terms.

$$0 = 2s^2 + 20s - 6{,}000$$ Subtract 6,000 from both sides.

$$0 = s^2 + 10s - 3{,}000$$ Divide both sides by 2.

$$0 = (s-50)(s+60)$$ Factor.

$$s - 50 = 0 \quad \text{or} \quad s + 60 = 0$$ Set each factor equal to 0.

$$s = 50 \quad \bigg| \quad s = -60$$

The solution $s = -60$ must be discarded. The man drove 50 mph to the convention and $50 + 10$, or 60, mph on the return trip.

These answers are correct, because a 600-mile trip at 50 mph would take $\frac{600}{50}$, or 12 hours. At 60 mph, the same trip would take only 10 hours, which is 2 hours less time. ∎

Self Check 3   Terrence drives his motorcycle 600 miles to Key West, Florida. On the return trip, he is able to increase his speed by 10 mph and save 3 hours of driving time. How fast did he drive in each direction?

Now Try Exercise 19.

3. Solve Falling Body Problems

EXAMPLE 4   Solving a Falling Body Problem

If an object is thrown straight up into the air with an initial velocity of 144 feet per second, its height is given by the formula $h = 144t - 16t^2$, where h represents its height (in feet) and t represents the time (in seconds) since it was thrown. How long will it take for the object to return to the point from which it was thrown?

SOLUTION   When the object returns to its starting point, its height is again 0. Thus, we can set h equal to 0 and solve for t.

$$h = 144t - 16t^2$$

$$0 = 144t - 16t^2$$ Let $h = 0$.

$$0 = 16t(9-t)$$ Factor.

$$16t = 0 \quad \text{or} \quad 9 - t = 0$$ Set each factor equal to 0.

$$t = 0 \quad \bigg| \quad t = 9$$

At $t = 0$, the object's height is 0, because it was just released. When $t = 9$, the height is again 0, and the object has returned to its starting point. ∎

Self Check 4   How long does it take the ball in Example 4 to reach a height of 324 feet?

Now Try Exercise 23.

4. Solve Business Problems

EXAMPLE 5   **Solving a Business Problem**

A bus company shuttles 1,120 passengers daily between Rockford, Illinois, and O'Hare Airport. The current one-way fare is $10. For each 25¢ increase in the fare, the company predicts that it will lose 48 passengers. What increase in fare will produce daily revenue of $10,208?

SOLUTION   Let q represent the number of quarters the fare will be increased. Then the new fare will be $(10 + 0.25q)$. Since the company will lose 48 passengers for each 25¢ increase, $48q$ passengers will be lost when the rate increases by q quarters. The passenger load will then be $(1,120 - 48q)$ passengers.

Since the daily revenue of $10,208 will be the product of the rate and the number of passengers, we have

$$(10 + 0.25q)(1,120 - 48q) = 10,208$$

$$11,200 - 480q + 280q - 12q^2 = 10,208 \qquad \text{Remove parentheses.}$$

$$-12q^2 - 200q + 992 = 0 \qquad \text{Combine like terms and subtract 10,208 from both sides.}$$

$$3q^2 + 50q - 248 = 0 \qquad \text{Divide both sides by } -4.$$

Since the left side is difficult to factor, we will solve this equation with the Quadratic Formula.

$$q = \frac{-b \pm \sqrt{b^2 - 4ac}}{2a}$$

$$q = \frac{-50 \pm \sqrt{50^2 - 4(3)(-248)}}{2(3)} \qquad \text{Substitute 3 for } a, \text{ 50 for } b, \text{ and } -248 \text{ for } c.$$

$$q = \frac{-50 \pm \sqrt{2,500 + 2,976}}{6}$$

$$q = \frac{-50 \pm \sqrt{5,476}}{6}$$

$$q = \frac{-50 \pm 74}{6}$$

$$q = \frac{-50 + 74}{6} \quad \text{or} \quad q = \frac{-50 - 74}{6}$$

$$= \frac{24}{6} \qquad\qquad\qquad = \frac{-124}{6}$$

$$= 4 \qquad\qquad\qquad = -\frac{62}{3}$$

Since the number of riders cannot be negative, the result of $-\frac{62}{3}$ must be discarded. To generate $10,208 in daily revenues, the company should raise the fare by 4 quarters, or $1, to $11.

Self Check 5   A rock band has been drawing average crowds of 400 people. It is projected that for every $1 increase in the $10 ticket price, the average attendance will decrease by 20. At what ticket price will nightly receipts be $4,500?

Now Try Exercise 31.

5. Solve Shared-Work Problems

EXAMPLE 6   Solving a Shared-Work Problem

One environmental company can clean up an oil spill on a beach in 2 days less time than its competitor. Working together they were able to clean up the spill in 10 days. How long would it have taken the first company to clean up the spill if it worked alone?

SOLUTION   Suppose the first company can clean up the spill in x days. Then the first company can do $\frac{1}{x}$ of the job each day. Because the first company can do the work in 2 days less than its competitor, it will take the competitor $(x + 2)$ days to clean up the spill. The competitor can do $\frac{1}{x+2}$ of the job each day.

Working together, they can clean up the spill in 10 days. So together they can do $\frac{1}{10}$ of the job each day. The sum of the work each can do in one day is equal to the work that they can do together in one day.

The part the first company can clean up in one day	plus	the part the second company can clean up in one day	equals	the part they can clean up together in one day.
$\dfrac{1}{x}$	$+$	$\dfrac{1}{x+2}$	$=$	$\dfrac{1}{10}$

We can solve this equation for x.

$$\frac{1}{x} + \frac{1}{x+2} = \frac{1}{10}$$

$$10x(x+2)\left(\frac{1}{x} + \frac{1}{x+2}\right) = 10x(x+2)\left(\frac{1}{10}\right) \qquad \text{Multiply both sides by } 10x(x+2) \text{ to eliminate the fractions.}$$

$$\frac{10x(x+2)}{x} + \frac{10x(x+2)}{x+2} = \frac{10x(x+2)}{10} \qquad \text{Distribute the multiplication by } 10x(x+2).$$

$$10(x+2) + 10x = x(x+2) \qquad \frac{x}{x} = 1, \frac{x+2}{x+2} = 1, \text{ and } \frac{10}{10} = 1.$$

$$10x + 20 + 10x = x^2 + 2x \qquad \text{Use the Distributive Property to remove parentheses.}$$

$$0 = x^2 - 18x - 20 \qquad \text{Subtract } 20x \text{ and } 20 \text{ from both sides.}$$

Since the right side cannot be factored over the integers, we will solve the equation with the Quadratic Formula.

$$x = \frac{-b \pm \sqrt{b^2 - 4ac}}{2a}$$

$$x = \frac{-(-18) \pm \sqrt{(-18)^2 - 4(1)(-20)}}{2(1)} \qquad \text{Substitute 1 for } a, -18 \text{ for } b, \text{ and } -20 \text{ for } c.$$

$$x = \frac{18 \pm \sqrt{324 + 80}}{2}$$

$$x = \frac{18 \pm \sqrt{404}}{2}$$

$$x = \frac{18 \pm 20.09975124}{2}$$

$$x = \frac{18 + 20.09975124}{2} \quad \text{or} \quad x = \frac{18 - 20.09975124}{2}$$

$$\approx 19.05 \qquad\qquad\qquad \approx -1.05$$

Since the work cannot be completed in a negative number of days, we discard the solution of -1.05. Thus, the first company can complete the job working alone in a little over 19 days. ∎

Self Check 6   A hose can fill a swimming pool in 7 hours. Another hose needs 2 more hours to fill the pool than the two hoses combined. How long would it take the second hose to fill the pool?

Now Try Exercise 39.

Self Check Answers   **1.** 15 feet by 25 feet   **2.** 36 feet   **3.** 40 mph to Key West and 50 mph returning   **4.** 4.5 seconds   **5.** $15   **6.** ≈4.9 hours

Exercises **1.4**

Getting Ready
You should be able to complete these vocabulary and concept statements before you proceed to the practice exercises.

Fill in the blanks.

1. The formula for the area of a rectangle is _____ .
2. The formula that relates distance, rate, and time is

   _____ .

Practice
Solve each problem.

3. **Geometric problem** A rectangle is 4 feet longer than it is wide. If its area is 32 square feet, find its dimensions.

4. **Geometric problem** A rectangle is 5 times as long as it is wide. If the area is 125 square feet, find its perimeter.

5. **Dallas Cowboys video screen** The Dallas Cowboys stadium has the world's largest video screen. The rectangular screen's length is 88 feet more than its width. If the video screen has an area of 11,520 square feet, find the dimensions of the screen.

Ken Durden/Shutterstock.com

6. **IMAX screen** A large movie screen is in the Panasonic IMAX theater at Darling Harbor, Sydney, Australia. The rectangular screen has an area of 11,349 square feet. Find the dimensions of the screen if it is 20 feet longer than it is wide.

7. **Geometric problem** The side of a square is 4 centimeters shorter than the side of a second square. If the sum of their areas is 106 square centimeters, find the length of one side of the larger square.

8. **Geometric problem** If two opposite sides of a square are increased by 10 meters and the other sides are decreased by 8 meters, the area of the rectangle that is formed is 63 square meters. Find the area of the original square.

9. **Flags** In 1912, an order by President Taft fixed the width and length of the U.S. flag in the ratio of 1 to 1.9. If 100 square feet of cloth are to be used to make a U.S. flag, estimate its dimensions to the nearest $\frac{1}{4}$ foot.

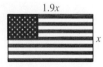

10. **Geometric problem** Find the dimensions of a rectangle whose area is 180 cm² and whose perimeter is 54 cm.

11. **Metal fabrication** A piece of tin, 12 inches on a side, is to have four equal squares cut from its corners, as in the illustration. If the edges are then to be folded up to make a box with a floor area of 64 square inches, find the depth of the box.

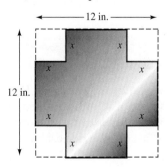

12. **Making gutters** A piece of sheet metal, 18 inches wide, is bent to form the gutter shown in the illustration. If the cross-sectional area is 36 square inches, find the depth of the gutter.

13. **Geometric problem** The base of a triangle is one-third as long as its height. If the area of the triangle is 24 square meters, how long is its base?

14. **Geometric problem** The base of a triangle is one-half as long as its height. If the area of the triangle is 100 square yards, find its height.

15. **Right triangle** If one leg of a right triangle is 14 meters shorter than the other leg, and the hypotenuse is 26 meters, find the length of the two legs.

16. **Right triangle** If one leg of a right triangle is five times the other leg, and the hypotenuse is $10\sqrt{26}$ centimeters, find the length of the two legs.

17. **Manufacturing** A manufacturer of television sets received an order for sets with a 46-inch screen (measured along the diagonal), as shown in the illustration. If the televisions are $17\frac{1}{2}$ inches wider than they are high, find the dimensions of the screen to the nearest tenth of an inch.

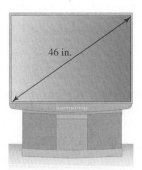

18. **Finding dimensions** An oriental rug is 2 feet longer than it is wide. If the diagonal of the rug is 12 feet, to the nearest tenth of a foot, find its dimensions.

19. **Cycling rates** A cyclist rides from DeKalb to Rockford, a distance of 40 miles. His return trip takes 2 hours longer, because his speed decreases by 10 mph. How fast does he ride each way?

20. **Travel times** Jake drives a tractor from one town to another, a distance of 120 kilometers. He drives 10 kilometers per hour faster on the return trip, cutting 1 hour off the time. How fast does he drive each way?

21. **Uniform motion problem** If the speed were increased by 10 mph, a 420-mile trip would take 1 hour less time. How long will the trip take at the slower speed?

22. **Uniform motion problem** By increasing her usual speed by 25 kilometers per hour, a bus driver decreases the time on a 25-kilometer trip by 10 minutes. Find the usual speed.

23. **Ballistics** The height of a projectile fired upward with an initial velocity of 400 feet per second is given by the formula $h = -16t^2 + 400t$, where h is the height in feet and t is the time in seconds. Find the time required for the projectile to return to earth.

24. **Ballistics** The height of an object tossed upward with an initial velocity of 104 feet per second is given by the formula $h = -16t^2 + 104t$, where h is the height in feet and t is the time in seconds. Find the time required for the object to return to its point of departure.

25. **Falling coins** An object will fall s feet in t seconds, where $s = 16t^2$. How long will it take for a penny to hit the ground if it is dropped from the top of the Sears Tower in Chicago? (*Hint:* The tower is 1,454 feet tall.)

26. **Movie stunts** According to the *Guinness Book of World Records, 1998,* stuntman Dan Koko fell a distance of 312 feet into an airbag after jumping from the Vegas World Hotel and Casino. The distance d in feet traveled by a free-falling object in t seconds is given by the formula $d = 16t^2$. To the nearest tenth of a second, how long did the fall last?

27. **Accidents** The height h (in feet) of an object that is dropped from a height of s feet is given by the formula $h = s - 16t^2$, where t is the time the object has been falling. A 5-foot-tall woman on a sidewalk looks directly overhead and sees a window washer drop a bottle from 4 stories up. How long does she have to get out of the way? Round to the nearest tenth. (A story is 12 feet.)

28. **Ballistics** The height of an object thrown upward with an initial velocity of 32 feet per second is given by the formula $h = -16t^2 + 32t$, where t is the time in seconds. How long will it take the object to reach a height of 16 feet?

29. Setting fares A bus company has 3,000 passengers daily, paying a 25¢ fare. For each nickel increase in fare, the company projects that it will lose 80 passengers. What fare increase will produce $994 in daily revenue?

30. Jazz concerts A jazz group on tour has been drawing average crowds of 500 persons. It is projected that for every $1 increase in the $12 ticket price, the average attendance will decrease by 50. At what ticket price will nightly receipts be $5,600?

31. Concert receipts Tickets for the annual symphony orchestra pops concert cost $15, and the average attendance at the concerts has been 1,200 persons. Management projects that for each 50¢ decrease in ticket price, 40 more patrons will attend. How many people attended the concert if the receipts were $17,280?

32. Projecting demand The *Vilas County News* earns a profit of $20 per year for each of its 3,000 subscribers. Management projects that the profit per subscriber would increase by 1¢ for each additional subscriber over the current 3,000. How many subscribers are needed to bring a total profit of $120,000?

33. Investment problems Morgan and Chloe each have a bank CD. Morgan's is $1,000 larger than Chloe's, but the interest rate is 1% less. Last year Morgan received interest of $280, and Chloe received $240. Find the rate of interest for each CD.

34. Investment problem Scott and Laura have both invested some money. Scott invested $3,000 more than Laura and at a 2% higher interest rate. If Scott received $800 annual interest and Laura received $400, how much did Scott invest?

35. Buying microwave ovens Some mathematics professors would like to purchase a $150 microwave oven for the department workroom. If four of the professors don't contribute, everyone's share will increase by $10. How many professors are in the department?

36. Digital cameras A merchant could sell one model of digital cameras at list price for $180. If he had three more cameras, he could sell each one for $10 less and still receive $180. Find the list price of each camera.

37. Filling storage tanks Two pipes are used to fill a water storage tank. The first pipe can fill the tank in 4 hours, and the two pipes together can fill the tank in 2 hours less time than the second pipe alone. How long would it take for the second pipe to fill the tank?

38. Filling swimming pools A hose can fill a swimming pool in 6 hours. Another hose needs 3 more hours to fill the pool than the two hoses combined. How long would it take the second hose to fill the pool?

39. Mowing lawns Kristy can mow a lawn in 1 hour less time than her brother Steven. Together they can finish the job in 5 hours. How long would it take Kristy if she worked alone?

40. Cleaning the garage Working together, Sarah and Heidi can clean the garage in 2 hours. If they work alone, it takes Heidi 3 hours longer than it takes Sarah. How long would it take Heidi to clean the garage alone?

41. Planting windscreens A farmer intends to construct a windscreen by planting trees in a quarter-mile row. His daughter points out that 44 fewer trees will be needed if they are planted 1 foot farther apart. If her dad takes her advice, how many trees will be needed? A row starts and ends with a tree. (*Hint:* 1 mile = 5,280 feet.)

42. Angle between spokes If a wagon wheel had 10 more spokes, the angle between spokes would decrease by 6°. How many spokes does the wheel have?

43. Architecture A **golden rectangle** is one of the most visually appealing of all geometric forms. The front of the Parthenon, built in Athens in the 5th century B.C. and shown in the illustration, is a golden rectangle. In a golden rectangle, the length l and the height h of the rectangle must satisfy the following equation. If a rectangular billboard is to have a height of 15 feet, how long should it be if it is to form a golden rectangle? Round to the nearest tenth of a foot.

$$\frac{l}{h} = \frac{h}{l - h}$$

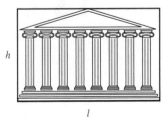

44. Golden ratio Rectangle $ABCD$, shown here, will be a **golden rectangle** if $\frac{AB}{AD} = \frac{BC}{BE}$ where $AE = AD$. Let $AE = 1$ and find the ratio of AB to AD.

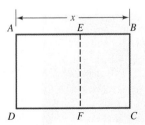

45. Automobile engines As the piston shown moves upward, it pushes a cylinder of a gasoline/air mixture that is ignited by the spark plug. The formula that gives the volume of a cylinder is $V = \pi r^2 h$, where r is the radius and h the height. Find the radius of the piston (to the nearest hundredth of an inch) if it displaces 47.75 cubic inches of gasoline/air mixture as it moves from its lowest to its highest point.

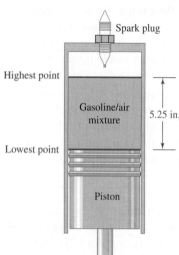

Spark plug

Highest point

Gasoline/air mixture

5.25 in.

Lowest point

Piston

46. History One of the important cities of the ancient world was Babylon. Greek historians wrote that the city was square-shaped. Its area numerically exceeded its perimeter by about 124. Find its dimensions in miles. (Round to the nearest tenth.)

Discovery and Writing

47. Is it possible for a rectangle to have a width that is 3 units shorter than its diagonal and a length that is 4 units longer than its diagonal?

48. Which of the preceding application problems did you find the hardest? Why?

Review
Perform the operations and simplify.

49. $\dfrac{2}{x} - \dfrac{1}{x-3}$

50. $\dfrac{1}{x} \cdot \dfrac{x^2 - 5x}{x-3}$

51. $\dfrac{x+3}{x^2-x-6} \div \dfrac{x^2+3x}{x^2-9}$

52. $\dfrac{\dfrac{1}{x} - \dfrac{1}{2}}{x-2}$

53. $\dfrac{\dfrac{1}{x} + \dfrac{1}{y}}{\dfrac{1}{x} - \dfrac{1}{y}}$

54. $\dfrac{x}{x+1} + \dfrac{x+1}{x} \cdot \dfrac{x^2-x}{3}$

1.5 Complex Numbers

In this section, we will learn to

1. Simplify imaginary numbers.

2. Perform operations on complex numbers.

3. Find powers of i.

4. Determine the absolute value of a complex number.

5. Solve quadratic equations with complex roots.

6. Factor the sum of two squares.

The Blue Man Group is a highly successful group of creative performers who cover themselves in blue grease paint, wear latex bald caps, and dress in black. They combine rock music, comedy, multimedia theatrics, and sophisticated lighting to entertain their audiences.

There are a variety of themes used in Blue Man performances, and the group uses fractals in their shows. Fractals are beautiful art designs that are computer generated from numbers called *complex numbers*. We will study these numbers in this section. A fractal is shown on the next page.

All of the quadratic equations that we considered in Section 1.3 had roots that were real numbers. However, the solutions of many quadratic equations are not real numbers. For example, if we use the Quadratic Formula to solve $x^2 + x + 2 = 0$, we get solutions that are not real.

$$x = \frac{-b \pm \sqrt{b^2 - 4ac}}{2a}$$

$$x = \frac{-1 \pm \sqrt{1^2 - 4(1)(2)}}{2(1)} \qquad \text{Substitute 1 for } a, \text{ 1 for } b, \text{ and 2 for } c.$$

$$x = \frac{-1 \pm \sqrt{1 - 8}}{2}$$

$$x = \frac{-1 \pm \sqrt{-7}}{2}$$

Each solution involves $\sqrt{-7}$. This is not a real number, because the square of no real number is -7.

1. Simplify Imaginary Numbers

For many years, mathematicians considered numbers such as $\sqrt{-1}$, $\sqrt{-5}$ and $\sqrt{-9}$ to make no sense. Sir Isaac Newton (1642–1727) called them "impossible numbers." In the 17th century, these symbols were called **imaginary numbers** by René Descartes. Today, they have important uses, such as describing the behavior of alternating current in electronics. The imaginary numbers are based on the imaginary unit, and the imaginary unit is denoted by the letter i.

Imaginary Unit i   The **imaginary unit** is denoted by the letter i and is defined as

$$i = \sqrt{-1}$$

From the definition it follows that $i^2 = -1$.

Because imaginary numbers follow the rules for exponents, we have

$$(3i)^2 = 3^2 i^2 = 9(-1) = -9$$

Since $(3i)^2 = -9$, $3i$ is the square root of -9, and we can write

$$\sqrt{-9} = \sqrt{(-1)(9)}$$
$$= \sqrt{-1}\sqrt{9} \qquad \sqrt{ab} = \sqrt{a}\sqrt{b}$$
$$= 3i \qquad\qquad \sqrt{-1} = i$$

The Multiplication Property of Radicals can be used to simplify imaginary numbers. Four examples are shown in the table below.

Comment

When $\sqrt{b}i$ and $\sqrt{bi}$ are handwritten, it is easy to confuse them. To avoid confusion, we will write i first so that it is clear that i is not under the radical symbol.

Imaginary Numbers Written in Terms of i
$\sqrt{-25} = \sqrt{(-1)(25)} = \sqrt{-1}\sqrt{25} = (i)(5) = 5i$
$-\sqrt{-169} = -\sqrt{(-1)(169)} = -\sqrt{-1}\sqrt{169} = -(i)(13) = -13i$
$\sqrt{-24} = \sqrt{(-1)(24)} = \sqrt{-1}\sqrt{24} = (i)(2\sqrt{6}) = 2\sqrt{6}i$ or $2i\sqrt{6}$
$\sqrt{-\dfrac{8}{49}} = \sqrt{(-1)\left(\dfrac{8}{49}\right)} = \sqrt{-1}\sqrt{\dfrac{8}{49}} = (i)\left(\dfrac{\sqrt{8}}{\sqrt{49}}\right) = (i)\left(\dfrac{2\sqrt{2}}{7}\right) = \dfrac{2\sqrt{2}}{7}i$ or $\dfrac{2i\sqrt{2}}{7}$

Caution

If a and b both are negative, then $\sqrt{ab} \neq \sqrt{a}\sqrt{b}$. For example, the correct simplification of $\sqrt{-16}\sqrt{-4}$ is

$$\sqrt{-16}\sqrt{-4} = (4i)(2i) = 8i^2 = 8(-1) = -8$$

The following simplification is incorrect. Since both numbers are negative, the Multiplication Property of Radicals does not apply.

$$\sqrt{-16}\sqrt{-4} = \sqrt{(-16)(-4)} = \sqrt{64} = 8$$

ACCENT ON TECHNOLOGY

Simplifying Imaginary Numbers

If we place our graphing calculator in $a + bi$ mode, it can be used to simplify many imaginary numbers. Please note that on a graphing calculator i is located above the decimal point key.

In Figure 1-9, we see the graphing calculator results for $\sqrt{-81}$, $\sqrt{-\frac{4}{25}}$, and $\sqrt{-108}$.

```
√(-81)
                  9i
√(-4/25)
                 .4i
√(-108)
         10.39230485i
```

FIGURE 1-9

2. Perform Operations on Complex Numbers

Numbers that are the sum or difference of a real number and an imaginary number, such as $3 + 4i$, $-5 + 7i$, and $-1 - 9i$, are called *complex numbers*.

Complex Numbers

A **complex number** is a number that can be written in the form $a + bi$, where a and b are real numbers and $i = \sqrt{-1}$.

The number a is called the **real part,** and b is called the **imaginary part.**

If $b = 0$, the complex number $a + bi$ is the real number a. If $a = 0$ and $b \neq 0$, the complex number $a + bi$ is the imaginary number bi. It follows that the set of real numbers and the set of imaginary numbers are subsets of the set of complex numbers. Figure 1-10 illustrates how the various sets of numbers are related.

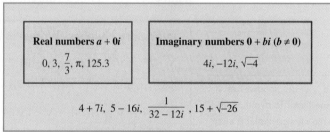

Complex numbers

Real numbers $a + 0i$	Imaginary numbers $0 + bi$ ($b \neq 0$)
$0, 3, \frac{7}{3}, \pi, 125.3$	$4i, -12i, \sqrt{-4}$

$$4 + 7i, \ 5 - 16i, \ \frac{1}{32 - 12i}, \ 15 + \sqrt{-26}$$

FIGURE 1-10

To determine whether two complex numbers are equal, we can use the following definition.

Equality of Complex Numbers   Two complex numbers are **equal** if their real parts are equal and their imaginary parts are equal. If $a + bi$ and $c + di$ are two complex numbers, then

$$a + bi = c + di \quad \text{if and only if} \quad a = c \quad \text{and} \quad b = d$$

EXAMPLE 1   **Using the Definition of Equality of Complex Numbers**

For what numbers x and y is $3x + 4i = (2y + x) + xi$?

SOLUTION   Since the numbers are equal, their imaginary parts must be equal: $x = 4$. Since their real parts are equal, $3x = 2y + x$. We can solve the system

$$\begin{cases} x = 4 \\ 3x = 2y + x \end{cases}$$

by substituting 4 for x in the second equation and solving for y. We find that $y = 4$. The solution is $x = 4$ and $y = 4$. ∎

Self Check 1   Find x: $a + (x + 3)i = a - (2x - 1)i$.

Now Try Exercise 19.

Complex numbers can be added and subtracted as if they were binomials.

Adding and Subtracting   Two complex numbers such as $a + bi$ and $c + di$ are **added** and **subtracted** as if
Complex Numbers   they were binomials:

$$(a + bi) + (c + di) = (a + c) + (b + d)i$$
$$(a + bi) - (c + di) = (a - c) + (b - d)i$$

Because of the preceding definition, the sum or difference of two complex numbers is another complex number.

EXAMPLE 2   **Adding and Subtracting Complex Numbers**

Simplify: **a.** $(3 + 4i) + (2 + 7i)$ **b.** $(-5 + 8i) - (2 - 12i)$

SOLUTION   **a.** $(3 + 4i) + (2 + 7i) = 3 + 4i + 2 + 7i$
$$= 3 + 2 + 4i + 7i$$
$$= 5 + 11i$$

b. $(-5 + 8i) - (2 - 12i) = -5 + 8i - 2 + 12i$
$$= -5 - 2 + 8i + 12i$$
$$= -7 + 20i$$ ∎

Self Check 2   Simplify: **a.** $(5 - 2i) + (-3 + 9i)$ **b.** $(2 + 5i) - (6 + 7i)$

Now Try Exercise 23.

Complex numbers can also be multiplied as if they were binomials.

Multiplying Complex Numbers	The numbers $a + bi$ and $c + di$ are **multiplied** as if they were binomials, with $i^2 = -1$:

$$(a + bi)(c + di) = (ac - bd) + (ad + bc)i$$

Because of this definition, the product of two complex numbers is another complex number.

EXAMPLE 3 **Multiplying Complex Numbers**

Multiply: **a.** $(3 + 4i)(2 + 7i)$ **b.** $(5 - 7i)(1 + 3i)$

SOLUTION **a.** $(3 + 4i)(2 + 7i) = 6 + 21i + 8i + 28i^2$

$$= 6 + 21i + 8i + 28(-1) \qquad i^2 = -1$$
$$= 6 - 28 + 29i$$
$$= -22 + 29i$$

b. $(5 - 7i)(1 + 3i) = 5 + 15i - 7i - 21i^2$

$$= 5 + 15i - 7i - 21(-1) \qquad i^2 = -1$$
$$= 5 + 21 + 8i$$
$$= 26 + 8i$$

Self Check 3 Multiply: $(2 - 5i)(3 + 2i)$.

Now Try Exercise 33.

To avoid errors in determining the sign of the result, always express numbers in $a + bi$ form before attempting any algebraic manipulations.

EXAMPLE 4 **Multiplying Complex Numbers**

Multiply: $\left(-2 + \sqrt{-16}\right)\left(4 - \sqrt{-9}\right)$.

SOLUTION We change each number to $a + bi$ form:

$$-2 + \sqrt{-16} = -2 + \sqrt{16}\sqrt{-1} = -2 + 4i$$
$$4 - \sqrt{-9} = 4 - \sqrt{9}\sqrt{-1} = 4 - 3i$$

and then find the product.

$$(-2 + 4i)(4 - 3i) = -8 + 6i + 16i - 12i^2$$
$$= -8 + 6i + 16i - 12(-1) \qquad i^2 = -1$$
$$= -8 + 12 + 22i$$
$$= 4 + 22i$$

Self Check 4 Multiply: $\left(3 + \sqrt{-25}\right)\left(2 - \sqrt{-9}\right)$.

Now Try Exercise 37.

Before we discuss the division of complex numbers, we introduce the concept of a complex conjugate.

Complex Conjugates   The complex numbers $a + bi$ and $a - bi$ are called **complex conjugates** of each other.

For example,

- $2 + 5i$ and $2 - 5i$ are complex conjugates.
- $-\frac{1}{2} + 4i$ and $-\frac{1}{2} - 4i$ are complex conjugates.
- $13i$ and $-13i$ are complex conjugates. That is, the complex conjugate of $0 + 13i$ is $0 - 13i$.

What makes this concept important is the fact that the product of two complex conjugates is always a real number. For example,

$$(2 + 5i)(2 - 5i) = 4 - 10i + 10i - 25i^2$$

$$= 4 - 25(-1) \qquad i^2 = -1$$

$$= 4 + 25$$

$$= 29$$

In general, we have

$$(a + bi)(a - bi) = a^2 - abi + abi - b^2i^2$$

$$= a^2 - b^2(-1) \qquad i^2 = -1$$

$$= a^2 + b^2$$

To divide complex numbers, we use the concept of complex conjugates to rationalize the denominator.

EXAMPLE 5   **Dividing Complex Numbers**

Divide and write the result in $a + bi$ form: $\dfrac{3}{2 + i}$.

SOLUTION   To divide, we rationalize the denominator and simplify.

$$\frac{3}{2 + i} = \frac{3(2 - i)}{(2 + i)(2 - i)}$$ To make the denominator a real number, multiply the numerator and denominator by the complex conjugate of $2 + i$, which is $2 - i$.

$$= \frac{6 - 3i}{4 - 2i + 2i - i^2}$$ Multiply.

$$= \frac{6 - 3i}{4 + 1}$$ Simplify the denominator.

$$= \frac{6 - 3i}{5}$$

$$= \frac{6}{5} - \frac{3}{5}i$$

It is common to accept $\frac{6}{5} - \frac{3}{5}i$ as a substitute for $\frac{6}{5} + \left(-\frac{3}{5}\right)i$. ∎

Self Check 5   Divide and write the result in $a + bi$ form: $\dfrac{3}{3 - i}$.

Now Try Exercise 45.

EXAMPLE 6   Dividing Complex Numbers

Divide and write the result in $a + bi$ form: $\dfrac{2 - \sqrt{-16}}{3 + \sqrt{-1}}$.

SOLUTION $\dfrac{2 - \sqrt{-16}}{3 + \sqrt{-1}} = \dfrac{2 - 4i}{3 + i}$ Change each number to $a + bi$ form.

$= \dfrac{(2 - 4i)(3 - i)}{(3 + i)(3 - i)}$ To make the denominator a real number, multiply the numerator and denominator by $3 - i$.

$= \dfrac{6 - 2i - 12i + 4i^2}{9 - 3i + 3i - i^2}$ Remove parentheses.

$= \dfrac{2 - 14i}{9 + 1}$ Combine like terms; $i^2 = -1$.

$= \dfrac{2}{10} - \dfrac{14i}{10}$

$= \dfrac{1}{5} - \dfrac{7}{5}i$ ■

Self Check 6   Divide and write the result in $a + bi$ form: $\dfrac{3 + \sqrt{-25}}{2 - \sqrt{-1}}$.

Now Try Exercise 53.

Examples 5 and 6 illustrate that the quotient of two complex numbers is another complex number.

3. Find Powers of i

The powers of i with natural number exponents produce an interesting pattern as shown in the table.

Comment

i^0 is 1 because any number except 0 raised to the 0th power is 1.

Powers of i	
$i^1 = \sqrt{-1} = i$	$i^5 = i^4 i = 1i = i$
$i^2 = \left(\sqrt{-1}\right)^2 = -1$	$i^6 = i^4 i^2 = 1(-1) = -1$
$i^3 = i^2 i = -1i = -i$	$i^7 = i^4 i^3 = 1(-i) = -i$
$i^4 = i^2 i^2 = (-1)(-1) = 1$	$i^8 = i^4 i^4 = 1(1) = 1$
The pattern continues: $i, -1, -i, 1, \ldots$	

EXAMPLE 7   Simplifying Powers of i

Simplify: i^{365}.

SOLUTION   Since $i^4 = 1$, each occurrence of i^4 is a factor of 1. To determine how many factors of i^4 are in i^{365}, we divide 365 by 4. The quotient is 91, and the remainder is 1.

$$i^{365} = (i^4)^{91} \cdot i^1$$

$$= 1^{91} \cdot i^1 \qquad i^4 = 1$$

$$= i \qquad\qquad 1^{91} = 1 \text{ and } 1 \cdot i = i$$

■

Self Check 7   Simplify: $i^{1,999}$.

Now Try Exercise 63.

The result of Example 7 illustrates the following theorem.

Powers of i   If n is a natural number that has a remainder of r when divided by 4, then

$$i^n = i^r$$

When n is divisible by 4, the remainder r is 0 and $i^0 = 1$.

We can also simplify powers of i that involve negative integer exponents.

$$i^{-1} = \frac{1}{i} = \frac{1 \cdot i}{i \cdot i} = \frac{i}{-1} = -i \qquad\qquad i^{-2} = \frac{1}{i^2} = \frac{1}{-1} = -1$$

$$i^{-3} = \frac{1}{i^3} = \frac{1 \cdot i}{i^3 \cdot i} = \frac{i}{i^4} = \frac{i}{1} = i \qquad\qquad i^{-4} = \frac{1}{i^4} = \frac{1}{1} = 1$$

ACCENT ON TECHNOLOGY   **Performing Operations on Complex Numbers**

If we place our graphing calculator in $a + bi$ mode we can use it to perform operations on complex numbers. Addition, subtraction, and multiplication of $7 - 2i$ and $2 + i$ are shown in Figure 1-11(a). Division of $7 - 2i$ and $2 + i$ is shown in Figure 1-11(b). We can also find powers of i using our graphing calculator. i^{24} and i^{25} are shown in Figure 1-11(c). Since 4E^-13i can be approximated as 0, $i^{24} = 1$. Since -5E^-13 can be approximated as 0, $i^{25} = i$.

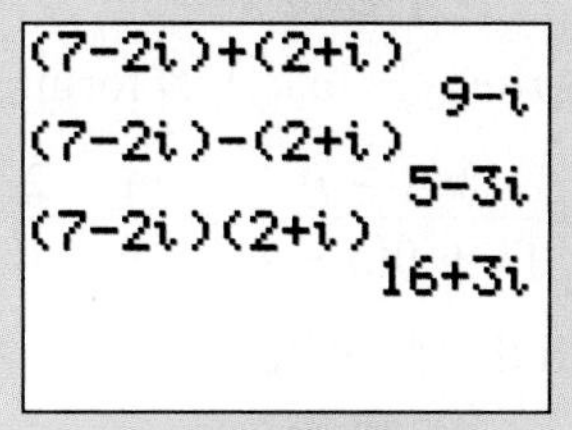

```
(7-2i)+(2+i)
            9-i
(7-2i)-(2+i)
           5-3i
(7-2i)(2+i)
          16+3i
```
(a)

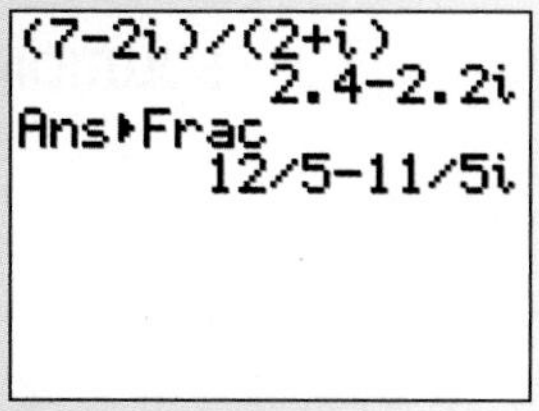

```
(7-2i)/(2+i)
         2.4-2.2i
Ans▶Frac
         12/5-11/5i
```
(b)

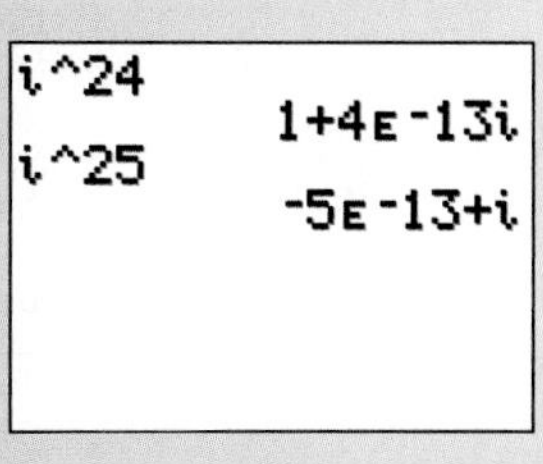

```
i^24
         1+4E-13i
i^25
         -5E-13+i
```
(c)

FIGURE 1-11

4. Determine the Absolute Value of a Complex Number

Absolute Value of a Complex Number	If $a + bi$ is a complex number, then
	$$\lvert a + bi \rvert = \sqrt{a^2 + b^2}$$

Because of the previous definition, the absolute value of a complex number is a real number. For this reason, i does not appear in the result.

EXAMPLE 8   **Determine the Absolute Value of a Complex Number**

Write without absolute value symbols: **a.** $\lvert 3 + 4i \rvert$ **b.** $\lvert 4 - 6i \rvert$

SOLUTION   In each case, we apply the definition of absolute value of a complex number.

a. $\lvert 3 + 4i \rvert = \sqrt{3^2 + 4^2}$

$\qquad\qquad = \sqrt{9 + 16}$

$\qquad\qquad = \sqrt{25}$

$\qquad\qquad = 5$

b. $\lvert 4 - 6i \rvert = \sqrt{4^2 + (-6)^2}$

$\qquad\qquad = \sqrt{16 + 36}$

$\qquad\qquad = \sqrt{52}$

$\qquad\qquad = \sqrt{4 \cdot 13}$

$\qquad\qquad = \sqrt{4}\sqrt{13}$

$\qquad\qquad = 2\sqrt{13}$

Self Check 8   Write without absolute value symbols: $\lvert 2 - 5i \rvert$.

Now Try Exercise 75.

EXAMPLE 9   **Determine the Absolute Value of a Complex Number**

Write without absolute value symbols: **a.** $\left\lvert \dfrac{2i}{3 + i} \right\rvert$ **b.** $\lvert a + 0i \rvert$

SOLUTION   **a.** We first write $\dfrac{2i}{3 + i}$ in $a + bi$ form:

$$\frac{2i}{3 + i} = \frac{2i(3 - i)}{(3 + i)(3 - i)} = \frac{6i - 2i^2}{9 - i^2} = \frac{6i + 2}{10} = \frac{1}{5} + \frac{3}{5}i$$

and then find the absolute value of $\frac{1}{5} + \frac{3}{5}i$.

$$\left\lvert \frac{2i}{3 + i} \right\rvert = \left\lvert \frac{1}{5} + \frac{3}{5}i \right\rvert = \sqrt{\left(\frac{1}{5}\right)^2 + \left(\frac{3}{5}\right)^2} = \sqrt{\frac{10}{25}} = \frac{\sqrt{10}}{5}$$

b. $\lvert a + 0i \rvert = \sqrt{a^2 + 0^2} = \sqrt{a^2} = \lvert a \rvert$

From part b, we see that $\lvert a \rvert = \sqrt{a^2}$.

Self Check 9   Write without absolute value symbols: $\left\lvert \dfrac{3i}{2 - i} \right\rvert$.

Now Try Exercise 85.

5. Solve Quadratic Equations with Complex Roots

The roots of many quadratic equations are complex numbers, as the following example shows.

EXAMPLE 10   **Solving a Quadratic Equation with Complex Roots**

Solve: $x^2 - 4x + 5 = 0$.

SOLUTION   In this equation, $a = 1$, $b = -4$, and $c = 5$.

$$x = \frac{-b \pm \sqrt{b^2 - 4ac}}{2a}$$

$$= \frac{-(-4) \pm \sqrt{(-4)^2 - 4(1)(5)}}{2(1)} \qquad \text{Substitute 1 for } a, -4 \text{ for } b, \text{ and } 5 \text{ for } c.$$

$$= \frac{4 \pm \sqrt{16 - 20}}{2}$$

$$= \frac{4 \pm \sqrt{-4}}{2}$$

$$= \frac{4 \pm 2i}{2} \qquad \sqrt{-4} = \sqrt{4}\sqrt{-1} = 2i$$

$$= 2 \pm i \qquad \frac{4 \pm 2i}{2} = \frac{2(2 \pm i)}{2} = 2 \pm i$$

The roots $x = 2 + i$ and $x = 2 - i$ both satisfy the equation. Note that the roots are complex conjugates.   ∎

Self Check 10   Solve: $x^2 + 3x + 4 = 0$.

Now Try Exercise 113.

6. Factor the Sum of Two Squares

We have seen that the sum of two squares cannot be factored over the set of integers. However, it is possible to factor the sum of two squares over the set of complex numbers. For example, to factor $9x^2 + 16y^2$ we proceed as follows:

$$9x^2 + 16y^2 = 9x^2 - (-1)16y^2$$

$$= 9x^2 - i^2(16y^2) \qquad i^2 = -1$$

$$= 9x^2 - 16y^2 i^2$$

$$= (3x + 4yi)(3x - 4yi) \qquad \text{Factor the difference of two squares.}$$

EXAMPLE 11   **Factoring the Sum of Two Squares**

Factor: $100x^2 + 144y^2$.

SOLUTION   We will rewrite $100x^2 + 144y^2$ as the difference of two squares and then factor.

$$100x^2 + 144y^2 = 100x^2 - (-1)144y^2$$

$$= 100x^2 - i^2 144y^2 \qquad i^2 = -1$$

$$= 100x^2 - 144y^2 i^2$$

$$= (10x + 12yi)(10x - 12yi) \qquad \text{Factor the difference of two squares.}$$

Self Check 11   Factor $x^2 + 225y^2$.

Now Try Exercise 121.

Self Check Answers   **1.** $-\dfrac{2}{3}$ **2. a.** $2 + 7i$ **b.** $-4 - 2i$ **3.** $16 - 11i$ **4.** $21 + i$

5. $\dfrac{9}{10} + \dfrac{3}{10}i$ **6.** $\dfrac{1}{5} + \dfrac{13}{5}i$ **7.** $-i$ **8.** $\sqrt{29}$ **9.** $\dfrac{3\sqrt{5}}{5}$

10. $-\dfrac{3}{2} \pm \dfrac{\sqrt{7}}{2}i$ **11.** $(x + 15yi)(x - 15yi)$

Exercises **1.5**

Getting Ready
You should be able to complete these vocabulary and concept statements before you proceed to the practice exercises.

Fill in the blanks.

1. $\sqrt{-3}$, $\sqrt{-9}$, and $\sqrt{-12}$ are examples of _____ numbers.

2. In the complex number $a + bi$, a is the _____ part, and b is the _____ part.

3. If $a = 0$ and $b \neq 0$ in the complex number $a + bi$, the number is an _____ number.

4. If $b = 0$ in the complex number $a + bi$, the number is a _____ number.

5. The complex conjugate of $2 + 5i$ is _____.

6. By definition, $|a + bi| = $ _____.

7. The absolute value of a complex number is a _____ number.

8. The product of two complex conjugates is a _____ number.

Practice
Simplify the imaginary numbers.

9. $\sqrt{-144}$

10. $-\sqrt{-225}$

11. $-2\sqrt{-24}$

12. $7\sqrt{-48}$

13. $\sqrt{-\dfrac{50}{9}}$

14. $-\sqrt{-\dfrac{72}{25}}$

15. $-7\sqrt{-\dfrac{3}{8}}$

16. $5\sqrt{-\dfrac{5}{27}}$

Find the values of x and y.

17. $x + (x + y)i = 3 + 8i$

18. $x + 5i = y - yi$

19. $3x - 2yi = 2 + (x + y)i$

20. $\begin{cases} 2 + (x + y)i = 2 - i \\ x + 3i = 2 + 3i \end{cases}$

Perform all operations. Give all answers in $a + bi$ form.

21. $(2 - 7i) + (3 + i)$

22. $(-7 + 2i) + (2 - 8i)$

23. $(5 - 6i) - (7 + 4i)$

24. $(11 + 2i) - (13 - 5i)$

25. $(14i + 2) + \left(2 - \sqrt{-16}\right)$

26. $\left(5 + \sqrt{-64}\right) - (23i - 32)$

27. $\left(3 + \sqrt{-4}\right) - \left(2 + \sqrt{-9}\right)$

28. $\left(7 - \sqrt{-25}\right) + \left(-8 + \sqrt{-1}\right)$

29. $-5(3 + 5i)$

30. $5(2 - i)$

31. $7i(4 - 8i)$

32. $-2i(3 - 7i)$

33. $(2 + 3i)(3 + 5i)$

34. $(5 - 7i)(2 + i)$

35. $(2 + 3i)^2$

36. $(3 - 4i)^2$

37. $\left(11 + \sqrt{-25}\right)\left(2 - \sqrt{-36}\right)$

38. $\left(6 + \sqrt{-49}\right)\left(6 - \sqrt{-49}\right)$

39. $\left(\sqrt{-16} + 3\right)\left(2 + \sqrt{-9}\right)$

40. $\left(12 - \sqrt{-4}\right)\left(-7 + \sqrt{-25}\right)$

41. $\dfrac{1}{-i}$

42. $\dfrac{3}{i}$

43. $\dfrac{-4}{3i}$

44. $\dfrac{10}{7i}$

45. $\dfrac{1}{2 + i}$

46. $\dfrac{-2}{3 - i}$

47. $\dfrac{2i}{7 + i}$

48. $\dfrac{-3i}{2 + 5i}$

49. $\dfrac{2 + i}{3 - i}$

50. $\dfrac{3 - i}{1 + i}$

51. $\dfrac{4 - 5i}{2 + 3i}$

52. $\dfrac{34 + 2i}{2 - 4i}$

53. $\dfrac{5 - \sqrt{-16}}{-8 + \sqrt{-4}}$

54. $\dfrac{3 - \sqrt{-9}}{2 - \sqrt{-1}}$

55. $\dfrac{2 + i\sqrt{3}}{3 + i}$

56. $\dfrac{3 + i}{4 - i\sqrt{2}}$

Simplify each expression.

57. i^9

58. i^{26}

59. i^{38}

60. i^{99}

61. i^{87}

62. i^{44}

63. i^{100}

64. i^{201}

65. i^{-6}

66. i^0

67. i^{-10}

68. i^{-31}

69. $\dfrac{1}{i^3}$

70. $\dfrac{3}{i^5}$

71. $\dfrac{-4}{i^{10}}$

72. $\dfrac{-10}{i^{24}}$

Write without absolute value symbols.

73. $|3 + 4i|$

74. $|5 + 12i|$

75. $|2 + 3i|$

76. $|5 - i|$

77. $\left| -7 + \sqrt{-49} \right|$

78. $\left| -2 - \sqrt{-16} \right|$

79. $\left| \dfrac{1}{2} + \dfrac{1}{2}i \right|$

80. $\left| \dfrac{1}{2} - \dfrac{1}{4}i \right|$

81. $|-6i|$

82. $|5i|$

83. $\left| \dfrac{2}{1 + i} \right|$

84. $\left| \dfrac{3}{3 + i} \right|$

85. $\left| \dfrac{-3i}{2 + i} \right|$

86. $\left| \dfrac{5i}{i - 2} \right|$

87. $\left| \dfrac{i + 2}{i - 2} \right|$

88. $\left| \dfrac{2 + i}{2 - i} \right|$

Solve each quadratic equation using the Square Root Property.

89. $x^2 = -169$

90. $x^2 = -81$

91. $y^2 + 54 = 0$

92. $x^2 + 125 = 0$

93. $2x^2 = -90$

94. $5x^2 = -400$

95. $9x^2 = -7$

96. $25x^2 = -11$

97. $2x^2 + 15 = 0$

98. $-5x^2 = 11$

99. $(x + 1)^2 + 12 = 0$

100. $(y + 2)^2 + 120 = 0$

101. $(5x + 1)^2 = -8$

102. $(7y + 2)^2 + 48 = 0$

Solve the following quadratic equations by completing the square. Simplify the solutions and write them in a + bi form.

103. $x^2 - 10x + 37 = 0$

104. $a^2 + 16a + 82 = 0$

105. $y^2 + 11y = -49$

106. $x^2 - 5x = -22$

107. $9x^2 = 18x - 14$

108. $7z^2 = -14z - 13$

109. $2x^2 = 14x - 30$

110. $5x^2 + x = -5$

Solve the following quadratic equations by the Quadratic Formula. Simplify the solutions and write them in a + bi form.

111. $x^2 + 2x + 2 = 0$

112. $a^2 + 4a + 8 = 0$

113. $y^2 + 4y + 5 = 0$

114. $x^2 + 2x + 5 = 0$

115. $x^2 - 2x = -5$

116. $z^2 - 3z = -8$

117. $x^2 - \dfrac{2}{3}x = -\dfrac{2}{9}$

118. $x^2 + \dfrac{5}{4} = x$

Factor each expression over the set of complex numbers.

119. $x^2 + 4$

120. $16a^2 + 9$

121. $25p^2 + 36q^2$

122. $100r^2 + 49s^2$

123. $2y^2 + 8z^2$

124. $12b^2 + 75c^2$

125. $50m^2 + 2n^2$

126. $64a^4 + 4b^2$

Applications

In electronics, the formula $V = IR$ is called Ohm's Law. It gives the relationship in a circuit between the voltage V (in volts), the current I (in amperes), and the resistance R (in ohms).

127. Electronics Find V when $I = 3 - 2i$ amperes and $R = 3 + 6i$ ohms.

128. Electronics Find R when $I = 2 - 3i$ amperes and $V = 21 + i$ volts.

129. Electronics The impedance Z in an AC (alternating current) circuit is a measure of how much the circuit impedes (hinders) the flow of current through it. The impedance is related to the voltage V and the current I by the following formula.

$$V = IZ$$

If a circuit has a current of $(0.5 + 2.0i)$ amps and an impedance of $(0.4 - 3.0i)$ ohms, find the voltage.

130. Fractals Complex numbers are fundamental in the creation of the intricate geometric shape shown below, called a *fractal*. The process of creating this image is based on the following sequence of steps, which begins by picking any complex number, which we will call z.

   1. Square z, and then add that result to z.

   2. Square the result from step 1, and then add it to z.

   3. Square the result from step 2, and then add it to z.

If we begin with the complex number i, what is the result after performing steps 1, 2, and 3?

Kostyantyn Ivanyshen/Shutterstock.com

Discovery and Writing

131. Show that the addition of two complex numbers is commutative by adding the complex numbers $a + bi$ and $c + di$ in both orders and observing that the sums are equal.

132. Show that the multiplication of two complex numbers is commutative by multiplying the complex numbers $a + bi$ and $c + di$ in both orders and observing that the products are equal.

133. Show that the addition of complex numbers is associative.

134. Find three examples of complex numbers that are reciprocals of their own conjugates.

135. Explain how to determine whether two complex numbers are equal.

136. Define the complex conjugate of a complex number.

Review

Simplify each expression. Assume that all variables represent nonnegative numbers.

137. $\sqrt{8x^3}\sqrt{4x}$

138. $\left(\sqrt{x} - 5\right)^2$

139. $\left(\sqrt{x + 1} - 2\right)^2$

140. $\left(-3\sqrt{2x + 1}\right)^2$

141. $\dfrac{4}{\sqrt{5} - 1}$

142. $\dfrac{x - 4}{\sqrt{x} + 2}$

1.6 Polynomial and Radical Equations

In this section, we will learn to

   1. Solve polynomial equations by factoring.
   2. Solve other equations by factoring.
   3. Solve radical equations.
   4. Solve applications of radical equations.

John Hoffman/Shutterstock.com

Pike's Peak is in the Rocky Mountain range, near Colorado Springs, Colorado. Standing at its summit, a person can see for miles.

The distance a person can see from the peak of the mountain is called the *horizon distance*. If this distance, d, is measured in miles and the height of the observer, h, is measured in feet, d and h are related by the formula $d = \sqrt{1.5h}$.

Since the height of Pike's Peak is approximately 14,000 feet, we can substitute 14,000 for h into the formula and simplify.

$$d = \sqrt{1.5h}$$

$$d = \sqrt{1.5(\mathbf{14,000})} \qquad \text{Substitute 14,000 for } h.$$

$$= \sqrt{21,000}$$

$$\approx 144.9137675$$

From the top of Pike's Peak, a person can see about 145 miles.

Since this equation contains a radical, it is called a *radical equation,* one of the topics of the section.

1. Solve Polynomial Equations by Factoring

The equation $ax^2 + bx + c = 0$ is a polynomial equation of second degree, because its left side contains a second-degree polynomial. Many polynomial equations of higher degree can be solved by factoring. A strategy for solving a polynomial equation is shown here.

Strategy for Solving Polynomial Equations

1. Write the polynomial equation in standard form.
   - Arrange the terms of the polynomial in descending order based on their degrees.
   - Set the polynomial equal to 0.
2. Use the Zero-Factor Theorem and solve by factoring.

EXAMPLE 1   **Solving Polynomial Equations by Factoring**

Solve: **a.** $6x^3 - x^2 - 2x = 0$ **b.** $x^4 - 5x^2 + 4 = 0$

SOLUTION   We will solve each equation by factoring.

a. $6x^3 - x^2 - 2x = 0$

$\qquad x(6x^2 - x - 2) = 0 \qquad \text{Factor out } x.$

$\qquad x(3x - 2)(2x + 1) = 0 \qquad \text{Factor } 6x^2 - x - 2.$

We set each factor equal to 0.

$$x = 0 \quad \text{or} \quad 3x - 2 = 0 \quad \text{or} \quad 2x + 1 = 0$$

$$x = \frac{2}{3} \qquad\qquad x = -\frac{1}{2}$$

Verify that each solution satisfies the original equation.

b. $\qquad\qquad x^4 - 5x^2 + 4 = 0$

$\qquad (x^2 - 4)(x^2 - 1) = 0 \qquad \text{Factor } x^4 - 5x^2 + 4.$

$(x + 2)(x - 2)(x + 1)(x - 1) = 0 \qquad \text{Factor each difference of two squares.}$

We set each factor equal to 0.

$$x + 2 = 0 \quad \text{or} \quad x - 2 = 0 \quad \text{or} \quad x + 1 = 0 \quad \text{or} \quad x - 1 = 0$$
$$x = -2 \quad \mid \quad x = 2 \quad \mid \quad x = -1 \quad \mid \quad x = 1$$

Verify that each solution satisfies the original equation.

Self Check 1   Solve: $2x^3 + 3x^2 - 2x = 0$.

Now Try Exercise 9.

2. Solve Other Equations by Factoring

To solve another type of equation by factoring, we use a property that states that equal powers of equal numbers are equal.

Power Property of Real Numbers   If a and b are numbers, n is an integer, and $a = b$, then
$$a^n = b^n$$

When we raise both sides of an equation to the same power, the resulting equation might not be equivalent to the original one. For example, if we raise both sides of

(1) $x = 4$ with a solution set of $\{4\}$

to the second power, we obtain

(2) $x^2 = 16$ with a solution set of $\{4, -4\}$

Equations 1 and 2 have different solution sets, and the solution -4 of Equation 2 does not satisfy Equation 1. Because raising both sides of an equation to the same power often introduces **extraneous solutions** (false solutions that don't satisfy the original equation), we must check all suspected roots to be certain that they satisfy the original equation.

The following equation has an extraneous solution.

$$x - x^{1/2} - 6 = 0$$
$$(x^{1/2} - 3)(x^{1/2} + 2) = 0 \qquad\qquad \text{Factor } x - x^{1/2} - 6.$$
$$x^{1/2} - 3 = 0 \quad \text{or} \quad x^{1/2} + 2 = 0 \qquad \text{Set each factor equal to 0.}$$
$$x^{1/2} = 3 \quad \mid \quad x^{1/2} = -2$$

Because equal powers of equal numbers are equal, we can square both sides of the previous equations to get

$$(x^{1/2})^2 = (3)^2 \quad \text{or} \quad (x^{1/2})^2 = (-2)^2$$
$$x = 9 \quad \mid \quad x = 4$$

The number 9 satisfies the equation $x - x^{1/2} - 6 = 0$ but 4 does not, as the following check shows:

If $x = 9$	**If $x = 4$**
$x - x^{1/2} - 6 = 0$	$x - x^{1/2} - 6 = 0$
$9 - 9^{1/2} - 6 \stackrel{?}{=} 0$	$4 - 4^{1/2} - 6 \stackrel{?}{=} 0$
$9 - 3 - 6 \stackrel{?}{=} 0$	$4 - 2 - 6 \stackrel{?}{=} 0$
$0 = 0$	$-4 \neq 0$

The number 9 is the only root.

EXAMPLE 2   **Solving Other Types of Equations by Factoring**

Solve: **a.** $2x^{2/5} - 5x^{1/5} - 3 = 0$ **b.** $3(3x - 2x^{1/2}) = -1$

SOLUTION   We will solve each equation by factoring.

a. $2x^{2/5} - 5x^{1/5} - 3 = 0$

$(2x^{1/5} + 1)(x^{1/5} - 3) = 0$ Factor $2x^{2/5} - 5x^{1/5} - 3$.

$2x^{1/5} + 1 = 0$ or $x^{1/5} - 3 = 0$ Set each factor equal to 0.

$2x^{1/5} = -1$ $x^{1/5} = 3$

$x^{1/5} = -\dfrac{1}{2}$

We can raise both sides of each of the previous equations to the fifth power to obtain

$$(x^{1/5})^5 = \left(-\frac{1}{2}\right)^5 \quad \text{or} \quad (x^{1/5})^5 = (3)^5$$

$$x = -\frac{1}{32} \qquad\qquad x = 243$$

Verify that each solution satisfies the original equation.

b. $3(3x - 2x^{1/2}) = -1$

$9x - 6x^{1/2} + 1 = 0$ Remove parentheses and add 1 to both sides.

$(3x^{1/2} - 1)(3x^{1/2} - 1) = 0$ Factor.

$3x^{1/2} - 1 = 0$ or $3x^{1/2} - 1 = 0$ Set each factor equal to 0.

$x^{1/2} = \dfrac{1}{3}$ $x^{1/2} = \dfrac{1}{3}$ Solve each equation.

We can square both sides of each of the previous equations to obtain

$$(x^{1/2})^2 = \left(\frac{1}{3}\right)^2 \qquad \text{or} \qquad (x^{1/2})^2 = \left(\frac{1}{3}\right)^2$$

$$x = \frac{1}{9} \qquad\qquad\qquad x = \frac{1}{9}$$

Here, the solutions are the same. Verify that $x = \frac{1}{9}$ satisfies the equation.   ∎

Self Check 2   Solve: $x^{2/5} - x^{1/5} - 2 = 0$.

Now Try Exercise 19.

3. Solve Radical Equations

Radical equations are equations containing radicals with variables in the radicand. To solve such equations, we use the Power Property of Real Numbers. A strategy for solving radical equations with square roots or cube roots is stated here.

Strategy for Solving Radical Equations with Square Roots or Cube Roots

1. Isolate one radical on the left side of the equation.
2. Raise both sides of the equation to the same power (the index of the radical).
   - If the radical is a square root, square both sides.
   - If the radical is a cube root, cube both sides.
3. If a radical remains, repeat Steps 1 and 2.
4. Solve the resulting equation and check your answers.

We can extend the strategy stated above to solve radical equations containing fourth roots, fifth roots, ..., nth roots.

EXAMPLE 3   Solving an Equation with One Radical

Solve: $\sqrt{x + 3} - 4 = 7$.

SOLUTION   We will isolate the radical on the left side and then square both sides.

$$\sqrt{x + 3} - 4 = 7$$

$$\sqrt{x + 3} = 11 \qquad \text{Add 4 to both sides to isolate the radical.}$$

$$\left(\sqrt{x + 3}\right)^2 = (11)^2 \qquad \text{Square both sides.}$$

$$x + 3 = 121 \qquad \text{Simplify.}$$

$$x = 118 \qquad \text{Subtract 3 from both sides.}$$

Since squaring both sides might introduce extraneous roots, we must check the result of 118.

$$\sqrt{x + 3} - 4 = 7$$

$$\sqrt{118 + 3} - 4 \stackrel{?}{=} 7 \qquad \text{Substitute 118 for } x.$$

$$\sqrt{121} - 4 \stackrel{?}{=} 7$$

$$11 - 4 \stackrel{?}{=} 7$$

$$7 = 7$$

Comment

Remember to check all roots when solving radical equations, because raising both sides of an equation to a power can introduce extraneous roots.

Because it checks, 118 is a root of the equation. ∎

Self Check 3   Solve: $\sqrt{x - 3} + 4 = 7$.

Now Try Exercise 25.

EXAMPLE 4   Solving an Equation with One Radical

Solve: $\sqrt{x + 3} = 3x - 1$.

SOLUTION   We will square both sides of the equation to eliminate the radical and solve the resulting equations by factoring.

$$\sqrt{x + 3} = 3x - 1$$

$$\left(\sqrt{x + 3}\right)^2 = (3x - 1)^2 \qquad \text{Square both sides.}$$

$$x + 3 = 9x^2 - 6x + 1 \qquad \text{Remove parentheses.}$$

$$0 = 9x^2 - 7x - 2 \qquad \text{Add } -x - 3 \text{ to both sides.}$$

$$0 = (9x + 2)(x - 1) \qquad \text{Factor } 9x^2 - 7x - 2.$$

$$9x + 2 = 0 \quad \text{or} \quad x - 1 = 0 \qquad \text{Set each factor equal to 0.}$$

$$x = -\frac{2}{9} \qquad \qquad x = 1$$

Since squaring both sides can introduce extraneous roots, we must check each result.

$$\sqrt{x + 3} = 3x - 1 \qquad \text{or} \qquad \sqrt{x + 3} = 3x - 1$$

$$\sqrt{-\frac{2}{9} + 3} \stackrel{?}{=} 3\left(-\frac{2}{9}\right) - 1 \qquad\qquad \sqrt{1 + 3} \stackrel{?}{=} 3(1) - 1$$

$$\sqrt{\frac{25}{9}} \stackrel{?}{=} -\frac{2}{3} - 1 \qquad\qquad\qquad \sqrt{4} \stackrel{?}{=} 3 - 1$$

$$\frac{5}{3} \neq -\frac{5}{3} \qquad\qquad\qquad\qquad\qquad 2 = 2$$

Since $-\frac{2}{9}$ does not satisfy the equation, it is extraneous. Since 1 checks, it is the only solution. ∎

Caution

> If we *forget* to check solution(s) to a radical equation, we risk having an *incorrect* solution set. *Failure* to check both $x = -\frac{2}{9}$ and $x = 1$ in Example 4 would have resulted in an *error*.

Self Check 4   Solve: $\sqrt{x - 2} = 2x - 10$.

Now Try Exercise 37.

EXAMPLE 5   **Solving a Radical Equation with a Cube Root**

Solve: $\sqrt[3]{x^3 + 56} = x + 2$.

SOLUTION   To eliminate the radical, we cube both sides of the equation.

$$\sqrt[3]{x^3 + 56} = x + 2$$

$$\left(\sqrt[3]{x^3 + 56}\right)^3 = (x + 2)^3 \qquad\qquad \text{Cube both sides.}$$

$$x^3 + 56 = x^3 + 6x^2 + 12x + 8 \qquad \text{Remove parentheses.}$$

$$0 = 6x^2 + 12x - 48 \qquad\qquad \text{Simplify.}$$

$$0 = x^2 + 2x - 8 \qquad\qquad\quad \text{Divide both sides by 6.}$$

$$0 = (x + 4)(x - 2) \qquad\qquad \text{Factor } x^2 + 2x - 8.$$

$$x + 4 = 0 \quad \text{or} \quad x - 2 = 0 \qquad \text{Set each factor equal to 0.}$$

$$x = -4 \qquad\qquad x = 2$$

We check each suspected solution to see whether either is extraneous.

For $x = -4$	*For $x = 2$*
$\sqrt[3]{x^3 + 56} = x + 2$	$\sqrt[3]{x^3 + 56} = x + 2$
$\sqrt[3]{(-4)^3 + 56} \stackrel{?}{=} -4 + 2$	$\sqrt[3]{2^3 + 56} \stackrel{?}{=} 2 + 2$
$\sqrt[3]{-64 + 56} \stackrel{?}{=} -2$	$\sqrt[3]{8 + 56} \stackrel{?}{=} 4$
$\sqrt[3]{-8} \stackrel{?}{=} -2$	$\sqrt[3]{64} \stackrel{?}{=} 4$
$-2 = -2$	$4 = 4$

Since both values satisfy the equation, -4 and 2 are roots. ∎

Self Check 5   Solve: $\sqrt[3]{x^3 + 7} = x + 1$.

Now Try Exercise 49.

EXAMPLE 6   **Solving an Equation with Two Radicals**

Solve: $\sqrt{2x + 3} + \sqrt{x - 2} = 4$.

SOLUTION   We can write the equation in the form

$$\sqrt{2x + 3} = 4 - \sqrt{x - 2} \qquad \text{Subtract } \sqrt{x - 2} \text{ from both sides.}$$

so that the left side contains one radical. We then square both sides to get

$$\left(\sqrt{2x + 3}\right)^2 = \left(4 - \sqrt{x - 2}\right)^2$$

$$2x + 3 = 16 - 8\sqrt{x - 2} + x - 2$$

$$2x + 3 = 14 - 8\sqrt{x - 2} + x \qquad \text{Combine like terms.}$$

$$x - 11 = -8\sqrt{x - 2} \qquad \text{Subtract 14 and } x \text{ from both sides.}$$

We then square both sides again to eliminate the radical.

$$(x - 11)^2 = \left(-8\sqrt{x - 2}\right)^2$$

$$x^2 - 22x + 121 = 64(x - 2)$$

$$x^2 - 22x + 121 = 64x - 128$$

$$x^2 - 86x + 249 = 0$$

$$(x - 3)(x - 83) = 0$$

$$x - 3 = 0 \quad \text{or} \quad x - 83 = 0 \qquad \text{Set each factor equal to 0.}$$

$$x = 3 \quad \Big| \quad x = 83$$

Substituting these results into the equation will show that 83 doesn't check; it is extraneous. However, 3 does satisfy the equation and is a root.   ∎

Self Check 6   Solve: $\sqrt{2x + 1} + \sqrt{x + 5} = 6$.

Now Try Exercise 53.

4. Solve Applications of Radical Equations

Many applications can be solved using radical equations.

EXAMPLE 7   **Solving an Application Problem**

A highway curve banked at 8° will accommodate traffic traveling s mph if the radius of the curve is r feet, according to the formula $s = 1.45\sqrt{r}$. Find what radius is necessary to accommodate 70-mph traffic. (See Figure 1-12.)

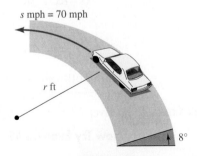

FIGURE 1-12

SOLUTION   We can substitute 70 for s in the formula and solve for r.

$$s = 1.45\sqrt{r}$$

$$70 = 1.45\sqrt{r} \qquad \text{Substitute 70 for } s.$$

$$\frac{70}{1.45} = \sqrt{r} \qquad \text{Divide both sides by 1.45.}$$

$$\left(\frac{70}{1.45}\right)^2 = r \qquad \text{Square both sides.}$$

$$2330.558859 = r$$

We can use a calculator to find r. The radius of the curve is approximately 2,331 feet.

Self Check 7   Find the radius necessary to accommodate 65-mph traffic.

Now Try Exercise 61.

Self Check Answers   **1.** $0, \dfrac{1}{2}, -2$ **2.** $-1, 32$ **3.** 12 **4.** 6 **5.** $1, -2$ **6.** 4
7. approximately 2,010 feet

Exercises **1.6**

Getting Ready

You should be able to complete these vocabulary and concept statements before you proceed to the practice exercises.

Fill in the blanks.

1. Equal powers of equal real numbers are _____.
2. If a and b are real numbers and $a = b$ then $a^2 =$ _____.
3. False solutions that don't satisfy the equation are called _____ solutions.
4. Radical equations contain radicals with variables in their _____.

Practice

Use factoring to solve each equation for real values of the variable.

5. $x^3 + 9x^2 + 20x = 0$
6. $x^3 + 4x^2 - 21x = 0$

7. $6a^3 - 5a^2 - 4a = 0$
8. $8b^3 - 10b^2 + 3b = 0$

9. $y^4 - 26y^2 + 25 = 0$
10. $y^4 - 13y^2 + 36 = 0$

11. $x^4 - 37x^2 + 36 = 0$
12. $x^4 - 50x^2 + 49 = 0$

13. $2y^4 - 46y^2 = -180$
14. $2x^4 - 102x^2 = -196$

15. $z^{3/2} - z^{1/2} = 0$
16. $r^{5/2} - r^{3/2} = 0$

17. $2m^{2/3} + 3m^{1/3} - 2 = 0$
18. $6t^{2/5} + 11t^{1/5} + 3 = 0$

19. $x - 13x^{1/2} + 12 = 0$
20. $p + p^{1/2} - 20 = 0$

21. $2t^{1/3} + 3t^{1/6} - 2 = 0$
22. $z^3 - 7z^{3/2} - 8 = 0$

23. $6p + p^{1/2} - 1 = 0$
24. $3r - r^{1/2} - 2 = 0$

Find all real solutions of each equation.

25. $\sqrt{x - 2} - 3 = 2$
26. $\sqrt{a - 3} - 5 = 0$

27. $3\sqrt{x + 1} = \sqrt{6}$
28. $\sqrt{x + 3} = 2\sqrt{x}$

29. $\sqrt{5a - 2} = \sqrt{a + 6}$
30. $\sqrt{16x + 4} = \sqrt{x + 4}$

31. $2\sqrt{x^2 + 3} = \sqrt{-16x - 3}$

32. $\sqrt{x^2 + 1} = \dfrac{\sqrt{-7x + 11}}{\sqrt{6}}$

33. $\sqrt[3]{7x + 1} = 4$
34. $\sqrt[3]{11a - 40} = 5$

35. $\sqrt[4]{30t + 25} = 5$
36. $\sqrt[4]{3z + 1} = 2$

37. $\sqrt{x^2 + 21} = x + 3$
38. $\sqrt{5 - x^2} = -(x + 1)$

39. $\sqrt{y + 2} = 4 - y$ **40.** $\sqrt{3z + 1} = z - 1$

41. $x - \sqrt{7x - 12} = 0$ **42.** $x - \sqrt{4x - 4} = 0$

43. $x + 4 = \sqrt{\dfrac{6x + 6}{5}} + 3$ **44.** $\sqrt{\dfrac{8x + 43}{3}} - 1 = x$

45. $\sqrt{\dfrac{x^2 - 1}{x - 2}} = 2\sqrt{2}$ **46.** $\dfrac{\sqrt{x^2 - 1}}{\sqrt{3x - 5}} = \sqrt{2}$

47. $\sqrt[3]{x^3 + 7} = x + 1$ **48.** $\sqrt[3]{x^3 - 7} + 1 = x$

49. $\sqrt[3]{8x^3 + 61} = 2x + 1$ **50.** $\sqrt[3]{8x^3 - 37} = 2x - 1$

51. $\sqrt{2p + 1} - 1 = \sqrt{p}$ **52.** $\sqrt{r} + \sqrt{r + 2} = 2$

53. $\sqrt{x + 3} = \sqrt{2x + 8} - 1$

54. $\sqrt{x + 2} + 1 = \sqrt{2x + 5}$

55. $\sqrt{y + 8} - \sqrt{y - 4} = -2$

56. $\sqrt{z + 5} - 2 = \sqrt{z - 3}$

57. $\sqrt{2b + 3} - \sqrt{b + 1} = \sqrt{b - 2}$

58. $\sqrt{a + 1} + \sqrt{3a} = \sqrt{5a + 1}$

59. $\sqrt{\sqrt{b} + \sqrt{b + 8}} = 2$

60. $\sqrt{\sqrt{x + 19} - \sqrt{x - 2}} = \sqrt{3}$

Applications

61. Height of a bridge The distance d (in feet) that an object will fall in t seconds is given by the following formula. To find the height of a bridge above a river, a man drops a stone into the water. (See the illustration.) If it takes the stone 5 seconds to hit the water, how high is the bridge?

$$t = \sqrt{\dfrac{d}{16}}$$

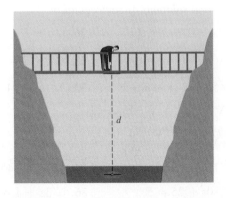

62. Horizon distance The higher a lookout tower, the farther an observer can see. (See the illustration.) The distance d (called the **horizon distance**, measured in miles) is related to the height h of the observer (measured in feet) by the following formula.

$$d = \sqrt{1.5h}$$

How tall must a tower be for the observer to see 30 miles?

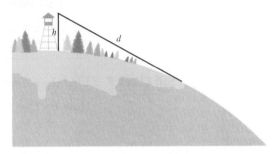

63. Carpentry During construction, carpenters often brace walls, as shown in the illustration. The appropriate length of the brace is given by the following formula.

$$l = \sqrt{f^2 + h^2}$$

If a carpenter nails a 10-foot brace to the wall 6 feet above the floor, how far from the base of the wall should he nail the brace to the floor?

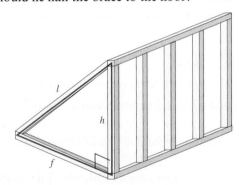

64. Windmills The power generated by a windmill is related to the velocity of the wind by the following formula where P is the power (in watts) and v is the velocity of the wind (in mph).

$$v = \sqrt[3]{\dfrac{P}{0.02}}$$

To the nearest 10 watts, find the power generated when the velocity of the wind is 31 mph.

65. Diamonds The *effective rate of interest r* earned by an investment is given by the following formula where P is the initial investment that grows to value A after n years.

$$r = \sqrt[n]{\frac{A}{P}} - 1$$

If a diamond buyer got $4,000 for a 1.03-carat diamond that he had purchased 4 years earlier and earned an annual rate of return of 6.5% on the investment, what did he originally pay for the diamond?

66. Theater productions The ropes, pulleys, and sandbags shown in the illustration are part of a mechanical system used to raise and lower scenery for a stage play. For the scenery to be in the proper position, the following formula must apply:

$$w_2 = \sqrt{w_1^2 + w_3^2}$$

If $w_2 = 12.5$ lb and $w_3 = 7.5$ lb, find w_1.

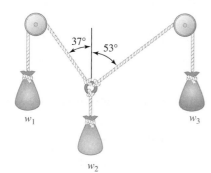

Discovery and Writing

67. Explain why squaring both sides of an equation might introduce extraneous roots.

68. Can cubing both sides of an equation introduce extraneous roots? Explain.

Review

Graph each subset of the real numbers on the number line.

69. The natural numbers between 0 and 4

70. The integers between -4 and 4

Write each inequality in interval notation and graph the interval.

71. $x \geq 3$ **72.** $x < -5$

73. $-2 \leq x < 1$ **74.** $-3 \leq x \leq 3$

75. $x < 1$ or $x \geq 2$ **76.** $x \leq 1$ or $x > 2$

1.7 Inequalities

In this section, we will learn to

1. Use the properties of inequalities.

2. Solve linear inequalities and applications.

3. Solve compound inequalities.

4. Solve quadratic inequalities.

5. Solve rational inequalities.

Studying abroad is a wonderful opportunity for students. Suppose you read about a program to study Spanish in Costa Rica for several weeks during the summer. The cost of the program includes $900 for round-trip airfare plus $350 per week, which covers tuition, meals per day, and living accommodations. If you have $3,000, how many weeks can you afford to spend in Costa Rica?

If x represents the number of weeks you can spend in Costa Rica, we can write an inequality that represents the cost.

Airfare	plus	cost of $350 per week for x weeks	is less than or equal to	$3,000.
900	+	$350x$	$\leq$	3,000

To solve this inequality, we can proceed as follows:

$$900 + 350x \leq 3,000$$

$$350x \leq 2,100 \qquad \text{Subtract 900 from both sides.}$$

$$x \leq 6 \qquad \text{Divide both sides by 350.}$$

There is enough money to spend up to 6 weeks in Costa Rica.

In this section, we will review the inequality symbols, learn inequality properties, and learn strategies that can be applied to solve several types of inequalities.

We previously introduced the following symbols.

Symbol	Read as	Examples
$\neq$	"is not equal to"	$8 \neq 10$ and $25 \neq 12$
$<$	"is less than"	$8 < 10$ and $12 < 25$
$>$	"is greater than"	$30 > 10$ and $100 > -5$
$\leq$	"is less than or equal to"	$-6 \leq 12$ and $-8 \leq -8$
$\geq$	"is greater than or equal to"	$12 \geq -5$ and $9 \geq 9$
$\approx$	"is approximately equal to"	$7.49 \approx 7.5$ and $\frac{1}{3} \approx 0.33$

Since the coordinates of points get larger as we move from left to right on the number line,

$a > b$ if point a lies to the right of point b on a number line.

$a < b$ if point a lies to the left of point b on a number line.

EXAMPLE 1   Using Inequality Symbols

a. $5 > -3$, because 5 lies to the right of -3 on the number line.

b. $-7 < -2$, because -7 lies to the left of -2 on the number line.

c. $3 \leq 3$, because $3 = 3$.

d. $2 \leq 3$, because $2 < 3$.

e. $x + 1 > x$, because $x + 1$ lies one unit to the right of x on the number line. ∎

Self Check 1   Write an inequality symbol to make a true statement:

a. $25 \,\square\, 12$ **b.** $-5 \,\square\, -5$ **c.** $-12 \,\square\, -20$

Now Try Exercise 1.

1. Use the Properties of Inequalities

The Trichotomy Property   For any real numbers a and b, one of the following statements is true.

$$a < b, \qquad a = b, \qquad \text{or} \qquad a > b$$

The Trichotomy Property indicates that one of the following statements is true about two real numbers. Either the first is less than the second, or the first is equal to the second, or the first is greater than the second.

The Transitive Property

If a, b, and c are real numbers, then

if $a < b$ and $b < c$, then $a < c$.

if $a > b$ and $b > c$, then $a > c$.

The first part of the Transitive Property indicates that if a first number is less than a second and the second number is less than a third, then the first number is less than the third.

The second part of the Transitive Property is similar, with the words "is greater than" substituted for "is less than."

Addition and Subtraction Properties of Inequality

Let a, b, and c represent real numbers.

If $a < b$, then $a + c < b + c$.

If $a < b$, then $a - c < b - c$.

Similar properties exist for $>$, $\leq$, and $\geq$.

This property states that *any real number can be added to (or subtracted from) both sides of an inequality to obtain another inequality with the same order (direction).*
For example, if we add 4 to (or subtract 4 from) both sides of $8 < 12$, we get

$$8 < 12 \qquad\qquad\qquad 8 < 12$$
$$8 + 4 < 12 + 4 \qquad\qquad 8 - 4 < 12 - 4$$
$$12 < 16 \qquad\qquad\qquad 4 < 8$$

and the $<$ symbol is unchanged.

Multiplication and Division Properties of Inequality

Let a, b, and c represent real numbers.

Part 1:   If $a < b$ and $c > 0$, then $ca < cb$.

If $a < b$ and $c > 0$, then $\frac{a}{c} < \frac{b}{c}$.

Part 2:   If $a < b$ and $c < 0$, then $ca > cb$.

If $a < b$ and $c < 0$, then $\frac{a}{c} > \frac{b}{c}$.

Similar properties exist for $>$, $\leq$, and $\geq$.

This property has two parts.

- Part 1 states that *both sides of an inequality can be multiplied (or divided) by the same positive number to obtain another inequality with the same order.*

For example, if we multiply (or divide) both sides of $8 < 12$ by 4, we get

$$8 < 12 \qquad\qquad\qquad 8 < 12$$

$$8(4) < 12(4) \qquad\qquad \frac{8}{4} < \frac{12}{4}$$

$$32 < 48 \qquad\qquad\qquad 2 < 3$$

and the $<$ symbol is unchanged.

- Part 2 states that *both sides of an inequality can be multiplied (or divided) by the same negative number to obtain another inequality with the opposite order.*

    For example, if we multiply (or divide) both sides of $8 < 12$ by -4, we get

$$8 < 12 \qquad\qquad\qquad 8 < 12$$

$$8(-4) > 12(-4) \qquad\qquad \frac{8}{-4} > \frac{12}{-4}$$

$$-32 > -48 \qquad\qquad\qquad -2 > -3$$

and the $<$ symbol is changed to a $>$ symbol.

Comment

The properties of inequalities are the same as the properties of equality, unless we are multiplying or dividing by a negative number.

2. Solve Linear Inequalities and Applications

Linear inequalities are inequalities such as $ax + c < 0$ or $ax - c \geq 0$ where $a \neq 0$. Numbers that make an inequality true when substituted for the variable are solutions of the inequality. Inequalities with the same solution set are called **equivalent inequalities.** Because of the previous properties, we can solve inequalities as we do equations. However, we must always remember to change the order of an inequality when multiplying (or dividing) both sides by a negative number.

EXAMPLE 2   Solving a Linear Inequality

Solve: $3(x + 2) < 8$.

SOLUTION   We proceed as with equations.

$$3(x + 2) < 8$$

$$3x + 6 < 8 \qquad \text{Remove parentheses.}$$

$$3x < 2 \qquad \text{Subtract 6 from both sides.}$$

$$x < \frac{2}{3} \qquad \text{Divide both sides by 3.}$$

All numbers that are less than $\frac{2}{3}$ are solutions of the inequality. The solution set can be expressed in interval notation as $\left(-\infty, \frac{2}{3}\right)$ and be graphed as in Figure 1-13.

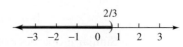

FIGURE 1-13

Self Check 2   Solve: $5(p - 4) > 25$.

Now Try Exercise 15.

EXAMPLE 3   **Solving a Linear Inequality**

Solve: $-5(x - 2) \le 20 + x$.

SOLUTION   We proceed as with equations.

$$-5(x - 2) \le 20 + x$$

$$-5x + 10 \le 20 + x \qquad \text{Remove parentheses.}$$

$$-6x + 10 \le 20 \qquad \text{Subtract } x \text{ from both sides.}$$

$$-6x \le 10 \qquad \text{Subtract 10 from both sides.}$$

We now divide both sides of the inequality by -6, which changes the order of the inequality.

$$x \ge \frac{10}{-6} \qquad \text{Divide both sides by } -6.$$

$$x \ge -\frac{5}{3} \qquad \text{Simplify the fraction.}$$

The graph of the solution set is shown in Figure 1-14. It is the interval $\left[-\frac{5}{3}, \infty\right)$.

−5/3

−3 −2 −1 0 1 2

FIGURE 1-14

Self Check 3   Solve: $-4(x + 3) \ge 16$.

Now Try Exercise 19.

Caution

> The most common error made when solving inequalities is forgetting to reverse the inequality symbol when multiplying or dividing both sides by a negative number. In Example 3, both sides were divided by -6. If the order of the inequality isn't reversed, the solution will be incorrect.

EXAMPLE 4   **Solving an Application of a Linear Inequality**

An empty truck with driver weighs 4,350 pounds. It is loaded with feed corn weighing 31 pounds per bushel. Between farm and market is a bridge with a 10,000-pound load limit. How many bushels can the truck legally carry?

Robert Kyllo/Shutterstock.com

SOLUTION   The empty truck with driver weighs 4,350 pounds, and corn weighs 31 pounds per bushel. If we let b represent the number of bushels in a legal load, the weight of the corn will be $31b$ pounds. Since the combined weight of the truck, driver, and cargo cannot exceed 10,000 pounds, we can form the following inequality.

The weight of the empty truck with driver	plus	the weight of the corn	must be less than or equal to	10,000 pounds.
4,350	+	$31b$	$\le$	10,000

We can solve the inequality as follows.

$$4{,}350 + 31b \le 10{,}000$$

$$31b \le 5{,}650 \qquad \text{Subtract 4,350 from each side.}$$

$$b \le 182.2580645 \qquad \text{Divide both sides by 31.}$$

The truck can legally carry $182\frac{1}{4}$ bushels or less. ∎

Self Check 4   If the empty truck and driver in example 3 weigh 3,800 pounds, how many bushels can the truck legally carry?

Now Try Exercise 87.

3. Solve Compound Inequalities

The statement that x is between 2 and 5 implies two inequalities,

$$x > 2 \text{ and } x < 5$$

Comment

Remember that $2 < x < 5$ means that $x > 2$ and $x < 5$. The word *and* indicates that both inequalities must be true at the same time.

It is customary to write both inequalities as one **compound inequality:**

$$2 < x < 5 \qquad \text{Read as "2 is less than } x \text{ and } x \text{ is less than 5."}$$

To express that x is not between 2 and 5, we must convey the idea that either x is greater than or equal to 5, or that x is less than or equal to 2. This is equivalent to the statement

Caution

It is incorrect to write $x \ge 5$ or $x \le 2$ as $2 \ge x \ge 5$ because this would means that $2 \ge 5$, which is false.

$$x \ge 5 \quad \text{or} \quad x \le 2$$

This inequality is satisfied by all numbers x that satisfy one or both of its parts.

EXAMPLE 5   **Solving a Compound Inequality by Isolating x between the Inequality Symbols**

Solve: $5 < 3x - 7 \le 8$.

SOLUTION   We can isolate x between the inequality symbols by adding 7 to each part of the inequality to get

$$5 + 7 < 3x - 7 + 7 \le 8 + 7 \qquad \text{Add 7 to each part.}$$

$$12 < 3x \le 15 \qquad \text{Do the additions.}$$

and dividing all parts by 3 to get

$$4 < x \le 5$$

The solution set is the interval $(4, 5]$, whose graph appears in Figure 1-15.

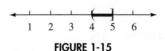

FIGURE 1-15

Self Check 5   Solve: $-5 \le 2x + 1 < 9$.

Now Try Exercise 33.

EXAMPLE 6   **Solving a Compound Inequality by Solving Each Inequality Separately**

Solve: $3 + x \le 3x + 1 < 7x - 2$.

SOLUTION Because it is impossible to isolate x between the inequality symbols, we must solve each inequality separately.

$$3 + x \le 3x + 1 \quad \text{and} \quad 3x + 1 < 7x - 2$$

$3 \le 2x + 1$	$1 < 4x - 2$
$2 \le 2x$	$3 < 4x$
$1 \le x$	$\dfrac{3}{4} < x$
$x \ge 1$	$x > \dfrac{3}{4}$

Since the connective in this inequality is *and*, the solution set is the intersection (or overlap) of the intervals $[1, \infty)$ and $\left(\frac{3}{4}, \infty\right)$, which is $[1, \infty)$. The graph is shown in Figure 1-16.

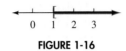

FIGURE 1-16

Self Check 6   Solve: $x + 1 < 2x - 3 \le 3x - 5$.

Now Try Exercise 47.

Comment   It is possible for an inequality to be true for all values of its variable. It is also possible for an inequality to have no solutions.

For example,

- $x < x + 1$ is true for all numbers x.
- $x > x + 1$ is true for no numbers x.

4. Solve Quadratic Inequalities

If $a \ne 0$, inequalities like $ax^2 + bx + c < 0$ and $ax^2 + bx + c \ge 0$ are called **quadratic inequalities.** We will begin by giving two methods for solving quadratic inequalities.

Strategy for Solving Quadratic Inequalities

Method 1: Constructing a Table and Testing Numbers

- Solve the quadratic equation and use the roots of the equation to establish intervals on a number line.
- Construct a table. To do so, write down each interval, select a number to test from each interval, test the selected value to determine if it satisfies the inequality, and then write the result.
- Use the results from the table and write the solution of the quadratic inequality.

Method 2: Constructing a Sign Graph

- Solve the quadratic equation and use the roots of the equation to establish intervals on the number line.
- Construct a sign graph by determining the sign of each factor on each interval. Write these signs on the number line.
- Use these results to write the solution to the quadratic inequality.

The roots of the quadratic equation *will be included* in the solution if the quadratic inequality involves $\le$ or $\ge$. The roots *will not be included* in the solution if the quadratic inequality involves $<$ or $>$.

In our first example, we will solve the quadratic inequality using both of the strategies or methods outlined earlier.

EXAMPLE 7   **Solving a Quadratic Inequality**

Solve: $x^2 - x - 6 > 0$.

SOLUTION   **Method 1:**   First we solve the equation $x^2 - x - 6 = 0$.

$$x^2 - x - 6 = 0$$

$$(x + 2)(x - 3) = 0$$

$$x + 2 = 0 \quad \text{or} \quad x - 3 = 0$$

$$x = -2 \qquad \qquad x = 3$$

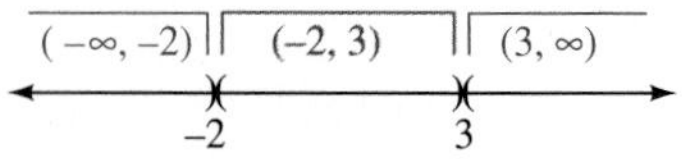

The graphs of these solutions establish the three intervals shown in Figure 1-17. To determine which intervals are solutions, we construct a table and test a number in each interval and see whether it satisfies the inequality.

Interval	Test Value	Inequality $x^2 - x - 6 > 0$	Result
$(-\infty, -2)$	-6	$(-6)^2 - (-6) - 6 \overset{?}{>} 0$ $36 > 0$ True	The numbers in this interval are solutions.
$(-2, 3)$	0	$0^2 - 0 - 6 \overset{?}{>} 0$ $-6 > 0$ False	The numbers in this interval are not solutions.
$(3, \infty)$	5	$5^2 - 5 - 6 \overset{?}{>} 0$ $14 > 0$ True	The numbers in this interval are solutions.

The solutions are in the intervals $(-\infty, -2)$ or $(3, \infty)$ as shown in Figure 1-18. Note that we write a parenthesis next to -2 and 3 because they do not satisfy the quadratic inequality. The quadratic inequality given contains $>$.

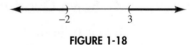

FIGURE 1-18

Method 2:   A second method relies on the number line and a notation that keeps track of the signs of the factors of $x^2 - x - 6$, which are $(x - 3)(x + 2)$.

- First, we consider the factor $x - 3$.
  If $x = 3$, then $x - 3 = 0$.
  If $x < 3$, then $x - 3$ is negative.
  If $x > 3$, then $x - 3$ is positive.

- Next, we consider the factor $x + 2$.
  If $x = -2$, then $x + 2 = 0$.
  If $x < -2$, then $x + 2$ is negative.
  If $x > -2$, then $x + 2$ is positive.

We construct a **sign graph** as shown in Figure 1-19 by using $+$ and $-$ signs and writing these above the number line.

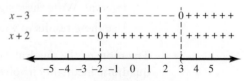

FIGURE 1-19

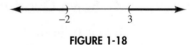

FIGURE 1-17

Only to the left of -2 and to the right of 3 do the signs of both factors agree. Only there is the product positive. The solutions are in the intervals $(-\infty, -2)$ or $(3, \infty)$. ∎

Self Check 7   Solve: $x(x + 1) - 6 \geq 0$.

Now Try Exercise 59.

EXAMPLE 8   **Solving a Quadratic Inequality**

Solve: $x(x + 3) \leq -2$.

SOLUTION   We remove parentheses and add 2 to both sides to make the right side of the equation equal to 0 and solve $x^2 + 3x + 2 \leq 0$.

Method 1:   First we solve the equation $x^2 + 3x + 2 = 0$.

$$x^2 + 3x + 2 = 0$$
$$(x + 2)(x + 1) = 0$$
$$x + 2 = 0 \quad \text{or} \quad x + 1 = 0$$
$$x = -2 \quad | \quad x = -1$$

These solutions establish the intervals shown in Figure 1-20.

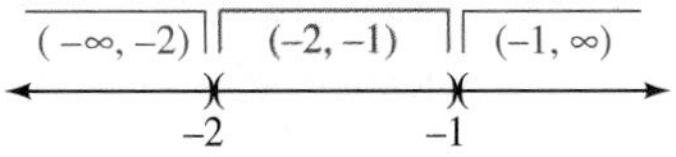

FIGURE 1-20

The solutions of $x^2 + 3x + 2 \leq 0$ will be the numbers in one or more of these intervals. To determine which intervals are solutions, we test a number in each interval to see whether it satisfies the inequality.

Interval	Test Value	Inequality $x^2 + 3x + 2 \leq 0$	Result
$(-\infty, -2)$	-7	$(-7)^2 + 3(-7) + 2 \overset{?}{\leq} 0$ $30 \leq 0$ False	The numbers in this interval are not solutions.
$(-2, -1)$	$-\dfrac{3}{2}$	$\left(-\dfrac{3}{2}\right)^2 + 3\left(-\dfrac{3}{2}\right) + 2 \overset{?}{\leq} 0$ $-\dfrac{1}{4} \leq 0$ True	The numbers in this interval are solutions.
$(-1, \infty)$	0	$0^2 - 0 + 2 \overset{?}{\leq} 0$ $2 \leq 0$ False	The numbers in this interval are not solutions.

The solution is the interval $[-2, -1]$ whose graph appears in Figure 1-21. Note that we write a bracket next to -2 and -1 because they satisfy the quadratic inequality. The quadratic inequality given contains $\leq$.

FIGURE 1-21

Method 2:   We construct a sign graph.

- First, we consider the factor $x + 1$.

    If $x = -1$, then $x + 1 = 0$

    If $x < -1$, then $x + 1$ is negative.

    If $x > -1$, then $x + 1$ is positive.

- Next, we consider the factor $x + 2$.

  If $x = -2$ then $x + 2 = 0$.

  If $x < -2$ then $x + 2$ is negative.

  If $x > -2$ then $x + 2$ is positive.

The sign graph is shown in Figure 1-22.

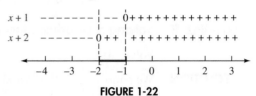

FIGURE 1-22

Only between -2 and -1 do the factors have opposite signs. Here, the product is negative. The solution is the interval $[-2, -1]$.

Self Check 8   Solve: $x^2 - 5x - 6 \leq 0$.

Now Try Exercise 61.

EXAMPLE 9   Solving a Quadratic Inequality

Solve: $x^2 - 5 \geq 0$.

SOLUTION   This equation will not factor using only integers. However, we can solve it by adding 5 to both sides and using the Square-Root Property. We will then use Method 1 and solve the inequality.

$$x^2 - 5 = 0$$

$$x^2 - 5 + 5 = 0 + 5 \qquad \text{Add 5 to both sides.}$$

$$x^2 = 5$$

$$x = \sqrt{5} \quad \text{or} \quad x = -\sqrt{5}$$

These solutions establish the intervals shown in the table and Figure 1-23. To decide which ones are solutions, we test a number in each interval to see whether it is a solution.

Interval	Test Value	Inequality $x^2 - 5 \geq 0$	Result
$\left(-\infty, -\sqrt{5}\right)$	-3	$(-3)^2 - 5 \overset{?}{\geq} 0$ $4 \geq 0$ \quad True	The numbers in this interval are solutions.
$\left(-\sqrt{5}, \sqrt{5}\right)$	0	$(0)^2 - 5 \overset{?}{\geq} 0$ $-5 \geq 0$ \quad False	The numbers in this interval are not solutions.
$\left(\sqrt{5}, \infty\right)$	3	$(3)^2 - 5 \overset{?}{\geq} 0$ $4 \geq 0$ \quad True	The numbers in this interval are solutions.

As shown in Figure 1-23, the solution is the union of two intervals:

$$\left(-\infty, -\sqrt{5}\right] \cup \left[\sqrt{5}, \infty\right).$$

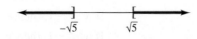

FIGURE 1-23

Self Check 9   Solve $x^2 - 7 \leq 0$.

Now Try Exercise 67.

4. Solve Rational Inequalities

Inequalities that contain fractions with polynomial numerators and denominators are called **rational inequalities.** To solve them, we can use the same strategies that we use to solve quadratic inequalities.

EXAMPLE 10   **Solving a Rational Inequality**

Solve the rational inequality: $\dfrac{x^2 - x - 2}{x^2 - 4x + 3} \leq 0$.

SOLUTION   **Method 1:** The intervals are found by solving $x^2 - x - 2 = 0$ and $x^2 - 4x + 3 = 0$. The solutions of the first equation are -1 and 2, and the solutions of the second equation are 1 and 3. These solutions establish the five intervals shown in Figure 1-24.

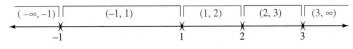

FIGURE 1-24

The solutions of $\dfrac{x^2 - x - 2}{x^2 - 4x + 3} \leq 0$ will be the numbers in one or more of these intervals. To determine which intervals are solutions, we test a number in each interval to see whether it satisfies the inequality.

Interval	Test Value	Inequality $\dfrac{x^2 - x - 2}{x^2 - 4x + 3} \leq 0$	Result
$(-\infty, -1)$	-2	$\dfrac{(-2)^2 - (-2) - 2}{(-2)^2 - 4(-2) + 3} \leq 0$ $\dfrac{4}{15} \leq 0$ False	The numbers in this interval are not solutions.
$(-1, 1)$	0	$\dfrac{(0)^2 - (0) - 2}{(0)^2 - 4(0) + 3} \leq 0$ $-\dfrac{2}{3} \leq 0$ True	The numbers in this interval are solutions.
$(1, 2)$	$\dfrac{3}{2}$	$\dfrac{\left(\dfrac{3}{2}\right)^2 - \left(\dfrac{3}{2}\right) - 2}{\left(\dfrac{3}{2}\right)^2 - 4\left(\dfrac{3}{2}\right) + 3} \leq 0$ $\dfrac{5}{3} \leq 0$ False	The numbers in this interval are not solutions.
$(2, 3)$	$\dfrac{5}{2}$	$\dfrac{\left(\dfrac{5}{2}\right)^2 - \left(\dfrac{5}{2}\right) - 2}{\left(\dfrac{5}{2}\right)^2 - 4\left(\dfrac{5}{2}\right) + 3} \leq 0$ $-\dfrac{7}{3} \leq 0$ True	The numbers in this interval are solutions.
$(3, \infty)$	4	$\dfrac{(4)^2 - (4) - 2}{(4)^2 - 4(4) + 3} \leq 0$ $\dfrac{10}{3} \leq 0$ False	The numbers in this interval are not solutions.

The numbers in the intervals $(-1, 1)$ and $(2, 3)$ satisfy the inequality, but the numbers in the intervals $(-\infty, -1)$, $(1, 2)$, and $(3, \infty)$ do not. The solution set is $[-1, 1) \cup [2, 3)$. The graph of the solution set is shown in Figure 1-25.

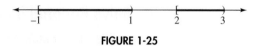

FIGURE 1-25

Because $x = -1$ and $x = 2$ make the numerator 0, they satisfy the inequality. Thus, their graphs are drawn with brackets to show that –1 and 2 are included. Because 1 and 3 give 0's in the denominator, the parentheses at $x = 1$ and $x = 3$ show that 1 and 3 are not in the solution set.

Method 2:   We factor each trinomial and write the inequality in the form

$$\frac{(x - 2)(x + 1)}{(x - 3)(x - 1)} \le 0$$

We then construct the sign graph shown in Figure 1-26. The value of the fraction will be 0 when $x = 2$ and $x = -1$. The value will be negative when there is an odd number of negative factors. This happens between -1 and 1 and between 2 and 3.

The graph of the solution set also appears in Figure 1-26. The brackets at -1 and 2 show that these numbers are in the solution set. The parentheses at 1 and 3 show that these numbers are not in the solution set.

Caution

A common error made when writing the solution set for a rational inequality that contains $\le$ or $\ge$ is to write a bracket next to each number in the solution set. The real numbers that make the denominator zero will *never* be included in the solution set, and we should always write a parenthesis next to them.

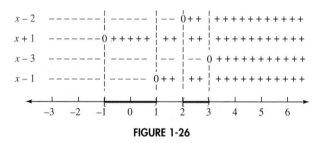

FIGURE 1-26

Self Check 10   Solve: $\dfrac{x^2 + 2x - 3}{x^2 + 4x + 3} > 0$.

Now Try Exercise 73.

EXAMPLE 11   Solving a Rational Inequality

Solve: $\dfrac{6}{x} > 2$.

SOLUTION   We will first rewrite the rational inequality and get a 0 on the right side. Then we can use either method to solve the inequality.

To get a 0 on the right side, we subtract 2 from both sides. We then combine like terms on the left side.

$$\frac{6}{x} > 2$$

$$\frac{6}{x} - 2 > 0 \qquad \text{Subtract 2 from both sides.}$$

$$\frac{6}{x} - \frac{2x}{x} > 0 \qquad \frac{x}{x} = 1$$

$$\frac{6 - 2x}{x} > 0 \qquad \text{Add the numerators and keep the common denominator.}$$

The inequality now has the form of a rational inequality. We can use either method to solve the inequality. The intervals are found by solving $6 - 2x = 0$ and $x = 0$. The solution of the first equation is 3, and the solution of the second equation is 0. This determines the intervals $(-\infty, 0)$, $(0, 3)$, and $(3, \infty)$. Because only the numbers in the interval $(0, 3)$ satisfy the original inequality, the solution set is $(0, 3)$. The graph is shown in Figure 1-27.

FIGURE 1-27

We could construct a sign graph as in Figure 1-28 and obtain the same solution set.

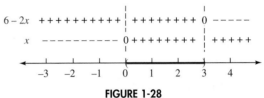

FIGURE 1-28

Self Check 11   Solve: $\dfrac{2}{x} < 4$.

Now Try Exercise 77.

Caution   It is tempting to solve Example 11 by multiplying both sides by x and solving the inequality $6 > 2x$. However, multiplying both sides by x gives $6 > 2x$ only when x is positive. If x is negative, multiplying both sides by x will reverse the direction of the $>$ symbol, and the inequality $\frac{6}{x} > 2$ will be equivalent to $6 < 2x$. If you fail to consider both cases, you will get a wrong answer.

Self Check Answers   **1. a.** $>$ or $\geq$ **b.** $\leq$ or $\geq$ **c.** $>$ or $\geq$ **2.** $(9, \infty)$

3. $(-\infty, -7]$ **4.** 200 bushels or less   **5.** $[-3, 4)$

6. $(4, \infty)$ **7.** $(-\infty, -3] \cup [2, \infty)$ **8.** $[-1, 6]$

9. $\left[-\sqrt{7}, \sqrt{7}\right]$ **10.** $(-\infty, -3) \cup (-3, -1) \cup (1, \infty)$

11. $(-\infty, 0) \cup \left(\dfrac{1}{2}, \infty\right)$

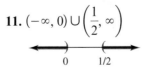

Exercises **1.7**

Fill in the blanks.

1. If $x > y$, then x lies to the _____ of y on a number line.
2. $a < b$, _____, or $a > b$.
3. If $a < b$ and $b < c$, then _____.
4. If $a < b$ then $a + c <$ _____.
5. If $a < b$ then $a - c <$ _____.
6. If $a < b$ and $c > 0$, then ac _____ bc.
7. If $a < b$ and $c < 0$, then ac _____ bc.
8. If $a < b$ and $c < 0$, then $\frac{a}{c}$ _____ $\frac{b}{c}$.
9. $3x - 5 < 12$ and $ax + c > 0$ $(a \neq 0)$ are examples of _____ inequalities.
10. $ax^2 + bx - c \geq 0$ and $3x^2 - 6x < 0$ are examples of _____ inequalities.
11. If two inequalities have the same solution set, they are called _____ inequalities.
12. An inequality that contains a fraction with a polynomial numerator and denominator is called a _____ inequality.

Practice
Solve each inequality, graph the solution set, and write the answer in interval notation. Do not worry about drawing your graphs exactly to scale.

13. $3x + 2 < 5$

14. $-2x + 4 < 6$

15. $3x + 2 \geq 5$

16. $-2x + 4 \geq 6$

17. $-5x + 3 > -2$

18. $4x - 3 > -4$

19. $-5x + 3 \leq -2$

20. $4x - 3 \leq -4$

21. $2(x - 3) \leq -2(x - 3)$

22. $3(x + 2) \leq 2(x + 5)$

23. $\frac{3}{5}x + 4 > 2$

24. $\frac{1}{4}x - 3 > 5$

25. $\frac{x + 3}{4} < \frac{2x - 4}{3}$

26. $\frac{x + 2}{5} > \frac{x - 1}{2}$

27. $\frac{6(x - 4)}{5} \geq \frac{3(x + 2)}{4}$

28. $\frac{3(x + 3)}{2} < \frac{2(x + 7)}{3}$

29. $\frac{5}{9}(a + 3) - a \geq \frac{4}{3}(a - 3) - 1$

30. $\frac{2}{3}y - y \leq -\frac{3}{2}(y - 5)$

31. $\frac{2}{3}a - \frac{3}{4}a < \frac{3}{5}\left(a + \frac{2}{3}\right) + \frac{1}{3}$

32. $\frac{1}{4}b + \frac{2}{3}b - \frac{1}{2} > \frac{1}{2}(b + 1) + b$

33. $4 < 2x - 8 \leq 10$

34. $3 \leq 2x + 2 < 6$

35. $9 \geq \frac{x - 4}{2} > 2$

36. $5 < \frac{x - 2}{6} < 6$

37. $0 \leq \frac{4 - x}{3} \leq 5$

38. $0 \geq \frac{5 - x}{2} \geq -10$

39. $-2 \geq \frac{1 - x}{2} \geq -10$

40. $-2 \leq \frac{1 - x}{2} < 10$

41. $-3x > -2x > -x$

42. $-3x < -2x < -x$

43. $x < 2x < 3x$

44. $x > 2x > 3x$

45. $2x + 1 < 3x - 2 < 12$

46. $2 - x < 3x + 5 < 18$

47. $2 + x < 3x - 2 < 5x + 2$

48. $x > 2x + 3 > 4x - 7$

49. $3 + x > 7x - 2 > 5x - 10$

50. $2 - x < 3x + 1 < 10x$

51. $x \le x + 1 \le 2x + 3$

52. $-x \ge -2x + 1 \ge -3x + 1$

53. $x^2 + 7x + 12 < 0$

54. $x^2 - 13x + 12 \le 0$

55. $x^2 - 5x + 6 \ge 0$

56. $6x^2 + 5x - 6 > 0$

57. $x^2 + 5x + 6 < 0$

58. $x^2 + 9x + 20 \ge 0$

59. $6x^2 + 5x + 1 \ge 0$

60. $x^2 + 9x + 20 < 0$

61. $6x^2 - 5x < -1$

62. $9x^2 + 24x > -16$

63. $2x^2 \ge 3 - x$

64. $9x^2 \le 24x - 16$

65. $x^2 - 3 \ge 0$

66. $x^2 - 7 \le 0$

67. $x^2 - 11 < 0$

68. $x^2 - 20 > 0$

69. $\dfrac{x + 3}{x - 2} < 0$

70. $\dfrac{x + 3}{x - 2} > 0$

71. $\dfrac{x^2 + x}{x^2 - 1} > 0$

72. $\dfrac{x^2 - 4}{x^2 - 9} < 0$

73. $\dfrac{x^2 + 5x + 6}{x^2 + x - 6} \ge 0$

74. $\dfrac{x^2 + 10x + 25}{x^2 - x - 12} \le 0$

75. $\dfrac{6x^2 - x - 1}{x^2 + 4x + 4} > 0$

76. $\dfrac{6x^2 - 3x - 3}{x^2 - 2x - 8} < 0$

77. $\dfrac{3}{x} > 2$

78. $\dfrac{3}{x} < 2$

79. $\dfrac{6}{x} < 4$

80. $\dfrac{6}{x} > 4$

81. $\dfrac{3}{x - 2} \le 5$

82. $\dfrac{3}{x + 2} \le 4$

83. $\dfrac{6}{x^2 - 1} < 1$

84. $\dfrac{6}{x^2 - 1} > 1$

Applications
Solve each problem.

85. Long distance A long-distance telephone call costs 40¢ for the first three minutes and 10¢ for each additional minute. At most how many minutes can a person talk and not exceed $2?

86. Buying a computer A student who can afford to spend up to $2,000 sees the ad shown in the illustration. If she buys a computer, how many games can she buy?

Big Sale!!!!

◀▐▌ **$1,695.95**

Games
$19.95

87. Buying albums Andy can spend up to $275 on an iPod and some albums. If he can buy an iPod for $150 and albums for $9.75, what is the greatest number of albums that he can buy?

88. Buying DVDs Mary wants to spend less than $600 for a DVD recorder and some DVDs. If the recorder of her choice costs $425 and DVDs cost $7.50 each, how many DVDs can she buy?

89. Buying a refrigerator A woman who has $1,200 to spend wants to buy a refrigerator. Refer to the following table and write an inequality that shows how much she can pay for the refrigerator.

State sales tax	6.5%
City sales tax	0.25%

90. Renting a rototiller The cost of renting a rototiller is $17.50 for the first hour and $8.95 for each additional hour. How long can a person have the rototiller if the cost must be less than $75?

iStockphoto.com/
Keith Webber Jr.

91. Real estate taxes A city council has proposed the following two methods of taxing real estate:

Method 1	$2,200 + 4% of assessed value
Method 2	$1,200 + 6% of assessed value

For what range of assessments *a* would the first method benefit the taxpayer?

92. Medical plans A college provides its employees with a choice of the two medical plans shown in the following table. For what size hospital bills is Plan 2 better for the employee than Plan 1? (*Hint:* The cost to the employee includes both the deductible payment and the employee's coinsurance payment.)

Plan 1	Plan 2
Employee pays $100	Employee pays $200
Plan pays 70% of the rest	Plan pays 80% of the rest

93. Medical plans To save costs, the college in Exercise 92 raised the employee deductible, as shown in the following table. For what size hospital bills is Plan 2 better for the employee than Plan 1? (*Hint:* The cost to the employee includes both the deductible payment and the employee's coinsurance payment.)

Plan 1	Plan 2
Employee pays $200	Employee pays $400
Plan pays 70% of the rest	Plan pays 80% of the rest

94. Geometry The perimeter of a rectangle is to be between 180 inches and 200 inches. Find the range of values for its length when its width is 40 inches.

95. Geometry The perimeter of an equilateral triangle is to be between 50 centimeters and 60 centimeters. Find the range of lengths of one side.

96. Geometry The perimeter of a square is to be from 25 meters to 60 meters. Find the range of values for its area.

Discovery and Writing

97. Express the relationship $20 < l < 30$ in terms of P, where $P = 2l + 2w$.

98. Express the relationship $10 < C < 20$ in terms of F, where $F = \frac{9}{5}C + 32$.

99. The techniques used for solving linear equations and linear inequalities are similar, yet different. Explain.

100. Explain why the relation $\geq$ is transitive.

Review

In Exercises 101–106,
$A = \left\{ -9, -\pi, -2, -\frac{1}{2}, 0, 1, 2, \sqrt{7}, \frac{21}{2} \right\}.$

101. Which numbers are even integers?

102. Which numbers are natural numbers?

103. Which numbers are prime numbers?

104. Which numbers are irrational numbers?

105. Which numbers are real numbers?

106. Which numbers are rational numbers?

1.8 Absolute Value

In this section, we will learn to

1. Define and use absolute value.
2. Solve equations of the form $|x| = k$.
3. Solve equations with two absolute values.
4. Solve inequalities of the forms $|x| < k$ and $|x| \leq k$.
5. Solve inequalities of the forms $|x| > k$ and $|x| \geq k$.
6. Solve compound inequalities with absolute value.
7. Solve inequalities with two absolute values.

Brocreative/Shutterstock.com

Cancun, Mexico, is a popular spring-break destination for college students. The beautiful beaches, exciting nightlife, and tropical weather make it a pleasurable experience.

The average annual temperature in Cancun is 78°F with fluctuations of approximately 7 degrees. We can represent this temperature range using absolute value notation. If we let x represent the temperature at a given time, the absolute value of the difference between x and 78 is less than or equal to 7. We can write this as $|x - 78| \leq 7$.

In this section, we will review absolute value and examine its consequences in greater detail.

1. Define and Use Absolute Value

Absolute Value

The **absolute value** of the real number x, denoted by $|x|$, is defined as follows:

If $x \geq 0$, then $|x| = x$.

If $x < 0$, then $|x| = -x$.

This definition provides a way to associate a nonnegative real number with any real number.

- If $x \geq 0$, then x (which is positive or 0) is its own absolute value.
- If $x < 0$, then $-x$ (which is positive) is the absolute value.

Either way, $|x|$ is positive or 0:

$$|x| \geq 0 \qquad \text{for all real numbers } x$$

EXAMPLE 1   **Using the Definition of Absolute Value**

Write each expression without using absolute value symbols:

a. $|7|$ **b.** $|-3|$ **c.** $-|-7|$ **d.** $|x - 2|$

SOLUTION   In each case, we will apply the definition of absolute value.

a. Because 7 is positive, $|7| = 7$.

b. Because -3 is negative, $|-3| = -(-3) = 3$.

c. The expression $-|-7|$ means "the negative of the absolute value of -7." Thus, $-|-7| = -(7) = -7$.

d. To denote the absolute value of a variable quantity, we must give a conditional answer.

If $x - 2 \geq 0$, then $|x - 2| = x - 2$.

If $x - 2 < 0$, then $|x - 2| = -(x - 2) = -x + 2$.

Self Check 1   Write each expression without using absolute value symbols:

a. $|0|$ **b.** $|-17|$ **c.** $|x + 5|$

Now Try Exercise 13.

2. Solve Equations of the Form $|x| = k$

In the equation $|x| = 8$, x can be either 8 or -8, because $|8| = 8$ and $|-8| = 8$. In general, the following is true.

Absolute Value Equations

If $k \geq 0$, then

$$|x| = k \quad \text{is equivalent to} \quad x = k \quad \text{or} \quad x = -k$$

The absolute value of a number represents the distance on the number line from a point to the origin. The solutions of $|x|$ are the coordinates of the two points that lie exactly k units from the origin. (See Figure 1-29.)

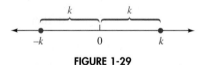

FIGURE 1-29

The equation $|x - 3| = 7$ indicates that a point on the number line with a coordinate of $x - 3$ is 7 units from the origin. Thus, $x - 3$ can be 7 or -7.

$$x - 3 = 7 \quad \text{or} \quad x - 3 = -7$$
$$x = 10 \quad \quad \quad x = -4$$

The solutions of 10 and -4 are shown in Figure 1-30. Both of these numbers satisfy the equation.

FIGURE 1-30

$$
\begin{array}{ccc}
|x - 3| = 7 & \text{and} & |x - 3| = 7 \\
|10 - 3| = 7 & & |-4 - 3| = 7 \\
|7| = 7 & & |-7| = 7 \\
7 = 7 & & 7 = 7
\end{array}
$$

Comment   In general, absolute value equations will have two solutions. Consider these exceptions:

- Note that if $k < 0$, then $|x| = k$ has no solution.

  For example, the equation $|x| = -5$ has no solution.

- If $k = 0$, then $|x| = k$ has only one solution.

  For example, the equation $|x| = 0$ has one solution, $x = 0$.

EXAMPLE 2   **Solving an Absolute Value Equation**

Solve: $|3x - 5| + 3 = 10$.

SOLUTION   We will first write the equation in the form $|x| = k$. To do so, we will subtract 3 from both sides of the equation.

$$|3x - 5| + 3 = 10$$
$$|3x - 5| + 3 - 3 = 10 - 3$$
$$|3x - 5| = 7$$

The equation $|3x - 5| = 7$ is equivalent to two equations

$$3x - 5 = 7 \quad \text{or} \quad 3x - 5 = -7$$

which can be solved separately:

$$
\begin{array}{c|c}
3x - 5 = 7 \quad \text{or} & 3x - 5 = -7 \\
3x = 12 & 3x = -2 \\
x = 4 & x = -\dfrac{2}{3}
\end{array}
$$

The solution set consists of the points shown in Figure 1-31.

FIGURE 1-31

Self Check 2   Solve: $|2x + 3| = 7$.

Now Try Exercise 23.

3. Solve Equations with Two Absolute Values

The equation $|a| = |b|$ is true when $a = b$ or when $a = -b$. For example,

$$|3| = |3| \quad \text{or} \quad |3| = |-3|$$
$$3 = 3 \quad \bigg| \quad 3 = 3$$

In general, the following is true.

Equations with Two Absolute Values   If a and b represent algebraic expressions, the equation $|a| = |b|$ is equivalent to

$$a = b \quad \text{or} \quad a = -b$$

EXAMPLE 3   **Solving an Equation with Two Absolute Values**

Solve: $|2x| = |x - 3|$.

SOLUTION   The equation $|2x| = |x - 3|$ will be true when $2x$ and $x - 3$ are equal or when they are negatives. This gives two equations, which can be solved separately:

$$2x = x - 3 \quad \text{or} \quad 2x = -(x - 3)$$
$$x = -3 \qquad\qquad 2x = -x + 3$$
$$3x = 3$$
$$x = 1$$

Verify that -3 and 1 satisfy the equation.   ∎

Self Check 3   Solve: $|3x + 1| = |5x - 3|$.

Now Try Exercise 39.

We will now turn our attention to inequalities and solve inequalities containing absolute values.

4. Solve Inequalities of the Forms $|x| < k$ and $|x| \le k$

The inequality $|x| < 5$ indicates that a point with coordinate x is less than 5 units from the origin. (See Figure 1-32.) Thus, x is between -5 and 5, and

$$|x| < 5 \quad \text{is equivalent to} \quad -5 < x < 5$$

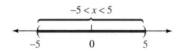

FIGURE 1-32

In general, the inequality $|x| < k \ (k > 0)$ indicates that a point with coordinate x is less than k units from the origin. (See Figure 1-33.)

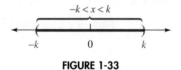

FIGURE 1-33

Similarly, the inequality $|x| \le k \ (k > 0)$ indicates that a point with coordinate x is less than or equal to k units from the origin.

Inequalities of the Forms   • If $k > 0$, then $|x| < k$ is equivalent to $-k < x < k$.
$|x| < k$ and $|x| \le k$ • If $k > 0$, then $|x| \le k$ is equivalent to $-k \le x \le k$.

EXAMPLE 4   **Solving an Absolute Value Inequality of the Form $|x| < k$**

Solve: $|x - 2| < 7$.

SOLUTION   The inequality $|x - 2| < 7$ is equivalent to

$$-7 < x - 2 < 7$$

We can add 2 to each part of this inequality to get

$$-5 < x < 9$$

The solution set is the interval $(-5, 9)$, shown in Figure 1-34.

FIGURE 1-34

Self Check 4   Solve: $|x + 3| < 9$.

Now Try Exercise 47.

5. Solve Inequalities of the Forms $|x| > k$ and $|x| \geq k$

The inequality $|x| > 5$ indicates that a point with coordinate x is more than 5 units from the origin. (See Figure 1-35.) Thus, $x < -5$ or $x > 5$.

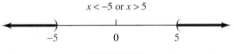

FIGURE 1-35

In general, the inequality $|x| > k$ $(k > 0)$ indicates that a point with coordinate x is more than k units from the origin. (See Figure 1-36.)

FIGURE 1-36

Similarly, the inequality $|x| \geq k$ $(k > 0)$ indicates that a point with coordinate x is k or more units from the origin and thus, $x \leq -5$ or $x \geq 5$.

Inequalities of the Forms $|x| > k$ and $|x| \geq k$	If $k > 0$, then • $|x| > k$ is equivalent to $x < -k$ or $x > k$ • $|x| \geq k$ is equivalent to $x \leq -k$ or $x \geq k$

EXAMPLE 5   **Solving an Absolute Value Inequality of the Form $|x| \geq k$**

Solve: $\left| \dfrac{2x + 3}{2} \right| + 7 \geq 12$.

SOLUTION   We begin by subtracting 7 from both sides of the inequality to isolate the absolute value on the left side.

$$\left| \frac{2x + 3}{2} \right| \geq 5$$

This result is equivalent to two inequalities that can be solved separately.

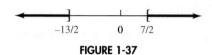

$$\frac{2x+3}{2} \le -5 \quad \text{or} \quad \frac{2x+3}{2} \ge 5$$
$$2x + 3 \le -10 \qquad\qquad 2x + 3 \ge 10$$
$$2x \le -13 \qquad\qquad 2x \ge 7$$
$$x \le -\frac{13}{2} \qquad\qquad x \ge \frac{7}{2}$$

The solution set is the union of the intervals $\left(-\infty, -\frac{13}{2}\right]$ and $\left[\frac{7}{2}, \infty\right)$. Its graph appears in Figure 1-37.

<!-- number line: -13/2   0   7/2 -->

FIGURE 1-37

Self Check 5   Solve: $\left|\dfrac{3x-6}{3}\right| + 2 \ge 12$.

Now Try Exercise 59.

6. Solve Compound Inequalities with Absolute Value

EXAMPLE 6   **Solving a Compound Inequality with Absolute Value**

Solve: $0 < |x - 5| \le 3$.

SOLUTION   The inequality $0 < |x - 5| \le 3$ consists of two inequalities that can be solved separately. The solution will be the intersection of the inequalities

$$0 < |x - 5| \quad \text{and} \quad |x - 5| \le 3$$

The inequality $0 < |x - 5|$ is true for all x except 5. The inequality $|x - 5| \le 3$ is equivalent to the inequality

$$-3 \le x - 5 \le 3$$
$$2 \le x \le 8 \qquad \text{Add 5 to each part.}$$

   The solution set is the intersection of these two solutions, which is the interval $[2, 8]$, except 5. This is the union of the intervals $[2, 5)$ and $(5, 8]$, as shown in Figure 1-38.

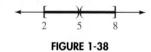

<!-- number line: 2   5   8 -->

FIGURE 1-38

Self Check 6   Solve: $0 < |x + 2| \le 5$.

Now Try Exercise 63.

7. Solve Inequalities with Two Absolute Values

In Example 9b of Section 1.5, we saw that $|a|$ could be defined as

$$|a| = \sqrt{a^2}$$

We will use this fact in the next example.

EXAMPLE 7   **Solving an Inequality with Two Absolute Values**

Solve $|x + 2| > |x + 1|$ and give the result in interval notation.

SOLUTION
$$|x + 2| > |x + 1|$$

$$\sqrt{(x + 2)^2} > \sqrt{(x + 1)^2} \qquad \text{Use } |a| = \sqrt{a^2}.$$

$$(x + 2)^2 > (x + 1)^2 \qquad \text{Square both sides.}$$

$$x^2 + 4x + 4 > x^2 + 2x + 1 \qquad \text{Expand each binomial.}$$

$$4x > 2x - 3 \qquad \text{Subtract } x^2 \text{ and 4 from both sides.}$$

$$2x > -3 \qquad \text{Subtract } 2x \text{ from both sides.}$$

$$x > -\frac{3}{2} \qquad \text{Divide both sides by 2.}$$

The solution set is the interval $\left(-\frac{3}{2}, \infty\right)$. Check several numbers in this interval to verify that this interval is the solution.   ■

Self Check 7   Solve $|x - 3| \le |x + 2|$ and give the result in interval notation.

Now Try Exercise 77.

Three other properties of absolute value are sometimes useful.

Properties of Absolute Value   If a and b are real numbers, then

1. $|ab| = |a||b|$ **2.** $\left|\dfrac{a}{b}\right| = \dfrac{|a|}{|b|}$ $(b \neq 0)$ **3.** $|a + b| \le |a| + |b|$

Properties 1 and 2 above indicate that the absolute value of a product (or a quotient) is the product (or the quotient) of the absolute values.

Property 3 indicates that the absolute value of a sum is either equal to or less than the sum of the absolute values.

Self Check Answers   **1. a.** 0   **b.** 17   **c.** if $x + 5 \ge 0$, $|x + 5| = x + 5$; if $x + 5 < 0$,

$|x + 5| = -x - 5$ **2.** $-5, 2$ **3.** $2, \dfrac{1}{4}$

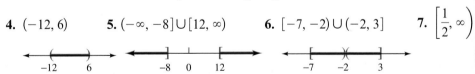

4. $(-12, 6)$ **5.** $(-\infty, -8] \cup [12, \infty)$ **6.** $[-7, -2) \cup (-2, 3]$ **7.** $\left[\dfrac{1}{2}, \infty\right)$

Exercises **1.8**

Getting Ready
You should be able to complete these vocabulary and concept statements before you proceed to the practice exercises.

Fill in the blanks.

1. If $x \ge 0$, then $|x| = $ ___.

2. If $x < 0$, then $|x| = $ ___.

3. $|x| = k$ is equivalent to _____.

4. $|a| = |b|$ is equivalent to $a = b$ or _____.

5. $|x| < k$ is equivalent to _____.

6. $|x| > k$ is equivalent to _____.

7. $|x| \ge k$ is equivalent to _____.

8. $\sqrt{a^2} = $ ____.

Practice

Write each expression without absolute value symbols.

9. $|7|$

10. $|-9|$

11. $|0|$

12. $|3 - 5|$

13. $|5| - |-3|$

14. $|-3| + |5|$

15. $|\pi - 2|$

16. $|\pi - 4|$

17. $|x - 5|$ and $x \geq 5$

18. $|x - 5|$ and $x \leq 5$

19. $|x^3|$

20. $|2x|$

Solve each absolute value equation for x.

21. $|x + 2| = 2$

22. $|2x + 5| = 3$

23. $|3x - 1| - 7 = -2$

24. $|7x - 5| + 5 = 8$

25. $\left|\dfrac{3x - 4}{2}\right| = 5$

26. $\left|\dfrac{10x + 1}{2}\right| = \dfrac{9}{2}$

27. $\left|\dfrac{2x - 4}{5}\right| + 6 = 8$

28. $\left|\dfrac{3x + 11}{7}\right| - 15 = -14$

29. $\left|\dfrac{x - 3}{4}\right| = -2$

30. $\left|\dfrac{x + 5}{2}\right| + 3 = 2$

31. $\left|\dfrac{x - 5}{3}\right| = 0$

32. $\left|\dfrac{x + 7}{9}\right| = 0$

33. $\left|\dfrac{4x - 2}{x}\right| = 3$

34. $\left|\dfrac{2(x - 3)}{3x}\right| = 6$

35. $|x| = x$

36. $|x| + x = 2$

37. $|x + 3| = |x|$

38. $|x + 5| = |5 - x|$

39. $|x - 3| = |2x + 3|$

40. $|x - 2| = |3x + 8|$

41. $|x + 2| = |x - 2|$

42. $|2x - 3| = |3x - 5|$

43. $\left|\dfrac{x + 3}{2}\right| = |2x - 3|$

44. $\left|\dfrac{x - 2}{3}\right| = |6 - x|$

45. $\left|\dfrac{3x - 1}{2}\right| = \left|\dfrac{2x + 3}{3}\right|$

46. $\left|\dfrac{5x + 2}{3}\right| = \left|\dfrac{x - 1}{4}\right|$

Solve each absolute value inequality. Express the solution set in interval notation, and graph it.

47. $|x - 3| < 6$

48. $|x - 2| \geq 4$

49. $|x + 3| > 6$

50. $|x + 2| \leq 4$

51. $|2x + 4| \geq 10$

52. $|5x - 2| < 7$

53. $|3x + 5| + 1 \leq 9$

54. $|2x - 7| - 3 > 2$

55. $|x + 3| > 0$

56. $|x - 3| \leq 0$

57. $\left|\dfrac{5x + 2}{3}\right| < 1$

58. $\left|\dfrac{3x + 2}{4}\right| > 2$

59. $3\left|\dfrac{3x - 1}{2}\right| > 5$

60. $2\left|\dfrac{8x + 2}{5}\right| \leq 1$

61. $\dfrac{|x - 1|}{-2} > -3$

62. $\dfrac{|2x - 3|}{-3} < -1$

Solve each compound inequality with absolute value. Express the solution set in interval notation, and graph it.

63. $0 < |2x + 1| < 3$

64. $0 < |2x - 3| < 1$

65. $8 > |3x - 1| > 3$

66. $8 > |4x - 1| > 5$

67. $2 < \left|\dfrac{x - 5}{3}\right| < 4$

68. $3 < \left|\dfrac{x - 3}{2}\right| < 5$

69. $10 > \left|\dfrac{x - 2}{2}\right| > 4$

70. $5 \geq \left|\dfrac{x + 2}{3}\right| > 1$

71. $2 \le \left| \dfrac{x + 1}{3} \right| < 3$

72. $8 > \left| \dfrac{3x + 1}{2} \right| > 2$

Solve each inequality and express the solution using interval notation.

73. $|x + 1| \ge |x|$

74. $|x + 1| < |x + 2|$

75. $|2x + 1| < |2x - 1|$

76. $|3x - 2| \ge |3x + 1|$

77. $|x + 1| < |x|$

78. $|x + 2| \le |x + 1|$

79. $|2x + 1| \ge |2x - 1|$

80. $|3x - 2| < |3x + 1|$

Applications

81. Finding temperature ranges The temperatures on a summer day satisfy the inequality $|t - 78°| \le 8°$, where t is the temperature in degrees Fahrenheit. Express this range without using absolute value symbols.

82. Finding operating temperatures A car CD player has an operating temperature of $|t - 40°| < 80°$, where t is the temperature in degrees Fahrenheit. Express this range without using absolute value symbols.

83. Range of camber angles The specifications for a certain car state that the camber angle c of its wheels should be $0.6° \pm 0.5°$. Express this range with an inequality containing an absolute value.

84. Tolerance of a sheet of steel A sheet of steel is to be 0.25 inch thick, with a tolerance of 0.015 inch. Express this specification with an inequality containing an absolute value.

85. Humidity level A Steinway piano should be placed in an environment where the relative humidity h is between 38% and 72%. Express this range with an inequality containing an absolute value.

© Istockphoto.com/iLexx

86. Light bulbs A light bulb is expected to last h hours, where $|h - 1,500| \le 200$. Express this range without using absolute value symbols.

87. Error analysis In a lab, students measured the percent of copper p in a sample of copper sulfate. The students know that copper sulfate is actually 25.46% copper by mass. They are to compare their results to the actual value and find the amount of *experimental error.*

   a. Which measurements shown in the illustration satisfy the absolute value inequality $|p - 25.46| \le 1.00$?

   b. What can be said about the amount of error for each of the trials listed in part a?

Lab 4	Section A
Title:	
"Percent copper (CU) in copper sulfate (CuSO₄·5H₂O)"	

$$\text{Title: "Percent copper (CU) in copper sulfate } (CuSO_4{\cdot}5H_2O)\text{"}$$

Results

	% Copper
Trial #1:	22.91%
Trial #2:	26.45%
Trial #3	26.49%
Trial #4:	24.76%

88. Error analysis See Exercise 87.

   a. Which measurements satisfy the absolute value inequality $|p - 25.46| > 1.00$?

   b. What can be said about the amount of error for each of the trials listed in part a?

Discovery and Writing

89. Explain how to find the absolute value of a number.

90. Explain why the equation $|x| + 9 = 0$ has no solution.

91. Explain the use of parentheses and brackets when graphing inequalities.

92. If $k > 0$, explain the differences between the solution sets of $|x| < k$ and $|x| > k$.

93. If $k < 0$, explain why the solution set of $|x| < k$ has no solution.

94. If $k < 0$, explain why the solution set of of $|x| > k$ is all real numbers.

Review

Write each number in scientific notation.

95. 37,250

96. 0.0003725

Write each number in standard notation.

97. 5.23×10^5

98. 7.9×10^{-4}

Simplify each expression.

99. $(x - y)^2 - (x + y)^2$

100. $(p + q)^2 + (p - q)^2$

CHAPTER REVIEW

SECTION 1.1   Linear Equations and Rational Equations

Definitions and Concepts	Examples
An **equation** is a statement indicating that two quantities are equal. There can be restrictions on the variable in an equation.	**Equations:** $2x - 5 = 10$, $\dfrac{2(x-2)}{x-3} = \dfrac{7x+3}{x+2}$ In the equation $2x - 5 = 10$, x can be any real number. In the equation $\dfrac{2(x-2)}{x-3} = \dfrac{7x+3}{x+2}$, x cannot be 3 or -2, because this would give a 0 in the denominator.
Properties of equality: If $a = b$ and c is a number, then $\quad a + c = b + c \quad$ and $\quad a - c = b - c$ $\quad ac = bc \quad\quad$ and $\quad \dfrac{a}{c} = \dfrac{b}{c} \quad (c \neq 0)$	If $a = b$, then $\quad a + 7 = b + 7 \quad$ and $\quad a - 7 = b - 7$ $\quad 7a = 7b \quad$ and $\quad \dfrac{a}{7} = \dfrac{b}{7}$
A **linear equation** is an equation that can be written in the form $ax + b = 0$ $(a \neq 0)$. To solve a linear equation, use the properties of equality to isolate x on one side of the equation.	Solve $3x - 5 = 4$. $\quad 3x - 5 = 4$ $\quad 3x - 5 + 5 = 4 + 5 \qquad$ Add 5 to both sides. $\quad\quad\quad 3x = 9 \qquad$ Combine like terms. $\quad\quad\quad \dfrac{3x}{3} = \dfrac{9}{3} \qquad$ Divide both sides by 3. $\quad\quad\quad x = 3$
An **identity** is an equation that is true for all acceptable replacements for its variable. A **contradiction** is an equation that is false for all acceptable replacements for its variable.	**Identities:** $x + x = 2x$, $\quad 2(x + 1) = 2x + 2$ **Contradictions:** $x + 1 = x$, $\quad 2(x + 1) = 2x + 3$
Rational equations are equations that contain rational expressions. To solve rational equations, multiply both sides of the equation by an expression that will remove the denominators and solve the resulting equation. Be sure to check the answers to identify any **extraneous solutions.**	Solve $\dfrac{2x}{x-3} = \dfrac{6}{x-3}$. $\quad \dfrac{2x}{x-3} = \dfrac{6}{x-3}$ $\quad (x-3)\left(\dfrac{2x}{x-3}\right) = (x-3)\left(\dfrac{6}{x-3}\right) \qquad$ Multiply both sides by $x - 3$. $\quad\quad\quad 2x = 6 \qquad$ Simplify. $\quad\quad\quad x = 3 \qquad$ Divide both sides by 2 and simplify. The result of 3 is extraneous because when you substitute 3 into the original equation, you get a denominator of 0.

Definitions and Concepts	Examples
Formulas can be solved for a specific variable.	Solve $A = \dfrac{1}{2}bh$ for h. $$A = \frac{1}{2}bh$$ $2A = bh$ Multiply both sides by 2. $\dfrac{2A}{b} = \dfrac{bh}{b}$ Divide both sides by b. $\dfrac{2A}{b} = h$ Simplify.

EXERCISES

Find the restrictions on x, if any.

1. $3x + 7 = 4$

2. $x + \dfrac{1}{x} = 2$

3. $\dfrac{1}{x-1} = 4$

4. $\dfrac{1}{x-2} = \dfrac{2}{x-3}$

Solve each equation and classify it as an identity, a conditional equation, or a contradiction.

5. $3(9x + 4) = 28$

6. $\dfrac{3}{2}a = 7(a + 11)$

7. $8(3x - 5) - 4(x + 3) = 12$

8. $\dfrac{x+3}{x+4} + \dfrac{x+3}{x+2} = 2$

9. $\dfrac{3}{x-1} = \dfrac{1}{2}$

10. $\dfrac{8x^2 + 72x}{9 + x} = 8x$

11. $\dfrac{3x}{x-1} - \dfrac{5}{x+3} = 3$

12. $x + \dfrac{1}{2x-3} = \dfrac{2x^2}{2x-3}$

13. $\dfrac{4}{x^2 - 13x - 48} - \dfrac{1}{x^2 + x - 6} = \dfrac{2}{x^2 - 18x + 32}$

14. $\dfrac{a-1}{a+3} + \dfrac{2a-1}{3-a} = \dfrac{2-a}{a-3}$

Solve each formula for the indicated variable.

15. $C = \dfrac{5}{9}(F - 32); F$

16. $P_n = l + \dfrac{si}{f}; f$

17. $\dfrac{1}{f} = \dfrac{1}{f_1} + \dfrac{1}{f_2}; f_1$

18. $S = \dfrac{a - lr}{1 - r}; l$

SECTION 1.2   Application of Linear Equations

Definitions and Concepts	Examples
Use the following steps to solve an application problem: **1.** Analyze the problem. **2.** Pick a variable to represent the quantity to be found. **3.** Form an equation. **4.** Solve the equation. **5.** Check the solution in the words of the problem.	Two students leave their dorm in two cars traveling in opposite directions. If one student drives at a rate of 55 mph and the other at a rate of 50 mph, how long will it take for them to be 210 miles apart? **Analyze the problem and pick a variable.** We can organize the facts of the problem in the following chart. Since each student drives the same amount of time, let t represent that time.

	d	$=$	r	$\cdot$	t
Student 1	$55t$		55		t
Student 2	$50t$		50		t

Definitions and Concepts	Examples
	Form and solve an equation. Since the students are driving in opposite directions, the distance they are apart in t hours is the sum of the distances they drive, a total of 210 miles. We can form and solve the following equation:
	$55t + 50t = 210$
	$\quad\quad 105t = 210$ Combine like terms.
	$\quad\quad\quad\quad t = 2$ Divide both sides by 105.
	Check. In 2 hours, student 1 drives $55(2)$ miles and student 2 drives $50(2)$ miles, or 110 miles plus 100 miles. At this time, they will be 210 miles apart.

EXERCISES

19. Test scores Carlos took four tests in an English class. On each successive test, his score improved by 4 points. If his mean score was 66%, what did he score on the first test?

20. Fencing a garden A homeowner has 100 ft of fencing to enclose a rectangular garden. If the garden is to be 5 ft longer than it is wide, find its dimensions.

21. Travel Two women leave a shopping center by car traveling in opposite directions. If one car averages 45 mph and the other 50 mph, how long will it take for the cars to be 285 miles apart?

22. Travel Two taxis leave an airport and travel in the same direction. If the average speed of one taxi is 40 mph and the average speed of the other taxi is 46 mph, how long will it take before the cars are 3 miles apart?

23. Preparing a solution A liter of fluid is 50% alcohol. How much water must be added to dilute it to a 20% solution?

24. Washing windows Scott can wash 37 windows in 3 hours, and Bill can wash 27 windows in 2 hours. How long will it take the two of them to wash 100 windows?

25. Filling a tank A tank can be filled in 9 hours by one pipe and in 12 hours by another. How long will it take both pipes to fill the empty tank?

26. Producing brass How many ounces of pure zinc must be alloyed with 20 ounces of brass that is 30% zinc and 70% copper to produce brass that is 40% zinc?

27. Lending money A bank lends $10,000, part of it at 11% annual interest and the rest at 14%. If the annual income is $1,265, how much was lent at each rate?

28. Producing oriental rugs An oriental rug manufacturer can use one loom with a setup cost of $750 that can weave a rug for $115. Another loom, with a setup cost of $950, can produce a rug for $95. How many rugs are produced if the costs are the same on each loom?

SECTION 1.3   Quadratic Equations

Definitions and Concepts	Examples
A **quadratic equation** is an equation that can be written in the form $ax^2 + bx + c = 0$, where a, b, and c are real numbers and $a \neq 0$.	$3x^2 - 5x - 7 = 0, \quad 5x^2 - 25 = 0, \quad 7x^2 + 14x = 0$

Definitions and Concepts	Examples
Zero-Factor Theorem: If $ab = 0$, then $a = 0$ or $b = 0$.	Solve $x^2 - x - 6 = 0$ using the Zero-Factor Theorem. $x^2 - x - 6 = 0$ $(x + 2)(x - 3) = 0$ Factor $x^2 - x - 6$. $x + 2 = 0$ or $x - 3 = 0$ $x = -2$ $x = 3$
Square Root Property: If $c > 0$, $x^2 = c$ has two real roots: $x = \sqrt{c}$ or $x = -\sqrt{c}$	If $x^2 = 32$, then $x = \sqrt{32}$ or $x = -\sqrt{32}$ $= \sqrt{16 \cdot 2}$ $= -\sqrt{16 \cdot 2}$ $= 4\sqrt{2}$ $= -4\sqrt{2}$
Steps to complete the square: 1. Make the coefficient of x^2 equal to 1. 2. Get the constant on the right side of the equation. 3. Complete the square on x. Take one-half the coefficient of x, square it, and add it to both sides of the equation. 4. Factor the resulting perfect-square trinomial and combine like terms. 5. Solve the resulting quadratic equation by using the Square Root Property.	Solve $x^2 - x - 6 = 0$ by completing the square: 1. Since the coefficient of $x^2 = 1$, we go to Step 2. 2. Add 6 to both sides to get the constant on the right side: $x^2 - x = 6$. 3. $x^2 - x + \left(-\dfrac{1}{2}\right)^2 = 6 + \left(-\dfrac{1}{2}\right)^2$ 4. $\left(x - \dfrac{1}{2}\right)^2 = \dfrac{25}{4}$ 5. $x - \dfrac{1}{2} = \pm\sqrt{\dfrac{25}{4}}$ $x = \dfrac{1}{2} \pm \dfrac{5}{2}$ $x = \dfrac{6}{2} = 3$ or $x = -\dfrac{4}{2} = -2$
Quadratic Formula: $x = \dfrac{-b \pm \sqrt{b^2 - 4ac}}{2a}$ $(a \neq 0)$	If $3x^2 - 5x + 1 = 0$, then $a = 3$, $b = -5$, and $c = 1$. So $x = \dfrac{-b \pm \sqrt{b^2 - 4ac}}{2a} = \dfrac{-(-5) \pm \sqrt{(-5)^2 - 4(3)(1)}}{2(3)}$ $= \dfrac{5 \pm \sqrt{25 - 12}}{6} = \dfrac{5 \pm \sqrt{13}}{6}$
Discriminant: The value $b^2 - 4ac$ is the **discriminant.** • If $b^2 - 4ac = 0$, the roots of $ax^2 + bx + c = 0$ are repeated rational numbers. • If $b^2 - 4ac > 0$ and a perfect square, the roots of $ax^2 + bx + c = 0$ are two different rational numbers.	In $4x^2 - 12x + 9 = 0$, the discriminant is $b^2 - 4ac = (-12)^2 - 4(4)(9) = 0$. So the roots are repeated rational numbers. In $x^2 - x - 6 = 0$, the discriminant is $b^2 - 4ac = (-1)^2 - 4(1)(-6) = 25$. Since 25 is positive and a perfect square, the roots are two different rational numbers.

Definitions and Concepts	Examples
• If $b^2 - 4ac > 0$ and not a perfect square, the roots of $ax^2 + bx + c = 0$ are two different irrrational numbers.	In $x^2 - x - 5 = 0$, the discriminant is $b^2 - 4ac = (-1)^2 - 4(1)(-5) = 21$. Since 21 is positive and not a perfect square, the roots are two different irrational numbers.
• If $b^2 - 4ac < 0$, the roots of $ax^2 + bx + c = 0$ are two different nonreal numbers.	In $3x^2 - 2x + 1 = 0$, the discriminant is $b^2 - 4ac = (-2)^2 - 4(3)(1) = -8 < 0$. So the roots are two different nonreal numbers.

EXERCISES

Solve each equation by factoring.

29. $2x^2 - x - 6 = 0$ **30.** $12x^2 + 13x = 4$

31. $5x^2 - 8x = 0$ **32.** $27x^2 = 30x - 8$

Solve each equation by using the Square Root Property.

33. $2x^2 = 16$ **34.** $12x^2 = 60$

35. $(4z - 5)^2 = 32$ **36.** $(5x - 7)^2 = 45$

Solve each equation by completing the square.

37. $x^2 - 8x + 15 = 0$ **38.** $3x^2 + 18x = -24$

39. $5x^2 - x - 1 = 0$ **40.** $5x^2 - x = 0$

Use the Quadratic Formula to solve each equation.

41. $x^2 + 5x - 14 = 0$ **42.** $3x^2 - 25x = 18$

43. $5x^2 = 1 - x$ **44.** $5 = a^2 + 2a$

45. Calculate the discriminant associated with the equation $6x^2 + 5x + 1 = 0$.

46. Determine the number and nature of the roots of the equation in Exercise 45.

47. Find the value of k that will make the roots of $kx^2 + 4x + 12 = 0$ equal.

48. Find the values of k that will make the roots of $4y^2 + (k + 2)y = 1 - k$ equal.

49. Solve: $\dfrac{1}{a} - \dfrac{1}{5} = \dfrac{3}{2a}$.

50. Solve: $\dfrac{4}{a - 4} + \dfrac{4}{a - 1} = 5$.

SECTION 1.4   Applications of Quadratic Equations

Definitions and Concepts	Examples
Many real-life problems are modeled by quadratic equations.	If a missile is launched straight up into the air with an initial velocity of 128 feet per second, its height will be given by the formula $h = -16t^2 + 128t$, where h represents its height (in feet) and t represents the time (in seconds) since it was launched. How long will it take the missile to return to its starting point?
	When the missile returns to its starting point, its height will again be 0. So we let $h = 0$ and solve for t.
	$$0 = -16t^2 + 128t$$
	$$0 = -16t^2(t - 8)$$
	$$-16t = 0 \quad \text{or} \quad t - 8 = 0$$
	$$t = 0 \quad | \quad t = 8$$
	The missile will leave its starting point at 0 seconds and return at 8 seconds.

EXERCISES

51. Fencing a field A farmer wishes to enclose a rectangular garden with 300 yards of fencing. A river runs along one side of the garden, so no fencing is needed there. Find the dimensions of the rectangle if the area is 10,450 square yards.

52. Flying rates A jet plane, flying 120 mph faster than a propeller-driven plane, travels 3,520 miles in 3 hours less time than the propeller plane requires to fly the same distance. How fast does each plane fly?

53. Flight of a ball A ball thrown into the air reaches a height h (in feet) according to the formula $h = -16t^2 + 64t$, where t is the time elapsed since the ball was thrown. Find the shortest time it will take the ball to reach a height of 48 feet.

54. Width of a walk A man built a walk of uniform width around a rectangular pool. If the area of the walk is 117 square feet and the dimensions of the pool are 16 feet by 20 feet, how wide is the walk?

SECTION 1.5   Complex Numbers

Definitions and Concepts	Examples
Complex numbers: Numbers that can be written in the form $a + bi$, where a and b are real numbers and $i = \sqrt{-1}$, are **complex numbers**.	$7 - 2i$, $9 + 5i$, and $2 + \sqrt{7}i$ are complex numbers. $\sqrt{-36} = \sqrt{(-1)(36)} = \sqrt{-1}\sqrt{36} = (i)(6) = 6i$
Equality of complex numbers: $a + bi = c + di$ if and only if $a = c$ and $b = d$	$3 + \sqrt{4}i = \frac{6}{2} + 2i$ because $3 = \frac{6}{2}$ and $\sqrt{4} = 2$.
Adding, subtracting, and multiplying complex numbers: • $(a + bi) + (c + di) = (a + c) + (b + d)i$ • $(a + bi) - (c + di) = (a - c) + (b - d)i$ • $(a + bi)(c + di) = (ac - bd) + (ad + bc)i$ or multiply them as if they were binomials.	$(-3 + 4i) + (2 + 7i) = (-3 + 2) + (4 + 7)i = -1 + 11i$ $(-3 + 4i) - (2 + 7i) = (-3 - 2) + (4 - 7)i = -5 - 3i$ $(-3 + 4i)(2 + 7i) = -3(2) - 3(7i) + 4i(2) + 4i(7i)$ $\qquad = -6 - 21i + 8i + 28i^2$ $\qquad = -6 - 21i + 8i - 28$ $\qquad = -34 - 13i$
The **complex conjugate** of $a + bi$ is $a - bi$.	$3 + 4i$ and $3 - 4i$ are complex conjugates.
Division of complex numbers: To divide complex numbers, rationalize the denominator.	Divide $2 + i$ by $2 - i$. $\dfrac{2 + i}{2 - i} = \dfrac{(2 + i)(2 + i)}{(2 - i)(2 + i)}$ $\qquad \dfrac{2+i}{2+i} = 1$ $= \dfrac{4 + 2i + 2i + i^2}{4 + 2i - 2i - i^2}$ $= \dfrac{4 + 4i - 1}{4 - (-1)}$ $= \dfrac{3 + 4i}{5}$ $= \dfrac{3}{5} + \dfrac{4}{5}i$

Definitions and Concepts	Examples
Powers of i: If n is a natural number that has a remainder of r when divided by 4, then $i^n = i^r$.	$i^2 = -1, i^3 = -i, i^4 = 1, i^5 = i, i^6 = -i, i^7 = -i,$ $i^8 = 1, \ldots$
Absolute value of a complex number: $\lvert a + bi \rvert = \sqrt{a^2 + b^2}$	$\lvert 5 - 7i \rvert = \sqrt{5^2 + (-7)^2} = \sqrt{25 + 49} = \sqrt{74}$

EXERCISES

Perform all operations and express all answers in $a + bi$ form.

55. $(2 - 3i) + (-4 + 2i)$　　**56.** $(2 - 3i) - (4 + 2i)$

57. $(3 - \sqrt{-36}) + (\sqrt{-16} + 2)$

58. $(3 + \sqrt{-9})(2 - \sqrt{-25})$

59. $\dfrac{3}{i}$　　　　　　**60.** $-\dfrac{2}{i^3}$

61. $\dfrac{3}{1 + i}$　　　　　**62.** $\dfrac{2i}{2 - i}$

63. $\dfrac{3 + i}{3 - i}$　　　　　**64.** $\dfrac{3 - 2i}{1 + i}$

65. Simplify: i^{53}.　　　**66.** Simplify: i^{103}.

67. $\lvert 3 - i \rvert$　　　　　**68.** $\left\lvert \dfrac{1 + i}{1 - i} \right\rvert$

69. Solve: $3x^2 - 2x + 1 = 0$.

70. Solve: $3x^2 + 4 = 2x$.

SECTION **1.6**　　Polynomials and Radical Equations

Definitions and Concepts	Examples
Polynomial equations: Many polynomial equations of higher degree can be solved by factoring.	Solve $x^3 - 5x^2 + 6x = 0$. $x^3 - 5x^2 + 6x = 0$ $x(x^2 - 5x + 6) = 0$　　Factor out x. $x(x - 2)(x - 3) = 0$　　Factor $x^2 - 5x + 6$. $x = 0$　or　$x - 2 = 0$　or　$x - 3 = 0$ 　　　　　　　$x = 2$　　　　$x = 3$ The solution set is $\{0, 2, 3\}$.
Power property of real numbers: If $a = b$, then $a^2 = b^2$.	If $x = 5$, then $x^2 = 5^2$ or $x^2 = 25$.
Factoring can be used to solve certain nonpolynomial equations. Check all solutions because extraneous roots can be introduced.	Solve $2x - 5x^{1/2} + 3 = 0$. $2x - 5x^{1/2} + 3 = 0$ $(2x^{1/2} - 3)(x^{1/2} - 1) = 0$ $2x^{1/2} - 3 = 0$　or　$x^{1/2} - 1 = 0$ 　　$x^{1/2} = \dfrac{3}{2}$　　　　$x^{1/2} = 1$ 　　$x = \dfrac{9}{4}$　　　　　$x = 1$ Both roots check.

Definitions and Concepts	Examples
Radical equations: To solve radical equations, use the Power Property of Real Numbers.	Solve $\sqrt{2x-3} = x-1$. $$\sqrt{2x-3} = x-1$$ $$\left(\sqrt{2x-3}\right)^2 = (x-1)^2 \qquad \text{Square both sides.}$$ $$2x-3 = x^2 - 2x + 1$$ $$0 = x^2 - 4x + 4$$ $$0 = (x-2)(x-2)$$ $$x-2 = 0 \quad \text{or} \quad x-2 = 0$$ $$x = 2 \qquad\qquad x = 2$$ Since 2 checks, it is a root.

EXERCISES

Solve each equation.

71. $\dfrac{3x}{2} - \dfrac{2x}{x-1} = x - 3$

72. $\dfrac{12}{x} - \dfrac{x}{2} = x - 3$

73. $x^4 - 2x^2 + 1 = 0$

74. $x^4 + 36 = 37x^2$

75. $a - a^{1/2} - 6 = 0$

76. $x^{2/3} + x^{1/3} - 6 = 0$

77. $\sqrt{x-1} + x = 7$

78. $\sqrt{a+9} - \sqrt{a} = 3$

79. $\sqrt{5-x} + \sqrt{5+x} = 4$

80. $\sqrt{y+5} + \sqrt{y} = 1$

SECTION 1.7   Inequalities

Definitions and Concepts	Examples
Addition, Subtraction, Multiplication, and Division Properties of Inequalities: If a, b, and c are real numbers: If $a < b$, $a + c < b + c$ and $a - c < b - c$. If $a < b$ and $c > 0$, $ac < bc$ and $\dfrac{a}{c} < \dfrac{b}{c}$. If $a < b$ and $c < 0$, $ac > bc$ and $\dfrac{a}{c} > \dfrac{b}{c}$.	 If $x < 10$, then $x + 6 < 10 + 6$ and $x - 6 < 10 - 6$. If $x < 12$, then $3x < 3(12)$ and $\dfrac{x}{3} < \dfrac{12}{3}$. If $x \le 12$, then $-3x \ge -3(12)$ and $\dfrac{x}{-3} \ge \dfrac{12}{-3}$.
Trichotomy Property: $a < b$, $a = b$, or $a > b$ **Transitive Property:** If $a < b$ and $b < c$, then $a < c$.	 Either $x < 3$, $x = 3$, or $x > 3$. If $x < 4$ and $4 < y$, then $x < y$.

Definitions and Concepts	Examples
Types of inequalities:	
• Linear inequality	$-4(x - 3) \geq 7$
• Compound inequality	$-5 \leq 2x + 1 < 3$
• Quadratic inequality	$x^2 - 2x + 3 \leq 0$
• Rational inequality	$\dfrac{x + 1}{x - 2} > 0$
Solving a linear inequality: Use the same steps to solve a linear inequality as you would use to solve a linear equation. However, remember to reverse the order of the inequality when you multiply (or divide) both sides of an inequality by a negative number.	Solve the *linear* inequality: $-4(x - 3) \geq 7$. $-4(x - 3) \geq 7$ $-4x + 12 \geq 7$ Remove parentheses. $-4x \geq -5$ Subtract 12 from both sides. $x \leq \dfrac{5}{4}$ Divide both sides by -4. In interval notation, the solution is $\left(-\infty, \frac{5}{4}\right]$.
Solving a compound inequality: For a compound inequality, isolate x in the middle, if possible. If not, solve each inequality separately. The intersection of the intervals is the solution.	Solve the *compound* inequality: $-5 \leq 2x + 1 < 3$. $-5 \leq 2x + 1 < 3$ $-6 \leq 2x < 2$ Subtract 1 from all three parts. $-3 \leq x < 1$ Divide each part by 2. In interval notation, the solution is $[-3, 1)$.
Solving quadratic and rational inequalities: Quadratic and rational inequalities can be solved by constructing a table and testing values or by constructing a sign graph.	Solve the *rational* inequality: $\frac{x + 1}{x - 2} > 0$. First note that the solution of $x + 1 = 0$ $(x = -1)$ and the solution of $x - 2 = 0$ $(x = 2)$ form three intervals: $(-\infty, -1)$, $(-1, 2)$, and $(2, \infty)$. To determine which intervals are solutions, we test a number in each interval to see whether it satisfies the inequality.

Definitions and Concepts	**Examples**

Interval	Test Value	Inequality $\dfrac{x+1}{x-2} > 0$	Result
$(-\infty, -1)$	-2	$\dfrac{-2+1}{-2-2} = \dfrac{1}{4} > 0$ True	The numbers in this interval are solutions.
$(-1, 2)$	0	$\dfrac{0+1}{0-2} = -\dfrac{1}{2} > 0$ False	The numbers in this interval are not solutions.
$(2, \infty)$	3	$\dfrac{3+1}{3-2} = 4 > 0$ True	The numbers in this interval are solutions.

The solution is the union of two intervals:

$(-\infty, -1) \cup (2, \infty)$

EXERCISES

Solve each inequality; graph the solution set and write the answer in interval notation.

81. $2x - 9 < 5$

82. $5x + 3 \geq 2$

83. $\dfrac{5(x-1)}{2} < x$

84. $\dfrac{1}{4}x + \dfrac{2}{3}x - x > \dfrac{1}{2} + \dfrac{1}{2}(x+1)$

85. $0 \leq \dfrac{3+x}{2} < 4$

86. $2 + a < 3a - 2 \leq 5a + 2$

87. $(x + 2)(x - 4) > 0$

88. $(x - 1)(x + 4) < 0$

89. $x^2 - 2x - 3 < 0$

90. $2x^2 + x - 3 > 0$

91. $\dfrac{x+2}{x-3} \geq 0$

92. $\dfrac{x-1}{x+4} \leq 0$

93. $\dfrac{x^2 + x - 2}{x - 3} \geq 0$

94. $\dfrac{5}{x} < 2$

SECTION **1.8**   Absolute Value

Definitions and Concepts	**Examples**
Definition of absolute value of x: $|x| = \begin{cases} x \text{ when } x \geq 0 \\ -x \text{ when } x < 0 \end{cases}$	$|5| = 5 \qquad |-7| = 7 \qquad -|-10| = -10$

Definitions and Concepts	Examples
Absolute value equations:	Solve: $\|x - 2\| = 6$.
• If $k \geq 0$, then $\|x\| = k$ is equivalent to $x = k$ or $x = -k$.	$x - 2 = 6 \quad$ or $\quad x - 2 = -6$ $\qquad x = 8 \qquad \quad \; x = -4$
• If a and b are algebraic expressions, $\|a\| = \|b\|$ is equivalent to $a = b$ or $a = -b$.	Solve: $\|3x\| = \|x - 2\|$. $3x = x - 2 \quad$ or $\quad 3x = -(x - 2)$ $2x = -2 \qquad\qquad 3x = -x + 2$ $\quad x = -1 \qquad\qquad\; 4x = 2$ $\qquad\qquad\qquad\qquad\quad x = \dfrac{1}{2}$ Both solutions check.
Absolute value inequality properties:	Solve: $\|x - 2\| < 6$.
• If $k > 0$, then $\|x\| < k$ is equivalent to $-k < x < k$.	$-6 < x - 2 < 6$
• If $k > 0$, then $\|x\| > k$ is equivalent to $x > k$ or $x < -k$.	$-4 < x < 8 \qquad$ Add 2 to all three parts. The solution in interval notation is $(-4, 8)$.
These two properties also hold for $\leq$ and $\geq$.	Solve: $\|x - 2\| > 6$. $x - 2 > 6 \quad$ or $\quad x - 2 < -6$ $\quad x > 8 \qquad\qquad\; x < -4 \qquad$ Add 2 to both parts. Thus, $x < -4 \quad$ or $\quad x > 8$ The solution in interval notation is $(-\infty, -4) \cup (8, \infty)$.
Properties of absolute value:	
1. $\|ab\| = \|a\|\|b\|$	$\|-3x\| = \|-3\|\|x\| = 3\|x\|$
2. $\left\|\dfrac{a}{b}\right\| = \dfrac{\|a\|}{\|b\|} \quad (b \neq 0)$	$\left\|\dfrac{-3}{x}\right\| = \dfrac{\|-3\|}{\|x\|} = \dfrac{3}{\|x\|} \; (x \neq 0)$
3. $\|a + b\| \leq \|a\| + \|b\|$	$\|x + 3\| \leq \|x\| + \|3\|$

EXERCISES

Solve each equation or inequality.

95. $\|x + 1\| = 6$

96. $\|2x - 1\| = \|2x + 1\|$

97. $\left\|\dfrac{3x + 11}{7}\right\| - 1 = 0$

98. $\left\|\dfrac{2a - 6}{3a}\right\| - 6 = 0$

99. $\|x + 3\| < 3$

100. $\|3x - 7\| \geq 1$

101. $\left\|\dfrac{x + 2}{3}\right\| < 1$

102. $\left\|\dfrac{x - 3}{4}\right\| > 8$

103. $1 < \|2x + 3\| < 4$

104. $0 < \|3x - 4\| < 7$

CHAPTER TEST

Find all restrictions on x.

1. $\dfrac{x}{x(x-1)} = 2$

2. $\dfrac{4}{3x-2} + 3 = 7$

Solve each equation.

3. $7(2a+5) - 7 = 6(a+8)$

4. $\dfrac{3}{x^2-5x-14} = \dfrac{4}{x^2+5x+6}$

5. Solve for x: $z = \dfrac{x-\mu}{\sigma}$.

6. Solve for a: $\dfrac{1}{a} = \dfrac{1}{b} + \dfrac{1}{c}$.

7. Test scores A student's average on three tests is 75. If the final is to count as two one-hour tests, what grade must the student make to bring the average up to 80?

8. Investment A woman invested part of $20,000 at 6% interest and the rest at 7%. If her annual interest is $1,260, how much did she invest at 6%?

Solve each equation.

9. $4x^2 - 8x + 3 = 0$

10. $2b^2 - 12 = -5b$

11. Write the Quadratic Formula.

12. Use the Quadratic Formula to solve $3x^2 - 5x - 9 = 0$.

13. Find k such that $x^2 + (k+1)x + k + 4 = 0$ will have two equal roots.

14. Height of a projectile The height of a projectile shot up into the air is given by the formula $h = -16t^2 + 128t$. Find the time t required for the projectile to return to its starting point.

Perform each operation and write all answers in $a + bi$ form.

15. $(4-5i) - (-3+7i)$

16. $(4-5i)(3-7i)$

17. $\dfrac{2}{2-i}$

18. $\dfrac{1+i}{1-i}$

Simplify each expression.

19. i^{13}

20. i^0

Find each absolute value.

21. $|5 - 12i|$

22. $\left|\dfrac{1}{3+i}\right|$

Solve each equation.

23. $z^4 - 13z^2 + 36 = 0$

24. $2p^{2/5} - p^{1/5} - 1 = 0$

25. $\sqrt{x+5} = 12$

26. $\sqrt{2z+3} = 1 - \sqrt{z+1}$

Solve each inequality; graph the solution set and write the answer using interval notation.

27. $5x - 3 \le 7$

28. $\dfrac{x+3}{4} > \dfrac{2x-4}{3}$

29. $5 \le 2x - 1 < 7$

30. $1 + x < 3x - 3 < 4x - 2$

31. $x^2 - 7x - 8 \ge 0$

32. $\dfrac{x+2}{x-1} \le 0$

Solve each equation.

33. $\left|\dfrac{3x+2}{2}\right| = 4$

34. $|x+3| = |x-3|$

Solve each inequality; graph the solution set and write the answer using interval notation.

35. $|2x - 5| > 2$

36. $\left|\dfrac{2x+3}{3}\right| \le 5$

CUMULATIVE REVIEW EXERCISES

Consider the set $\{-5, -3, -2, 0, 1, \sqrt{2}, 2, \frac{5}{2}, 5, 6, 11\}$.

1. Which numbers are even integers?

2. Which numbers are prime numbers?

Write each inequality as an interval and graph it.

3. $-4 \le x < 7$

4. $x \ge 2$ or $x < 0$

Determine which property of the real numbers justifies each expression.

5. $(a + b) + c = c + (a + b)$

6. If $x < 3$ and $3 < y$, then $x < y$.

Simplify each expression. Assume that all variables represent positive numbers. Give all answers with positive exponents.

7. $(81a^4)^{1/2}$

8. $81(a^4)^{1/2}$

9. $(a^{-3}b^{-2})^{-2}$

10. $\left(\dfrac{4x^4}{12x^2y}\right)^{-2}$

11. $\left(\dfrac{4x^0y^2}{x^2y}\right)^{-2}$

12. $\left(\dfrac{4x^{-5}y^2}{6x^{-2}y^{-3}}\right)^2$

13. $(a^{1/2}b)^2(ab^{1/2})^2$

14. $(a^{1/2}b^{1/2}c)^2$

Rationalize each denominator and simplify.

15. $\dfrac{3}{\sqrt{3}}$

16. $\dfrac{2}{\sqrt[3]{4x}}$

17. $\dfrac{3}{y - \sqrt{3}}$

18. $\dfrac{3x}{\sqrt{x} - 1}$

Simplify each expression and combine like terms.

19. $\sqrt{75} - 3\sqrt{5}$

20. $\sqrt{18} + \sqrt{8} - 2\sqrt{2}$

21. $\left(\sqrt{2} - \sqrt{3}\right)^2$

22. $\left(3 - \sqrt{5}\right)\left(3 + \sqrt{5}\right)$

Perform the operations and simplify when necessary.

23. $(3x^2 - 2x + 5) - 3(x^2 + 2x - 1)$

24. $5x^2(2x^2 - x) + x(x^2 - x^3)$

25. $(3x - 5)(2x + 7)$

26. $(z + 2)(z^2 - z + 2)$

27. $3x + 2\overline{)6x^3 + x^2 + x + 2}$

28. $x^2 + 2\overline{)3x^4 + 7x^2 - x + 2}$

Factor each polynomial.

29. $3t^2 - 6t$

30. $3x^2 - 10x - 8$

31. $x^8 - 2x^4 + 1$

32. $x^6 - 1$

Perform the operations and simplify.

33. $\dfrac{x^2 - 4}{x^2 + 5x + 6} \cdot \dfrac{x^2 - 2x - 15}{x^2 + 3x - 10}$

34. $\dfrac{6x^3 + x^2 - x}{x + 2} \div \dfrac{3x^2 - x}{x^2 + 4x + 4}$

35. $\dfrac{2}{x + 3} + \dfrac{5x}{x - 3}$

36. $\dfrac{x - 2}{x + 3}\left(\dfrac{x + 3}{x^2 - 4} - 1\right)$

37. $\dfrac{\dfrac{1}{a} + \dfrac{1}{b}}{\dfrac{1}{ab}}$

38. $\dfrac{x^{-1} - y^{-1}}{x - y}$

Solve each equation.

39. $\dfrac{3x}{x + 5} = \dfrac{x}{x - 5}$

40. $8(2x - 3) - 3(5x + 2) = 4$

Solve each formula for the indicated variable.

41. $\dfrac{1}{R} = \dfrac{1}{R_1} + \dfrac{1}{R_2}$; R

42. $S = \dfrac{a - lr}{1 - r}$; r

43. Gardening A gardener wishes to enclose her rectangular raspberry patch with 40 feet of fencing. The raspberry bushes are planted along the garage, so no fencing is needed on that side. Find the dimensions if the total area is to be 192 square feet.

44. Financial planning A college student invested part of a $25,000 inheritance at 7% interest and the rest at 6%. If his annual interest is $1,670, how much did he invest at 6%?

*Perform the operations. If the result is not real, express
the answer in a + bi form.*

45. $\dfrac{2 + i}{2 - i}$

46. $\dfrac{i(3 - i)}{(1 + i)(1 + i)}$

47. $|3 + 4i|$

48. $\dfrac{5}{i^7} + 5i$

Solve each equation.

49. $\dfrac{x + 3}{x - 1} - \dfrac{6}{x} = 1$

50. $x^4 + 36 = 13x^2$

51. $\sqrt{y + 2} + \sqrt{11 - y} = 5$

52. $z^{2/3} - 13z^{1/3} + 36 = 0$

*Graph the solution set of each inequality and write the
answer using interval notation.*

53. $5x - 7 \le 4$

54. $x^2 - 8x + 15 > 0$

55. $\dfrac{x^2 + 4x + 3}{x - 2} \ge 0$

56. $\dfrac{9}{x} > x$

57. $|2x - 3| \ge 5$

58. $\left|\dfrac{3x - 5}{2}\right| < 2$

The Rectangular Coordinate System and Graphs of Equations

2

CAREERS AND MATHEMATICS: Cartographer

© Stock Connection Blue/Alamy

Cartographers collect, analyze, interpret, and map geographic information about the earth's surface. Their work involves geographical research and compiling data to produce maps. They analyze latitude and longitude, elevation and distance, as well as population density, land-use patterns, precipitation levels, and demographic characteristics. Their maps are prepared in either digital or graphic form, using information provided by geodetic surveys, aerial cameras, satellites, light-imaging detection, and Geographic Information Systems (GIS).

Education and Mathematics Required
- Cartographers usually have a bachelor's degree in cartography, geography, surveying, engineering, forestry, computer science, or a physical science.
- College Algebra, Trigonometry, Calculus I and II, Elementary Statistics, and Spatial Statistics are usually required.

How Cartographers Use Math and Who Employs Them
- Math helps cartographers with map scale, coordinate systems, and map projection. Map scale is the relationship between distances on a map and the corresponding distances on the earth's surface. Coordinate systems are numeric methods of representing locations on the earth's surface. Map projection is a function or transformation that relates coordinates of points on a curved surface to coordinates of points on a plane.
- Most cartographers work with engineering, architectural, and surveying firms. Government agencies hire cartographers to work in highway departments and in areas such as land management, natural resources planning, and national defense.

Career Outlook and Salary
- Employment of cartographers is expected to grow 19 percent from 2008 to 2018.
- The median annual wage in May 2008 was $51,180.

For more information see: www.bls.gov/oco

Mathematical expressions often indicate relationships between two variables. To visualize these relationships, we draw graphs of their equations.

2.1 The Rectangular Coordinate System

In this section, we will learn to

1. Plot points in the rectangular coordinate system.
2. Graph linear equations.
3. Graph vertical and horizontal lines.
4. Solve applications using linear equations.
5. Find the distance between two points.
6. Find the midpoint of a line segment.

We often say that a picture is worth a thousand words. In fact, pictures and graphs are an effective way to present information. For this reason, they appear frequently in newspapers and magazines. For example, the graph shown in Figure 2-1 provides a visual representation of the manatee population in Florida after the year 2003.

From the graph we can note many facts about manatees. Among them are:

- The number of manatees in Florida declined from 2005 to 2007.
- In 2005, the population was about 3,100 animals.
- In this time period, the lowest population of manatees occurred in 2004.

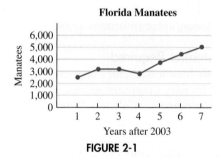

Florida Manatees

FIGURE 2-1

In mathematics, graphs are also an effective way to present information. In this chapter, we will draw graphs of equations containing two variables and then discuss the information that we can derive from graphs.

The solutions of an equation with variables x and y such as $y = -\frac{1}{2}x + 4$ are ordered pairs of real numbers (x, y) that satisfy the equation. To find some ordered pairs that satisfy the equation, we substitute **input values** of x into the equation and find the corresponding **output values** of y. For example, if we substitute 2 for x, we obtain

$$y = -\frac{1}{2}x + 4$$

$$y = -\frac{1}{2}(2) + 4 \qquad \text{Substitute 2 for } x.$$

$$= -1 + 4$$

$$= 3$$

Since $y = 3$ when $x = 2$, the ordered pair $(2, 3)$ is a solution of the equation. The first coordinate, 2, of the ordered pair is usually called the **x-coordinate**. The second coordinate, 3, is usually called the **y-coordinate**. The solution $(2, 3)$ and several other solutions are listed in the table of values shown in Figure 2-2.

Comment

To complete the table of solutions, we first pick values for x. Next we compute each y-value. Then we write each solution as an ordered pair. Note that we choose x-values that are multiples of the denominator, 2. This makes the computations easier when multiplying the x-value by $-\frac{1}{2}$ to find the corresponding y-value.

$y = -\dfrac{1}{2}x + 4$		
x	y	(x, y)
-4	6	$(-4, 6)$
-2	5	$(-2, 5)$
0	4	$(0, 4)$
2	3	$(2, 3)$
4	2	$(4, 2)$

Pick values for x. ⟶   Compute each y-value.   ⟵ Write each solution as an ordered pair.

FIGURE 2-2

ACCENT ON TECHNOLOGY   **Generating Table Values**

© Texas Instruments images used with permission

If an equation in x and y is solved for y, we can use a graphing calculator to generate a table of solutions. The instructions in this discussion are for a TI-84 Plus graphing calculator. For details about other brands, please consult the owner's manual.

To construct a table of solutions for $x + 2y = 8$, we first solve the equation for y.

$$x + 2y = 8$$

$$2y = -x + 8 \qquad \text{Subtract } x \text{ from both sides.}$$

$$y = -\frac{1}{2}x + 4 \qquad \text{Divide both sides by 2 and simplify.}$$

- To construct a table of values for $y = -\frac{1}{2}x + 4$, we first enter the equation. We press ▭ and enter $-(1/2)x + 4$, as shown in Figure 2-3(a).
- Next, we press ▭ ▭ and enter one value for x on the line labeled TblStart=. In Figure 2-3(b), -4 has been entered on this line. Other values for x that will appear in the table are determined by setting an **increment value** on the line labeled ∆Tbl=. In Figure 2-3(b), an increment value of 2 has been entered. This means that each x-value in the table will be 2 units larger than the previous one.
- Finally, we press ▭ ▭ to obtain the table of values shown in Figure 2-3(c). This table contains all of the solutions listed in Figure 2-2, plus the two additional solutions $(6, 1)$ and $(8, 0)$.

To see other values, we simply scroll up and down the screen by pressing the up and down arrow keys.

(a)

(b)

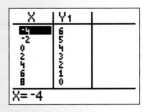

(c)

FIGURE 2-3

Before we can present the table of solutions shown in Figure 2-2 in graphical form, we need to discuss the rectangular coordinate system.

1. Plot Points in the Rectangular Coordinate System

The **rectangular coordinate system** consists of two perpendicular number lines that divide the plane into four **quadrants,** numbered as shown in Figure 2-4. The horizontal number line is called the **x-axis,** and the vertical number line is called the **y-axis.** These axes intersect at a point called the **origin,** which is 0 on each axis. The positive direction on the x-axis is to the right, the positive direction on the y-axis is upward, and the same unit distance is used on both axes, unless otherwise indicated.

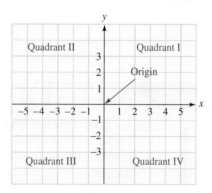

FIGURE 2-4

To plot (or graph) the point associated with the pair $x = 2$ and $y = 3$, denoted as $(2, 3)$, we start at the origin, count 2 units to the right, and then count 3 units up. (See Figure 2-5.) Point P (which lies in the first quadrant) is the graph of the ordered pair $(2, 3)$. The ordered pair $(2, 3)$ gives the **coordinates** of point P.

To plot point Q with coordinates $(-4, 6)$, we start at the origin, count 4 units to the left, and then count 6 units up. Point Q lies in the second quadrant. Point R with coordinates $(6, -4)$ lies in the fourth quadrant.

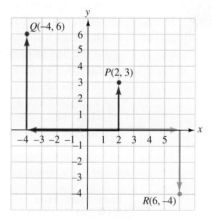

FIGURE 2-5

Caution

The ordered pairs $(-4, 6)$ and $(6, -4)$ represent **different** points. $(-4, 6)$ is in the second quadrant and $(6, -4)$ is in the fourth quadrant.

2. Graph Linear Equations

The **graph of the equation** $y = -\frac{1}{2}x + 4$ is the graph of all points (x, y) on the rectangular coordinate system whose coordinates satisfy the equation. To graph $y = -\frac{1}{2}x + 4$, we plot the pairs listed in the table of solutions shown in Figure 2-6. These points lie on the line shown in the figure. This line is the graph of the equation.

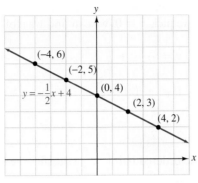

FIGURE 2-6

Comment   When we say that the graph of an equation is a line, we imply two things:

1. Every point with coordinates that satisfy the equation will lie on the line.
2. Every point on the line will have coordinates that satisfy the equation.

When the graph of an equation is a line, we call the equation a **linear equation.** These equations are often written in **standard form** as $Ax + By = C$, where A, B, and C are specific numbers (called **constants**) and x and y are variables. Either A or B can be 0, but A and B cannot both be 0.

Standard Form of an Equation of a Line

The **standard form of an equation of a line** is

$$Ax + By = C$$

where A, B, and C are real numbers and A and B are not both 0.

Here are four examples of linear equations written in standard form.

Linear Equation	Values of A, B, and C
$3x + 2y = 6$	$A = 3, B = 2, C = 6$
$5x - 2y = -10$	$A = 5, B = -2, C = -10$
$2y = 7$	$A = 0, B = 2, C = 7$
$x = -4$	$A = 1, B = 0, C = -4$

EXAMPLE 1   Graphing a Linear Equation

Graph: $x + 2y = 5$.

SOLUTION   We will solve the equation for y and form a table of solutions by picking values for x, substituting them into the equation, and solving for the other variable y. We will plot the points represented in the table of solutions and draw a line through the points.

Solve the equation for y.

$$x + 2y = 5$$

$$x - x + 2y = 5 - x \qquad \text{Subtract } x \text{ from both sides.}$$

$$2y = -x + 5 \qquad \text{Simplify.}$$

$$y = -\frac{1}{2}x + \frac{5}{2} \qquad \text{Divide both sides by 2.}$$

Pick values for x and solve for y.
If we pick $x = 0$, we can find y as follows:

$$y = -\frac{1}{2}x + \frac{5}{2}$$

$$y = -\frac{1}{2}(0) + \frac{5}{2} \qquad \text{Substitute 0 in for } x.$$

$$y = \frac{5}{2} \qquad \text{Simplify.}$$

The ordered pair $\left(0, \dfrac{5}{2}\right)$ satisfies the equation.

To find another ordered pair, we pick $x = 1$ and find y.

$$y = -\frac{1}{2}x + \frac{5}{2}$$

$$y = -\frac{1}{2}(1) + \frac{5}{2} \qquad \text{Substitute 1 in for } x.$$

$$y = 2 \qquad \text{Simplify.}$$

The ordered pair $(1, 2)$ satisfies the equation.

These pairs and others that satisfy the equation are shown in Figure 2-7. We plot the points and join them with a line to get the graph of the equation.

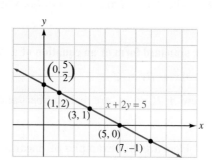

$x + 2y = 5$		
x	y	(x, y)
0	$\dfrac{5}{2}$	$\left(0, \dfrac{5}{2}\right)$
1	2	$(1, 2)$
3	1	$(3, 1)$
5	0	$(5, 0)$
7	-1	$(7, -1)$

FIGURE 2-7

Comment

Even though there are infinitely many points that lie on a line, only two are required to graph a line. However, it is a good idea to find a third point as a check.

Self Check 1   Graph: $3x - 2y = 6$.

Now Try Exercise 35.

EXAMPLE 2   **Graphing a Linear Equation**

Graph: $3(y + 2) = 2x - 3$.

SOLUTION   We will solve the equation for y and find ordered pairs (x, y) that satisfy the equation. Then, we will plot the points and graph the line.

Solve the equation for y.

$$3(y + 2) = 2x - 3$$

$$3y + 6 = 2x - 3 \qquad \text{Use the Distributive Property to remove parentheses.}$$

$$3y = 2x - 9 \qquad \text{Subtract 6 from both sides.}$$

$$y = \frac{2}{3}x - 3 \qquad \text{Divide both sides by 3.}$$

Pick values for *x*, solve for *y*.
We now substitute numbers for *x* to find the corresponding values of *y*. If we let
$x = 0$ and find *y*, we get

$$y = \frac{2}{3}x - 3$$

$$y = \frac{2}{3}(0) - 3 \qquad \text{Substitute 0 for } x.$$

$$y = -3 \qquad\qquad \text{Simplify.}$$

The point $(0, -3)$ lies on the graph.
If we let $x = 3$, we get

$$y = \frac{2}{3}x - 3$$

$$y = \frac{2}{3}(3) - 3 \qquad \text{Substitute 3 for } x.$$

$$y = 2 - 3 \qquad\qquad \text{Simplify.}$$

$$y = -1$$

The point $(3, -1)$ lies on the graph.
We plot these points and others, as in Figure 2-8, and draw the line that passes
through the points.

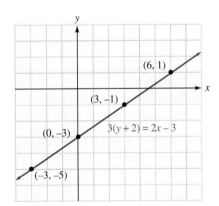

$3(y + 2) = 2x - 3$		
x	y	(x, y)
-3	-5	$(-3, -5)$
0	-3	$(0, -3)$
3	-1	$(3, -1)$
6	1	$(6, 1)$

FIGURE 2-8

Self Check 2   Graph: $2(x - 1) = 6 - 8y$.

Now Try Exercise 39.

ACCENT ON TECHNOLOGY   **Graphing Equations**

We can graph equations with a graphing calculator. To see a graph, we must
choose the minimum and maximum values of the *x*- and *y*-coordinates that will
appear on the calculator's window. A window with standard settings of

Xmin $= -10$ Xmax $= 10$ Ymin $= -10$ Ymax $= 10$

will produce a graph where the value of *x* is in the interval $[-10, 10]$, and the
value of *y* is in the interval $[-10, 10]$.

- To use a graphing calculator to graph $3x + 2y = 12$, we first solve the equation for y.

$$2y = -3x + 12 \qquad \text{Subtract } 3x \text{ from both sides.}$$

$$y = -\frac{3}{2}x + 6 \qquad \text{Divide both sides by 2.}$$

- Next, we press [Y=] and enter the right side of the equation. The screen is shown in Figure 2-9(a).
- We then press [GRAPH] to obtain the graph shown in Figure 2-9(b).

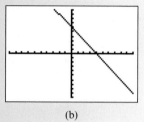

(a)   (b)

FIGURE 2-9

In Figure 2-9, the graph intersects the y-axis at the point $(0, 6)$, which is called the **y-intercept.** It intersects the x-axis at the point $(4, 0)$, which is called the **x-intercept.**

Intercepts of a Line

The **y-intercept** of a line is the point $(0, b)$, where the line intersects the y-axis. To find b, substitute 0 for x in the equation of the line and solve for y.

The **x-intercept** of a line is the point $(a, 0)$, where the line intersects the x-axis. To find a, substitute 0 for y in the equation of the line and solve for x.

EXAMPLE 3   **Graphing a Line by Finding the Intercepts**

Use the x- and y-intercepts to graph the equation $3x + 2y = 12$.

SOLUTION   To find the y-intercept, we substitute 0 for x and solve for y. To find the x-intercept, we substitute 0 for y and solve for x. We will also find a third point as a check and then plot the points and draw the graph.

Find the y-intercept.

To find the y-intercept, we substitute 0 for x and solve for y.

$$3x + 2y = 12$$

$$3(0) + 2y = 12 \qquad \text{Substitute 0 for } x.$$

$$2y = 12 \qquad \text{Simplify.}$$

$$y = 6 \qquad \text{Divide both sides by 2.}$$

The y-intercept is the point $(0, 6)$.

Find the x-intercept.

To find the x-intercept, we substitute 0 for y and solve for x.

$$3x + 2y = 12$$

$$3x + 2(0) = 12 \qquad \text{Substitute 0 for } y.$$

$$3x = 12 \qquad \text{Simplify.}$$

$$x = 4 \qquad \text{Divide both sides by 3.}$$

The x-intercept is the point $(4, 0)$.

Find a third point as a check.

If we let $x = 2$, we will find that $y = 3$.

$$3x + 2y = 12$$

$$3(2) + 2y = 12 \qquad \text{Substitute 2 for } x.$$

$$6 + 2y = 12 \qquad \text{Simplify.}$$

$$2y = 6 \qquad \text{Subtract 6 from both sides.}$$

$$y = 3 \qquad \text{Divide both sides by 2.}$$

The point $(2, 3)$ satisfies the equation.

We plot each pair (as in Figure 2-10) and join them with a line to get the graph of the equation.

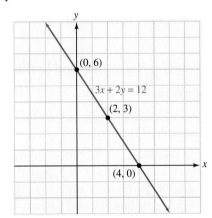

$3x + 2y = 12$		
x	y	(x, y)
0	6	$(0, 6)$
2	3	$(2, 3)$
4	0	$(4, 0)$

FIGURE 2-10

Self Check 3   Graph: $2x - 3y = 12$.

Now Try Exercise 47.

ACCENT ON TECHNOLOGY   **Finding Intercepts Using Zoom and Trace**

We can use the trace feature on a graphing calculator to find the approximate coordinates of any point on a graph. When we press ⟨TRACE⟩, a flashing cursor will appear on the screen. The coordinates of the cursor will also appear at the bottom of the screen.

- To find the y-intercept of the graph of $2y = -5x - 7$ (or $y = -\frac{5}{2}x - \frac{7}{2}$), we graph the equation, using $[-10, 10]$ for x and $[-10, 10]$ for y, and press ⟨TRACE⟩ to get Figure 2-11(a). We see from the figure that the y-intercept is $(0, -3.5)$.

- We can approximate the x-intercept by using the left arrow key and moving the cursor toward the *x*-intercept until we arrive at a point with the coordinates shown in Figure 2-11(b). We see from the graph that the *x*-coordinate of the *x*-intercept is approximately −1.489362.

To get better results, we can zoom in to get a magnified picture, trace again, and move the cursor to the point with coordinates shown in Figure 2-11(c). Since the *y*-coordinate is almost 0, we now have a good approximation for the *x*-intercept. The *x*-coordinate of the *x*-intercept is approximately −1.382979. We can achieve better results with repeated zooms.

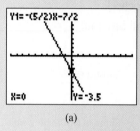 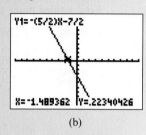 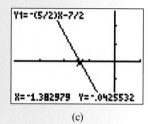

(a)   (b)   (c)

FIGURE 2-11

3. Graph Horizontal and Vertical Lines

In the next example, we will graph a horizontal and a vertical line.

EXAMPLE 4   **Graphing Horizontal and Vertical Lines**

Graph: **a.** $y = 2$ **b.** $x = -3$

SOLUTION   In each case, we will plot a few ordered pairs that satisfy the equation and then draw the graph of the line.

a. In the equation $y = 2$, the value of y is always 2. Any value can be used for x. If we pick *x*-values of −3, 0, 2, and 4, we get the ordered pairs: $(-3, 2)$, $(0, 2)$, $(2, 2)$, and $(4, 2)$. Plotting the pairs shown in Figure 2-12, we see that the graph is a horizontal line, parallel to the *x*-axis and having a *y*-intercept of $(0, 2)$. The line has no *x*-intercept.

b. In the equation $x = -3$, the value of x is always −3. Any value can be used for y. If we pick *y*-values of −2, 0, 2 and 3, we get the ordered pairs: $(-3, -2)$, $(-3, 0)$, $(-3, 2)$ and $(-3, 3)$. After plotting the pairs shown in Figure 2-12, we see that the graph is a vertical line, parallel to the *y*-axis and having an *x*-intercept of $(-3, 0)$. The line has no *y*-intercept.

$y = 2$		
x	y	(x, y)
−3	2	$(-3, 2)$
0	2	$(0, 2)$
2	2	$(2, 2)$
4	2	$(4, 2)$

$x = -3$		
x	y	(x, y)
−3	−2	$(-3, -2)$
−3	0	$(-3, 0)$
−3	2	$(-3, 2)$
−3	3	$(-3, 3)$

FIGURE 2-12

Self Check 4   Graph: **a.** $x = 2$ **b.** $y = -3$

Now Try Exercise 51.

Example 4 suggests the following facts.

Equations of Vertical and Horizontal Lines   If a and b are real numbers, then

- The graph of the equation $x = a$ is a vertical line with x-intercept of $(a, 0)$. If $a = 0$, the line $x = 0$ is the y-axis.
- The graph of the equation $y = b$ is a horizontal line with y-intercept of $(0, b)$. If $b = 0$, the line $y = 0$ is the x-axis.

---

4. Solve Applications Using Linear Equations

EXAMPLE 5   **Solving an Application Problem**

A computer purchased for \$2,750 is expected to depreciate according to the formula $y = -550x + \$2,750$, where y is the value of the computer after x years. When will the computer be worth nothing?

SOLUTION   The computer will have no value when its value (y) is 0. To find x when $y = 0$, we substitute 0 for y and solve for x.

$$y = -550x + 2,750$$

$$0 = -550x + 2,750$$

$$-2,750 = -550x \qquad \text{Subtract 2,750 from both sides.}$$

$$5 = x \qquad \text{Divide both sides by } -550.$$

The computer will have no value in 5 years.

Self Check 5   When will the value of the computer be \$1,650?

Now Try Exercise 97.

5. Find the Distance between Two Points

To derive the formula used to find the distance between two points on a rectangular coordinate system, we use **subscript notation** and denote the points as

$P(x_1, y_1)$ Read as "point P with coordinates of x sub 1 and y sub 1."

$Q(x_2, y_2)$ Read as "point Q with coordinates of x sub 2 and y sub 2."

If $P(x_1, y_1)$ and $Q(x_2, y_2)$ are two points in Figure 2-13 and point R has coordinates (x_2, y_1), triangle PQR is a right triangle. By the Pythagorean Theorem, the square of the hypotenuse of right triangle PQR is equal to the sum of the squares of the two legs. Because leg RQ is vertical, the square of its length is $(y_2 - y_1)^2$. Since leg PR is horizontal, the square of its length is $(x_2 - x_1)^2$. Thus, we have

$$(1) \qquad d^2 = (x_2 - x_1)^2 + (y_2 - y_1)^2$$

Because equal positive numbers have equal positive square roots, we can take the positive square root of both sides of Equation 1 to obtain the **Distance Formula**.

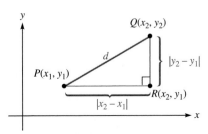

FIGURE 2-13

The Distance Formula   The distance d between points (x_1, y_1) and (x_2, y_2) is given by

$$d = \sqrt{(x_2 - x_1)^2 + (y_2 - y_1)^2}$$

EXAMPLE 6   Finding the Distance between Two Points

Find the distance between $P(-1, -2)$ and $Q(-7, 8)$.

SOLUTION   We use the Distance Formula, $d = \sqrt{(x_2 - x_1)^2 + (y_2 - y_1)^2}$, to find the distance between $P(-1, -2)$ and $Q(-7, 8)$.

If we let $P(-1, -2) = P(x_1, y_1)$ and $Q(-7, 8) = Q(x_2, y_2)$, we can substitute -1 for x_1, -2 for y_1, -7 for x_2, and 8 for y_2 into the formula and simplify.

$$d(PQ) = \sqrt{(x_2 - x_1)^2 + (y_2 - y_1)^2} \qquad \text{Read } d(PQ) \text{ as "the length of segment } PQ.\text{"}$$

$$d(PQ) = \sqrt{[-7 - (-1)]^2 + [8 - (-2)]^2}$$

$$= \sqrt{(-6)^2 + (10)^2}$$

$$= \sqrt{36 + 100}$$

$$= \sqrt{136}$$

$$= \sqrt{4 \cdot 34}$$

$$= 2\sqrt{34} \qquad\qquad \sqrt{4 \cdot 34} = \sqrt{4}\sqrt{34} = 2\sqrt{34}$$

Self Check 6   Find the distance between $P(-2, -5)$ and $Q(3, 7)$.

Now Try Exercise 73.

6. Find the Midpoint of a Line Segment

If point M in Figure 2-14 lies midway between points $P(x_1, y_1)$ and $Q(x_2, y_2)$, point M is called the **midpoint** of segment PQ. To find the coordinates of M, we find the average of the x-coordinates and the average of the y-coordinates of P and Q.

The Midpoint Formula   The midpoint of the line segment with endpoints at $P(x_1, y_1)$ and $Q(x_2, y_2)$ is the point M with coordinates of

$$M = \left(\frac{x_1 + x_2}{2}, \frac{y_1 + y_2}{2} \right)$$

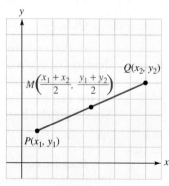

FIGURE 2-14

You will be asked to prove this formula in Exercise 107 by using the Distance Formula to show that $d(PM) + d(MQ) = d(PQ)$.

EXAMPLE 7   **Finding the Midpoint of a Line Segment**

Find the midpoint of the segment joining $P(-7, 2)$ and $Q(1, -4)$.

SOLUTION   We use the Midpoint Formula, $M = \left(\dfrac{x_1 + x_2}{2}, \dfrac{y_1 + y_2}{2}\right)$ to find the midpoint of the line segment joining $P(-7, 2)$ and $Q(1, -4)$. To do so, we substitute $P(-7, 2)$ for $P(x_1, y_1)$ and $Q(1, -4)$ for $Q(x_2, y_2)$ into the Midpoint Formula to get

$$x_M = \frac{x_1 + x_2}{2} \quad \text{and} \quad y_M = \frac{y_1 + y_2}{2}$$
$$= \frac{-7 + 1}{2} \qquad\qquad = \frac{2 + (-4)}{2}$$
$$= \frac{-6}{2} \qquad\qquad = \frac{-2}{2}$$
$$= -3 \qquad\qquad = -1$$

The midpoint is $M(-3, -1)$.

Self Check 7   Find the midpoint of the segment joining $P(-7, -8)$ and $Q(-2, 10)$.

Now Try Exercise 83.

EXAMPLE 8   **Using Midpoint Formula to Find Coordinates**

The midpoint of the segment joining $P(-3, 2)$ and $Q(x_2, y_2)$ is $M(1, 4)$. Find the coordinates of Q.

SOLUTION   We can let $P(x_1, y_1) = P(-3, 2)$ and $M(x_M, y_M) = M(1, 4)$, and then find the coordinates x_2 and y_2 of point $Q(x_2, y_2)$.

$$x_M = \frac{x_1 + x_2}{2} \quad \text{and} \quad y_M = \frac{y_1 + y_2}{2}$$
$$1 = \frac{-3 + x_2}{2} \qquad\qquad 4 = \frac{2 + y_2}{2}$$
$$2 = -3 + x_2 \qquad\qquad 8 = 2 + y_2 \qquad \text{Multiply both sides by 2.}$$
$$5 = x_2 \qquad\qquad 6 = y_2$$

The coordinates of point Q are $(5, 6)$.

Self Check 8   If the midpoint of a segment PQ is $M(2, -5)$ and one endpoint is $Q(6, 9)$, find P.

Now Try Exercise 87.

Self Check Answers   1.   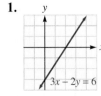   2.   3.   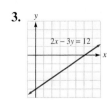   4.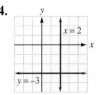

5. 2 years   6. 13   7. $M\left(-\dfrac{9}{2}, 1\right)$ 8. $(-2, -19)$

Exercises **2.1**

Getting Ready

You should be able to complete these vocabulary and concept statements before you proceed to the practice exercises.

Fill in the blanks.

1. The coordinate axes divide the plane into four _____.

2. The coordinate axes intersect at the _____.

3. The positive direction on the x-axis is _____.

4. The positive direction on the y-axis is _____.

5. The x-coordinate is the ____ coordinate in an ordered pair.

6. The y-coordinate is the _____ coordinate in an ordered pair.

7. A _____ equation is an equation whose graph is a line.

8. The point where a line intersects the _____ is called the y-intercept.

9. The point where a line intersects the x-axis is called the _____.

10. The graph of the equation $x = a$ will be a _____ line.

11. The graph of the equation $y = b$ will be a _____ line.

12. Complete the Distance Formula:

    $d =$ _____.

13. If a point divides a segment into two equal segments, the point is called the _____ of the segment.

14. The midpoint of the segment joining $P(x_1, y_1)$ and

    $Q(x_2, y_2)$ is _____.

Practice

Refer to the illustration and determine the coordinates of each point.

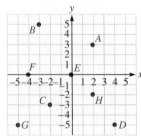

15. A
16. B
17. C
18. D
19. E
20. F
21. G
22. H

Graph each point. Indicate the quadrant in which the point lies, or the axis on which it lies.

23. $(2, 5)$ 24. $(-3, 4)$
25. $(-4, -5)$ 26. $(6, 2)$
27. $(5, 2)$ 28. $(3, -4)$
29. $(4, 0)$ 30. $(0, 2)$

Solve each equation for y and graph the equation. Then check your graph with a graphing calculator.

31. $y - 2x = 7$ 32. $y + 3 = -4x$

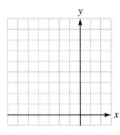   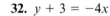

33. $y + 5x = 5$ 34. $y - 3x = 6$

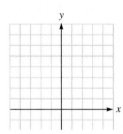   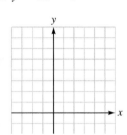

35. $6x - 3y = 10$ 36. $4x + 8y - 1 = 0$

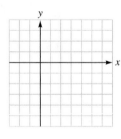   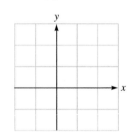

37. $3x = 6y - 1$ 38. $2x + 1 = 4y$

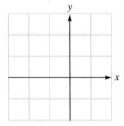   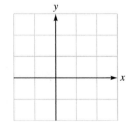

39. $2(x + y + 1) = x + 2$ **40.** $5(x + 2) = 3y - x$

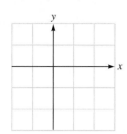   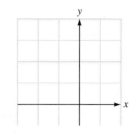

Find the x- and y-intercepts and use them to graph each equation.

41. $x + y = 5$ **42.** $x - y = 3$

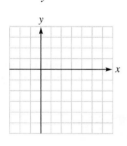   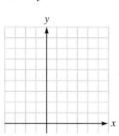

43. $2x - y = 4$ **44.** $3x + y = 9$

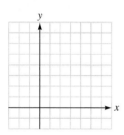   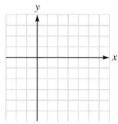

45. $3x + 2y = 6$ **46.** $2x - 3y = 6$

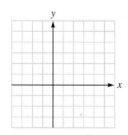   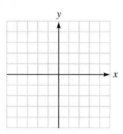

47. $4x - 5y = 20$ **48.** $3x - 5y = 15$

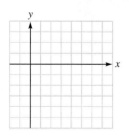   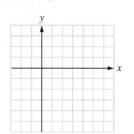

Graph each equation.

49. $y = 3$ **50.** $x = -4$

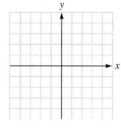   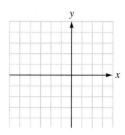

51. $3x + 5 = -1$ **52.** $7y - 1 = 6$

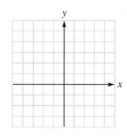   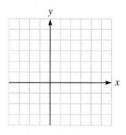

53. $3(y + 2) = y$ **54.** $4 + 3y = 3(x + y)$

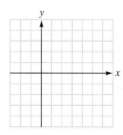   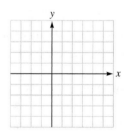

55. $3(y + 2x) = 6x + y$ **56.** $5(y - x) = x + 5y$

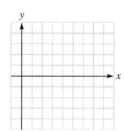   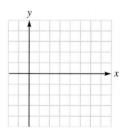

Use a graphing calculator to graph each equation and then find the x-coordinate of the x-intercept to the nearest hundredth.

57. $y = 3.7x - 4.5$ **58.** $y = \dfrac{3}{5}x + \dfrac{5}{4}$

59. $1.5x - 3y = 7$ **60.** $0.3x + y = 7.5$

Find the distance between P and O(0, 0).

61. $P(4, -3)$ **62.** $P(-5, 12)$

63. $P(-3, 2)$ **64.** $P(5, 0)$

65. $P(1, 1)$ **66.** $P(6, -8)$

67. $P\left(\sqrt{3}, 1\right)$ **68.** $P\left(\sqrt{7}, \sqrt{2}\right)$

Find the distance between P and Q.

69. $P(3, 7)$; $Q(6, 3)$ **70.** $P(4, 9)$; $Q(9, 21)$

71. $P(4, -6)$; $Q(-1, 6)$ **72.** $P(0, 5)$; $Q(6, -3)$

73. $P(-2, -15)$; $Q(-6, -21)$ **74.** $P(-7, 11)$; $Q(-11, 7)$

75. $P(3, -3)$; $Q(-5, 5)$ **76.** $P(6, -3)$; $Q(-3, 2)$

77. $P(\pi, -2)$; $Q(\pi, 5)$ **78.** $P\left(\sqrt{5}, 0\right)$; $Q(0, 2)$

Find the midpoint of the line segment PQ.

79. $P(2, 4)$; $Q(6, 8)$ **80.** $P(3, -6)$; $Q(-1, -6)$

81. $P(2, -5)$; $Q(-2, 7)$ **82.** $P(0, 3)$; $Q(-10, -13)$

83. $P(-8, 5)$; $Q(6, -4)$ **84.** $P(3, -2)$; $Q(2, -3)$

85. $P(0, 0)$; $Q\left(\sqrt{5}, \sqrt{5}\right)$ **86.** $P\left(\sqrt{3}, 0\right)$; $Q\left(0, -\sqrt{5}\right)$

One endpoint P and the midpoint M of line segment PQ are given. Find the coordinates of the other endpoint, Q.

87. $P(1, 4)$; $M(3, 5)$ **88.** $P(2, -7)$; $M(-5, 6)$

89. $P(5, -5)$; $M(5, 5)$ **90.** $P(-7, 3)$; $M(0, 0)$

91. Show that a triangle with vertices at $(13, -2)$, $(9, -8)$, and $(5, -2)$ is isosceles.

92. Show that a triangle with vertices at $(-1, 2)$, $(3, 1)$, and $(4, 5)$ is isosceles.

93. In the illustration, points M and N are the midpoints of AC and BC, respectively. Find the length of MN.

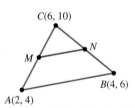

94. In the illustration, points M and N are the mid-points of AC and BC, respectively. Show that $d(MN) = \frac{1}{2}[d(AB)]$.

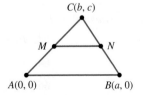

95. In the illustration, point M is the midpoint of the hypotenuse of right triangle AOB. Show that the area of rectangle $OLMN$ is one-half of the area of triangle AOB.

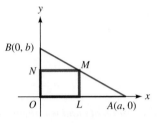

96. Rectangle $ABCD$ in the illustration is twice as long as it is wide, and its sides are parallel to the coordinate axes. If the perimeter is 42, find the coordinates of point C.

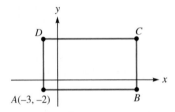

Applications

97. **House appreciation** A house purchased for \$225,000 is expected to appreciate according to the formula $y = 17{,}500x + 225{,}000$, where y is the value of the house after x years. Find the value of the house 5 years later.

98. **Car depreciation** A Chevy Cruze car purchased for \$17,000 is expected to depreciate according to the formula $y = -1{,}360x + 17{,}000$, where y is the value of the car after x years. When will the car be worthless?

99. Demand equations The number of photo scanners that consumers buy depends on price. The higher the price, the fewer photo scanners people will buy. The equation that relates price to the number of photo scanners sold at that price is called a **demand equation.** If the demand equation for a photo scanner is $p = -\frac{1}{10}q + 170$, where p is the price and q is the number of photo scanners sold at that price, how many photo scanners will be sold at a price of $150?

100. Supply equations The number of television sets that manufacturers produce depends on price. The higher the price, the more TVs manufacturers will produce. The equation that relates price to the number of TVs produced at that price is called a **supply equation.** If the supply equation for a 25-inch TV is $p = \frac{1}{10}q + 130$, where p is the price and q is the number of TVs produced for sale at that price, how many TVs will be produced if the price is $150?

101. Meshing gears The rotational speed V of a large gear (with N teeth) is related to the speed v of the smaller gear (with n teeth) by the equation $V = \frac{nv}{N}$. If the larger gear in the illustration is making 60 revolutions per minute, how fast is the smaller gear spinning?

102. Crime prevention The number n of incidents of family violence requiring police response appears to be related to d, the money spent on crisis intervention, by the equation $n = 430 - 0.005d$. What expenditure would reduce the number of incidents to 350?

103. Navigation See the illustration. An ocean liner is located 23 miles east and 72 miles north of Pigeon Cove Lighthouse, and its home port is 47 miles west and 84 miles south of the lighthouse. How far is the ship from port?

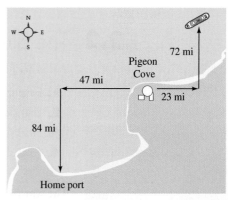

104. Engineering Two holes are to be drilled at locations specified by the engineering drawing shown in the illustration. Find the distance between the centers of the holes.

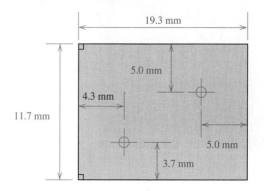

Discovery and Writing

105. Explain how to graph a line using the intercept method.

106. Explain how to determine the quadrant in which the point $P(a, b)$ lies.

107. In Figure 2-14, show that $d(PM) + d(MQ) = d(PQ)$.

108. Use the result of Exercise 107 to explain why point M is the midpoint of segment PQ.

Review

Graph each interval on the number line.

109. $[-3, 2) \cup (-2, 3]$

110. $(-1, 4) \cap [-2, 2]$

111. $[-3, -2) \cap (2, 3]$

112. $[-4, -3) \cup (2, 3]$

Solve each equation.

113. $\dfrac{3}{y + 6} = \dfrac{4}{y + 4}$

114. $\dfrac{z + 4}{z^2 + z} - \dfrac{z + 1}{z^2 + 2z} = \dfrac{8}{z^2 + 3z + 2}$

2.2 The Slope of a Nonvertical Line

In this section, we will learn to

1. Find the slope of a line.
2. Use slope to solve applications.
3. Find slopes of horizontal and vertical lines.
4. Find slopes of parallel and perpendicular lines.

© dk/Alamy

The world's steepest passenger railway is the Lookout Mountain Incline Railway in Chattanooga, Tennessee. Passengers experience breathtaking views of the city and surrounding mountains as the trolley-style railcars travel up Lookout Mountain. The grade or steepness of the track is 72.7% near the top.

Mathematicians use the term **slope** to represent the measure of the steepness of a line. We will explore the topic of slope in this section because it has many real-life applications.

1. Find the Slope of a Line

Suppose that a college student rents a room for $300 per month, plus a $200 nonrefundable deposit. The table shown in Figure 2-15(b) gives the cost (y) for different numbers of months (x). If we construct a graph from these data, we get the line shown in Figure 2-15(a).

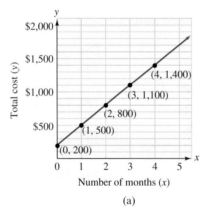

(a)

Time in Months	Total Cost
x	y
0	200
1	500
2	800
3	1,100
4	1,400

(b)

FIGURE 2-15

From the graph, we can see that if x changes from 0 to 1, y changes from 200 to 500. As x changes from 1 to 2, y changes from 500 to 800, and so on. The ratio of the change in y divided by the change in x is the constant 300.

$$\frac{\text{Change in } y}{\text{Change in } x} = \frac{500 - 200}{1 - 0} = \frac{800 - 500}{2 - 1} = \frac{1,100 - 800}{3 - 2} = \frac{1,400 - 1,100}{4 - 3} = \frac{300}{1} = 300$$

The ratio of the change in y divided by the change in x between any two points on any line is always a constant. This constant rate of change is called the **slope** of the line.

The Slope of a Nonvertical Line   The **slope of the nonvertical line** (see Figure 2-16) passing through points $P(x_1, y_1)$ and $Q(x_2, y_2)$ is

$$m = \frac{\text{change in } y}{\text{change in } x} = \frac{y_2 - y_1}{x_2 - x_1} \qquad (x_2 \neq x_1)$$

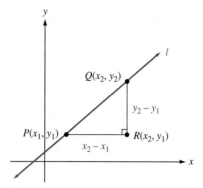

FIGURE 2-16

Comment

Slope is often considered to be a measure of the steepness or tilt of a line. Note that we can use the coordinates of any two points on a line to compute the slope of the line.

EXAMPLE 1   **Finding the Slope of a Line Given Two Points**

Find the slope of the line passing through $P(-1, -2)$ and $Q(7, 8)$.

SOLUTION   We will substitute the points $P(-1, -2)$ and $Q(7, 8)$ into the slope formula, $m = \dfrac{\text{change in } y}{\text{change in } x} = \dfrac{y_2 - y_1}{x_2 - x_1}$, to find the slope of the line.

Let $P(x_1, y_1) = P(-1, -2)$ and $Q(x_2, y_2) = Q(7, 8)$. Then we substitute -1 for x_1, -2 for y_1, 7 for x_2, and 8 for y_2 to get

$$m = \frac{\text{change in } y}{\text{change in } x}$$

$$m = \frac{y_2 - y_1}{x_2 - x_1}$$

$$= \frac{8 - (-2)}{7 - (-1)}$$

$$= \frac{10}{8}$$

$$= \frac{5}{4}$$

The slope of the line is $\frac{5}{4}$. See Figure 2-17.

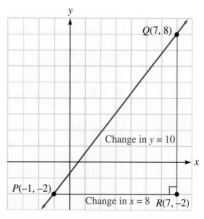

FIGURE 2-17

We would have obtained the same result if we had let $P(x_1, y_1) = P(7, 8)$ and $Q(x_2, y_2) = Q(-1, -2)$.

Self Check 1   Find the slope of the line passing through $P(-3, -4)$ and $Q(5, 9)$.

Now Try Exercise 13.

Caution

> When calculating slope, always subtract the y-values and the x-values in the same order.
>
> $$m = \frac{y_2 - y_1}{x_2 - x_1} \quad \text{or} \quad m = \frac{y_1 - y_2}{x_1 - x_2}$$
>
> Otherwise, we will obtain an *incorrect* result.

A slope can be a positive real number, 0, or a negative real number. If the denominator of the slope formula is 0, slope is not defined.

The change in y (often denoted as Δy) is the **rise** of the line between points P and Q. The change in x (often denoted as Δx) is the **run.** Using this terminology, we can define slope to be the ratio of the rise to the run:

$$m = \frac{y_2 - y_1}{x_2 - x_1} = \frac{\Delta y}{\Delta x} = \frac{\text{rise}}{\text{run}} \qquad (\Delta x \neq 0)$$

EXAMPLE 2   **Finding the Slope of a Line Given Its Equation in Standard Form**

Find the slope of the line determined by $5x + 2y = 10$. (See Figure 2-18.)

SOLUTION   We will find the coordinates of the x- and y-intercepts and substitute into slope formula, $m = \dfrac{\text{change in } y}{\text{change in } x} = \dfrac{y_2 - y_1}{x_2 - x_1}$, to find the slope of the line

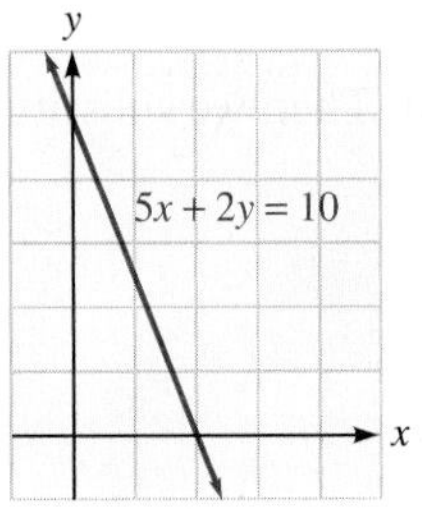

- If $y = 0$, then $x = 2$, and the point $(2, 0)$ lies on the line.
- If $x = 0$, then $y = 5$, and the point $(0, 5)$ lies on the line.

We then find the slope of the line between $P(2, 0)$ and $Q(0, 5)$.

$5x + 2y = 10$

FIGURE 2-18

$$m = \frac{\text{change in } y}{\text{change in } x}$$

$$m = \frac{y_2 - y_1}{x_2 - x_1}$$

$$= \frac{5 - 0}{0 - 2}$$

$$= -\frac{5}{2}$$

The slope is $-\frac{5}{2}$.

Self Check 2   Find the slope of the line determined by $3x - 2y = 9$.

Now Try Exercise 27.

2. Use Slope to Solve Applications

EXAMPLE 3   **Using Slope to Solve an Application Problem**

If carpet costs $25 per square yard plus a delivery charge of $30, the total cost C of n square yards is given by the formula

Total cost	equals	cost per square yard	times	the number of square yards purchased	plus	the delivery charge.
C	$=$	25	$\cdot$	n	$+$	30

Graph the equation $C = 25n + 30$ and interpret the slope of the line.

SOLUTION   We will complete a table of solutions and graph the equation on a coordinate system with a vertical C-axis and a horizontal n-axis. Figure 2-19 shows a table of ordered pairs and the graph.

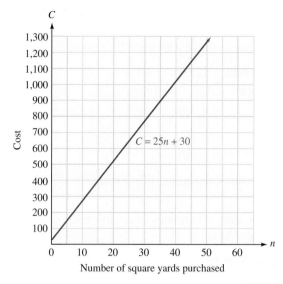

FIGURE 2-19

If we pick the points (30, 780) and (50, 1,280) to find the slope, we have

$$m = \frac{\Delta C}{\Delta n}$$

$$= \frac{C_2 - C_1}{n_2 - n_1}$$

$$= \frac{1{,}280 - 780}{50 - 30} \qquad \text{Substitute 1,280 for } C_2, \text{ 780 for } C_1, \text{ 50 for } n_2, \text{ and 30 for } n_1.$$

$$= \frac{500}{20}$$

$$= 25$$

The slope of 25 (in dollars/square yard) is the cost per square yard of the carpet.

Self Check 3   If the cost of the carpet in Example 3 increases to $35 per square yard, find the slope of the line.

Now Try Exercise 81.

EXAMPLE 4   **Solving an Application Problem**

It takes a skier 25 minutes to complete the course shown in Figure 2-20. Find his average rate of descent in feet per minute.

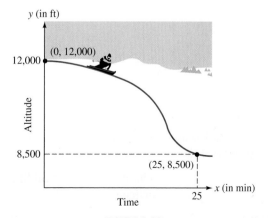

FIGURE 2-20

SOLUTION   To find the average rate of descent, we will find the ratio of the change in altitude to the change in time. To find this ratio, we will calculate the slope of the line passing through the points (0, 12,000) and (25, 8,500).

$$\begin{array}{c} \text{Average rate} \\ \text{of descent} \end{array} = \frac{12,000 - 8,500}{0 - 25}$$

$$= \frac{3,500}{-25}$$

$$= -140$$

The average rate of descent is 140 ft/min.   ■

Self Check 4   If it takes the skier 20 minutes to complete the course, find his average rate of descent in feet per minute.

Now Try Exercise 83.

3. Find Slopes of Horizontal and Vertical Lines

If $P(x_1, y_1)$ and $Q(x_2, y_2)$ are points on the horizontal line shown in Figure 2-21(a), then $y_1 = y_2$, and the numerator of the fraction is 0.

$$\frac{y_2 - y_1}{x_2 - x_1} \qquad \text{On a horizontal line, } x_2 \neq x_1.$$

Thus, the value of the fraction is 0, and the slope of the horizontal line is 0.

If $P(x_1, y_1)$ and $Q(x_2, y_2)$ are points on the vertical line shown in Figure 2-21(b), then $x_1 = x_2$, and the denominator of the fraction is 0.

$$\frac{y_2 - y_1}{x_2 - x_1} \qquad \text{On a vertical line, } y_2 \neq y_1.$$

Since the denominator of a fraction cannot be 0, the slope of a vertical line is not defined.

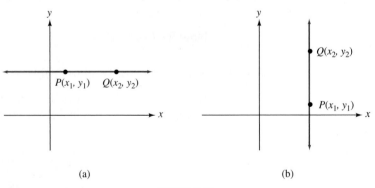

(a)   (b)

FIGURE 2-21

Slopes of Horizontal and Vertical Lines   The slope of a horizontal line (a line with an equation of the form $y = b$) is 0.

The slope of a vertical line (a line with an equation of the form $x = a$) is not defined.

Here are a few facts about slope. See Figure 2-22.

- If a line rises as we follow it from left to right, as in Figure 2-22(a), its slope is positive.
- If a line drops as we follow it from left to right, as in Figure 2-22(b), its slope is negative.
- If a line is horizontal, as in Figure 2-22(c), its slope is 0.
- If a line is vertical, as in Figure 2-22(d), it has no defined slope.

Slope Concepts

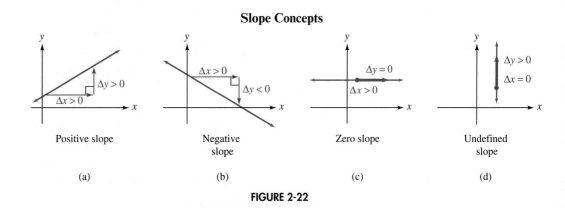

Positive slope	Negative slope	Zero slope	Undefined slope
(a)	(b)	(c)	(d)

FIGURE 2-22

4. Find Slopes of Parallel and Perpendicular Lines

To see a relationship between parallel lines and their slopes, we refer to the parallel lines l_1 and l_2 shown in Figure 2-23, with slopes of m_1 and m_2, respectively. Because right triangles ABC and DEF are similar, it follows that

$$m_1 = \frac{\Delta y \text{ of } l_1}{\Delta x \text{ of } l_1} = \frac{\Delta y \text{ of } l_2}{\Delta x \text{ of } l_2} = m_2$$

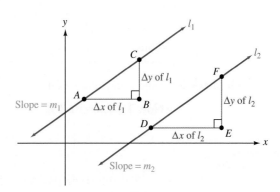

FIGURE 2-23

This shows that if two nonvertical lines are parallel, they have the same slope. It is also true that when two lines have the same slope, they are parallel.

Slopes of Parallel Lines   Nonvertical parallel lines have the same slope, and lines having the same slope are parallel.

Since vertical lines are parallel, lines with undefined slopes are parallel.

EXAMPLE 5   **Solving a Slope Problem Involving Parallel Lines**

The lines in Figure 2-24 are parallel. Find y.

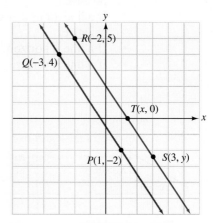

FIGURE 2-24

SOLUTION   Since the lines are parallel, their slopes are equal. To find y, we will find the slope of each line, set them equal, and solve the resulting equation.

Slope of PQ = slope of RS

$$\frac{-2 - 4}{1 - (-3)} = \frac{y - 5}{3 - (-2)}$$

$$\frac{-6}{4} = \frac{y - 5}{5} \qquad \text{Simplify.}$$

$$-30 = 4(y - 5) \qquad \text{Multiply both sides by 20.}$$

$$-30 = 4y - 20 \qquad \text{Remove parentheses and simplify.}$$

$$-10 = 4y \qquad \text{Add 20 to both sides.}$$

$$-\frac{5}{2} = y \qquad \text{Divide both sides by 4 and simplify.}$$

Thus, $y = -\frac{5}{2}$. ∎

Self Check 5   Find x in Figure 2-24.

Now Try Exercise 59.

Comment

If the product of two numbers is -1, the numbers are called **negative reciprocals.**

The following theorem relates perpendicular lines and their slopes.

Slopes of Perpendicular Lines   If two nonvertical lines are perpendicular, the product of their slopes is -1.

If the product of the slopes of two lines is -1, the lines are perpendicular.

PROOF   Suppose l_1 and l_2 are lines with slopes of m_1 and m_2 that intersect at some point. See Figure 2-25. Then superimpose a coordinate system over the lines so that the intersection point is the origin. Let $P(a, b)$ be a point on l_1, and let $Q(c, d)$ be a point on l_2. Neither point P nor point Q can be the origin.

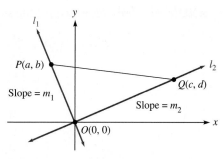

FIGURE 2-25

First, we suppose that l_1 and l_2 are perpendicular. Then triangle POQ is a right triangle with its right angle at O. By the Pythagorean Theorem,

$$d(OP)^2 + d(OQ)^2 = d(PQ)^2$$

$$(a - 0)^2 + (b - 0)^2 + (c - 0)^2 + (d - 0)^2 = (a - c)^2 + (b - d)^2$$

$$a^2 + b^2 + c^2 + d^2 = a^2 - 2ac + c^2 + b^2 - 2bd + d^2$$

$$0 = -2ac - 2bd$$

$$bd = -ac$$

(1) $$\frac{b}{a} \cdot \frac{d}{c} = -1 \qquad \text{Divide both sides by } ac.$$

The coordinates of P are (a, b), and the coordinates of O are $(0, 0)$. Using the definition of slope, we have

$$m_1 = \frac{b - 0}{a - 0} = \frac{b}{a}$$

Similarly, we have

$$m_2 = \frac{d}{c}$$

We substitute m_1 for $\frac{b}{a}$ and m_2 for $\frac{d}{c}$ in Equation 1 to obtain

$$m_1 m_2 = -1$$

Hence, if lines l_1 and l_2 are perpendicular, the product of their slopes is -1.

Conversely, we suppose that the product of the slopes of lines l_1 and l_2 is -1. Because the steps in the previous discussion are reversible, we have $d(OP)^2 + d(OQ)^2 = d(PQ)^2$. By the Pythagorean Theorem, triangle POQ is a right triangle. Thus, l_1 and l_2 are perpendicular. ∎

Comment

It is also true that a horizontal line is perpendicular to a vertical line.

EXAMPLE 6   Solving a Slope Problem Involving Perpendicular Lines

Are the lines shown in Figure 2-26 perpendicular?

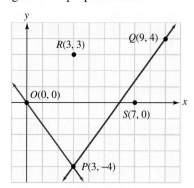

FIGURE 2-26

SOLUTION   We will determine the slopes of the lines and see whether their product is –1.

$$\text{Slope of } OP = \frac{\Delta y}{\Delta x}$$

$$= \frac{y_2 - y_1}{x_2 - y_1}$$

$$= \frac{-4 - 0}{3 - 0}$$

$$= -\frac{4}{3}$$

$$\text{Slope of } PQ = \frac{\Delta y}{\Delta x}$$

$$= \frac{y_2 - y_1}{x_2 - y_1}$$

$$= \frac{4 - (-4)}{9 - 3}$$

$$= \frac{8}{6}$$

$$= \frac{4}{3}$$

Since the product of the slopes is $-\frac{16}{9}$ and not -1, the lines are not perpendicular.

Self Check 6   Is either line in Figure 2-26 perpendicular to the line passing through R and S?

Now Try Exercise 55.

Self Check Answers   1. $\dfrac{13}{8}$ **2.** $\dfrac{3}{2}$ **3.** 35   **4.** 175 ft/min   **5.** $\dfrac{4}{3}$ **6.** yes

Exercises **2.2**

Getting Ready
You should be able to complete these vocabulary and concept statements before you proceed to the practice exercises.

Fill in the blanks.

1. The slope of a nonvertical line is defined to be the change in y _____ by the change in x.
2. The change in ___ is often called the rise.
3. The change in x is often called the ____.
4. When computing the slope from the coordinates of two points, always subtract the y-values and the x-values in the _____.
5. The symbol Δy means _____ y.
6. The slope of a _____ line is 0.
7. The slope of a _____ line is undefined.
8. If the slopes of two lines are equal, the lines are _____.
9. If the product of the slopes of two lines is -1, the lines are _____.
10. If two lines are perpendicular, the product of their slopes is ____.

Practice
Find the slope of the line passing through each pair of points, if possible.

11. $P(2, 2)$; $Q(-1, -1)$ **12.** $P(3, -1)$; $Q(5, 3)$

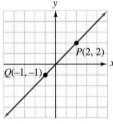

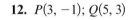

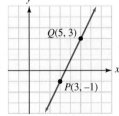

13. $P(-6, 3)$; $Q(6, -2)$ **14.** $P(2, 5)$; $Q(3, 10)$

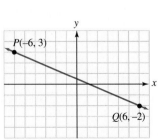

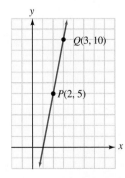

15. $P(3, -2)$; $Q(-1, 5)$

16. $P(3, 7)$; $Q(6, 16)$

17. $P(8, -7)$; $Q(4, 1)$

18. $P(5, 17)$; $Q(17, 17)$

19. $P(-4, 3)$; $Q(-4, -3)$

20. $P\left(2, \sqrt{7}\right)$; $Q\left(\sqrt{7}, 2\right)$

21. $P\left(\dfrac{3}{2}, \dfrac{2}{3}\right)$; $Q\left(\dfrac{5}{2}, \dfrac{7}{3}\right)$

22. $P\left(-\dfrac{2}{5}, \dfrac{1}{3}\right)$; $Q\left(\dfrac{3}{5}, -\dfrac{5}{3}\right)$

23. $P(a + b, c)$; $Q(b + c, a)$ assume $c \neq a$

24. $P(b, 0)$; $Q(a + b, a)$ assume $a \neq 0$

Find two points on the line and use slope formula to find the slope of the line.

25. $y = 3x + 2$ **26.** $y = 5x - 8$

27. $5x - 10y = 3$ **28.** $8y + 2x = 5$

29. $3(y + 2) = 2x - 3$ **30.** $4(x - 2) = 3y + 2$

31. $3(y + x) = 3(x - 1)$ **32.** $2x + 5 = 2(y + x)$

Find the slope of the line, if possible.

33. $y = 7$ **34.** $2y = 5$

35. $x = -\dfrac{1}{2}$ **36.** $x - 7 = 0$

Determine whether the slope of the line is positive, negative, 0, or undefined.

37. 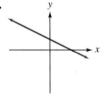   **38.**

39. 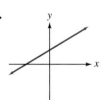   **40.**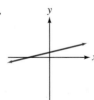

41. 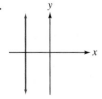   **42.**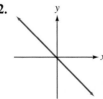

Determine whether the lines with the given slopes are parallel, perpendicular, or neither.

43. $m_1 = 3$; $m_2 = -\dfrac{1}{3}$ **44.** $m_1 = \dfrac{2}{3}$; $m_2 = \dfrac{3}{2}$

45. $m_1 = \sqrt{8}$; $m_2 = 2\sqrt{2}$ **46.** $m_1 = 1$; $m_2 = -1$

47. $m_1 = -\sqrt{2}$; $m_2 = \dfrac{\sqrt{2}}{2}$

48. $m_1 = 2\sqrt{7}$; $m_2 = \sqrt{28}$

49. $m_1 = -0.125$; $m_2 = 8$

50. $m_1 = 0.125$; $m_2 = \dfrac{1}{8}$

51. $m_1 = ab^{-1}$; $m_2 = -a^{-1}b$ $(a \neq 0, b \neq 0)$

52. $m_1 = \left(\dfrac{a}{b}\right)^{-1}$; $m_2 = -\dfrac{b}{a}$ $(a \neq 0, b \neq 0, a \neq b)$

Determine whether the line through the given points and the line through $R(-3, 5)$ and $S(2, 7)$ are parallel, perpendicular, or neither.

53. $P(2, 4)$; $Q(7, 6)$ **54.** $P(-3, 8)$; $Q(-13, 4)$

55. $P(-4, 6)$; $Q(-2, 1)$ **56.** $P(0, -9)$; $Q(4, 1)$

57. $P(a, a)$; $Q(3a, 6a)$ $(a \neq 0)$

58. $P(b, b)$; $Q(-b, 6b)$ $(b \neq 0)$

Lines PQ and RS are either parallel or perpendicular. Find x or y.

59. Parallel: $P(-3, 7)$; $Q(2, 9)$; $R(10, -4)$; $S(x, -6)$

60. Parallel: $P(2, -3)$; $Q(5, 7)$; $R(3, -1)$; $S(6, y)$

61. Perpendicular: $P(2, -7)$; $Q(1, 0)$; $R(-9, 5)$; $S(-2, y)$

62. Perpendicular: $P(1, -2)$; $Q(3, 4)$; $R(x, 6)$; $S(6, 5)$

Find the slopes of lines PQ and PR, and determine whether points P, Q, and R lie on the same line.

63. $P(-2, 8)$; $Q(-6, 9)$; $R(2, 5)$

64. $P(1, -1)$; $Q(3, -2)$; $R(-3, 0)$

65. $P(-a, a)$; $Q(0, 0)$; $R(a, -a)$

66. $P(a, a + b)$; $Q(a + b, b)$; $R(a - b, a)$

Determine which, if any, of the three lines PQ, PR, and QR are perpendicular.

67. $P(5, 4)$; $Q(2, -5)$; $R(8, -3)$

68. $P(8, -2)$; $Q(4, 6)$; $R(6, 7)$

69. $P(1, 3)$; $Q(1, 9)$; $R(7, 3)$

70. $P(2, -3)$; $Q(-3, 2)$; $R(3, 8)$

71. $P(0, 0)$; $Q(a, b)$; $R(-b, a)$

72. $P(a, b)$; $Q(-b, a)$; $R(a - b, a + b)$

73. **Right triangles** Show that the points $A(-1, -1)$, $B(-3, 4)$, and $C(4, 1)$ are the vertices of a right triangle.

74. **Right triangles** Show that the points $D(0, 1)$, $E(-1, 3)$, and $F(3, 5)$ are the vertices of a right triangle.

75. **Squares** Show that the points $A(1, -1)$, $B(3, 0)$, $C(2, 2)$, and $D(0, 1)$ are the vertices of a square.

76. Squares Show that the points $E(-1, -1)$, $F(3, 0)$, $G(2, 4)$, and $H(-2, 3)$ are the vertices of a square.

77. Parallelograms Show that the points $A(-2, -2)$, $B(3, 3)$, $C(2, 6)$, and $D(-3, 1)$ are the vertices of a parallelogram. (Show that both pairs of opposite sides are parallel.)

78. Trapezoids Show that points $E(1, -2)$, $F(5, 1)$, $G(3, 4)$, and $H(-3, 4)$ are the vertices of a trapezoid. (Show that only one pair of opposite sides is parallel.)

79. Geometry In the illustration, points M and N are midpoints of CB and BA, respectively. Show that MN is parallel to AC.

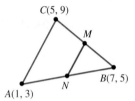

80. Geometry In the illustration, $d(AB) = d(AC)$. Show that AD is perpendicular to BC.

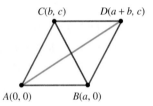

Applications

81. Rate of growth When a college started an aviation program, the administration agreed to predict enrollments using a straight-line method. If the enrollment during the first year was 12, and the enrollment during the fifth year was 26, find the rate of growth per year (the slope of the line). See the illustration.

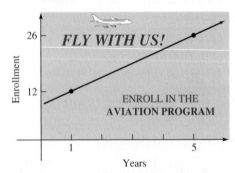

82. Rate of growth A small business predicts sales according to a straight-line method. If sales were $50,000 in the first year and $110,000 in the third year, find the rate of growth in dollars per year (the slope of the line).

83. Rate of decrease The price of computers has been dropping steadily for the past ten years. If a desktop PC cost $6,700 ten years ago, and the same computing power cost $2,200 three years ago, find the rate of decrease per year. (Assume a straight-line model.)

84. Hospital costs The table shows the changing mean daily cost for a hospital room. For the ten-year period, find the rate of change per year of the portion of the room cost that is absorbed by the hospital.

Year	Total Cost to the Hospital	Amount Passed on to Patient
2000	$459	$212
2005	$670	$295
2010	$812	$307

85. Charting temperature changes The following Fahrenheit temperature readings were recorded over a four-hour period.

Time	12:00	1:00	2:00	3:00	4:00
Temperature	47°	53°	59°	65°	71°

Let t represent the time (in hours), with 12:00 corresponding to $t = 0$. Let T represent the temperature. Plot the points (t, T), and draw the line through those points. Explain the meaning of $\frac{\Delta T}{\Delta t}$.

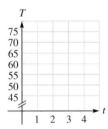

86. Tracking the Dow The Dow Jones Industrial Averages at the close of trade on three consecutive days were as follows:

Day	Monday	Tuesday	Wednesday
Close	12,981	12,964	12,947

Let d represent the day, with $d = 0$ corresponding to Monday, and let D represent the Dow Jones average. Plot the points (d, D), and draw the graph. Explain the meaning of $\frac{\Delta D}{\Delta d}$.

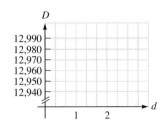

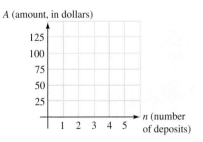

87. Speed of an airplane A pilot files a flight plan indicating her intention to fly at a constant speed of 590 mph. Write an equation that expresses the distance traveled in terms of the flying time. Then graph the equation and interpret the slope of the line. (*Hint: d = rt.*)

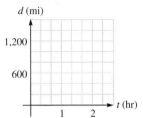

88. Growth of savings A student deposits $25 each month in a Holiday Club account at her bank. The account pays no interest. Write an equation that expresses the amount A in her account in terms of the number of deposits n. Then graph the line, and interpret the slope of the line.

Discovery and Writing
89. Explain why the slope of a vertical line is undefined.
90. Explain how to determine whether two lines are parallel, perpendicular, or neither.

Review
Solve each equation for y and simplify.
91. $3x + 7y = 21$ **92.** $y - 3 = 5(x + 2)$

93. $\dfrac{x}{5} + \dfrac{y}{2} = 1$ **94.** $x - 5y = 15$

Factor each expression.
95. $6p^2 + p - 12$
96. $b^3 - 27$
97. $mp + mq + np + nq$
98. $x^4 + x^2 - 2$

2.3 Writing Equations of Lines

In this section, we will learn to

1. Use point-slope form to write an equation of a line.
2. Use slope-intercept form to write an equation of a line.
3. Graph linear equations using the slope and *y*-intercept.
4. Determine whether linear equations represent lines that are parallel, perpendicular, or neither.
5. Write equations of parallel and perpendicular lines.
6. Recognize and use the standard form of the equation of a line.
7. Write an equation of a line that models a real-life problem.
8. Use linear curve fitting to solve problems.

Suppose we purchase a Kawasaki motorcycle for $10,000 and know that it depreciates $1,200 in value each year.

We can use the facts given to write a linear equation that represents the value y of the motorcycle x years after it was purchased. Because the motorcycle's value decreases $1,200 each year, the slope of the line's graph is $-1,200$. Because the purchase price is $10,000, we know that when we let $x = 0$, the value of y will equal 10,000. A linear equation that satisfies these two conditions is $y = -1,200x + 10,000$. This equation represents the straight-line depreciation of the motorcycle.

In this section, we will write equations of lines given specific characteristics or features of the line.

1. Use Point-Slope Form to Write an Equation of a Line

Suppose that line l in Figure 2-27 has a slope of m and passes through the point $P(x_1, y_1)$. If $Q(x, y)$ is any other point on line l, we have

$$m = \frac{y - y_1}{x - x_1}$$

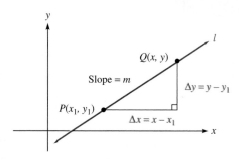

FIGURE 2-27

Comment

$y - y_1 = m(x - x_1)$ is often referred to as **point-slope formula** because we substitute a point and slope into it to write an equation of a line.

If we multiply both sides by $x - x_1$, we have

(1) $y - y_1 = m(x - x_1)$

Since Equation 1 displays the coordinates of the point (x_1, y_1) on the line and the slope m of the line, it is called the **point-slope form** of the equation of a line.

Point-Slope Form of an Equation of a Line

The equation of a line passing through $P(x_1, y_1)$ and with slope m is

$$y - y_1 = m(x - x_1).$$

EXAMPLE 1

Writing an Equation of the Line with a Given Slope Passing through a Given Point

Write an equation of the line with slope $-\frac{5}{3}$ and passing through $P(3, -1)$.

SOLUTION   We will substitute $-\frac{5}{3}$ for m, 3 for x_1, and -1 for y_1 in the point-slope form and simplify.

$$y - y_1 = \mathbf{m}(x - x_1) \qquad \text{This is the point-slope form.}$$

$$y - (-1) = -\frac{5}{3}(x - 3) \qquad \text{Substitute } -\frac{5}{3} \text{ for } m, 3 \text{ for } x_1, \text{ and } -1 \text{ for } y_1.$$

$$y + 1 = -\frac{5}{3}x + 5 \qquad \text{Remove parentheses.}$$

$$y = -\frac{5}{3}x + 4 \qquad \text{Subtract 1 from both sides.}$$

An equation of the line is $y = -\frac{5}{3}x + 4$.

Self Check 1   Write an equation of the line with slope $-\frac{2}{3}$ and passing through $P(-4, 5)$.

Now Try Exercise 11.

EXAMPLE 2   **Writing an Equation of the Line that Passes through Two Given Points**

Find an equation of the line passing through $P(3, 7)$ and $Q(-5, 3)$.

SOLUTION   We will find the slope of the line and then choose either point P or point Q and substitute both the slope and coordinates of the point into the point-slope form. First we find the slope of the line.

$$m = \frac{y_2 - y_1}{x_2 - x_1} \qquad \text{This is the slope formula.}$$

$$= \frac{3 - 7}{-5 - 3} \qquad \text{Substitute 3 for } y_2, 7 \text{ for } y_1, -5 \text{ for } x_2, \text{ and 3 for } x_1.$$

$$= \frac{-4}{-8}$$

$$= \frac{1}{2}$$

We can choose either point P or point Q and substitute its coordinates into the point-slope form. If we choose $P(3, 7)$, we substitute $\frac{1}{2}$ for m, 3 for x_1, and 7 for y_1.

$$y - y_1 = m(x - x_1) \qquad \text{This is the point-slope form.}$$

$$y - 7 = \frac{1}{2}(x - 3) \qquad \text{Substitute } \frac{1}{2} \text{ for } m, 3 \text{ for } x_1, \text{ and 7 for } y_1.$$

$$y = \frac{1}{2}x - \frac{3}{2} + 7 \qquad \text{Remove parentheses and add 7 to both sides.}$$

$$y = \frac{1}{2}x + \frac{11}{2} \qquad -\frac{3}{2} + 7 = -\frac{3}{2} + \frac{14}{2} = \frac{11}{2}$$

An equation of the line is $y = \frac{1}{2}x + \frac{11}{2}$.

Self Check 2   Find an equation of the line passing through $P(-5, 4)$ and $Q(8, -6)$.

Now Try Exercise 23.

2. Use Slope-Intercept Form to Write an Equation of a Line

Since the y-intercept of the line shown in Figure 2-28 is the point $P(0, b)$, we can write an equation of the line by substituting 0 for x_1 and b for y_1 into the point-slope form and simplifying.

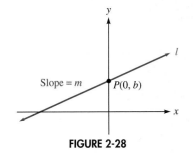

FIGURE 2-28

$$y - y_1 = m(x - x_1) \qquad \text{This is the point-slope form.}$$

$$y - b = m(x - 0) \qquad \text{Substitute 0 for } x_1 \text{ and } b \text{ for } y_1.$$

$$y - b = mx \qquad x - 0 = x$$

$$(2) \qquad y = mx + b \qquad \text{Add } b \text{ to both sides.}$$

Because Equation 2 displays the slope m and the y-coordinate b of the y-intercept, it is called the **slope-intercept form** of the equation of a line.

Slope-Intercept Form of an Equation of a line   An equation of the line with slope m and y-intercept $(0, b)$ is

$$y = mx + b$$

Three examples of linear equations written in slope-intercept form are shown below.

Comment

Note that when a line is written in slope-intercept form the coefficient of x is the slope and the constant term is the y-coordinate of the y-intercept.

Example	Slope	y-Intercept
$y = 2x + 7$	$m = 2$	$b = 7; (0, 7)$
$y = \dfrac{2}{3}x - 5$	$m = \dfrac{2}{3}$	$b = -5; (0, -5)$
$y = -4x + \dfrac{1}{5}$	$m = -4$	$b = \dfrac{1}{5}; \left(0, \dfrac{1}{5}\right)$

EXAMPLE 3   **Using Slope-Intercept Form to Write an Equation of a Line**

Use slope-intercept form to write an equation of the line with slope 4 that passes through $P(5, 9)$.

SOLUTION   Since we know that $m = 4$ and that the ordered pair $(5, 9)$ satisfies the equation, we substitute 4 for m, 5 for x, and 9 for y in the equation $y = mx + b$ and solve for b.

$$y = \boldsymbol{mx} + b \qquad \text{This is the slope-intercept form.}$$

$$9 = 4(5) + b \qquad \text{Substitute 4 for } m, 5 \text{ for } x, \text{ and 9 for } y.$$

$$9 = 20 + b \qquad \text{Simplify.}$$

$$-11 = b \qquad \text{Subtract 20 from both sides.}$$

Comment

When we are given a point and slope we can determine an equation of the line by substituting into point-slope form or slope-intercept form.

Because $m = 4$ and $b = -11$, an equation is $y = 4x - 11$. ∎

Self Check 3   Use slope-intercept form to write an equation of the line with slope $\frac{7}{3}$ and passing through $(3, 1)$.

Now Try Exercise 37.

3. Graph Linear Equations Using the Slope and y-Intercept.

It is easy to graph a linear equation when it is written in slope-intercept form. For example, to graph $y = \frac{4}{3}x - 2$ we note that $b = -2$ and that the y-intercept is $(0, b) = (0, -2)$. (See Figure 2-29.)

Because the slope is $\frac{\Delta y}{\Delta x} = \frac{4}{3}$, we can locate another point Q on the line by starting at point P and counting 3 units to the right and 4 units up. The change in x from point P to point Q is $\Delta x = 3$, and the corresponding change in y is $\Delta y = 4$. The line joining points P and Q is the graph of the equation.

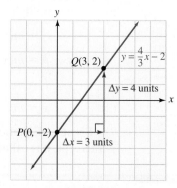

FIGURE 2-29

EXAMPLE 4   **Finding the Slope and y-Intercept of a Line and Graphing the Line**

Find the slope and the y-intercept of the line with equation $3(y + 2) = 6x - 1$ and graph it.

SOLUTION   We will write the equation in the form $y = mx + b$ to find the slope m and the y-intercept $(0, b)$. Then we will use m and b to graph the line.

$$3(y + 2) = 6x - 1$$

$$3y + 6 = 6x - 1 \qquad \text{Remove parentheses.}$$

$$3y = 6x - 7 \qquad \text{Subtract 6 from both sides.}$$

$$y = 2x - \frac{7}{3} \qquad \text{Divide both sides by 3.}$$

The slope of the graph is 2, and the y-intercept is $\left(0, -\frac{7}{3}\right)$. We plot the y-intercept. Then we find a second point on the line by moving 1 unit to the right and 2 units up to the point $\left(1, -\frac{1}{3}\right)$. To get the graph, we draw a line through the two points, as shown in Figure 2-30.

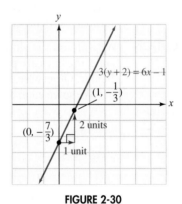

FIGURE 2-30

Self Check 4   Find the slope and the y-intercept of the line with equation $2(x - 3) = -3(y + 5)$. Then graph it.

Now Try Exercise 45.

4. Determine Whether Linear Equations Represent Lines that Are Parallel, Perpendicular, or Neither

EXAMPLE 5   **Determining Whether Lines are Parallel, Perpendicular, or Neither**

Determine whether the lines represented by $4x + 8y = 10$ and $2x = 12 - 4y$ are parallel, perpendicular, or neither.

SOLUTION   We will find the slope of each line and compare their slopes. If their slopes are the same, the lines are parallel. If the product of their slopes is -1, the lines are perpendicular. Otherwise, they are neither parallel nor perpendicular.

We solve each equation for y and write each equation in slope-intercept form.

$$4x + 8y = 10 \qquad\qquad\qquad 2x = 12 - 4y$$

$$8y = -4x + 10 \qquad\qquad\qquad 4y = -2x + 12$$

$$y = \frac{-4x}{8} + \frac{10}{8} \qquad\qquad\qquad y = \frac{-2x}{4} + \frac{12}{4}$$

$$y = -\frac{1}{2}x + \frac{5}{4} \qquad\qquad\qquad y = -\frac{1}{2}x + 3$$

Since the values of b are different, the lines are distinct. Since each slope is $-\frac{1}{2}$, the lines are parallel.

Self Check 5   Are the lines represented by $y = 3x + 2$ and $6x - 2y = 5$ parallel, perpendicular, or neither?

Now Try Exercise 57.

EXAMPLE 6   **Determining Whether Lines are Parallel, Perpendicular, or Neither**

Determine whether the lines represented by $4x + 8y = 10$ and $4x - 2y = 21$ are parallel, perpendicular, or neither.

SOLUTION   We will find the slope of each line and compare their slopes. If their slopes are the same, the lines are parallel. If the product of their slopes is -1, the lines are perpendicular. Otherwise, they are neither parallel nor perpendicular.

We solve each equation for y and write each equation in slope-intercept form.

$$4x + 8y = 10 \qquad\qquad\qquad 4x - 2y = 21$$

$$8y = -4x + 10 \qquad\qquad\qquad -2y = -4x + 21$$

$$y = \frac{-4x}{8} + \frac{10}{8} \qquad\qquad\qquad y = \frac{-4x}{-2} + \frac{21}{-2}$$

$$y = -\frac{1}{2}x + \frac{5}{4} \qquad\qquad\qquad y = 2x - \frac{21}{2}$$

Since the product of their slopes $\left(-\frac{1}{2} \text{ and } 2\right)$ is -1, the lines are perpendicular.

Self Check 6   Are the lines represented by $3x + 2y = 7$ and $y = \frac{2}{3}x + 3$ parallel, perpendicular, or neither?

Now Try Exercise 59.

---

5. Write Equations of Parallel and Perpendicular Lines

EXAMPLE 7   **Writing an Equation of a Parallel Line**

Write an equation of the line passing through $P(-2, 5)$ and parallel to the line $y = 8x - 3$.

SOLUTION   We will substitute the coordinates of $P(-2, 5)$ and the slope of the line parallel to $y = 8x - 3$ into point-slope form and simplify the results to write an equation of the parallel line.

The slope of the line given by $y = 8x - 3$ is 8, the coefficient of x. Since the graph of the desired equation is to be parallel to the graph of $y = 8x - 3$, its slope must also be 8.

We will substitute -2 for x_1, 5 for y_1, and 8 for m into the point-slope form and simplify.

$$y - y_1 = m(x - x_1)$$

$$y - 5 = 8[x - (-2)] \qquad \text{Substitute 5 for } y_1, \text{ 8 for } m, \text{ and } -2 \text{ for } x_1.$$

$$y - 5 = 8(x + 2) \qquad -(-2) = 2$$

$$y - 5 = 8x + 16 \qquad \text{Use the Distributive Property to remove parentheses.}$$

$$y = 8x + 21 \qquad \text{Add 5 to both sides.}$$

An equation of the desired line is $y = 8x + 21$. ■

Self Check 7　Write an equation of the line passing through $Q(1, 2)$ and parallel to the line $y = 8x - 3$.

Now Try Exercise 67.

EXAMPLE 8　**Writing an Equation of a Perpendicular Line**

Write an equation of the line passing through $P(-2, 5)$ and perpendicular to the line $y = 8x - 3$.

SOLUTION　We will substitute the coordinates of $P(-2, 5)$ and the slope of the line perpendicular to $y = 8x - 3$ into point-slope form and simplify to write an equation of the parallel line.

Because the slope of the given line is 8, the slope of the desired perpendicular line must be $-\frac{1}{8}$.

We substitute -2 for x_1, 5 for y_1, and $-\frac{1}{8}$ for m into the point-slope form and simplify.

$$y - y_1 = m(x - x_1)$$

$$y - 5 = -\frac{1}{8}[x - (-2)] \qquad \text{Substitute 5 for } y_1, -\frac{1}{8} \text{ for } m, \text{ and } -2 \text{ for } x_1.$$

$$y - 5 = -\frac{1}{8}(x + 2) \qquad -(-2) = 2$$

$$y = -\frac{1}{8}x - \frac{1}{4} + 5 \qquad \text{Remove parentheses and add 5 to both sides.}$$

$$y = -\frac{1}{8}x + \frac{19}{4} \qquad -\frac{1}{4} + 5 = -\frac{1}{4} + \frac{20}{4} = \frac{19}{4}$$

An equation of the line is $y = -\frac{1}{8}x + \frac{19}{4}$. ■

Self Check 8　Write an equation of the line passing through $Q(1, 2)$ and perpendicular to $y = 8x - 3$.

Now Try Exercise 73.

6. Recognize and Use the Standard Form of the Equation of a Line

We have shown that the graph of any equation of the form $y = mx + b$ is a line with slope m and y-intercept $(0, b)$. In Section 2.1, we saw that the graph of any equation of the form $Ax + By = C$ (where A and B are not *both* zero) is also a line. We consider three possibilities.

- If $A \neq 0$ and $B \neq 0$, the equation $Ax + By = C$ can be written in slope-intercept form.

$$Ax + By = C$$

$$By = -Ax + C \qquad \text{Subtract } Ax \text{ from both sides.}$$

$$y = -\frac{A}{B}x + \frac{C}{B} \qquad \text{Divide both sides by } B.$$

This is an equation of a line with slope $-\frac{A}{B}$ and y-intercept $\left(0, \frac{C}{B}\right)$.

- If $A = 0$ and $B \neq 0$, the equation $Ax + By = C$ can be written in the form $y = \frac{C}{B}$. This is an equation of a horizontal line with y-intercept $\left(0, \frac{C}{B}\right)$.

- If $A \neq 0$ and $B = 0$, the equation $Ax + By = C$ can be written in the form $x = \frac{C}{A}$. This is an equation of a vertical line with x-intercept at $\left(\frac{C}{A}, 0\right)$.

Recall that $Ax + By = C$ is called the **standard form of the equation of a line.**

Standard Form of an Equation on a Line

The **standard form of an equation of a line** is

$$Ax + By = C$$

where A, B, and C are real numbers, and both A and B are not zero.

- If $B \neq 0$, the slope of the nonvertical line is $-\frac{A}{B}$ and the y-intercept is $\left(0, \frac{C}{B}\right)$.
- If $B = 0$ the graph is a vertical line with x-intercept of $\left(\frac{C}{A}, 0\right)$.

Comment When writing equations in $Ax + By = C$ form, we usually clear the equation of fractions and make A positive. For example, the equation $-x + \frac{5}{2}y = 2$ can be changed to $2x - 5y = -4$ by multiplying both sides by -2. We will also divide out any common integer factors of A, B, and C. For example, we would write $4x + 8y = 12$ as $x + 2y = 3$.

Comment Sometimes linear equations are written with a constant term of 0 on the right side of the equation. This form is referred to as *general form*. The line $2x - 5y + 3 = 0$ is written in general form.

EXAMPLE 9 **Using Formulas to Find the Slope and y-Intercept**

Find the slope and the y-intercept of the graph of $3x - 2y = 5$.

SOLUTION The equation $3x - 2y = 5$ is in standard form, with $A = 3$, $B = -2$, and $C = 5$. We will use $-\frac{A}{B}$ to determine the slope of the line and $\left(0, \frac{C}{B}\right)$ to determine the y-intercept.

The slope of the graph is

$$m = -\frac{A}{B} = -\frac{3}{-2} = \frac{3}{2}$$

and the y-intercept is

$$\left(0, \frac{C}{B}\right) = \left(0, \frac{5}{-2}\right)$$

The slope is $\frac{3}{2}$ and the y-intercept is $\left(0, -\frac{5}{2}\right)$.

Self Check 9   Find the slope and the y-intercept of the graph of $3x - 4y = 12$.

Now Try Exercise 77.

We summarize the various forms of an equation of a line as follows.

Forms of an Equation of a Line	
Standard form	$Ax + By = C$ A and B cannot both be 0.
Slope-intercept form	$y = mx + b$ The slope is m, and the y-intercept is $(0, b)$.
Point-slope form	$y - y_1 = m(x - x_1)$ The slope is m, and the line passes through (x_1, y_1).
A horizontal line	$y = b$ The slope is 0, and the y-intercept is $(0, b)$.
A vertical line	$x = a$ There is no defined slope, and the x-intercept is $(a, 0)$.

7. Write an Equation of the Line that Models a Real-Life Problem

For tax purposes, many businesses use *straight-line depreciation* to find the declining value of aging equipment.

EXAMPLE 10   **Solving an Application Problem**

mpanch/Shutterstock.com

A business purchases a digital multimedia projector for $1,970 and expects it to last for ten years. It can then be sold as scrap for a *salvage value* of $270.

If y is the value of the projector after x years of use, and y and x are related by the equation of a line,

a. Find an equation of the line.

b. Find the value of the projector after $2\frac{1}{2}$ years.

c. Find the economic meaning of the y-intercept of the line.

d. Find the economic meaning of the slope of the line.

SOLUTION   a. We will find the slope and use the point-slope form to write an equation of the line. (See Figure 2-31.)

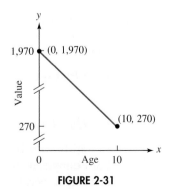

FIGURE 2-31

When the projector is new, its age x is 0, and its value y is \$1,970. When the projector is 10 years old, $x = 10$ and $y = \$270$. Since the line passes through the points (0, 1,970) and (10, 270) the slope of the line is

$$m = \frac{y_2 - y_1}{x_2 - x_1} \qquad \text{This is the slope formula.}$$

$$= \frac{270 - 1{,}970}{10 - 0} \qquad \text{Substitute 270 for } y_2, 1{,}970 \text{ for } y_1, 10 \text{ for } x_2, \text{ and 0 for } x_1.$$

$$= \frac{-1{,}700}{10}$$

$$= -170$$

To find an equation of the line, we substitute -170 for m, 0 for x_1, and 1,970 for y_1 in the point-slope form and simplify.

$$y - y_1 = m(x - x_1)$$

$$y - 1{,}970 = -170(x - 0)$$

(3) $$y = -170x + 1{,}970$$

The value y of the projector is related to its age x by the equation $y = -170x + 1{,}970$.

b. To find the value after $2\frac{1}{2}$ years, we will substitute 2.5 for x in Equation 3 and solve for y.

$$y = -170x + 1{,}970$$

$$= -170(\mathbf{2.5}) + 1{,}970 \qquad \text{Substitute 2.5 for } x.$$

$$= -425 + 1{,}970$$

$$= 1{,}545$$

In $2\frac{1}{2}$ years, the projector will be worth \$1,545.

c. The y-intercept of the graph is $(0, b)$, where b is the value of y when $x = 0$.

$$y = -170x + 1{,}970$$

$$y = -170(\mathbf{0}) + 1{,}970 \qquad \text{Substitute 0 for } x.$$

$$y = 1{,}970$$

The y-coordinate b of the y-intercept is the value of a 0-year-old projector, which is the projector's original cost, \$1,970.

d. Each year, the value decreases by \$170, because the slope of the line is -170. The slope of the depreciation line is called the *annual depreciation rate*. ∎

Problems that have an annual appreciation rate can be worked similarly.

Self Check 10   A business purchases a Canon copier for \$2,700 and expects it to last for ten years. It can then be sold for \$300. Write a straight-line depreciation line for the copier.

Now Try Exercise 91.

8. Use Linear Curve Fitting to Solve Problems

In statistics, the process of using one variable to predict another is called **regression.** For example, if we know a woman's height, we can make a good prediction about her weight, because taller women usually weigh more than shorter women.

Figure 2-32 shows the result of sampling ten women and finding their heights and weights. The graph of the ordered pairs (h, w) is called a **scattergram.**

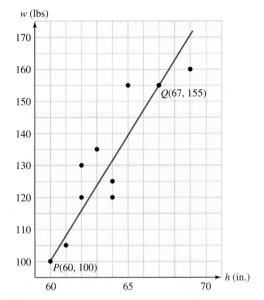

Woman	Height (h) in inches	Weight (w) in pounds
1	60	100
2	61	105
3	62	120
4	62	130
5	63	135
6	64	120
7	64	125
8	65	155
9	67	155
10	69	160

FIGURE 2-32

To write a **prediction equation** (sometimes called a **regression equation**), we must find an equation of the line that comes closer to all of the points in the scattergram than any other possible line. There are statistical methods to find this equation, but we can only approximate it here.

To write an approximation of the regression equation, we place a straightedge on the scattergram shown in Figure 2-32 and draw the line joining two points that seems to best fit all the points. In the figure, line PQ is drawn, where point P has coordinates of $(60, 100)$ and point Q has coordinates of $(67, 155)$.

Our approximation of the regression equation will be an equation of the line passing through points P and Q. To find an equation of this line, we first find its slope.

$$m = \frac{y_2 - y_1}{x_2 - x_1}$$ This is the slope formula.

$$= \frac{155 - 100}{67 - 60}$$ Substitute 155 for y_2, 100 for y_1, 67 for x_2, and 60 for x_1.

$$= \frac{55}{7}$$

We can then use point-slope form to find an equation of the line.

$$y - y_1 = m(x - x_1)$$ This is the point-slope form.

$$y - 100 = \frac{55}{7}(x - 60)$$ Choose $(60, 100)$ for (x_1, y_1).

$$y = \frac{55}{7}x - \frac{3,300}{7} + 100$$ Remove parentheses and add 100 to both sides.

(4) $$y = \frac{55}{7}x - \frac{2,600}{7}$$ Simplify.

Our approximation of the regression equation is $y = \frac{55}{7}x - \frac{2,600}{7}$.

To predict the weight of a woman who is 66 inches tall, for example, we substitute 66 for x in Equation 4 and simplify.

$$y = \frac{55}{7}x - \frac{2,600}{7}$$

$$y = \frac{55}{7}(66) - \frac{2,600}{7}$$

$$y \approx 147.1428571$$

We would predict that a 66-inch-tall woman chosen at random will weigh about 147 pounds.

ACCENT ON TECHNOLOGY   Linear Regression

We can use the linear regression feature on a graphing calculator to find an equation of the line that best fits a given set of data points. We will do that for the data given in Figure 2-32.

• First, we press $\boxed{\text{STAT}}$ to enter the statistics menu on the calculator. This screen is shown in Figure 2-33(a). Next we press $\boxed{\text{ENTER}}$ to input our data. We can input our heights into the L1 column and our weights into the L2 column. This is shown in Figure 2-33(b).

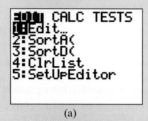

(a)   (b)

FIGURE 2-33

• To obtain an equation of the regression line, we press $\boxed{\text{STAT}}$ and then the right-arrow key once to access the calculate menu. This screen is shown in Figure 2-34(a). To calculate a linear regression equation, we press $\boxed{4}$ to select LinReg $(ax + b)$ and then $\boxed{\text{ENTER}}$ to obtain an equation. The screen is shown in Figure 2-34(b).

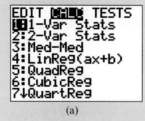   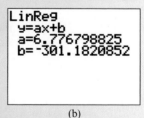

(a)   (b)

FIGURE 2-34

Note that the regression line is of the form $y = ax + b$, where a is the slope and b is the y-coordinate of the y-intercept. If we substitute the values shown in Figure 2-34(b) for a and b and round to hundredths, we can write an equation of the line that best fits the data. The regression line is $y = 6.78x - 301.18$.

Self Check Answers   **1.** $y = -\dfrac{2}{3}x + \dfrac{7}{3}$ **2.** $y = -\dfrac{10}{13}x + \dfrac{2}{13}$ **3.** $y = \dfrac{7}{3}x - 6$ **4.** $-\dfrac{2}{3}, (0, -3)$

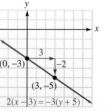

5. parallel   **6.** perpendicular   **7.** $y = 8x - 6$ **8.** $y = -\dfrac{1}{8}x + \dfrac{17}{8}$

9. $\dfrac{3}{4}$, $(0, -3)$ **10.** $y = -240x + 2{,}700$

Exercises **2.3**

Getting Ready

You should be able to complete these vocabulary and concept statements before you proceed to the practice exercises.

Fill in the blanks.

1. The formula for the point-slope form of a line is _____.

2. In the equation $y = mx + b$, ___ is the slope of the graph of the line, and $(0, b)$ is the _____.

3. The equation $y = mx + b$ is called the _____ form of the equation of a line.

4. The standard form of an equation of a line is _____.

5. The slope of the graph of $Ax + By = C$ is ____.

6. The y-intercept of the graph of $Ax + By = C$ is ____.

Practice

Write an equation of the line with the given properties. Your answer should be written in standard form.

7. $m = 2$ passing through $P(2, 4)$

8. $m = -3$ passing through $P(3, 5)$

9. $m = 2$ passing through $P\left(-\frac{3}{2}, \frac{1}{2}\right)$

10. $m = -6$ passing through $P\left(\frac{1}{4}, -2\right)$

11. $m = \frac{2}{5}$ passing through $P(-1, 1)$

12. $m = -\frac{1}{5}$ passing through $P(-2, -3)$

13. $m = 0$ passing through $P(-6, -3)$

14. $m = 0$ passing through $P(7, 5)$

15. m is undefined passing through $P(-6, -3)$

16. m is undefined passing through $P(6, -1)$

17. $m = \pi$ passing through $P(\pi, 0)$

18. $m = \pi$ passing through $P(0, \pi)$

Find an equation of each line shown. Your answer should be written in standard form.

19.

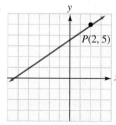

20.

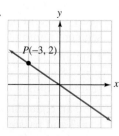

Write an equation of a line that passes through the two given points. Your answer should be written in slope-intercept form.

21. $P(0, 0)$, $Q(4, 4)$ **22.** $P(-5, -5)$, $Q(0, 0)$

23. $P(3, 4)$, $Q(0, -3)$ **24.** $P(4, 0)$, $Q(6, -8)$

Write an equation in slope-intercept form of each line shown.

25.

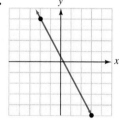

26.

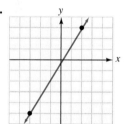

Use slope-intercept form to write an equation of the line with the given properties.

27. $m = 3$; $b = -2$ **28.** $m = -\dfrac{1}{3}$; $b = \dfrac{2}{3}$

29. $m = 5$; $b = -\dfrac{1}{5}$ **30.** $m = \sqrt{2}$; $b = \sqrt{2}$

31. $m = a$; $b = \dfrac{1}{a}$ **32.** $m = a$; $b = 2a$

33. $m = a$; $b = a$ **34.** $m = \dfrac{1}{a}$; $b = a$

Use slope-intercept form to write an equation of a line passing through the given point and having the given slope. Express the answer in standard form.

35. $P(0, 0)$; $m = \dfrac{3}{2}$ **36.** $P(-3, -7)$; $m = -\dfrac{2}{3}$

37. $P(-3, 5)$; $m = -3$ **38.** $P(-5, 1)$; $m = 1$

39. $P\left(0, \sqrt{2}\right)$; $m = \sqrt{2}$

40. $P\left(-\sqrt{3}, 0\right)$; $m = 2\sqrt{3}$

Write each equation in slope-intercept form to determine the slope and y-intercept. Then use the slope and y-intercept to graph the line.

41. $x - y = 1$

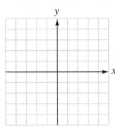

42. $x + y = 2$

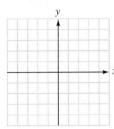

43. $x = \dfrac{3}{2}y - 3$

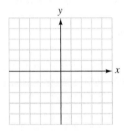

44. $x = -\dfrac{4}{5}y + 2$

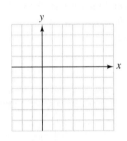

45. $3(y - 4) = -2(x - 3)$

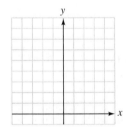

46. $-4(2x + 3) = 3(3y + 8)$

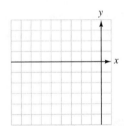

Find the slope and the y-intercept of the lines determined by the given equations.

47. $3x - 2y = 8$

48. $-2x + 4y = 12$

49. $-2(x + 3y) = 5$

50. $5(2x - 3y) = 4$

51. $x = \dfrac{2y - 4}{7}$

52. $3x + 4 = -\dfrac{2(y - 3)}{5}$

Determine whether the graphs of each pair of equations are parallel, perpendicular, or neither.

53. $y = 3x + 4, y = 3x - 7$

54. $y = 4x - 13, y = \dfrac{1}{4}x + 13$

55. $x + y = 2, y = x + 5$

56. $x = y + 2, y = x + 3$

57. $y = 3x + 7, 2y = 6x - 9$

58. $2x + 3y = 9, 3x - 2y = 5$

59. $x = 3y + 4, y = -3x + 7$

60. $3x + 6y = 1, y = \dfrac{1}{2}x$

61. $y = 3, x = 4$

62. $y = -3, y = -7$

63. $x = \dfrac{y - 2}{3}, 3(y - 3) + x = 0$

64. $2y = 8, 3(2 + x) = 3(y + 2)$

Write an equation of the line that passes through the given point and is parallel to the given line. Your answer should be written in slope-intercept form.

65. $P(0, 0), y = 4x - 7$

66. $P(0, 0), x = -3y - 12$

67. $P(2, 5), 4x - y = 7$

68. $P(-6, 3), y + 3x = -12$

69. $P(4, -2), x = \dfrac{5}{4}y - 2$

70. $P(1, -5), x = -\dfrac{3}{4}y + 5$

Write an equation of the line that passes through the given point and is perpendicular to the given line. Your answer should be written in slope-intercept form.

71. $P(0, 0), y = 4x - 7$

72. $P(0, 0), x = -3y - 12$

73. $P(2, 5), 4x - y = 7$

74. $P(-6, 3), y + 3x = -12$

75. $P(4, -2), x = \dfrac{5}{4}y - 2$

76. $P(1, -5), x = -\dfrac{3}{4}y + 5$

Use the method of Example 9 to find the slope and the y-intercept of the graph of each equation.

77. $4x + 5y = 20$ **78.** $9x - 12y = 17$

79. $2x + 3y = 12$ **80.** $5x + 6y = 30$

81. Find an equation of the line perpendicular to the line $y = 3$ and passing through the midpoint of the segment joining $(2, 4)$ and $(-6, 10)$.

82. Find an equation of the line parallel to the line $y = -8$ and passing through the midpoint of the segment joining $(-4, 2)$ and $(-2, 8)$.

83. Find an equation of the line parallel to the line $x = 3$ and passing through the midpoint of the segment joining $(2, -4)$ and $(8, 12)$.

84. Find an equation of the line perpendicular to the line $x = 3$ and passing through the midpoint of the segment joining $(-2, 2)$ and $(4, -8)$.

Applications

In Exercises 85–95, assume straight-line depreciation or straight-line appreciation.

85. Depreciation A Toyota Tundra truck was purchased for $24,300. Its salvage value at the end of its 7-year useful life is expected to be $1,900. Find a depreciation equation.

86. Depreciation A small business purchases the laptop computer shown. It will be depreciated over a 4-year period, when its salvage value will be $300. Find a depreciation equation.

$2,700

87. Appreciation A condominium in San Diego was purchased for $475,000. The owners expect the condominium to double in value in 10 years. Find an appreciation equation.

88. Appreciation A house purchased for $112,000 is expected to double in value in 12 years. Find an appreciation equation.

89. Depreciation Find a depreciation equation for the TV in the following want ad.

> *For Sale*: 3-year-old 54-inch TV, $1,900 new. Asking $1,190. Call 875-5555. Ask for Mike.

90. Depreciation A Bose Wave Radio cost $555 when new and is expected to be worth $80 after 5 years. What will it be worth after 3 years?

91. Salvage value A copier cost $1,050 when new and will be depreciated at the rate of $120 per year. If the useful life of the copier is 8 years, find its salvage value.

92. Rate of depreciation A ski boat that cost $27,600 when new will have no salvage value after 12 years. Find its annual rate of depreciation.

93. Value of an antique An antique table is expected to appreciate $40 each year. If the table will be worth $450 in 2 years, what will it be worth in 13 years?

94. Value of an antique An antique clock is expected to be worth $350 after 2 years and $530 after 5 years. What will the clock be worth after 7 years?

95. Purchase price of real estate A cottage that was purchased 3 years ago is now appraised at $47,700. If the property has been appreciating $3,500 per year, find its original purchase price.

96. Computer repair A computer repair company charges a fixed amount, plus an hourly rate, for a service call. Use the information in the illustration to find the hourly rate

AAA Computer Repair	
Typical Charges	
2 hours	$ 70
4 hours	$105

97. Automobile repair An auto repair shop charges an hourly rate, plus the cost of parts. If the cost of labor for a $1\frac{1}{2}$-hour radiator repair is $69, find the cost of labor for a 5-hour transmission overhaul.

98. Printer charges A printer charges a fixed setup cost, plus $1 for every 100 copies. If 700 copies cost $52, how much will it cost to print 1,000 copies?

99. Predicting fires A local fire department recognizes that city growth and the number of reported fires are related by a linear equation. City records show that 300 fires were reported in a year when the local population was 57,000 persons, and 325 fires were reported in a year when the population was 59,000 persons. How many fires can be expected in the year when the population reaches 100,000 persons?

100. Estimating the cost of rain gutter A neighbor tells you that an installer of rain gutter charges $60, plus a dollar amount per foot. If the neighbor paid $435 for the installation of 250 feet of gutter, how much will it cost you to have 300 feet installed?

We can use the Distance Formula to find the length of CP, which is r:

$$r = \sqrt{(x - h)^2 + (y - k)^2}$$

After squaring both sides, we get

$$r^2 = (x - h)^2 + (y - k)^2$$

This equation is called the **standard form of an equation of a circle.**

The Standard Form of an Equation of a Circle with Center at (h, k) and Radius r

The graph of any equation that can be written in the standard form

$$(x - h)^2 + (y - k)^2 = r^2$$

is a circle with radius r and center at point (h, k).

If $r = 0$, the circle is a single point called a **point circle.** If the center of a circle is the origin, then $(h, k) = (0, 0)$ and we have the following result.

The Standard Form of an Equation of a Circle with Center at $(0, 0)$ and Radius r

The graph of any equation that can be written in the standard form

$$x^2 + y^2 = r^2$$

is a circle with radius r and center at the origin.

The equations of four circles written in standard form and their graphs are shown in Figure 2-48.

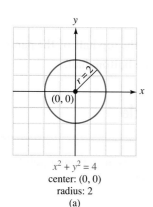

$x^2 + y^2 = 4$
center: $(0, 0)$
radius: 2
(a)

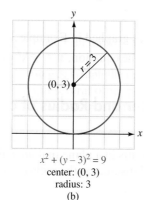

$x^2 + (y - 3)^2 = 9$
center: $(0, 3)$
radius: 3
(b)

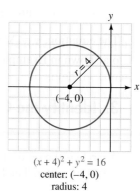

$(x + 4)^2 + y^2 = 16$
center: $(-4, 0)$
radius: 4
(c)

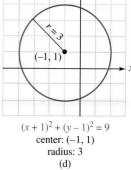

$(x + 1)^2 + (y - 1)^2 = 9$
center: $(-1, 1)$
radius: 3
(d)

FIGURE 2-48

If we are given an equation of a circle in standard form, we can easily determine the coordinates of its center and the length of its radius.

EXAMPLE 5   Finding the Center and Radius of a Circle in Standard Form

Find the center and radius of the circle with the equation $(x - 3)^2 + (y + 2)^2 = 36$.

SOLUTION   Since the equation is written in standard form, we can identify h, k, and r by comparing the equation to the equation $(x - h)^2 + (y - k)^2 = r^2$.

Standard Form: $(x - h)^2 + (y - k)^2 = r^2$

Given Form: $(x - 3)^2 + [y - (-2)]^2 = 6^2$

We see that $h = 3$, $k = -2$, and $r = 6$. The center (h, k) of the circle is at $(3, -2)$ and the radius is 6.

Self Check 5   Identify the center and radius of the circle with equation of $(x - 4)^2 + y^2 = 49$.

Now Try Exercise 69.

4. Write Equations of Circles

EXAMPLE 6   **Writing an Equation of a Circle Given the Center and Radius**

Write an equation of the circle with center at $(-2, 4)$ and radius 1.

SOLUTION   We will substitute the coordinates of the center and the radius into the standard form $(x - h)^2 + (y - k)^2 = r^2$.

$$(x - h)^2 + (y - k)^2 = r^2$$
$$[x - (-2)]^2 + (y - 4)^2 = 1^2 \qquad \text{Substitute } -2 \text{ for } h, 4 \text{ for } k, \text{ and } 1 \text{ for } r.$$
$$(x + 2)^2 + (y - 4)^2 = 1 \qquad \text{Simplify.}$$

An equation of the circle with center at $(-2, 4)$ and radius 1 is $(x + 2)^2 + (y - 4)^2 = 1$.

Self Check 6   Write an equation in standard form of the circle with center $(2, -4)$ and radius $\sqrt{5}$.

Now Try Exercise 79.

If we square the binomials in the equation of the circle found in Example 6, we get another form of the equation, called the **general equation of a circle.**

$$(x + 2)^2 + (y - 4)^2 = 1 \qquad \text{This is the equation of Example 6.}$$
$$x^2 + 4x + 4 + y^2 - 8y + 16 = 1 \qquad \text{Remove parentheses.}$$
$$x^2 + y^2 + 4x - 8y + 19 = 0 \qquad \text{Subtract 1 from both sides and simplify.}$$

The general form is $x^2 + y^2 + 4x - 8y + 19 = 0$.

Comment

We can easily recognize the equation of a circle. A circle's equation will always contain both x^2 and y^2 terms, and the coefficients of both terms will be equal.

The General Form of an Equation of a Circle

The **general form** of an equation of a circle is

$$x^2 + y^2 + cx + dy + e = 0$$

where c, d, and e are real numbers.

EXAMPLE 7   **Finding the General Form of an Equation of a Circle Given the Center and Radius**

Find the general form of an equation of the circle with radius 5 and center at $(3, 2)$.

SOLUTION   We will substitute 5 for r, 3 for h, and 2 for k in the standard form of the equation of a circle and simplify.

$$(x - h)^2 + (y - k)^2 = r^2 \qquad \text{This is the standard equation.}$$
$$(x - 3)^2 + (y - 2)^2 = 5^2 \qquad \text{Substitute.}$$
$$x^2 - 6x + 9 + y^2 - 4y + 4 = 25 \qquad \text{Remove parentheses.}$$
$$x^2 + y^2 - 6x - 4y - 12 = 0 \qquad \text{Subtract 25 from both sides and simplify.}$$

The general form is $x^2 + y^2 - 6x - 4y - 12 = 0$.

Self Check 7   Find the general form of an equation of a circle with radius 6 and center at $(-2, 5)$.

Now Try Exercise 83.

EXAMPLE 8   **Finding the General Form of an Equation of a Circle Given the Endpoints of Its Diameter**

Find the general form of an equation of the circle with endpoints of its diameter at $(8, -3)$ and $(-4, 13)$.

SOLUTION   We will find the center and the radius of the circle, substitute into the standard equation, and simplify. To determine the center, we will find the midpoint of the diameter. To find the radius, we will find the distance from the center to one end-point of the diameter.

Step 1: Find the center of the circle.   We will find the midpoint of its diameter. Since $(x_1, y_1) = (8, -3)$ and $(x_2, y_2) = (-4, 13)$, we know that $x_1 = 8$, $x_2 = -4$, $y_1 = -3$, and $y_2 = 13$.

$$h = \frac{x_1 + x_2}{2} \qquad\qquad k = \frac{y_1 + y_2}{2} \qquad \text{Use the Midpoint Formula.}$$

$$h = \frac{8 + (-4)}{2} \qquad\qquad k = \frac{-3 + 13}{2}$$

$$= \frac{4}{2} \qquad\qquad\qquad = \frac{10}{2}$$

$$= 2 \qquad\qquad\qquad\quad = 5$$

The center of the circle is at $(h, k) = (2, 5)$.

Step 2: Find the radius of the circle.   To find the radius, we find the distance between the center and one endpoint of the diameter. The center is at $(2, 5)$ and one end-point is $(8, -3)$.

$$r = \sqrt{(x_2 - x_1)^2 + (y_2 - y_1)^2} \qquad \text{Use the Distance Formula.}$$

$$r = \sqrt{(2 - 8)^2 + [5 - (-3)]^2} \qquad \text{Substitute 8 for } x_1, -3 \text{ for } y_1, 2 \text{ for } x_2, \text{ and 5 for } y_2.$$

$$= \sqrt{(-6)^2 + (8)^2}$$

$$= \sqrt{36 + 64}$$

$$= 10 \qquad\qquad\qquad\qquad \sqrt{36 + 64} = \sqrt{100} = 10$$

The radius of the circle is 10 units.

Step 3: Substitute and simplify.   To find an equation of a circle with center at $(2, 5)$ and radius 10, we substitute 2 for h, 5 for k, and 10 for r in the standard equation of the circle and simplify:

$$(x - h)^2 + (y - k)^2 = r^2 \qquad\qquad \text{This is the standard equation.}$$

$$(x - 2)^2 + (y - 5)^2 = 10^2$$

$$x^2 - 4x + 4 + y^2 - 10y + 25 = 100 \qquad \text{Remove parentheses.}$$

$$x^2 + y^2 - 4x - 10y - 71 = 0 \qquad\qquad \text{Subtract 100 from both sides and simplify.}$$

The general form of the equation is $x^2 + y^2 - 4x - 10y - 71 = 0$. ∎

Self Check 8   Find an equation of a circle with endpoints of its diameter at $(-2, 2)$ and $(6, 8)$.

Now Try Exercise 87.

5. Graph Circles

We can convert the general form of the equation of a circle into standard form by completing the square on x and y.

EXAMPLE 9   **Graphing a Circle Whose Equation is in General Form**

Graph the circle whose equation is $2x^2 + 2y^2 - 8x + 4y - 40 = 0$.

SOLUTION   We will convert the general equation of the circle into standard form by completing the square on x and y. We will then use the coordinates of the center of the circle and its radius to draw its graph.

Step 1: Complete the square on x and y.   First, we divide both sides of the equation by 2 to make the coefficients of x^2 and y^2 equal to 1.

$$2x^2 + 2y^2 - 8x + 4y - 40 = 0$$

$$x^2 + y^2 - 4x + 2y - 20 = 0$$

To find the coordinates of the center and the radius, we add 20 to both sides and write the equation in standard form by completing the square on both x and y:

$$x^2 + y^2 - 4x + 2y = 20$$

$$x^2 - 4x + y^2 + 2y = 20$$

$$x^2 - 4x + \mathbf{4} + y^2 + 2y + \mathbf{1} = 20 + \mathbf{4} + \mathbf{1}$$ Add 4 and 1 to both sides to complete the square.

$$(x - 2)^2 + (y + 1)^2 = 25$$ Factor $x^2 - 4x + 4$ and $y^2 + 2y + 1$.

$$(x - 2)^2 + [y - (-1)]^2 = 5^2$$

Step 2: Graph the circle.   From the standard form of the equation of the circle, we see that its radius is 5 and that the coordinates of its center are $h = 2$ and $k = -1$. Thus, the center of the circle is at is $(2, -1)$.

To graph the circle, we plot its center $(2, -1)$ and locate points on the circle that are 5 units from its center. The graph of the circle is shown in Figure 2-49.

Because the radius of the circle is 5, the easiest points to locate on the circle to draw the graph are the points that are 5 units to the right, 5 units to the left, 5 units above, and 5 units below the center.

Those points are $(7, -1)$, $(-3, -1)$, $(2, 4)$, and $(2, -6)$.

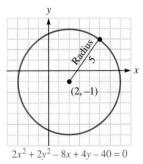

$$2x^2 + 2y^2 - 8x + 4y - 40 = 0$$

FIGURE 2-49

Self Check 9   Graph: $2x^2 + 2y^2 + 4x - 8y + 2 = 0$.

Now Try Exercise 107.

ACCENT ON TECHNOLOGY **Graphing a Circle**

When we graphed $y^2 = 2x$ using a graphing calculator, we saw that it is necessary to solve the equation for y in terms of x. The same is true when we graph a circle on a graphing calculator.

• To graph $(x - 2)^2 + (y + 1)^2 = 25$, we must solve the equation for y:

$$(x - 2)^2 + (y + 1)^2 = 25$$

$$(y + 1)^2 = 25 - (x - 2)^2$$

$$y + 1 = \pm\sqrt{25 - (x - 2)^2}$$

$$y = -1 \pm \sqrt{25 - (x - 2)^2}$$

This last expression represents two equations: $y = -1 + \sqrt{25 - (x - 2)^2}$ and $y = -1 - \sqrt{25 - (x - 2)^2}$.

• We graph both of these equations separately on the same coordinate axes by entering the first equation as Y_1 and the second as Y_2, as shown in Figure 2-50(a). Depending on the window setting of the maximum and minimum values of x and y, the graph may not appear to be a circle. However, if we use the ZOOM 5: ZSquare window, the graph will be circular. See Figures 2-50(b) and (c). If it appears that there are gaps in the graph, that is due to the way the calculator draws graphs by darkening pixels.

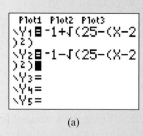

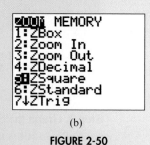

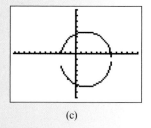

(a)   (b)   (c)

FIGURE 2-50

Note that in this example, it is easier to graph the circle by hand than with a calculator.

6. Solve Equations by Using a Graphing Calculator

We can solve many equations using the graphing concepts discussed in this chapter and a graphing calculator. For example, the solutions of $x^2 - x - 3 = 0$ will be the numbers x that will make $y = 0$ in the equation $y = x^2 - x - 3$. These numbers will be the x-coordinates of the x-intercepts of the graph of $y = x^2 - x - 3$.

ACCENT ON TECHNOLOGY **Solving Equations**

To use a graphing calculator to solve $x^2 - x - 3 = 0$, we graph the equation $y = x^2 - x - 3$, and find the x-coordinates of the x-intercepts. These will be the solutions of the equation. The graph of $y = x^2 - x - 3$ is shown in Figure 2-51(c).

• Use the ZERO command shown earlier in this section to solve the equation. Notice the screen in Figures 2-51(h) and (i) looks different than in Figures 2-51(e) and (f). On some calculators we can just enter the value we want as the left and right bound rather than using the cursor to get to a bound. We see that the two solutions to the equation are $x \approx -1.302776$ and $x \approx 2.3027756$.

- Every equation does not have a solution over the real numbers. For example, to try and solve the equation $x^2 + 2x + 2 = 0$, graph the equation $y = x^2 + 2x + 2$. We can see that it does not intersect the x-axis, and there is no solution to the equation. See Figure 2-51(k).

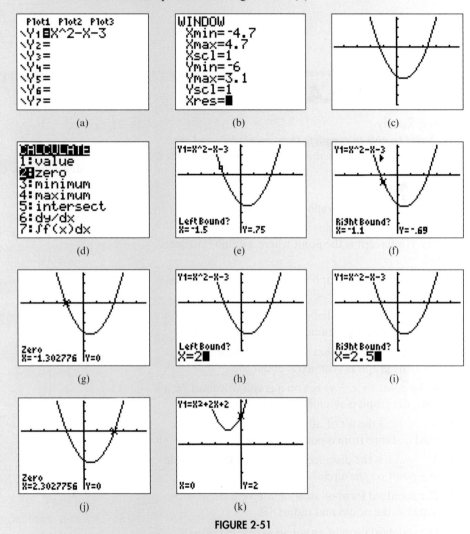

FIGURE 2-51

Self Check Answers

1.
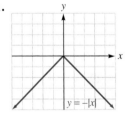
$y = -|x|$

2.
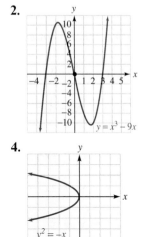
$y = x^3 - 9x$

3.
$y = -\sqrt{x}$

4.
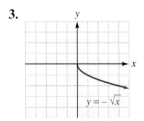
$y^2 = -x$

5. $(4, 0)$, 7   **6.** $(x - 2)^2 + (y + 4)^2 = 5$ **7.** $x^2 + y^2 + 4x - 10y - 7 = 0$

8. $x^2 + y^2 - 4x - 10y + 4 = 0$ **9.**

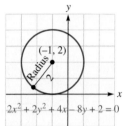

Exercises **2.4**

Getting Ready

You should be able to complete these vocabulary and concept statements before you proceed to the practice exercises.

Fill in the blanks.

1. The point where a graph intersects the x-axis is called the _____.

2. The y-intercept is the point where a graph intersects the _____.

3. If a line divides a graph into two congruent halves, we call the line an _____.

4. If the point $(-x, y)$ lies on a graph whenever (x, y) does, the graph is symmetric about the _____.

5. If the point $(x, -y)$ lies on a graph whenever (x, y) does, the graph is symmetric about the _____.

6. If the point $(-x, -y)$ lies on a graph whenever (x, y) does, the graph is symmetric about the _____.

7. A _____ is the set of all points in a plane that are fixed distance from a point called its _____.

8. A _____ is the distance from the center of a circle to a point on the circle.

9. The standard form of an equation of a circle with center at the origin and radius r is _____.

10. The standard form of an equation of a circle with center at (h, k) and radius r is _____.

Practice

Find the x- and y-intercepts of each graph. Do not graph the equation.

11. $y = x^2 - 4$ **12.** $y = x^2 - 9$

13. $y = 4x^2 - 2x$ **14.** $y = 2x - 4x^2$

15. $y = x^2 - 4x - 5$ **16.** $y = x^2 - 10x + 21$

17. $y = x^2 + x - 2$ **18.** $y = x^2 + 2x - 3$

19. $y = x^3 - 9x$ **20.** $y = x^3 + x$

21. $y = x^4 - 1$ **22.** $y = x^4 - 25x^2$

Graph each equation. Check your graph with a graphing calculator.

23. $y = x^2$

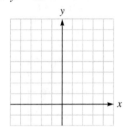

24. $y = -x^2$

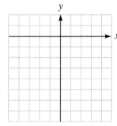

25. $y = -x^2 + 2$

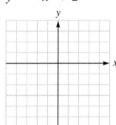

26. $y = x^2 - 1$

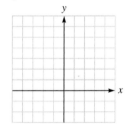

27. $y = x^2 - 4x$

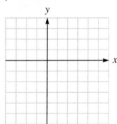

28. $y = x^2 + 2x$

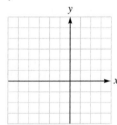

29. $y = \frac{1}{2}x^2 - 2x$

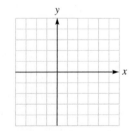

30. $y = \frac{1}{2}x^2 + 3$

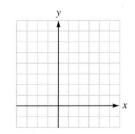

Find the symmetries, if any, of the graph of each equation. Do not graph the equation.

31. $y = x^2 + 2$

32. $y = 3x + 2$

33. $y^2 + 1 = x$

34. $y^2 + y = x$

35. $y^2 = x^2$

36. $y = 3x + 7$

37. $y = 3x^2 + 7$

38. $x^2 + y^2 = 1$

39. $y = 3x^3 + 7$

40. $y = 3x^3 + 7x$

41. $y^2 = 3x$

42. $y = 3x^4 + 7$

43. $y = |x|$

44. $y = |x + 1|$

45. $|y| = x$

46. $|y| = |x|$

Graph each equation. Be sure to find any intercepts and symmetries. Check your graph with a graphing calculator.

47. $y = x^2 + 4x$

48. $y = x^2 - 6x$

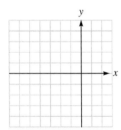
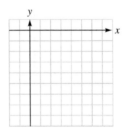

49. $y = x^3$

50. $y = x^3 + x$

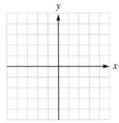
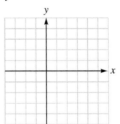

51. $y = |x - 2|$

52. $y = |x| - 2$

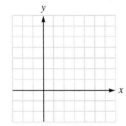
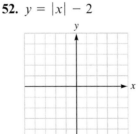

53. $y = 3 - |x|$

54. $y = 3|x|$

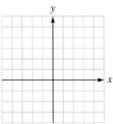
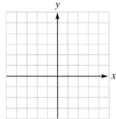

55. $y^2 = -x$

56. $y^2 = 4x$

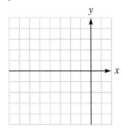
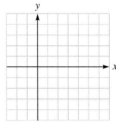

57. $y^2 = 9x$

58. $y^2 = -4x$

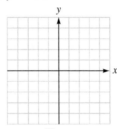
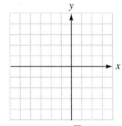

59. $y = \sqrt{x} - 1$

60. $y = 1 - \sqrt{x}$

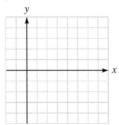
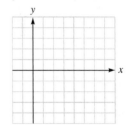

61. $xy = 4$

62. $xy = -9$

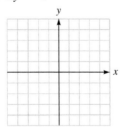
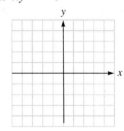

Identify the center and radius of each circle written in standard form.

63. $x^2 + y^2 = 100$

64. $x^2 + y^2 = 81$

65. $x^2 + (y - 5)^2 = 49$

66. $x^2 + (y + 3)^2 = 8$

67. $(x + 6)^2 + y^2 = \dfrac{1}{4}$

68. $(x - 5)^2 + y^2 = \dfrac{16}{25}$

69. $(x - 4)^2 + (y - 1)^2 = 9$

70. $(x + 11)^2 + (x + 7)^2 = 121$

71. $\left(x - \dfrac{1}{4}\right)^2 + (y + 2)^2 = 45$

72. $\left(x + \sqrt{5}\right)^2 + (y - 3)^2 = 1$

Write an equation in standard form of the circle with the given properties.

73. Center at the origin; $r = 5$

74. Center at the origin; $r = \sqrt{3}$

75. Center at $(0, -6)$; $r = 6$

76. Center at $(0, 7)$; $r = 9$

77. Center at $(8, 0)$; $r = \dfrac{1}{5}$

78. Center at $(-10, 0)$; $r = \sqrt{11}$

79. Center at $(-2, 12)$, $r = 13$

80. Center at $\left(\dfrac{2}{7}, -5\right)$; $r = 7$

Write an equation in general form of the circle with the given properties.

81. Center at the origin; $r = 1$

82. Center at the origin; $r = 4$

83. Center at $(6, 8)$; $r = 4$

84. Center at $(5, 3)$; $r = 2$

85. Center at $(3, -4)$; $r = \sqrt{2}$

86. Center at $(-9, 8)$; $r = 2\sqrt{3}$

87. Ends of diameter at $(3, -2)$ and $(3, 8)$

88. Ends of diameter at $(5, 9)$ and $(-5, -9)$

89. Center at $(-3, 4)$ and passing through the origin

90. Center at $(-2, 6)$ and passing through the origin

Convert the general form of each circle given into standard form.

91. $x^2 + y^2 - 6x + 4y + 4 = 0$

92. $x^2 + y^2 + 4x - 8y - 5 = 0$

93. $x^2 + y^2 - 10x - 12y + 57 = 0$

94. $x^2 + y^2 + 2x + 18y + 57 = 0$

95. $2x^2 + 2y^2 - 8x - 16y + 22 = 0$

96. $3x^2 + 3y^2 + 6x - 30y + 3 = 0$

Graph each circle.

97. $x^2 + y^2 - 25 = 0$

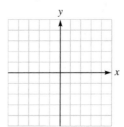

98. $x^2 + y^2 - 8 = 0$

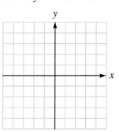

99. $(x - 1)^2 + (y + 2)^2 = 4$

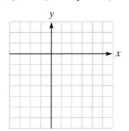

100. $(x + 1)^2 + (y - 2)^2 = 9$

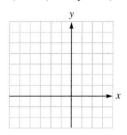

101. $x^2 + y^2 + 2x - 24 = 0$

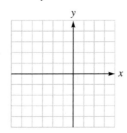

102. $x^2 + y^2 - 4y = 12$

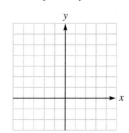

103. $x^2 + y^2 + 4x + 2y - 11 = 0$

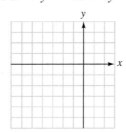

104. $x^2 + y^2 - 6x + 2y + 1 = 0$

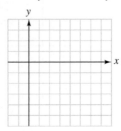

105. $9x^2 + 9y^2 - 12y = 5$ **106.** $4x^2 + 4y^2 + 4y = 15$

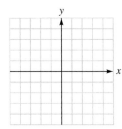   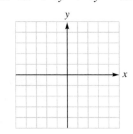

107. $4x^2 + 4y^2 - 4x + 8y + 1 = 0$

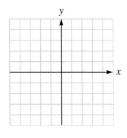

108. $9x^2 + 9y^2 - 6x + 18y + 1 = 0$

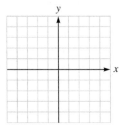

 Use a graphing calculator to graph each equation. Then find the coordinates of the vertex of the parabola to the nearest hundredth.

109. $y = 2x^2 - x + 1$ **110.** $y = x^2 + 5x - 6$

111. $y = 7 + x - x^2$ **112.** $y = 2x^2 - 3x + 2$

 Use a graphing calculator to solve each equation. Round to the nearest hundredth.

113. $x^2 - 7 = 0$ **114.** $x^2 - 3x + 2 = 0$

115. $x^3 - 3 = 0$ **116.** $3x^3 - x^2 - x = 0$

Applications

117. Golfing Phil Mickelson's tee shot follows a path given by $y = 64t - 16t^2$, where y is the height of the ball (in feet) after t seconds of flight. How long will it take for the ball to strike the ground?

118. Golfing Halfway through its flight, the golf ball of Exercise 117 reaches the highest point of its trajectory. How high is that?

 119. Stopping distances The stopping distance D (in feet) for a Ford Fusion car moving V miles per hour is given by $D = 0.08V^2 + 0.9V$. Graph the equation for velocities between 0 and 60 mph.

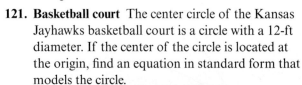

 120. Stopping distances See Exercise 119. How much farther does it take to stop at 60 mph than at 30 mph?

121. Basketball court The center circle of the Kansas Jayhawks basketball court is a circle with a 12-ft diameter. If the center of the circle is located at the origin, find an equation in standard form that models the circle.

ulisse/Shutterstock.com

122. Oil spill Oil spills from a tanker in the Gulf of Mexico and surfaces continuously at coordinates $(0, 0)$. If oil spreads in a circular pattern for ten hours and the circle's radius increases at a rate of 2 inches per hour, write an equation of the circle that models the range of the spill's effect.

123. Super Loop The Fire Ball Super Loop is a rollercoaster ride that is shaped like a circle. Find an equation of the loop in standard form if it is positioned 5 feet off of the ground, has a diameter of 60 feet, and its center is at coordinates $(0, 35)$.

Kenneth William Caleno/Shutterstock.com

124. Hurricane As a hurricane strengthens an eye begins to form at the center of the storm. At a wind speed of 80 mph the eye of a hurricane is circular when viewed from above and is 30 miles in diameter. If the eye is located at map coordinates $(5, 10)$, find an equation, in standard form, of the circle that models the eye of the hurricane.

125. CB radios The CB radio of a trucker covers the circular area shown in the illustration. Find an equation of that circle, in general form.

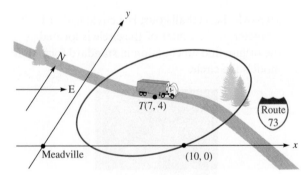

126. Firestone tires Two 24-inch-diameter Firestone tires stand against a wall, as shown in the illustration. Find equations in general form of the circular boundaries of the tires.

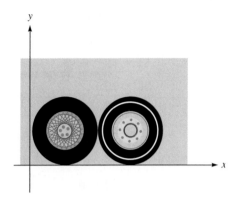

Discovery and Writing

The solution of the inequality $y < 0$ consists of those numbers x for which the graph of y lies below the x-axis. To solve $y < 0$, we graph y and trace to find numbers x that produce negative values of y. Solve each inequality.

127. $x^2 + x - 6 < 0$ **128.** $x^2 - 3x - 10 > 0$

When converting a circle's equation from general to standard form, it is possible to obtain a constant term on the right side that is zero or negative. If the constant term is zero the graph is a single point. If the constant term is negative the graph is nonexistent. Determine whether the graph of the equation is a single point or nonexistent.

129. $x^2 - 4x + y^2 - 6y + 13 = 0$

130. $x^2 - 12x + y^2 + 4y + 43 = 0$

Review
Solve each equation.

131. $3(x + 2) + x = 5x$

132. $12b + 6(3 - b) = b + 3$

133. $\dfrac{5(2 - x)}{3} - 1 = x + 5$

134. $\dfrac{r - 1}{3} = \dfrac{r + 2}{6} + 2$

135. Mixing an alloy In 60 ounces of alloy for watch cases, there are 20 ounces of gold. How much copper must be added to the alloy so that a watch case weighing 4 ounces, made from the new alloy, will contain exactly 1 ounce of gold?

136. Mixing coffee To make a mixture of 80 pounds of coffee worth $272, a grocer mixes coffee worth $3.25 a pound with coffee worth $3.85 a pound. How many pounds of cheaper coffee should the grocer use?

2.5 Proportion and Variation

In this section, we will learn to

1. Solve proportions.
2. Use direct variation to solve problems.
3. Use inverse variation to solve problems.
4. Use joint variation to solve problems.
5. Use combined variation to solve problems.

Skydiving is an adventurous sport. Jumping out of an aircraft at a height of 14,000 feet and free falling delivers a rush of adrenaline that is an exhilarating experience and one that skydivers never forget.

The unique experience allows skydivers to fall approximately 1,300 feet every five seconds and reach speeds between 120 and 150 miles per hour.

For a free-falling object, we can calculate the distance fallen given the amount of time. In fact, the distance of the fall is proportional to the square of the time. In this section we will consider how variables are related. These ways include direct variation, inverse variation, and joint variation and combinations of these.

1. Solve Proportions

The quotient of two numbers is often called a **ratio.** For example, the fraction $\frac{3}{2}$ (or the expression 3:2) can be read as "the ratio of 3 to 2." Some examples are

$$\frac{3}{5}, \qquad 7:9, \qquad \frac{x+1}{9}, \qquad \frac{a}{b}, \qquad \text{and} \qquad \frac{x^2-4}{x+5}$$

An equation indicating that two ratios are equal is called a **proportion.** Some examples of proportions are

$$\frac{2}{3} = \frac{4}{6}, \qquad \frac{x}{y} = \frac{3}{5}, \qquad \text{and} \qquad \frac{x^2+8}{2(x+3)} = \frac{17(x+3)}{2}$$

In the proportion $\frac{a}{b} = \frac{c}{d}$, the numbers a and d are called the **extremes,** and the numbers b and c are called the **means.**

To develop an important property of proportions, we suppose that

$$\frac{a}{b} = \frac{c}{d}$$

and multiply both sides by bd to get

$$bd\left(\frac{a}{b}\right) = bd\left(\frac{c}{d}\right)$$

$$\frac{bda}{b} = \frac{bdc}{d}$$

$$da = bc$$

Thus, if $\frac{a}{b} = \frac{c}{d}$ then $ad = bc$. This proves the following statement.

Property of Proportions   In any proportion, the product of the extremes is equal to the product of the means.

We can use this property to solve proportions.

EXAMPLE 1 **Solving a Proportion**

Solve the proportion: $\dfrac{x}{5} = \dfrac{2}{x + 3}$.

SOLUTION We will use the property of proportions to solve the proportion.

$$\frac{x}{5} = \frac{2}{x + 3}$$

$$x(x + 3) = 5 \cdot 2 \qquad \text{The product of the extremes equals the product of the means.}$$

$$x^2 + 3x = 10 \qquad \text{Remove parentheses and simplify.}$$

$$x^2 + 3x - 10 = 0 \qquad \text{Subtract 10 from both sides.}$$

$$(x - 2)(x + 5) = 0 \qquad \text{Factor the trinomial.}$$

$$x - 2 = 0 \quad \text{or} \quad x + 5 = 0 \qquad \text{Set each factor equal to 0.}$$

$$x = 2 \qquad\qquad x = -5$$

Thus, $x = 2$ or $x = -5$. Verify each solution. ∎

Self Check 1 Solve: $\dfrac{2}{5} = \dfrac{3}{x - 4}$.

Now Try Exercise 13.

EXAMPLE 2 **Solving an Application Problem Involving a Proportion**

Gasoline and oil for a Nissan outboard boat motor are to be mixed in a 50:1 ratio. How many ounces of oil should be mixed with 6 gallons of gasoline?

SOLUTION We will first express 6 gallons in terms of ounces.

$$6 \text{ gallons} = 6 \cdot 128 \text{ ounces} = 768 \text{ ounces}$$

Then, we let x represent the number of ounces of oil needed, set up the proportion, and solve it.

$$\frac{50}{1} = \frac{768}{x}$$

$$50x = 768 \qquad \text{The product of the extremes equals the product of the means.}$$

$$x = \frac{768}{50} \qquad \text{Divide both sides by 50.}$$

$$x = 15.36$$

Approximately 15 ounces of oil should be added to 6 gallons of gasoline. ∎

Self Check 2 How many ounces of oil should be mixed with 6 gallons of gas if the ratio is to be 40 parts of gas to 1 part of oil?

Now Try Exercise 15.

2. Use Direct Variation to Solve Problems

Two variables are said to **vary directly** or be **directly proportional** if their ratio is a constant. The variables x and y vary directly when

$$\frac{y}{x} = k \quad \text{or, equivalently,} \quad y = kx \quad (k \text{ is a constant})$$

Direct Variation   The words **"y varies directly with x,"** or **"y is directly proportional to x,"** mean that $y = kx$ for some real-number constant k.

The number k is called the **constant of proportionality.**

EXAMPLE 3   **Solving a Direct Variation Problem**

Distance traveled in a given time varies directly with the speed. If a car travels 70 miles at 30 mph, how far will it travel in the same time at 45 mph?

SOLUTION   We will use direct variation to solve the problem.

The phrase *distance varies directly with speed* translates into the formula $d = ks$, where d represents the distance traveled and s represents the speed. The constant of proportionality k can be found by substituting 70 for d and 30 for s in the equation $d = ks$.

$$d = ks$$

$$70 = k(30)$$

$$k = \frac{7}{3}$$

To evaluate the distance d traveled at 45 mph, we substitute $\frac{7}{3}$ for k and 45 for s into the formula $d = ks$.

$$d = \frac{7}{3}s$$

$$= \frac{7}{3}(45)$$

$$= 105$$

In the time it takes to go 70 miles at 30 mph, the car could travel 105 miles at 45 mph. ∎

Self Check 3   How far will the car travel in the same time if its speed is 60 mph?

Now Try Exercise 39.

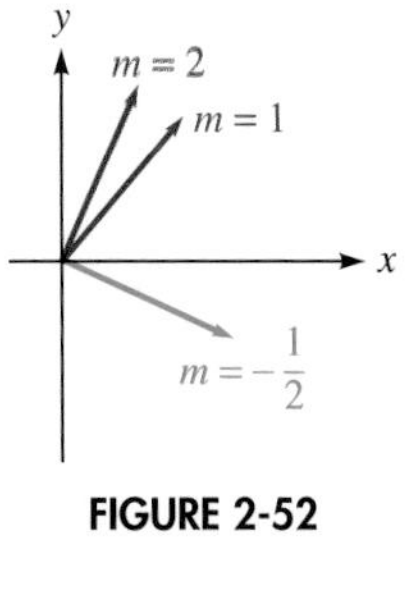

FIGURE 2-52

The statement *y varies directly with x* is equivalent to the equation $y = mx$, where m is the constant of proportionality. Because the equation is in the form $y = mx + b$ with $b = 0$, its graph is a line with slope m and y-intercept at $(0, 0)$. The graphs of $y = mx$ for several values of m are shown in Figure 2-52.

The graph of the relationship of direct variation is always a line that passes through the origin.

3. Use Inverse Variation to Solve Problems

Two variables are said to **vary inversely** or be **inversely proportional** if their product is a constant.

$$xy = k \quad \text{or, equivalently,} \quad y = \frac{k}{x} \quad (k \text{ is a constant})$$

Inverse Variation   The words **"y varies inversely with x,"** or **"y is inversely proportional to x,"** mean that $y = \frac{k}{x}$ for some real-number constant k.

EXAMPLE 4   Solving an Inverse Variation Problem

Intensity of illumination from a light source varies inversely with the square of the distance from the source. If the intensity of a light source is 100 lumens at a distance of 20 feet, find the intensity at 30 feet.

SOLUTION   We will use inverse variation to solve the problem.

If I is the intensity and d is the distance from the light source, the phrase *intensity varies inversely with the square of the distance* translates into the formula

$$I = \frac{k}{d^2}$$

We can evaluate k by substituting 100 for I and 20 for d in the formula and solving for k.

$$I = \frac{k}{d^2}$$

$$100 = \frac{k}{20^2}$$

$$k = 40{,}000$$

To find the intensity at a distance of 30 feet, we substitute 40,000 for k and 30 for d in the formula

$$I = \frac{k}{d^2}$$

$$I = \frac{40{,}000}{30^2}$$

$$= \frac{400}{9}$$

At 30 feet, the intensity of light would be $\frac{400}{9}$ lumens per square centimeter. ■

Self Check 4   Find the intensity at 50 feet.

Now Try Exercise 43.

The statement *y is inversely proportional to x* is equivalent to the equation $y = \frac{k}{x}$, where k is a constant. Figure 2-53 shows the graphs of $y = \frac{k}{x}$ $(x > 0)$ for three values of k. In each case, the equation determines one branch of a curve called a **hyperbola.** Verify these graphs with a graphing calculator.

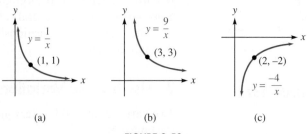

(a)   (b)   (c)

FIGURE 2-53

4. Use Joint Variation to Solve Problems

Joint Variation   The words **"y varies jointly with w and x"** mean that $y = kwx$ for some real-number constant k.

EXAMPLE 5   **Solving a Joint Variation Problem**

Kinetic energy of an object varies jointly with its mass and the square of its velocity. A 25-gram mass moving at the rate of 30 centimeters per second has a kinetic energy of 11,250 dyne-centimeters. Find the kinetic energy of a 10-gram mass that is moving at 40 centimeters per second.

SOLUTION   We will use joint variation to solve the problem. If we let E, m, and v represent the kinetic energy, mass, and velocity, respectively, the phrase *energy varies jointly with the mass and the square of its velocity* translates into the formula

$$E = kmv^2$$

The constant k can be evaluated by substituting 11,250 for E, 25 for m, and 30 for v in the formula.

$$E = kmv^2$$
$$11{,}250 = k(25)(30)^2$$
$$11{,}250 = 22{,}500k$$
$$k = \frac{1}{2}$$

We can now substitute $\frac{1}{2}$ for k, 10 for m, and 40 for v in the formula and evaluate E.

$$E = kmv^2$$
$$= \frac{1}{2}(10)(40)^2$$
$$= 8{,}000$$

A 10-gram mass that is moving at 40 centimeters per second has a kinetic energy of 8,000 dyne-centimeters. ∎

Self Check 5   Find the kinetic energy of a 25-gram mass that is moving at 100 centimeters per second.

Now Try Exercise 45.

5. Use Combined Variation to Solve Problems

The preceding terminology can be used in various combinations. In each of the following statements shown in the following table, the formula on the left translates into the words on the right.

Formula	Words
$y = \dfrac{kx}{z}$	y varies directly with x and inversely with z.
$y = kx^2\sqrt[3]{z}$	y varies jointly with the square of x and the cube root of z.
$y = \dfrac{kx\sqrt{z}}{\sqrt[3]{t}}$	y varies jointly with x and the square root of z and inversely with the cube root of t.
$y = \dfrac{k}{xz}$	y varies inversely with the product of x and z.

EXAMPLE 6   **Using Combined Variation to Solve a Problem**

The time it takes to build a highway varies directly with the length of the road but inversely with the number of workers. If it takes 100 workers 4 weeks to build 2 miles of the George Washington Memorial Parkway, how long will it take 80 workers to build 10 miles of the parkway?

SOLUTION   We will use combined variation to solve the problem.

We can let t represent the time in weeks, l represent the length in miles, and w represent the number of workers. Because time varies directly with the length of the parkway but inversely with the number of workers, the relationship between these variables can be expressed by the equation

$$t = \frac{kl}{w}$$

We substitute 4 for t, 100 for w, and 2 for l to find k:

$$4 = \frac{k(2)}{100}$$

$$400 = 2k \qquad \text{Multiply both sides by 100.}$$

$$200 = k \qquad \text{Divide both sides by 2.}$$

We now substitute 80 for w, 10 for l, and 200 for k in the equation $t = \frac{kl}{w}$ and simplify:

$$t = \frac{kl}{w}$$

$$t = \frac{200(10)}{80}$$

$$= 25$$

It will take 25 weeks for 80 workers to build 10 miles of parkway.   ■

Self Check 6   How long will it take 100 workers to build 20 miles of parkway?

Now Try Exercise 47.

Self Check Answers   **1.** $\dfrac{23}{2}$ **2.** 19.2 oz   **3.** 140 mi   **4.** 16 lumens per cm^2

5. 125,000 dyne-centimeters   **6.** 40 weeks

Exercises **2.5**

Getting Ready

You should be able to complete these vocabulary and concept statements before you proceed to the practice exercises.

Fill in the blanks.

1. A ratio is the _____ of two numbers.
2. A proportion is a statement that two _____ are equal.
3. In the proportion $\frac{a}{b} = \frac{c}{d}$, b and c are called the _____.
4. In the proportion $\frac{a}{b} = \frac{c}{d}$, a and d are called the _____.
5. In a proportion, the product of the _____ is equal to the product of the _____.
6. Direct variation translates into the equation _____.
7. The equation $y = \frac{k}{x}$ indicates _____ variation.
8. In the equation $y = kx$, k is called the _____ of proportionality.
9. The equation $y = kxz$ represents _____ variation.
10. In the equation $y = \frac{kx^2}{z}$, y varies directly with ___ and inversely with __.

Practice

Solve each proportion.

11. $\dfrac{4}{x} = \dfrac{2}{7}$
12. $\dfrac{5}{2} = \dfrac{x}{6}$
13. $\dfrac{x}{2} = \dfrac{3}{x+1}$
14. $\dfrac{x+5}{6} = \dfrac{7}{8-x}$

Set up and solve a proportion to answer each question.

15. The ratio of women to men in a mathematics class is 3:5. How many women are in the class if there are 30 men?
16. The ratio of lime to sand in mortar is 3:7. How much lime must be mixed with 21 bags of sand to make mortar?

Find the constant of proportionality.

17. y is directly proportional to x. If $x = 30$, then $y = 15$.
18. z is directly proportional to t. If $t = 7$, then $z = 21$.
19. I is inversely proportional to R. If $R = 20$, then $I = 50$.
20. R is inversely proportional to the square of I. If $I = 25$, then $R = 100$.
21. E varies jointly with I and R. If $R = 25$ and $I = 5$, then $E = 125$.

22. z is directly proportional to the sum of x and y. If $x = 2$ and $y = 5$, then $z = 28$.

Solve each problem.

23. y is directly proportional to x. If $y = 15$ when $x = 4$, find y when $x = \frac{7}{5}$.
24. w is directly proportional to z. If $w = -6$ when $z = 2$, find w when $z = -3$.
25. w is inversely proportional to z. If $w = 10$ when $z = 3$, find w when $z = 5$.
26. y is inversely proportional to x. If $y = 100$ when $x = 2$, find y when $x = 50$.
27. P varies jointly with r and s. If $P = 16$ when $r = 5$ and $s = -8$, find P when $r = 2$ and $s = 10$.
28. m varies jointly with the square of n and the square root of q. If $m = 24$ when $n = 2$ and $q = 4$, find m when $n = 5$ and $q = 9$.

Determine whether the graph could represent direct variation, inverse variation, or neither.

29.

30.

31.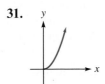

32.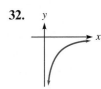

Applications

Set up and solve the required proportion.

33. **Caffeine** Many convenience stores sell supersize 44-ounce soft drinks in refillable cups. For each of the products listed in the table, find the amount of caffeine contained in one of the large cups. Round to the nearest milligram.

Soft drink, 12 oz	Caffeine (mg)
Mountain Dew	55
Coca-Cola Classic	47
Pepsi	37

Based on data from the *Los Angeles Times*

Joe Raedle/Getty Images

34. Cellphones A country has 221 mobile cellular telephones per 250 inhabitants. If the country's population is about 280,000, how many mobile cellular telephones does the country have?

35. Wallpapering Read the instructions on the label of wallpaper adhesive. Estimate the amount of adhesive needed to paper 500 square feet of kitchen walls if a heavy wallpaper will be used.

> COVERAGE: One-half gallon will hang approximately 4 single rolls (140 sq ft), depending on the weight of the wall covering and the condition of the wall

36. Recommended dosages The recommended child's dose of the sedative hydroxine is 0.006 gram per kilogram of body mass. Find the dosage for a 30-kg child in milligrams.

37. Gas laws The volume of a gas varies directly with the temperature and inversely with the pressure. When the temperature of a certain gas is 330°C, the pressure is 40 pounds per square inch and the volume is 20 cubic feet. Find the volume when the pressure increases 10 pounds per square inch and the temperature decreases to 300°C.

38. Hooke's Law The force f required to stretch a spring a distance d is directly proportional to d. A force of 5 newtons stretches a spring 0.2 meter. What force will stretch the spring 0.35 meter?

39. Free-falling objects The distance that an object will fall in t seconds varies directly with the square of t. An object falls 16 feet in 1 second. How long will it take the object to fall 144 feet?

40. Heat dissipation The power, in watts, dissipated as heat in a resistor varies directly with the square of the voltage and inversely with the resistance. If 20 volts are placed across a 20-ohm resistor, it will dissipate 20 watts. What voltage across a 10-ohm resistor will dissipate 40 watts?

41. Period of a pendulum The time required for one complete swing of a pendulum is called the **period** of the pendulum. The period varies directly with the square of its length. If a 1-meter pendulum has a period of 1 second, find the length of a pendulum with a period of 2 seconds.

42. Frequency of vibration The **pitch,** or **frequency,** of a vibrating string varies directly with the square root of the tension. If a string vibrates at a frequency of 144 hertz due to a tension of 2 pounds, find the frequency when the tension is 18 pounds.

43. Illumination Intensity of illumination from a light source varies inversely with the square of the distance from the source. If the intensity of a light source is 60 lumens at a distance of 10 feet, find the intensity at 20 feet.

44. Illumination Intensity of illumination from a light source varies inversely with the square of the distance from the source. If the intensity of a light source is 100 lumens at a distance of 15 feet, find the intensity at 25 feet.

45. Kinetic energy The kinetic energy of an object varies jointly with its mass and the square of its velocity. What happens to the energy when the mass is doubled and the velocity is tripled?

46. Heat dissipation The power, in watts, dissipated as heat in a resistor varies jointly with the resistance, in ohms, and the square of the current, in amperes. A 10-ohm resistor carrying a current of 1 ampere dissipates 10 watts. How much power is dissipated in a 5-ohm resistor carrying a current of 3 amperes?

47. Gravitational attraction The gravitational attraction between two massive objects varies jointly with their masses and inversely with the square of the distance between them. What happens to this force if each mass is tripled and the distance between them is doubled?

48. Gravitational attraction In Problem 47, what happens to the force if one mass is doubled and the other tripled and the distance between them is halved?

49. Plane geometry The area of an equilateral triangle varies directly with the square of the length of a side. Find the constant of proportionality.

50. Solid geometry The diagonal of a cube varies directly with the length of a side. Find the constant of proportionality.

Discovery and Writing

51. Explain the terms *extremes* and *means*.

52. Distinguish between a *ratio* and a *proportion*.

53. Explain the term *joint variation*.

54. Explain why $\frac{y}{x} = k$ indicates that y varies directly with x.

55. Explain why $xy = k$ indicates that y varies inversely with x.

56. As temperature increases on the Fahrenheit scale, it also increases on the Celsius scale. Is this direct variation? Explain.

Review
Perform each operation and simplify.

57. $\dfrac{1}{x+2} + \dfrac{2}{x+1}$

58. $\dfrac{x^2-1}{x+1} \cdot \dfrac{x-1}{x^2-2x+1}$

59. $\dfrac{x^2 + 3x - 4}{x^2 - 5x + 4} \div \dfrac{x - 1}{x^2 - 3x - 4}$

60. $\dfrac{x + 2}{3x - 3} \div (2x + 4)$

61. $\dfrac{x^2 + 4 - (x + 2)^2}{4x^2}$

62. $\dfrac{\dfrac{1}{x} - \dfrac{1}{3}}{\dfrac{1}{x} - 1}$

CHAPTER REVIEW

SECTION 2.1   The Rectangular Coordinate System

Definitions and Concepts	Examples
The rectangular coordinate system divides the plane into four quadrants.	

The **graph** of an equation in x and y is the set of all points (x, y) that satisfy the equation.	Use the x- and y-intercepts to graph the equation $6x + 4y = 24$.
The **y-intercept** of a line is the point $(0, b)$, where the line intersects the y-axis. To find b, substitute 0 for x in the equation of the line and solve for y.	**Find the y-intercept.** To find the y-intercept, we substitute 0 for x and solve for y.

$$6x + 4y = 24$$
$$6(0) + 4y = 24 \quad \text{Substitute 0 in for } x.$$
$$4y = 24 \quad \text{Simplify.}$$
$$y = 6 \quad \text{Divide both sides by 4.}$$

The y-intercept is the point $(0, 6)$.

The **x-intercept** of a line is the point $(a, 0)$, where the line intersects the x-axis. To find a, substitute 0 for y in the equation of the line and solve for x.	**Find the x-intercept.** To find the x-intercept, we substitute 0 for y and solve for x.

$$6x + 4y = 24$$
$$6x + 4(0) = 24 \quad \text{Substitute 0 for } y.$$
$$6x = 24 \quad \text{Simplify.}$$
$$x = 4 \quad \text{Divide both sides by 6.}$$

The x-intercept is the point $(4, 0)$.

Definitions and Concepts	Examples
	Find a third point as a check. If we let $x = 2$, we will find that $y = 3$.

$$6x + 4y = 24$$

$$6(2) + 4y = 24 \qquad \text{Substitute 2 for } x.$$

$$12 + 4y = 24 \qquad \text{Simplify.}$$

$$4y = 12 \qquad \text{Subtract 12 from both sides.}$$

$$y = 3 \qquad \text{Divide both sides by 4.}$$

The point $(2, 3)$ satisfies the equation.

We plot each pair and join them with a line to get the graph of the equation.

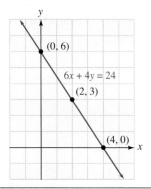

Equation of a vertical line through (a, b):

$\quad x = a$

Equation of a horizontal line through (a, b):

$\quad y = b$

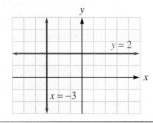

The Distance Formula:
The distance d between points (x_1, y_1) and (x_2, y_2) is given by

$$d = \sqrt{(x_2 - x_1)^2 + (y_2 - y_1)^2}$$

Find the distance between $P(-5, 2)$ and $Q(-3, 4)$.

We can use the Distance Formula
$d = \sqrt{(x_2 - x_1)^2 + (y_2 - y_1)^2}$ to find the distance between $P(-5, 2)$ and $Q(-3, 4)$. If we let $P(-5, 2) = P(x_1, y_1)$ and $Q(-3, 4) = Q(x_2, y_2)$, we can substitute -5 for x_1, 2 for y_1, -3 for x_2, and 4 for y_2 into the formula and simplify.

$$d(PQ) = \sqrt{(x_2 - x_1)^2 + (y_2 - y_1)^2}$$

$$d(PQ) = \sqrt{[-3 - (-5)]^2 + [4 - (2)]^2}$$

$$= \sqrt{(2)^2 + (2)^2}$$

$$= \sqrt{4 + 4} = \sqrt{8} = \sqrt{4 \cdot 2} = 2\sqrt{2}$$

The distance between the two points is $2\sqrt{2}$.

The Midpoint Formula:
The midpoint of the line segment joining (x_1, y_1) and (x_2, y_2) is the point M with coordinates

$$M = \left(\frac{x_1 + x_2}{2}, \frac{y_1 + y_2}{2} \right)$$

To find the midpoint of the segment with endpoints at $(-4, 5)$ and $(6, 7)$, average the x-coordinates and average the y-coordinates:

The midpoint is $\left(\frac{-4 + 6}{2}, \frac{5 + 7}{2} \right) = \left(\frac{2}{2}, \frac{12}{2} \right) = (1, 6)$.

EXERCISES

Refer to the illustration and find the coordinates of each point.

1. *A*   **2.** B

3. *C*   **4.** *D*

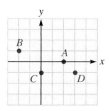

Graph each point. Indicate the quadrant in which the point lies or the axis on which it lies.

5. $(-3, 5)$ **6.** $(5, -3)$

7. $(0, -7)$ **8.** $\left(-\frac{1}{2}, 0\right)$

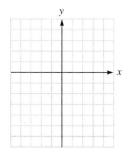

Solve each equation for y and graph the equation. Then check your graph with a graphing calculator.

9. $2x - y = 6$ **10.** $2x + 5y = -10$

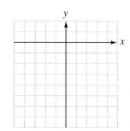   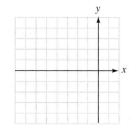

Use the x- and the y-intercepts to graph each equation.

11. $3x - 5y = 15$ **12.** $x + y = 7$

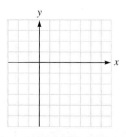   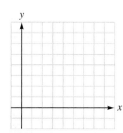

13. $x + y = -7$ **14.** $x - 5y = 5$

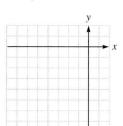   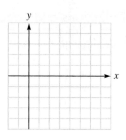

Graph each equation.

15. $y = 4$ **16.** $x = -2$

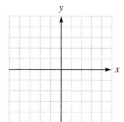   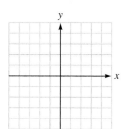

17. Depreciation A Ford Mustang purchased for $18,750 is expected to depreciate according to the formula $y = -2,200x + 18,750$. Find its value after 3 years.

18. House appreciation A house purchased for $250,000 is expected to appreciate according to the formula $y = 16,500x + 250,000$, where y is the value of the house after x years. Find the value of the house 5 years later.

Find the length of the segment PQ.

19. $P(-3, 7)$; $Q(3, -1)$ **20.** $P(-8, 6)$; $Q(-12, 10)$

21. $P\left(\sqrt{3}, 9\right)$; $Q\left(\sqrt{3}, 7\right)$ **22.** $P(a, -a)$; $Q(-a, a)$

Find the midpoint of the segment PQ.

23. $P(-3, 7)$; $Q(3, -1)$ **24.** $P(0, 5)$; $Q(-12, 10)$

25. $P\left(\sqrt{3}, 9\right)$; $Q\left(\sqrt{3}, 7\right)$ **26.** $P(a, -a)$; $Q(-a, a)$

SECTION **2.2**   The Slope of a Nonvertical Line

Definitions and Concepts	Examples

The slope of a nonvertical line passing through points $P(x_1, y_1)$ and $Q(x_2, y_2)$ is
 	
 $$m = \frac{\text{change in } y}{\text{change in } x} = \frac{y_2 - y_1}{x_2 - x_1} \quad (x_2 \neq x_1)$$
	Find the slope of the line passing through $P(-1, -3)$ and $Q(7, 9)$.
	      We will substitute the points $P(-1, -3)$ and $Q(7, 9)$ into the slope formula,
	
	 $$m = \frac{\text{change in } y}{\text{change in } x} = \frac{y_2 - y_1}{x_2 - x_1}$$
	 to find the slope of the line.
	      Let $P(x_1, y_1) = P(-1, -3)$ and $Q(x_2, y_2) = Q(7, 9)$. Then we substitute -1 for x_1, -3 for y_1, 7 for x_2, and 9 for y_2 to get
	
	 $$m = \frac{\text{change in } y}{\text{change in } x}$$
	 $$m = \frac{y_2 - y_1}{x_2 - x_1}$$
	 $$= \frac{9 - (-3)}{7 - (-1)} = \frac{12}{8} = \frac{3}{2}$$
	 The slope of the line is $\frac{3}{2}$.

Slopes of horizontal and vertical lines: The slope of a horizontal line (a line with an equation of the form $y = b$) is 0. The slope of a vertical line (a line with an equation of the form $x = a$) is not defined.	The slope of the graph of $y = 7$ is 0. The slope of the line $x = 6$ is not defined.
Slopes of parallel lines: Nonvertical parallel lines have the same slope. **Slopes of perpendicular lines:** The product of the slopes of two perpendicular lines is -1, provided neither line is vertical.	Determine whether the lines with the given slopes are parallel, perpendicular, or neither. $$m_1 = -6; m_2 = \frac{1}{6}$$ The product of the slopes $m_1 = -6$ and $m_2 = \frac{1}{6}$ is -1. The lines are perpendicular.

EXERCISES

Find the slope of the line PQ, if possible.

27. $P(3, -5)$; $Q(1, 7)$ **28.** $P(2,7)$; $Q(-5,-7)$

29. $P(b, a)$; $Q(a, b)$ **30.** $P(a + b, b)$; $Q(b, b - a)$

Find two points on the line and find the slope of the line.

31. $y = 3x + 6$ **32.** $y = 5x - 6$

Determine whether the slope of each line is 0 or undefined.

33.

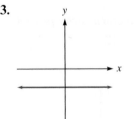

34.

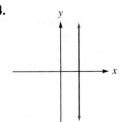

Determine whether the slope of each line is positive or negative.

35.

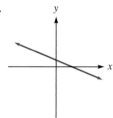

36.

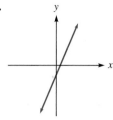

Determine whether the lines with the given slopes are parallel, perpendicular, or neither.

37. $m_1 = 5; m_2 = -\dfrac{1}{5}$

38. $m_1 = \dfrac{2}{7}; m_2 = \dfrac{7}{2}$

39. A line passes through $(-2, 5)$ and $(6, 10)$. A line parallel to it passes through $(2, 2)$ and $(10, y)$. Find y.

40. A line passes through $(-2, 5)$ and $(6, 10)$. A line perpendicular to it passes through $(-2, 5)$ and $(x, -3)$. Find x.

41. Rate of descent If an airplane descends 3,000 feet in 15 minutes, what is the average rate of descent in feet per minute?

42. Rate of growth A small business predicts sales according to a straight-line method. If sales were $50,000 in the first year and $147,500 in the third year, find the rate of growth in dollars per year (the slope of the line).

SECTION **2.3**   Writing Equations of Lines

Definitions and Concepts	Examples
Point-slope form: An equation of the line passing through $P(x_1, y_1)$ and with slope m is $$y - y_1 = m(x - x_1)$$	Write an equation of the line with slope $-\frac{4}{3}$ and passing through $P(3, -2)$. We will substitute $-\frac{4}{3}$ for m, 3 for x_1, and -2 for y_1 in the point-slope form $y - y_1 = m(x - x_1)$ and simplify. $$y - y_1 = m(x - x_1)$$ $y - (-2) = -\dfrac{4}{3}(x - 3)$ Substitute $-\frac{4}{3}$ for m, 3 for x_1, and -2 for y_1. $y + 2 = -\dfrac{4}{3}x + 4$ Remove parentheses. $y = -\dfrac{4}{3}x + 2$ Subtract 2 from both sides.
Slope-intercept form: An equation of the line with slope m and y-intercept $(0, b)$ is $y = mx + b$.	The equation of the line $y = -\frac{5}{3}x + 4$ is written in slope-intercept form.
Standard form of an equation of a line: $$Ax + By = C$$	The above equation written in standard form $Ax + By = C$ is $5x + 3y = 12$.
Slope-intercept form can be used to find the slope and the y-intercept from the equation of a line.	Find the slope and the y-intercept of the line with equation $2x + 5y = -10$. We will write the equation in the form $y = mx + b$ to find the slope m and the y-intercept $(0, b)$. $2x + 5y = -10$ $5y = -2x - 10$ Subtract $2x$ from both sides. $y = -\dfrac{2}{5}x - 2$ Divide both sides by 5. The slope of the graph is $-\frac{2}{5}$, and the y-intercept is $(0, -2)$.

Definitions and Concepts	Examples
Horizontal line: $y = b$ The slope is 0, and the y-intercept is $(0, b)$.	The equation of the horizontal line with slope 0 and y-intercept $(0, 7)$ is $y = 7$
Vertical line: $x = a$ There is no defined slope, and the x-intercept is $(a, 0)$.	The equation of the vertical line with no defined slope and the x-intercept $(-6, 0)$ is $x = -6$.

EXERCISES

Use point-slope form to write an equation of each line. Write each equation in standard form.

43. The line passes through the origin and the point $(-5, 7)$.

44. The line passes through $(-2, 1)$ and has a slope of -4.

45. The line passes through $(2, -1)$ and has a slope of $-\frac{1}{5}$.

46. The line passes through $(7, -5)$ and $(4, 1)$.

Use slope-intercept form to write an equation of each line.

47. The line has a slope of $\frac{2}{3}$ and a y-intercept of 3.

48. The slope is $-\frac{3}{2}$ and the line passes through $(0, -5)$.

Use slope-intercept form to graph each equation.

49. $y = \dfrac{3}{5}x - 2$

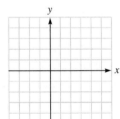

50. $y = -\dfrac{4}{3}x + 3$

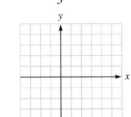

Find the slope and the y-intercept of the graph of each line.

51. $3x - 2y = 10$

52. $2x + 4y = -8$

53. $-2y = -3x + 10$

54. $2x = -4y - 8$

55. $5x + 2y = 7$

56. $3x - 4y = 14$

Write an equation of each line.

57. The line has a slope of 0 and passes through $(-5, 17)$.

58. The line has no defined slope and passes through $(-5, 17)$.

Write an equation of each line. Write the answer in slope-intercept form.

59. The line is parallel to $3x - 4y = 7$ and passes through $(2, 0)$.

60. The line passes through $(7, -2)$ and is parallel to the line segment joining $(2, 4)$ and $(4, -10)$.

61. The line passes through $(0, 5)$ and is perpendicular to the line $x + 3y = 4$.

62. The line passes through $(7, -2)$ and is perpendicular to the line segment joining $(2, 4)$ and $(4, -10)$.

Determine whether the graphs of each pair of equations are parallel, perpendicular, or neither.

63. $y = 3x + 8, 2y = 6x - 19$

64. $2x + 3y = 6, 3x - 2y = 15$

SECTION **2.4**　　　　Graphs of Equations

Definitions and Concepts	Examples

To graph an equation:

1. Find the x- and y-intercepts.
2. Find the symmetries of the graph.
3. Plot some additional points, if necessary, and draw the graph.

Intercepts of a graph:
To find the x-intercepts, let $y = 0$ and solve for x.
To find the y-intercepts, let $x = 0$ and solve for y.

Graph: $y = x^3 - 4x$.

To graph $y = x^3 - 4x$, we will find the x- and y-intercepts, test for symmetries, plot points, and join the points with a smooth curve.

Step 1: Find the x- and y-intercepts. To find the x-intercepts, we let $y = 0$ and solve for x.

$$y = x^3 - 4x$$

$$0 = x^3 - 4x \qquad \text{Substitute 0 for } y.$$

$$0 = x(x^2 - 4) \qquad \text{Factor out } x.$$

$$0 = x(x + 2)(x - 2) \qquad \text{Factor } x^2 - 4.$$

$$x = 0 \quad \text{or} \quad x + 2 = 0 \quad \text{or} \quad x - 2 = 0 \quad \text{Set each factor}$$
$$x = -2 \qquad\qquad x = 2 \qquad \text{equal to 0.}$$

The x-intercepts are $(0, 0)$, $(-2, 0)$, and $(2, 0)$.
　　To find the y-intercepts, we let $x = 0$ and solve for y.

$$y = x^3 - 4x$$

$$y = 0^3 - 4(0) \qquad \text{Substitute 0 in for } x.$$

$$y = 0$$

The y-intercept is $(0, 0)$.

Test for x-axis symmetry:
To test for x-axis symmetry, replace y with $-y$. If the resulting equation is equivalent to the original one, the graph is symmetric about the x-axis.

Step 2: Test for symmetries. We test for symmetry about the x-axis by replacing y with $-y$.

(1)　　$y = x^3 - 4x$ 　　This is the original equation.

　　　　$-y = x^3 - 4x$ 　　Replace y with $-y$.

(2)　　$y = -x^3 + 4x$ 　　Multiply both sides by -1.

Since Equations 1 and 2 are different, the graph is not symmetric about the x-axis.
　　To test for y-axis symmetry, we replace x with $-x$.

Test for y-axis symmetry:
To test for y-axis symmetry, replace x with $-x$. If the resulting equation is equivalent to the original one, the graph is symmetric about the y-axis.

---

(1)　$y = x^3 - 4x$ 　　　This is the original equation.

　　　$y = (-x)^3 - 4(-x)$ 　　Replace x with $-x$.

(3)　$y = -x^3 + 4x$ 　　　Simplify.

Since Equations 1 and 3 are different, the graph is not symmetric about the y-axis.

Definitions and Concepts	Examples

Test for origin symmetry:
To test for symmetry about the origin, replace x with $-x$ and y with $-y$. If the resulting equation is equivalent to the original one, the graph is symmetric about the origin.

To test for symmetry about the origin, we replace x with $-x$ and y with $-y$.

(1) $y = x^3 - 4x$ This is the original equation.

     $-y = (-x)^3 - 4(-x)$ Replace x with $-x$ and y with $-y$.

     $-y = -x^3 + 4x$ Simplify.

(4) $y = x^3 - 4x$ Multiply both sides by -1.

Since Equations 1 and 4 are the same, the graph is symmetric about the origin.

Plot some additional points and draw the graph.

Step 3: Graph the equation.   To graph the equation, we plot the x- and y-intercepts and several other pairs (x, y) with positive values of x. We can use the property of symmetry about the origin to draw the graph for negative values of x.

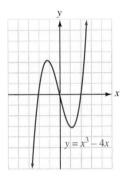

$y = x^3 - 4x$		
x	y	(x, y)
-2	0	$(-2, 0)$
0	0	$(0, 0)$
1	-3	$(1, -3)$
2	0	$(2, 0)$
3	15	$(3, 15)$

Circles:
A **circle** is the set of all points in a plane that are a fixed distance from a point called its **center.** The fixed distance is the **radius** of the circle.

The standard equation of a circle with center (h, k) and radius r:
The graph of any equation that can be written in the form

$$(x - h)^2 + (y - k)^2 = r^2$$

is a circle with radius r and center at point (h, k).

The standard equation of a circle with center $(0, 0)$ and radius r:
The graph of any equation that can be written in the form

$$x^2 + y^2 = r^2$$

is a circle with radius r and center at the origin.

Find the center and radius of the circle with the equation $(x - 6)^2 + (y + 5)^2 = 4$.

Standard Form: $(x - h)^2 + (y - k)^2 = r^2$

Given Form: $(x - 6)^2 + [y - (-5)]^2 = 2^2$

We see that $h = 6$, $k = -5$, and $r = 2$. The center (h, k) of the circle is at $(6, -5)$ and the radius is 2.

Definitions and Concepts	Examples
The general form of an equation of a circle: The general form of an equation of a circle is $$x^2 + y^2 + cd + dy + e = 0$$ where c, d, and e are real numbers.	To convert the general form of the equation of the circle $x^2 + y^2 + 4x - 2y - 20 = 0$ into standard form, we must find the coordinates of the center and the radius. To do so, we will complete the square on both x and y: $$x^2 + y^2 + 4x - 2y = 20$$ $$x^2 + 4x + y^2 - 2y = 20$$ $$x^2 + 4x + 4 + y^2 - 2y + 1 = 20 + 4 + 1 \quad \text{Add 4 and 1 to both sides to complete the square.}$$ $$(x + 2)^2 + (y - 1)^2 = 25 \quad \text{Factor } x^2 - 4x + 4 \text{ and } y^2 - 2y + 1.$$

EXERCISES

Find the x- and y-intercepts of each graph. Do not graph the equation.

65. $y = 4x - 8x^2$

66. $y = x^2 - 10x - 24$

Find the symmetries, if any, of the graph of each equation. Do not graph the equation.

67. $y^2 = 8x$

68. $y = 3y^4 + 6$

69. $y = -2|x|$

70. $y = |x + 2|$

Graph each equation. Find all intercepts and symmetries.

71. $y = x^2 + 2$

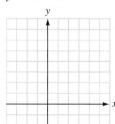

72. $y = x^3 - 2$

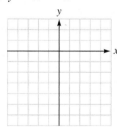

73. $y = \dfrac{1}{2}|x|$

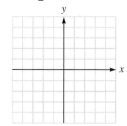

74. $y = -\sqrt{x - 4}$

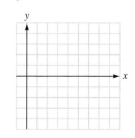

75. $y = \sqrt{x} + 2$

76. $y = |x + 1| + 2$

Use a graphing calculator to graph each equation.

77. $y = |x - 4| + 2$

78. $y = -\sqrt{x + 2} + 3$

79. $y = x + 2|x|$

80. $y^2 = x - 3$

Identify the center and radius of each circle written in standard form.

81. $x^2 + y^2 = 64$

82. $x^2 + (y - 6)^2 = 100$

83. $(x + 7)^2 + y^2 = \dfrac{1}{4}$

84. $(x - 5)^2 + (y + 1)^2 = 9$

Write an equation of each circle in standard form.

85. Center at $(0, 0)$; $r = 7$

86. Center at $(3, 0)$; $r = \dfrac{1}{5}$

87. Center at $(-2, 12)$, $r = 5$

88. Center at $\left(\dfrac{2}{7}, 5\right)$; $r = 9$

Write an equation of each circle in standard form and general form.

89. Center at $(-3, 4)$; radius 12

90. Ends of diameter at $(-6, -3)$ and $(5, 8)$

Convert the general form of each circle given into standard form.

91. $x^2 + y^2 + 6x - 4y + 4 = 0$

92. $2x^2 + 2y^2 - 8x - 16y - 10 = 0$

Graph each circle.

93. $x^2 + y^2 - 16 = 0$

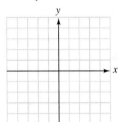

94. $x^2 + y^2 - 4x = 5$

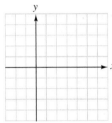

95. $x^2 + y^2 - 2y = 15$

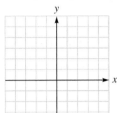

96. $x^2 + y^2 - 4x + 2y = 4$

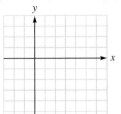

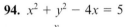 *Use a graphing calculator to solve each equation. If an answer is not exact, round to the nearest hundredth.*

97. $x^2 - 11 = 0$

98. $x^3 - x = 0$

99. $|x^2 - 2| - 1 = 0$

100. $x^2 - 3x = 5$

SECTION **2.5**   Proportion and Variation

Definitions and Concepts	Examples
Proportion: An equation indicating that two ratios are equal is called a **proportion.** In the proportion $\frac{a}{b} = \frac{c}{d}$, the numbers a and d are called the **extremes,** and the numbers b and c are called the **means.**	In the proportion $\frac{3}{5} = \frac{9}{15}$, 5 and 9 are called the means, and 3 and 15 are called the extremes.

Definitions and Concepts	Examples
Property of proportions: In any proportion, the product of the extremes is equal to the product of the means.	In the proportion on the previous page, the product of the means is equal to the product of the extremes: $5(9) = 3(15) = 45$

Use the property of proportions to solve a proportion for a variable.

Solve the proportion: $\dfrac{x}{6} = \dfrac{2}{x + 11}$ for x.

We will use the property of proportions to solve the proportion.

$$\frac{x}{6} = \frac{2}{x + 11}$$

$$x(x + 11) = 6 \cdot 2 \qquad \text{The product of the extremes equals the product of the means.}$$

$$x^2 + 11x = 12 \qquad \text{Remove parentheses and simplify.}$$

$$x^2 + 11x - 12 = 0 \qquad \text{Subtract 12 from both sides.}$$

$$(x + 12)(x - 1) = 0 \qquad \text{Factor the trinomial.}$$

$$x + 12 = 0 \quad \text{or} \quad x - 1 = 0 \qquad \text{Set each factor equal to 0.}$$

$$x = -12 \qquad\qquad x = 1$$

Thus, $x = -12$ or $x = 1$.

Direct variation:

The words **"y varies directly with x,"** or **"y is directly proportional to x,"** mean that $y = kx$ for some real-number constant k. The number k is called the **constant of proportionality.**

In the direct variation formula, $y = 3x$, find y when $x = 5$.

$$y = 3x = 3(5) = 15$$

Inverse variation:

The words **"y varies inversely with x,"** or **"y is inversely proportional to x,"** mean that $y = \frac{k}{x}$ for some real-number constant k.

In the inverse variation formula $y = \frac{k}{x}$, find the constant of variation if $y = 5$ when $x = 20$.

$$y = \frac{k}{x}$$

$$5 = \frac{k}{20} \qquad \text{Substitute 5 for } y \text{ and 20 for } x.$$

$$100 = k$$

Definitions and Concepts	Examples
Joint variation: The words **"y varies jointly with w and x"** mean that $y = kwx$ for some real-number constant k.	Kinetic energy of an object varies jointly with its mass and the square of its velocity. A 50-gram mass moving at the rate of 20 centimeters per second has a kinetic energy of 40,000 dyne-centimeters. Find the kinetic energy of a 10-gram mass that is moving at 60 centimeters per second. We will use joint variation to solve the problem. If we let E, m, and v represent the kinetic energy, mass, and velocity, respectively, the phrase *energy varies jointly with the mass and the square of its velocity* translates into the formula $$E = kmv^2.$$ The constant k can be evaluated by substituting 40,000 for E, 50 for m, and 20 for v in the formula $$E = kmv^2$$ $$40{,}000 = k(50)(20)^2$$ $$40{,}000 = 20{,}000k$$ $$k = 2$$ We can now substitute 2 for k, 10 for m, and 60 for v in the formula and evaluate E. $$E = kmv^2$$ $$= 2(10)(60)^2$$ $$= 72{,}000$$ A 10-gram mass that is moving at 60 centimeters per second has a kinetic energy of 72,000 dyne-centimeters.

EXERCISES

Solve each proportion.

101. $\dfrac{x + 3}{10} = \dfrac{x - 1}{x}$

102. $\dfrac{x - 1}{2} = \dfrac{12}{x + 1}$

103. Hooke's Law The force required to stretch a spring is directly proportional to the amount of stretch. If a 3-pound force stretches a spring 5 inches, what force would stretch the spring 3 inches?

104. Kinetic energy A moving body has a kinetic energy directly proportional to the square of its velocity. By what factor does the kinetic energy of an automobile increase if its speed increases from 30 mph to 50 mph?

105. Gas laws The volume of gas in a balloon varies directly as the temperature and inversely as the pressure. If the volume is 400 cubic centimeters when the temperature is 300 K and the pressure is 25 dynes per square centimeter, find the volume when the temperature is 200 K and the pressure is 20 dynes per square centimeter.

106. The area of a rectangle varies jointly with its length and width. Find the constant of proportionality.

107. Electrical resistance The resistance of a wire varies directly as the length of the wire and inversely as the square of its diameter. A 1,000-foot length of wire, 0.05 inch in diameter, has a resistance of 200 ohms. What would be the resistance of a 1,500-foot length of wire that is 0.08 inch in diameter?

108. **Billing for services** Angie's Painting and Decorating Service charges a fixed amount for accepting a wallpapering job and adds a fixed dollar amount for each roll hung. If the company bills a customer $177 to hang 11 rolls and $294 to hang 20 rolls, find the cost to hang 27 rolls.

109. **Paying for college** Rolf must earn $5,040 for next semester's tuition. Assume he works x hours tutoring algebra at $14 per hour and y hours tutoring Spanish at $18 per hour and makes his goal. Write an equation expressing the relationship between x and y, and graph the equation. If Rolf tutors algebra for 180 hours, how long must he tutor Spanish?

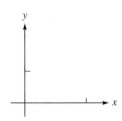

CHAPTER TEST

Indicate the quadrant in which the point lies or the axis on which it lies.

1. $(-3, \pi)$

2. $(0, -8)$

Find the x- and y-intercepts and use them to graph the equation.

3. $x + 3y = 6$

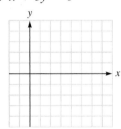

4. $2x - 5y = 10$

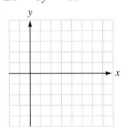

Graph each equation.

5. $2(x + y) = 3x + 5$

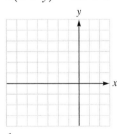

6. $3x - 5y = 3(x - 5)$

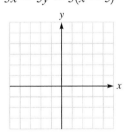

7. $\frac{1}{2}(x - 2y) = y - 1$

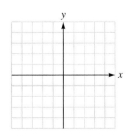

8. $\frac{x + y - 5}{7} = 3x$

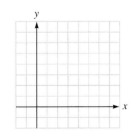

Find the distance between points P and Q.

9. $P(1, -1)$; $Q(-3, 4)$ 10. $P(0, \pi)$; $Q(-\pi, 0)$

Find the midpoint of the line segment PQ.

11. $P(3, -7)$; $Q(-3, 7)$

12. $P\left(0, \sqrt{2}\right)$; $Q\left(\sqrt{8}, \sqrt{18}\right)$

Find the slope of the line PQ.

13. $P(3, -9)$; $Q(-5, 1)$

14. $P\left(\sqrt{3}, 3\right)$; $Q\left(-\sqrt{12}, 0\right)$

Determine whether the two lines are parallel, perpendicular, or neither.

15. $y = 3x - 2$; $y = 2x - 3$
16. $2x - 3y = 5$; $3x + 2y = 7$

Write an equation of the line with the given properties. Your answers should be written in slope-intercept form, if possible.

17. Passing through $(3, -5)$; $m = 2$

18. $m = 3$; $b = \frac{1}{2}$

19. Parallel to $2x - y = 3$; $b = 5$

20. Perpendicular to $2x - y = 3$; $b = 5$

21. Passing through $\left(2, -\frac{3}{2}\right)$ and $\left(3, \frac{1}{2}\right)$

22. Parallel to the y-axis and passing through $(3, -4)$

Find the x- and y-intercepts of each graph.

23. $y = x^3 - 16x$

24. $y = |x - 4|$

Find the symmetries of each graph.

25. $y^2 = x - 1$

26. $y = x^4 + 1$

Graph each equation. Find all intercepts and symmetries.

27. $y = x^2 - 9$

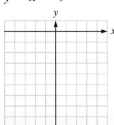

28. $x = |y|$

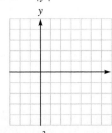

29. $y = 2\sqrt{x}$

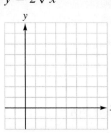

30. $x = y^3$

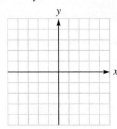

Write an equation of each circle in standard form.

31. Center at $(5, 7)$; radius of 8

32. Center at $(2, 4)$; passing through $(6, 8)$

Graph each equation.

33. $x^2 + y^2 = 9$

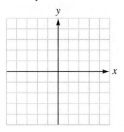

34. $x^2 - 4x + y^2 + 3 = 0$

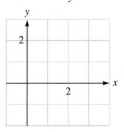

Write each statement as an equation.

35. y varies directly as the square of z.

36. w varies jointly with r and the square of s.

37. P varies directly with Q. $P = 7$ when $Q = 2$. Find P when $Q = 5$.

38. y is directly proportional to x and inversely proportional to the square of z, and $y = 16$ when $x = 3$ and $z = 2$. Find x when $y = 2$ and $z = 3$.

Use a graphing calculator to find the positive root of each equation. Round to two decimal places.

39. $x^2 - 7 = 0$

40. $x^2 - 5x - 5 = 0$

Functions

3

CAREERS AND MATHEMATICS: Computer Scientist

Benis Arapovic/Shutterstock.com

Computer scientists are highly trained innovative workers who design and invent new technology. They solve complex business, scientific, and general computing problems. Computer scientists conduct research on a variety of topics, including computer hardware, virtual reality, and robotics.

Education and Mathematics Required
- Most are required to possess a Ph.D. in computer science, computer engineering, or a closely related discipline. An aptitude for math is important.
- College Algebra, Trigonometry, Calculus, Linear Algebra, Ordinary Differential Equations, Theory of Analysis, Abstract Algebra, Graph Theory, Numerical Methods, and Combinatorics are math courses required.

How Computer Scientists Use Math and Who Employs Them
- Computer scientists use mathematics as they span a range of topics from theoretical studies of algorithms to the computation of implementing computing systems in hardware and software.
- Many computer scientists are employed by Internet service providers; Web search portals; and data processing, hosting, and related services firms. Others work for government, manufacturers of computer and electronic products, insurance companies, financial institutions, and universities.

Career Outlook and Earnings
- Employment of computer scientists is expected to grow by 24 percent through 2018, which is much faster than the average for all occupations.
- The median annual wages of computer and information scientists is approximately $98,000. Some earn more than $150,000 a year.

For more information see: www.bls.gov/oco

In this chapter, we will discuss one of the most important concepts in mathematics—the concept of a function.

3.1 Functions and Function Notation

In this section, we will learn to

1. Understand the concept of a function.
2. Determine whether an equation represents a function.
3. Find the domain of a function.
4. Evaluate a function.
5. Evaluate the difference quotient for a function.
6. Graph a function by plotting points.
7. Use the Vertical Line Test to identify functions.
8. Use linear functions to model applications.

Correspondences between the elements of two sets is a common occurrence in everyday life. For example,

- To every Motorola cell phone, there corresponds exactly one phone number.
- To every Honda Civic car, there corresponds exactly one vehicle identification number.
- To every case on the television show "Deal or No Deal," there corresponds exactly one amount of money.
- To every item's barcode at the Target store, there corresponds exactly one price.

This table shows the four highest-grossing movies of all time and the year each movie was released.

Movie	Year
Avatar	2009
The Lord of the Rings: The Return of the King	2003
Pirates of the Caribbean: Dead Man's Chest	2006
The Dark Knight	2008

The information shown in the table sets up a correspondence between a movie and the year it was released. Note that for each of the movies, there corresponds exactly one year in which it was released. The phase *there corresponds exactly one* is extremely important in mathematics, and we will use this idea to solve problems.

1. Understand the Concept of a Function

Correspondences in which exactly one quantity corresponds to (or depends on) another quantity according to some specific rule are called **functions**. Equations are frequently used in mathematics to represent functions. For example, the equation $y = x^2 - 1$ sets up a correspondence between two infinite sets of real numbers, x and y, according to the rule *square x and subtract 1*.

Equation: $y = x^2 - 1$

Correspondence: Each real number x determines exactly one real number y.

Rule: Square x and subtract 1.

Since the value of y depends on the number x, we call y the **dependent variable** and x the **independent variable.**

The equation $y = x^2 - 1$ determines what **output value** y will result from each **input value** x. This idea of inputs and outputs is shown in Figure 3-1(a). In the equation $y = x^2 - 1$, if the input x is 2, the output y is

$$y = 2^2 - 1 = 3$$

This is illustrated in Figure 3-1(b).

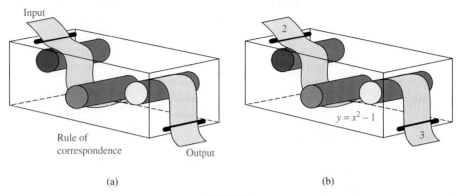

FIGURE 3-1

Comment

Correspondences can be set up by a verbal expression, by a table, by a set of ordered pairs, by an equation, and by a graph.

The equation $y = x^2 - 1$ also determines the table of ordered pairs, and the graph shown in Figure 3-2. To see how the table determines the correspondence, we find an input in the x-column and read across to find the corresponding output in the y-column. If we select $x = 2$ as an input, we get $y = 3$ for the output.

To see how the graph of $y = x^2 - 1$ determines the correspondence, we draw a vertical and horizontal line through any point (say, point P) on the graph shown in Figure 3-2. Because these lines intersect the x-axis at 2 and the y-axis at 3, the point $P(2, 3)$ associates 3 on the y-axis with 2 on the x-axis.

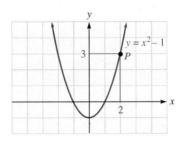

$y = x^2 - 1$		
x	y	(x, y)
-2	3	$(-2, 3)$
-1	0	$(-1, 0)$
0	-1	$(0, -1)$
1	0	$(1, 0)$
2	3	$(2, 3)$

FIGURE 3-2

Any correspondence that assigns exactly one value of y to each number x is called a **function**. We refer to the set of inputs as the **domain** of the function and the set of outputs as the **range** of the function.

Function	A **function** f is a correspondence between a set of input values x and a set of output values y, where to each x-value there corresponds exactly one y-value.
Domain	The set of input values x is called the **domain** of the function.
Range	The set of output values y is called the **range** of the function.

2. Determine Whether an Equation Represents a Function

EXAMPLE 1   **Determining Whether an Equation Represents a Function**

Determine whether the following equations define y to be a function of x.
a. $7x + y = 5$ **b.** $y = |x| + 1$ **c.** $y^2 = x + 2$

SOLUTION   In each case, we will solve each equation for y, if necessary, and determine whether each input value x determines exactly one output value y.

a. First, we solve $7x + y = 5$ for y to get

$$y = 5 - 7x$$

From this equation we see that for each input value x there corresponds exactly one output value y. For example, if $x = 3$, then $y = 5 - 7(3) = -16$. The equation defines y to be a function of x.

b. Since each input number x that we substitute for x determines exactly one output y, the equation $y = |x| + 1$ defines y to be a function of x.

c. First, we solve $y^2 = x + 2$ for y using the Square Root Property covered in Section 1.3 to get $y = \pm\sqrt{x + 2}$. Note that each input value x for which the function is defined (except -2) gives two outputs y. For example, if $x = 7$, then $y = \pm\sqrt{7 + 2} = \pm\sqrt{9} = \pm 3$. For this reason, the equation does not define y to be a function of x. ■

Caution

Some equations represent functions and some do not. If to some input x in an equation there corresponds more than one output y, the equation will not represent a function.

Self Check 1   Determine whether each equation defines y to be a function of x.
a. $y = |x|$ **b.** $y = \sqrt{x}$ **c.** $y^2 = 2x$

Now Try Exercise 17.

3. Find the Domain of a Function

The **domain** of a function is the set of all real numbers x for which the function is defined. Thus, to find the domain of a function we must find the set of numbers that are permissible inputs for x. It is often helpful to determine any restrictions on the input values for x. For example, we consider the information in the following table.

Equation	Restriction	Domain
$y = x^3 - x^2 + x - 5$	There are no restrictions on x because any real number can be cubed, squared and combined with constants.	all real numbers, $(-\infty, \infty)$
$y = \sqrt{x}$	The restriction that x cannot be a negative real number is placed on x because the square root of a negative number is an imaginary number.	$x \geq 0$, the interval $[0, \infty)$
$y = \dfrac{1}{x}$	The restriction that x cannot be 0 is placed on x because division by 0 isn't defined.	$x \neq 0$, $(-\infty, 0) \cup (0, \infty)$

EXAMPLE 2   **Finding the Domain of a Function**

Find the domain of the function defined by each equation:

a. $y = 2x - 5$ **b.** $y = \sqrt{3x - 2}$ **c.** $y = \dfrac{3}{x + 2}$

SOLUTION   We must determine what numbers are permissible inputs for x. This set of numbers is the domain.

a. Any real number that we input for x can be multiplied by 2 and then 5 can be subtracted from the result. Thus, the domain is the interval $(-\infty, \infty)$.

b. Since the radicand must be nonnegative, we have

$$3x - 2 \geq 0$$

$$3x \geq 2 \qquad \text{Add 2 to both sides.}$$

$$x \geq \frac{2}{3} \qquad \text{Divide both sides by 3.}$$

Thus, the domain is the interval $\left[\frac{2}{3}, \infty\right)$.

c. Since the fraction $\frac{3}{x+2}$ is undefined when $x = -2$, -2 is not a permissible input value. Since all other values of x are permissible inputs, the domain is $(-\infty, -2) \cup (-2, \infty)$.

Self Check 2   Find the domain of each function: **a.** $y = |x| + 2$ **b.** $y = \sqrt[3]{x + 2}$

Now Try Exercise 27.

EXAMPLE 3   **Finding the Domain of a Function**

Find the domain of the function defined by the equation $y = \dfrac{1}{x^2 - 5x - 6}$.

SOLUTION   We can factor the denominator to see what values of x will give 0's in the denominator. These values are not in the domain.

$$x^2 - 5x - 6 = 0$$

$$(x - 6)(x + 1) = 0$$

$$x - 6 = 0 \quad \text{or} \quad x + 1 = 0$$

$$x = 6 \qquad\qquad x = -1$$

The domain is $(-\infty, -1) \cup (-1, 6) \cup (6, \infty)$.

Self Check 3   Find the domain of the function defined by the equation $y = \dfrac{2}{x^2 - 16}$.

Now Try Exercise 41.

4. Evaluate a Function

To indicate that y is a function of x, we often use **function notation** and write

$$y = f(x) \qquad \text{Read as "}y\text{ is a function of }x\text{."}$$

The notation $y = f(x)$ provides a way of denoting the value of y (the dependent variable) that corresponds to some input number x (the independent variable). For example, if $y = f(x)$, the value of y that is determined when $x = 2$ is denoted by $f(2)$, read as "f of 2." If $f(x) = 5 - 7x$, we can evaluate $f(2)$ by substituting 2 for x.

$$f(x) = 5 - 7x$$

$$f(2) = 5 - 7(2) \qquad \text{Substitute the input 2 for } x.$$

$$= -9$$

If $x = 2$, then $y = f(2) = -9$.
To evaluate $f(-5)$, we substitute -5 for x.

$$f(x) = 5 - 7x$$

$$f(-5) = 5 - 7(-5) \qquad \text{Substitute the input } -5 \text{ for } x.$$

$$= 40$$

If $x = -5$, then $y = f(-5) = 40$.

Comment

To see why function notation is helpful, consider the following sentences. Note that the second sentence is much more concise.

1. In the function $y = 3x^2 + x - 4$, find the value of y when $x = -3$.

2. In the function $f(x) = 3x^2 + x - 4$, find $f(-3)$.

In this context, the notations y and $f(x)$ both represent the output of a function and can be used interchangeably, but function notation is more concise.

Sometimes functions are denoted by letters other than f. The notations $y = g(x)$ and $y = h(x)$ also denote functions involving the independent variable x.

EXAMPLE 4   **Evaluating a Function**

Let $g(x) = 3x^2 + x - 4$. Find **a.** $g(-3)$ **b.** $g(k)$ **c.** $g(-t^3)$ **d.** $g(k + 1)$

SOLUTION   In each case, we will substitute the input value into the function and simplify.

a. $g(x) = 3x^2 + x - 4$ **b.** $g(x) = 3x^2 + x - 4$

$g(-3) = 3(-3)^2 + (-3) - 4$ $g(k) = 3k^2 + k - 4$

$\quad = 3(9) - 3 - 4$

$\quad = 20$

c. $g(x) = 3x^2 + x - 4$ **d.** $g(x) = 3x^2 + x - 4$

$g(-t^3) = 3(-t^3)^2 + (-t^3) - 4$ $g(k + 1) = 3(k + 1)^2 + (k + 1) - 4$

$\quad = 3t^6 - t^3 - 4$ $\quad = 3(k^2 + 2k + 1) + k + 1 - 4$

$\qquad\qquad\qquad\qquad\qquad\qquad\qquad = 3k^2 + 6k + 3 + k + 1 - 4$

$\qquad\qquad\qquad\qquad\qquad\qquad\qquad = 3k^2 + 7k$

Self Check 4   Evaluate: **a.** $g(0)$ **b.** $g(2)$ **c.** $g(k - 1)$

Now Try Exercise 49.

ACCENT ON TECHNOLOGY   **Evaluating a Function**

Functions can be easily evaluated on a graphing calculator. There are several ways to do this, but one of the easiest is to use the graph editor and the table. Press [Y=] and input $3x^2 + x - 4$ in the graph editor. Next, press [2nd] [WINDOW] and use the table setup shown on the next page. Then press [2nd] [GRAPH] and enter values for x. The function values will appear in the table.

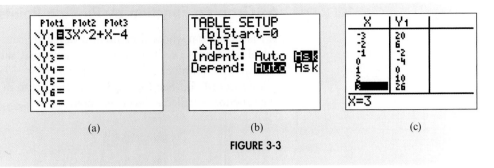

(a)   (b)   (c)

FIGURE 3-3

5. Evaluate the Difference Quotient for a Function

The fraction $\frac{f(x + h) - f(x)}{h}$ is called the **difference quotient** and is important in calculus. The difference quotient can be used to find quantities such as the velocity of a guided missile or the rate of change of a company's profit.

EXAMPLE 5   **Finding the Difference Quotient**

If $f(x) = x^2 - 2x - 5$, evaluate $\dfrac{f(x + h) - f(x)}{h}$.

SOLUTION   We will evaluate the difference quotient in three steps. Find $f(x + h)$. Then subtract $f(x)$. Then divide by h.

Step 1: Find $f(x + h)$.

$$f(x) = x^2 - 2x - 5$$

$$f(x + h) = (x + h)^2 - 2(x + h) - 5 \qquad \text{Substitute } x + h \text{ for } x.$$

$$= x^2 + 2xh + h^2 - 2x - 2h - 5$$

Step 2: Find $f(x + h) - f(x)$. We can use the result from Step 1.

$$f(x + h) - f(x) = x^2 + 2xh + h^2 - 2x - 2h - 5 - (x^2 - 2x - 5) \qquad \text{Subtract } f(x).$$

$$= x^2 + 2xh + h^2 - 2x - 2h - 5 - x^2 + 2x + 5 \qquad \text{Remove parentheses.}$$

$$= 2xh + h^2 - 2h \qquad \text{Combine like terms.}$$

Step 3: Find the difference quotient $\frac{f(x + h) - f(x)}{h}$. We can use the result from Step 2.

$$\frac{f(x + h) - f(x)}{h} = \frac{2xh + h^2 - 2h}{h} \qquad \text{Divide both sides by } h.$$

$$= \frac{h(2x + h - 2)}{h} \qquad \text{In the numerator, factor out } h.$$

$$= 2x + h - 2 \qquad \text{Divide out } h: \frac{h}{h} = 1.$$

Comment

After the completion of Step 2, in a polynomial function, when $f(x + h) - f(x)$ is simplified, each term will always include an h. Use that fact to check your work as you progress through the problem.

Self Check 5   If $f(x) = x^2 + 2$, evaluate $\dfrac{f(x + h) - f(x)}{h}$.

Now Try Exercise 65.

6. Graph a Function by Plotting Points

If f is a function whose domain and range are sets of real numbers, its graph is the set of all points $(x, f(x))$ in the xy-plane that satisfy the equation $y = f(x)$. For example, the graph of the function $y = f(x) = -7x + 5$ is a line with slope -7 and y-intercept $(0, 5)$. (See Figure 3-4.) If the graph of a function is a nonvertical line, the function is called a **linear function.**

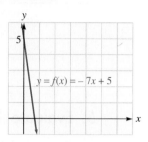

FIGURE 3-4

| Graph of a Function | The **graph** of a function f in the xy-plane is the set of all points (x, y) where x is in the domain of f, y is in the range of f, and $y = f(x)$. |

EXAMPLE 6   Graphing a Function by Plotting Points

Graph the functions. **a.** $f(x) = -2|x| + 3$ **b.** $f(x) = \sqrt{x - 2}$

SOLUTION   In each case, we will make a table of solutions and plot the points given by the table. Then we will connect the points by drawing a smooth curve through them and obtain the graph of the function.

a. $f(x) = -2|x| + 3$

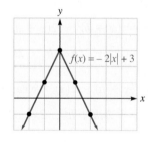

$f(x) = -2|x| + 3$		
x	$f(x)$	$(x, f(x))$
-2	-1	$(-2, -1)$
-1	1	$(-1, 1)$
0	3	$(0, 3)$
1	1	$(1, 1)$
2	-1	$(2, -1)$

FIGURE 3-5

b. $f(x) = \sqrt{x - 2}$

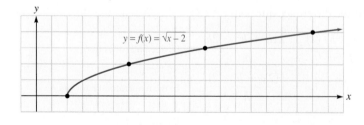

$f(x) = \sqrt{x - 2}$		
x	$f(x)$	$(x, f(x))$
2	0	$(2, 0)$
6	2	$(6, 2)$
11	3	$(11, 3)$
18	4	$(18, 4)$

FIGURE 3-6

Self Check 6   Graph: $f(x) = |x + 3|$.

Now Try Exercise 83.

Both the **domain** and the **range** of a function can be identified by viewing the graph of the function. The inputs or x-values that correspond to points on the graph of the function can be identified on the x-axis and used to state the domain of the function. The outputs or $f(x)$ values that correspond to points on the graph of the function can be identified on the y-axis and used to state the range of the function. (See Figure 3-7).

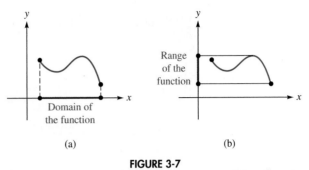

(a)   (b)

FIGURE 3-7

From Figure 3-6, we can see that the domain of the function $f(x) = -2|x| + 3$ is the set of all real numbers, and that the range is the set of real numbers that are less than or equal to 3.

From Figure 3-6, we can see that the domain of the function $f(x) = \sqrt{x - 2}$ is the set of all real numbers greater than or equal to 2, and that the range is the set of all real numbers greater than or equal to 0.

In Figure 3-8, we see the graphs of several basic functions.

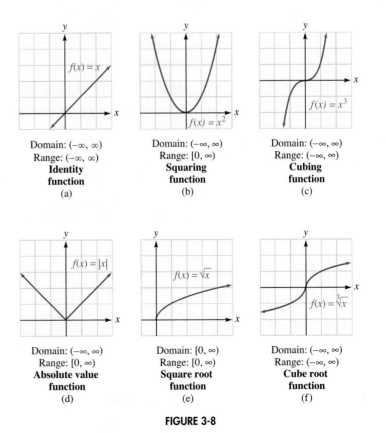

Domain: $(-\infty, \infty)$
Range: $(-\infty, \infty)$
Identity function
(a)

Domain: $(-\infty, \infty)$
Range: $[0, \infty)$
Squaring function
(b)

Domain: $(-\infty, \infty)$
Range: $(-\infty, \infty)$
Cubing function
(c)

Domain: $(-\infty, \infty)$
Range: $[0, \infty)$
Absolute value function
(d)

Domain: $[0, \infty)$
Range: $[0, \infty)$
Square root function
(e)

Domain: $(-\infty, \infty)$
Range: $(-\infty, \infty)$
Cube root function
(f)

FIGURE 3-8

ACCENT ON TECHNOLOGY

Graphing the Absolute Value Function

To graph the absolute value function on a graphing calculator, we must call up the function abs(into the graphing window. We press [MATH], then scroll right to NUM as shown below in Figure 3-9.

Press [Y=]. Go to [MATH] NUM and press [1]. Input [x] and [GRAPH]. Use [ZOOM] [4] as the window.

 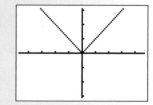  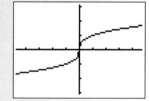

FIGURE 3-9

ACCENT ON TECHNOLOGY

Graphing the Cube Root Function

We can graph this function on a graphing calculator using the cube root key by selecting [MATH] [4] from the graph window as shown in Figure 3-10. Input [x] and [GRAPH]. Use [ZOOM] [4] as the window.

FIGURE 3-10

7. Use the Vertical Line Test to Identify Functions

We can use a **Vertical Line Test** to determine whether a graph represents a function.

Vertical Line Test
- If every vertical line that can be drawn intersects the graph in no more than one point, the graph represents a function. See Figure 3-11(a).

- If a vertical line can be drawn that intersects the graph at more than one point, the graph does not represent a function. See Figure 3-11(b).

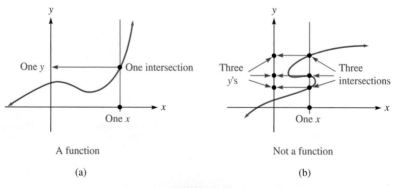

FIGURE 3-11

EXAMPLE 7   **Using the Vertical Line Test to Identify Functions**

Determine which of the following graphs represent functions.

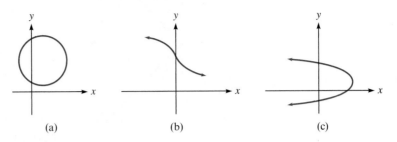

(a)   (b)   (c)

SOLUTION   We will use the Vertical Line Test by drawing several vertical lines through each graph. If every vertical line that intersects the graph does so exactly once, the graph represents a function. Otherwise, the graph does not represent a function.

a. This graph fails the Vertical Line Test, so it does not represent a function.

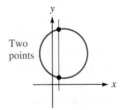

b. This graph passes the Vertical Line Test, so it does represent a function.

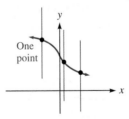

c. This graph fails the Vertical Line Test, so it does not represent a function.

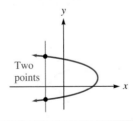

Self Check 7   Does the graph shown in Figure 3-2 represent a function?

Now Try Exercise 97.

Comment

Some graphs represent functions and some do not. Graphs that pass the Vertical Line Test are functions.

Not all equations define functions. For example, the equation $x = |y|$ does not define a function, because two values of y can correspond to one number x. For example, if $x = 2$, then y can be either 2 or -2. The graph of the equation is shown in Figure 3-12. Since the graph does not pass the Vertical Line Test, it does not represent a function.

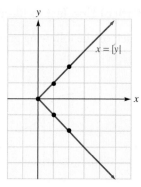

$x = |y|$		
x	y	(x, y)
2	-2	$(2, -2)$
1	-1	$(1, -1)$
0	0	$(0, 0)$
1	1	$(1, 1)$
2	2	$(2, 2)$

FIGURE 3-12

Correspondences between a set of input values x (called the domain) and a set of output values y (called the range), where to each x-value in the domain there corresponds one or more y-values in the range, are called **relations.** Although the graph in Figure 3-12 does not represent a function, it does represent a relation. We note that all functions are relations, but not all relations are functions.

Another way to visualize the definition of function is to consider the diagram shown in Figure 3-13(a). The function f that assigns the element y to the element x is represented by an arrow leaving x and pointing to y. The set of elements in **X** from which arrows originate is the domain of the function. The set of elements in **Y** to which arrows point is the range.

To constitute a function, each element of the domain must determine exactly one y-value in the range. However, the same value of y could correspond to several numbers x. In the function shown in Figure 3-13(b), the single value y corresponds to the three numbers x_1, x_2, and x_3 in the domain.

The correspondence shown in Figure 3-13(c) is not a function, because two values of y correspond to the same number x. However, it is still a relation.

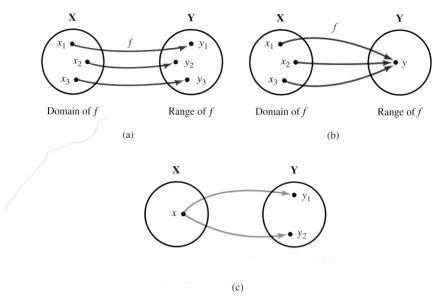

FIGURE 3-13

8. Use Linear Functions to Model Applications

We have seen that the equation of a nonvertical line defines a linear function—an important function in mathematics and its applications.

Linear Functions   A **linear function** is a function determined by an equation of the form

$$f(x) = mx + b \quad \text{or} \quad y = mx + b$$

EXAMPLE 8   Using a Linear Function to Model an Application

Cost of a Fraternity Dance   The cost associated with a fraternity dance is $300 for Country Club rental and $24 for each couple that attends.
a. Write the cost C of the dance in terms of the number of couples x attending.
b. Find the cost if 45 couples attend the dance.

SOLUTION   Use the information stated in the problem to write a linear function that models the application.

a. The cost C for the dance is $24 per couple plus the $300 rental fee. If x couples attend, the cost is $24x$ plus 300. Therefore, the linear cost function is

$$C(x) = 24x + 300$$

b. To find the cost when 45 couples attend, we find C(45).

$$C(x) = 24x + 300$$
$$C(\mathbf{45}) = 24(\mathbf{45}) + 300 \qquad \text{Substitute 45 for } x.$$
$$= 1{,}080 + 300$$
$$= 1{,}380$$

If 45 couples attend, the cost of the dance is $1,380.

Self Check 8   Find the cost when 60 couples attend.

Now Try Exercise 105.

EXAMPLE 9   Writing a Linear Function that Models an Application

Heart Rates   The target heart rate at which a person should train to get an effective workout is a linear function of his age. For a 20-year-old starting an exercise program, the target heart rate should be 120 beats per minute. For a 40-year-old, it should be 108 beats per minute. Express the target heart rate R as a function of age A.

SOLUTION   We will use the information stated in the problem to find the slope of the linear function and write a linear function that models the application.

Since the target heart rate (R) is given to be a linear function of age (A), there are constants m and b such that

$$R = mA + b$$

Since $R = 120$ when $A = 20$, the point $(A_1, R_1) = (20, 120)$ lies on the straight-line graph of this function. Since $R = 108$ when $A = 40$, the point $(A_2, R_2) = (40, 108)$ also lies on that line. The slope of the line is

$$m = \frac{R_2 - R_1}{A_2 - A_1}$$
$$= \frac{\mathbf{108} - \mathbf{120}}{\mathbf{40} - \mathbf{20}} \qquad \text{Substitute 108 for } R_2\text{, 120 for } R_1\text{, 40 for } A_2\text{, and 20 for } A_1.$$
$$= -\frac{12}{20}$$
$$= -\frac{3}{5}$$
$$= -0.6$$

Thus, $m = -0.6$. To determine b, we can substitute -0.6 for m and the coordinates of one point, say $P(20, 120)$, in the equation $R = mA + b$ and solve for b.

$$R = mA + b$$

$$120 = -0.6(20) + b \qquad \text{Substitute 120 for } R, \text{ 20 for } b, \text{ and } -0.6 \text{ for } m.$$

$$120 = -12 + b$$

$$132 = b \qquad\qquad\qquad \text{Add 12 to both sides.}$$

If we substitute -0.6 for m and 132 for b in $R = mA + b$, we obtain $R = -0.6A + 132$.

Self Check 9   Find the target heart rate for a 30-year-old person.

Now Try Exercise 111.

Self Check Answers
1. **a.** a function   **b.** a function   **c.** not a function   2. **a.** $(-\infty, \infty)$
b. $(-\infty, \infty)$ 3. $(-\infty, -4) \cup (-4, 4) \cup (4, \infty)$ 4. **a.** -4 **b.** 10
c. $3k^2 - 5k - 2$ 5. $2x + h$ 6.   7. yes
8. \$1,740   9. 114 beats per minute

$f(x) = |x + 3|$

Exercises **3.1**

Getting Ready
You should be able to complete these vocabulary and concept statements before you proceed to the practice exercises.

Fill in the blanks.

1. A correspondence that assigns exactly one value of y to any number x is called a _____.

2. A correspondence that assigns one or more values of y to any number x is called a _____.

3. The set of input numbers x in a function is called the _____ of the function.

4. The set of all output values y in a function is called the _____ of the function.

5. The statement "y is a function of x" can be written as the equation _____.

6. The graph of a function $y = f(x)$ in the xy-plane is the set of all points _____ that satisfy the equation, where x is in the _____ of f and y is in the _____ of f.

7. In the function of Exercise 5, ___ is called the independent variable.

8. In the function of Exercise 5, y is called the _____ variable.

9. If every _____ line that intersects a graph does so _____, the graph represents a function.

10. A function that can be written in the form $y = mx + b$ is called a _____ function.

Practice
Assume that all variables represent real numbers. Determine whether each equation determines y to be a function of x.

11. $y = x$

12. $y - 2x = 0$

13. $y^2 = x$

14. $|y| = x$

15. $y = x^2$

16. $y - 7 = 7$

17. $y^2 - 4x = 1$

18. $|x - 2| = y$

19. $|x| = |y|$

20. $x = 7$

21. $y = 7$

22. $|x + y| = 7$

Let the function f be defined by the equation $y = f(x)$, where x and $f(x)$ are real numbers. Find the domain of each function.

23. $f(x) = 3x + 5$

24. $f(x) = -5x + 2$

25. $f(x) = x^2 - x + 1$

26. $f(x) = x^3 - 3x + 2$

27. $f(x) = \sqrt{x - 2}$

28. $f(x) = \sqrt{2x + 3}$

29. $f(x) = \sqrt{4 - x}$

30. $f(x) = 3\sqrt{2 - x}$

31. $f(x) = \sqrt{x^2 - 1}$

32. $f(x) = \sqrt{x^2 - 2x - 3}$

33. $f(x) = \sqrt[3]{x + 1}$

34. $f(x) = \sqrt[3]{5 - x}$

35. $f(x) = \dfrac{3}{x + 1}$

36. $f(x) = \dfrac{-7}{x + 3}$

37. $f(x) = \dfrac{x}{x - 3}$

38. $f(x) = \dfrac{x + 2}{x - 1}$

39. $f(x) = \dfrac{x}{x^2 - 4}$

40. $f(x) = \dfrac{2x}{x^2 - 9}$

41. $f(x) = \dfrac{1}{x^2 - 4x - 5}$

42. $f(x) = \dfrac{x}{2x^2 - 16x + 30}$

Let the function f be defined by $y = f(x)$*, where x and* $f(x)$
are real numbers. Find $f(2)$*,* $f(-3)$*,* $f(k)$*, and* $f(k^2 - 1)$*.*

43. $f(x) = 3x - 2$

44. $f(x) = 5x + 7$

45. $f(x) = \dfrac{1}{2}x + 3$

46. $f(x) = \dfrac{2}{3}x + 5$

47. $f(x) = x^2$

48. $f(x) = 3 - x^2$

49. $f(x) = x^2 + 3x - 1$

50. $f(x) = -x^2 - 2x + 1$

51. $f(x) = |x^2 + 1|$

52. $f(x) = |x^2 + x + 4|$

53. $f(x) = \dfrac{2}{x + 4}$

54. $f(x) = \dfrac{3}{x - 5}$

55. $f(x) = \dfrac{1}{x^2 - 1}$

56. $f(x) = \dfrac{3}{x^2 + 3}$

57. $f(x) = \sqrt{x^2 + 1}$

58. $f(x) = \sqrt{x^2 - 1}$

Evaluate the difference quotient for each function $f(x)$*.*

59. $f(x) = 3x + 1$

60. $f(x) = 5x - 1$

61. $f(x) = x^2 + 1$

62. $f(x) = x^2 - 3$

63. $f(x) = 4x^2 - 6$

64. $f(x) = 5x^2 + 3$

65. $f(x) = x^2 + 3x - 7$

66. $f(x) = x^2 - 5x + 1$

67. $f(x) = 2x^2 - 4x + 2$

68. $f(x) = 3x^2 + 2x - 3$

69. $f(x) = x^3$

70. $f(x) = \dfrac{1}{x}$

*Graph each function. Use the graph to identify the domain
and range of each function.*

71. $f(x) = 2x + 3$

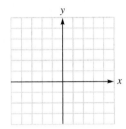

72. $f(x) = 3x + 2$

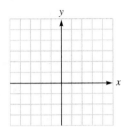

73. $f(x) = -\dfrac{3}{4}x + 4$

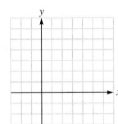

74. $f(x) = \dfrac{1}{2}x - 3$

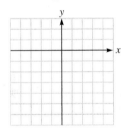

75. $2x = 3y - 3$

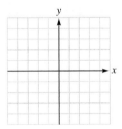

76. $3x = 2(y + 1)$

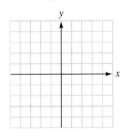

77. $f(x) = x^2 - 4$

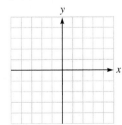

78. $f(x) = -x^2 + 3$

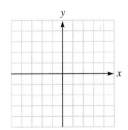

79. $f(x) = -x^3 + 2$

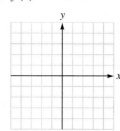

80. $f(x) = -x^3 + 1$

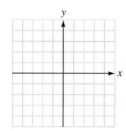

89. $f(x) = \sqrt{2x - 4}$

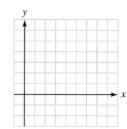

90. $f(x) = -\sqrt{2x - 4}$

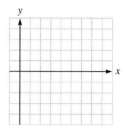

81. $f(x) = -|x|$

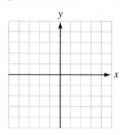

82. $f(x) = -|x| - 3$

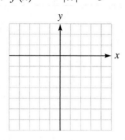

91. $f(x) = \sqrt[3]{x} + 2$

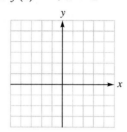

92. $f(x) = -\sqrt[3]{x} + 1$

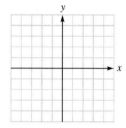

83. $f(x) = |x - 2|$

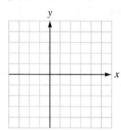

84. $f(x) = -|x - 2|$

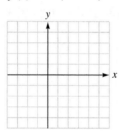

Draw lines to indicate the domain and range of each function as intervals on the x- and y-axes.

93.

94.

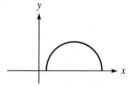

85. $f(x) = \left| \frac{1}{2}x + 3 \right|$

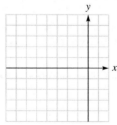

86. $f(x) = -\left| \frac{1}{2}x + 3 \right|$

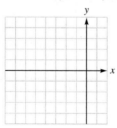

Use the Vertical Line Test to determine whether each graph represents a function.

95.

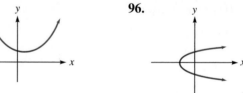

96.

97.

98.

87. $f(x) = -\sqrt{x + 1}$

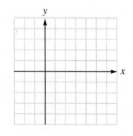

88. $f(x) = \sqrt{x} + 2$

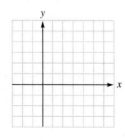

99.

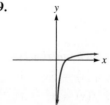

100.

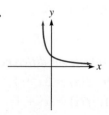

Use a graphing calculator to graph each function. Then determine the domain and range of the function.

101. $f(x) = |3x + 2|$ **102.** $f(x) = \sqrt{2x - 5}$

103. $f(x) = \sqrt[3]{5x - 1}$ **104.** $f(x) = -\sqrt[3]{3x + 2}$

Applications

105. Cost of *t*-shirts A chapter of Phi Theta Kappa, an honors society for two-year college students, is purchasing *t*-shirts for each of its members. A local company has agreed to make the shirts for $8 each plus a graphic arts fee of $75.

a. Write a linear function that describes the cost C for the shirts in terms of x, the number of *t*-shirts ordered.

b. Find the total cost of 85 *t*-shirts.

106. Service projects The Circle "K" Club is planning a service project for children at a local children's home. They plan to rent a "Dora the Explorer Moonwalk" for the event. The cost of the moonwalk will include a $60 delivery fee and $45 for each hour it is used. Express the total bill b in terms of the hours used h.

107. Cell phone plans A grandmother agrees to purchase a cell phone for emergency use only. AT&T now offers such a plan for $9.99 per month and $0.07 for each minute t the phone is used.

a. Write a linear function that describes the monthly cost C in terms of the time in minutes t the phone is used.

b. If the grandmother uses her phone for 20 minutes during the first month, what was her bill?

108. Concessions A concessionaire at a football game pays a vendor $40 per game for selling hot dogs at $2.50 each.

a. Write a linear function that describes the income I the vendor earns for the concessionaire during the game if the vendor sells h hot dogs.

b. Find the income if the vendor sells 175 hot dogs.

109. Home construction In a proposal to prospective clients, a contractor listed the following costs:

| 1. Fees, permits, site preparation | $14,000 |
| 2. Construction, per square foot | $95 |

a. Write a linear function the clients can use to determine the cost C of building a house having f square feet.

b. Find the cost to build a 2,600-square-foot house.

110. Temperature conversion The Fahrenheit temperature reading (F) is a linear function of the Celsius reading (C). If $C = 0$ when $F = 32$ and the readings are the same at $-40°$, express F as a function of C.

111. Cost of electricity The cost C of electricity in Eagle River is a linear function of x, the number of kilowatt-hours (kwh) used. If the cost of 100 kwh is $17 and the cost of 500 kwh is $57, find an equation that expresses C in terms of x.

112. Water billing The cost C of water is a linear function of n, the number of gallons used. If 1,000 gallons cost $4.70 and 9,000 gallons cost $14.30, express C as a function of n.

113. Coffee locations Suppose that in 2008 there were approximately 6,400 of your coffee company locations. Suppose that in 2012 this number had grown to approximately 13,168. Write a linear function that represents the number of coffee locations n as a function of time t. Let $t = 0$ represent 2008.

114. Cliff divers The cliff divers of Acapulco amaze tourists with their diving skills. The velocity v of a diver is a function of the time t the diver has fallen. If the initial velocity of the diver is 2 feet per second and $v = -66$ feet per second when $t = 2$ seconds, express v as a function t.

115. Exchange rates If fifty U.S. dollars can be exchanged for 69.5550 Euros and 125 U.S. dollars can be exchanged for 173.8875 Euros, write a linear function that represents the number of Euros E in terms of U.S. dollars D.

116. Exchange rates If fifty U.S. dollars can be exchanged for 600.1100 Mexican pesos and 125 U.S. dollars can be exchanged for 1500.275 Mexican pesos, write a linear function that represents the number of Mexican pesos P in terms of U.S. dollars D.

Discovery and Writing

Find all values of x that will make f(x) = 0.

117. $f(x) = 3x + 2$ **118.** $f(x) = -2x - 5$

119. Write a paragraph explaining how to find the domain of a function.

120. Write a paragraph explaining how to find the range of a function.

121. Explain why all functions are relations, but not all relations are functions.

 122. Use a graphing calculator to graph the function $f(x) = \sqrt{x}$, and use [TRACE] and [ZOOM] to find $\sqrt{5}$ to three decimal places.

Review

Consider this set: $\{-3, -1, 0, 0.5, \frac{3}{4}, 1, \pi, 7, 8\}$

123. Which numbers are natural numbers?

124. Which numbers are rational numbers?

125. Which numbers are prime numbers?

126. Which numbers are even numbers?

Write each in interval notation.

127.
      -4 7

128.
      -3 5

Graph each union of two intervals.

129. $(-3, 5) \cup [6, \infty)$ **130.** $(-\infty, 0) \cup (0, \infty)$

3.2 Quadratic Functions

In this section, we will learn to

1. Recognize the characteristics of a quadratic function.
2. Find the vertex of a parabola whose equation is in standard form.
3. Graph a quadratic function.
4. Find the vertex of a parabola whose equation is in general form.
5. Use a quadratic function to solve maximum and minimum problems.

Quadratic functions are important because we can use them to model many real-life problems. For example, the path of a basketball jump shot by Shaquille O'Neal and the path of a guided missile can be modeled with quadratic functions. Businesses like Coca Cola and Best Buy can use quadratic functions to help maximize the profit and revenue for the products they produce and sell.

1. Recognize the Characteristics of a Quadratic Function

The linear function $f(x) = mx + b$ $(m \neq 0)$ is a first-degree polynomial function, because its right side is a first-degree polynomial in the variable x. A function defined by a polynomial of second degree is called a **quadratic function.**

Quadratic Function   A **quadratic function** is a second-degree polynomial function in one variable of the form

$$f(x) = ax^2 + bx + c \quad \text{or} \quad y = ax^2 + bx + c,$$

where a, b, and c are real numbers and $a \neq 0$.

Some examples of quadratic functions are

$$f(x) = x^2 - 2x - 3 \quad \text{and} \quad f(x) = -2x^2 - 8x - 3.$$

Quadratic functions can be graphed by plotting points. For example, to graph the function $f(x) = x^2 - 2x - 3$, we plot several points with coordinates that satisfy the equation. We then join them with a smooth curve to obtain the graph shown in Figure 3-14.

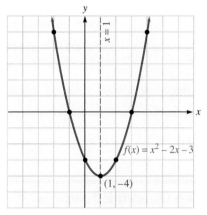

$f(x) = x^2 - 2x - 3$		
x	$f(x)$	$(x, f(x))$
-2	5	$(-2, 5)$
-1	0	$(-1, 0)$
0	-3	$(0, -3)$
1	-4	$(1, -4)$
2	-3	$(2, -3)$
3	0	$(3, 0)$
4	5	$(4, 5)$

Domain: $(-\infty, \infty)$, **Range:** $[-4, \infty)$

FIGURE 3-14

A table of values and the graph of $f(x) = -2x^2 - 8x - 3$ are shown in Figure 3-15.

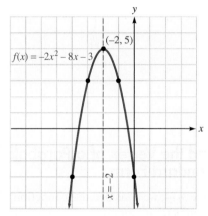

$f(x) = -2x^2 - 8x - 3$		
x	$f(x)$	$(x, f(x))$
-4	-3	$(-4, -3)$
-3	3	$(-3, 3)$
-2	5	$(-2, 5)$
-1	3	$(-1, 3)$
0	-3	$(0, -3)$

Domain: $(-\infty, \infty)$, **Range:** $(-\infty, 5]$

FIGURE 3-15

The graph of a quadratic function is called a **parabola**, a cup-shaped curve that either opens upward $\cup$ or downward $\cap$. The graphs in Figures 3-14 and 3-15 suggest that the graph of a quadratic function has the following characteristics.

Characteristics of Quadratic Functions

Characteristics	Examples	
Equation of a quadratic function $f(x) = ax^2 + bx + c$	$f(x) = x^2 - 2x - 3$	$f(x) = -2x^2 - 8x - 3$
If $a > 0$, the parabola **opens up**. If $a < 0$, the parabola **opens down**.	$a = 1$, opens up	$a = -2$, opens down
The **vertex** is the turning point of the parabola.	Vertex is $(1, -4)$	Vertex is $(-2, 5)$
The **minimum** or **maximum** **point** occurs at the vertex.	$(1, -4)$ is the minimum or lowest point on the graph.	$(-2, 5)$ is the maximum or highest point on the graph.
The **axis of symmetry** is the vertical line that intersects the parabola at the vertex. The parabola is symmetric about this vertical line.	The graph of $x = 1$ is the axis of symmetry.	The graph of $x = -2$ is the axis of symmetry.

ACCENT ON TECHNOLOGY / Graphing Quadratic Functions

We can use a graphing calculator to graph quadratic functions. If we use window settings of $[-10, 10]$ for x and $[-10, 10]$ for y, the graph of $f(x) = x^2 - 2x - 3$ will look like Figure 3-16(a). The graph of $f(x) = -2x^2 - 8x - 3$ will look like Figure 3-16(b).

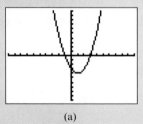
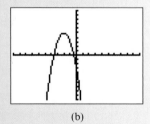

(a)   (b)

FIGURE 3-16

2. Find the Vertex of a Parabola Whose Equation Is in Standard Form

In Figure 3-14, we considered the function $f(x) = x^2 - 2x - 3$ with vertex $(1, -4)$. If we complete the square on the right side of the equation we will obtain

$$f(x) = (x^2 - 2x + 1) - 3 - 1 \qquad \text{One-half of } -2 \text{ is } -1, \text{ and } (-1)^2 \text{ is } 1.$$
$$\text{Add 1 and subtract 1 on the right side of the equation.}$$

$$f(x) = (x - 1)(x - 1) - 4 \qquad \text{Factor } x^2 - 2x + 1.$$

$$f(x) = (x - 1)^2 - 4$$

In this factored form, the coordinates of the vertex of the parabola can be read from the equation. The vertex of $f(x) = (x - 1)^2 - 4$ is $(1, -4)$. We call this factored form the **standard form** of an equation of a quadratic function.

Standard Form of an Equation of a Quadratic Function	The graph of a quadratic function $$y = f(x) = a(x - h)^2 + k \ (a \neq 0)$$ is a parabola with vertex at (h, k). The parabola opens upward when $a > 0$ and downward when $a < 0$. The axis of symmetry of the parabola is the vertical line graph of the equation $x = h$.

EXAMPLE 1   **Finding the Vertex of a Parabola in Standard Form**

Find the vertex of the graph of each quadratic function:
a. $f(x) = 2(x - 3)^2 + 5$ **b.** $f(x) = -3(x + 2)^2 - 4$

SOLUTION   In each case, the equation of the quadratic function is given in standard form. From the equations, we can identify h and k, because the vertex is the point with coordinates (h, k).

a. We identify the values of h and k.

Standard Form: $f(x) = a(x - h)^2 + k$

Given Function: $f(x) = 2(x - 3)^2 + 5$ $h = 3$ and $k = 5$.

Since $h = 3$ and $k = 5$, the vertex is the point with coordinates of $(3, 5)$.

b. We identify the values of h and k.

Standard Form: $f(x) = a(x - h)^2 + k$

Given Function: $f(x) = -3(x + 2)^2 - 4$

$\qquad\qquad\qquad f(x) = -3[x - (-2)]^2 + (-4)$ $h = -2$ and $k = -4$.

Since $h = -2$ and $k = -4$, the vertex is the point with coordinates of $(-2, -4)$.

Self Check 1   Find the vertex of the graph of the quadratic function $f(x) = 2(x + 5)^2 - 4$.

Now Try Exercise 19.

3. Graph a Quadratic Function

The easiest way to graph a quadratic function is to follow these steps.

Strategy for Graphing a Quadratic Function	To graph a quadratic function 1. Determine whether the parabola opens upward or downward. 2. Find the vertex of the parabola. 3. Find the x-intercept(s). 4. Find the y-intercept. 5. Identity one additional point on the graph. 6. Draw a smooth curve through the points found in Steps 2–5.

EXAMPLE 2 **Graphing a Quadratic Function Written in Standard Form**

Graph the quadratic function $f(x) = 2(x + 1)^2 - 8$.

SOLUTION We first determine whether the parabola opens upward or downward. Then we will find the vertex and the x- and y-intercepts. Finally, we will find one additional point and draw a smooth curve through the plotted points.

Step 1: Determine whether the parabola opens upward or downward.

$$\text{Standard Form: } f(x) = a(x - h)^2 + k$$

$$\text{Given Form: } \quad f(x) = 2(x + 1)^2 - 8$$

Since $a = 2$ and 2 is positive, the parabola opens upward.

Step 2: Find the vertex of the parabola.

$$\text{Standard Form: } f(x) = a(x - h)^2 + k$$

$$\text{Given Form: } \quad f(x) = 2(x + 1)^2 - 8$$

$$f(x) = 2[x - (-1)]^2 + (-8)$$

Since $h = -1$ and $k = -8$ the vertex is the point with coordinates of $(-1, -8)$.

Step 3: Find the x-intercept(s).
To find the x-intercepts, we substitute 0 for $f(x)$ and solve for x.

$f(x) = 2(x + 1)^2 - 8$	
$0 = 2(x + 1)^2 - 8$	Substitute 0 for $f(x)$.
$8 = 2(x + 1)^2$	Add 8 to both sides of the equation.
$4 = (x + 1)^2$	Divide both sides by 2.
$x + 1 = \pm 2$	Write $(x + 1)^2$ on the left side and use the Square Root Property.
$x = -1 \pm 2$	Subtract 1 from both sides.
$x = 1 \quad \text{or} \quad x = -3$	

The x-intercepts are the points with coordinates of $(1, 0)$ and $(-3, 0)$.

Step 4: Find the y-intercept.
To find the y-intercept, we substitute 0 in for x and solve for y.

$f(x) = 2(x + 1)^2 - 8$	
$y = 2(x + 1)^2 - 8$	Substitute y for $f(x)$.
$y = 2(0 + 1)^2 - 8$	Substitute 0 in for x.
$y = 2(1)^2 - 8$	
$y = 2 - 8$	
$y = -6$	

The y-intercept is the point with coordinates of $(0, -6)$.

Step 5: Identify one additional point on the graph.
Because of symmetry, the point $(-2, -6)$ is on the graph.

Step 6: Draw a smooth curve through the points found in Steps 2–5.
We can now draw the graph of the function as shown in Figure 3-17.

Comment

Sometimes the vertex of the graph of a quadratic function occurs at the origin, $(0, 0)$. In this case, both $h = 0$ and $k = 0$, and the equation of the parabola is of the form $f(x) = a(x - 0)^2 + 0$ or $f(x) = ax^2$ and $a \neq 0$.

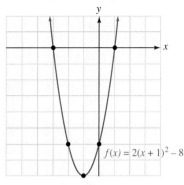

$$f(x) = 2(x + 1)^2 - 8$$

FIGURE 3-17

Self Check 2   Graph the function: $f(x) = -(x - 2)^2 + 4$.

Now Try Exercise 41.

Comment

The graph of a quadratic function can have one, two, or no x-intercepts. However, it will always have one y-intercept.

4. Find the Vertex of a Parabola Whose Equation Is in General Form

To graph a quadratic function given in the form $f(x) = ax^2 + bx + c$ $(a \neq 0)$, called **general form,** we must find the coordinates of the vertex of the parabola. To do so, we can complete the square on $ax^2 + bx$ to change the equation into standard form $y = a(x - h)^2 + k$. As we have seen, we can read the coordinates (h, k) of the vertex from this form. Once the quadratic function is in standard form, we can graph the function following the steps stated earlier.

EXAMPLE 3   **Finding the Vertex of a Parabola Written in General Form**

Find the vertex of the parabola whose equation is $f(x) = -2x^2 + 12x - 16$.

SOLUTION   We will complete the square on x, write the equation in standard form, and identify h and k, the coordinates of the vertex.
   We begin by completing the square on $-2x^2 + 12x$.

$f(x) = -2x^2 + 12x - 16$	Identify a: $a = -2$.
$f(x) = -2(x^2 - 6x) - 16$	Factor $a = -2$ from $-2x^2 + 12x$.
$f(x) = -2(x^2 - 6x + 9 - 9) - 16$	One-half of -6 is -3 and $(-3)^2 = 9$. Add and subtract 9 within the parentheses.
$f(x) = -2(x^2 - 6x + 9) - 2(-9) - 16$	Distribute the multiplication by -2.
$f(x) = -2(x - 3)^2 + 18 - 16$	Factor $x^2 - 6x + 9$ and multiply.
$f(x) = -2(x - 3)^2 + 2$	Simplify.

The equation is now in standard form with $h = 3$ and $k = 2$. Therefore, the vertex is the point with coordinates $(h, k) = (3, 2)$.

Self Check 3   Find the vertex of the graph of $y = 4x^2 - 16x + 19$.

Now Try Exercise 27.

To find formulas for the coordinates of the vertex of a parabola defined by $y = ax^2 + bx + c$ $(a \neq 0)$, we can complete the square on x to write the equation in standard form $(y = a(x - h)^2 + k)$:

$$y = ax^2 + bx + c$$

$$y = a\left(x^2 + \frac{b}{a}x\right) + c \qquad \text{Factor } a \text{ from } ax^2 + bx.$$

$$y = a\left(x^2 + \frac{b}{a}x + \frac{b^2}{4a^2} - \frac{b^2}{4a^2}\right) + c \qquad \text{Add and subtract } \frac{b^2}{4a^2} \text{ within the parentheses.}$$

$$y = a\left(x^2 + \frac{b}{a}x + \frac{b^2}{4a^2}\right) - a\left(\frac{b^2}{4a^2}\right) + c \qquad \text{Distribute the multiplication of } a.$$

$$y = a\left(x + \frac{b}{2a}\right)^2 + c - \frac{b^2}{4a} \qquad \text{Factor } x^2 + \frac{b}{a}x + \frac{b^2}{4a^2} \text{ and simplify } a\left(\frac{b^2}{4a^2}\right).$$

$$y = a\left[x - \left(-\frac{b}{2a}\right)\right]^2 + c - \frac{b^2}{4a} \qquad -\left(-\frac{b}{2a}\right) = \frac{b}{2a}$$

Comment

You don't need to memorize the formula for the y-coordinate of the vertex of a parabola. It is usually convenient to find the y-coordinate by substituting $-\frac{b}{2a}$ for x in the function and solving for y.

If we compare the last equation to the form $y = a(x - h)^2 + k$, we see that $h = -\frac{b}{2a}$ and $k = c - \frac{b^2}{4a}$. This result gives the following fact.

Vertex of a Parabola The graph of the function $y = f(x) = ax^2 + bx + c$ $(a \neq 0)$ is a parabola with vertex at $\left(-\frac{b}{2a}, c - \frac{b^2}{4a}\right)$.

EXAMPLE 4 **Graphing a Quadratic Function Written in General Form**

Graph the function: $y = f(x) = -2x^2 - 5x + 3$.

SOLUTION We begin by determining whether the parabola opens upward or downward. Then we find the vertex by using the formula $h = -\frac{b}{2a}$. Next we find the x- and y-intercepts and one additional point and then draw a smooth curve through the plotted points.

Step 1: Determine whether the parabola opens up or downward.
The equation has the form $y = ax^2 + bx + c$, where $a = -2$, $b = -5$, and $c = 3$. Since $a < 0$, the parabola opens downward.

Step 2: Find the vertex.
To find the x-coordinate of the vertex, we substitute the values of a and b into the formula $x = -\frac{b}{2a}$.

$$x = -\frac{b}{2a} = -\frac{-5}{2(-2)} = -\frac{5}{4}$$

The x-coordinate of the vertex is $-\frac{5}{4}$. To find the y-coordinate, we substitute $-\frac{5}{4}$ for x in the equation and solve for y.

$$y = -2x^2 - 5x + 3$$

$$y = -2\left(-\frac{5}{4}\right)^2 - 5\left(-\frac{5}{4}\right) + 3 \qquad \text{Substitute } -\frac{5}{4} \text{ for } x.$$

$$= -2\left(\frac{25}{16}\right) + \frac{25}{4} + 3$$

$$= -\frac{25}{8} + \frac{50}{8} + \frac{24}{8}$$

$$= \frac{49}{8}$$

Since the vertex is the point $\left(-\frac{5}{4}, \frac{49}{8}\right)$, we can plot it on the coordinate system in Figure 3-18(a) and draw the axis of symmetry.

Step 3: Find the x-intercept(s).
To find the x-intercepts, we substitute 0 for y and solve for x.

$$y = -2x^2 - 5x + 3$$

$0 = -2x^2 - 5x + 3$ Substitute 0 for y.

$0 = 2x^2 + 5x - 3$ Divide both sides by -1 to make the leading coefficient positive.

$0 = (2x - 1)(x + 3)$ Factor the trinomial.

$2x - 1 = 0$ or $x + 3 = 0$ Set each factor equal to 0.

$x = \frac{1}{2}$ | $x = -3$ Solve each linear equation.

The x-intercepts are $\left(\frac{1}{2}, 0\right)$ and $(-3, 0)$. We plot these intercepts as shown in Figure 3-18(a).

Step 4: Find the y-intercept.
To find the y-intercept, we let $x = 0$ and solve for y.

$$y = -2x^2 - 5x + 3$$

$= -2(0)^2 - 5(0) + 3$ Substitute 0 for x.

$= 0 - 0 + 3$

$= 3$

The y-intercept is $(0, 3)$. We plot the intercept as shown in Figure 3-18(a).

Step 5: Plot one additional point.
Because of symmetry, we know that the point $\left(-2\frac{1}{2}, 3\right)$ is on the graph. We plot this point on the coordinate system in Figure 3-18(a).

Step 6: We can now draw the graph of the function, as shown in Figure 3-18(b).

Comment

The y-intercept of a parabola written in the general form $f(x) = ax^2 + bx + c$ $(a \neq 0)$ is the point $(0, c)$. This is because when we substitute 0 for x, y is always c.

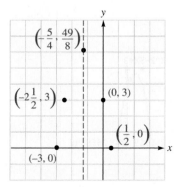

(a)

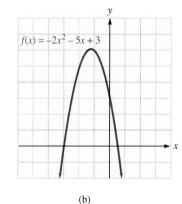

(b)

FIGURE 3-18

Self Check 4   Graph the function: $f(x) = -3x^2 + 7x - 2$.

Now Try Exercise 53.

ACCENT ON TECHNOLOGY   **Finding the Maximum Point or Minimum Point (Vertex) of a Parabola**

We can use a graphing calculator to find the maximum point or minimum point (vertex) of a parabola. Consider the $f(x) = -2x^2 - 5x + 3$ given in Example 4. We will apply the following steps to determine the maximum point or vertex of the parabola.

1. Enter the function.

2. Set a window that shows the function.

3. Graph the function.

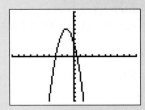

4. Go to the CALC menu by pressing [2nd] [TRACE] and select maximum.

5. Use the [TRACE] key to get a left bound.

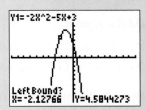

6. Use the [TRACE] key to get a right bound.

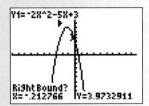

7. Use the [TRACE] key to make a guess.

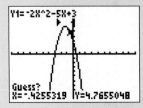

8. Press [ENTER] and find the maximum point.

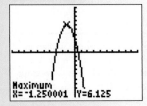

FIGURE 3-19

We see that the maximum point or the vertex is $(-1.250001, 6.125)$.

5. Use a Quadratic Function to Solve Maximum and Minimum Problems

EXAMPLE 5   **Using a Quadratic Function to Solve a Maximum Area Problem**

The Montana Dude Rancher's Association has 400 feet of fencing to enclose a rectangular corral. To save money and fencing, the association intends to use the bank of a river as one boundary of the corral, as in Figure 3-20. Find the dimensions that will enclose the largest area.

SOLUTION   We will represent the fenced area with a quadratic function. Since its parabolic graph opens downward, the largest or maximum area will occur at the vertex. We can use the vertex formula to find the vertex.

Step 1: Represent the area with a quadratic function.
Let x represent the width of the fenced area. Then $400 - 2x$ represents the length. Because the area A of a rectangle is the product of the length and the width, we have

$$A = (400 - 2x)x \quad \text{or} \quad A(x) = -2x^2 + 400x$$

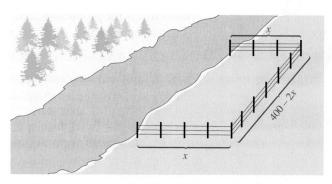

FIGURE 3-20

The graph of this area function is a parabola. Since the coefficient of x^2 is negative, the parabola opens downward and its vertex is its highest point. The A-coordinate of the vertex (x, A) represents the maximum area, and the x-coordinate represents the width of the corral that will give the maximum area.

Step 2: Find the vertex of the parabola.
We compare the equations

$$A(x) = -2x^2 + 400x \quad \text{and} \quad y = ax^2 + bx + c$$

to see that $a = -2, b = 400$, and $c = 0$. Using the vertex formula, the vertex of the parabola is the point with coordinates

$$\left(-\frac{b}{2a}, c - \frac{b^2}{4a}\right) = \left(-\frac{400}{2(-2)}, 0 - \frac{400^2}{4(-2)}\right) = (100, 20{,}000)$$

Comment   Note that we could have determined the y-coordinate of 20,000 by finding $A(100)$.

$$A(x) = -2x^2 + 400x$$
$$A(100) = -2(100)^2 + 400(100) \qquad \text{Substitute 100 for } x.$$
$$= -2(10{,}000) + 40{,}000$$
$$= -20{,}000 + 40{,}000$$
$$= 20{,}000$$

If the fence runs 100 feet out from the river, 200 feet parallel to the river, and 100 feet back to the river, it will enclose the largest possible area, which is 20,000 square feet. ■

Self Check 5   Find the largest area possible if the association has 1,200 feet of fencing available.

Now Try Exercise 55.

EXAMPLE 6   Using a Quadratic Function to Solve a Minimum Cost Problem

A company that makes and sells water skis has found that the total weekly cost C of producing x water skis is given by the function $C(x) = 0.5x^2 - 210x + 26{,}250$. Find the production level that minimizes the weekly cost and find that weekly minimum cost.

SOLUTION   The weekly cost function $C(x)$ is a quadratic function whose graph is a parabola that opens upward. The minumum value of $C(x)$ occurs at the vertex of the parabola. We will use the vertex formula to find the vertex of the parabola.

Since the coefficient of x^2 is 0.5 (a positive real number), the x-coordinate of the vertex is the production level that will minimize the cost, and the y-coordinate is that minimum cost. We compare the equations

$$C(x) = 0.5x^2 - 210x + 26{,}250 \quad \text{and} \quad y = ax^2 + bx + c$$

to see that $a = 0.5$, $b = -210$, and $c = 26{,}250$. Using the vertex formula, we see that the vertex of the parabola is the point with coordinates

$$\left(-\frac{b}{2a}, \, c - \frac{b^2}{4a}\right) = \left(-\frac{-210}{2(0.5)}, \, 26{,}250 - \frac{(-210)^2}{4(0.5)}\right) = (210, \, 4{,}200)$$

Comment   We can also determine the y-coordinate of 4,200 by finding $C(210)$.

$$C(x) = 0.5x^2 - 210x + 26{,}250$$

$$C(210) = 0.5(210)^2 - 210(210) + 26{,}250 \qquad \text{Substitute 210 for } x.$$

$$= 0.5(44{,}100) - 44{,}100 + 26{,}250$$

$$= 22{,}050 - 17{,}850$$

$$= 4{,}200$$

If the company makes 210 water skis each week, it will minimize its production cost. The minimum weekly cost will be $4,200. ■

Self Check 6   A company that makes and sells baseball caps has found that the total monthly cost C of producing x caps is given by the function $C(x) = 0.2x^2 - 80x + 9{,}000$. Find the production level that will minimize the monthly cost and find the minimum cost.

Now Try Exercise 67.

Self Check Answers   1. $(-5, -4)$ 2. [graph with $f(x) = -(x-2)^2 + 4$]   3. $(2, 3)$ 4. [graph with $f(x) = -3x^2 + 7x - 2$]
5. 180,000 square feet
6. 200, $1,000

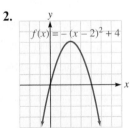

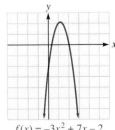

Exercises **3.2**

Getting Ready

You should be able to complete these vocabulary and concept statements before you proceed to the practice exercises.

Fill in the blanks.

1. A quadratic function is defined by the equation _____ $(a \neq 0)$.

2. The standard form for the equation of a parabola is _____ $(a \neq 0)$.

3. The vertex of the parabolic graph of the equation $y = 2(x - 3)^2 + 5$ will be at _____.

4. The vertical line that intersects the parabola at its vertex is the _____.

5. If the parabola opens _____ the vertex will be a minimum point.

6. If the parabola opens _____ the vertex will be a maximum point.

7. The x-coordinate of the vertex of the parabolic graph of $f(x) = ax^2 + bx + c$ is _____.

8. The y-coordinate of the vertex of the parabolic graph of $f(x) = ax^2 + bx + c$ is _____.

Practice

Determine whether the graph of each quadratic function opens upward or downward. State whether a maximum or minimum point occurs at the vertex of the parabola.

9. $f(x) = \frac{1}{2}x^2 + 3$ 10. $f(x) = 2x^2 - 3x$

11. $f(x) = -3(x + 1)^2 + 2$ 12. $f(x) = -5(x - 1)^2 - 1$

13. $f(x) = -2x^2 + 5x - 1$ 14. $f(x) = 2x^2 - 3x + 1$

Find the vertex of each parabola.

15. $y = x^2 - 1$ 16. $y = -x^2 + 2$
17. $f(x) = (x - 3)^2 + 5$ 18. $f(x) = -2(x - 3)^2 + 4$

19. $f(x) = -2(x + 6)^2 - 4$ 20. $f(x) = \frac{1}{3}(x + 1)^2 - 5$

21. $f(x) = \frac{2}{3}(x - 3)^2$ 22. $f(x) = 7(x + 2)^2 + 8$

23. $f(x) = x^2 - 4x + 4$ 24. $y = x^2 - 10x + 25$

25. $y = x^2 + 6x - 3$ 26. $y = -x^2 + 9x - 2$

27. $y = -2x^2 + 12x - 17$ 28. $y = 2x^2 + 16x + 33$

29. $y = 3x^2 - 4x + 5$ 30. $y = -4x^2 + 3x + 4$

31. $y = \frac{1}{2}x^2 + 4x - 3$ 32. $y = -\frac{2}{3}x^2 + 3x - 5$

Graph each quadratic function given in standard form.

33. $f(x) = x^2 - 4$ 34. $f(x) = x^2 + 1$

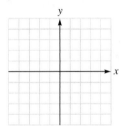   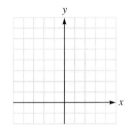

35. $f(x) = -3x^2 + 6$ 36. $f(x) = -4x^2 + 4$

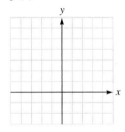   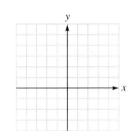

37. $f(x) = -\frac{1}{2}x^2 + 8$ 38. $f(x) = \frac{1}{2}x^2 - 2$

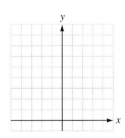   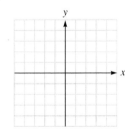

39. $f(x) = (x - 3)^2 - 1$ 40. $f(x) = (x + 3)^2 - 1$

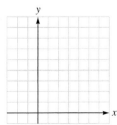   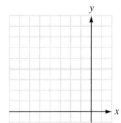

41. $f(x) = 2(x + 1)^2 - 2$

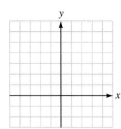

42. $f(x) = -\dfrac{3}{4}(x - 2)^2$

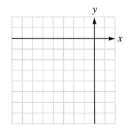

43. $f(x) = -(x + 4)^2 + 1$

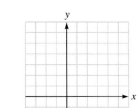

44. $f(x) = -3(x - 4)^2 + 3$

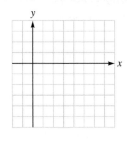

45. $f(x) = -3(x - 2)^2 + 6$

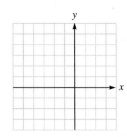

46. $f(x) = 2(x - 3)^2 - 4$

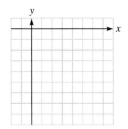

Graph each quadratic function given in general form.

47. $f(x) = x^2 + 2x$

48. $f(x) = x^2 - 6x$

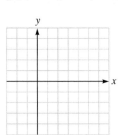

49. $f(x) = x^2 - 4x + 1$

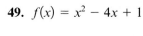

50. $f(x) = x^2 - 6x - 7$

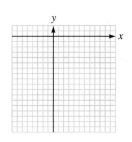

51. $f(x) = 2x^2 - 12x + 10$

52. $f(x) = -x^2 - 4x + 1$

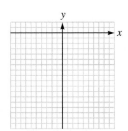
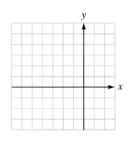

53. $f(x) = -3x^2 - 6x - 9$

54. $f(x) = -3x^2 - 3x + 18$

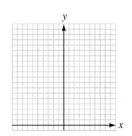

Applications

55. Police investigations A police officer seals off the scene of an accident using a roll of yellow tape that is 300 feet long. What dimensions should be used to seal off the maximum rectangular area around the collision? Find the maximum area.

56. Maximizing area The rectangular garden shown has a width of x and a perimeter of 100 feet. Find x such that the area of the rectangle is maximum.

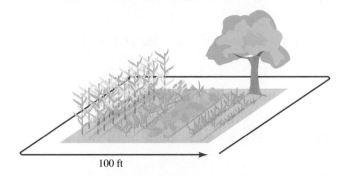

100 ft

57. Maximizing storage area A farmer wants to partition a rectangular feed storage area in a corner of his barn, as shown in the illustration. The barn walls form two sides of the stall, and the farmer has 50 feet of partition for the remaining two sides. What dimensions will maximize the area?

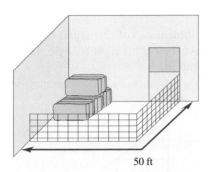

50 ft

58. Maximizing grazing area A rancher wishes to enclose a rectangular partitioned corral with 1,800 feet of fencing. (See the illustration.) What dimensions of the corral would enclose the largest possible area? Find the maximum area.

59. Sheet metal fabrication A 24-inch-wide sheet of metal is to be bent into a rectangular trough with the cross section shown in the illustration. Find the dimensions that will maximize the amount of water the trough can hold. That is, find the dimensions that will maximize the cross-sectional area.

Depth
24 in.
Width

60. Landscape design A gardener will use D feet of edging to border a rectangular plot of ground. Show that the maximum area will be enclosed if the rectangle is a square.

61. Architecture A parabolic arch has an equation of $x^2 + 20y - 400 = 0$, where x is measured in feet. Find the maximum height of the arch.

62. Path of a guided missile A guided missile is propelled from the origin of a coordinate system with the x-axis along the ground and the y-axis vertical. Its path, or **trajectory,** is given by the equation $y = 400x - 16x^2$. Find the object's maximum height.

63. Height of a basketball The path of a basketball thrown from the free throw line can be modeled by the quadratic function $f(x) = -0.06x^2 + 1.5x + 6$, where x is the horizontal distance (in feet) from the free throw line and $f(x)$ is the height (in feet) of the ball. Find the maximum height of the basketball.

64. Ballistics A child throws a ball up a hill that makes an angle of 45° with the horizontal. The ball lands 100 feet up the hill. Its trajectory is a parabola with equation $y = -x^2 + ax$ for some number a. Find a.

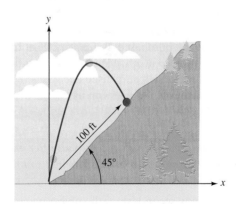

65. Maximizing height A ball is thrown straight up from the top of a building 144 ft. tall with an initial velocity of 64 ft per second. The distance $s(t)$ (in feet) of the ball from the ground is given by $s(t) = 144 + 64t - 16t^2$. Find the maximum height attained by the ball.

66. Flat-panel television sets A wholesaler of appliances finds that she can sell $(2,400 - p)$ flat-panel television sets each week when the price is p dollars. What price will maximize revenue?

67. Digital camera A company that produces and sells digital cameras has determined that the total weekly cost C of producing x digital cameras is given by the function $C(x) = 1.5x^2 - 144x + 5{,}856$. Determine the production level that minimizes the weekly cost for producing the digital cameras and find that weekly minimum cost.

68. Finding mass transit fares The Municipal Transit Authority serves 150,000 commuters daily when the fare is $1.80. Market research has determined that every penny decrease in the fare will result in 1,000 new riders. What fare will maximize revenue?

69. Finding hotel rates A 300-room hotel is two-thirds filled when the nightly room rate is $90. Experience has shown that each $5 increase in cost results in 10 fewer occupied rooms. Find the nightly rate that will maximize income.

70. Selling concert tickets Tickets for a concert are cheaper when purchased in quantity. The first 100 tickets are priced at $10 each, but each additional block of 100 tickets purchased decreases the cost of each ticket by 50¢. How many blocks of tickets should be sold to maximize the revenue?

Use this information: At a time t seconds after an object is tossed vertically upward, it reaches a height s in feet given by the equation $s = 80t - 16t^2$.

71. In how many seconds does the object reach its maximum height?

72. In how many seconds does the object return to the point from which it was thrown?

73. What is the maximum height reached by the object?

74. Show that it takes the same amount of time for the object to reach its maximum height as it does to return from that height to the point from which it was thrown.

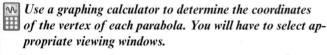 *Use a graphing calculator to determine the coordinates of the vertex of each parabola. You will have to select appropriate viewing windows.*

75. $y = 2x^2 + 9x - 56$ **76.** $y = 14x - \dfrac{x^2}{5}$

77. $y = (x - 7)(5x + 2)$ **78.** $y = -x(0.2 + 0.1x)$

Discovery and Writing
Find all values of x that will make $f(x) = 0$.

79. $f(x) = x^2 - 5x + 6$ **80.** $f(x) = 6x^2 + x - 2$

81. Find the dimensions of the largest rectangle that can be inscribed in the right triangle ABC shown in the illustration.

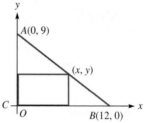

82. Point P lies in the first quadrant and on the line $x + y = 1$ in such a position that the area of triangle OPA is maximum. Find the coordinates of P. (See the illustration.)

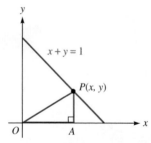

83. The sum of two numbers is 6, and the sum of the squares of those two numbers is as small as possible. What are the numbers?

84. What number most exceeds its square?

The maximum or minimum value of a quadratic function can be found automatically by using a computer program, such as Excel.

85. Find the minimum value of the function $f(x) = 2x^2 - 3x - 4$ by using the Solver in Excel. Give the value of x that minimizes the function as well as the minimum value of the function. See Problem 119 in Section 1.3.

86. Find the maximum value of the function $f(x) = -2x^2 + 3x + 4$ by using the Solver in Excel. Give the value of x that maximizes the function as well as the maximum value of the function. See Problem 119 in Section 1.3.

Review
Find $f(a)$ and $f(-a)$.

87. $f(x) = x^2 - 3x$

88. $f(x) = x^3 - 3x$

89. $f(x) = (5 - x)^2$

90. $f(x) = \dfrac{1}{x^2 - 4}$

91. $f(x) = 7$

92. $f(x) = -|x|$

3.3 Polynomial and Other Functions

In this section, we will learn to

1. Understand the characteristics of polynomial functions.

2. Graph polynomial functions.

3. Determine whether a function is even, odd, or neither.

4. Identify the open intervals on which a function is increasing, decreasing, or constant.

5. Graph piecewise-defined functions.

6. Evaluate and graph the greatest-integer function.

© Istockphoto.com/Steve Maehl

So far, we have discussed two types of polynomial functions—first-degree (or linear) functions, and second-degree (or quadratic) functions. In this section, we will discuss polynomial functions of higher degree.

Polynomial functions can be used to model the path of a roller coaster or to model the fluctuation of gasoline prices over the past few months. Goliath, a hypercoaster, is located in Atlanta at Six Flags over Georgia. It climbs to a height of 200 feet and reaches speeds of nearly 70 mph. It has over 4,400 feet of steel track. Portions of Goliath's tracks can be modeled with a polynomial function.

Polynomial Functions

A **polynomial function in one variable (say, x)** is a function of the form

$$f(x) = a_n x^n + a_{n-1} x^{n-1} + \cdots + a_1 x + a_0$$

where $a_n, a_{n-1}, \ldots, a_1$, and a_0 are real numbers and n is a whole number.

The **degree of a polynomial function** is the largest power of x that appears in the polynomial.

The table shows three basic polynomial functions that we have already covered:

Name	Function	Degree	Graph
Constant function	$f(x) = 5$	0	Horizontal line
Linear function	$f(x) = 2x - 7$	1	Nonvertical line
Quadratic function	$f(x) = -2x^2 + 4x - 5$	2	Parabola

Here are two examples of higher-degree polynomial functions, along with their graphs.

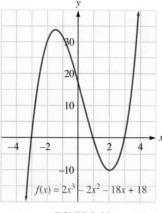

| FIGURE 3-21 | FIGURE 3-22 |

1. Understand the Characteristics of Polynomial Functions

There are several basic characteristics common to all polynomial functions. We will list several of them.

1: The graphs of polynomial functions are smooth and continuous curves.

Like the graphs of linear and quadratic functions, the graphs of higher-degree polynomial functions are smooth continuous curves. Because their graphs are smooth, they have no cusps or corners. Because they are continuous, their graphs have no breaks or holes. They can always be drawn without lifting the pencil from the paper.

- The graph of a polynomial function

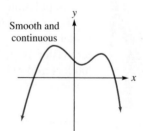

Graph of a polynomial function

FIGURE 3-23

- The graphs of two functions that are **not** polynomial functions

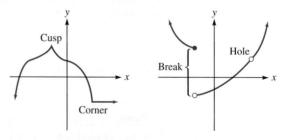

Not the graph of a polynomial function   Not the graph of a polynomial function

FIGURE 3-24

2: Many polynomial functions have graphs similar to the graphs of $f(x) = x$, $f(x) = x^2$, and $f(x) = x^3$.

Many polynomial functions are of the form $f(x) = x^n$, and several of their graphs are shown in Figure 3-25. Note that when n is even, the graph has the same general shape as $y = x^2$. When n is odd and greater than 1, the graph has the same general shape as $y = x^3$. However, the graphs are flatter at the origin and steeper as n becomes large.

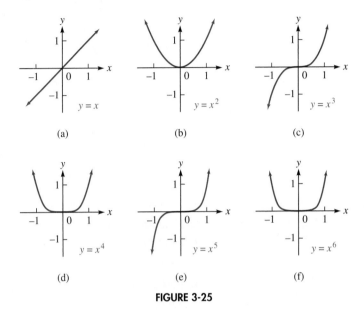

FIGURE 3-25

3: The end behavior of the graph of a polynomial function is similar to the graph of its term with highest degree.

The ends of the graph of any polynomial function will be similar to the graph of its term with the highest power of x, because when n becomes large, the other terms become relatively insignificant.

Consider the polynomial function $f(x) = x^3 + x^2 - 2x$. The end behavior of its graph will be similar to the ends of the graph of its leading term x^3. In Figure 3-25 (c), we see that the graph of $y = x^3$ falls on the far left and rises on the far right. Therefore, the graph of $f(x)$ will also fall on the far left and rise on the far right. See Figure 3-26.

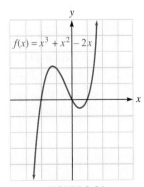

FIGURE 3-26

4: Polynomial functions can be symmetric about the y-axis or the origin.

In Section 2.4, we learned that a graph is symmetric about the y-axis if the graph of $y = f(x)$ has the same y-coordinate when the function is evaluated at x or at $-x$.

Thus, a function is symmetric about the y-axis if $f(x) = f(-x)$ for all values of x that are in the domain of the function. See Figure 3-27(a). Also recall that a graph is symmetric about the origin if the point $(-x, -f(x))$ lies on the graph whenever $(x, f(x))$ does. In this case, $f(-x) = -f(x)$. See Figure 3-27(b).

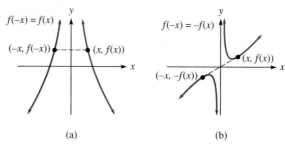

(a)   (b)

FIGURE 3-27

2. Graph Polynomial Functions

To graph polynomial functions, we can use the following steps.

Strategy for Graphing Polynomial Functions

1. Find any symmetries of the graph.
2. Find the x- and y-intercepts of the graph.
3. Determine where the graph is above and below the x-axis.
4. Plot a few points, if necessary, and draw the graph as a smooth continuous curve.

EXAMPLE 1   **Graphing a Polynomial Function of Degree 3**

Graph the function: $f(x) = x^3 - 4x$.

SOLUTION   We will use the four steps stated above to graph the polynomial function.

Step 1: Find any symmetries of the graph.
To test for symmetry about the y-axis, we check to see whether $f(x) = f(-x)$. To test for symmetry about the origin, we check to see whether $f(-x) = -f(x)$.

$$f(x) = x^3 - 4x$$

$$f(-x) = (-x)^3 - 4(-x) \qquad \text{Substitute } -x \text{ for } x.$$

$$f(-x) = -x^3 + 4x \qquad \text{Simplify.}$$

Since $f(x) \neq f(-x)$, there is no symmetry about the y-axis. However, since $f(-x) = -f(x)$, there is symmetry about the origin.

Step 2: Find the x- and y-intercepts of the graph.
To find the x-intercepts, we let $f(x) = 0$ and solve for x.

$$x^3 - 4x = 0$$

$$x(x^2 - 4) = 0 \qquad \text{Factor out } x.$$

$$x(x + 2)(x - 2) = 0 \qquad \text{Factor } x^2 - 4.$$

$$x = 0 \quad \text{or} \quad x + 2 = 0 \quad \text{or} \quad x - 2 = 0 \qquad \text{Set each factor equal to 0.}$$

$$x = -2 \qquad x = 2$$

The x-intercepts are $(0, 0)$, $(-2, 0)$, and $(2, 0)$. If we let $x = 0$ and solve for $f(x)$, we see that the y-intercept is also $(0, 0)$.

Step 3: Determine where the graph is above or below the x-axis.

To determine where the graph is above or below the x-axis, we plot the solutions of $x^3 - 4x = 0$ (the x-coordinates of the x-intercepts) on a number line and establish the four intervals shown in Figure 3-28. We then test a number from each interval to determine the sign of $f(x)$. (For a review of this process, see Example 7 in Section 1.7.)

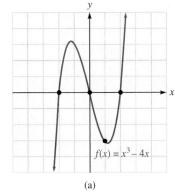

Sign of $f(x) = x^3 - 4x$

	$-$	$+$	$-$	$+$
	$(-\infty, -2)$	$(-2, 0)$	$(0, 2)$	$(2, \infty)$
Test point	$f(-3) = -15$ -2	$f(-1) = 3$ 0	$f(1) = -3$ 2	$f(3) = 15$
Graph of $f(x)$	below the x-axis	above the x-axis	below the x-axis	above the x-axis

FIGURE 3-28

Comment

Note that the far right and far left ends of the graph are similar to the ends of the graph of $f(x) = x^3$, which is the leading term of the function $f(x) = x^3 - 4x$ or the term with highest degree. On the far right the graph rises and on the far left the graph falls.

Step 4: Plot a few points and draw the graph as a smooth continuous curve.

We now plot the intercepts and one additional point. In the previous step we found that $f(1) = -3$. This will be the additional point we plot $(1, -3)$. Making use of our knowledge of symmetry and where the graph is above and below the x-axis we now draw the graph as shown in Figure 3-29(a). A calculator graph, using the standard viewing window, is shown in Figure 3-29(b).

$f(x) = x^3 - 4x$		
x	$f(x)$	$(x, f(x))$
-2	0	$(-2, 0)$
0	0	$(0, 0)$
1	-3	$(1, -3)$
2	0	$(2, 0)$

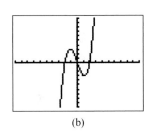

(a)

(b)

FIGURE 3-29

Self Check 1   Graph: $f(x) = x^3 - 9x$.

Now Try Exercise 13.

The peak and valley of the graph shown in Figure 3-29 are called **turning points.** In calculus, such points are called **local minima** and **local maxima.** Although we cannot find these points without using calculus, we can approximate them by plotting points or by using the TRACE feature on a graphing calculator. We can also use the CALC Minimum and CALC Maximum features on a graphing calculator to find these local extrema.

Note that Figure 3-29 shows the graph of a third-degree polynomial and the graph has two turning points. This suggests the following result from calculus that helps us understand the shape of many polynomial graphs.

Number of Turning Points   If $f(x)$ is a polynomial function of nth degree, then the graph of $f(x)$ will have $n - 1$, or fewer, turning points.

EXAMPLE 2   **Graphing a Polynomial Function of Degree 4**

Graph the function: $f(x) = x^4 - 5x^2 + 4$.

SOLUTION   We will use the four steps for graphing a polynomial function.

Step 1: Find any symmetries of the graph.
Because x appears with only even exponents, $f(x) = f(-x)$, and the graph is symmetric about the y-axis. The graph is not symmetric about the origin.

Step 2: Find the x- and y-intercepts of the graph.
To find the x-intercepts, we let $f(x) = 0$ and solve for x.

$$x^4 - 5x^2 + 4 = 0$$
$$(x^2 - 4)(x^2 - 1) = 0$$
$$(x + 2)(x - 2)(x + 1)(x - 1) = 0$$

$$x + 2 = 0 \quad \text{or} \quad x - 2 = 0 \quad \text{or} \quad x + 1 = 0 \quad \text{or} \quad x - 1 = 0$$
$$x = -2 \quad | \quad x = 2 \quad | \quad x = -1 \quad | \quad x = 1$$

The x-intercepts are $(-2, 0)$, $(2, 0)$, $(-1, 0)$, and $(1, 0)$. To find the y-intercept, we let $x = 0$ and see that the y-intercept is $(0, 4)$.

Step 3: Determine where the graph is above or below the x-axis.
To determine where the graph is above or below the x-axis, we plot the x-coordinates of the x-intercepts on a number line and establish the five intervals shown in Figure 3-30. We then test a number from each interval to determine the sign of $f(x)$.

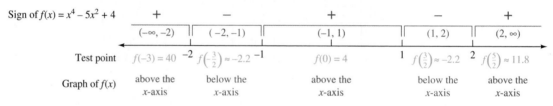

FIGURE 3-30

Comment

Note that the far right and far left ends of the graph are similar to the ends of the graph of $f(x) = x^4$, which is the leading term of the function $f(x) = x^4 - 5x^2 + 4$ or the term with highest degree. On the far right and far left the graph rises. Also note that the graph has three turning points.

Step 4: Plot a few points and draw the graph as a smooth continuous curve.
We can now plot the intercepts and use our knowledge of symmetry and where the graph is above and below the x-axis to draw the graph, as shown in Figure 3-31(a). A calculator graph, using the standard viewing window, is shown in Figure 3-31(b).

$f(x) = x^4 - 5x^2 + 4$		
x	$f(x)$	$(x, f(x))$
-2	0	$(-2, 0)$
-1	0	$(-1, 0)$
0	4	$(0, 4)$
1	0	$(1, 0)$
$\dfrac{3}{2}$	$-\dfrac{35}{16}$	$\left(\dfrac{3}{2}, -\dfrac{35}{16}\right)$
2	0	$(2, 0)$

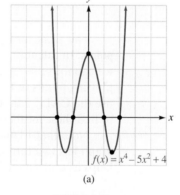

(a)

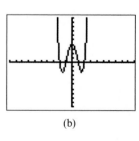

(b)

FIGURE 3-31

Self Check 2   Graph: $f(x) = x^4 - 10x^2 + 9$.

Now Try Exercise 19.

3. Determine Whether a Function Is Even, Odd, or Neither

Functions can be classified as being even, odd, or neither even nor odd. These definitions are given next.

Even Function   A function is an **even function** if $f(-x) = f(x)$ for all x in the domain of f.

Even functions are symmetric about the y-axis.

Odd Function   A function is an **odd function** if $f(-x) = -f(x)$ for all x in the domain of f.

Odd functions are symmetric about the origin.

Since the graph of the function $f(x) = x^3 - 4x$ in Example 1 is symmetric about the origin, it represents an odd function. Since the graph of the function $f(x) = x^4 - 5x + 4$ in Example 2 is symmetric about the y-axis, it represents an even function. If $f(x)$ does not have either of these symmetries, it is neither even nor odd.

EXAMPLE 3   **Determining Whether a Function Is Even, Odd, or Neither**

Determine whether each function is even, odd, or neither:
a. $f(x) = x^2 + 1$ **b.** $f(x) = x^3$

SOLUTION   To check whether the function is an even function, we find $f(-x)$ and see whether $f(-x) = f(x)$. To check whether the function is an odd function, we find $f(-x)$ and see whether $f(-x) = -f(x)$.

a. $f(-x) = (-x)^2 + 1 = x^2 + 1 = f(x)$

Since $f(-x) = f(x)$, the function is an even function.

The graph of $f(x) = x^2 + 1$ is shown in Figure 3-32. We see that the graph is symmetric about the y-axis.

b. $f(-x) = (-x)^3 = -x^3 = -f(x)$

Since $f(-x) \neq f(x)$, the function is not an even function.

However, the function is an odd function, because $f(-x) = -f(x)$.

The graph of $f(x) = x^3$ is shown in Figure 3-33. Note that the graph is symmetric about the origin.

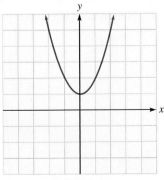

FIGURE 3-32

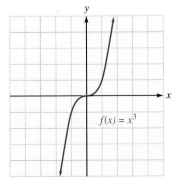

FIGURE 3-33

Self Check 3 Classify each function as even, odd, or neither: **a.** $f(x) = x^3 + x$
b. $f(x) = x^2 + 4$

Now Try Exercise 23.

4. Identify the Open Intervals on Which a Function Is Increasing, Decreasing, or Constant

If we trace the graph of a function from left to right and the values $f(x)$ increase as shown in Figure 3-34(a), we say that the function is an **increasing function.** If the values $f(x)$ decrease as in Figure 3-34(b), we say that the function is a **decreasing function.** If the values $f(x)$ remain unchanged as x increases, we say that the function is a **constant function.**

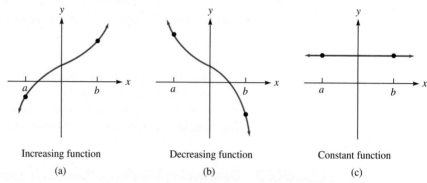

Increasing function	Decreasing function	Constant function
(a)	(b)	(c)

FIGURE 3-34

As we would expect, when we trace the graphs of some functions from left to right, there will be intervals where the function increases, intervals where the function decreases, and intervals where the function is constant. We will always state these intervals as open intervals because of the following definition.

Increasing on an Open Interval A function f is **increasing on an open interval** (a, b) if for any x_1 and x_2 in (a, b), where $x_1 < x_2$, then $f(x_1) < f(x_2)$.

Decreasing on an Open Interval A function f is **decreasing on an open interval** (a, b) if for any x_1 and x_2 in (a, b), where $x_1 < x_2$, then $f(x_1) > f(x_2)$.

Constant on an Open Interval A function f is **constant on an open interval** (a, b) if for any x_1 and x_2 in (a, b), $f(x_1) = f(x_2)$.

EXAMPLE 4 **Identify the Open Intervals Where a Function Increases or Decreases**

Use the graph of the polynomial function shown to determine the intervals on which the function is increasing or decreasing.

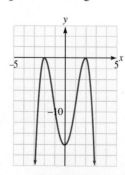

FIGURE 3-35

SOLUTION   We trace the graph of the function from left to right and identify the open intervals on which $f(x)$ values increase and where they decrease. We see from the graph shown in Figure 3-35 that the values of $f(x)$ increase on the open intervals $(-\infty, -2)$ and $(0, 2)$. They decrease on the open intervals $(-2, 0)$ and $(2, \infty)$. Hence, the function is increasing on $(-\infty, -2) \cup (0, 2)$ and is decreasing on $(-2, 0) \cup (2, \infty)$. ∎

Self Check 4   Use the graph shown in Figure 3-35 to identify the open intervals, if any, on which the function is constant.

Now Try Exercise 41.

Comment

Intervals on which a function is increasing, decreasing, or constant are open intervals. We use interval notation and x-values to write the open intervals. Note that y-values are not used.

5. Graph Piecewise-Defined Functions

Some functions, called **piecewise-defined functions**, are defined by using different equations for different intervals in their domains. To illustrate, we will graph the piecewise-defined function f given by

$$f(x) = \begin{cases} -2 & \text{if } x \leq 0 \\ x + 1 & \text{if } x > 0 \end{cases}$$

To evaluate this piecewise-defined function, we must determine which part of the function's definition to use.

*   If $x \leq 0$, we use the top part of the definition and the corresponding value of $f(x)$ is -2. Some examples are:

    $$f(-2) = -2, \quad f(-1) = -2, \quad \text{and} \quad f(0) = -2$$

    In the interval $(-\infty, 0]$, the function value is always -2 and the graph is the horizontal line $y = -2$.

*   If $x > 0$, we use the bottom part of the definition and the corresponding value of $f(x)$ is $x + 1$. Some examples are:

    $$f(1) = 1 + 1 = 2 \text{ and } f(2) = 2 + 1 = 3$$

    In the interval $(0, \infty)$, the graph of the function is the line $f(x) = x + 1$. This is a linear function with slope $m = 1$ and y-intercept $(0, b) = (0, 1)$.

The graph of this piecewise-defined function appears in Figure 3-36.

Comment

Note that when $x = 0$, we have $f(x) = -2$. For this reason, the point $(0, -2)$ is shown as a closed point and the point $(0, 1)$ is shown as an open point.

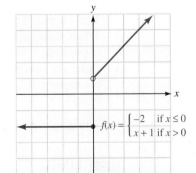

$$f(x) = \begin{cases} -2 & \text{if } x \leq 0 \\ x + 1 & \text{if } x > 0 \end{cases}$$

FIGURE 3-36

EXAMPLE 5   **Graphing a Piecewise-Defined Function**

Graph the function: $f(x) = \begin{cases} -x & \text{if } x < 0 \\ x^2 & \text{if } 0 \leq x \leq 1. \\ 1 & \text{if } x > 1 \end{cases}$

SOLUTION   This piecewise-defined function is defined in three parts. We will graph each part of the function. That is, we will graph $f(x) = -x$ in the interval $(-\infty, 0)$, $f(x) = x^2$ in the interval $[0, 1]$, and $f(x) = 1$ in the interval $(1, \infty)$.

If $x < 0$, the value of $f(x)$ is determined by the equation $f(x) = -x$. We graph the line with slope $m = -1$ and y-intercept $(0, b) = (0, 0)$ in the interval $(-\infty, 0)$. In the interval $(-\infty, 0)$, the function is decreasing.

If $0 \le x \le 1$, the value of $f(x)$ is x^2. We graph the parabola in the interval $[0, 1]$. In the open interval $(0, 1)$ the function is increasing.

If $x > 1$, the value of $f(x)$ is 1. In the open interval $(1, \infty)$, the function is constant and its graph is the same as the graph of $y = 1$.

The graph of the piecewise function appears in Figure 3-37.

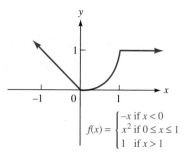

$$f(x) = \begin{cases} -x & \text{if } x < 0 \\ x^2 & \text{if } 0 \le x \le 1 \\ 1 & \text{if } x > 1 \end{cases}$$

FIGURE 3-37

Self Check 5   Graph: $f(x) = \begin{cases} 2x & \text{if } x \le 0 \\ x - 1 & \text{if } x > 0 \end{cases}$.

Now Try Exercise 57.

ACCENT ON TECHNOLOGY   **Piecewise-Defined Functions**

Piecewise-defined functions can be graphed on a calculator. We will graph the piecewise-defined function $f(x) = \begin{cases} -x & \text{if } x < 0 \\ x^2 & \text{if } 0 \le x \le 1 \\ 1 & \text{if } x > 1 \end{cases}$ given in Example 5.

- The inequality symbols are found by pressing [2nd] [MATH] (TEST) shown in Figure 3-38(a).
- The "and" command is found in the LOGIC menu which is accessed via the TEST menu. See Figure 3-38(b).
- When graphing these functions, place the calculator in DOT mode. See Figure 3-38(c).
- Enter the function as shown in Figure 3-38(d). Note that the use of parentheses is essential. A viewing window of $[-5, 5]$ for x and $[-3, 3]$ for y is used here.

The graph of the function is shown in Figure 3-38(e).

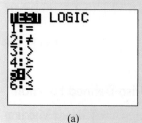

(a)

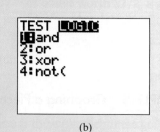

(b)

(c)

FIGURE 3-38

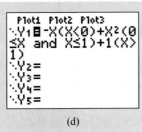

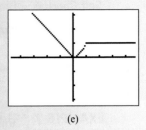

(d)   (e)

FIGURE 3-38 (Continued)

6. Evaluate and Graph the Greatest-Integer Function

The **greatest-integer function** is important in many business applications and in the field of computer science. This function is determined by the equation $f(x) = [\![x]\!]$, where the value of $f(x)$ that corresponds to x is the greatest integer that is less than or equal to x.

Comment

Recall that the set of integers is $\{ \dots , -2, -1, 0, 1, 2, \dots \}$.

For example,

$$f(2.71) = [\![2.71]\!] = 2$$

$$f(23.5) = [\![23.5]\!] = 23$$

$$f(10) = [\![10]\!] = 10$$

$$f(\pi) = [\![\pi]\!] = 3$$

$$f(-2.5) = [\![-2.5]\!] = -3$$

EXAMPLE 6 **Graphing the Greatest-Integer Function**

Graph: $f(x) = [\![x]\!]$.

SOLUTION We will list several intervals and determine the corresponding values of the greatest-integer function. Then we will use these values to graph the function.

$[0, 1)$ $f(x) = [\![x]\!] = 0$ For numbers from 0 to 1 (not including 1), the greatest integer in the interval is 0.

$[1, 2)$ $f(x) = [\![x]\!] = 1$ For numbers from 1 to 2 (not including 2), the greatest integer in the interval is 1.

$[2, 3)$ $f(x) = [\![x]\!] = 2$ For numbers from 2 to 3 (not including 3), the greatest integer in the interval is 2.

Within each interval, the values of y are constant, but they jump by 1 at integer values of x. The graph is shown in Figure 3-39. From the graph, we can see that the domain of the greatest integer function is the interval $(-\infty, \infty)$. The range is the set of integers.

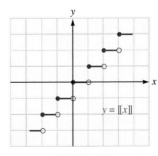

FIGURE 3-39

Self Check 6   Find **a.** $[\![7.61]\!]$ and **b.** $[\![-3.75]\!]$.

Now Try Exercise 63.

Since the greatest-integer function is made up of a series of horizontal line segments, it is an example of a group of functions called **step functions.**

EXAMPLE 7   Graphing a Step Function Occurring in an Application

To print business forms, a printing company charges customers $10 for the order, plus $20 for each box containing 200 forms. The printing company counts any portion of a box as a full box. Graph this step function.

SOLUTION   To graph the step function, we will determine the cost for printing various amounts of boxes of forms. Then we will graph our result.

If we order the forms and then change our minds before the forms are printed, the cost will be $10. Thus, the ordered pair (0, 10) will be on the graph.

If we purchase up to one full box, the cost will be $10 for the order and $20 for the printing, for a total of $30. Thus, the ordered pair (1, 30) will be on the graph.

The cost for $1\frac{1}{2}$ boxes will be the same as the cost for 2 full boxes, or $50. Thus, the ordered pairs (1.5, 50) and (2, 50) are on the graph.

The complete graph is shown in Figure 3-40.

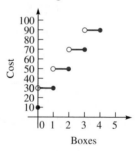

FIGURE 3-40

Self Check 7   Find the cost of $4\frac{1}{2}$ boxes.

Now Try Exercise 69.

Self Check Answers   **1.**

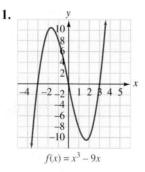

$f(x) = x^3 - 9x$

2.

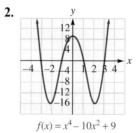

$f(x) = x^4 - 10x^2 + 9$

3. a. odd   **b.** even

4. no intervals   **5.**

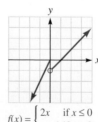

$f(x) = \begin{cases} 2x & \text{if } x \le 0 \\ x - 1 & \text{if } x > 0 \end{cases}$

6. a. 7   **b.** −4   **7.** $110

Exercises **3.3**

Getting Ready
You should be able to complete these vocabulary and concept statements before you proceed to the practice exercises.

Fill in the blanks.

1. The degree of the function $y = f(x) = x^4 - 3$ is ___.
2. Peaks and valleys on a polynomial graph are called _____ points.
3. The graph of a nth degree polynomial function can have at most _____ turning points.
4. If the graph of a function is symmetric about the _____, it is called an even function.
5. If the graph of a function is symmetric about the origin, it is called an _____ function.
6. If the values of $f(x)$ get larger as x increases on an interval, we say that the function is _____ on the open interval.
7. _____ functions are defined by different equations for different intervals in their domains.
8. If the values of $f(x)$ get smaller as x increases on an interval, we say that the function is _____ on the open interval.
9. $\llbracket 3.69 \rrbracket =$ ___.
10. If the values of $f(x)$ do not change as x increases on an interval, we say that the function is _____ on the open interval.

Practice
Graph each polynomial function.

11. $f(x) = x^3 - 9x$

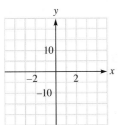

12. $f(x) = x^3 - 16x$

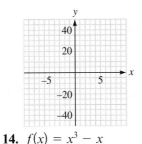

13. $f(x) = -x^3 - 4x^2$

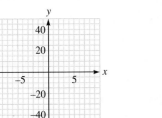

14. $f(x) = x^3 - x$

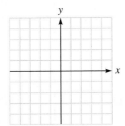

15. $f(x) = x^3 + x^2$

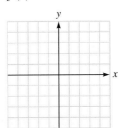

16. $f(x) = -x^3 + 1$

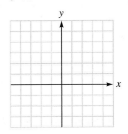

17. $f(x) = x^3 - x^2 - 4x + 4$

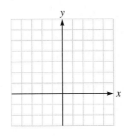

18. $f(x) = 4x^3 - 4x^2 - x + 1$

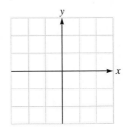

19. $f(x) = x^4 - 2x^2 + 1$

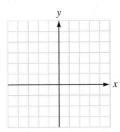

20. $f(x) = x^4 - 5x^2 + 4$

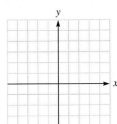

21. $f(x) = -x^4 + 5x^2 - 4$

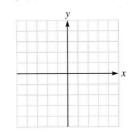

22. $f(x) = x(x - 3)(x - 2)(x + 1)$

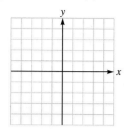

Determine whether each function is even, odd, or neither.

23. $f(x) = x^4 + x^2$

24. $f(x) = x^3 - 2x$

25. $f(x) = x^3 + x^2$

26. $f(x) = x^6 - x^2$

27. $f(x) = x^5 + x^3$

28. $f(x) = x^3 - x^2$

29. $f(x) = 2x^3 - 3x$

30. $f(x) = 4x^2 - 5$

Determine whether each function is even, odd, or neither.

31.

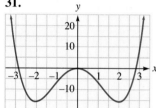

32.

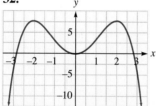

33.

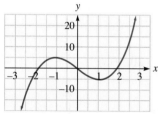

34.

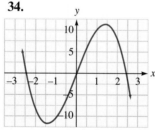

35.

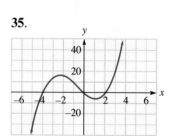

36.

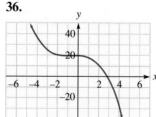

State the open intervals where each function is increasing, decreasing, or constant.

37.

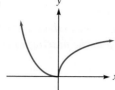

38.

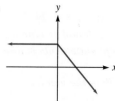

39.

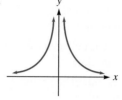

40.

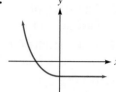

41.

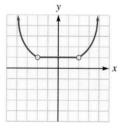

42.

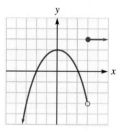

43. $f(x) = x^2 - 4x + 4$

44. $f(x) = 4 - x^2$

Evaluate each piecewise-defined function.

45. $f(x) = \begin{cases} 2x + 2 & \text{if } x < 0 \\ 3 & \text{if } x \geq 0 \end{cases}$

    a. $f(-2)$ **b.** $f(0)$

46. $f(x) = \begin{cases} x - 2 & \text{if } x < 1 \\ x^2 & \text{if } x \geq 1 \end{cases}$

    a. $f(1)$ **b.** $f(5)$

47. $f(x) = \begin{cases} 2 & \text{if } x < 0 \\ 2 - x & \text{if } 0 \leq x < 2 \\ x + 1 & \text{if } x \geq 2 \end{cases}$

    a. $f(-1)$ **b.** $f(1)$ **c.** $f(2)$

48. $f(x) = \begin{cases} 2x & \text{if } x < 0 \\ 3 - x & \text{if } 0 \leq x < 2 \\ |x| & \text{if } x \geq 2 \end{cases}$

    a. $f(-0.5)$ **b.** $f(0)$ **c.** $f(2)$

Graph each piecewise-defined function.

49. $f(x) = \begin{cases} x + 2 & \text{if } x < 0 \\ 2 & \text{if } x \geq 0 \end{cases}$

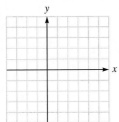

50. $f(x) = \begin{cases} 2x & \text{if } x < 0 \\ -2x & \text{if } x \geq 0 \end{cases}$

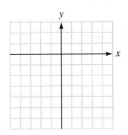

51. $f(x) = \begin{cases} x & \text{if } x \leq 0 \\ 2 & \text{if } x > 0 \end{cases}$

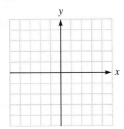

52. $f(x) = \begin{cases} -x & \text{if } x < 0 \\ \frac{1}{2}x & \text{if } x > 0 \end{cases}$

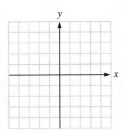

53. $f(x) = \begin{cases} -4 - x & \text{if } x < 1 \\ 3 & \text{if } x \geq 1 \end{cases}$

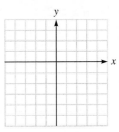

54. $f(x) = \begin{cases} -5 - x & \text{if } x < 1 \\ -3 & \text{if } x \geq 1 \end{cases}$

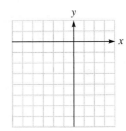

55. $f(x) = \begin{cases} -x & \text{if } x < 0 \\ x^2 & \text{if } x \geq 0 \end{cases}$

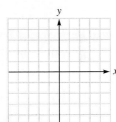

56. $f(x) = \begin{cases} |x| & \text{if } x < 0 \\ \sqrt{x} & \text{if } x \geq 0 \end{cases}$

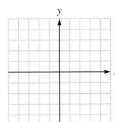

57. $f(x) = \begin{cases} 0 & \text{if } x < 0 \\ x^2 & \text{if } 0 \leq x \leq 2 \\ 4 - 2x & \text{if } x > 2 \end{cases}$

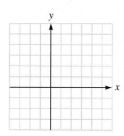

58. $f(x) = \begin{cases} 2 & \text{if } x < 0 \\ 2 - x & \text{if } 0 \leq x < 2 \\ x & \text{if } x \geq 2 \end{cases}$

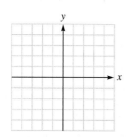

Evalute each function at the indicated x-values.

59. $f(x) = [\![x]\!]$ **a.** $f(3)$ **b.** $f(-4)$ **c.** $f(-2.3)$

60. $f(x) = [\![3x]\!]$ **a.** $f(4)$ **b.** $f(-2)$ **c.** $f(-1.2)$

61. $f(x) = [\![x + 3]\!]$ **a.** $f(-1)$ **b.** $f\left(\dfrac{2}{3}\right)$ **c.** $f(1.3)$

62. $f(x) = [\![4x]\!] - 1$ **a.** $f(-3)$ **b.** $f(0)$ **c.** $f(\pi)$

Graph each function.

63. $y = [\![2x]\!]$

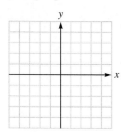

64. $y = \left[\!\!\left[\dfrac{1}{3}x + 3\right]\!\!\right]$

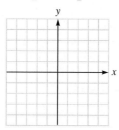

65. $y = [\![x]\!] - 1$

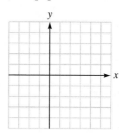

66. $y = [\![x + 2]\!]$

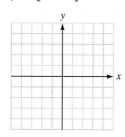

Applications

67. Grading scales A mathematics instructor assigns letter grades according to the following scale.

From	Up to but less than	Grade
60%	70%	D
70%	80%	C
80%	90%	B
90%	100% (including 100%)	A

Graph the ordered pairs (p, g), where p represents the percent and g represents the grade. Find the final semester grade of a student who has test scores of 67%, 73%, 84%, 87%, and 93%.

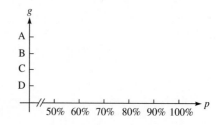

68. Calculating grades See Exercise 67 and find the final semester grade of a student who has test scores of 53%, 65%, 64%, 73%, 89%, and 82%.

69. Renting a jeep A rental company charges $20 to rent a Jeep Wrangler for one day, plus $4 for every 100 miles (or portion of 100 miles) that it is driven. Graph the ordered pairs (m, C), where m represents the miles driven and C represents the cost. Find the cost if the Jeep is driven 275 miles in one day.

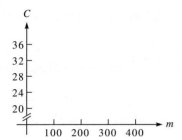

70. Riding in a taxi A taxicab company charges $3 for a trip up to 1 mile, and $2 for every extra mile (or portion of a mile). Graph the ordered pairs (m, C), where m represents the miles traveled and C represents the cost. Find the cost to ride $10\frac{1}{4}$ miles.

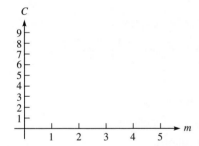

71. Computer communications An on-line information service charges for connect time at a rate of $12 per hour, computed for every minute or fraction of a minute. Graph the points (t, C), where C is the cost of t minutes of connect time. Find the cost of $7\frac{1}{2}$ minutes.

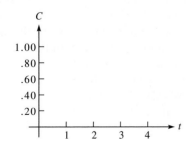

72. iPad repair There is a charge of $30, plus $40 per hour (or fraction of an hour), to repair an iPad. Graph the points (t, C), where t is the time it takes to do the job and C is the cost. If it takes 4 hours to repair the iPad, how much did it cost?

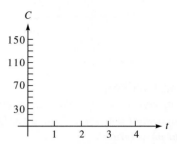

73. Rounding numbers Measurements are rarely exact; they are often *rounded* to an appropriate precision. Graph the points (x, y), where y is the result of rounding the number x to the nearest ten.

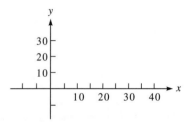

74. Signum function Computer programmers often use the following function, denoted by $y = \text{sgn } x$. Graph this function and find its domain and range.

$$y = \begin{cases} -1 & \text{if } x < 0 \\ 0 & \text{if } x = 0 \\ 1 & \text{if } x > 0 \end{cases}$$

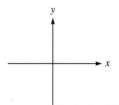

75. Graph the function defined by $y = \frac{|x|}{x}$ and compare it to the graph in Exercise 74. Are the graphs the same?

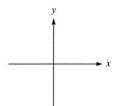

76. Graph: $y = x + |x|$.

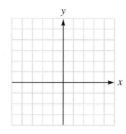

Discovery and Writing

Use a graphing calculator to explore the properties of graphs of polynomial functions. Write a paragraph summarizing your observations.

77. Graph the function $y = x^2 + ax$ for several values of a. How does the graph change?

78. Graph the function $y = x^3 + ax$ for several values of a. How does the graph change?

79. Graph the function $y = (x - a)(x - b)$ for several values of a and b. What is the relationship between the x-intercepts and the equation?

80. Use the insight you gained in Exercise 79 to factor $x^3 - 3x^2 - 4x + 12$.

Review

81. If $f(x) = 3x + 2$, find $f(x + 1)$ and $f(x) + 1$.

82. If $f(x) = x^2$, find $f(x - 2)$ and $f(x) - 2$.

83. If $f(x) = \dfrac{3x + 1}{5}$, find $f(x - 3)$ and $f(x) - 3$.

84. If $f(x) = 8$, find $f(x + 8)$ and $f(x) + 8$.

85. Solve: $2x^2 - 3 = x$.

86. Solve: $4x^2 = 24x - 37$.

3.4 Transformations of the Graphs of Functions

In this section, we will learn to

1. Use vertical translations to graph functions.
2. Use horizontal translations to graph functions.
3. Graph functions using two translations.
4. Use reflections about the x- and y-axes to graph functions.
5. Use vertical stretching and shrinking to graph functions.
6. Use horizontal stretching and shrinking to graph functions.
7. Graph functions using a combination of transformations.

We can often transform the graph of a function into the graph of another function by shifting the graph vertically or horizontally. Also, we can reflect a graph about the x- or y-axis, and stretch or shrink a graph horizontally or vertically to transform the graph of a function into the graph of another function. In this section, we will graph new functions from known ones using these methods.

Consider a white water rafting trip on the Ocoee River in Tennessee. Suppose one company charges a group of students $20 for each hour on the river, plus $100 for a guide and equipment. The cost of the rafting trip can be represented by the function

$$C_1(t) = 20t + 100$$

where $C_1(t)$ represents the cost in dollars to raft t hours on the river.

If the company increases its charge for the guide and equipment to $150, the new cost function can be represented by

$$C_2(t) = 20t + 150$$

The graphs of the two cost functions are shown in Figure 3-41.

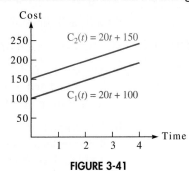

FIGURE 3-41

Note that if we shift the graph of $C_1(t)$ 50 units vertically, we obtain the graph of $C_2(t)$. This shift is called a *translation*.

As we continue our study of translations, it will be helpful to review the graphs of the basic functions that are shown in Figure 3-8 in Section 3.1. In this section, the graphs of $f(x) = x^2$, $f(x) = x^3$, $f(x) = |x|$, $f(x) = \sqrt{x}$, and $f(x) = \sqrt[3]{x}$ will be translated and stretched in various ways.

1. Use Vertical Translations to Graph Functions

Comment

The graphs of the functions shown in Figure 3-42 are exactly the graphs we would expect based on our previous study of quadratic functions.

The graphs of functions can be identical except for their position in the xy-plane. For example, Figure 3-42 shows the graph of $y = x^2 + k$ for three values of k. If $k = 0$, we have the graph of $y = x^2$. The graph of $y = x^2 + 2$ is identical to the graph of $y = x^2$, except that it is shifted 2 units upward. The graph of $y = x^2 - 3$ is identical to the graph of $y = x^2$, except that it is shifted 3 units downward. These shifts are called **vertical translations**.

$y = x^2$		
x	y	(x, y)
-2	4	$(-2, 4)$
-1	1	$(-1, 1)$
0	0	$(0, 0)$
1	1	$(1, 1)$
2	4	$(2, 4)$

$y = x^2 + 2$		
x	y	(x, y)
-2	6	$(-2, 6)$
-1	3	$(-1, 3)$
0	2	$(0, 2)$
1	3	$(1, 3)$
2	6	$(2, 6)$

$y = x^2 - 3$		
x	y	(x, y)
-2	1	$(-2, 1)$
-1	-2	$(-1, -2)$
0	-3	$(0, -3)$
1	-2	$(1, -2)$
2	1	$(2, 1)$

FIGURE 3-42

In general, we can make the following observations:

Vertical Translations   If f is a function and k is a positive number, then

- The graph of $y = f(x) + k$ is identical to the graph of $y = f(x)$ except that it is translated k units upward.

- The graph of $y = f(x) - k$ is identical to the graph of $y = f(x)$ except that it is translated k units downward.

EXAMPLE 1   **Using Vertical Translations to Graph Functions**

Graph each function: **a.** $g(x) = |x| - 2$ **b.** $h(x) = |x| + 3$

SOLUTION   We will use vertical translations of $f(x) = |x|$ to graph each function.

a. The graph of $g(x) = |x| - 2$ is identical to the graph of $f(x) = |x|$, except that it is translated 2 units downward. It is translated downward because 2 is subtracted from $|x|$. The graph of $g(x)$ is shown in Figure 3-43(a).

b. The graph of $h(x) = |x| + 3$ is identical to the graph of $f(x) = |x|$, except that it is translated 3 units upward. It is translated upward because 3 is added to $|x|$. The graph of $h(x)$ is shown in Figure 3-43(b).

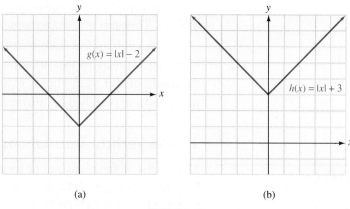

(a)   (b)

FIGURE 3-43

Self Check 1   *Fill in the blanks:* The graph of $g(x) = x^2 + 3$ is identical to the graph of $f(x) = x^2$, except that it is translated ___ units _____. The graph of $h(x) = x^2 - 4$ is identical to the graph of $f(x) = x^2$, except that it is translated ___ units _____.

Now Try Exercise 27.

2. Use Horizontal Translations to Graph Functions

Comment

The graphs of the functions shown in Figure 3-44 are exactly the graphs we would expect based on our previous knowledge of quadratic functions.

Figure 3-44 shows the graph of $y = (x + k)^2$ for three values of k. If $k = 0$, we have the graph of $y = x^2$. The graph of $y = (x - 2)^2$ is identical to the graph of $y = x^2$, except that it is shifted 2 units to the right. The graph of $y = (x + 3)^2$ is identical to the graph of $y = x^2$, except that it is shifted 3 units to the left. These shifts are called **horizontal translations.**

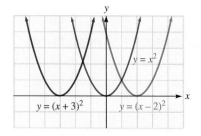

$y = x^2$		
x	y	(x, y)
-2	4	$(-2, 4)$
-1	1	$(-1, 1)$
0	0	$(0, 0)$
1	1	$(1, 1)$
2	4	$(2, 4)$

$y = (x - 2)^2$		
x	y	(x, y)
0	4	$(0, 4)$
1	1	$(1, 1)$
2	0	$(2, 0)$
3	1	$(3, 1)$
4	4	$(4, 4)$

$y = (x + 3)^2$		
x	y	(x, y)
-5	4	$(-5, 4)$
-4	1	$(-4, 1)$
-3	0	$(-3, 0)$
-2	1	$(-2, 1)$
-1	4	$(-1, 4)$

FIGURE 3-44

In general, we can make the following observations.

Horizontal Translations

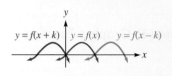

If f is a function and k is a positive number, then

- The graph of $y = f(x - k)$ is identical to the graph of $y = f(x)$ except that it is translated k units to the right.

- The graph of $y = f(x + k)$ is identical to the graph of $y = f(x)$ except that it is translated k units to the left.

EXAMPLE 2   **Using Horizontal Translations to Graph Functions**

Graph each function: **a.** $g(x) = |x - 4|$ **b.** $h(x) = |x + 2|$

SOLUTION   We will use horizontal translations of the graph of $f(x) = |x|$ to graph each function.

a. The graph of $g(x) = |x - 4|$ is identical to the graph of $f(x) = |x|$, except that it is translated 4 units to the right. It is translated 4 units to the right because within the absolute value symbols 4 is subtracted from x. The graph of $g(x)$ is shown in Figure 3-45(a).

b. The graph of $h(x) = |x + 2|$ is identical to the graph of $f(x) = |x|$, except that it is translated 2 units to the left. It is translated 2 units to the left because within the absolute value symbols 2 is added to x. The graph of $h(x)$ is shown in Figure 3-45(b).

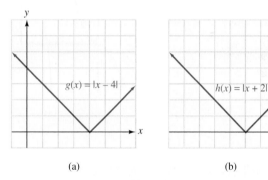

(a)   (b)

FIGURE 3-45

Self Check 2   *Fill in the blanks:* The graph of $g(x) = (x - 3)^2$ is identical to the graph of $f(x) = x^2$ except that it is translated ___ units to the _____. The graph of $h(x) = (x + 2)^2$ is identical to the graph of $f(x) = x^2$ except that it is translated ___ units to the

___ .

Now Try Exercise 29.

Caution

When using horizontal translations to graph a function, it is easy to shift the function in the *wrong* direction.

- If we see a positive constant subtracted from x, we have the tendency to shift the graph left. This is **incorrect.**

- If we see a positive constant added to x, we have the tendency to shift the graph right. This too is **incorrect.**

We should avoid making these common errors.

3. Graph Functions Using Two Translations

Sometimes we can obtain a graph by using both a horizontal and a vertical translation.

EXAMPLE 3   **Using a Horizontal and a Vertical Translation to Graph a Function**

Graph each function:
a. $g(x) = (x - 5)^3 + 4$ **b.** $h(x) = (x + 2)^2 - 2$

SOLUTION   By inspection, we see that the function in part **a** involves two translations of $f(x) = x^3$ and the function in part **b** involves two translations of $f(x) = x^2$. We will perform the horizontal translation first, followed by a vertical translation to obtain the graph of each function.

a. The graph of $g(x) = (x - 5)^3 + 4$ is identical to the graph of $f(x) = x^3$, except that it is translated 5 units to the right and 4 units upward, as shown in Figure 3-46(a).

b. The graph of $h(x) = (x + 2)^2 - 2$ is identical to the graph of $f(x) = x^2$, except that it is translated 2 units to the left and 2 units downward, as shown in Figure 3-46(b).

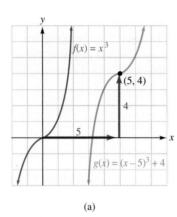

(a)

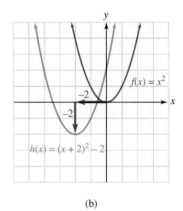

(b)

FIGURE 3-46

Self Check 3   *Fill in the blanks:* The graph of $g(x) = |x - 4| + 5$ is identical to the graph of $f(x) = |x|$, except that it is translated ___ units to the _____ and ___ units _____.

Now Try Exercise 31.

ACCENT ON TECHNOLOGY   **Translations of Functions**

We can use a graphing calculator to help understand translations of functions. Compare the graph of each function listed to the graph of $f(x) = \sqrt{x}$.

   a. $g(x) = \sqrt{x} - 4$ **b.** $h(x) = \sqrt{x + 2}$ **c.** $k(x) = \sqrt{x - 2} + 3$

a. Graph the functions $f(x) = \sqrt{x}$ and $g(x) = \sqrt{x} - 4$ on the same screen using the window shown in Figure 3-47(a). From these graphs, we can see that the graph of g is obtained by shifting the graph of f downward 4 units.

b. Graph the functions $f(x) = \sqrt{x}$ and $h(x) = \sqrt{x + 2}$ on the same screen using the window shown in Figure 3-47(b). From these graphs, we can see that the graph of h is obtained by shifting the graph of f to the left 2 units.

c. Graph the functions $f(x) = \sqrt{x}$ and $k(x) = \sqrt{x - 2} + 3$ on the same screen using the window shown in Figure 3-47(c). From these graphs, we can see that the graph of k is obtained by shifting the graph of f right 2 units and upward 3 units.

Comment

To make the graph of $f(x) = \sqrt{x}$ appear bold, scroll to the left of Y₁ and press [ENTER].

(a) Vertical shift downward 4 units. The graph of $f(x) = \sqrt{x}$ is the bold graph.

 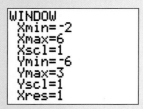 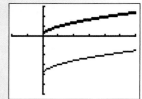

(b) Horizontal shift left 2 units. The graph of $f(x) = \sqrt{x}$ is the bold graph.

 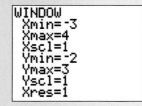 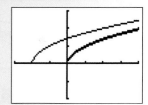

(c) Horizontal shift 2 units right and vertical shift 3 units upward. The graph of $f(x) = \sqrt{x}$ is the bold graph.

 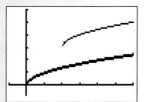

FIGURE 3-47

4. Use Reflections about the x- and y- Axes to Graph Functions

Figure 3-48(a) shows that the graph of $y = -\sqrt{x}$ is identical to the graph of $y = \sqrt{x}$ except that it is reflected about the x-axis. Figure 3-48(b) shows that the graph of $y = \sqrt{-x}$ is identical to the graph of $y = \sqrt{x}$ except that it is reflected about the y-axis.

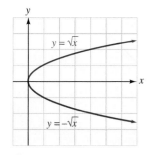

$y = -\sqrt{x}$		
x	y	(x, y)
0	0	$(0, 0)$
1	-1	$(1, -1)$
4	-2	$(4, -2)$

(a)

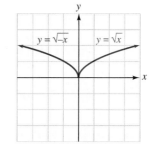

$y = \sqrt{-x}$		
x	y	(x, y)
0	0	$(0, 0)$
-1	1	$(-1, 1)$
-4	2	$(-4, 2)$

(b)

FIGURE 3-48

In general, we can make the following observations.

Reflections If f is a function, then

- The graph of $y = -f(x)$ is identical to the graph of $f(x)$ except that it is reflected about the x-axis.

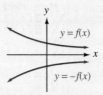

- The graph of $y = f(-x)$ is identical to the graph of $f(x)$ except that it is reflected about the y-axis.

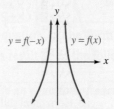

EXAMPLE 4 **Using Reflections to Graph Functions**

Graph each function: **a.** $g(x) = -|x + 1|$ **b.** $h(x) = |-x + 1|$

SOLUTION By inspection, we see that the function given in part **a** involves a reflection about the x-axis and the function given in part **b** involves a reflection about the y-axis. We will use reflections to draw the graph of each function.

a. The graph of $g(x) = -|x + 1|$ is identical to the graph of $f(x) = |x + 1|$, except that it is reflected about the x-axis. This is because $g(x) = -f(x)$. The graphs of both functions are shown in Figure 3-49(a).

b. The graph of $h(x) = |-x + 1|$ is identical to the graph of $f(x) = |x + 1|$, except that it is reflected about the y-axis. This is because $f(-x) = h(x)$. The graphs of both functions are shown in Figure 3-49(b).

Comment

It is often helpful to think of a reflection as a mirror image of the graph about the x- or y-axis.

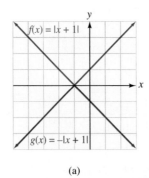

(a)

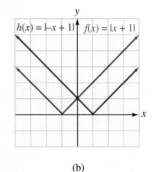

(b)

FIGURE 3-49

Self Check 4   *Fill in the blanks:* The graph of $g(x) = -\sqrt[3]{x}$ is identical to the graph of $f(x) = \sqrt[3]{x}$, except that it is reflected about the ___ axis. The graph of $h(x) = \sqrt{-x-4}$ is identical to the graph of $f(x) = \sqrt{x-4}$, except that it is reflected about the ___ axis.

Now Try Exercise 51.

5. Use Vertical Stretching and Shrinking to Graph Functions

Figure 3-50 shows the graphs of $y = x^2$, $y = 3x^2$, and $y = \frac{1}{3}x^2$.

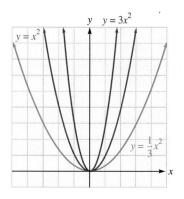

$y = x^2$		
x	y	(x, y)
-2	4	$(-2, 4)$
-1	1	$(-1, 1)$
0	0	$(0, 0)$
1	1	$(1, 1)$
2	4	$(2, 4)$

$y = 3x^2$		
x	y	(x, y)
-2	12	$(-2, 12)$
-1	3	$(-1, 3)$
0	0	$(0, 0)$
1	3	$(1, 3)$
2	12	$(2, 12)$

$y = \dfrac{1}{3}x^2$		
x	y	(x, y)
-2	$\dfrac{4}{3}$	$\left(-2, \dfrac{4}{3}\right)$
-1	$\dfrac{1}{3}$	$\left(-1, \dfrac{1}{3}\right)$
0	0	$(0, 0)$
1	$\dfrac{1}{3}$	$\left(1, \dfrac{1}{3}\right)$
2	$\dfrac{4}{3}$	$\left(2, \dfrac{4}{3}\right)$

FIGURE 3-50

Because each value of $y = 3x^2$ is 3 times greater than the corresponding value of $y = x^2$, its graph is stretched vertically by a factor of 3. Because each value of $y = \frac{1}{3}x^2$ is 3 times smaller than the corresponding value of $y = x^2$, its graph shrinks vertically by a factor of $\frac{1}{3}$.

In general, we can make the following observations.

Vertical Stretching and Shrinking

If f is a function and $k > 1$, then

- The graph of $y = kf(x)$ can be obtained by stretching the graph of $y = f(x)$ vertically by multiplying each value of $f(x)$ by k.

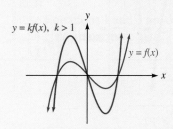

If f is a function and $0 < k < 1$, then

• The graph of $y = kf(x)$ can be obtained by shrinking the graph of $y = f(x)$ vertically by multiplying each value of $f(x)$ by k.

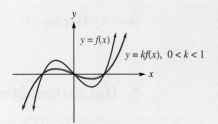

EXAMPLE 5   Graphing a Function by Vertical Stretching or Shrinking

Graph each function: **a.** $g(x) = 2|x|$ **b.** $h(x) = \dfrac{1}{2}|x|$

SOLUTION   We will vertically stretch or vertically shrink the basic function $f(x) = |x|$ to graph each of the given functions.

a. The graph of $g(x) = 2|x|$ is identical to the graph of $f(x) = |x|$ except that it is vertically stretched by a factor of 2. This is because each value of $|x|$ is multiplied by 2. The graphs of both functions are shown in Figure 3-51.

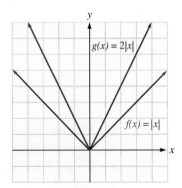

FIGURE 3-51

b. The graph of $g(x) = \frac{1}{2}|x|$ is identical to the graph of $f(x) = |x|$ except that it is vertically shrunk by a factor of $\frac{1}{2}$. This is because each value of $|x|$ is multiplied by $\frac{1}{2}$. The graphs of both functions are shown in Figure 3-52.

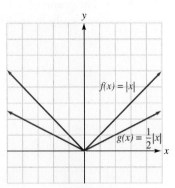

FIGURE 3-52

Comment

Note that vertically stretching the graph of a function narrows the graph of the function. Vertically shrinking the graph of a function widens the graph.

Self Check 5   *Fill in the blanks:* The graph of $g(x) = 5x^3$ is identical to the graph of $f(x) = x^3$, except that it is vertically _____ by a factor of ___. The graph of $h(x) = \frac{1}{5}x^3$ is identical to the graph $f(x) = x^3$, except that it is vertically _____ by a factor of ___.

Now Try Exercise 57.

6. Use Horizontal Stretching and Shrinking to Graph Functions

Functions can also be graphed by using horizontal stretchings and shrinkings.

Horizontal Stretching and Shrinking

If f is a function and $k > 1$, then

- The graph of $y = f(kx)$ can be obtained by shrinking the graph of $y = f(x)$ horizontally by multiplying each x-value of $f(x)$ by $\frac{1}{k}$.

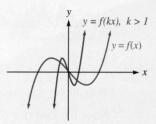

If f is a function and $0 < k < 1$, then

- The graph of $y = f(kx)$ can be obtained by stretching the graph of $y = f(x)$ horizontally by multiplying each x-value of $f(x)$ by $\frac{1}{k}$.

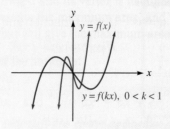

Because we multiply each x-value of $f(x)$ by k to obtain the graph of $y = f(kx)$, the graphs are horizontally stretched or shrunk by a factor of $\frac{1}{k}$.

EXAMPLE 6   **Graphing a Function by Horizontally Shrinking or Stretching**

Graph $y = (3x)^2 - 1$ using the graph of the function $y = x^2 - 1$.

SOLUTION   Since $3 > 1$, the graph of $y = (3x)^2 - 1$ can be obtained by shrinking the graph of $y = x^2 - 1$ horizontally by multiplying each x-coordinate of $y = x^2 - 1$ by $\frac{1}{3}$.

First, we complete the table of solutions for $y = x^2 - 1$, shown in Figure 3-53. Next, we multiply each x-value in the table by $\frac{1}{3}$ to obtain the table of solutions for $y = (3x)^2 - 1$, also shown in Figure 3-53.

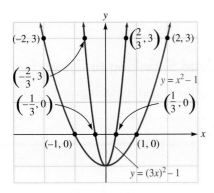

$y = x^2 - 1$		
x	y	(x, y)
-2	3	$(-2, 3)$
-1	0	$(-1, 0)$
0	-1	$(0, -1)$
1	0	$(1, 0)$
2	3	$(2, 3)$

$y = (3x)^2 - 1$		
x	y	(x, y)
$-\dfrac{2}{3}$	3	$\left(-\dfrac{2}{3}, 3\right)$
$-\dfrac{1}{3}$	0	$\left(-\dfrac{1}{3}, 0\right)$
0	-1	$(0, -1)$
$\dfrac{1}{3}$	0	$\left(\dfrac{1}{3}, 0\right)$
$\dfrac{2}{3}$	3	$\left(\dfrac{2}{3}, 3\right)$

FIGURE 3-53

We now draw each graph as shown in Figure 3-53. ∎

Self Check 6   *Fill in the blank:* The graph of $g(x) = \left(\frac{1}{3}x\right)^2 - 1$ is identical to the graph of $f(x) = x^2 - 1$, except that it is horizontally _____ by a factor of ___.

Now Try Exercise 63.

We can summarize the ideas in this section as follows.

Summary of Transformations   If f is a function and k represents a positive number, then

The graph of	***can be obtained by graphing $y = f(x)$ and***
$y = f(x) + k$	translating the graph k units upward.
$y = f(x) - k$	translating the graph k units downward.
$y = f(x + k)$	translating the graph k units to the left.
$y = f(x - k)$	translating the graph k units to the right.
$y = -f(x)$	reflecting the graph about the x-axis.
$y = f(-x)$	reflecting the graph about the y-axis.
$y = kf(x) \quad k > 1$	stretching the graph vertically by multiplying each value of $f(x)$ by k.
$y = kf(x) \quad 0 < k < 1$	shrinking the graph vertically by multiplying each value $f(x)$ by k.
$y = f(kx) \quad k > 1$	shrinking the graph horizontally by multiplying each x-value of $f(x)$ by $\frac{1}{k}$.
$y = f(kx) \quad 0 < k < 1$	stretching the graph horizontally by multiplying each x-value of $f(x)$ by $\frac{1}{k}$.

EXAMPLE 7   **Applying Transformations of Graphs**

Figure 3-54 shows the graph of $y = f(x)$. Use this graph and a translation to sketch the graph of **a.** $y = f(x) + 2$ **b.** $y = f(x - 2)$ **c.** $y = 2f(x)$

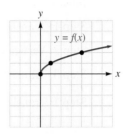

FIGURE 3-54

SOLUTION   We will use the summary of transformations in the section given earlier to graph each function.

a. The graph of $y = f(x) + 2$ is identical to the graph of $y = f(x)$, except that it is translated 2 units upward. See Figure 3-55(a).

b. The graph of $y = f(x - 2)$ is identical to the graph of $y = f(x)$, except that it is translated 2 units to the right. See Figure 3-55(b).

c. The graph of $y = 2f(x)$ is identical to the graph of $y = f(x)$, except that it is stretched vertically by multiplying each y-value of $f(x)$ by 2. See Figure 3-55(c).

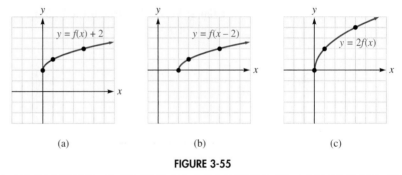

(a)   (b)   (c)

FIGURE 3-55

Self Check 7   Use Figure 3-50 and a reflection to sketch the graph of **a.** $y = -f(x)$
b. $y = f(-x)$

Now Try Exercise 79.

7. Graph Functions Using a Combination of Transformations

To graph functions involving a combination of transformations, we must apply each translation or stretching to the function.

Strategy for Graphing Using a Sequence of Transformations

To graph a function using a combination of transformations, perform the transformations in the following order:

1. horizontal translation
2. stretching or shrinking
3. reflection
4. vertical translation

EXAMPLE 8 **Graphing a Function by Using a Combination of Transformations**

Use the function $f(x) = |x|$ to graph $g(x) = 3|x - 2| + 4$.

SOLUTION We will graph $g(x) = 3|x - 2| + 4$ by applying three transformations to the basic function $f(x) = |x|$:

Step 1: Translate $f(x) = |x|$ horizontally 2 units to the right.

Step 2: Vertically stretch the graph by a factor of 3.

Step 3: Translate the graph vertically 4 units upward.

Step 1: Translate $f(x) = |x|$ horizontally 2 units to the right to obtain the graph of $y = |x - 2|$.

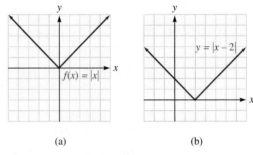

(a)   (b)

FIGURE 3-56

Step 2: Vertically stretch the graph of $y = |x - 2|$ by a factor of 3 to obtain the graph of $y = 3|x - 2|$.

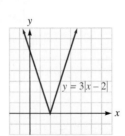

FIGURE 3-57

Step 3: Translate the graph of $y = 3|x - 2|$ vertically 4 units upward to obtain the graph of $g(x) = 3|x - 2| + 4$.

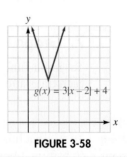

FIGURE 3-58

■

Self Check 8 Use the graph of the function $f(x) = x^3$ to graph $g(x) = \frac{1}{3}(x + 1)^3 - 2$.

Now Try Exercise 69.

Self Check Answers　**1.** 3, upward; 4, downward　**2.** 3, right; 2, left　**3.** 4, right; 5, upward

4. x; y　**5.** stretched, 5; shrunk, $\dfrac{1}{5}$　**6.** stretched, 3

7. a. 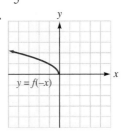　**b.** 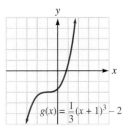　**8.**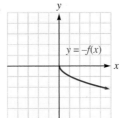

Exercises **3.4**

Getting Ready

You should be able to complete these vocabulary and concept statements before you proceed to the practice exercises.

Fill in the blanks.

1. The graph of $y = f(x) + 5$ is identical to the graph of $y = f(x)$ except that it is translated 5 units

　　　.

2. The graph of _____ is identical to the graph of $y = f(x)$ except that it is translated 7 units downward.

3. The graph of $y = f(x - 3)$ is identical to the graph of $y = f(x)$ except that it is translated 3 units

　　　.

4. The graph of $y = f(x + 2)$ is identical to the graph of $y = f(x)$ except that it is translated 2 units

　　　.

5. To draw the graph of $y = (x + 2)^2 - 3$, translate the graph of $y = x^2$ ___ units to the left and 3 units

　　　.

6. To draw the graph of $y = (x - 3)^3 + 1$, translate the graph of $y = x^3$ 3 units to the _____ and 1 unit

　　　.

7. The graph of $y = f(-x)$ is a reflection of the graph of $y = f(x)$ about the _____.

8. The graph of _____ is a reflection of the graph of $y = f(x)$ about the x-axis.

9. The graph of $y = f(4x)$ shrinks the graph of $y = f(x)$ _____ by multiplying each x-value of $f(x)$ by $\frac{1}{4}$.

10. The graph of $y = 8f(x)$ stretches the graph of $y = f(x)$ _____ by a factor of 8.

Practice

The graph of each function is a translation of the graph of $f(x) = x^2$. Graph each function.

11. $g(x) = x^2 - 2$　　**12.** $g(x) = (x - 2)^2$

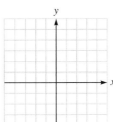

13. $g(x) = (x + 3)^2$　　**14.** $g(x) = x^2 + 3$

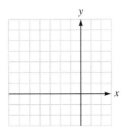

15. $h(x) = (x + 1)^2 + 2$　　**16.** $h(x) = (x - 3)^2 - 1$

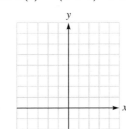

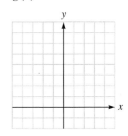

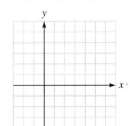

17. $h(x) = \left(x + \dfrac{1}{2}\right)^2 - \dfrac{1}{2}$ **18.** $h(x) = \left(x - \dfrac{3}{2}\right)^2 + \dfrac{5}{2}$

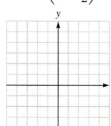   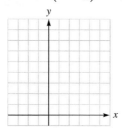

The graph of each function is a translation of the graph of $f(x) = x^3$. Graph each function.

19. $g(x) = x^3 + 1$ **20.** $g(x) = x^3 - 3$

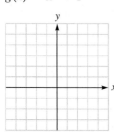   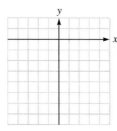

21. $g(x) = (x - 2)^3$ **22.** $g(x) = (x + 3)^3$

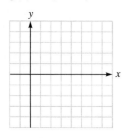   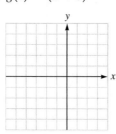

23. $h(x) = (x - 2)^3 - 3$ **24.** $h(x) = (x + 1)^3 + 4$

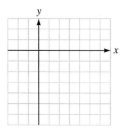   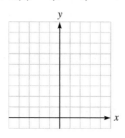

25. $y + 2 = x^3$ **26.** $y - 7 = (x - 5)^3$

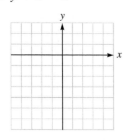   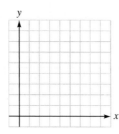

The graph of each function is a translation of the graph of $f(x) = |x|$. Graph each function.

27. $g(x) = |x| + 2$ **28.** $g(x) = |x| - 2$

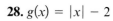

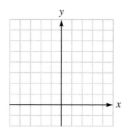   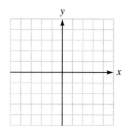

29. $g(x) = |x - 5|$ **30.** $g(x) = |x + 4|$

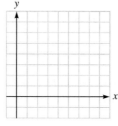   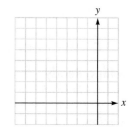

31. $f(x) = |x + 2| - 1$ **32.** $h(x) = |x - 3| + 3$

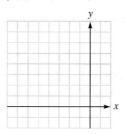   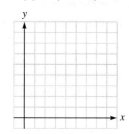

The graph of each function is a translation of the graph of $f(x) = \sqrt{x}$. Graph each function.

33. $g(x) = \sqrt{x} + 1$ **34.** $g(x) = \sqrt{x} - 3$

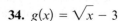

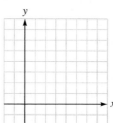   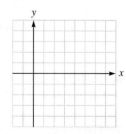

35. $g(x) = \sqrt{x + 2}$ **36.** $g(x) = \sqrt{x - 4}$

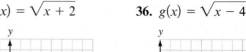

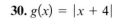

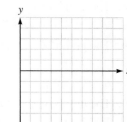

37. $h(x) = \sqrt{x - 2} - 1$

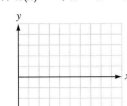

38. $h(x) = \sqrt{x + 2} + 3$

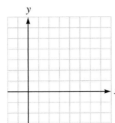

47. $h(x) = -x^3$

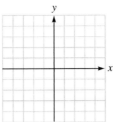

48. $f(x) = -|x|$

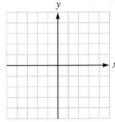

The graph of each function is a translation of the graph of $f(x) = \sqrt[3]{x}$. *Graph each function.*

39. $g(x) = \sqrt[3]{x} - 4$

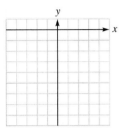

40. $g(x) = \sqrt[3]{x} + 3$

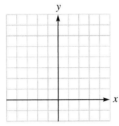

49. $f(x) = -\sqrt{x}$

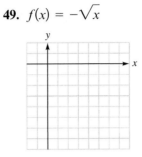

50. $g(x) = \sqrt[3]{-x}$

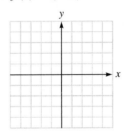

41. $g(x) = \sqrt[3]{x} - 2$

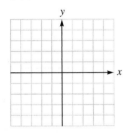

42. $g(x) = \sqrt[3]{x} + 5$

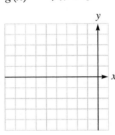

51. $f(x) = |-x|$

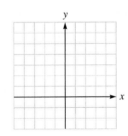

52. $g(x) = (-x)^2$

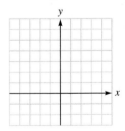

43. $h(x) = \sqrt[3]{x + 1} - 1$

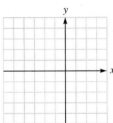

44. $h(x) = \sqrt[3]{x - 1} - 1$

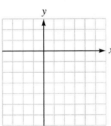

The graph of each function is a vertical stretching or shrinking of the graph of $y = x^2$, $y = x^3$, *or* $y = |x|$. *Graph each function.*

53. $f(x) = 2x^2$

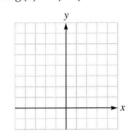

54. $g(x) = \frac{1}{2}x^2$

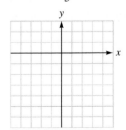

The graph of each function is a reflection of the graph of $y = x^2$, $y = x^3$, $y = |x|$, $y = \sqrt{x}$, *or* $y = \sqrt[3]{x}$. *Graph each function.*

45. $f(x) = -x^2$

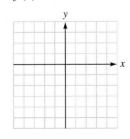

46. $g(x) = (-x)^3$

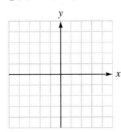

55. $h(x) = -3x^2$

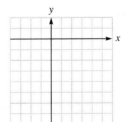

56. $f(x) = -\frac{1}{3}x^2$

57. $f(x) = \dfrac{1}{2}x^3$

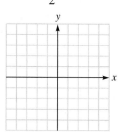

58. $g(x) = 2x^3$

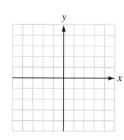

67. $h(x) = -2|x| + 3$

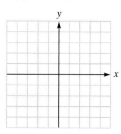

68. $f(x) = -2|x + 3|$

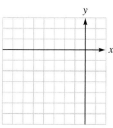

59. $h(x) = -3|x|$

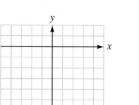

60. $f(x) = \dfrac{1}{3}|x|$

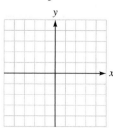

69. $f(x) = 2|x - 2| + 1$

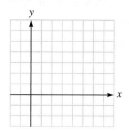

70. $f(x) = -3|x + 5| - 2$

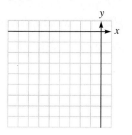

The graph of each function is a horizontal stretching or shrinking of the graph of $y = x^2$ or $y = x^3$. Graph each function.

61. $f(x) = \left(\dfrac{1}{2}x\right)^3$

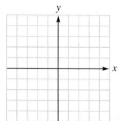

62. $f(x) = (2x)^3$

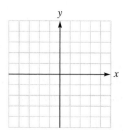

71. $f(x) = 2\sqrt{x} + 3$
$(x \geq 0)$

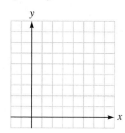

72. $g(x) = 2\sqrt{x + 3}$
$(x \geq -3)$

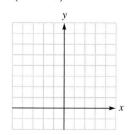

63. $f(x) = (2x)^2$

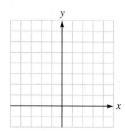

64. $f(x) = (-2x)^3$

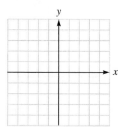

73. $h(x) = 2\sqrt{x - 2} + 1$
$(x \geq 2)$

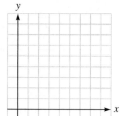

74. $h(x) = \dfrac{1}{2}\sqrt{x + 5} - 2$
$(x \geq -5)$

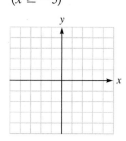

Graph each function using a combination of transformations applied to the graph of a basic function.

65. $g(x) = 3(x + 2)^2 - 1$

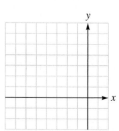

66. $g(x) = -\dfrac{1}{3}(x + 1)^2 + 1$

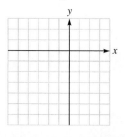

75. $g(x) = -2(x + 2)^3 - 1$

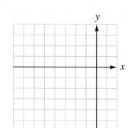

76. $g(x) = \dfrac{1}{3}(x + 1)^3 - 1$

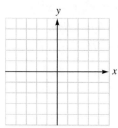

77. $f(x) = 2\sqrt[3]{x} + 4$

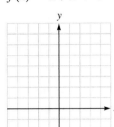

78. $f(x) = -2\sqrt[3]{x + 1}$

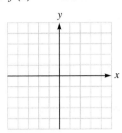

85. $y = 2f(-x)$

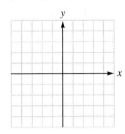

86. $y = f(x + 1) - 2$

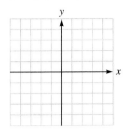

Use the following graph and a translation, stretching, or reflection to sketch the graph of each function.

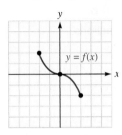

79. $y = f(x) + 1$

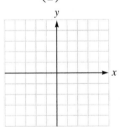

80. $y = f(x + 1)$

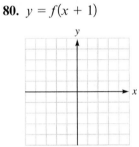

81. $y = 2f(x)$

82. $y = f\left(\dfrac{x}{2}\right)$

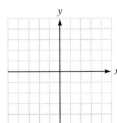

83. $y = f(x - 2) + 1$

84. $y = -f(x)$

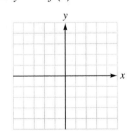

Discovery and Writing

Use a graphing calculator to perform each experiment. Write a brief paragraph describing your findings.

87. Investigate the translations of the graph of a function by graphing the parabola $y = (x - k)^2 + k$ for several values of k. What do you observe about successive positions of the vertex?

88. Investigate the translations of the graph of a function by graphing the parabola $y = (x - k)^2 + k^2$ for several values of k. What do you observe about successive positions of the vertex?

89. Investigate the horizontal stretching of the graph of a function by graphing $y = \sqrt{ax}$ for several values of a. What do you observe?

90. Investigate the vertical stretching of the graph of a function by graphing $y = b\sqrt{x}$ for several values of b. What do you observe? Are these graphs different from the graphs in Exercise 89?

Write a paragraph using your own words.

91. Explain why the effect of vertically stretching a graph by a factor of -1 is to reflect the graph about the x-axis.

92. Explain why the effect of horizontally stretching a graph by a factor of -1 is to reflect the graph about the y-axis.

Review

Simplify each function.

93. $\dfrac{x^2 + x - 6}{x^2 + 5x + 6}$

94. $\dfrac{2x^2 + 3x}{2x^2 + x - 3}$

Find the domain of each function. Use interval notation to write each answer.

95. $f(x) = \dfrac{x + 7}{x - 3}$

96. $f(x) = \dfrac{x^2 + 1}{x^2 + 3x + 2}$

Perform each division and write the answer in quotient $+ \frac{remainder}{divider}$ form.

97. $\dfrac{x^2 + 3x}{x + 1}$

98. $\dfrac{x^2 + 3}{x + 1}$

3.5 Rational Functions

In this section, we will learn to

1. Find the domain of a rational function.
2. Understand the characteristics of rational functions and their graphs.
3. Find vertical asymptotes of rational functions.
4. Find horizontal asymptotes of rational functions.
5. Identify slant asymptotes of rational functions.
6. Graph rational functions.
7. Understand when a graph has a missing point.
8. Solve problems modeled by rational functions.

Walter G Arce/Shutterstock.com

We have discussed polynomial functions and now focus on another class of functions called **rational functions**. Rational functions are defined by rational expressions that are quotients of polynomials.

For example, consider the time t it takes a Nascar driver to drive the 500 miles in the Daytona 500 race. The time t can be defined as a function of the average rate of the driver's speed r. That is

$$t = f(r) = \frac{500}{r}$$

If a driver averages a speed of 170 mph, we can evaluate $f(170)$ to determine the time in hours driven.

$$f(\mathbf{170}) = \frac{500}{\mathbf{170}} = \frac{50}{17} \qquad \text{This is approximately 2.94 hours.}$$

We will discuss this type of function in this section.

Rational Functions A **rational function** is a function defined by an equation of the form

$$y = \frac{P(x)}{Q(x)}$$

where $P(x)$ and $Q(x)$ are polynomials and $Q(x) \neq 0$.

1. Find the Domain of a Rational Function

Because rational functions are quotients of polynomials and $Q(x)$ is the denominator of a fraction, $Q(x)$ cannot equal 0. Thus, the domain of a rational function must exclude all values of x for which $Q(x) = 0$.

Here are some examples of rational functions and their domains:

Function	Domain
$f(x) = \dfrac{3}{x + 7}$	$(-\infty, -7) \cup (-7, \infty)$, x cannot equal -7
$f(x) = \dfrac{5x + 2}{x^2 - 4}$	$(-\infty, -2) \cup (-2, 2) \cup (2, \infty)$, x cannot equal 2 or -2
$f(x) = \dfrac{2x}{x^2 + 3}$	$(-\infty, \infty)$, x can equal any real number

EXAMPLE 1   **Finding the Domain of a Rational Function**

Find the domain of $f(x) = \dfrac{3x + 2}{x^2 - 7x + 12}$.

SOLUTION   We can factor the denominator to see what values of x will give 0's in the denominator. These values are not in the domain.

To find the numbers x that make the denominator 0, we set $x^2 - 7x - 12$ equal to 0 and solve for x.

$$x^2 - 7x + 12 = 0$$

$$(x - 4)(x - 3) = 0 \qquad \text{Factor } x^2 - 7x + 12.$$

$$x - 4 = 0 \quad \text{or} \quad x - 3 = 0 \qquad \text{Set each factor equal to 0.}$$

$$x = 4 \quad \Big| \quad x = 3 \qquad \text{Solve each linear equation.}$$

Since 4 and 3 make the denominator 0, the domain is the set of all real numbers except $x = 4$ and $x = 3$. In interval notation, we have $(-\infty, 3) \cup (3, 4) \cup (4, \infty)$. ∎

Caution

> Do not exclude values of the variable from the domain that make the numerator equal to zero, unless the denominator has the same factor. We will discuss this situation later in this section.

Self Check 1   Find the domain of $f(x) = \dfrac{2x - 3}{x^2 - x - 2}$.

Now Try Exercise 23.

ACCENT ON TECHNOLOGY

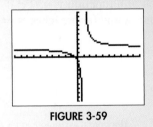

FIGURE 3-59

Domain and Range of a Rational Function

We can use a graphing calculator to find domains and ranges of rational functions. If we use settings of $[-10, 10]$ for x and $[-10, 10]$ for y and graph $f(x) = \frac{2x + 1}{x - 1}$, we will obtain Figure 3-59.

From the graph, we can see that every real number x except 1 gives a value of y. Thus, the domain of the function is $(-\infty, 1) \cup (1, \infty)$. We can also see that y can be any value except 2. The range of the function is $(-\infty, 2) \cup (2, \infty)$.

Comment

> When using a graphing calculator to graph a rational function, make sure that the function is entered properly. To avoid making an error, place both the numerator and denominator in parentheses.

2. Understand the Characteristics of Rational Functions and Their Graphs

Consider the rational function $t = f(r) = \frac{500}{r}$ given at the beginning of the section. A graph of this rational function is shown in Figure 3-60.

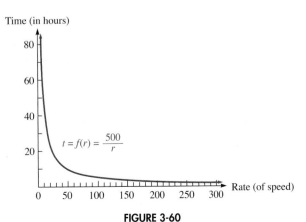

FIGURE 3-60

We see from the graph that as the rate of speed increases, the time it takes to complete the race decreases. In fact, if we drive at rocket speed, we will arrive in almost no time at all. We can express this by saying that "as the rate increases without bound (or approaches ∞), the time it takes to complete the race approaches 0 hours." When a graph approaches a line as shown in the figure, we call the line an **asymptote.** The horizontal line representing the rate axis shown in the graph is a **horizontal asymptote.**

We also see from the graph that as the rate of speed decreases, the time it takes to complete the race increases. In fact, if the car goes at turtle speed, it will take almost forever to finish the race. We can express this by saying that "as the rate gets slower and slower (or approaches 0 mph), the time approaches ∞." The vertical line representing the time axis shown on the graph is a **vertical asymptote.** A vertical asymptote is a vertical line that the graph approaches, but never touches.

One important characteristic of the graph of a rational function is that asymptotes often occur. We will first consider vertical and horizontal asymptotes. Later in the section, we will discuss slant asymptotes.

Vertical Asymptote   The line $x = a$ is a **vertical asymptote** of the graph of a function $y = f(x)$ if $f(x)$ either increases or decreases without bound (approaches ∞ or $-∞$) as x approaches a.

Horizontal Asymptote   The line $y = b$ is a **horizontal asymptote** of the graph of a function $y = f(x)$ if $f(x)$ approaches b as x increases or decreases without bound (approaches ∞ or $-∞$).

Figure 3-61 shows a typical vertical and typical horizontal asymptote.

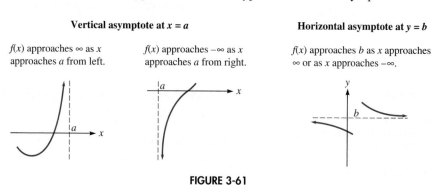

FIGURE 3-61

The graph of the rational function $f(x) = \frac{1}{x}$, called the **reciprocal function,** is shown in Figure 3-62. The domain of the function is $(-∞, 0) \cup (0, ∞)$ and the range is $(-∞, 0) \cup (0, ∞)$.

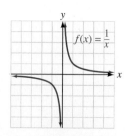

FIGURE 3-62

The reciprocal function has a vertical asymptote of $x = 0$ because

- We see from the graph that as x approaches 0 from the right that y or $f(x)$ approaches ∞.
- We see from the graph that as x approaches 0 from the left that y or $f(x)$ approaches $-\infty$.

The reciprocal function has a horizontal asymptote of $y = 0$ because

- We see from the graph that as x approaches ∞ that y or $f(x)$ approaches 0.
- We see from the graph that as x approaches $-\infty$ that y or $f(x)$ approaches 0.

3. Find Vertical Asymptotes of Rational Functions

To find the vertical asymptotes of a rational function written in simplest form, we must find the values of x for which the denominator of the rational function is 0 and the function is undefined. For example, since the denominator of $f(x) = \frac{2x-1}{x+2}$ is 0 when $x = -2$, there are no corresponding values of y and the line $x = -2$ is a vertical asymptote. We note that when x approaches -2 from the right or from the left, $f(x)$ approaches $-\infty$ and ∞ respectively. A graph of the function appears in Figure 3-63.

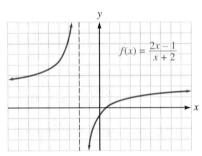

FIGURE 3-63

Strategy for Locating Vertical Asymptotes

To locate the vertical asymptotes of the rational function $f(x) = \frac{P(x)}{Q(x)}$, we follow these steps:

Step 1: Factor $P(x)$ and $Q(x)$ and remove any common factors.

Step 2: Set the denominator equal to 0 and solve the equation.

If a is a solution of the equation found in Step 2, $x = a$ is a vertical asymptote.

EXAMPLE 2   **Finding Vertical Asymptotes of Rational Functions**

Find the vertical asymptotes, if any, of each function:

a. $f(x) = \dfrac{2x}{x^2 - 16}$ **b.** $g(x) = \dfrac{x - 4}{x^2 - 16}$ **c.** $h(x) = \dfrac{5x}{x^2 + 16}$

SOLUTION   We will locate the vertical asymptotes of the graph of each function by factoring its numerator and/or denominator and removing any common factors. Then we will set the resulting denominator equal to zero, solve the equation, and identify the vertical asymptotes.

a. $f(x) = \dfrac{2x}{x^2 - 16}$

$f(x) = \dfrac{2x}{(x + 4)(x - 4)}$ Factor the denominator completely.

Practice

The time t it takes to travel 600 miles is a function of the mean rate of speed r:

$$t = f(r) = \frac{600}{r}$$

Find t for the given values of r.

13. 30 mph   **14.** 40 mph

15. 50 mph   **16.** 60 mph

Suppose the cost (in dollars) of removing p% of the pollution in a river is given by the function

$$C = f(p) = \frac{50,000p}{100 - p} \quad (0 \le p < 100)$$

Find the cost of removing each percent of pollution.

17. 10%   **18.** 30%

19. 50%   **20.** 80%

Find the domain of each rational function. Do not graph the function.

21. $f(x) = \dfrac{x^2}{x - 2}$

22. $f(x) = \dfrac{x^3 - 3x^2 + 1}{x + 3}$

23. $f(x) = \dfrac{2x^2 + 7x - 2}{x^2 - 25}$

24. $f(x) = \dfrac{5x^2 + 1}{x^2 + 5}$

25. $f(x) = \dfrac{x - 1}{x^3 - x}$

26. $f(x) = \dfrac{x + 2}{2x^2 - 9x + 9}$

27. $f(x) = \dfrac{3x^2 + 5}{x^2 + 1}$

28. $f(x) = \dfrac{7x^2 - x + 2}{x^4 + 4}$

Find the vertical asymptotes, if any, of each rational function. Do not graph the function.

29. $f(x) = \dfrac{x}{x - 3}$

30. $f(x) = \dfrac{2x}{2x + 5}$

31. $f(x) = \dfrac{x + 2}{x^2 - 1}$

32. $f(x) = \dfrac{x - 4}{x^2 - 16}$

33. $f(x) = \dfrac{1}{x^2 - x - 6}$

34. $f(x) = \dfrac{x + 2}{2x^2 - 6x - 8}$

35. $f(x) = \dfrac{x^2}{x^2 + 5}$

36. $f(x) = \dfrac{x^3 - 3x^2 + 1}{2x^2 + 3}$

Find the horizontal asymptotes, if any, of each rational function. Do not graph the function.

37. $f(x) = \dfrac{2x - 1}{x}$

38. $f(x) = \dfrac{x^2 + 1}{3x^2 - 5}$

39. $f(x) = \dfrac{x^2 + x - 2}{2x^2 - 4}$

40. $f(x) = \dfrac{5x^2 + 1}{5 - x^2}$

41. $f(x) = \dfrac{x + 1}{x^3 - 4x}$

42. $f(x) = \dfrac{x}{2x^2 - x + 11}$

43. $f(x) = \dfrac{x^2}{x - 2}$

44. $f(x) = \dfrac{x^4 + 1}{x - 3}$

Find the slant asymptote, if any, of each rational function. Do not graph the function.

45. $f(x) = \dfrac{x^2 - 5x - 6}{x - 2}$

46. $f(x) = \dfrac{x^2 - 2x + 11}{x + 3}$

47. $f(x) = \dfrac{2x^2 - 5x + 1}{x - 4}$

48. $f(x) = \dfrac{5x^3 + 1}{x + 5}$

49. $f(x) = \dfrac{x^3 + 2x^2 - x - 1}{x^2 - 1}$

50. $f(x) = \dfrac{-x^3 + 3x^2 - x + 1}{x^2 + 1}$

Find all vertical, horizontal, and slant asymptotes, x- and y-intercepts, and symmetries, and then graph each function. Check your work with a graphing calculator.

51. $y = \dfrac{1}{x - 2}$

52. $y = \dfrac{3}{x + 3}$

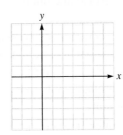

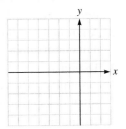

53. $y = \dfrac{x}{x - 1}$

54. $y = \dfrac{x}{x + 2}$

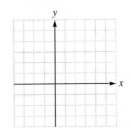

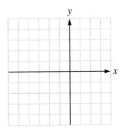

55. $f(x) = \dfrac{x + 1}{x + 2}$

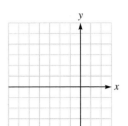

56. $f(x) = \dfrac{x - 1}{x - 2}$

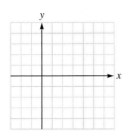

63. $y = \dfrac{x^2 + 2x - 3}{x^3 - 4x}$

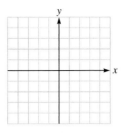

64. $y = \dfrac{3x^2 - 4x + 1}{2x^3 + 3x^2 + x}$

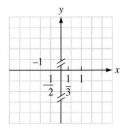

57. $f(x) = \dfrac{2x - 1}{x - 1}$

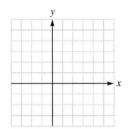

58. $f(x) = \dfrac{3x + 2}{x^2 - 4}$

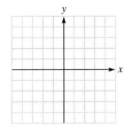

65. $y = \dfrac{x^2 - 9}{x^2}$

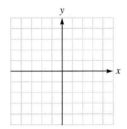

66. $y = \dfrac{3x^2 - 12}{x^2}$

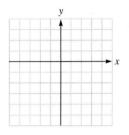

67. $f(x) = \dfrac{x}{(x + 3)^2}$

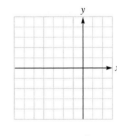

68. $f(x) = \dfrac{x}{(x - 1)^2}$

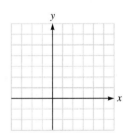

59. $g(x) = \dfrac{x^2 - 9}{x^2 - 4}$

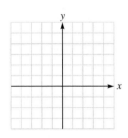

60. $g(x) = \dfrac{x^2 - 4}{x^2 - 9}$

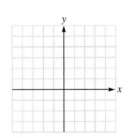

69. $f(x) = \dfrac{x + 1}{x^2(x - 2)}$

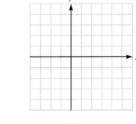

70. $f(x) = \dfrac{x - 1}{x^2(x + 2)^2}$

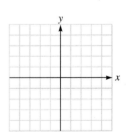

61. $g(x) = \dfrac{x^2 - x - 2}{x^2 - 4x + 3}$

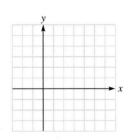

62. $g(x) = \dfrac{x^2 + 7x + 12}{x^2 - 7x + 12}$

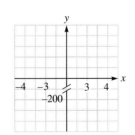

71. $y = \dfrac{x}{x^2 + 1}$

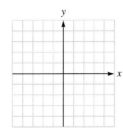

72. $y = \dfrac{x - 1}{x^2 + 2}$

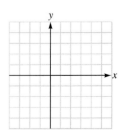

73. $y = \dfrac{3x^2}{x^2 + 1}$

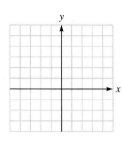

74. $y = \dfrac{x^2 - 9}{2x^2 + 1}$

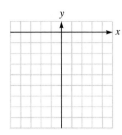

Graph each rational function. Note that the numerator and denominator of the fraction share a common factor.

79. $f(x) = \dfrac{x^2}{x}$

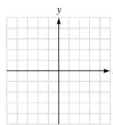

80. $f(x) = \dfrac{x^2 - 1}{x - 1}$

75. $h(x) = \dfrac{x^2 - 2x - 8}{x - 1}$

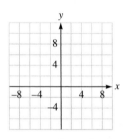

76. $h(x) = \dfrac{x^2 + x - 6}{x + 2}$

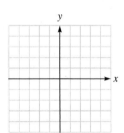

81. $f(x) = \dfrac{x^3 + x}{x}$

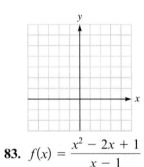

82. $f(x) = \dfrac{x^3 - x^2}{x - 1}$

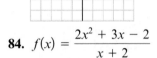

77. $f(x) = \dfrac{x^3 + x^2 + 6x}{x^2 - 1}$

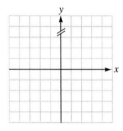

83. $f(x) = \dfrac{x^2 - 2x + 1}{x - 1}$

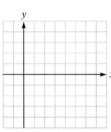

84. $f(x) = \dfrac{2x^2 + 3x - 2}{x + 2}$

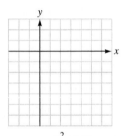

78. $f(x) = \dfrac{x^3 - 2x^2 + x}{x^2 - 4}$

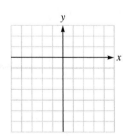

85. $f(x) = \dfrac{x^3 - 1}{x - 1}$

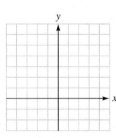

86. $f(x) = \dfrac{x^2 - x}{x^2}$

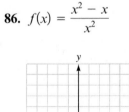

Applications

A service club wants to publish a directory of its members. Some investigation shows that the cost of typesetting and photography will be $700, and the cost of printing each directory will be $3.25.

87. a. Find a function that gives the total cost C of printing x directories.

   b. Find the total cost of printing 500 directories.

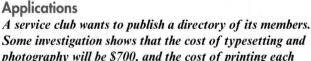

c. Find a function that gives the mean cost per directory $\overline{C}$ of printing x directories.

d. Find the mean cost per directory if 500 directories are printed.

e. Find the mean cost per directory if 1,000 directories are printed.

f. Find the mean cost per directory if 2,000 directories are printed.

An electric company charges $10 per month plus 20¢ for each kilowatt-hour (kwh) of electricity used.

88. a. Find a function that gives the total cost C of n kwh of electricity.

b. Find the total cost for using 775 kwh.

c. Find a function that gives the mean cost per kwh, $\overline{C}$, when using n kwh.

d. Find the mean cost per kwh when 775 kwh are used. Round to the nearest hundredth.

e. Find the mean cost per kwh when 3,200 kwh are used. Round to the nearest hundredth.

89. Utility costs An overseas electric company charges $8.50 per month plus 9.5¢ for each kilowatt-hour (kwh) of electricity used.

a. Find a linear function that gives the total cost C of n kwh of electricity.

b. Find a rational function that gives the average cost per kwh when using n kwh.

c. Find the average cost per kwh when 850 kwh are used.

90. Scheduling work crews The following rational function gives the number of days it would take two construction crews, working together, to frame a house that crew 1 (working alone) could complete in t days and crew 2 (working alone) could complete in $(t + 3)$ days.

$$f(t) = \frac{t^2 + 3t}{2t + 3}$$

a. If crew 1 could frame a certain house in 21 days, how long would it take both crews working together?

b. If crew 2 could frame a certain house in 25 days, how long would it take both crews working together?

Discovery and Writing

91. Can a rational function have two horizontal asymptotes? Explain.

92. Can a rational function have two slant asymptotes? Explain.

In Exercises 93–96, a, b, c, and d are nonzero constants.

93. Show that $y = 0$ is a horizontal asymptote of the graph of $y = \frac{ax + b}{cx^2 + d}$.

94. Show that $y = \frac{a}{c}x$ is a slant asymptote of the graph of $y = \frac{ax^3 + b}{cx^2 + d}$.

95. Show that $y = \frac{a}{c}$ is a horizontal asymptote of the graph of $y = \frac{ax^2 + b}{cx^2 + d}$.

96. Graph the rational function $y = \frac{x^3 + 1}{x}$ and explain why the curve is said to have a *parabolic asymptote*.

Use a graphing calculator to perform each experiment. Write a brief paragraph describing your findings.

97. Investigate the positioning of the vertical asymptotes of a rational function by graphing $y = \frac{x}{x - k}$ for several values of k. What do you observe?

98. Investigate the positioning of the vertical asymptotes of a rational function by graphing $y = \frac{x}{x^2 - k}$ for $k = 4, 1, -1$, and 0. What do you observe?

99. Find the range of the rational function $y = \frac{kx^2}{x^2 + 1}$ for several values of k. What do you observe?

100. Investigate the positioning of the x-intercepts of a rational function by graphing $y = \frac{x^2 - k}{x}$ for $k = 1, -1$, and 0. What do you observe?

Review

Perform each operation.

101. $(2x^2 + 3x) + (x^2 - 2x)$　**102.** $(3x + 2) - (x^2 + 2)$

103. $(5x + 2)(2x + 5)$　**104.** $\dfrac{2x^2 + 3x + 1}{x + 1}$

105. If $f(x) = 3x + 2$, find $f(x + 1)$.

106. If $f(x) = x^2 + x$, find $f(2x + 1)$.

3.6 Operations on Functions

In this section, we will learn to

1. Add, subtract, multiply, and divide functions, specifying domains.
2. Write functions as sums, differences, products, or quotients of other functions.
3. Evaluate composite functions.
4. Determine domains for composite functions.
5. Write functions as compositions.
6. Use operations on functions to solve problems.

© Istockphoto.com/Ian McDonnell

Functions can be combined by addition, subtraction, multiplication, and division. In this section, we will explore these operations on functions and give careful attention to their domains and ranges.

Suppose that the functions $R(x) = 140x$ and $C(x) = 120{,}000 + 40x$ model a company's yearly revenue and cost for producing and selling surfboards. By subtracting the functions, $R(x) - C(x)$, we would arrive at a new function represented by

$$(R - C)(x) = 140x - (120{,}000 + 40x)$$
$$= 140x - 120{,}000 - 40x \qquad \text{Remove parentheses.}$$
$$= 100x - 120{,}000$$

This function represents the profit made by the company when it sells x surfboards. We will now discuss how to add, subtract, multiply, and divide functions.

1. Add, Subtract, Multiply, and Divide Functions, Specifying Domains

With the following definitions, it is possible to perform arithmetic operations on algebraic functions.

Adding, Subtracting, Multiplying, and Dividing Functions

If the ranges of functions f and g are subsets of the real numbers, then

1. The **sum** of f and g, denoted as $f + g$, is defined by

$$(f + g)(x) = f(x) + g(x)$$

2. The **difference** of f and g, denoted as $f - g$, is defined by

$$(f - g)(x) = f(x) - g(x)$$

3. The **product** of f and g, denoted as $f \cdot g$, is defined by

$$(f \cdot g)(x) = f(x)g(x)$$

4. The **quotient** of f and g, denoted as f/g, is defined by

$$(f/g)(x) = \frac{f(x)}{g(x)} \quad (g(x) \neq 0)$$

The **domain** of each function, unless otherwise restricted, is the set of real numbers x that are in the domains of both f and g. In the case of the quotient f/g, there is the restriction that $g(x) \neq 0$.

EXAMPLE 1   **Finding the Sum and Difference of Two Functions and Specifying Domains**

Let $f(x) = 3x + 1$ and $g(x) = 2x - 3$. Find each function and its domain:
a. $f + g$ **b.** $f - g$

SOLUTION   We will find the sum of f and g by using the definition

$$(f + g)(x) = f(x) + g(x)$$

We will find the difference of f and g by using the definition

$$(f - g)(x) = f(x) - g(x).$$

To find the domain of each result, we will consider the domains of both f and g.

a. $(f + g)(x) = f(x) + g(x)$

$$= (3x + 1) + (2x - 3)$$

$$= 5x - 2$$

Since the domain of both f and g is the set of real numbers, the domain of $f + g$ is the interval $(-\infty, \infty)$.

b. $(f - g)(x) = f(x) - g(x)$

$$= (3x + 1) - (2x - 3)$$

$$= 3x + 1 - 2x + 3 \qquad \text{Remove parentheses.}$$

$$= x + 4$$

Since the domain of both f and g is the set of real numbers, the domain of $f - g$ is the interval $(-\infty, \infty)$.

Self Check 1   Find $g - f$.

Now Try Exercise 15.

EXAMPLE 2   **Finding the Product and Quotient of Two Functions and Specifying Domains**

Let $f(x) = 3x + 1$ and $g(x) = 2x - 3$. Find each function and its domain:
a. $f \cdot g$ **b.** f/g

SOLUTION   We will multiply f and g by using the definition

$$(f \cdot g)(x) = f(x)g(x)$$

We will divide f by g by using the definition

$$(f/g)(x) = \frac{f(x)}{g(x)} \qquad (g(x) \neq 0)$$

We will find the domains of each result by considering the domains of both f and g.

a. $(f \cdot g)(x) = f(x) \cdot g(x)$

$$= (3x + 1)(2x - 3)$$

$$= 6x^2 - 7x - 3$$

Since the domain of both f and g is the set of real numbers, the domain of $f \cdot g$ is the interval $(-\infty, \infty)$.

b. $(f/g)(x) = \dfrac{f(x)}{g(x)}$ $(g(x) \neq 0)$

$ = \dfrac{3x + 1}{2x - 3}$ $(2x - 3 \neq 0)$

Since $\frac{3}{2}$ will make $2x - 3$ equal to 0, the domain of f/g is the set of all real numbers except $\frac{3}{2}$. This is $\left(-\infty, \frac{3}{2}\right) \cup \left(\frac{3}{2}, \infty\right)$.   ∎

Self Check 2   Find g/f and its domain.

Now Try Exercise 17.

EXAMPLE 3   **Using Operations on Functions and Specifying Domains**

Let $f(x) = x^2 - 4$ and $g(x) = \sqrt{x}$. Find each function and its domain: **a.** $f + g$
b. $f \cdot g$ **c.** f/g **d.** g/f

SOLUTION   To determine each result, we use the definitions of the sum, product, and quotient of two functions. We will determine the domain of each result by considering the domains of both f and g.

First, we find the domains of f and g. Because 4 can be subtracted from any real number squared, the domain of f is the interval $(-\infty, \infty)$. Because $\sqrt{x}$ is to be a real number, the domain of g is the interval $[0, \infty)$.

a. $(f + g)(x) = f(x) + g(x)$

$ = x^2 - 4 + \sqrt{x}$

The domain of $f + g$ consists of the numbers x that are in the domain of both f and g. This is $(-\infty, \infty) \cap [0, \infty)$, which is $[0, \infty)$. The domain of $f + g$ is $[0, \infty)$.

b. $(f \cdot g)(x) = f(x)g(x)$

$ = (x^2 - 4)\sqrt{x}$

$ = x^2\sqrt{x} - 4\sqrt{x}$ Distribute the multiplication of $\sqrt{x}$.

The domain of $f \cdot g$ consists of the numbers x that are in the domain of both f and g. The domain of $f \cdot g$ is $[0, \infty)$.

c. $(f/g)(x) = \dfrac{f(x)}{g(x)}$ $(g(x) \neq 0)$

$ = \dfrac{x^2 - 4}{\sqrt{x}}$

The domain of f/g consists of the numbers x that are in the domain of both f and g, except 0 (because division by 0 is undefined). The domain of f/g is $(0, \infty)$.
To write $(f/g)(x)$ in a different form, we can rationalize the denominator of $\frac{x^2 - 4}{\sqrt{x}}$ and simplify.

$(f/g)(x) = \dfrac{x^2 - 4}{\sqrt{x}}$

$ = \dfrac{(x^2 - 4)\sqrt{x}}{\sqrt{x} \cdot \sqrt{x}}$ Multiply numerator and denominator by $\sqrt{x}$.

$ = \dfrac{x^2\sqrt{x} - 4\sqrt{x}}{x}$

d. $(g/f)(x) = \dfrac{g(x)}{f(x)} \quad (f(x) \neq 0)$

$\qquad\quad = \dfrac{\sqrt{x}}{x^2 - 4}$

The domain of g/f consists of the numbers x that are in $[0, \infty)$, the domain of both f and g, except 2 (because division by 0 is undefined). The domain of g/f is $[0, 2) \cup (2, \infty)$. ∎

Self Check 3   Find $g - f$ and its domain.

Now Try Exercise 19.

We can perform these operations using a graphing calculator.

ACCENT ON TECHNOLOGY   **Operations on Functions**

A graphing calculator can be used to graph operations on function. Consider Example 3.

- We will use a bold setting to indicate the function $f + g$. In Figure 3-80, we see in the left graph that the graph of $f + g$ appears to be the same as f. However, if we zoom in to a small part of the graph as shown in the right graph, we can see that the functions are different. To input Y_1 and Y_2, press **VARS**, scroll right to Y-VARS, press **ENTER**, and select the desired function.

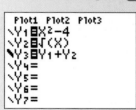 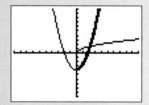 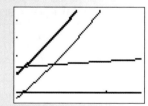

FIGURE 3-80

- We use the bold setting to indicate the function $f \cdot g$. Again, by zooming in to just part of the window, we can see the graphs are different. See Figure 3-81.

 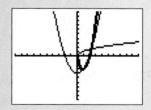 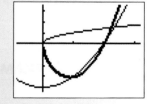

FIGURE 3-81

- We demonstrate the graph of f/g using a graphing calculator. See Figure 3-82. f/g appears bold.

 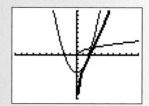

FIGURE 3-82

• We demonstrate the graph of g/f using a graphing calculator. See Figure 3-83. g/f appears bold.

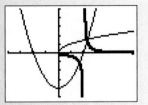

FIGURE 3-83

EXAMPLE 4   **Evaluating the Sum of Two Functions**

Find $(f + g)(3)$ when $f(x) = x^2 + 1$ and $g(x) = 2x + 1$.

SOLUTION   We will first find $(f + g)(x)$ and then find $(f + g)(3)$. To find $(f + g)(x)$, we proceed as follows:

$$(f + g)(x) = f(x) + g(x)$$

$$= x^2 + 1 + 2x + 1$$

$$= x^2 + 2x + 2$$

Comment

We can also determine $(f + g)(3)$ by finding $f(3) + g(3)$. Note that $f(3) = 10$, $g(3) = 7$, and $f(3) + g(3) = 10 + 7 = 17$.

To find $(f + g)(3)$, we proceed as follows:

$$(f + g)(x) = x^2 + 2(x) + 2$$

$$(f + g)(3) = 3^2 + 2(3) + 2 \qquad \text{Substitute 3 for } x.$$

$$= 9 + 6 + 2$$

$$= 17$$

Self Check 4   Find $(f \cdot g)(-2)$.

Now Try Exercise 25.

2. Write Functions as Sums, Differences, Products, or Quotients of Other Functions

EXAMPLE 5   **Writing Functions as Combinations of Other Functions**

Let $h(x) = x^2 + 3x + 2$. Find two functions f and g such that
a. $f + g = h$ **b.** $f \cdot g = h$

SOLUTION   For part **a**, we must find two functions f and g whose sum is h. For part **b**, we must find two functions f and g whose product is h.

a. There are many possibilities. One is $f(x) = x^2$ and $g(x) = 3x + 2$, then

$$(f + g)(x) = f(x) + g(x)$$

$$= (x^2) + (3x + 2)$$

$$= x^2 + 3x + 2$$

$$= h(x)$$

Another possibility is $f(x) = x^2 + 2x$ and $g(x) = x + 2$.

b. Again, there are many possibilities. One is suggested by factoring $x^2 + 3x + 2$.

$$x^2 + 3x + 2 = (x + 1)(x + 2)$$

If we let $f(x) = x + 1$ and $g(x) = x + 2$, then

$$(f \cdot g)(x) = f(x) \cdot g(x)$$
$$= (x + 1)(x + 2)$$
$$= x^2 + 3x + 2$$
$$= h(x)$$

Another possibility is $f(x) = 3$ and $g(x) = \frac{x^2}{3} + x + \frac{2}{3}$.

Self Check 5   Find two functions f and g such that $f - g = h$.

Now Try Exercise 35.

3. Evaluate Composite Functions

Often one quantity is a function of a second quantity that depends, in turn, on a third quantity. For example, the cost of a car trip is a function of the gasoline consumed. The amount of gasoline consumed, in turn, is a function of the number of miles driven. Such chains of dependence are analyzed mathematically as *composition of functions.*

Suppose that $y = f(x)$ and $y = g(x)$ define two functions. Any number x in the domain of g will produce a corresponding value $g(x)$ in the range of g. If $g(x)$ is in the domain of function f, then $g(x)$ can be substituted into f, and a corresponding value $f(g(x))$ will be determined. This two-step process defines a new function, called a **composite function,** denoted by $f \circ g$. (See Figure 3-84.)

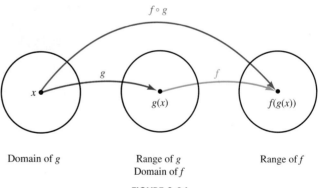

FIGURE 3-84

Composite Function   The **composite function** $f \circ g$ is defined by

$$(f \circ g)(x) = f(g(x))$$

The **domain** of $f \circ g$ consists of all those numbers in the domain of g for which $g(x)$ is in the domain of f.

To illustrate the previous definition, we consider the functions $f(x) = 5x + 1$ and $g(x) = 4x - 3$ and find $(f \circ g)(x)$ and $(g \circ f)(x)$.

$$(f \circ g)(x) = f(g(x)) \qquad\qquad (g \circ f)(x) = g(f(x))$$
$$= f(4x - 3) \qquad\qquad\qquad = g(5x + 1)$$
$$= 5(4x - 3) + 1 \qquad\qquad = 4(5x + 1) - 3$$
$$= 20x - 14 \qquad\qquad\qquad = 20x + 1$$

Since we get different results, the composition of functions is not commutative.

We have seen that a function can be represented by a machine. If we put a number from the domain into the machine (the input), a number from the range comes out (the output). For example, if we put 2 into the machine shown in Figure 3-85(a), the number $f(2) = 5(2) - 2 = 8$ comes out. In general, if we put x into the machine shown in Figure 3-85(b), the value $f(x)$ comes out.

Caution

Note that in the previous example $(f \circ g)(x) \neq (g \circ f)(x)$.

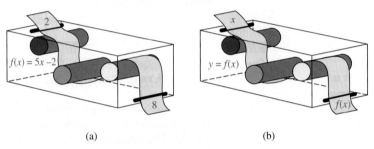

(a)   (b)

FIGURE 3-85

The function machines shown in Figure 3-86 illustrate the composition $f \circ g$. When we put a number x into the function g, the value $g(x)$ comes out. The value $g(x)$ then goes into function f, and $f(g(x))$ comes out.

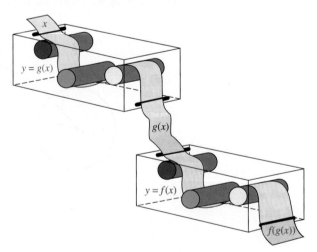

FIGURE 3-86

To further illustrate these ideas, we let $f(x) = 2x + 1$ and $g(x) = x - 4$.

- $(f \circ g)(9)$ means $f(g(9))$. In Figure 3-87(a), function g receives the number 9 and subtracts 4, and the number $g(x) = 5$ comes out. The 5 goes into the f function, which doubles it and adds 1. The final result, 11, is the output of the composite function $f \circ g$:

$$(f \circ g)(9) = f(g(9)) = f(5) = 2(5) + 1 = 11$$

• $(f \circ g)(x)$ means $f(g(x))$. In Figure 3-87(a), function g receives the number x and subtracts 4, and the number $x - 4$ comes out. The $x - 4$ goes into the f function, which doubles it and adds 1. The final result, $2x - 7$, is the output of the composite function $f \circ g$.

$$(f \circ g)(x) = f(g(x)) = f(x - 4) = 2(x - 4) + 1 = 2x - 7$$

• $(g \circ f)(-2)$ means $g(f(-2))$. In Figure 3-87(b), function f receives the number -2, doubles it and adds 1, and releases -3 into the g function. Function g subtracts 4 from -3 and releases a final output of -7. Thus,

$$(g \circ f)(-2) = g(f(-2)) = g(-3) = -3 - 4 = -7$$

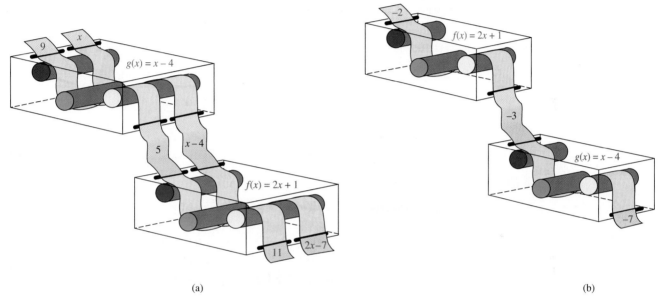

(a)   (b)

FIGURE 3-87

EXAMPLE 6   **Finding the Composition of Two Functions**

If $f(x) = 2x + 7$ and $g(x) = x^2 - 1$, find   **a.** $(f \circ g)(x)$ **b.** $(g \circ f)(x)$

SOLUTION   In part **a**, because $(f \circ g)(x)$ means $f(g(x))$, we will replace x in $f(x) = 2x + 7$ with $g(x)$. In part **b**, because $(g \circ f)(x)$ means $g(f(x))$ we will replace x in $g(x) = x^2 - 1$ with $f(x)$.

a. $(f \circ g)(x) = f(g(x))$

$\qquad\qquad = f(x^2 - 1)$ Substitute $x^2 - 1$ for $g(x)$.

$\qquad\qquad = 2(x^2 - 1) + 7$ Evaluate $f(x^2 - 1)$.

$\qquad\qquad = 2x^2 - 2 + 7$ Remove parentheses.

$\qquad\qquad = 2x^2 + 5$

b. $(g \circ f)(x) = g(f(x))$

$\qquad\qquad = g(2x + 7)$ Substitute $2x + 7$ for $f(x)$.

$\qquad\qquad = (2x + 7)^2 - 1$ Evaluate $g(2x + 7)$.

$\qquad\qquad = 4x^2 + 28x + 49 - 1$ Square the binomial.

$\qquad\qquad = 4x^2 + 28x + 48$

Self Check 6   If $f(x) = 2x + 7$ and $h(x) = x + 1$, find $(f \circ h)(x)$.

Now Try Exercise 51.

EXAMPLE 7   **Evaluating the Composition of Two Functions**

If $f(x) = 3x - 2$ and $g(x) = 3x^2 + 6x - 5$, find $(f \circ g)(-2)$.

SOLUTION   Because $(f \circ g)(-2)$ means $f(g(-2))$, we first find $g(-2)$. We then find $f(g(-2))$.

$$g(x) = 3x^2 + 6x - 5$$
$$g(-2) = 3(-2)^2 + 6(-2) - 5$$
$$= 3(4) - 12 - 5$$
$$= -5$$
$$(f \circ g)(-2) = f(g(-2))$$
$$= f(-5)$$
$$= 3(-5) - 2$$
$$= -17$$

Self Check 7   Find $(g \circ f)(-1)$.

Now Try Exercise 43.

4. Determine Domains of Composite Functions

To be in the domain of the composite function $f \circ g$, a number x has to be in the domain of g, and the output of g must be in the domain of f. Thus, the domain of $f \circ g$ consists of those inputs x that are in the domain of g and for which $g(x)$ is in the domain of f.

Strategy to Determine the Domain of $f \circ g$ To determine the domain of $(f \circ g)(x) = f(g(x))$, apply the following restrictions to the composition:

1. If x is not in the domain of g, it will not be in the domain of $f \circ g$.
2. Any x that has an output $g(x)$ that is not in the domain of f will not be in the domain of $f \circ g$.

EXAMPLE 8   **Finding the Domain of Composite Functions**

Let $f(x) = \sqrt{x}$ and $g(x) = x - 3$. Find the domain of **a.** $f \circ g$ **b.** $g \circ f$

SOLUTION   We will first find the domains of $f(x)$ and $g(x)$. Then, we will find the domain of $f \circ g$ and $g \circ f$ by applying the restrictions stated above.

For $\sqrt{x}$ to be a real number, x must be a nonnegative real number. Thus, the domain of f is the interval $[0, \infty)$. Since any real number x can be an input into g, the domain of g is the interval $(-\infty, \infty)$.

a. The domain of $f \circ g$ is the set of real numbers x such that x is in the domain of g and $g(x)$ is in the domain of f. We have seen that all values of x are in the domain of g. However, $g(x)$ must be nonnegative, because $g(x)$ must be in the domain of f. So we must find the values of x such that $g(x)$ is greater than or equal to 0.

$$g(x) \geq 0 \qquad \text{$g(x)$ must be nonnegative.}$$

$$x - 3 \geq 0 \qquad \text{Substitute } x - 3 \text{ for } g(x).$$

$$x \geq 3 \qquad \text{Add 3 to both sides.}$$

Since $x \geq 3$, the domain of $f \circ g$ is the interval $[3, \infty)$.

b. The domain of $g \circ f$ is the set of real numbers x such that x is in the domain of f and $f(x)$ is in the domain of g. We have seen that only nonnegative values of x are in the domain of f. Because all values of $f(x)$ are in the domain of g, the domain of $g \circ f$ is the domain of f, which is the interval $[0, \infty)$.

Self Check 8   Find the domain of $f \circ f$.

Now Try Exercise 55.

Composition of Functions and Domain

Consider the two functions given in Example 8(a). We can use a graphing calculator to graph the composite function $f \circ g$. To enter the composition, we enter $y_1(y_2)$ using **VARS**. See Figure 3-88. From the graph, we can determine the domain of $f \circ g$. We see that the domain is $[3, \infty)$.

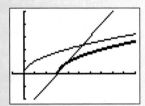

FIGURE 3-88

EXAMPLE 9   **Finding the Domain of Composite Functions**

Let $f(x) = \dfrac{x + 3}{x - 2}$ and $g(x) = \dfrac{1}{x}$.

a. Find the domain of $f \circ g$.   **b.** Find $f \circ g$.

SOLUTION   We will first find the domains of $f(x)$ and $g(x)$. Then, we will find the domain of $f \circ g$ and $g \circ f$ by applying the restrictions stated earlier.

Since any real number except 2 can be an input into f, the domain of f is $(-\infty, 2) \cup (2, \infty)$.

For $\frac{1}{x}$ to be a real number, x cannot be 0. Thus, the domain of g is $(-\infty, 0) \cup (0, \infty)$.

a. The domain of $f \circ g$ is the set of real numbers x such that x is in the domain of g and $g(x)$ is in the domain of f. We have seen that all values of x but 0 are in the domain of g and that all values of $g(x)$ but 2 are in the domain of f. So we must exclude 0 from the domain of $f \circ g$ and all values of x where $g(x) = 2$. To find the excluded values, we proceed as follows:

$$g(x) = 2$$

$$\frac{1}{x} = 2 \qquad \text{Substitute } \frac{1}{x} \text{ for } g(x).$$

$$1 = 2x \qquad \text{Since } x \neq 0, \text{ we can multiply both sides by } x.$$

$$x = \frac{1}{2} \qquad \text{Divide both sides by 2.}$$

The domain of $f \circ g$ is the set of all real numbers except 0 and $\frac{1}{2}$, which is $(-\infty, 0) \cup \left(0, \frac{1}{2}\right) \cup \left(\frac{1}{2}, \infty\right)$.

b. To find $f \circ g$, we proceed as follows.

$$(f \circ g)(x) = f(g(x))$$

$$= f\left(\frac{1}{x}\right) \qquad \text{Substitute } \frac{1}{x} \text{ for } g(x).$$

$$= \frac{\dfrac{1}{x} + 3}{\dfrac{1}{x} - 2} \qquad \text{Substitute } \frac{1}{x} \text{ for } x \text{ in } f.$$

$$= \frac{1 + 3x}{1 - 2x} \qquad \text{Multiply numerator and denominator by } x.$$

Thus, $(f \circ g)(x) = \frac{1 + 3x}{1 - 2x}$. ∎

Self Check 9 Let $f(x) = \dfrac{x}{x - 1}$ and $g(x) = \dfrac{1}{x}$. **a.** Find the domain of $f \circ g$ **b.** Find $f \circ g$

Now Try Exercise 63.

Caution

> Example 9 illustrates that the domain of the composite function $f \circ g$ **cannot** always be found by finding $f \circ g$ and analyzing it. In the example, we found that the domain is all real numbers, except for 0 and $\frac{1}{2}$. If we only analyze the form $f \circ g$, we would state the domain incorrectly as all real numbers except $\frac{1}{2}$.

5. Write Functions as Compositions

Comment

The result of a decomposition is not unique. There are several possibilities. In the example to the right, another possibility is $f(x) = x^2$ and $g(x) = (2x^2 - 6x + 3)^2$.

When we form a composite function $f \circ g$, we obtain a function h. It is possible to reverse this composition process and begin with a function h and express it as a composition of two functions. This process is called **decomposition.**

For example, consider $h(x) = (2x^2 - 6x + 3)^4$. The function h takes $2x^2 - 6x + 3$ and raises it to the fourth power. To write the function h as a composition of functions f and g, we can let

$$f(x) = x^4 \quad \text{and} \quad g(x) = 2x^2 - 6x + 3$$

Then

$$(f \circ g)(x) = f(g(x))$$

$$= f(2x^2 - 6x + 3)$$

$$= (2x^2 - 6x + 3)^4$$

$$= h(x)$$

EXAMPLE 10   Writing a Function as a Composition of Two Functions

Let $h(x) = \sqrt{x + 1}$. Find two functions f and g such that $f \circ g = h$.

SOLUTION   Because h takes the square root of the algebraic function $x + 1$, we let $f(x) = \sqrt{x}$ and $g(x) = x + 1$.

$$f(x) = \sqrt{x} \quad \text{and} \quad g(x) = x + 1$$

We can check the composition $f \circ g$ to see that it gives the original function h.

$$(f \circ g)(x) = f(g(x))$$

$$= f(x + 1)$$

$$= \sqrt{x + 1}$$

$$= h(x)$$

Self Check 10   Let $h(x) = \sqrt[3]{x^2 - 5}$. Find functions f and g such that $f \circ g = h$.

Now Try Exercise 73.

6. Use Operations on Functions to Solve Problems

EXAMPLE 11   Solving an Application Using Composition of Functions

A laboratory sample is removed from a cooler at a temperature of $15°$ F. Technicians then warm the sample at the rate of $3°$F per hour. Express the sample's temperature in degrees Celsius as a function of the time t (in hours) since it was removed from the cooler.

SOLUTION   We first write a Fahrenheit temperature function that represents the warming of the sample. We then use composition of functions to write degrees Celsius as a function of degrees Fahrenheit.

The temperature of the sample is $15°$ F when $t = 0$. Because the sample warms at $3°$ F per hour, it warms $3t°$ after t hours. Thus, the Fahrenheit temperature after t hours is given by the function

$$F(t) = 3t + 15 \qquad \text{\small $F(t)$ is the Fahrenheit temperature and t represents the time in hours.}$$

The Celsius temperature is a function of the Fahrenheit temperature $F(t)$, given by the formula

$$C(F(t)) = \frac{5}{9}(F(t) - 32)$$

To express the sample's Celsius temperature as a function of time, we find the composition function $C \circ F$.

$$(C \circ F)(t) = C(F(t))$$

$$= C(3t + 15) \qquad \text{Substitute for } F(t).$$

$$= \frac{5}{9}[(3t + 15) - 32] \qquad \text{Substitute } 3t + 15 \text{ for } F(t) \text{ in } C(F(t)).$$

$$= \frac{5}{9}(3t - 17) \qquad \text{Simplify.}$$

$$= \frac{15}{9}t - \frac{85}{9}$$

$$= \frac{5}{3}t - \frac{85}{9}$$

Self Check 11   Find the $(C \circ F)(10)$.

Now Try Exercise 93.

Self Check Answers   1. $(g - f)(x) = -x - 4$ 2. $(g/f)(x) = \dfrac{2x - 3}{3x + 1}, \left(-\infty, -\dfrac{1}{3}\right) \cup \left(-\dfrac{1}{3}, \infty\right)$

3. $(g - f)(x) = \sqrt{x} - x^2 + 4, [0, \infty)$ 4. $(f \cdot g)(2) = -15$
5. One possibility is $f(x) = 2x^2$ and $g(x) = x^2 - 3x - 2$.
6. $(f \circ h)(x) = 2x + 9$ 7. $(g \circ f)(-1) = 40$ 8. $[0, \infty)$
9. $(-\infty, 0) \cup (0, 1) \cup (1, \infty); (f \circ g)(x) = \dfrac{1}{1 - x}$
10. One possibility is $f(x) = \sqrt[3]{x}, g(x) = x^2 - 5$ 11. approximately $7.2°C$

Exercises **3.6**

Getting Ready
You should be able to complete these vocabulary and concept statements before you proceed to the practice exercises.

Fill in the blanks.

1. $(f + g)(x) =$ _____
2. $(f - g)(x) =$ _____
3. $(f \cdot g)(x) =$ _____
4. $(f/g)(x) =$ _____, where $g(x) \neq 0$
5. The domain of $f + g$ is the _____ of the domains of f and g.
6. $(f \circ g)(x) =$ _____
7. $(g \circ f)(x) =$ _____
8. To determine $(f \circ g)(-5)$, first find _____.
9. Composition of functions is not _____.
10. To be in the domain of the composite function $f \circ g$, a number x has to be in the _____ of g, and the output of g must be in the _____ of f.

Practice
Let $f(x) = 2x + 1$ and $g(x) = 3x - 2$. Find each function and its domain.

11. $f + g$ 12. $f - g$

13. $f \cdot g$ 14. f/g

Let $f(x) = x^2 + x$ and $g(x) = x^2 - 1$. Find each function and its domain.

15. $f - g$ 16. $f + g$

17. f/g **18.** $f \cdot g$

Let $f(x) = x^2 - 7$ and $g(x) = \sqrt{x}$. Find each function and its domain.

19. $f + g$ **20.** $f - g$

21. f/g **22.** $f \cdot g$

Let $f(x) = x^2 - 1$ and $g(x) = 3x - 2$. Find each value, if possible.

23. $(f + g)(2)$ **24.** $(f + g)(-3)$

25. $(f - g)(0)$ **26.** $(f - g)(-5)$

27. $(f \cdot g)(2)$ **28.** $(f \cdot g)(-1)$

29. $(f/g)\left(\dfrac{2}{3}\right)$ **30.** $(f/g)(t)$

Find two functions f and g such that $h(x)$ can be expressed as the function indicated. Several answers are possible.

31. $h(x) = 3x^2 + 2x; \ f + g$

32. $h(x) = 3x^2; \ f \cdot g$

33. $h(x) = \dfrac{3x^2}{x^2 - 1}; \ f/g$

34. $h(x) = 5x + x^2; \ f - g$

35. $h(x) = x(3x^2 + 1); \ f - g$

36. $h(x) = (3x - 2)(3x + 2); \ f + g$

37. $h(x) = x^2 + 7x - 18; \ f \cdot g$

38. $h(x) = 5x^5; \ f/g$

Let $f(x) = 2x - 5$ and $g(x) = 5x - 2$. Find each value.

39. $(f \circ g)(2)$ **40.** $(g \circ f)(-3)$

41. $(f \circ f)\left(-\dfrac{1}{2}\right)$ **42.** $(g \circ g)\left(\dfrac{3}{5}\right)$

Let $f(x) = 3x^2 - 2$ and $g(x) = 4x + 4$. Find each value.

43. $(f \circ g)(-3)$ **44.** $(g \circ f)(3)$

45. $(f \circ f)\left(\sqrt{3}\right)$ **46.** $(g \circ g)(-4)$

Let $f(x) = 3x$ and $g(x) = x + 1$. Determine the domain of each composite function and then find the composite function.

47. $f \circ g$ **48.** $g \circ f$

49. $f \circ f$ **50.** $g \circ g$

Let $f(x) = x^2$ and $g(x) = 2x$. Determine the domain of each composite function and then find the composite function.

51. $g \circ f$ **52.** $f \circ g$

53. $g \circ g$ **54.** $f \circ f$

Let $f(x) = \sqrt{x}$ and $g(x) = x + 1$. Determine the domain of each composite function and then find the composite function.

55. $f \circ g$ **56.** $g \circ f$

57. $f \circ f$ **58.** $g \circ g$

Let $f(x) = \sqrt{x + 1}$ and $g(x) = x^2 - 1$. Determine the domain of each composite function and then find the composite function.

59. $g \circ f$ **60.** $f \circ g$

61. $g \circ g$ **62.** $f \circ f$

Let $f(x) = \dfrac{1}{x - 1}$ and $g(x) = \dfrac{1}{x - 2}$. Determine the domain of each composite function and then find the composite function.

63. $f \circ g$ **64.** $g \circ f$

65. $f \circ f$ **66.** $g \circ g$

Find two functions f and g such that the composition $f \circ g = h$ expresses the given correspondence. Several answers are possible.

67. $h(x) = 3x - 2$ **68.** $h(x) = 7x - 5$

69. $h(x) = x^2 - 2$ **70.** $h(x) = x^3 - 3$

71. $h(x) = (x - 2)^2$ **72.** $h(x) = (x - 3)^3$

73. $h(x) = \sqrt{x + 2}$ **74.** $h(x) = \dfrac{1}{x - 5}$

75. $h(x) = \sqrt{x} + 2$ **76.** $h(x) = \dfrac{1}{x} - 5$

77. $h(x) = x$ **78.** $f(x) = 3$

Use the graphs of functions f and g to answer each problem.

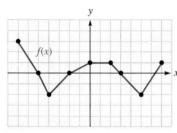

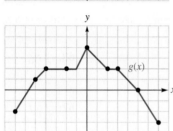

79. $(f + g)(-4)$ **80.** $(f - g)(1)$
81. $(f \cdot g)(5)$ **82.** $(f/g)(-1)$
83. $(f \circ g)(3)$ **84.** $(g \circ f)(2)$
85. $(f \circ f)(-2)$ **86.** $(g \circ g)(-5)$

Use the tables of values of f and g to answer each problem.

x	$f(x)$
2	4
4	9
6	13
13	17

x	$g(x)$
0	0
2	4
3	9
4	16

87. $(f + g)(2)$ **88.** $(f/g)(4)$
89. $(f \circ g)(2)$ **90.** $(g \circ f)(2)$

Applications

91. DVD camcorder Suppose that the functions $R(x) = 300x$ and $C(x) = 60{,}000 + 40x$ model a company's monthly revenue and cost for producing and selling DVD camcorders.

   a. Find $(R - C)(x)$, the function that models the monthly profit, $P(x)$.

   b. Find the company's profit if 500 camcorders are produced and sold in one month.

92. TV screen The height of the television screen shown is 13 inches.

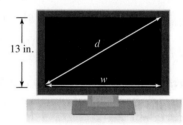

   a. Write a formula to find the area of the viewing screen.

   b. Use the Pythagorean Theorem to write a formula to find the width w of the screen.

   c. Write a formula to find the area of the screen as a function of the diagonal d.

93. Area of an oil spill Suppose an oil spill from a tanker is spreading in the shape of a circular ripple. If the function $d(t) = 3t$ represents the diameter of the spill in inches at time t minutes, express the area, A, of the oil spill as a function of time. Find the area of the oil spill after 2 hours. Round to one decimal place.

94. Area of a square Write a formula for the area A of a square in terms of its perimeter P.

95. Perimeter of a square Write a formula for the perimeter P of a square in terms of its area A.

96. Ceramics When the temperature of a pot in a kiln is $1{,}200°$ F, an artist turns off the heat and leaves the pot to cool at a controlled rate of $81°$ F per hour. Express the temperature of the pot in degrees Celsius as a function of the time t (in hours) since the kiln was turned off.

Discovery and Writing

97. Let $f(x) = 3x$. Show that $(f + f)(x) = f(x + x)$.

98. Let $g(x) = x^2$. Show that $(g + g)(x) \neq g(x + x)$.

99. Let $f(x) = \frac{x - 1}{x + 1}$. Find $(f \circ f)(x)$.

100. Let $g(x) = \frac{x}{x - 1}$. Find $(g \circ g)(x)$.

Let $f(x) = x^2 - x$, $g(x) = x - 3$, and $h(x) = 3x$. Use a graphing calculator to graph both functions on the same axes. Write a brief paragraph summarizing your observations.

101. f and $f \circ g$

102. f and $g \circ f$

103. f and $f \circ h$

104. f and $h \circ f$

Review

Solve each equation for y.

105. $x = 3y - 7$

106. $x = \dfrac{7}{y}$

107. $x = \dfrac{y}{y + 3}$

108. $x = \dfrac{y - 1}{y}$

3.7 Inverse Functions

In this section, we will learn to

1. Understand the definition of a one-to-one function.

2. Determine whether a function is one-to-one.

3. Verify inverse functions.

4. Find the inverse of a one-to-one function.

5. Understand the relationship between the graphs of f and f^{-1}.

contax66/Shutterstock.com

In this section, we will discuss inverse functions. A function and its inverse do opposite things.

Suppose we climb the Great Wall of China on a summer day when the temperature reaches a high of 35°C.

The linear function defined by $F = \frac{9}{5}C + 32$ gives a formula to convert degrees Celsius to degrees Fahrenheit. If we substitute a Celsius reading into the formula, a Fahrenheit reading comes out. For example, if we substitute 35 for C, we obtain a Fahrenheit reading of 95°:

$$F = \frac{9}{5}C + 32$$
$$= \frac{9}{5}(35) + 32$$
$$= 63 + 32$$
$$= 95$$

If we want to find a Celsius reading from a Fahrenheit reading, we need a formula into which we can substitute a Fahrenheit reading and have a Celsius reading come out. Such a formula is $C = \frac{5}{9}(F - 32)$, which takes the Fahrenheit reading of 95° and turns it back into a Celsius reading of 35°.

$$C = \frac{5}{9}(F - 32)$$
$$= \frac{5}{9}(95 - 32)$$
$$= \frac{5}{9}(63)$$
$$= 35$$

The functions defined by these two formulas do opposite things. The first turns $35°C$ into $95°$ Fahrenheit, and the second turns $95°$ Fahrenheit back into $35°C$. Such functions are called *inverse functions.*

Some functions have inverses that are functions and some do not. To guarantee that the inverse of a function will also be a function, we must know that the function is *one-to-one.*

1. Understand the Definition of a One-to-One Function

In this section, we will find inverses of functions that are one-to-one. *One-to-one functions* are functions whose inverses are also functions.

We now examine what it means for a function to be one-to-one. Consider the following two functions:

Function 1: To each student, there corresponds exactly one eye color

Function 2: To each student, there corresponds exactly one college identification number

Function 1 **is not a one-to-one function** because two different students can have the same eye color.

Function 2 **is a one-to-one function** because two different students will always have two different ID numbers.

Recall that each element x in the domain of a function has a single output y. For some functions, different numbers x in the domain can have the same output. See Figure 3-89(a). For other functions, called **one-to-one functions,** different numbers x have different outputs. See Figure 3-89(b).

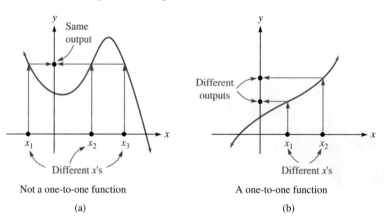

Not a one-to-one function

(a)

A one-to-one function

(b)

FIGURE 3-89

One-to-One Functions A function f from a set **X** to a set **Y** is called a **one-to-one function** if and only if different numbers in the domain of f have different outputs in the range of f.

The previous definition implies that if x_1 and x_2 are two numbers in the domain of f and $x_1 \neq x_2$, then $f(x_1) \neq f(x_2)$.

2. Determine Whether a Function Is One-to-One

EXAMPLE 1 **Determining Whether a Function Is One-to-One**

Determine whether each function is one-to-one.
a. $f(x) = x^4 + x^2$ **b.** $f(x) = x^3$

SOLUTION   We will examine the functions and determine whether the definition of a one-to-one function applies. If different *x*-values always produce different *y*-values, the function is one-to-one.

a. The function $f(x) = x^4 + x^2$ is not one-to-one, because different numbers in the domain have the same output. For example, 2 and –2 have the same output: $f(2) = f(-2) = 20$.

b. The function $f(x) = x^3$ is one-to-one, because different numbers *x* produce different outputs $f(x)$. This is because different numbers have different cubes.

Self Check 1   Determine whether $f(x) = \sqrt{x}$ is one-to-one.

Now Try Exercise 11.

A **Horizontal Line Test** can be used to determine whether the graph of a function represents a one-to-one function. If every horizontal line that intersects the graph of a function does so exactly once, the function passes the Horizontal Line Test and is one-to-one. See Figure 3-90(a). If any horizontal line intersects the graph of a function more than once, the function fails the Horizontal Line Test and is not one-to-one. See Figure 3-90(b).

Comment

A one-to-one function satisfies both the Horizontal and Vertical Line Tests.

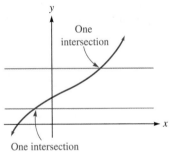

A one-to-one function

(a)

Not a one-to-one function

(b)

FIGURE 3-90

EXAMPLE 2   **Using the Horizontal Line Test**

Use the Horizontal Line Test to determine whether each graph represents a one-to-one function.

a.

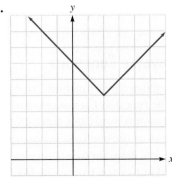

b.

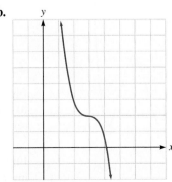

SOLUTION   We will use the Horizontal Line Test and draw many horizontal lines. If every horizontal line that intersects the graph does so exactly once, the function is one-to-one. If any horizontal line intersects the graph more than once, the function is not one-to-one.

a. Because the horizontal line drawn in Figure 3-91 intersects the graph in two places, we know that the function fails the Horizontal Line Test and is not a one-to-one function.

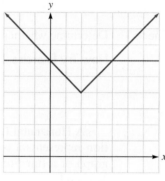

FIGURE 3-91

b. Several horizontal lines are drawn in Figure 3-92, and each one intersects the graph exactly once. We conclude that the graph passes the Horizontal Line Test and represents a one-to-one function.

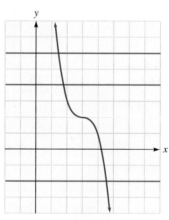

FIGURE 3-92

Self Check 2   Determine whether the graph in the margin represents a one-to-one function.

Now Try Exercise 19.

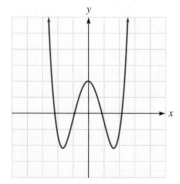

3. Verify Inverse Functions

Figure 3-93(a) illustrates a function f from set **X** to set **Y**. Since three arrows point to a single y, the function f is not one-to-one. If the arrows in Figure 3-93(a) were reversed, the diagram would not represent a function.

If the arrows of the one-to-one function f in Figure 3-93(b) were reversed, as in Figure 3-93(c), the diagram would represent a function. This function is called the **inverse of function** f and is denoted by the symbol f^{-1}.

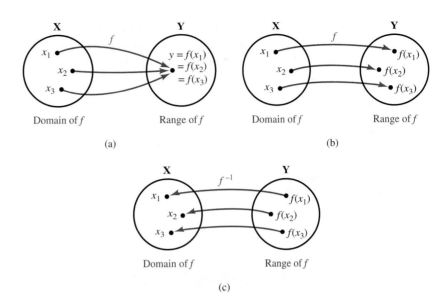

FIGURE 3-93

Caution

The -1 in the notation for inverse function is **not** an exponent. Remember that

$$f^{-1}(x) \neq \frac{1}{f(x)}$$

Consider the functions:

$$f(x) = 4x \quad \text{and} \quad g(x) = \frac{x}{4}$$

These functions are inverses of each other because f multiplies any input x by **4** and function g will take the result and divide it by 4. The final result will be the original input x.

We can show that the composition of these functions (in either order) is the identity function.

$$(f \circ g)(x) = f(g(x)) = f\left(\frac{x}{4}\right) = 4\left(\frac{x}{4}\right) = x \quad \text{and} \quad (g \circ f)(x) = g(f(x)) = g(4x) = \frac{4x}{4} = x$$

Since $g(x)$ is the inverse of $f(x)$, we can write $g(x)$ using inverse notation as $f^{-1}(x) = \frac{x}{4}$. Thus, $(f \circ f^{-1}) = x$ and $(f^{-1} \circ f)(x) = x$.

We can now define inverse functions.

Inverse Functions   If f and g are two one-to-one functions such that $(f \circ g)(x) = x$ for every x in the domain of g and $(g \circ f)(x) = x$ for every x in the domain of f, then f and g are **inverse functions**. Function g can be denoted as f^{-1} and is called the **inverse function of f**.

We can also list two important properties of one-to-one functions.

Properties of a   **Property 1:** If f is a one-to-one function, there is a one-to-one function $f^{-1}(x)$
One-to-One Function   such that

$$(f^{-1} \circ f)(x) = x \quad \text{and} \quad (f \circ f^{-1})(x) = x.$$

Property 2: The domain of f is the range of f^{-1} and the range of f is the domain of f^{-1}.

Figure 3-94 shows a one-to-one function f and its inverse f^{-1}. To the number x in the domain of f, there corresponds an output $f(x)$ in the range of f. Since $f(x)$ is in the domain of f^{-1}, the output for $f(x)$ under the function f^{-1} is $f^{-1}(f(x)) = x$. Thus, $(f^{-1} \circ f)(x) = f^{-1}(f(x)) = x$.

Comment

To show that one function is the inverse of another, we must show that their compositions are the **identity function,** x.

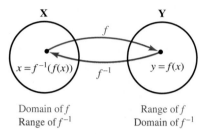

Domain of f
Range of f^{-1}

Range of f
Domain of f^{-1}

FIGURE 3-94

EXAMPLE 3 **Verifying that Two Functions are Inverses**

Verify that $f(x) = x^3$ and $g(x) = \sqrt[3]{x}$ are inverse functions.

SOLUTION To show that f and g are inverse functions, we must show that $f \circ g$ and $g \circ f$ are x, the identity function.

$$(f \circ g)(x) = f(g(x)) = f(\sqrt[3]{x}) = (\sqrt[3]{x})^3 = x$$
$$(g \circ f)(x) = g(f(x)) = g(x^3) = \sqrt[3]{x^3} = x$$

Because g is the inverse of f, we can use inverse notation and write $g(x) = \sqrt[3]{x}$ as $f^{-1}(x) = \sqrt[3]{x}$. Because f is the inverse of g, we can use inverse notation and write $f(x) = x^3$ as $g^{-1}(x) = x^3$. ∎

Self Check 3 If $x \geq 0$, are $f(x) = x^2$ and $g(x) = \sqrt{x}$ inverse functions?

Now Try Exercise 21.

4. Find the Inverse of a One-to-One Function

If f is the one-to-one function $y = f(x)$, then f^{-1} reverses the correspondence of f. That is, if $f(a) = b$, then $f^{-1}(b) = a$. To determine f^{-1}, we follow these steps.

Strategy for Finding f^{-1} from a Given Function $f(x)$

Step 1: Replace $f(x)$ with y.
Step 2: Interchange the variables x and y.
Step 3: Solve the resulting equation for y.
Step 4: Replace y with $f^{-1}(x)$.

Once $f^{-1}(x)$ is determined, it should be verified by showing that $(f \circ f^{-1})(x) = x$ and $(f^{-1} \circ f)(x) = x$.

EXAMPLE 4 **Finding the Inverse of a One-to-One Function**

Find the inverse of $f(x) = \dfrac{3}{2}x + 2$ and verify the result.

SOLUTION We will use the strategy given above to find f^{-1}. We then will verify the result by showing that $(f \circ f^{-1})(x) = x$ and $(f^{-1} \circ f)(x) = x$.

To find f^{-1}, we use the following steps.

Step 1: Replace $f(x)$ with y.

$$f(x) = \frac{3}{2}x + 2$$

$$y = \frac{3}{2}x + 2$$

Step 2: Interchange the variables x and y.

$$x = \frac{3}{2}y + 2$$

Step 3: Solve the resulting equation for y.

$$x = \frac{3}{2}y + 2$$

$$2x = 3y + 4 \qquad \text{Multiply both sides by 2.}$$

$$2x - 4 = 3y \qquad \text{Subtract 4 from both sides.}$$

$$y = \frac{2x - 4}{3} \qquad \text{Divide both sides by 3.}$$

Step 4: Replace y with $f^{-1}(x)$.

$$y = \frac{2x - 4}{3}$$

$$f^{-1}(x) = \frac{2x - 4}{3}$$

The inverse of $f(x) = \frac{3}{2}x + 2$ is $f^{-1}(x) = \frac{2x-4}{3}$.

To verify the result, we will use $f(x) = \frac{3}{2}x + 2$ and $f^{-1}(x) = \frac{2x-4}{3}$ and show that $(f \circ f^{-1})(x) = x$ and $(f^{-1} \circ f)(x) = x$.

$$(f \circ f^{-1})(x) = f(f^{-1}(x))$$

$$= f\left(\frac{2x - 4}{3}\right)$$

$$= \frac{3}{2}\left(\frac{2x - 4}{3}\right) + 2$$

$$= x - 2 + 2$$

$$= x$$

$$(f^{-1} \circ f)(x) = f^{-1}(f(x))$$

$$= f^{-1}\left(\frac{3}{2}x + 2\right)$$

$$= \frac{2\left(\frac{3}{2}x + 2\right) - 4}{3}$$

$$= \frac{3x + 4 - 4}{3}$$

$$= x$$

Caution

After completing Step 3, if y does not represent a function of x, the process ends and f does not have an inverse.

Self Check 4   Find $f(2)$. Then find $f^{-1}(5)$. Explain the significance of the results.

Now Try Exercise 27.

5. Understand the Relationship Between the Graphs of f and f^{-1}

Because we interchange the positions of x and y to find the inverse of a function, the point (b, a) lies on the graph of $y = f^{-1}(x)$ whenever the point (a, b) lies on the graph of $y = f(x)$. Thus, the graph of a function and its inverse are reflections of each other about the line $y = x$.

EXAMPLE 5   **Finding f^{-1} and Graphing Both f and f^{-1}**

Find the inverse of $f(x) = x^3 + 3$. Graph the function and its inverse on the same set of coordinate axes.

SOLUTION   We will find the inverse of the function $f(x)$ using the strategy given in the section. We will use translations to graph both f and f^{-1}.

We first find f^{-1} and proceed as follows:

Step 1: Replace $f(x)$ with y.

$$f(x) = x^3 + 3$$

$$y = x^3 + 3$$

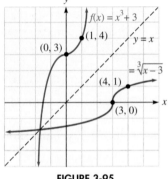

FIGURE 3-95

Step 2: Interchange the variables x and y.

$$x = y^3 + 3$$

Step 3: Solve the resulting equation for y.

$$x - 3 = y^3$$

$$y = \sqrt[3]{x - 3}$$

Step 4: Replace y with $f^{-1}(x)$.

$$f^{-1}(x) = \sqrt[3]{x - 3}$$

Comment

We can also graph f and f^{-1} by completing a table of solutions. The x and y columns of f can be reversed to obtain the table of solutions for f^{-1}.

We now graph f and f^{-1}. To graph $f(x) = x^3 + 3$, we translate the graph of $y = x^3$ vertically upward 3 units. To graph $f^{-1}(x) = \sqrt[3]{x - 3}$ we translate the graph of $y = \sqrt[3]{x}$ horizontally 3 units to the right. The graphs of f and f^{-1} are shown in Figure 3-95. In the graph that appears in Figure 3-95, the line $y = x$ is the axis of symmetry.

Self Check 5   Find $f(2)$. Then find $f^{-1}(11)$. Explain the significance of the result.

Now Try Exercise 45.

In the next example, we will consider a function that is not one-to-one but becomes so when we restrict its domain. By restricting the domain of the function and making it one-to-one, we are able to find its inverse and examine the function and its inverse graphically.

EXAMPLE 6   **Restricting the Domain of f to Make It One-to-One; Finding f^{-1}; Graphing f and f^{-1}; Stating Domain and Range**

The function $y = f(x) = x^2 + 3$ is not one-to-one. However, it becomes one-to-one when we restrict its domain to the interval $(-\infty, 0]$. Under this restriction,
a. Find the inverse of f.
b. Graph each function and state each one's domain and range.

SOLUTION   We will find f^{-1} by using the four-step strategy given in the section. We will then graph f and f^{-1} by using translations and then identify the domain and range from the graphs of each.

a. We first find f^{-1} and follow these steps:

Step 1: Replace $f(x)$ with y.

$$f(x) = x^2 + 3 \quad (x \leq 0)$$

$$y = x^2 + 3$$

Step 2: Interchange the variables x and y.

$$x = y^2 + 3 \quad (y \leq 0) \qquad \text{Interchange } x \text{ and } y.$$

Step 3: Solve the resulting equation for y.

$$x - 3 = y^2$$

To solve this equation for y, we take the square root of both sides. Because $y \leq 0$, we have $-\sqrt{x - 3} = y \quad (y \leq 0)$

Step 4: Replace y with $f^{-1}(x)$.

The inverse of f is defined by $f^{-1}(x) = -\sqrt{x - 3}$.

b. We graph the function $f(x) = x^2 + 3$ with domain $(-\infty, 0]$ by translating the graph of the parabola $y = x^2$ with domain $(-\infty, 0]$ vertically upward 3 units. From the graph, we see that the y coordinates are 3 and above and thus the range is the interval $[3, \infty)$. (See Figure 3-96.)

We graph the function $f^{-1}(x) = -\sqrt{x - 3}$ by translating the graph of $y = \sqrt{x}$ horizontally to the right 3 units and then reflecting the graph about the x-axis. It has domain $[3, \infty)$ and range $(-\infty, 0]$. (See Figure 3-96.) Note that the line of symmetry is shown and is $y = x$.

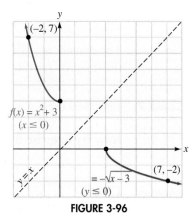

FIGURE 3-96

Domain of f and **Range** of f^{-1}: $(-\infty, 0]$
Range of f and **Domain** of f^{-1}: $[3, \infty)$

Self Check 6   Find the inverse of f when its domain is restricted to the interval $[0, \infty)$.

Now Try Exercise 53.

If a function is defined by the equation $y = f(x)$, we can often find the domain of f by inspection. Finding the range can be more difficult. One way to find the range of f is to find the domain of f^{-1}.

EXAMPLE 7   **Using the Domain of $f^{-1}(x)$ to Find the Range of $f(x)$**

Find the domain and range of $f(x) = \frac{2}{x} + 3$. Find its range by finding the domain of $f^{-1}(x)$.

SOLUTION   We will find the domain of $f(x) = \frac{2}{x} + 3$ by identifying the values of x that make the function undefined. We will then find $f^{-1}(x)$ and find its domain. The domain of $f^{-1}(x)$ will be the range of $f(x)$.

Because x cannot be 0, the domain of f is $(-\infty, 0) \cup (0, \infty)$. Next, we find $f^{-1}(x)$.

Step 1: Replace $f(x)$ with y.

$$f(x) = \frac{2}{x} + 3$$

$$y = \frac{2}{x} + 3$$

Step 2: Interchange the variables x and y.

$$x = \frac{2}{y} + 3 \qquad \text{Interchange } x \text{ and } y.$$

Step 3: Solve the resulting equation for y.

$$xy = 2 + 3y \qquad \text{Multiply both sides by } y.$$

$$xy - 3y = 2 \qquad \text{Subtract 3y from both sides.}$$

$$y(x - 3) = 2 \qquad \text{Factor out } y.$$

$$y = \frac{2}{x - 3} \qquad \text{Divide both sides by } x - 3.$$

Step 4: Replace y with $f^{-1}(x)$.

$$f^{-1}(x) = \frac{2}{x - 3}$$

The domain of $f^{-1}(x) = \frac{2}{x-3}$ is $(-\infty, 3) \cup (3, \infty)$ because x cannot be 3. Because the range of f is the domain of f^{-1}, the range of f is $(-\infty, 3) \cup (3, \infty)$. ■

Self Check 7   Find the range of $y = f(x) = \frac{3}{x} - 1$.

Now Try Exercise 61.

Self Check Answers   1. yes   2. no   3. yes   4. 5, 2   5. 11, 2   6. $f^{-1}(x) = \sqrt{x - 3}$
7. $(-\infty, -1) \cup (-1, \infty)$

Exercises **3.7**

Getting Ready
You should be able to complete these vocabulary and concept statements before you proceed to the practice exercises.

Fill in the blanks.

1. If different numbers in the domain of a function have different outputs, the function is called a _____ function.

2. If every _____ line intersects the graph of a function only once, the function is one-to-one.

3. Two functions f and g are inverses if their composition in either order is the _____ function.

4. The graph of a function and its inverse are reflections of each other about the line _____.

Practice

Determine whether each function is one-to-one.

5. $f(x) = 3x$

6. $f(x) = \dfrac{1}{2}x$

7. $f(x) = x^2 + 3$

8. $f(x) = x^4 - x^2$

9. $f(x) = x^3 - x$

10. $f(x) = x^2 - x$

11. $f(x) = |x|$

12. $f(x) = |x - 3|$

13. $f(x) = 5$

14. $f(x) = \sqrt{x - 5}$

15. $f(x) = (x - 2)^2; x \geq 2$

16. $f(x) = \dfrac{1}{x}$

Use the Horizontal Line Test to determine whether each graph represents a one-to-one function.

17.

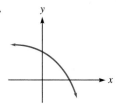

18.

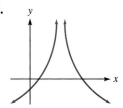

19.

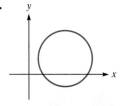

20.

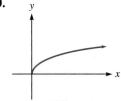

Verify that the functions are inverses by showing that $f \circ g$ and $g \circ f$ are the identity function.

21. $f(x) = 5x$ and $g(x) = \dfrac{1}{5}x$

22. $f(x) = 4x + 5$ and $g(x) = \dfrac{x - 5}{4}$

23. $f(x) = \dfrac{x + 1}{x}$ and $g(x) = \dfrac{1}{x - 1}$

24. $f(x) = \dfrac{x + 1}{x - 1}$ and $g(x) = \dfrac{x + 1}{x - 1}$

Each equation defines a one-to-one function f. Determine f^{-1} and verify that $f \circ f^{-1}$ and $f^{-1} \circ f$ are both the identity function.

25. $f(x) = 3x$

26. $f(x) = \dfrac{1}{3}x$

27. $f(x) = 3x + 2$

28. $f(x) = 2x - 5$

29. $f(x) = x^3 + 2$

30. $f(x) = (x + 2)^3$

31. $f(x) = \sqrt[5]{x}$

32. $f(x) = \sqrt[5]{x} + 4$

33. $f(x) = \dfrac{1}{x + 3}$

34. $f(x) = \dfrac{1}{x - 2}$

35. $f(x) = \dfrac{1}{2x}$

36. $f(x) = \dfrac{1}{x^3}$

Find the inverse of each one-to-one function and graph both the function and its inverse on the same set of coordinate axes.

37. $y = 5x$

38. $y = \dfrac{3}{2}x$

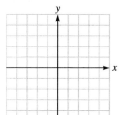

39. $y = 2x - 4$

40. $y = \dfrac{3}{2}x - 2$

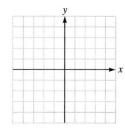

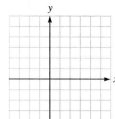

41. $x - y = 2$

42. $x + y = 0$

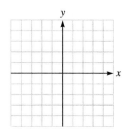

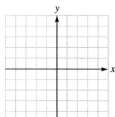

43. $2x + y = 4$

44. $3x + 2y = 6$

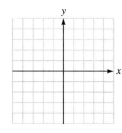

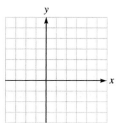

45. $f(x) = \sqrt[3]{x} - 4$

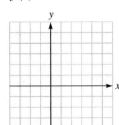

46. $f(x) = \sqrt[3]{x} + 3$

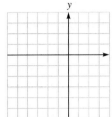

54. $f(x) = \dfrac{1}{x^2}$ $(x > 0)$

55. $f(x) = x^4 - 8$ $(x \geq 0)$

56. $f(x) = \dfrac{-1}{x^4}$ $(x < 0)$

47. $f(x) = (x - 6)^3$

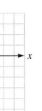

48. $f(x) = x^3 + 2$

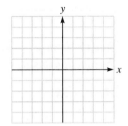

57. $f(x) = \sqrt{4 - x^2}$ $(0 \leq x \leq 2)$

58. $f(x) = \sqrt{x^2 - 1}$ $(x \leq -1)$

Find the domain and the range of f. Find the range by finding the domain of f^{-1}.

59. $f(x) = \dfrac{x}{x - 2}$

60. $f(x) = \dfrac{x - 2}{x + 3}$

49. $f(x) = \dfrac{1}{2x}$

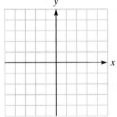

61. $f(x) = \dfrac{1}{x} - 2$

62. $f(x) = \dfrac{3}{x} - \dfrac{1}{2}$

50. $f(x) = \dfrac{1}{x - 3}$

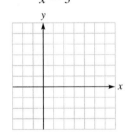

Applications

63. Buying pizza A pizzeria charges \$8.50 plus 75¢ per topping for a medium pizza.

51. $f(x) = \dfrac{x + 1}{x - 1}$

52. $f(x) = \dfrac{x - 1}{x}$

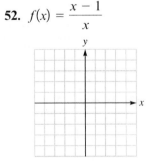

a. Find a linear function that expresses the cost $f(x)$ of a medium pizza in terms of the number of toppings x.

b. Find the cost of a pizza that has four toppings.

c. Find the inverse of the function found in part (a) to find a formula that gives the number of toppings $f^{-1}(x)$ in terms of the cost x.

d. If Josh has \$10, how many toppings can he afford?

The function f defined by the given equation is one-to-one on the given domain. Find $f^{-1}(x)$.

53. $f(x) = x^2 - 3$ $(x \leq 0)$

64. Cell phone bills A phone company charges $11 per month plus a nickel per call.

    a. Find a rational function that expresses the average cost $f(x)$ of a call in a month when x calls were made.

    b. To the nearest tenth of a cent, find the average cost of a call in a month when 68 calls were made.

    c. Find the inverse of the function found in part (a) to find a formula that gives the number of calls $f^{-1}(x)$ that can be made for an average cost x.

    d. How many calls need to be made for an average cost of 15¢ per call?

Discovery and Writing

65. Write a brief paragraph to explain why the range of f is the domain of f^{-1}.

66. Write a brief paragraph to explain why the graphs of a function and its inverse are reflections about the line $y = x$.

67. Let $f(x) = x^5 + x^3 + x + 3$. Find $f^{-1}(3)$. (Hint: Do not find $f^{-1}(x)$. Use observation and the fact that if $f(a) = b$, then $f^{-1}(b) = a$.)

68. Let $f(x) = x^5 + x^3 + x - 3$. Find $f^{-1}(-3)$. (Hint: Do not find $f^{-1}(x)$. Use the fact that if $f(a) = b$, then $f^{-1}(b) = a$.)

Use a graphing calculator to graph each function for various values of a.

69. For what values of a is $f(x) = x^3 + ax$ a one-to-one function?

70. For what values of a is $f(x) = x^3 + ax^2$ a one-to-one function?

Review
Simplify each expression.

71. $16^{3/4}$ **72.** $25^{-1/2}$

73. $(-8)^{2/3}$ **74.** $-8^{2/3}$

75. $\left(\dfrac{64}{125}\right)^{-1/3}$ **76.** $49^{3/2}$

77. $49^{-1/2}$ **78.** $\left(\dfrac{9}{25}\right)^{-3/2}$

CHAPTER REVIEW

SECTION **3.1**   Functions and Function Notation

Definitions and Concepts	Examples
A **function** f is a correspondence between a set of input values x and a set of output values y, where to each x-value there corresponds exactly one y-value.	Determine whether the equation $y^2 = 4x$ defines y to be a function of x. First we solve for y. $$y = \pm\sqrt{4x}$$ $$y = \pm 2\sqrt{x}$$ Since for each real number input for x (except 0) there corresponds an output of two y-values, y is not a function of x.
The set of input values x is called the **domain** of a function. The set of output values y is called the **range** of the function.	Find the domain of $g(x) = \sqrt{x - 4}$. Since $x - 4$ must be non-negative $$x - 4 \geq 0$$ $$x \geq 4$$ The domain is $[4, \infty)$.

Definitions and Concepts	Examples
To evaluate a function $f(x)$ at a given input value x, we substitute the input value for x.	Let $f(x) = \frac{6}{x-6}$. Find $f(-3)$. $f(-3) = \dfrac{6}{(-3)-6} = \dfrac{6}{-9} = -\dfrac{2}{3}$
The fraction $\frac{f(x+h)-f(x)}{h}$ is called the **difference quotient** and is important in calculus.	See Example 5 in Section 3.1.
The **graph** of a function f in the xy-plane is the set of all points (x, y) where x is in the domain of f, y is in the range of f, and $y = f(x)$.	Graph the function $f(x) = -2\lvert x\rvert + 5$ and determine the domain and range of the function. To graph the function, we make a table of values and plot the points by drawing a smooth curve through them. $$f(x) = -2\lvert x\rvert + 5$$ 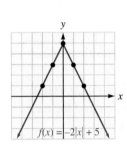<table><tr><th colspan="3">$f(x) = -2\lvert x\rvert + 5$</th></tr><tr><th>$x$</th><th>$f(x)$</th><th>$(x, f(x))$</th></tr><tr><td>$-2$</td><td>1</td><td>$(-2, 1)$</td></tr><tr><td>$-1$</td><td>3</td><td>$(-1, 3)$</td></tr><tr><td>0</td><td>5</td><td>$(0, 5)$</td></tr><tr><td>1</td><td>3</td><td>$(1, 3)$</td></tr><tr><td>2</td><td>1</td><td>$(2, 1)$</td></tr></table>
The domain and the range of a function can be identified by viewing the graph of the function. The inputs or x-values that correspond to points on the graph of the function can be identified on the x-axis and used to state the domain of the function. The outputs or $f(x)$ values that correspond to points on the graph of the function can be identified on the y-axis and used to state the range of the function. **Vertical line test:** If every vertical line that intersects a graph does so exactly once, every number x determines exactly one value of y, and the graph represents a function.	The domain of $f(x) = -2\lvert x\rvert + 5$ is $(-\infty, \infty)$. The range is $(-\infty, 5]$. Note that the graph of the function passes the Vertical Line Test.
A **linear function** is a function determined by an equation of the form $f(x) = mx + b$ or $y = mx + b$.	$f(x) = 3x + 2 \quad f(x) = -\dfrac{1}{2}x - 7$

EXERCISES

Determine whether each equation defines y to be a function of x. Assume that all variables represent real numbers.

1. $y = 3$

2. $y + 5x^2 = 2$

3. $y^2 - x = 5$

4. $y = \lvert x\rvert + x$

Find the domain of each function. Write each answer using interval notation.

5. $f(x) = 3x^2 - 5$

6. $f(x) = \dfrac{3x}{x-5}$

7. $f(x) = \sqrt{x-1}$

8. $f(x) = \sqrt{x^2+1}$

Find $f(2)$, $f(-3)$, and $f(k)$.

9. $f(x) = 5x - 2$

10. $f(x) = \dfrac{6}{x - 5}$

11. $f(x) = |x - 2|$

12. $f(x) = \dfrac{x^2 - 3}{x^2 + 3}$

Evaluate the difference quotient for each function $f(x)$.

13. $f(x) = 5x - 6$

14. $f(x) = 2x^2 - 7x + 3$

Graph each function. Use the graph to identify the domain and range of each function.

15. $f(x) = -x^2 + 4$

16. $f(x) = 3|x - 2|$

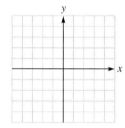

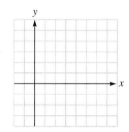

Use the Vertical Line Test to determine whether each graph represents a function.

17.

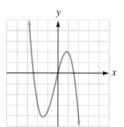

18.

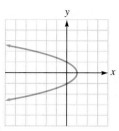

19. Concessions A concessionaire at a basketball game pays a vendor $50 per game for selling hamburgers at $3.50 each.

   a. Write a linear function that describes the income I the vendor earns for the concessionaire during the game if the vendor sells h hamburgers.

   b. Find the income if the vendor sells 200 hamburgers.

20. Cost of electricity The cost C of electricity in Boston is a linear function of x, the number of kilowatt-hours (kwh) used. If the cost of 100 kwh is $20 and the cost of 500 kwh is $80, find a linear function that expresses C in terms of x.

SECTION **3.2**   Quadratic Functions

Definitions and Concepts

A **quadratic function** is a second-degree polynomial function in one variable of the form

$$f(x) = ax^2 + bx + c \quad \text{or} \quad y = ax^2 + bx + c$$

where a, b, and c are real numbers and $a \neq 0$.

The graph of a quadratic function of the form $f(x) = ax^2 + bx + c \ (a \neq 0)$ is a **parabola** with vertex at $\left(-\frac{b}{2a}, c - \frac{b^2}{4a}\right)$.

- If $a > 0$, the parabola **opens upward**.
- If $a < 0$, the parabola **opens downward**.

Examples

$$f(x) = 3x^2 - 2x + 1$$

- $a = 3$, the parabola opens upward.

$$y = -\frac{1}{2}x^2 - 4$$

- $a = -\frac{1}{2}$, the parabola opens downward.

Definitions and Concepts	Examples
The **standard form of an equation of a quadratic function** is: $$y = f(x) = a(x - h)^2 + k \quad (a \neq 0)$$ The vertex is at (h, k). • The parabola opens upward when $a > 0$ and downward when $a < 0$. • The axis of symmetry of the parabola is the vertical line graph of the equation $x = h$.	$$f(x) = 4(x - 2)^2 - 8$$ • $a = 4$, the parabola opens upward. The vertex is $(2, -8)$. The axis of symmetry is $x = 2$. $$y = -\frac{1}{3}(x + 2)^2 + 5$$ • $a = -\frac{1}{3}$, the parabola opens downward. The vertex is $(-2, 5)$. The axis of symmetry is $x = -2$.

Graphing a quadratic function:

To graph a quadratic function:

1. Determine whether the parabola opens upward or downward.
2. Find the vertex of the parabola.
3. Find the x-intercept(s).
4. Find the y-intercept.
5. Identify one additional point on the graph.
6. Draw a smooth curve through the points found in Steps 2–5.

Graph the quadratic function $f(x) = 3(x + 2)^2 - 3$.

Step 1: Determine whether the parabola opens upward or downward.

Standard Form: $f(x) = a(x - h)^2 + k$

Given Form: $f(x) = 3(x + 2)^2 - 3$

Since $a = 3$ and 3 is positive, the parabola opens upward.

Step 2: Find the vertex of the parabola.

Standard Form: $f(x) = a(x - h)^2 + k$

Given Form: $f(x) = 3(x + 2)^2 - 3$

$f(x) = 3[x - (-2)]^2 + (-3)$

Since $h = -2$ and $k = -3$ the vertex is the point $(-2, -3)$.

Step 3: Find the x-intercept(s).
To find the x-intercepts, we substitute 0 for $f(x)$ and solve for x.

$$f(x) = 3(x + 2)^2 - 3$$
$$0 = 3(x + 2)^2 - 3$$
$$3 = 3(x + 2)^2$$
$$1 = (x + 2)^2$$
$$x + 2 = \pm 1$$
$$x = -2 \pm 1$$
$$x = -1 \quad \text{or} \quad x = -3$$

The x-intercepts are the points $(-3, 0)$ and $(-1, 0)$.

Definitions and Concepts	Examples
	Step 4: Find the *y*-intercept. To find the *y*-intercept, we substitute 0 in for *x* and solve for *y*. $$f(x) = 3(x + 2)^2 - 3$$ $$y = 3(x + 2)^2 - 3$$ $$y = 3(0 + 2)^2 - 3$$ $$y = 3(2)^2 - 3$$ $$y = 12 - 3$$ $$y = 9$$ The *y*-intercept is the point $(0, 9)$.
	Step 5: Identify one additional point on the graph. Because of symmetry, the point $(-4, -9)$ is on the graph.
	Step 6: Draw a smooth curve through the points found in Steps 2–5. 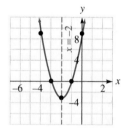 The axis of symmetry of the parabola is $x = -2$ because $h = -2$.
If the graph of the parabola opens downward, then the vertex is the **maximum point** on the graph of the parabola. If the graph of the parabola opens upward, then the vetex is the **minimum point** on the graph of the parabola.	**Minimum cost** A company has found that the total monthly cost C of producing x air-hockey tables is given by $C(x) = 1.5x^2 - 270x + 28{,}665$. Find the production level that minimizes the monthly cost and find that monthly minimum cost. The function $C(x)$ is a quadratic function whose graph is a parabola that opens upward. The minumum value of $C(x)$ occurs at the vertex of the parabola. We will use the vertex formula to find the vertex of the parabola. We compare the equations $C(x) = 1.5x^2 - 270x + 28{,}665$ and $y = ax^2 + bx + c$ to see that $a = 1.5$, $b = -270$, and $c = 28{,}665$. Using the vertex formula, we see that the vertex of the parabola is the point with coordinates $$\left(-\frac{b}{2a},\ c - \frac{b^2}{4a}\right) = \left(-\frac{-270}{2(1.5)},\ 28{,}665 - \frac{(-270)^2}{4(1.5)}\right)$$ $$= (90,\ 16{,}515)$$ If the company makes 90 air-hockey tables each month, it will minimize its production cost. The minimum monthly cost will be \$16,515.

EXERCISES

Determine whether the graph of each quadratic function opens upward or downward. State whether a maximum or minimum point occurs at the vertex of the parabola.

21. $f(x) = \dfrac{1}{2}x^2 + 4$ **22.** $f(x) = -4(x + 1)^2 + 5$

Find the vertex of each parabola.

23. $f(x) = 2(x - 1)^2 + 6$ **24.** $y = -2(x + 4)^2 - 5$

25. $y = x^2 + 6x - 4$ **26.** $y = -4x^2 + 4x - 9$

Graph each quadratic function and find its vertex.

27. $f(x) = (x - 2)^2 - 3$ **28.** $f(x) = -(x - 4)^2 + 4$

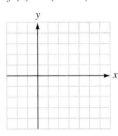   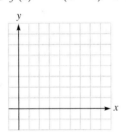

29. $y = x^2 - x$ **30.** $y = x - x^2$

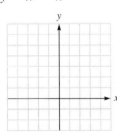   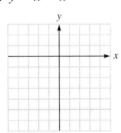

31. $y = x^2 - 3x - 4$ **32.** $y = 3x^2 - 8x - 3$

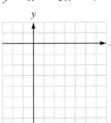   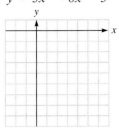

33. Architecture A parabolic arch has an equation of $3x^2 + y - 300 = 0$. Find the maximum height of the arch.

34. Puzzle problem The sum of two numbers is 1, and their product is as large as possible. Find the numbers.

35. Maximizing area A rancher wishes to enclose a rectangular corral with 1,400 feet of fencing. What dimensions of the corral will maximize the area? Find the maximum area.

36. Digital camera A company that produces and sells digital cameras has determined that the total weekly cost C of producing x digital cameras is given by the function $C(x) = 1.5x^2 - 150x + 4,850$. Determine the production level that minimizes the weekly cost for producing the digital cameras and find that weekly minimum cost.

SECTION **3.3**   Polynomial and Other Functions

Definitions and Concepts	Examples
A **polynomial function in one variable (say, x)** is a function of the form $$f(x) = a_n x^n + a_{n-1}x^{n-1} + \cdots + a_1 x + a_0$$ where a_n, a_{n-1}, ..., a_1, and a_0 are real numbers and n is a whole number. The **degree of a polynomial function** is the largest power of x that appears in the polynomial.	$f(x) = 3x^2 + 4x - 7$, degree of 2 $f(x) = -17x^4 + 3x^3 - 2x^2 + 13$, degree of 4

Definitions and Concepts	Examples
Graphing polynomial functions: 1. Find any symmetries of the graph. 2. Find the x- and y-intercepts of the graph. 3. Determine where the graph is above and below the x-axis. 4. Plot a few points, if necessary, and draw the graph as a smooth continuous curve. If $f(-x) = f(x)$ for all x in the domain of f, the graph of the function is symmetric about the y-axis, and the function is called an **even function.** If $f(-x) = -f(x)$ for all x in the domain of f, the function is symmetric about the origin, and the function is called an **odd function.**	Graph the function $f(x) = -x^3 + 9x$ and determine whether it is even, odd, or neither.

Graph the function $f(x) = -x^3 + 9x$ and determine whether it is even, odd, or neither.

Step 1: Find any symmetries of the graph.
To test for symmetry about the y-axis, we check to see whether $f(x) = f(-x)$. To test for symmetry about the origin, we check to see whether $f(x) = -f(x)$.

$$f(x) = -x^3 + 9x$$
$$f(-x) = -(-x)^3 + 9(-x)$$
$$f(-x) = x^3 - 9x$$

Since $f(x) \neq f(-x)$, there is no symmetry about the y-axis. However, since $f(-x) = -f(x)$, there is symmetry about the origin.

Step 2: Find the x- and y-intercepts of the graph. To find the x-intercepts, we let $f(x) = 0$ and solve for x.

$$-x^3 + 9x = 0$$
$$-x(x^2 - 9) = 0$$
$$-x(x + 3)(x - 3) = 0$$

$$-x = 0 \quad \text{or} \quad x + 3 = 0 \quad \text{or} \quad x - 3 = 0$$
$$x = 0 \quad \quad \quad x = -3 \quad \quad \quad x = 3$$

The x-intercepts are $(0, 0)$, $(-3, 0)$, and $(3, 0)$.

If we let $x = 0$ and solve for $f(x)$, we see that the y-intercept is also $(0, 0)$.

Step 3: Determine where the graph is above or below the x-axis. To determine where the graph is above or below the x-axis, we plot the solutions of $-x^3 + 9x = 0$ on a number line and establish the four intervals shown in the figure. We then test a number from each interval to determine the sign of $f(x)$.

$+$	$-$	$+$	$-$
$(-\infty, -3)$	$(-3, 0)$	$(0, 3)$	$(3, \infty)$

$f(-4) = 28$ -3 $f(-1) = -8$ 0 $f(1) = 8$ 3 $f(4) = -28$

| above the x-axis | below the x-axis | above the x-axis | below the x-axis |

Definitions and Concepts	Examples
	Step 4: Plot a few points and draw the graph as a smooth continuous curve. We now plot the intercepts and one additional point. In the previous step we found that $f(1) = 8$. This will be the additional point we plot $(1, 8)$. Making use of our knowledge of symmetry about the origin and where the graph is above and below the x-axis, we now draw the graph as shown. 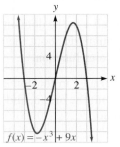 The function $f(x) = -x^3 + 9x$ is an odd function because the function is symmetric about the origin.
Some functions, called **piecewise-defined functions,** are defined by using different equations for different intervals in their domains.	$f(x) = \begin{cases} x - 2 & \text{if } x < 3 \\ x^2 & \text{if } x \geq 3 \end{cases}$ • $f(2) = 2 - 2 = 0$ • $f(4) = (4)^2 = 16$
The **greatest-integer function** is important in many business applications and in the field of computer science. This function is determined by the equation $f(x) = [\![x]\!]$, where the value of $f(x)$ that corresponds to x is the greatest integer that is less than or equal to x.	Let $f(x) = [\![x - 3]\!]$. Find $f(2.2)$. $\quad f(x) = [\![x - 3]\!]$ $\quad f(2.2) = [\![2.2 - 3]\!] = [\![-0.8]\!] = -1$ -1 is the greatest integer less than or equal to -0.8.

EXERCISES

Graph each polynomial function and determine whether it is even, odd, or neither.

37. $y = x^3 - x$

38. $y = x^2 - 4x$

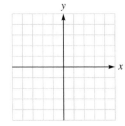

39. $y = x^3 - x^2$

40. $y = 1 - x^4$

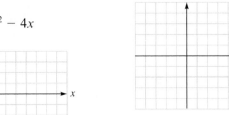

Evaluate each piecewise-defined function.

41. $f(x) = \begin{cases} x - 2 & \text{if } x < 3 \\ x^2 & \text{if } x \geq 3 \end{cases}$

    a. $f(-2)$ **b.** $f(3)$

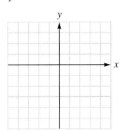

42. $f(x) = \begin{cases} 2 & \text{if } x < 0 \\ 2 - x & \text{if } 0 \le x < 2 \\ x + 1 & \text{if } x \ge 2 \end{cases}$

  a. $f\left(\dfrac{3}{2}\right)$ **b.** $f(2)$

Graph each piecewise-defined function and determine the open intervals on which it is increasing, decreasing, or constant.

43. $y = f(x) = \begin{cases} x + 5 & \text{if } x \le 0 \\ 5 - x & \text{if } x > 0 \end{cases}$

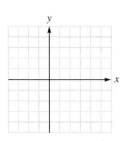

44. $y = f(x) = \begin{cases} x + 3 & \text{if } x \le 0 \\ 3 & \text{if } x > 0 \end{cases}$

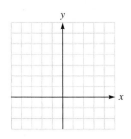

Evaluate each function at the indicated x-values.

45. $f(x) = [\![2x]\!]$ Find $f(1.7)$.

46. $f(x) = [\![x - 5]\!]$ Find $f(4.99)$.

Graph each function.

47. $f(x) = [\![x]\!] + 2$

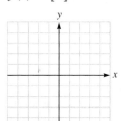

48. $f(x) = [\![x - 1]\!]$

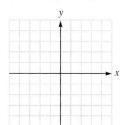

49. Renting a Jeep A rental company charges $20 to rent a Jeep Wrangler for one day, plus $8 for every 100 miles (or portion of 100 miles) that it is driven. Find the cost if the Jeep is driven 295 miles in one day.

50. Riding in a taxi A taxicab company charges $4 for a trip up to 1 mile, and $2 for every extra mile (or portion of a mile). Find the cost to ride $11\frac{1}{2}$ miles.

SECTION **3.4**   Transformations of the Graphs of Functions

Definitions and Concepts	Examples
Vertical translations: If $k > 0$, the graph of $\begin{cases} y = f(x) + k \\ y = f(x) - k \end{cases}$ is identical to the graph of $y = f(x)$, except that it is translated k units $\begin{cases} \text{upward} \\ \text{downward} \end{cases}$.	The function $g(x) = \sqrt{x + 3} - 2$ is a translation of the graph of $f(x) = \sqrt{x}$. Graph both on one set of coordinate axes.

Definitions and Concepts	Examples

Horizontal translations:

If $k > 0$, the graph of $\begin{cases} y = f(x - k) \\ y = f(x + k) \end{cases}$

is identical to the graph of $y = f(x)$, except that it is

translated k units to the $\begin{cases} \text{right} \\ \text{left} \end{cases}$.

Vertical stretchings:

If f is a function and $k > 1$, then
- The graph of $y = kf(x)$ can be obtained by stretching the graph of $y = f(x)$ vertically by multiplying each value of $f(x)$ by k.

If f is a function and $0 < k < 1$, then
- The graph of $y = kf(x)$ can be obtained by shrinking the graph of $y = f(x)$ vertically by multiplying each value of $f(x)$ by k.

Horizontal stretchings:

If f is a function and $k > 1$, then
- The graph of $y = f(kx)$ can be obtained by shrinking the graph of $y = f(x)$ horizontally by multiplying each x-value of $f(x)$ by $\frac{1}{k}$.

If f is a function and $0 < k < 1$,
- The graph of $y = f(kx)$ can be obtained by stretching the graph of $y = f(x)$ horizontally by multiplying each x-value of $f(x)$ by $\frac{1}{k}$.

Reflections:

If f is a function, then
- The graph of $y = -f(x)$ is identical to the graph of $y = f(x)$ except that it is reflected about the x-axis.
- The graph of $y = f(-x)$ is identical to the graph of $y = f(x)$ except that it is reflected about the y-axis.

To graph functions involving a combination of transformations, we must apply each transformation to the basic function. We will apply these transformations in the following order:

1. Horizontal translation
2. Stretching or shrinking
3. Reflection
4. Vertical translation

By inspection, we see that the function $g(x) = \sqrt{x + 3} - 2$ involves two translations of $f(x) = \sqrt{x}$. The graph of $g(x) = \sqrt{x + 3} - 2$ is identical to the graph of $f(x) = \sqrt{x}$ except it is translated 3 units to the left and 2 units downward as shown in the figure.

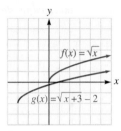

Graph: $g(x) = -\frac{1}{3}|x|$.

The graph of $g(x) = -\frac{1}{3}|x|$ is identical to the graph of $f(x) = |x|$ except that it is vertically shrunk by a factor of $\frac{1}{3}$ and reflected about the x-axis. This is because each value of $|x|$ is multiplied by $-\frac{1}{3}$. The graphs of both functions are shown in the figure.

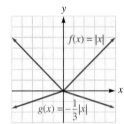

Graph $g(x) = 2(x - 4)^3 + 1$.

We will graph $g(x) = 2(x - 4)^3 + 1$ by applying three translations to the basic function $f(x) = x^3$: translate $f(x) = x^3$ horizontally 4 units to the right, stretch the graph by a factor of 2, and translate the graph vertically 1 unit upward. The graphs of both functions are shown in the figure.

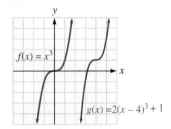

EXERCISES

Each function is a translation of a basic function. Graph both on one set of coordinate axes.

51. $g(x) = x^2 + 5$

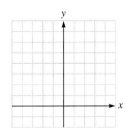

52. $g(x) = (x - 7)^3$

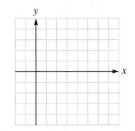

53. $g(x) = \sqrt{x + 2} + 3$

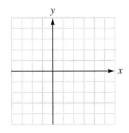

54. $g(x) = |x - 4| + 2$

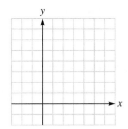

Each function is a stretching of $f(x) = x^3$. Graph both on one set of coordinate axes.

55. $g(x) = \dfrac{1}{3}x^3$

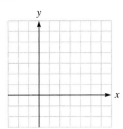

56. $g(x) = (-5x)^3$

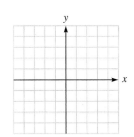

Graph each function using a combination of translations and stretchings.

57. $g(x) = -|x - 4| + 3$

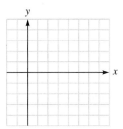

58. $g(x) = \dfrac{1}{4}|x - 4| + 1$

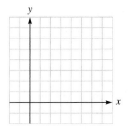

59. $g(x) = 3\sqrt{x + 3} + 2$

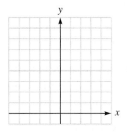

60. $g(x) = \dfrac{1}{3}(x + 3)^3 + 2$

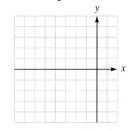

61. $f(x) = \sqrt{-x} + 3$

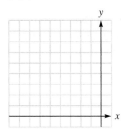

62. $g(x) = 2\sqrt[3]{x} - 5$

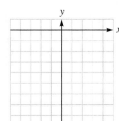

SECTION **3.5**   Rational Functions

Definitions and Concepts	Examples
A **rational function** is a function defined by an equation of the form $y = \frac{P(x)}{Q(x)}$, where $P(x)$ and $Q(x)$ are polynomials and $Q(x) \neq 0$.	$f(x) = \dfrac{3}{x + 2}$ $f(x) = \dfrac{3x + 4}{x^2 - 3x + 4}$

Definitions and Concepts	Examples

Locating vertical asymptotes:

To locate the vertical asymptotes of rational function $f(x) = \frac{P(x)}{Q(x)}$, we follow these steps:

Step 1: Factor $P(x)$ and $Q(x)$ and remove any common factors.

Step 2: Set the denominator equal to 0 and solve the equation.

If a is a solution of the equation found in Step 2, $x = a$ is a vertical asymptote.

Locating horizontal asymptotes:

To locate the horizontal asymptote of the rational function $f(x) = \frac{P(x)}{Q(x)}$, we consider three cases:

Case 1: If the degree of $P(x)$ *is less than* the degree of $Q(x)$, the line $y = 0$ is the horizontal asymptote.

Case 2: If the degree of $P(x)$ and $Q(x)$ are equal, the line $y = \frac{p}{q}$, where p and q are the leading coefficients of $P(x)$ and $Q(x)$, is the horizontal asymptote.

Case 3: If the degree of $P(x)$ is greater than the degree of $Q(x)$, there is no horizontal asymptote.

A third type of asymptote is called a **slant asymptote.** These asymptotes occur when the degree of the numerator of a rational function is 1 more than the degree of the denominator. As the name implies, it is a slanted line, neither vertical nor horizontal.

Locating slant asymptotes:

If the degree of $P(x)$ is 1 greater than the degree of $Q(x)$ for the rational function $f(x) = \frac{P(x)}{Q(x)}$, there is a slant asymptote. To find it, divide $P(x)$ by $Q(x)$ and ignore the remainder.

Graph: $y = f(x) = \dfrac{4x}{x - 2}$.

We will use the steps outlined to graph the function.

Step 1: Symmetry
We find $f(-x)$.

$$f(-x) = \frac{4(-x)}{(-x) - 2} = \frac{-4x}{-x - 2} = \frac{4x}{x + 2}$$

Because $f(-x) \neq f(x)$ and $f(-x) \neq -f(x)$, there is no symmetry about the y-axis or the origin.

Step 2: Vertical asymptotes
We first note that $f(x)$ is in simplest form. We then set the denominator equal to 0 and solve for x. Since the solution is 2, there will be a vertical asymptote at $x = 2$.

Step 3: y- and x-intercepts
We can find the y-intercept by finding $f(0)$.

$$f(0) = \frac{4(0)}{0 - 2} = \frac{0}{-2} = 0$$

The y-intercept is $(0, 0)$.

We can find the x-intercepts by setting the numerator equal to 0 and solving for x:

$$4x = 0$$
$$x = 0$$

The x-intercept is $(0, 0)$.

Step 4: Horizontal asymptotes
Since the degrees of the numerator and denominator of the polynomials are the same, the line

$$y = \frac{4}{1} = 4 \qquad \begin{array}{l}\text{The leading coefficient of the numerator is 4.}\\ \text{The leading coefficient of the denominator is 1.}\end{array}$$

is a horizontal asymptote.

Step 5: Slant asymptotes
Since the degree of the numerator is not 1 greater than the degree of the denominator, there are no slant asymptotes.

Definitions and Concepts	Examples
Steps to graph a rational function: We will use the following steps to graph the rational function $f(x) = \frac{P(x)}{Q(x)}$, where $\frac{P(x)}{Q(x)}$ is in simplest form (no common factors). 1. Check symmetries. 2. Look for vertical asymptotes. 3. Look for the y- and x-intercepts. 4. Look for horizontal asymptotes. 5. Look for slant asymptotes. 6. Graph the function.	**Step 6: Graph** First, we plot the intercept $(0, 0)$ and draw the asymptotes. We then find one additional point on our graph to see what happens when x is greater than 2. To do so, we choose 3, a value of x that is greater than 2, and evaluate $f(3)$. $$f(3) = \frac{4(3)}{3 - 2} = \frac{12}{1} = 12$$ Since $f(3) = 12$, the point $(3, 12)$ lies on our graph. We sketch the graph as shown in the figure. 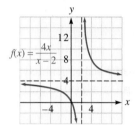

EXERCISES

Find the domain of each rational function.

63. $f(x) = \dfrac{3x^2 + x - 2}{x^2 - 25}$

64. $f(x) = \dfrac{2x^2 + 1}{x^2 + 7}$

Find the vertical asymptotes, if any, of each rational function.

65. $f(x) = \dfrac{x + 5}{x^2 - 1}$

66. $f(x) = \dfrac{x - 7}{x^2 - 49}$

67. $f(x) = \dfrac{x}{x^2 + x - 6}$

68. $f(x) = \dfrac{5x + 2}{2x^2 - 6x - 8}$

Find the horizontal asymptotes, if any, of each rational function.

69. $f(x) = \dfrac{2x^2 + x - 2}{4x^2 - 4}$

70. $f(x) = \dfrac{5x^2 + 4}{4 - x^2}$

71. $f(x) = \dfrac{x + 1}{x^3 - 4x}$

72. $f(x) = \dfrac{x^3}{2x^2 - x + 11}$

Find the slant asymptote, if any, for each rational function.

73. $f(x) = \dfrac{2x^2 - 5x + 1}{x - 4}$

74. $f(x) = \dfrac{5x^3 + 1}{x + 5}$

Graph each rational function.

75. $f(x) = \dfrac{2x}{x - 4}$

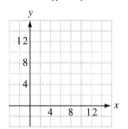

76. $f(x) = \dfrac{-4x}{x + 4}$

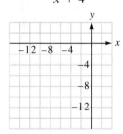

77. $f(x) = \dfrac{x}{(x - 1)^2}$

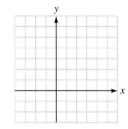

78. $f(x) = \dfrac{(x - 1)^2}{x}$

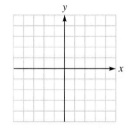

79. $f(x) = \dfrac{x^2 - x - 2}{x^2 + x - 2}$

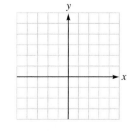

80. $f(x) = \dfrac{x^3 + x}{x^2 - 4}$

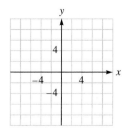

SECTION **3.6**   Operations on Functions

Definitions and Concepts	Examples
Adding, subtracting, multiplying, and dividing functions: If the ranges of functions f and g are subsets of the real numbers, then	Let $f(x) = 3x + 5$ and $g(x) = 4x - 7$. Find each function and its domain: **a.** $f + g$ **b.** $f - g$ **c.** $f \cdot g$ **d.** f/g
1. The **sum of f and g**, denoted as $f + g$, is defined by $(f + g)(x) = f(x) + g(x)$.	**a.** $(f + g)(x) = f(x) + g(x)$ $= (3x + 5) + (4x - 7)$ $= 7x - 2$ Since the domain of both f and g is the set of real numbers, the domain of $f + g$ is the interval $(-\infty, \infty)$.
2. The **difference of f and g**, denoted as $f - g$, is defined by $(f - g)(x) = f(x) - g(x)$.	**b.** $(f - g)(x) = f(x) - g(x)$ $= (3x + 5) - (4x - 7)$ $= 3x + 5 - 4x + 7$ $= -x + 12$ Since the domain of both f and g is the set of real numbers, the domain of $f - g$ is the interval $(-\infty, \infty)$.
3. The **product of f and g**, denoted as $f \cdot g$, is defined by $(f \cdot g)(x) = f(x) \cdot g(x)$.	**c.** $(f \cdot g)(x) = f(x) \cdot g(x)$ $= (3x + 5)(4x - 7)$ $= 12x^2 - x - 35$ Since the domain of both f and g is the set of real numbers, the domain of $f \cdot g$ is the interval $(-\infty, \infty)$.
4. The **quotient of f and g**, denoted as f/g, is defined by $(f/g)(x) = \frac{f(x)}{g(x)}$, $g(x) \neq 0$. The domain of each function, unless otherwise restricted, is the set of real numbers x that are in the domains of both f and g. In the case of the quotient f/g, there is the restriction that $g(x) \neq 0$.	**d.** $(f/g)(x) = \frac{f(x)}{g(x)}$ $= \frac{3x + 5}{4x - 7} \quad (4x - 7 \neq 0)$ Since $\frac{7}{4}$ will make $4x - 7$ equal to 0, the domain of f/g is the set of all real numbers except $\frac{7}{4}$. This is $\left(-\infty, \frac{7}{4}\right) \cup \left(\frac{7}{4}, \infty\right)$.
The **composite function** $f \circ g$ is defined by $(f \circ g)(x) = f(g(x))$.	If $f(x) = 2x + 7$ and $g(x) = x^2 + 1$, find $(f \circ g)(x)$ and its domain.

Definitions and Concepts	Examples
The domain of $f \circ g$ consists of all those numbers in the domain of g for which $g(x)$ is in the domain of f.	Because $(f \circ g)(x)$ means $f(g(x))$, we will replace x in $f(x) = 2x + 7$ with $g(x)$. $$(f \circ g)(x) = f(g(x))$$ $$= f(x^2 + 1)$$ $$= 2(x^2 + 1) + 7$$ $$= 2x^2 + 9$$ The domain of $(f \circ g)(x)$ is the interval $(-\infty, \infty)$ because the domain of both f and g consists of all real numbers.

EXERCISES

Let $f(x) = x^2 - 1$ and $g(x) = 2x + 1$. Find each function and its domain.

81. $f + g$ **82.** $f \cdot g$

83. $f - g$ **84.** f/g

Let $f(x) = 2x^2 - 1$ and $g(x) = 2x - 1$. Find each value, if possible.

85. $(f + g)(-3)$ **86.** $(f - g)(-5)$

87. $(f \cdot g)(2)$ **88.** $(f/g)\left(\dfrac{1}{2}\right)$

Let $f(x) = x^2 - 1$ and $g(x) = 2x + 1$. Find each function and its domain.

89. $f \circ g$ **90.** $g \circ f$

Let $f(x) = x^2 - 5$ and $g(x) = 3x + 1$. Find each value.

91. $(f \circ g)(-2)$ **92.** $(g \circ f)(-2)$

Find two functions f and g such that the composition $f \circ g = h$ expresses the given correspondence. Several answers are possible.

93. $h(x) = (x - 5)^2$ **94.** $h(x) = (x + 6)^3$

SECTION **3.7**   Inverse Functions

Definitions and Concepts	Examples
A function f from a set **X** to a set **Y** is called a **one-to-one function** if and only if different numbers in the domain of f have different outputs in the range of f. A **Horizontal Line Test** can be used to determine whether the graph of a function represents a one-to-one function. If every horizontal line that intersects the graph of a function does so exactly once, the function passes the Horizontal Line Test and is one-to-one.	Determine whether the function $f(x) = x^4 - 2x^2$ is one-to-one. The function $f(x) = x^4 - 2x^2$ is not one-to-one, because different numbers in the domain have the same output. For example, 2 and –2 have the same output: $f(2) = f(-2) = 8$.

Definitions and Concepts	Examples
Inverse functions: If f and g are two one-to-one functions such that $(f \circ g)(x) = x$ for every x in the domain of g and $(g \circ f)(x) = x$ for every x in the domain of f, then f and g are **inverse functions.** Function g is denoted as f^{-1} and is called the **inverse function of f.** **Properties of a one-to-one function:** **Property 1:** If f is a one-to-one function, there is a one-to-one function $f^{-1}(x)$ such that $(f^{-1} \circ f)(x) = x$ and $(f \circ f^{-1})(x) = x$. **Property 2:** The domain of f is the range of f^{-1}, and the range of f is the domain of f^{-1}.	Verify that $f(x) = x^5$ and $g(x) = \sqrt[5]{x}$ are inverse functions. To show that f and g are inverse functions, we must show that $f \circ g$ and $g \circ f$ are x, the identity function. $$(f \circ g)(x) = f(g(x)) = f\left(\sqrt[5]{x}\right) = \left(\sqrt[5]{x}\right)^5 = x$$ $$(g \circ f)(x) = g(f(x)) = g(x^5) = \sqrt[5]{x^5} = x$$ Because g is the inverse of f, we can use inverse notation and write $f(x) = x^5$ and $f^{-1}(x) = \sqrt[5]{x}$. Because f is the inverse of g, we can use inverse notation and write $g(x) = \sqrt[5]{x}$ and $g^{-1}(x) = x^5$.
Strategy for finding f^{-1}: **Step 1:** Replace $f(x)$ with y. **Step 2:** Interchange the variables x and y. **Step 3:** Solve the resulting equation for y. **Step 4:** Replace y with $f^{-1}(x)$. The graph of a function and its inverse are reflections of each other about the line $y = x$.	Find the inverse of $f(x) = x^3 + 5$. We will find the inverse of the function using the strategy given in the section. **Step 1:** Replace $f(x)$ with y. $$f(x) = x^3 + 5$$ $$y = x^3 + 5$$ **Step 2:** Interchange the variables x and y. $$x = y^3 + 5$$ **Step 3:** Solve the resulting equation for y. $$x - 5 = y^3$$ $$y = \sqrt[3]{x - 5}$$ **Step 4:** Replace y with $f^{-1}(x)$. $$f^{-1}(x) = \sqrt[3]{x - 5}$$

EXERCISES

Determine whether each function is one-to-one.

95. $f(x) = x^2 + 7$ **96.** $f(x) = x^3$

Use the Horizontal Line Test to determine whether each graph represents a one-to-one function.

97.
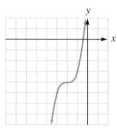

98.
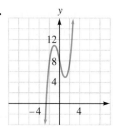

Verify that the functions are inverses by showing that $f \circ g$ and $g \circ f$ are the identity function.

99. $f(x) = 8x - 3$ **100.** $f(x) = \dfrac{1}{2 - x}$

Each equation defines a one-to-one function. Find f^{-1} and verify that $f \circ f^{-1}$ and $f^{-1} \circ f$ are the identity function.

101. $y = 7x - 1$ **102.** $y = \dfrac{1}{2 - x}$

103. $y = \dfrac{x}{1 - x}$ **104.** $y = \dfrac{3}{x^3}$

105. Find the inverse of the one-to-one function $f(x) = 2x - 5$ and graph both the function and its inverse on the same set of coordinate axes.

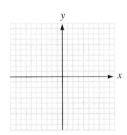

106. Find the range of $y = \frac{2x + 3}{5x - 10}$ by finding the domain of f^{-1}.

CHAPTER TEST

Find the domain of each function. Write each answer using interval notation.

1. $f(x) = \dfrac{3}{x - 5}$

2. $f(x) = \sqrt{x + 3}$

Find $f(-1)$ and $f(2)$.

3. $f(x) = \dfrac{x}{x - 1}$

4. $f(x) = \sqrt{x + 7}$

Find the vertex of each parabola.

5. $y = 3(x - 7)^2 - 3$

6. $y = x^2 - 2x - 3$

7. $f(x) = 3x^2 - 24x + 38$

8. $f(x) = 5 - 4x - x^2$

Graph each function.

9. $f(x) = x^4 - x^2$

10. $f(x) = x^5 - x^3$

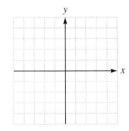
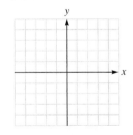

Assume that an object tossed vertically upward reaches a height of h feet after t seconds, where $h = 100t - 16t^2$.

11. In how many seconds does the object reach its maximum height?

12. What is that maximum height?

13. Suspension bridges The cable of a suspension bridge is in the shape of the parabola $x^2 - 2{,}500y + 25{,}000 = 0$ in the coordinate system shown in the illustration. (Distances are in feet.) How far above the roadway is the cable's lowest point?

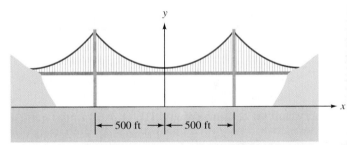

14. Refer to Question 13. How far above the roadway does the cable attach to the vertical pillars?

Graph each function.

15. $f(x) = (x - 3)^2 + 1$

16. $f(x) = \sqrt{x - 1} + 5$

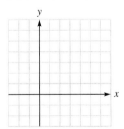
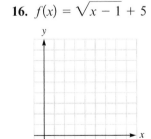

Find all asymptotes of the graph of each rational function. Do not graph the function.

17. $y = \dfrac{x - 1}{x^2 - 9}$

18. $y = \dfrac{x^2 - 5x - 14}{x - 3}$

Graph each rational function. Check for asymptotes, intercepts, and symmetry.

19. $y = \dfrac{x^2}{x^2 - 9}$

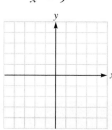

20. $y = \dfrac{x}{x^2 + 1}$

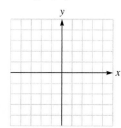

Graph each rational function. The numerator and denominator share a common factor.

21. $y = \dfrac{2x^2 - 3x - 2}{x - 2}$

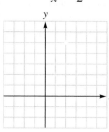

22. $y = \dfrac{x}{x^2 - x}$

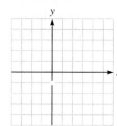

Let $f(x) = 3x$ and $g(x) = x^2 + 2$. Find each function.

23. $f + g$

24. $g \circ f$

25. f/g

26. $f \circ g$

Assume that $f(x)$ is one-to-one. Find f^{-1}.

27. $f(x) = \dfrac{x + 1}{x - 1}$

28. $f(x) = x^3 - 3$

Find the range of f by finding the domain of f^{-1}.

29. $y = \dfrac{3}{x} - 2$

30. $y = \dfrac{3x - 1}{x - 3}$

CUMULATIVE REVIEW EXERCISES

Use the x- and y-intercepts to graph each equation.

1. $5x - 3y = 15$

2. $3x + 2y = 12$

Find the length, the midpoint, and the slope of the line segment PQ.

3. $P\left(-2, \dfrac{7}{2}\right)$; $Q\left(3, -\dfrac{1}{2}\right)$

4. $P(3, 7)$; $Q(-7, 3)$

Write the equation of the line with the given properties. Give the answer in slope-intercept form.

5. The line passes through $(-3, 5)$ and $(3, -7)$.

6. The line passes through $\left(\dfrac{3}{2}, \dfrac{5}{2}\right)$ and has a slope of $\dfrac{7}{2}$.

7. The line is parallel to $3x - 5y = 7$ and passes through $(-5, 3)$.

8. The line is perpendicular to $x - 4y = 12$ and passes through the origin.

Graph each equation. Make use of intercepts and symmetries.

9. $x^2 = y - 2$

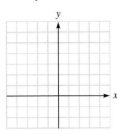

10. $y^2 = x - 2$

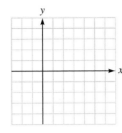

11. $x^2 + y^2 = 100$

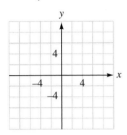

12. $x^2 - 2x + y^2 = 8$

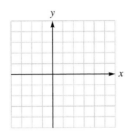

Solve each proportion.

13. $\dfrac{x - 2}{x} = \dfrac{x - 6}{5}$

14. $\dfrac{x + 2}{x - 6} = \dfrac{3x + 1}{2x - 11}$

15. Dental billing The billing schedule for dental X-rays specifies a fixed amount for the office visit plus a fixed amount for each X-ray exposure. If 2 X-rays cost $37 and 4 cost $54, find the cost of 5 exposures.

16. Automobile collisions The energy dissipated in an automobile collision varies directly with the square of the speed. By what factor does the energy increase in a 50-mph collision compared with a 20-mph collision?

Determine whether each equation defines a function.

17. $y = 3x - 1$

18. $y = x^2 + 3$

19. $y = \dfrac{1}{x - 2}$

20. $y^2 = 4x$

Find the domain of each function.

21. $f(x) = x^2 + 5$

22. $f(x) = \dfrac{7}{x + 2}$

23. $y = -\sqrt{x - 2}$

24. $f(x) = \sqrt{x + 4}$

Find the vertex of each parabola.

25. $y = x^2 + 5x - 6$

26. $f(x) = -x^2 + 5x + 6$

Graph each function.

27. $f(x) = x^2 - 4$

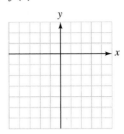

28. $f(x) = -x^2 + 4$

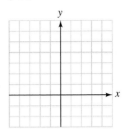

29. $f(x) = x^3 + x$

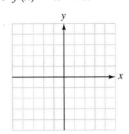

30. $f(x) = -x^4 + 2x^2 + 1$

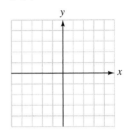

Graph each function. Show all asymptotes.

31. $f(x) = \dfrac{x}{x - 3}$

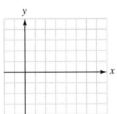

32. $f(x) = \dfrac{x^2 - 1}{x^2 - 9}$

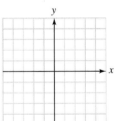

Let $f(x) = 3x - 4$ and $g(x) = x^2 + 1$. Find each function and its domain.

33. $(f + g)(x)$

34. $(f - g)(x)$

35. $(f \cdot g)(x)$

36. $(f/g)(x)$

Let $f(x) = 3x - 4$ and $g(x) = x^2 + 1$. Find each value.

37. $(f \circ g)(2)$

38. $(g \circ f)(2)$

39. $(f \circ g)(x)$

40. $(g \circ f)(x)$

Find the inverse of the function defined by each equation.

41. $y = 3x + 2$

42. $y = \dfrac{1}{x - 3}$

43. $y = x^2 + 5 \; (x \geq 0)$ **44.** $3x - y = 1$

Write each sentence as an equation.

45. y varies directly with the product of w and z.

46. y varies directly with x and inversely with the square of t.

Exponential and Logarithmic Functions

CAREERS AND MATHEMATICS: Epidemiologists

Kiselev Andrey Valerevich/Shutterstock.com

Epidemiologists investigate and describe the determinants and distribution of disease, disability, and other health outcomes. They also develop means for prevention and control. *Applied epidemiologists* typically work for state health agencies and are responsible for responding to disease outbreaks and determining the cause and method of containment. *Research epidemiologists* work in laboratories studying ways to prevent future outbreaks. This career can be quite rewarding, both mentally and financially. Epidemiologists spend a lot of time saving lives and finding solutions for better health.

Education and Mathematics Required

- Applied epidemiologists are generally required to have a master's degree from a school of public health. Research epidemiologists may need a Ph.D. or medical degree, depending on their area of work.
- College Algebra, Trigonometry, Calculus, Applied Data Analysis, Survey and Research Methods, Mathematical Statistics, and Biostatistics are required math courses.

How Epidemiologists Use Math and Who Employs Them

- Epidemiologists use mathematical models when they are tracking the progress of an infectious disease. The SIR model consists of three variables: S (for susceptible), I (for infectious), and R (for recovered). It is used for infectious diseases such as measles, mumps, and rubella.
- Government agencies employ 57%; hospitals employ 12%; colleges and universities employ 11%, and 9% are employed in areas of scientific research and developmental services like the American Cancer Society.

Career Outlook and Earnings

- Employment growth is projected to be 15% over the 2008–2018 decade, which is faster than average. This is due to an increased threat of bioterrorism and rare but infectious diseases, such as West Nile Virus or Avian flu.
- The median annual income is $61,700, with the top 10% of salaries at $92,610.

For more information see: www.bls.gov/oco

In this chapter, we will discuss exponential functions, which are often used in banking, ecology, and science. We will also discuss logarithmic functions, which are applied in chemistry, geology, and environmental science.

4.1 Exponential Functions and Their Graphs

In this section, we will learn to

1. Approximate and simplify exponential expressions.

2. Graph exponential functions.

3. Solve compound interest problems.

4. Define e and graph base-e exponential functions.

5. Use transformations to graph exponential functions.

Extreme water slides, called "plunge" or "plummet" slides, are fearsome water slides because of their heights. With near vertical drops, the slides are designed to allow riders to reach the greatest possible speeds. Summit Plummet at Blizzard Beach, a part of Walt Disney World Resort in Florida, stands 120 ft tall. On this slide, riders can achieve speeds up to 55 mph.

The shapes of extreme water slides can be modeled using *exponential functions*, the topic of this section.

Exponential functions are also important in business. Consider the graph shown in Figure 4-1. It shows the balance in a bank account in which $5,000 was invested in 1990 at 8%, compounded monthly. The graph shows that in the year 2015, the value of the account will be approximately $38,000, and in the year 2030, the value will be approximately $121,000.

The curve in Figure 4-1 is the graph of an exponential function. From the graph, we can see that the longer the money is kept on deposit, the more rapidly it will grow.

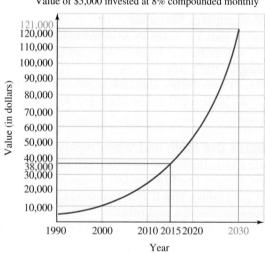

Value of $5,000 invested at 8% compounded monthly

FIGURE 4-1

Before we can discuss exponential functions, we must define irrational exponents.

1. Approximate and Simplify Exponential Expressions

We have discussed expressions of the form b^x, where x is a rational number.

- 5^2 means "the square of 5."
- $4^{1/3}$ means "the cube root of 4."
- $6^{-2/5} = \frac{1}{6^{2/5}}$ means "the reciprocal of the fifth root of 6^2."

To understand exponential functions and their graphs requires that we give meaning to b^x when x is an irrational number. Consider the expression

$3^{\sqrt{2}}$ where $\sqrt{2}$ is the irrational number 1.414213562 ...

We can use closer and closer approximations as shown below. Since $\sqrt{2}$ is an irrational number, we will use a calculator to find the approximations.

$$3^{\sqrt{2}} \approx 3^{1.4} \approx 4.655536722$$

$$3^{\sqrt{2}} \approx 3^{1.41} \approx 4.706965002$$

$$3^{\sqrt{2}} \approx 3^{1.414} \approx 4.727695035$$

$$3^{\sqrt{2}} \approx 3^{1.4142} \approx 4.72873393$$

Since the exponents of the expressions in the list are getting closer to $\sqrt{2}$, the values of the expressions are getting closer to the value of $3^{\sqrt{2}}$.

On a scientific calculator, there is an exponential key, usually y^x. On a graphing calculator, we can use the ⬛^⬛ key.

EXAMPLE 1   Approximating Exponential Expressions

Approximate each expression correct to 4 decimals.

a. $4^{2/3}$ **b.** $5^{-\sqrt{3}}$ **c.** $\left(\dfrac{4}{7}\right)^{\pi}$

SOLUTION   We will use a calculator.

a. $4\text{^}(2/3) \approx 2.5198$ Enter 2/3 in parentheses.

b. $5\text{^}\left(-\sqrt{3}\right) \approx 0.0616$ Enter $-\sqrt{3}$ in parentheses.

c. $(4/7)\text{^}\pi \approx 0.1724$ Enter the base, 4/7, in parentheses.

Figure 4-2 shows the graphing calculator screen that was used to find the values.

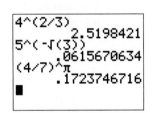

FIGURE 4-2

Self Check 1   Approximate each expression correct to 4 decimals.

a. $5^{3/5}$ **b.** $-3^{\sqrt{6}}$ **c.** $7^{-2.356}$

Now Try Exercise 15.

If b is a positive number and x is a real number, the expression b^x always represents a positive number. It is also true that the familiar properties of exponents hold for irrational exponents.

EXAMPLE 2   Simplifying Expressions with Irrational Exponents

Simplify each expression:

a. $\left(3^{\sqrt{2}}\right)^{\sqrt{2}}$

b. $a^{\sqrt{8}} \cdot a^{\sqrt{2}}$

SOLUTION   We will use properties of exponents to simplify each expression.

   a. $\left(3^{\sqrt{2}}\right)^{\sqrt{2}} = 3^{\sqrt{2}\sqrt{2}}$ Keep the base and multiply the exponents.

   $= 3^2$ $\sqrt{2}\sqrt{2} = \sqrt{4} = 2$

   $= 9$

   b. $a^{\sqrt{8}} \cdot a^{\sqrt{2}} = a^{\sqrt{8}+\sqrt{2}}$ Keep the base and add the exponents.

   $= a^{2\sqrt{2}+\sqrt{2}}$ $\sqrt{8} = \sqrt{4}\sqrt{2} = 2\sqrt{2}$

   $= a^{3\sqrt{2}}$ $2\sqrt{2} + \sqrt{2} = 3\sqrt{2}$

   ∎

Self Check 2   Simplify: **a.** $\left(2^{\sqrt{3}}\right)^{\sqrt{12}}$ **b.** $x^{\sqrt{20}} \cdot x^{\sqrt{5}}$

Now Try Exercise 19.

2. Graph Exponential Functions

If $b > 0$ and $b \neq 1$, the function $y = b^x$ defines a function, because for each input x, there is exactly one output y. Since x can be any real number, the domain of the function is the set of real numbers. Since the base b of the expression b^x is positive, y is always positive, and the range is the set of positive numbers. Since b^x is an exponential expression, the function is called an **exponential function.**

   We make the restriction that $b > 0$ to exclude any imaginary numbers that might result from taking even roots of negative numbers. The restriction that $b \neq 1$ excludes the constant function $f(x) = 1^x$, in which $f(x) = 1$ for every real number x.

Exponential Functions   An **exponential function with base b** is defined by the equation

$$f(x) = b^x \quad \text{or} \quad y = b^x \qquad (b > 0, b \neq 1, \text{ and } x \text{ is a real number})$$

The **domain of any exponential function** is the interval $(-\infty, \infty)$. The **range** is the interval $(0, \infty)$.

   Since the domain and range of $f(x) = b^x$ are sets of real numbers, we can graph exponential functions. For example, to graph

$$f(x) = 2^x$$

we find several points $(x, f(x))$ whose coordinates satisfy the equation, plot the points, and join them with a smooth curve, as in Figure 4-3(a). To graph the function

$$f(x) = \left(\frac{1}{2}\right)^x$$

we find several points $(x, f(x))$ whose coordinates satisfy the equation, plot the points, and join them with a smooth curve, as shown in Figure 4-3(b) on the next page.

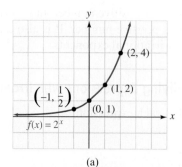

$f(x) = 2^x$		
x	$f(x)$	$(x, f(x))$
-1	$\dfrac{1}{2}$	$\left(-1, \dfrac{1}{2}\right)$
0	1	$(0, 1)$
1	2	$(1, 2)$
2	4	$(2, 4)$

(a)

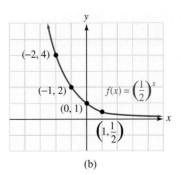

FIGURE 4-3

By looking at the graphs in Figure 4-3, we can see that the domain of each function is the interval $(-\infty, \infty)$ and that the range is the interval $(0, \infty)$.

EXAMPLE 3   **Graphing an Exponential Function**

Graph: $f(x) = 4^x$.

SOLUTION   We will find several points (x, y) that satisfy the equation, plot the points, and join them with a smooth curve, as in Figure 4-4.

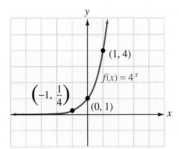

FIGURE 4-4

Self Check 3   Graph: $f(x) = \left(\dfrac{1}{4}\right)^x$.

Now Try Exercise 29.

The graph of $f(x) = 4^x$ in Example 3 has the following properties:

1. It passes through the point $(0, 1)$.
2. It passes through the point $(1, 4)$.
3. It approaches the x-axis. The x-axis is a horizontal asymptote.
4. The domain is the interval $(-\infty, \infty)$, and the range is the interval $(0, \infty)$.

This example illustrates the following properties of exponential functions.

Properties of Exponential Functions

- The **domain of the exponential function** $f(x) = b^x$ is $(-\infty, \infty)$, the set of real numbers.

- The **range** is $(0, \infty)$, the set of positive real numbers.

- The graph has a y-intercept at $(0, 1)$.

- The x-axis is a horizontal asymptote of the graph.

- The graph of $f(x) = b^x$ passes through the point $(1, b)$.

EXAMPLE 4   **Determining the Base of an Exponential Function**

The graph of an exponential function of the form $f(x) = b^x$ is shown in Figure 4-5. Find the value of b.

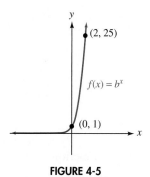

FIGURE 4-5

SOLUTION   We note that the graph passes through the point $(0, 1)$, a property of exponential functions of this form. Since the graph also passes through the point $(2, 25)$, we can find the base b by substituting 2 for x and 25 for $f(2)$ in the equation $f(x) = b^x$ and solving for b.

$$f(x) = b^x$$
$$f(2) = b^2$$
$$25 = b^2$$
$$5 = b \qquad b \text{ must be positive.}$$

The base b is 5. Note that the points $(0, 1)$ and $(2, 25)$ satisfy the equation $f(x) = 5^x$.

Self Check 4   Can a graph passing through $(0, 2)$ and $\left(1, \dfrac{3}{2}\right)$ be the graph of $f(x) = b^x$?

Now Try Exercise 45.

In Figure 4-3(a) (where $b = 2$ and $2 > 1$), the values of y increase as the values of x increase. Since the graph rises as we move to the right, the function is an increasing function. Such a function is said to model *exponential growth*.

In Figure 4-3(b) where $\left(b = \frac{1}{2} \text{ and } 0 < \frac{1}{2} < 1\right)$, the values of y decrease as the values of x increase. Since the graph drops as we move to the right, the function is a decreasing function. Such a function is said to model *exponential decay*. Two additional properties are stated here.

Additional Properties of Exponential Functions   If $b > 1$, then $f(x) = b^x$ is an **increasing function**. This function models **exponential growth**.

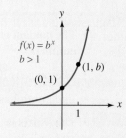

Increasing function

If $0 < b < 1$, then $f(x) = b^x$ is a **decreasing function.** This function models **exponential decay.**

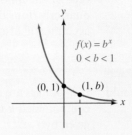

Decreasing function

Comment   Recall that $b^{-x} = \frac{1}{b^x} = \left(\frac{1}{b}\right)^x$. If $b > 1$, any function of the form $f(x) = b^{-x}$ models exponential decay, because $0 < \frac{1}{b} < 1$.

An exponential function $f(x) = b^x$ is either increasing (for $b > 1$) or decreasing (for $0 < b < 1$). Because different real numbers x always determine different values of b^x, an exponential function is one-to-one.

One-to-One Property of Exponential Functions   An exponential function defined by $f(x) = b^x$ or $y = b^x$, where $b > 0$ and $b \neq 1$, is one-to-one. This implies that

1. If $b^r = b^s$, then $r = s$.
2. If $r \neq s$, then $b^r \neq b^s$.

ACCENT ON TECHNOLOGY   **Graphing Exponential Functions**

When using a graphing calculator to draw the graphs of exponential functions, use care entering the function. Proper use of parentheses is critical to obtaining the correct graph. Since we know the shape of an exponential function, it is usually easy to find a good window for the graph. For any function in the form $f(x) = b^x$, where $b > 1$, the graph will be steeper as the base increases. The opposite is true if the base is between 0 and 1. Several graphs are shown below along with a window that works well with these basic functions.

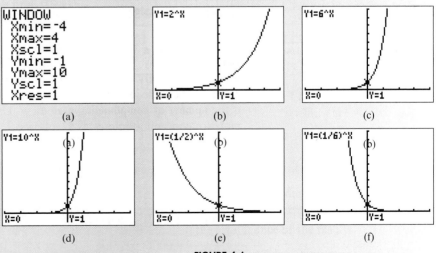

FIGURE 4-6

3. Solve Compound Interest Problems

Banks pay **interest** for using their customers' money. Interest is calculated as a percent of the amount on deposit in an account and is paid annually (once a year), quarterly (four times per year), monthly, or daily. Interest left on deposit in a bank account will also earn interest. Such accounts are said to earn **compound interest.**

Compound Interest Formula   If P dollars are deposited in an account earning interest at an annual rate r, compounded n times each year, the amount A in the account after t years is given by

$$A = P\left(1 + \frac{r}{n}\right)^{nt}$$

EXAMPLE 5   **Solving a Compound Interest Problem**

The parents of a newborn child invest $8,000 in a plan that earns 9% interest, compounded quarterly. If the money is left untouched, how much will the child have in the account in 55 years?

SOLUTION   We will substitute 8,000 for P, 0.09 for r, and 55 for t in the formula for compound interest. Because quarterly compounding means four times per year, we will substitute 4 for n.

$$A = P\left(1 + \frac{r}{n}\right)^{nt}$$

$$A = 8,000\left(1 + \frac{0.09}{4}\right)^{4 \cdot 55}$$

$$= 8,000(1.0225)^{220}$$

$$\approx 1,069,103.266 \qquad \text{Use a calculator.}$$

In 55 years, the account will be worth $1,069,103.27.

Self Check 5   Would $20,000 invested at 7% interest, compounded monthly, have provided more income at age 55?

Now Try Exercise 81.

In financial calculations, the initial amount deposited is often called the **present value,** denoted by PV. The amount to which the account will grow is called the **future value,** denoted by FV. The interest rate for each compounding period is called the **periodic interest rate,** i, and the number of times interest is compounded is the **number of compounding periods,** n. Using these definitions, an alternate formula for compound interest is as follows.

$$FV = PV(1 + i)^n$$

To use this formula to solve Example 5, we proceed as follows:

$$FV = PV(1 + i)^n$$

$$FV = 8,000(1 + 0.0225)^{220} \qquad i = \frac{0.09}{4} = 0.0225 \text{ and } n = 4(55) = 220.$$

$$= 8,000(1.0225)^{220}$$

$$\approx 1,069,103.266 \qquad \text{Use a calculator.}$$

4. Define e and Graph Base-e Exponential Functions

In mathematical models of natural events, the number

$$e = 2.71828182845904 \ldots$$

often appears as the base of an exponential function. We can introduce this number by considering the compound interest formula

$$A = P\left(1 + \frac{r}{n}\right)^{nt}$$ A is the amount, P is the initial deposit, r is the annual rate, n is the number of compoundings per year, and t is the time in years.

and allowing n to become very large. To see what happens, we let $n = rx$, where x is another variable.

$$A = P\left(1 + \frac{r}{n}\right)^{nt}$$

$$A = P\left(1 + \frac{r}{rx}\right)^{rxt}$$ Substitute rx for n.

$$A = P\left(1 + \frac{1}{x}\right)^{rxt}$$ Simplify $\frac{r}{rx}$.

$$A = P\left[\left(1 + \frac{1}{x}\right)^{x}\right]^{rt}$$ Remember that $(a^m)^n = a^{mn}$.

Since all variables in this formula are positive, r is a constant rate, and $n = rx$, it follows that as n becomes large, so does x. What happens to the value of A as n becomes large will depend on the value of $\left(1 + \frac{1}{x}\right)^{x}$ as x becomes large. Some results calculated for increasing values of x appear in the table shown.

x	$\left(1 + \dfrac{1}{x}\right)^{x}$
1	2
10	2.5937425
100	2.7048138
1,000	2.7169239
1,000,000	2.7182805
1,000,000,000	2.7182818

From the table, we can see that as x increases, the value of $\left(1 + \frac{1}{x}\right)^{x}$ approaches the value of e, and the formula

$$A = P\left[\left(1 + \frac{1}{x}\right)^{x}\right]^{rt}$$

becomes

$$A = Pe^{rt}$$ Substitute e for $\left(1 + \frac{1}{x}\right)^{x}$.

When the amount invested grows exponentially according to the formula $A = Pe^{rt}$, we say that interest is **compounded continuously.**

Continuous Compound Interest Formula   If P dollars are deposited in an account earning interest at an annual rate r, compounded continuously, the amount A after t years is given by the formula

$$A = Pe^{rt}$$

EXAMPLE 6   **Solving a Continuous Compound Interest Problem**

If the parents of the newborn child in Example 5 had invested $8,000 at an annual rate of 9%, compounded continuously, how much would the child have in the account in 55 years?

SOLUTION   We will substitute $8,000 for P, 0.09 for r, and 55 for t in the continuous compound interest formula $A = Pe^{rt}$.

$$A = Pe^{rt}$$

$$A = 8{,}000e^{(0.09)(55)}$$

$$= 8{,}000e^{4.95}$$

$$\approx 1{,}129{,}399.711 \qquad \text{Use a calculator.}$$

In 55 years, the balance will be $1,129,399.71, which is $60,296.44 more than the amount earned with quarterly compounding.   ∎

Self Check 6   Find the balance in 60 years.

Now Try Exercise 89.

To graph the exponential function $f(x) = e^x$, we plot several points and join them with a smooth curve [as in Figure 4-7(a)] or use a graphing calculator [as in Figure 4-7(b)].

$f(x) = e^x$		
x	$f(x)$	$(x, f(x))$
-1	0.37	$(-1, 0.37)$
0	1	$(0, 1)$
1	2.72	$(1, 2.72)$
2	7.39	$(2, 7.39)$

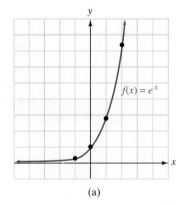

(a)

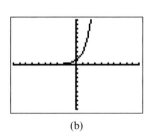

(b)

FIGURE 4-7

Comment   The graph of $f(x) = e^x$ is very important in mathematics and should be memorized.

5. Use Transformations to Graph Exponential Functions

In Section 3.4, translations, reflections, and stretchings were applied to functions. These may also be applied to the graphs of exponential functions. A summary of these transformations, when $k > 0$, is shown in the table.

Equation	Transformations of the Graph of $f(x) = b^x$
$y = b^x + k$	Translates the graph of $f(x) = b^x$ upward k units
$y = b^x - k$	Translates the graph of $f(x) = b^x$ downward k units
$y = b^{x-k}$	Translates the graph of $f(x) = b^x$ to the right k units
$y = b^{x+k}$	Translates the graph of $f(x) = b^x$ to the left k units
$y = -b^x$	Reflects the graph of $f(x) = b^x$ about the x-axis
$y = b^{-x}$	Reflects the graph of $f(x) = b^x$ about the y-axis
$y = kb^x$	• Vertically stretches the graph of $f(x) = b^x$ if $k > 1$ • Vertically shrinks the graph of $f(x) = b^x$ if $0 < k < 1$
$y = b^{kx}$	• Horizontally stretches the graph of $f(x) = b^x$ if $0 < k < 1$ • Horizontally shrinks the graph of $f(x) = b^x$ if $k > 1$

EXAMPLE 7 Using Translations to Graph an Exponential Function

On one set of axes, graph $f(x) = 2^x$ and $g(x) = 2^x + 3$.

SOLUTION The graph of $g(x) = 2^x + 3$ is identical to the graph of $f(x) = 2^x$, except that it is translated 3 units upward. (See Figure 4-8.)

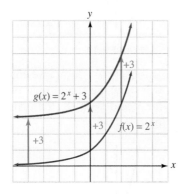

$f(x) = 2^x$		
x	$f(x)$	$(x, f(x))$
-4	$\dfrac{1}{16}$	$\left(-4, \dfrac{1}{16}\right)$
0	1	$(0, 1)$
2	4	$(2, 4)$

$g(x) = 2^x + 3$		
x	$g(x)$	$(x, g(x))$
-4	$3\dfrac{1}{16}$	$\left(-4, 3\dfrac{1}{16}\right)$
0	4	$(0, 4)$
2	7	$(2, 7)$

FIGURE 4-8

Self Check 7 On one set of axes, graph $f(x) = 2^x$ and $g(x) = 2^x - 2$.

Now Try Exercise 53.

EXAMPLE 8 Using Translations to Graph an Exponential Function with Base e

On one set of axes, graph $f(x) = e^x$ and $g(x) = e^{x-3}$.

SOLUTION The graph of $g(x) = e^{x-3}$ is identical to the graph of $f(x) = e^x$, except that it is translated 3 units to the right. (See Figure 4-9.)

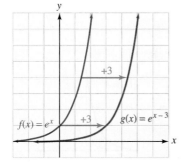

$f(x) = e^x$		
x	$f(x)$	$(x, f(x))$
-1	0.37	$(-1, 0.37)$
0	1	$(0, 1)$
1	2.72	$(1, 2.72)$
2	7.39	$(2, 7.39)$

$g(x) = e^{x-3}$		
x	$g(x)$	$(x, g(x))$
2	0.37	$(2, 0.37)$
3	1	$(3, 1)$
4	2.72	$(4, 2.72)$
5	7.39	$(5, 7.39)$

FIGURE 4-9

Self Check 8 On one set of axes, graph $f(x) = e^x$ and $g(x) = e^{x+2}$.

Now Try Exercise 61.

We can use a graphing calculator to graph exponential functions that are vertically or horizontally stretched or shrunk.

ACCENT ON TECHNOLOGY / **Graphing Exponential Functions**

Some exponential functions are very difficult to graph by hand. A graphing calculator can be used.

• To graph the exponential function $f(x) = 2(3^{x/2})$, we enter the right side of the equation after $Y_1 =$. The display will show the equation $Y_1 = 2(3^\wedge(X/2))$. If we use WINDOW settings of $[-10, 10]$ for x and $[-2, 18]$ for y and press GRAPH, we will obtain the graph shown in Figure 4-10.

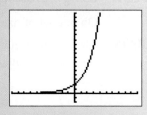

FIGURE 4-10

• To graph the exponential function $f(x) = 3e^{-x/2}$, we enter the right side of the equation $Y_1 =$. The display will show the equation $Y_1 = 3e^\wedge(-x/2)$. If we use WINDOW settings of $[-10, 10]$ for x and $[-2, 18]$ for y and press GRAPH, we will obtain the graph shown in Figure 4-11.

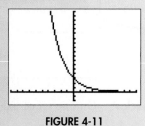

FIGURE 4-11

Self Check Answers   **1. a.** 2.6265   **b.** -14.7470 **c.** 0.0102   **2. a.** 64   **b.** $x^{3\sqrt{5}}$

3.

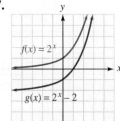

4. no   **5.** no   **6.** $1,771,251.33

7.

8.

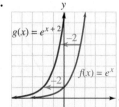

Exercises **4.1**

Getting Ready

You should be able to complete these vocabulary and con-
cept statements before you proceed to the practice exercises.

Fill in the blanks.

1. If $b > 0$ and $b \neq 1$, $y = b^x$ represents an _____ function.

2. If $f(x) = b^x$ represents an increasing function, then $b > $ __.

3. In interval notation, the domain of the exponential function $f(x) = b^x$ is _____.

4. The number b is called the _____ of the exponential function $y = b^x$.

5. The range of the exponential function $f(x) = b^x$ is _____.

6. The graphs of all exponential functions $y = b^x$ have the same ___-intercept, the point _____.

7. If $b > 0$ and $b \neq 1$, the graph of $y = b^x$ approaches the x-axis, which is called a horizontal _____ of the curve.

8. If $f(x) = b^x$ represents a decreasing function, then __ $< b <$ __.

9. If $b > 1$, then $y = b^x$ defines a (an) _____ function.

10. The graph of an exponential function $y = b^x$ always passes through the points $(0, 1)$ and _____.

11. To two decimal places, the value of e is ____.

12. The continuous compound interest formula is $A = $ ____.

13. Since $e > 1$, the base-e exponential function is a (an) _____ function.

14. The graph of the exponential function $y = e^x$ passes through the points $(0, 1)$ and _____.

Practice

 Use a calculator to find each value to four decimal places.

15. $4^{\sqrt{3}}$

16. $5^{\sqrt{2}}$

17. 7^{π}

18. $3^{-\pi}$

Simplify each expression.

19. $5^{\sqrt{2}}5^{\sqrt{2}}$

20. $\left(5^{\sqrt{2}}\right)^{\sqrt{2}}$

21. $\left(a^{\sqrt{8}}\right)^{\sqrt{2}}$

22. $a^{\sqrt{12}}a^{\sqrt{3}}$

Find $f(0)$ and $f(2)$ for each of the given exponential functions.

23. $f(x) = 5^x$

24. $f(x) = 4^{-x}$

25. $f(x) = \left(\dfrac{1}{3}\right)^{-x}$

26. $f(x) = \left(\dfrac{1}{4}\right)^{x}$

Graph each exponential function.

27. $f(x) = 3^x$

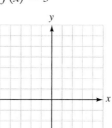

28. $f(x) = 5^x$

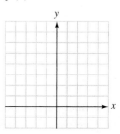

29. $f(x) = \left(\dfrac{1}{5}\right)^x$

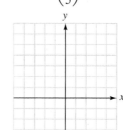

30. $f(x) = \left(\dfrac{1}{3}\right)^x$

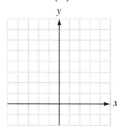

31. $f(x) = \left(\dfrac{3}{4}\right)^x$

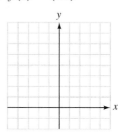

32. $f(x) = \left(\dfrac{4}{3}\right)^x$

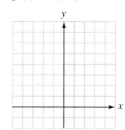

33. $f(x) = (1.5)^x$

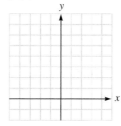

34. $f(x) = (0.3)^x$

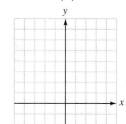

35. $f(x) = 3^{-x}$

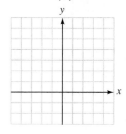

36. $f(x) = -5^x$

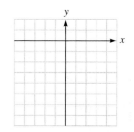

37. $f(x) = -\left(\dfrac{1}{5}\right)^{x}$

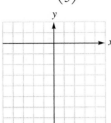

38. $f(x) = \left(\dfrac{1}{3}\right)^{-x}$

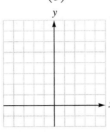

Determine whether the graph could represent an exponential function of the form $f(x) = b^{x}$.

39.

40.

41.

42.

Find the value of b, if any, that would cause the graph of $y = b^{x}$ *to look like the graph indicated.*

43.

44.

45.

46.

47.

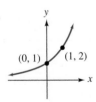

48.

49.

50.

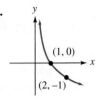

Graph each function by using transformations. **Do not use a graphing calculator.**

51. $f(x) = 3^{x} - 1$

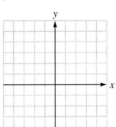

52. $f(x) = 2^{x} + 3$

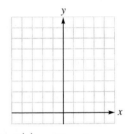

53. $f(x) = 2^{x} + 1$

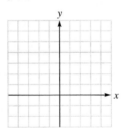

54. $f(x) = 4^{x} - 4$

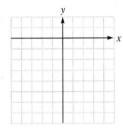

55. $f(x) = 3^{x-1}$

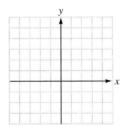

56. $f(x) = 2^{x+3}$

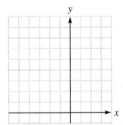

57. $f(x) = 3^{x+1}$

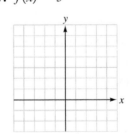

58. $f(x) = 2^{x-3}$

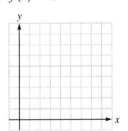

59. $f(x) = e^x - 4$

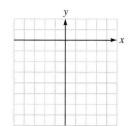

60. $f(x) = e^x + 2$

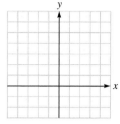

69. $f(x) = 2^{-x} - 3$

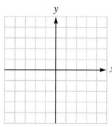

70. $f(x) = 4^{-x} + 4$

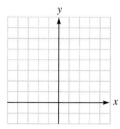

61. $f(x) = e^{x-2}$

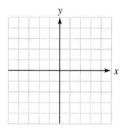

62. $f(x) = e^{x+3}$

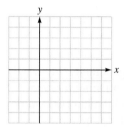

71. $f(x) = -e^x + 2$

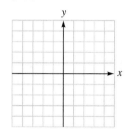

72. $f(x) = e^{-x} + 3$

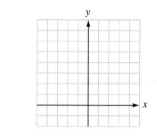

63. $f(x) = 2^{x+1} - 2$

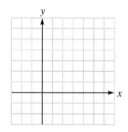

64. $f(x) = 3^{x-1} + 2$

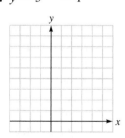

Use a graphing calculator to graph each function.

73. $f(x) = 5(2^x)$

74. $f(x) = 2(5^x)$

75. $f(x) = 3^{-x}$

76. $f(x) = 2^{-x}$

65. $y = 3^{x-2} + 1$

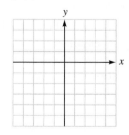

66. $y = 3^{x+2} - 1$

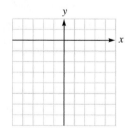

77. $f(x) = 2e^x$

78. $f(x) = 3e^{-x}$

79. $f(x) = 5e^{-0.5x}$

80. $f(x) = -3e^{2x}$

67. $f(x) = -3^x + 1$

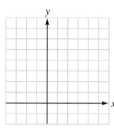

68. $f(x) = -2^x - 3$

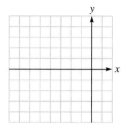

Applications

In Exercises 81–84, assume that there are no deposits or withdrawals.

81. Compound interest An initial deposit of $10,000 earns 8% interest, compounded quarterly. How much will be in the account in 10 years?

82. Compound interest An initial deposit of $1,000 earns 9% interest, compounded monthly. How much will be in the account in $4\frac{1}{2}$ years?

83. **Comparing interest rates** How much more interest could $500 earn in 5 years, compounded semi-annually (two times a year), if the annual interest rate were $5\frac{1}{2}\%$ instead of 5%?

84. **Comparing savings plans** Which institution in the ads provides the better investment?

> ### *Fidelity Savings & Loan*
> Earn 5.25%
> compounded monthly

> ### **Union Trust**
> Money Market Account
> paying 5.35%
> compounded annually

85. **Compound interest** If $1 had been invested on July 4, 1776, at 5% interest, compounded annually, what would it be worth on July 4, 2076?

86. **360/365 method** Some financial institutions pay daily interest, compounded by the 360/365 method, using the following formula. Using this method, what will an initial investment of $1,000 be worth in 5 years, assuming a 7% annual interest rate?

$$A = A_0\left(1 + \frac{r}{360}\right)^{365t} \quad (t \text{ is in years})$$

87. **Carrying charges** A college student takes advantage of the ad shown and buys a bedroom set for $1,100. He plans to pay the $1,100 plus interest when his income tax refund comes in 8 months. At that time, what will he need to pay?

> ### BUY NOW,
> ### PAY LATER!
>
> Only $1\frac{3}{4}\%$ interest per month.

88. **Credit card interest** A bank credit card charges interest at the rate of 21% per year, compounded monthly. If a senior in college charges $1,500 to pay for college expenses and intends to pay it in one year, what will she have to pay?

Mark Stout Photography/Shutterstock.com

89. **Continuous compound interest** An initial investment of $5,000 earns 8.2% interest, compounded continuously. What will the investment be worth in 12 years?

90. **Continuous compound interest** An initial investment of $2,000 earns 8% interest, compounded continuously. What will the investment be worth in 15 years?

91. **Comparison of compounding methods** An initial deposit of $5,000 grows at an annual rate of 8.5% for 5 years. Compare the final balances resulting from continuous compounding and annual compounding.

92. **Comparison of compounding methods** An initial deposit of $30,000 grows at an annual rate of 8% for 20 years. Compare the final balances resulting from continuous compounding and annual compounding.

93. **Frequency of compounding** $10,000 is invested in each of two accounts, both paying 6% annual interest. In the first account, interest compounds quarterly, and in the second account, interest compounds daily. Find the difference between the accounts after 20 years.

94. **Determining an initial deposit** An account now contains $11,180 and has been accumulating interest at a 7% annual rate, compounded continuously, for 7 years. Find the initial deposit.

95. **Saving for college** In 20 years, a father wants to accumulate $40,000 to pay for his daughter's college expenses. If he can get 6% interest, compounded quarterly, how much must he invest now to achieve his goal?

96. **Saving for college** In Problem 95, how much should he invest to achieve his goal if he can get 6% interest, compounded continuously?

97. **Population of a city** The population $P(t)$ of a small city can be approximated by the exponential function, $P(t) = 1,200e^{0.2t}$, where t represents time in years. What will be the population of the city in 12 years? Round to a whole number.

98. **Amount of drug present** The amount of a drug $A(t)$, in mg, present in the bloodstream t hours after being intravenously administered can be approximated by the exponential function, $A(t) = -1,000e^{-0.3t} + 1,250$. How much of the drug is present in the bloodstream after 14 hours? Round to a whole number.

Discovery and Writing

99. Financial planning To have P available in n years, A can be invested now in an account paying interest at an annual rate r, compounded annually. Show that

$$A = P(1 + r)^{-n}$$

100. If $2^{t+4} = k2^t$, find k.

101. If $5^{3t} = k^t$, find k.

102. a. If $e^{t+3} = ke^t$, find k.

   b. If $e^{3t} = k^t$, find k.

Review

Factor each expression completely.

103. $x^2 + 9x^4$

104. $x^2 - 9x^4$

105. $x^2 + x - 12$

106. $x^3 + 27$

4.2 Applications of Exponential Functions

In this section, we will learn to

1. Solve radioactive decay problems.

2. Solve oceanography problems.

3. Solve Malthusian population growth problems.

4. Solve epidemiology problems.

Flu kills an estimated 36,000 Americans each year and results is a much larger number of hospitalizations. The influenza virus replicates quickly and can rapidly infect a population. The most effective method of preventing the virus infection and its severe complications is a flu vaccination.

An event that changes with time, such as the spread of the influenza virus, can by modeled by an exponential function. In this section, we will see several important applications of these functions: radioactive decay, oceanography, population growth, and epidemiology.

A mathematical description of an observed event is called a **model** of that event. Many real-world occurrences change with time and can be modeled by exponential functions of the form

$$y = f(t) = ab^{kt} \qquad \text{Remember that } ab^{kt} \text{ means } a(b^{kt}).$$

where a, b, and k are constants and t represents time. If f is an increasing function, we say that y *grows exponentially.* If f is a decreasing function, we say that y *decays exponentially.*

1. Solve Radioactive Decay Problems

The atomic structure of a radioactive material changes as the material emits radiation. Uranium, for example, changes (decays) into thorium, then into radium, and eventually into lead.

Experiments have determined the time it takes for one-half of a sample of a given radioactive element to decompose. That time is a constant, called the element's **half-life.** The amount present decays exponentially according to this formula.

Radioactive Decay Formula	The amount A of radioactive material present at time t is given by

$$A = A_0 2^{-t/h}$$

where A_0 is the amount that was present initially (at $t = 0$) and h is the material's half-life.

EXAMPLE 1   **Solving a Radioactive Decay Problem**

The half-life of radium is approximately 1,600 years. How much of a 1-gram sample will remain after 1,000 years?

SOLUTION   In this example, $A_0 = 1$, $h = 1,600$, and $t = 1,000$. We substitute these values into the formula for radioactive decay and simplify.

$$A = A_0 2^{-t/h}$$

$$A = 1 \cdot 2^{-1,000/1,600}$$

$$\approx 0.648419777 \qquad \text{Use a calculator.}$$

After 1,000 years, approximately 0.65 gram of radium will remain.

Self Check 1   After 800 years, how much radium will remain?

Now Try Exercise 3.

2. Solve Oceanography Problems

Intensity of Light Formula	The intensity I of light (in lumens) at a distance x meters below the surface of a body of water decreases exponentially according to the formula

$$I = I_0 k^x$$

where I_0 is the intensity of light above the water and k is a constant that depends on the clarity of the water.

Comment   A lumen is a unit of standard measurement that describes how much light is contained in a certain area. The lumen is part of the photometry group which measures different aspects of light.

EXAMPLE 2   **Solving an Intensity of Light Problem**

At one location in the Atlantic Ocean, the intensity of light above water I_0 is 12 lumens and $k = 0.6$. Find the intensity of light at a depth of 5 meters.

SOLUTION   We will substitute 12 for I_0, 0.6 for k, and 5 for x into the formula for light intensity and then simplify.

$$I = I_0 k^x$$

$$I = 12(0.6)^5$$

$$I = 0.93312$$

At a depth of 5 meters, the intensity of the light is slightly less than 1 lumen.

Self Check 2   Find the intensity at a depth of 10 meters.

Now Try Exercise 11.

3. Solve Malthusian Population Growth Problems

An equation based on the exponential function provides a model for **population growth.** One such model, called the **Malthusian model of population growth,** assumes a constant birth rate and a constant death rate. In this model, the population P grows exponentially according to the following formula.

Malthusian Model of Population Growth

If b is the annual birth rate, d is the annual death rate, t is the time (in years), P_0 is the initial population at $t = 0$, and P is the current population, then

$$P = P_0 e^{kt}$$

where $k = b - d$ is the **annual growth rate,** the difference between the annual birth rate and death rate.

EXAMPLE 3   Using the Malthusian Model to Predict Population Growth

The population of the United States is approximately 300 million people. Assuming that the annual birth rate is 19 per 1,000 and the annual death rate is 7 per 1,000, what does the Malthusian model predict the U.S. population will be in 50 years?

SOLUTION   We can use the stated information to write the Malthusian model for U.S. population. We will then substitute into the model to predict the population in 50 years.

Since k is the difference between the birth and death rates, we have

$$k = b - d$$

$$k = \frac{19}{1,000} - \frac{7}{1,000} \qquad \text{Substitute } \frac{19}{1,000} \text{ for } b \text{ and } \frac{7}{1,000} \text{ for } d.$$

$$k = 0.019 - 0.007$$

$$= 0.012$$

We can now substitute 300,000,000 for P_0, 50 for t, and 0.012 for k in the formula for the Malthusian model of population growth and simplify.

$$P = P_0 e^{kt}$$

$$P = (300,000,000) e^{(0.012)(50)}$$

$$= (300,000,000) e^{0.6}$$

$$\approx 546,635,640.1 \qquad \text{Use a calculator.}$$

After 50 years, the U.S. population will exceed 546 million people. ∎

Self Check 3   Find the population in 100 years.

Now Try Exercise 23.

The English economist Thomas Robert Malthus (1766–1834) pioneered in population study. He believed that poverty and starvation were unavoidable, because the human population tends to grow exponentially, whereas the food supply tends to grow linearly.

EXAMPLE 4   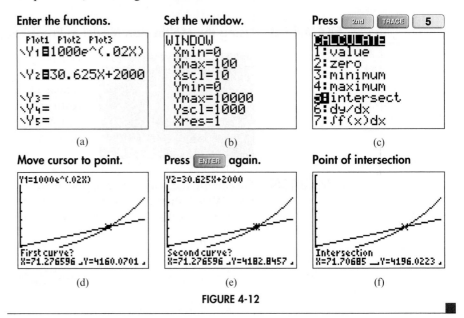 **Using a Graphing Calculator to Solve a Population Problem**

Suppose that a country with a population of 1,000 people is growing exponentially according to the formula

$$P = 1,000e^{0.02t}$$

where t is in years. Furthermore, assume that the food supply, measured in adequate food per day per person, is growing linearly according to the formula

$$y = 30.625x + 2,000$$

In how many years will the population outstrip the food supply?

SOLUTION   We can use a graphing calculator with WINDOW settings of $[0,100]$ for x and $[0,10,000]$ for y. After graphing the functions as shown in Figure 4-12, we find the point where the two graphs intersect. We can see that the food supply will be adequate for about 72 years. At that time, the population of approximately 4,200 people will begin to have problems, assuming all conditions remain static.

Enter the functions.	Set the window.	Press 2nd TRACE 5
Plot1 Plot2 Plot3 \Y₁■1000e^(.02X) \Y₂■30.625X+2000 \Y₃= \Y₄= \Y₅=	WINDOW Xmin=0 Xmax=100 Xscl=10 Ymin=0 Ymax=10000 Yscl=1000 Xres=1	CALCULATE 1:value 2:zero 3:minimum 4:maximum 5:intersect 6:dy/dx 7:∫f(x)dx
(a)	(b)	(c)

Move cursor to point.	Press ENTER again.	Point of intersection
Y1=1000e^(.02X) First curve? X=71.276596 ⌐Y=4160.0701	Y2=30.625X+2000 Second curve? X=71.276596 ⌐Y=4182.8457	Intersection X=71.70685 ⌐Y=4196.0223
(d)	(e)	(f)

FIGURE 4-12

Self Check 4   In 80 years, what is the approximate number of people per day that will not have adequate food?

Now Try Exercise 43.

4. Solve Epidemiology Problems

Many infectious diseases, including some caused by viruses, spread most rapidly when they first infect a population, but then more slowly as the number of uninfected individuals decreases. These situations are often modeled by a function, called a *logistic function*.

Logistic Epidemiology Model   The size P of an infected population at any time t in years is given by the logistic function

$$P = \frac{M}{1 + \left(\dfrac{M}{P_0} - 1\right)e^{-kt}}$$

where P_0 is the infected population size at $t = 0$, k is a constant determined by how contagious the virus is in a given environment, and M is the theoretical maximum size of the population P.

EXAMPLE 5   **Solving an Epidemiology Problem**

In a city with a population of 1,200,000, there are currently 1,000 cases of infection with HIV. If the spread of the disease is projected by the formula

$$P = \frac{1,200,000}{1 + (1,200 - 1)e^{-0.4t}}$$

how many people will be infected in 3 years?

SOLUTION   We can substitute 3 for t in the logistic formula and calculate P.

$$P = \frac{1,200,000}{1 + (1,200 - 1)e^{-0.4t}}$$

$$P = \frac{1,200,000}{1 + (1,199)e^{-0.4(3)}}$$

$$\approx 3,313.710094$$

In 3 years, approximately 3,300 people are expected to be infected.

Self Check 5   How many will be infected in 10 years?

Now Try Exercise 35.

ACCENT ON TECHNOLOGY   **Exponential Regression**

The table below shows the cooling temperatures of a hot cup of coffee after it is made.

Time in minutes	Temperature °F
0	179.5
5	168.7
8	158.1
11	149.2
15	141.7
18	134.6
22	125.4
25	123.5
30	116.3
34	113.2
38	109.1
42	105.7
45	102.2
50	100.5

graphx/Shutterstock.com

We can use a graphing calculator to determine an exponential regression model function that best fits or represents the given data.

- **Step 1:** Enter the data into lists in the calculator. Press ⌜STAT⌝ and then choose 1:Edit to access the list window and then enter the values. Enter the *x*-values (time in minutes) into L1 and the *y*-values (temperature) into LIST 2. See Figure 4-13(a).
- **Step 2:** Set a window to fit the data. See Figure 4-13(b).
- **Step 3:** Create a scatter plot of the data. To do so, select STAT PLOT, located above the ⌜Y=⌝ button, and turn Plot 1 on. See Figure 4-14(c). Press ⌜GRAPH⌝ to see the scatter plot. See Figure 4-13(d).

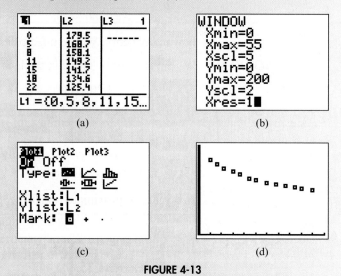

(a) (b)

(c) (d)

FIGURE 4-13

- **Step 4:** To determine the exponential regression function, we Press ⌜STAT⌝ and scroll right to the CALC menu. We scroll downward to 0:ExpReg and press ⌜ENTER⌝ twice. This gives the equation. See Figure 4-14 (a) and (b). The function is $y = 171.4617283 \cdot (0.9882469577)^x$.
- **Step 5:** To graph the exponential function, we first press ⌜Y=⌝, then we press ⌜VARS⌝ and scroll downward to 5: Statistics and press ⌜ENTER⌝. Finally, we select EQ, then press ⌜ENTER⌝ and ⌜GRAPH⌝. See Figure 4-14(c), (d), (e) and (f).

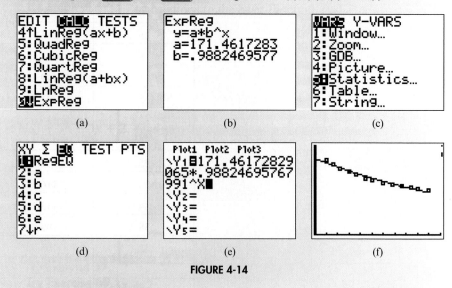

(a) (b) (c)

(d) (e) (f)

FIGURE 4-14

Self Check Answers **1.** about 0.71 g **2.** 0.073 lumen **3.** over 996 million **4.** About 503 **5.** about 52,000

Exercises **4.2**

Getting Ready

You should be able to complete these vocabulary and concept statements before you proceed to the practice exercises.

Fill in the blanks.

1. The Malthusian model assumes a constant _____ rate and a constant _____ rate.

2. The Malthusian prediction is pessimistic, because a _____ grows exponentially, but food supplies grow _____ .

Applications

Use a calculator to help solve each problem.

3. **Tritium decay** Tritium, a radioactive isotope of hydrogen, has a half-life of 12.4 years. Of an initial sample of 50 grams, how much will remain after 100 years?

4. **Chernobyl** In April 1986, the world's worst nuclear power disaster occurred at Chernobyl in the former USSR. An explosion released about 1,000 kilograms of radioactive cesium-137 (^{137}Cs) into the atmosphere. If the half-life of ^{137}Cs is 30.17 years, how much will remain in the atmosphere in 100 years?

5. **Chernobyl** Refer to Exercise 4. How much ^{137}Cs will remain in 200 years?

6. **Carbon-14 decay** The half-life of radioactive carbon-14 is 5,700 years. How much of an initial sample will remain after 3,000 years?

7. **Plutonium decay** One of the isotopes of plutonium, ^{237}Pu, decays with a half-life of 40 days. How much of an initial sample will remain after 60 days?

8. **Comparing radioactive decay** One isotope of holmium, ^{162}Ho, has a half-life of 22 minutes. The half-life of a second isotope, ^{164}Ho, is 37 minutes. Starting with a sample containing equal amounts, find the ratio of the amounts of ^{162}Ho to ^{164}Ho after one hour.

9. **Drug absorption in smokers** The biological half-life of the asthma medication theophylline is 4.5 hours for smokers. Find the amount of the drug retained in a smoker's system 12 hours after a dose of 1 unit is taken.

10. **Drug absorption in nonsmokers** For a nonsmoker, the biological half-life of theophylline is 8 hours. Find the amount of the drug retained in a nonsmoker's system 12 hours after taking a one-unit dose.

11. **Oceanography** The intensity I of light (in lumens) at a distance x meters below the surface is given by $I = I_0 k^x$, where I_0 is the intensity at the surface and k depends on the clarity of the water. At one location in the Arctic Ocean, $I_0 = 8$ lumens and $k = 0.5$. Find the intensity at a depth of 2 meters.

12. **Oceanography** At one location in the Atlantic Ocean, $I_0 = 14$ lumens and $k = 0.7$. Find the intensity of light at a depth of 12 meters. (See Exercise 11.)

13. **Oceanography** At a depth of 3 meters at one location in the Pacific Ocean, the intensity I of light is 1 lumen and $k = 0.5$. Find the intensity I_0 of light at the surface.

14. **Oceanography** At a depth of 2 meters at one location off the coast of Belize, the intensity I of light is 2 lumens and $k = 0.2$. Find the intensity I_0 of light at the surface.

15. **Bluegill population** A Wisconsin lake is stocked with 10,000 bluegill. The population is expected to grow exponentially according to the model $P = P_0 2^{t/2}$. How many bluegill will be in the lake in 5 years?

© Istockphoto.com/Dirk Nelson

16. **Community growth** The population of Eagle River is growing exponentially according to the model $P = 375(1.3)^t$, where t is measured in years from the present date. Find the population in 3 years.

17. **Newton's Law of Cooling** Some hot water, initially at 100°C, is placed in a room with a temperature of 40°C. The temperature T of the water after t hours is given by $T = 40 + 60(0.75)^t$. Find the temperature in $3\frac{1}{2}$ hours.

18. **Bacterial cultures** A colony of 6 million bacteria is growing in a culture medium. The population P after t hours is given by the formula $P = (6 \times 10^6)(2.3)^t$. Find the population after 4 hours.

19. **Population growth** The growth of a town's population is modeled by $P = 173e^{0.03t}$. How large will the population be when $t = 20$?

20. **Population decline** The decline of a city's population is modeled by $P = 1.2 \times 10^6 e^{-0.008t}$. How large will the population be when $t = 30$?

21. **Epidemics** The spread of hoof and mouth disease through a herd of cattle can be modeled by the formula $P = P_0 e^{0.27t}$, where P is the size of the infected population, P_0 is the infected population size at $t = 0$, and t is in days. If a rancher does not act quickly to treat two cases, how many cattle will have the disease in one week?

22. **Alcohol absorption** In one individual, the percent of alcohol absorbed into the bloodstream after drinking two glasses of wine is given by the following formula. Find the percent of alcohol absorbed into the blood after $\frac{1}{2}$ hour.

$$P = 0.3(1 - e^{-0.05t}) \text{ where } t \text{ is in minutes}$$

23. **World population growth** The population of the Earth is approximately 6 billion people and is growing at an annual rate of 1.9%. Assuming a Malthusian growth model, find the world population in 30 years.

24. **World population growth** See Exercise 23. Assuming a Malthusian growth model, find the world population in 40 years.

25. **World population growth** See Exercise 23. By what factor will the current population of the Earth increase in 50 years?

26. **World population growth** See Exercise 23. By what factor will the current population of the Earth increase in 100 years?

27. **Drug absorption** The percent P of the drug triazolam (a drug for treating insomnia) remaining in a person's bloodstream after t hours is given by $P = e^{-0.3t}$. What percent will remain in the bloodstream after 24 hours?

28. **Medicine** The concentration x of a certain drug in an organ after t minutes is given by $x = 0.08(1 - e^{-0.1t})$. Find the concentration of the drug in $\frac{1}{2}$ hour.

29. **Medicine** Refer to Exercise 28. Find the initial concentration of the drug (*Hint:* when $t = 0$).

30. **Spreading the news** Suppose the function

$$N = P(1 - e^{-0.1t})$$

is used to model the length of time t (in hours) it takes for N people living in a town with population P to hear a news flash. How many people in a town of 50,000 will hear the news between 1 and 2 hours after it happened?

31. **Spreading the news** How many people in the town described in Problem 30 will not have heard the news after 10 hours?

32. **Epidemics** Refer to Example 5. How many people will be infected with HIV in 5 years?

33. **Epidemics** Refer to Example 5. How many people will be infected with HIV in 8 years?

34. **Epidemics** In a city with a population of 450,000, there are currently 1,000 cases of hepatitis. If the spread of the disease is projected by the following logistic function, how many people will contract the hepatitis virus after 6 years?

$$P = \frac{450,000}{1 + (450 - 1)e^{-0.2t}}$$

35. **Epidemics** In an Indonesian city with a population of 55,000, there are currently 100 cases of the avian bird flu. If the spread of the disease is projected by the following formula, how many people will contract the bird flu after 2 years?

$$P = \frac{55,000}{1 + (550 - 1)e^{-0.8t}}$$

36. **Life expectancy** The life expectancy l of white females can be estimated by using the function $l = 78.5(1.001)^x$, where x is the current age. Find the life expectancy of a white female who is currently 50 years old. Give the answer to the nearest tenth.

37. **Oceanography** The width w (in millimeters) of successive growth spirals of the sea shell *Catapulus voluto,* shown in the illustration, is given by the function $w = 1.54e^{0.503n}$, where n is the spiral number. To the nearest tenth of a millimeter, find the width of the fifth spiral.

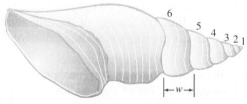

38. **Skydiving** Before the parachute opens, the velocity v (in meters per second) of a skydiver is given by $v = 50(1 - e^{-0.2t})$. Find the initial velocity.

39. Skydiving Refer to Exercise 38 and find the velocity after 20 seconds.

40. Free-falling objects After t seconds, a certain falling object has a velocity v given by $v = 50(1 - e^{-0.3t})$. Which is falling faster after 2 seconds, this object or the skydiver in Exercise 38?

41. Population growth In 1999, the male population of the United States was about 133 million, and the female population was about 139 million. Assuming a Malthusian growth model with a 1% annual growth rate, how many more females than males will there be in 20 years?

42. Population growth See Exercise 41. How many more females than males will there be in 50 years?

 Use a graphing calculator to solve each problem.

43. In Example 4, suppose that better farming methods change the formula for food growth to $y = 31x + 2{,}000$. How long will the food supply be adequate?

44. In Example 4, suppose that a birth control program changed the formula for population growth to $P = 1{,}000e^{0.01t}$. How long will the food supply be adequate?

Discovery and Writing

 45. The value of e can be calculated to any degree of accuracy by adding the first several terms of the following list. The more terms that are added, the closer the sum will be to e. Add the first six numbers in the following list. To how many decimal places is the sum accurate?

$$1, 1, \frac{1}{2}, \frac{1}{2 \cdot 3}, \frac{1}{2 \cdot 3 \cdot 4}, \frac{1}{2 \cdot 3 \cdot 4 \cdot 5}, \dots$$

 46. Graph the function defined by the equation $f(x) = \dfrac{e^x + e^{-x}}{2}$ from $x = -2$ to $x = 2$. The graph will look like a parabola, but it is not. The graph, called a **catenary,** is important in the design of power distribution networks, because it represents the shape of a uniform flexible cable whose ends are suspended from the same height. The function is called the **hyperbolic cosine function.**

 47. Graph the following logistic function, first discussed in Example 5. Use WINDOW settings of $[0, 20]$ for x and $[0, 1{,}500{,}000]$ for y.

$$P = \frac{1{,}200{,}000}{1 + (1{,}199)e^{-0.4t}}$$

 48. Use the TRACE capabilities of your graphing calculator to explore the logistic function of Example 5 and Exercise 47. As time passes, what value does P approach? How many years does it take for 20% of the population to become infected? For 80%?

Review
Find the value of x that makes each statement true.

49. $2^3 = x$ **50.** $3^x = 9$

51. $x^3 = 27$ **52.** $3^{-2} = x$

53. $x^{-3} = \dfrac{1}{8}$ **54.** $3^x = \dfrac{1}{3} - 1$

55. $9^{1/2} = x$ **56.** $x^{1/3} = 3$

4.3 Logarithmic Functions and Their Graphs

In this section, we will learn to

1. Evaluate logarithms.

2. Evaluate common logarithms.

3. Evaluate natural logarithms.

4. Graph logarithmic functions.

5. Use transformations to graph logarithmic functions.

Guests aboard the Royal Caribbean's cruise ship *Freedom of the Seas* can now "hang ten" while out to sea. The flowrider surf simulator allows riders to body board surf against a wave-like water flow of 34,000 gallons per minute.

It is important that water in the flowrider has the proper pH value. For example, if it is too acidic, the water will make our eyes and nose burn, and it will make our skin get dry and itchy. The pH of the water is one of the most important factors in pool water balance and should be tested frequently. To calculate pH, we need to understand *logarithms*, the topic of this section. As we continue through this chapter, we will see several real-life applications of logarithms.

Since exponential functions are one-to-one functions, each one has an inverse. For example, to find the inverse of the function $y = 3^x$, we interchange the positions of x and y to obtain $x = 3^y$. The graphs of these two functions are shown in Figure 4-15(a).

To find the inverse of the function $y = \left(\frac{1}{3}\right)^x$, we again interchange the positions of x and y to obtain $x = \left(\frac{1}{3}\right)^y$. The graphs of these two functions are shown in Figure 4-15(b).

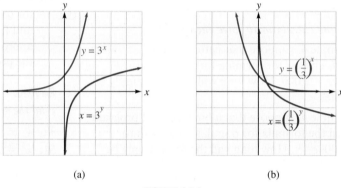

(a)   (b)

FIGURE 4-15

Comment   The graphs of the functions are inverses and therefore symmetric about the line $y = x$.

In general, the inverse of the function $y = b^x$ is $x = b^y$. When $b > 1$, their graphs appear as shown in Figure 4-16(a). When $0 < b < 1$, their graphs appear as shown in Figure 4-16(b).

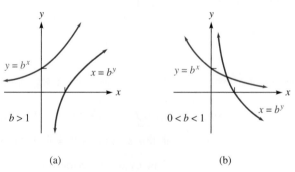

(a)   (b)

FIGURE 4-16

1. Evaluate Logarithms

Since an exponential function defined by $y = b^x$ is one-to-one, it has an inverse function that is defined by the equation $x = b^y$. To express this inverse function in the form $y = f^{-1}(x)$, we must solve the equation $x = b^y$ for y. To do this, we need the following definition.

Logarithmic Functions   If $b > 0$ and $b \neq 1$, the **logarithmic function with base b** is defined by

$$y = \log_b x \qquad \text{if and only if} \qquad x = b^y$$

The **domain of the logarithmic function** is the interval $(0, \infty)$. The **range** is the interval $(-\infty, \infty)$. The logarithmic function is also denoted as $f(x) = \log_b x$.

Caution

Since the domain of the logarithmic function is the set of positive numbers, the logarithm of 0 and the logarithm of a negative number are **undefined** in the set of real numbers.

The range of the logarithmic function is the set of real numbers, because the value of y in the equation $x = b^y$ can be any real number. The domain is the set of positive numbers, because the value of x in the equation $x = b^y$ ($b > 0$) is always positive.

Since the function $y = \log_b x$ is the inverse of the one-to-one exponential function $y = b^x$, the logarithmic function is also one-to-one.

The expression $x = b^y$ is said to be written in *exponential form*. The equivalent expression $y = \log_b x$ is said to be written in *logarithmic form*. To translate from one form to the other, it is helpful to keep track of the base and the exponent.

Exponential Form	*Logarithmic Form*
$x = b^y$	$y = \log_b x$
Base Exponent	Exponent Base

Comment   In general, we usually write $b^y = x$ instead of writing $x = b^y$. Both are acceptable. Also, many students prefer writing $\log_b x = y$ instead of writing $y = \log_b x$. This too is acceptable.

EXAMPLE 1   Converting from Exponential Form to Logarithmic Form

Write each equation in logarithmic form.

a. $2^6 = 64$ **b.** $27^{\frac{1}{3}} = \sqrt[3]{27} = 3$ **c.** $6^{-2} = \dfrac{1}{36}$

SOLUTION   For each part, we can use the definition of the logarithm of a number.

a. $2^6 = 64$ is equivalent to $\log_2 64 = 6$

b. $27^{\frac{1}{3}} = \sqrt[3]{27} = 3$ is equivalent to $\log_{27} 3 = \dfrac{1}{3}$

c. $6^{-2} = \dfrac{1}{36}$ is equivalent to $\log_6 \dfrac{1}{36} = -2$

Self Check 1   Write $13^{-2} = \dfrac{1}{169}$ in logarithmic form.

Now Try Exercise 21.

EXAMPLE 2   Converting from Logarithmic Form to Exponential Form

Write each equation in exponential form.

a. $\log_5 125 = 3$ **b.** $\log_{64} 8 = \dfrac{1}{2}$ **c.** $\log_{\frac{1}{4}} 16 = -2$

SOLUTION   For each part, we can use the definition of the logarithm of a number.

a. $\log_5 125 = 3$ is equivalent to $5^3 = 125$

b. $\log_{64} 8 = \dfrac{1}{2}$ is equivalent to $64^{\frac{1}{2}} = 8$

c. $\log_{\frac{1}{4}} 16 = -2$ is equivalent to $\left(\dfrac{1}{4}\right)^{-2} = 16$

Self Check 2   Write $\log_5 625 = 4$ in exponential form.

Now Try Exercise 29.

The definition of a logarithm can be used to find the values of many logarithms. Several examples are shown in the table.

Examples of the Values of Logarithms
$\log_5 25 = 2$ because **2** is the exponent to which 5 is raised to get 25: $5^2 = 25$
$\log_7 1 = 0$ because **0** is the exponent to which 7 is raised to get 1: $7^0 = 1$
$\log_{16} 4 = \dfrac{1}{2}$ because $\dfrac{1}{2}$ is the exponent to which 16 is raised to get 4: $16^{1/2} = \sqrt{16} = 4$
$\log_2 \dfrac{1}{8} = -3$ because -3 is the exponent to which 2 is raised to get $\dfrac{1}{8}$: $2^{-3} = \dfrac{1}{8}$

In each of these examples, the logarithm of a number is an exponent. In fact,

> $\log_b x$ *is the exponent to which b is raised to get x.*

To express this as an equation, we write $b^{\log_b x} = x$

EXAMPLE 3   **Finding an Unknown Term in a Logarithmic Equation**

Find y in each equation: **a.** $\log_2 8 = y$ **b.** $\log_5 1 = y$ **c.** $\log_7 \dfrac{1}{49} = y$

SOLUTION   For each part, we can use the definition of the logarithm of a number to find y.

a. $\log_2 8 = y$ is equivalent to $2^y = 8$. Since $2^3 = 8$, we have $2^y = 2^3$ and $y = 3$.

b. $\log_5 1 = y$ is equivalent to $5^y = 1$. Since $5^0 = 1$, we have $5^y = 5^0$ and $y = 0$.

c. $\log_7 \frac{1}{49} = y$ is equivalent to $7^y = \frac{1}{49}$. Since $7^{-2} = \frac{1}{49}$, we have $7^y = 7^{-2}$ and $y = -2$.

Self Check 3   Find y in each equation:

a. $\log_3 9 = y$ **b.** $\log_2 16 = y$ **c.** $\log_5 \dfrac{1}{25} = y$

Now Try Exercise 35.

EXAMPLE 4   **Finding an Unknown Term in a Logarithmic Equation**

Find a in each equation: **a.** $\log_a 32 = 5$ **b.** $\log_9 a = -\dfrac{1}{2}$ **c.** $\log_9 3 = a$.

SOLUTION   For each part, we will use the definition of the logarithm of a number to find a.

a. $\log_a 32 = 5$ is equivalent to $a^5 = 32$. Since $2^5 = 32$, we have $a^5 = 2^5$ and $a = 2$.

b. $\log_9 a = -\frac{1}{2}$ is equivalent to $9^{-1/2} = a$. Since $9^{-1/2} = \frac{1}{3}$, it follows that $a = \frac{1}{3}$.

c. $\log_9 3 = a$ is equivalent to $9^a = 3$. Since $3 = 9^{1/2}$, we have $9^a = 9^{1/2}$ and $a = \frac{1}{2}$.

Self Check 4   Find d in each equation:

a. $\log_4 \dfrac{1}{16} = d$ **b.** $\log_d 36 = 2$ **c.** $\log_8 d = -\dfrac{1}{3}$

Now Try Exercise 49.

2. Evaluate Common Logarithms

Many applications use base-10 logarithms (also called **common logarithms**). When the base b is not indicated in the notation $\log x$, we assume that $b = 10$:

> $\log x$ means $\log_{10} x$

Because base-10 logarithms appear so often, you should become familiar with the following base-10 logarithms:

Examples of Common Logarithms		
$\log_{10} \dfrac{1}{100} = -2$	because	$10^{-2} = \dfrac{1}{100}$
$\log_{10} \dfrac{1}{10} = -1$	because	$10^{-1} = \dfrac{1}{10}$
$\log_{10} 1 = 0$	because	$10^{0} = 1$
$\log_{10} 10 = 1$	because	$10^{1} = 10$
$\log_{10} 100 = 2$	because	$10^{2} = 100$
$\log_{10} 1,000 = 3$	because	$10^{3} = 1,000$

In general, we have

$$\log_{10} 10^{x} = x$$

ACCENT ON TECHNOLOGY   **Approximating Base-10 Logarithms**

We can use the [LOG] key on a graphing calculator to evaluate base-10 logarithms. Figure 4-17 shows the evaluation of $\log 2.34$ and demonstrates that **a logarithm is an exponent**. Notice that the logarithm is evaluated, and then the answer is used as the exponent of 10 to get 2.34. Remember: $\log 2.34$ is the exponent of 10 that will yield 2.34.

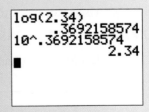

```
log(2.34)
          .3692158574
10^.3692158574
                 2.34
■
```

FIGURE 4-17

EXAMPLE 5   **Using a Calculator to Solve a Common Logarithmic Equation**

Find x in the equation $\log x = 0.7482$ to four decimal places.

SOLUTION   We can use the definition of the common logarithm and a calculator to find x.

The equation $\log x = 0.7482$ is equivalent to $10^{0.7482} = x$.

Therefore we have $x = 10^{0.7482} \approx 5.6002$.   ■

Self Check 5   Solve: $\log x = 1.87737$. Give the result to four decimal places.

Now Try Exercise 81.

3. Evaluate Natural Logarithms

We have seen the importance of the number e in mathematical models of events in nature. Base-e logarithms are just as important. They are called **natural logarithms** or **Napierian logarithms** after John Napier (1550–1617). They are usually written as $\ln x$, rather than $\log_e x$:

$$\ln x \qquad \text{means} \qquad \log_e x$$

Like all logarithmic functions, the domain of $f(x) = \ln x$ is the interval $(0, \infty)$, and the range is the interval $(-\infty, \infty)$.

To estimate the base-e logarithms of numbers, we can use a calculator. Scientific and graphing calculators have a natural logarithm key [LN], which is used like the [LOG] key.

ACCENT ON TECHNOLOGY   **Approximating Natural Logarithms**

We can use the [LN] key on a graphing calculator to evaluate base-e logarithms. Figure 4-18 shows the evaluation of $\ln 2.34$ and demonstrates that **a logarithm is an exponent**. Notice that the logarithm is evaluated, and then the answer is used as the exponent of e to get 2.34. Remember: $\ln 2.34$ is the exponent of e that will yield 2.34.

FIGURE 4-18

EXAMPLE 6   **Using a Calculator to Approximate Natural Logarithms**

Use a calculator to find **a.** $\ln 17.32$ **b.** $\ln(\log 0.05)$

SOLUTION   **a.** $\ln 17.32 \approx 2.851861903$, which means $e^{2.851861903} \approx 17.32$

b. $\ln(\log 0.05)$ has no value because $\log(0.05) < 0$. Our calculator gives an error message.

Self Check 6   Find each value to four decimal places:

a. $\ln \pi$ **b.** $\ln\left(\log \dfrac{1}{3}\right)$

Now Try Exercise 75.

EXAMPLE 7   **Using a Calculator to Solve Natural Logarithmic Equations**

Solve each equation and give the result to four decimal places:

a. $\ln x = 1.335$ **b.** $\ln x = \log 5.5$

SOLUTION   We will write the equation in exponential form and use our calculator to find x.

a. The equation $\ln x = 1.335$ is equivalent to $e^{1.335} = x$. We will use a calculator to evaluate the expression. $x = e^{1.335} \approx 3.8000$.

b. The equation $\ln x = \log 5.5$ is equivalent to $e^{\log 5.5} = x$. Using a calculator, we find $x = e^{\log 5.5} \approx 2.0967$.

Self Check 7   Solve each equation and give each result to four decimal places:

a. $\ln x = 1.9344$ b. $\log x = \ln 3.2$

Now Try Exercise 85.

4. Graph Logarithmic Functions

To draw the graph of a logarithmic function, we will use the fact that a logarithmic function is the inverse of an exponential function.

To graph the logarithmic function $y = f(x) = \log_2 x$, we calculate and plot several points with coordinates (x, y) that satisfy the equivalent equation $x = 2^y$. After joining these points with a smooth curve, we have the graph shown in Figure 4-19.

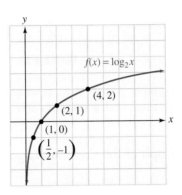

$f(x) = \log_2 x$		
x	$f(x)$	$(x, f(x))$
$\dfrac{1}{4}$	-2	$\left(\dfrac{1}{4}, -2\right)$
$\dfrac{1}{2}$	-1	$\left(\dfrac{1}{2}, -1\right)$
1	0	$(1, 0)$
2	1	$(2, 1)$
4	2	$(4, 2)$
8	3	$(8, 3)$

FIGURE 4-19

EXAMPLE 8   **Graphing a Logarithmic Function**

Graph: $y = f(x) = \log_{1/2} x$.

SOLUTION   To graph $y = f(x) = \log_{1/2} x$. we calculate and plot several points with coordinates (x, y) that satisfy the equation $x = \left(\dfrac{1}{2}\right)^y$. After joining these points with a smooth curve, we have the graph shown in Figure 4-20.

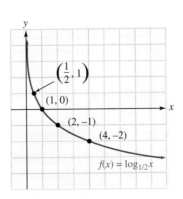

$f(x) = \log_{1/2} x$		
x	$f(x)$	$(x, f(x))$
$\dfrac{1}{4}$	2	$\left(\dfrac{1}{4}, 2\right)$
$\dfrac{1}{2}$	1	$\left(\dfrac{1}{2}, 1\right)$
1	0	$(1, 0)$
2	-1	$(2, -1)$
4	-2	$(4, -2)$
8	-3	$(8, -3)$

FIGURE 4-20

Self Check 8   Graph: $y = f(x) = \log_6 x$.

Now Try Exercise 105.

The graphs of all logarithmic functions are similar to those in Figure 4-21. If $b > 1$, the logarithmic function is an increasing function, as in Figure 4-21(a). If $0 < b < 1$, the logarithmic function is a decreasing function, as in Figure 4-21(b). We also know that the exponential function $f(x) = b^x$ and the logarithmic function $f(x) = \log_b x$ are inverse functions and therefore symmetric about the line $y = x$, as in Figure 4-21(c) and (d).

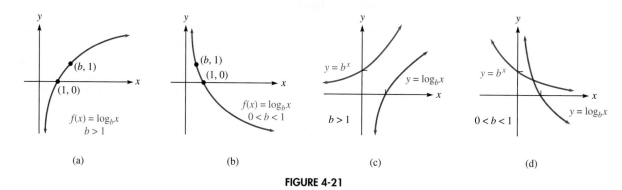

(a) (b) (c) (d)

FIGURE 4-21

To graph, $f(x) = \log_{10} x$, we can plot points that satisfy the equation $x = 10^y$ and join them with a smooth curve, as shown in Figure 4-22.

The Graph of the Common Logarithm Function $f(x) = \log_{10} x$

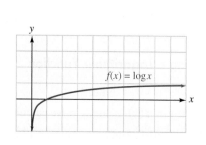

$f(x) = \log x$		
x	$f(x)$	$(x, f(x))$
$\dfrac{1}{100}$	-2	$\left(\dfrac{1}{100}, -2\right)$
$\dfrac{1}{10}$	-1	$\left(\dfrac{1}{10}, -1\right)$
1	0	$(1, 0)$
10	1	$(10, 1)$
100	2	$(100, 2)$

FIGURE 4-22

Comment   The function $f(x) = \log_{10} x$ is very important and its graph should be memorized.

As Figure 4-22 shows, the graph of $f(x) = \log_b x$ has these properties:

Properties of the Graph of $f(x) = \log_b x$
1. It passes through the point $(1, 0)$.
2. It passes through the point $(b, 1)$.
3. The y-axis is a vertical asymptote.
4. The domain is $(0, \infty)$, and the range is $(-\infty, \infty)$.

To graph $f(x) = \ln x$, we can plot points that satisfy the equation $x = e^y$ and join them with a smooth curve, as shown in Figure 4-23.

The Graph of the Natural Logarithm Function $f(x) = \ln x$

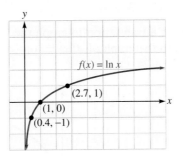

$f(x) = \ln x$		
x	$f(x)$	$(x, f(x))$
$\dfrac{1}{e} \approx 0.4$	-1	$(0.4, -1)$
1.0	0	$(1, 0)$
$e \approx 2.7$	1	$(2.7, 1)$
$e^2 \approx 7.4$	2	$(7.4, 2)$

FIGURE 4-23

Comment

The function $y = \ln x$ is very important and its graph should be memorized. The functions $y = e^x$ and $y = \ln x$ are inverses.

From the definition of logarithm, we know that

$$y = \ln x \qquad \text{if and only if} \qquad x = e^y \qquad \text{Remember that } \ln x = \log_e x.$$

Thus, $\ln e = 1$ because $e^1 = e$. This is an important property of the natural logarithm function. Two other important properties follow from the fact that $y = \ln x$ and $y = e^x$ are inverses:

$$\ln e^x = x \qquad \text{and} \qquad e^{\ln x} = x.$$

These two properties and examples are shown in the table:

Properties Involving e and ln	Examples
$\ln e^x = x$	$\ln e^8 = 8$
	$\ln e^{-5x} = -5x$
	$\ln e^{3x+2} = 3x + 2$
$e^{\ln x} = x$	$e^{\ln 4} = 4$
	$e^{\ln(7x)} = 7x$
	$e^{\ln(9x-2)} = 9x - 2$

5. Use Transformations to Graph Logarithmic Functions

The graphs of many functions involving logarithms are transformations of the basic logarithmic graphs. A summary of these are shown in the table for $k > 0$.

Equation	Transformations of the Graph of $f(x) = \log_b x$
$y = \log_b x + k$	Translates the graph of $f(x) = \log_b x$ upward k units
$y = \log_b x - k$	Translates the graph of $f(x) = \log_b x$ downward k units
$y = \log_b(x - k)$	Translates the graph of $f(x) = \log_b x$ to the right k units
$y = \log_b(x + k)$	Translates the graph of $f(x) = \log_b x$ to the left k units
$y = -\log_b x$	Reflects the graph of $f(x) = \log_b x$ about the x-axis
$y = \log_b(-x)$	Reflects the graph of $f(x) = \log_b x$ about the y-axis
$y = k \log_b x$	• Vertically stretches the graph of $y = \log_b x$ if $k > 1$ • Vertically shrinks the graph of $y = \log_b x$ if $0 < k < 1$
$y = \log_b(kx)$	• Horizontally stretches the graph of $y = \log_b x$ if $0 < k < 1$ • Horizontally shrinks the graph of $y = \log_b x$ if $k > 1$

EXAMPLE 9   **Using a Translation to Graph a Logarithmic Function**

Graph: $g(x) = 3 + \log_2 x$.

SOLUTION   We can use a translation to graph the function. The graph of $g(x) = 3 + \log_2 x$ is identical to the graph of $f(x) = \log_2 x$, except that it is translated 3 units upward. See Figure 4-24.

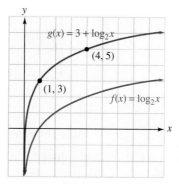

FIGURE 4-24

Self Check 9   Graph: $g(x) = -2 + \log_3 x$.

Now Try Exercise 111.

EXAMPLE 10   **Using a Translation to Graph a Logarithmic Function**

Graph: $g(x) = \log_{1/2}(x - 1)$.

SOLUTION   We can use a translation to graph the function. The graph of $g(x) = \log_{1/2}(x - 1)$ is identical to the graph of $f(x) = \log_{1/2} x$, except that it is translated 1 unit to the right. See Figure 4-25.

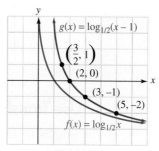

FIGURE 4-25

Self Check 10   Graph: $g(x) = \log_{1/3}(x + 2)$.

Now Try Exercise 113.

Many graphs of logarithmic functions involve translations of the graph of $f(x) = \ln x$. Consider the calculator graphs of the functions $f(x) = \ln x$, $g(x) = 2 + \ln x$, and $h(x) = -2 + \ln x$.

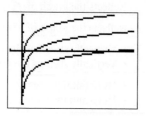

FIGURE 4-26

The graph of $f(x) = \ln x$ is the middle graph shown.

The graph of $g(x) = 2 + \ln x$ is 2 units above the graph of $f(x) = \ln x$.

The graph of $h(x) = -2 + \ln x$ is 2 units below the graph of $f(x) = \ln x$.

Figure 4-27 shows a calculator graph of the functions $f(x) = \ln x$, $g(x) = \ln(x + 2)$, and $h(x) = \ln(x - 2)$.

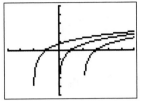

The graph of $f(x) = \ln x$ is the middle graph shown.

The graph of $g(x) = \ln(x + 2)$ is 2 units to the left of the graph of $f(x) = \ln x$.

The graph of $h(x) = \ln(x - 2)$ is 2 units to the right of the graph of $f(x) = \ln x$.

FIGURE 4-27

ACCENT ON TECHNOLOGY **Graphing Logarithmic Functions**

Graphing calculators can be used to graph logarithmic functions. However, the only bases that are built in are 10 and e. There is a way to graph logarithmic functions in other bases, but it involves changing the base. We will cover that in Section 4.5. In most applications either common logarithms or natural logarithms are used. When we are graphing logarithmic functions on a graphing calculator, we need to be aware of the domain so that we can set a proper window.

- Graph: $f(x) = -2 + \log_{10}\left(\frac{1}{2}x\right)$.

  To graph the function, we enter the right-hand side of the equation and use WINDOW settings of $[-1, 5]$ for x and $[-5, 1]$ for y. The graph is shown in Figure 4-28.

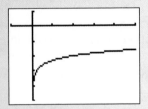

FIGURE 4-28

- Notice that the graph in Figure 4-28 appears to stop on the left and not continue toward negative infinity. That is a result of how the calculator plots points as it approaches the vertical asymptote.

Self Check Answers   **1.** $\log_{13}\dfrac{1}{169} = -2$ **2.** $5^4 = 625$ **3. a.** 2   **b.** 4   **c.** -2

4. a. -2 **b.** 6   **c.** $\dfrac{1}{2}$ **5.** 75.3998   **6. a.** 1.1447   **b.** no value

7. a. 6.9199   **b.** 14.5596

8.

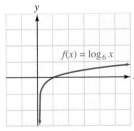

9.

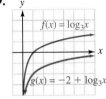

10.

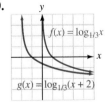

Exercises **4.3**

Getting Ready

You should be able to complete these vocabulary and concept statements before you proceed to the practice exercises.

Fill in the blanks.

1. The equation $y = \log_b x$ is equivalent to _____.
2. The domain of a logarithmic function is the interval _____.
3. The _____ of a logarithmic function is the interval $(-\infty, \infty)$.
4. $b^{\log_b x} = $ ___.
5. Because the exponential function is one-to-one, it has an _____ function.
6. The inverse of an exponential function is called a _____ function.
7. $\log_b x$ is the _____ to which b is raised to get x.
8. The y-axis is an _____ of a graph of $f(x) = \log_b x$.
9. The graph of $f(x) = \log_b x$ passes through the points _____ and _____.
10. $\log_{10} 10^x = $ ___.
11. ln x means _____.
12. The domain of the function $f(x) = \ln x$ is the interval _____.
13. The range of the function $f(x) = \ln x$ is the interval _____.
14. The graph of $f(x) = \ln x$ has the _____ as an asymptote
15. In the expression log x, the base is understood to be ___.
16. In the expression ln x, the base is understood to be ___.

Practice

Write each equation in logarithmic form.

17. $8^2 = 64$

18. $10^3 = 1,000$

19. $4^{-2} = \dfrac{1}{16}$

20. $3^{-4} = \dfrac{1}{81}$

21. $\left(\dfrac{1}{2}\right)^{-5} = 32$

22. $\left(\dfrac{1}{3}\right)^{-3} = 27$

23. $x^y = z$

24. $m^n = p$

Write each equation in exponential form.

25. $\log_3 81 = 4$

26. $\log_7 7 = 1$

27. $\log_{1/2} \dfrac{1}{8} = 3$

28. $\log_{1/5} 1 = 0$

29. $\log_4 \dfrac{1}{64} = -3$

30. $\log_6 \dfrac{1}{36} = -2$

31. $\log_\pi \pi = 1$

32. $\log_7 \dfrac{1}{49} = -2$

Find each value of x.

33. $\log_2 8 = x$

34. $\log_3 9 = x$

35. $\log_4 \dfrac{1}{64} = x$

36. $\log_6 216 = x$

37. $\log_{1/2} \dfrac{1}{8} = x$

38. $\log_{1/3} \dfrac{1}{81} = x$

39. $\log_9 3 = x$

40. $\log_{125} 5 = x$

41. $\log_{1/2} 8 = x$

42. $\log_{1/2} 16 = x$

43. $\log_8 x = 2$

44. $\log_7 x = 0$

45. $\log_7 x = 1$

46. $\log_2 x = 8$

47. $\log_{25} x = \dfrac{1}{2}$

48. $\log_4 x = \dfrac{1}{2}$

49. $\log_5 x = -2$

50. $\log_3 x = -4$

51. $\log_{36} x = -\dfrac{1}{2}$

52. $\log_{27} x = -\dfrac{1}{3}$

53. $\log_x 5^3 = 3$

54. $\log_x 5 = 1$

55. $\log_x \dfrac{9}{4} = 2$

56. $\log_x \dfrac{\sqrt{3}}{3} = \dfrac{1}{2}$

57. $\log_x \dfrac{1}{64} = -3$

58. $\log_x \dfrac{1}{100} = -2$

59. $\log_x \dfrac{9}{4} = -2$

60. $\log_x \dfrac{\sqrt{3}}{3} = -\dfrac{1}{2}$

61. $2^{\log_2 5} = x$

62. $3^{\log_3 4} = x$

63. $x^{\log_4 6} = 6$

64. $x^{\log_3 8} = 8$

Use a calculator to find each value to four decimal places.

65. log 3.25

66. log 0.57

67. log 0.00467

68. log 375.876

69. ln 45.7

70. ln 0.005

71. $\ln \dfrac{2}{3}$

72. $\ln \dfrac{12}{7}$

73. ln 35.15

74. ln 0.675

75. ln 7.896

76. ln 0.00465

77. log(ln 1.7)

78. ln(log 9.8)

79. ln(log 0.1)

80. log(ln 0.01)

Use a calculator to find y to four decimal places, if possible.

81. $\log y = 1.4023$

82. $\log y = 0.926$

83. $\log y = -3.71$

84. $\log y = \log \pi$

85. $\ln y = 1.4023$

86. $\ln y = 2.6490$

87. $\ln y = 4.24$

88. $\ln y = 0.926$

89. $\ln y = -3.71$

90. $\ln y = -0.28$

91. $\log y = \ln 8$

92. $\ln y = \log 7$

Find each value without using a calculator.

93. $\log 10{,}000$

94. $\log 1{,}000{,}000$

95. $\log 0.001$

96. $\log \dfrac{1}{100{,}000}$

97. $e^{(\ln 7)}$

98. $e^{(\ln 9)}$

99. $\ln(e^4)$

100. $\ln(e^{-6})$

Find the value of b, if any, that would cause the graph of $y = \log_b x$ to look like the graph shown.

101.

102.

103.

104.

Graph each function.

105. $f(x) = \log_3 x$

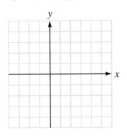

106. $f(x) = \log_4 x$

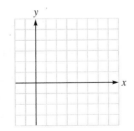

107. $f(x) = \log_{1/3} x$

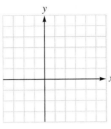

108. $f(x) = \log_{1/4} x$

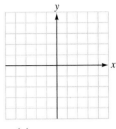

109. $f(x) = -\log_5 x$

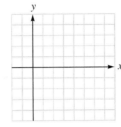

110. $f(x) = -\log_2 x$

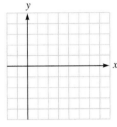

111. $f(x) = 2 + \log_2 x$

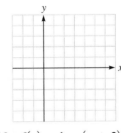

112. $f(x) = \log_2(x - 1)$

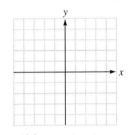

113. $f(x) = \log_3(x + 2)$

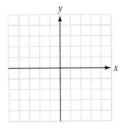

114. $f(x) = -3 + \log_3 x$

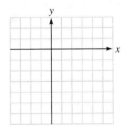

115. $f(x) = 3 + \log_3(x + 1)$

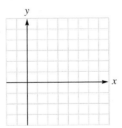

116. $f(x) = -3 + \log_3(x + 1)$

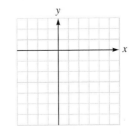

117. $f(x) = -3 + \ln x$

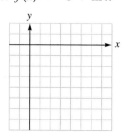

118. $f(x) = \ln(x + 1)$

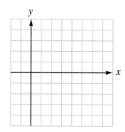

119. $f(x) = \ln(x - 4)$

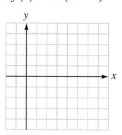

120. $f(x) = 2 + \ln x$

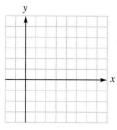

121. $f(x) = 1 - \ln x$

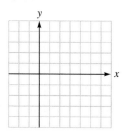

122. $f(x) = 2 - \ln x$

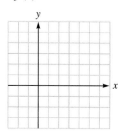

Use a graphing calculator to graph each function.

123. $f(x) = \log(3x)$

124. $f(x) = \log\left(\dfrac{x}{3}\right)$

125. $f(x) = \log(-x)$

126. $f(x) = -\log x$

127. $f(x) = \ln\left(\dfrac{1}{2}x\right)$

128. $f(x) = \ln x^2$

129. $f(x) = \ln(-x)$

130. $f(x) = \ln(3x)$

Discovery and Writing

131. Consider the following graphs. Which is larger, a or b, and why?

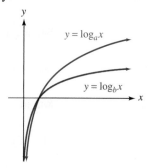

132. Consider the following graphs. Which is larger, a or b, and why?

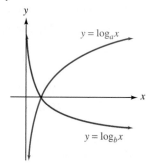

133. Pick two numbers and add their logarithms. Then find the logarithm of the product of those two numbers. What do you observe? Does it work for three numbers?

134. If $\log_a b = 7$, find $\log_b a$.

Review
Find the vertex of each parabola.

135. $y = x^2 + 7x + 3$

136. $y = 3x^2 - 8x - 1$

137. Fencing a pasture A farmer will use 3,400 feet of fencing to enclose and divide the pasture shown in the illustration. What dimensions will enclose the greatest area?

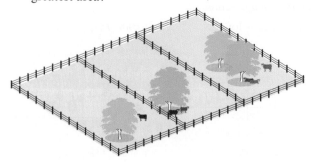

138. Selling appliances When the price is p dollars, an appliance dealer can sell $(2,200 - p)$ refrigerators. What price will maximize his revenue?

139. Write an equation of a line passing through the origin and parallel to the line $y = 5x - 8$.

140. Write an equation of a line passing through $(3, 2)$ and perpendicular to the line $y = \frac{2}{3}x - 12$.

4.4 Applications of Logarithmic Functions

In this section, we will learn to

1. Use logarithms to solve electrical engineering problems.
2. Use logarithms to solve geology problems.
3. Use logarithms to solve charging battery problems.
4. Use logarithms to solve population growth problems.
5. Use logarithms to solve isothermal expansion problems.

Earthquakes are one of the deadliest natural disasters that can occur. In 2004, an earthquake in the Indian Ocean triggered a series of tsunamis killing more than 225,000 people in eleven countries. Indonesia, Sri Lanka, Thailand, and India were hit the hardest. The earthquake was the second largest ever recorded and had a magnitude between 9.1 and 9.3 on the Richter scale. In 2010, a 7.0 catastrophic earthquake hit 16 miles west of Port-au-Prince, Haiti's capital, affecting more than 3 million people.

© Istockphoto.com/Claudia Dewald

In 2011, a massive earthquake of magnitude 9.0 hit off the coast of Japan, killing thousands and affecting Japan's nuclear reactors and triggering a nuclear crisis.

The intensity of an earthquake is based on a logarithmic function, the topic of this section. We will also use logarithmic functions to solve problems in engineering, geology, social science, and physics.

1. Use Logarithms to Solve Electrical Engineering Problems

Electronic engineers use common logarithms to measure the voltage gain of devices such as amplifiers or the length of a transmission line. The unit of gain, called the **decibel**, is defined by a logarithmic function.

© Istockphoto.com/Dragan Trifunovic

Decibel Voltage Gain   If E_O is the output voltage of a device and E_I is the input voltage, the **decibel voltage gain** is given by

$$dB \text{ gain} = 20 \log \frac{E_O}{E_I}$$

EXAMPLE 1   **Finding the Decibel Gain of an Amplifier**

Find the dB gain of an amplifier if its input is 0.5 volt and its output is 40 volts.

SOLUTION   We can find the decibel voltage gain by substituting the given values into the dB gain formula.

$$dB \text{ voltage gain} = 20 \log \frac{E_O}{E_I}$$

$$dB \text{ voltage gain} = 20 \log \frac{40}{0.5} \qquad \text{Substitute 40 for } E_O \text{ and 0.5 for } E_I.$$

$$= 20 \log 80$$

$$\approx 38.06179974 \qquad \text{Use a calculator.}$$

To the nearest decibel, the db gain is 38 decibels.

Self Check 1   Find the dB gain if the input is 0.7 volt.

Now Try Exercise 7.

2. Use Logarithms to Solve Geology Problems

Seismologists measure the intensity of earthquakes on the **Richter scale,** which is based on a logarithmic function.

Haiti's Presidential Palace after
2010 earthquake

Richter Scale   If R is the intensity of an earthquake, A is the amplitude (measured in micrometers), and P is the period (the time of one oscillation of the Earth's surface, measured in seconds), then

$$R = \log \frac{A}{P}$$

EXAMPLE 2   **Finding the Intensity of an Earthquake**

Find the intensity of an earthquake with amplitude of 5,000 micrometers $\left(\frac{1}{2} \text{ centimeter}\right)$ and a period of 0.07 second.

SOLUTION   We substitute 5,000 for A and 0.07 for P in the Richter scale formula and simplify.

$$R = \log \frac{A}{P}$$

$$R = \log \frac{5,000}{0.07}$$

$$\approx \log 71{,}428.57143 \qquad \text{Use a calculator.}$$

$$\approx 4.853871964$$

To the nearest tenth, the earthquake measures 4.9 on the Richter scale.

Self Check 2   Find the intensity of an aftershock with the same period but one-half of the amplitude.

Now Try Exercise 13.

3. Use Logarithms to Solve Charging Battery Problems

A battery charges at a rate that depends on how close it is to being fully charged—it charges fastest when it is most discharged. The formula that determines the time required to charge a battery to a certain level is based on a natural logarithmic function.

Charging Batteries   If M is the theoretical maximum charge that a battery can hold and k is a positive constant that depends on the battery and the charger, the length of time t (in minutes) required to charge the battery to a given level C is given by

$$t = -\frac{1}{k} \ln\left(1 - \frac{C}{M}\right)$$

EXAMPLE 3   **Solving a Charging Battery Problem**

How long will it take to bring a fully discharged battery to 90% of full charge? Assume that $k = 0.025$ and that time is measured in minutes.

SOLUTION   90% of full charge means 90% of M. We can substitute $0.90M$ for C and 0.025 for k in the formula for charging batteries to find t.

$$t = -\frac{1}{k} \ln\left(1 - \frac{C}{M}\right)$$

$$t = -\frac{1}{0.025} \ln\left(1 - \frac{0.90M}{M}\right)$$

$$= -40 \ln(1 - 0.9)$$

$$= -40 \ln(0.1)$$

$$\approx 92.10340372 \qquad \text{Use a calculator.}$$

The battery will reach 90% charge in about 92 minutes.

Self Check 3   How long will it take this battery to reach 80% of full charge?

Now Try Exercise 19.

4. Use Logarithms to Solve Population Growth Problems

If a population grows exponentially at a certain annual rate, the time required for the population to double is called the **doubling time** and is given by the following formula. You will be asked to prove this formula in Exercise 112 in Section 4.6.

Population Doubling Time	If r is the annual growth rate and t is the time (in years) required for a population to double, then

$$t = \frac{\ln 2}{r}$$

EXAMPLE 4 **Finding the Doubling Time of a Population**

The population of the Earth is growing at the approximate rate of 2% per year. If this rate continues, how long will it take the population to double?

SOLUTION Because the population is growing at the rate of 2% per year, we can substitute 0.02 for r in the formula for doubling time and simplify.

$$t = \frac{\ln 2}{r}$$

$$t = \frac{\ln 2}{0.02}$$

$$\approx 34.65735903$$

It will take about 35 years for the Earth's population to double. ∎

Self Check 4 If the world population's annual growth rate could be reduced to 1.5% per year, what would be the doubling time?

Now Try Exercise 21.

5. Use Logarithms to Solve Isothermal Expansion Problems

When energy is added to a gas, its temperature and volume could increase. In **isothermal expansion,** the temperature remains constant—only the volume changes. The energy required is calculated as follows.

Isothermal Expansion	If the temperature T is constant, the energy E required to increase the volume of 1 mole of gas from an initial volume V_i to a final volume V_f is given by

$$E = RT \ln\left(\frac{V_f}{V_i}\right)$$

E is measured in joules and T in Kelvins. R is the universal gas constant, which is 8.314 joules/mole/K.

EXAMPLE 5   **Solving an Isothermal Expansion Problem**

Find the amount of energy that must be supplied to triple the volume of 1 mole of gas at a constant temperature of 300 K.

SOLUTION   We substitute 8.314 for R and 300 for T in the formula. Since the final volume is to be three times the initial volume, we also substitute $3V_i$ for V_f.

$$E = RT \ln\left(\frac{V_f}{V_i}\right)$$

$$E = (8.314)(300)\ln\left(\frac{3V_i}{V_i}\right)$$

$$= 2{,}492.2 \ln 3$$

$$\approx 2{,}740.15877$$

Approximately 2,740 joules of energy must be added to triple the volume.

Self Check 5   What energy is required to double the volume?

Now Try Exercise 25.

Self Check Answers   **1.** about 35 decibels   **2.** about 4.6   **3.** about 64 min   **4.** about 46 years
5. 1,729 joules

Exercises **4.4**

Getting Ready
You should be able to complete these vocabulary and concept statements before you proceed to the practice exercises.

Fill in the blanks.

1. dB gain = _____

2. The intensity of an earthquake is measured by the formula R = _____ .

3. The formula for charging batteries is

_____ .

4. If a population grows exponentially at a rate r, the time it will take for the population to double is given by the formula t = _____ .

5. The formula for isothermal expansion is

_____ .

6. The logarithm of a negative number is _____ .

Applications
Use a calculator to solve each problem.

7. **Gain of an amplifier** An amplifier produces an output of 17 volts when the input signal is 0.03 volt. Find the decibel voltage gain.

8. **Transmission lines** A 4.9-volt input to a long transmission line decreases to 4.7 volts at the other end. Find the decibel voltage loss.

9. **Gain of an amplifier** Find the dB gain of an amplifier whose input voltage is 0.71 volt and whose output voltage is 20 volts.

10. **Gain of an amplifier** Find the dB gain of an amplifier whose output voltage is 2.8 volts and whose input voltage is 0.05 volt.

11. **dB gain** Find the dB gain of the amplifier shown below.

12. **dB gain** Find the dB gain of the amplifier shown below.

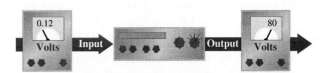

13. **Earthquakes** An earthquake has an amplitude of 5,000 micrometers and a period of 0.2 second. Find its measure on the Richter scale.

14. Earthquakes An earthquake has an amplitude of 8,000 micrometers and a period of 0.008 second. Find its measure on the Richter scale.

15. Earthquakes An earthquake with a period of $\frac{1}{4}$ second has an amplitude of 2,500 micrometers. Find its measure on the Richter scale.

16. Earthquakes An earthquake has a period of $\frac{1}{2}$ second and an amplitude of 5 cm. Find its measure on the Richter scale. (*Hint:* 1 cm = 10,000 micrometers)

17. Earthquakes An earthquake measuring between 3.5 and 5.4 on the Richter scale is often felt but rarely causes damage. Suppose an earthquake in Northern California has an amplitude of 6,000 micrometers and a period of 0.3 second. Is it likely to cause damage?

18. Earthquakes An earthquake measuring between 7 and 7.9 on the Richter scale is a major earthquake and can cause serious damage over larger areas. Suppose an earthquake in Chile has an amplitude of 198.5 cm and a period of 0.1 second. Would it cause serious damage over large areas? (*Hint:* 1 cm = 10,000 micrometers)

19. Battery charge If $k = 0.116$, how long will it take a battery to reach a 90% charge? Assume that the battery was fully discharged when it began charging.

20. Battery charge If $k = 0.201$, how long will it take a battery to reach a 40% charge? Assume that the battery was fully discharged when it began charging.

21. Population growth A town's population grows at the rate of 12% per year. If this growth rate remains constant, how long will it take the population to double?

22. Fish population growth One thousand bass were stocked in Catfish Lake in Eagle River, Wisconsin, a lake with no bass population. If the population of bass is expected to grow at a rate of 25% per year, how long will it take the population to double?

23. Population growth A population growing at an annual rate r will triple in a time t given by the formula $t = \frac{\ln 3}{r}$. How long will it take the population of the town in Exercise 21 to triple?

24. Fish population growth How long would it take the fish population in Exercise 22 to triple?

25. Isothermal expansion One mole of gas expands isothermically to triple its volume. If the gas temperature is 400 K, what energy is absorbed?

26. Isothermal expansion One mole of gas expands isothermically to double its volume. If the gas temperature is 300K, what energy is absorbed?

If an investment is growing continuously for t years, its annual growth rate r is given by the following formula, where P is the current value and P_0 is the amount originally invested.

$$r = \frac{1}{t} \ln \frac{P}{P_0}$$

27. Investing AOL's stock grew continuously from 1992–1999. An investment of $10,400 in the stock in 1992 was worth $10,400,000 in 1999. Find AOL's average annual growth rate during this period.

28. Investing A $5,000 investment in Dell Computer in 1995 was worth $237,000 in 1999. It grew continuously during this time period. Find the average annual growth rate of the stock.

29. Depreciation In business, equipment is often depreciated using the double declining-balance method. In this method, a piece of equipment with a life expectancy of N years, costing $C, will depreciate to a value of $V in n years, where n is given by the following formula.

$$n = \frac{\log V - \log C}{\log\left(1 - \dfrac{2}{N}\right)}$$

If a computer that cost $37,000 has a life expectancy of 5 years and has depreciated to a value of $8,000, how old is it?

30. Depreciation A word processor worth $470 when new had a life expectancy of 12 years. If it is now worth $189, how old is it? (See Exercise 29.)

31. Annuities If $P is invested at the end of each year in an annuity earning interest at an annual rate r, the amount in the account will be $A after n years, where

$$n = \frac{\log\left[\dfrac{Ar}{P} + 1\right]}{\log(1 + r)}$$

If $1,000 is invested each year in an annuity earning 12% annual interest, when will the account be worth $20,000?

32. Annuities If $5,000 is invested each year in an annuity earning 8% annual interest, when will the account be worth $50,000? (See Exercise 31.)

33. Breakdown voltage The coaxial power cable shown has a central wire with radius $R_1 = 0.25$ centimeter. It is insulated from a surrounding shield with inside radius $R_2 = 2$ centimeters. The maximum voltage the cable can withstand is called the **breakdown voltage** V of the insulation. V is given by the formula

$$V = ER_1 \ln \frac{R_2}{R_1}$$

where E is the **dielectric strength** of the insulation. If $E = 400,000$ volts/centimeter, find V.

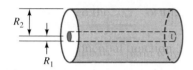

34. Breakdown voltage In Exercise 33, if the inside diameter of the shield were doubled, what voltage could the cable withstand?

35. Suppose you graph the function $f(x) = \ln x$ on a coordinate grid with a unit distance of 1 centimeter on the x- and y-axes. How far out must you go on the x-axis so that $f(x) = 12$? Give your result to the nearest mile.

36. Suppose you graph the function $f(x) = \log x$ on a coordinate grid with a unit distance of 1 centimeter on the x- and y-axes. How far out must you go on the x-axis so that $f(x) = 12$? Give the result to the nearest mile. Why is this result so much larger than the result in Exercise 35?

Discovery and Writing

37. One form of the logistic function is given by the following equation. Explain how you would find the y-intercept of its graph.

$$y = \frac{1}{1 + e^{-2x}}$$

38. Graph the function $y = \ln|x|$. Explain why the graph looks the way it does.

Review

Write an equation of the required line.

39. Having a slope of 7 and a y-intercept of 3

40. Parallel to the line $3x + 2y = 9$ and passing through the point $(-3, 5)$

41. A vertical line passing through $(2, 3)$

42. A horizontal line passing through $(2, 3)$

Simplify each expression.

43. $\dfrac{2(x + 2) - 1}{4x^2 - 9}$

44. $\dfrac{x + 1}{x} + \dfrac{x - 1}{x + 1}$

45. $\dfrac{x^2 + 3x + 2}{3x + 9} \cdot \dfrac{x + 3}{x^2 - 4}$

46. $\dfrac{1 + \dfrac{y}{x}}{\dfrac{y}{x} - 1}$

4.5 Properties of Logarithms

In this section, we will learn to

1. Use properties of logarithms to simplify expressions.

2. Use the Change-of-Base Formula.

3. Use logarithms to solve pH problems.

4. Use logarithms to solve problems in electronics.

5. Use logarithms to solve physiology problems.

Rolling Stone magazine lists the rock band U2 at number 22 in their list of the top 100 artists of all time. The Irish band is noted for lead singer Bono's vocals and their anthem-like sound.

Attending a U2 rock concert or cranking up a car stereo and playing U2's song "With or Without You" is an enjoyable activity. Music is an important part of our pop culture. Since many students prefer their music loud, the loudness of sound and its intensity are interesting concepts. In this section, we will see that loudness of sound and the intensity of sound are related by a formula involving the natural logarithmic function.

1. Use Properties of Logarithms to Simplify Expressions

Since logarithms are exponents, the properties of exponents have counterparts in the theory of logarithms. We begin with four basic properties.

Properties of Logarithms   If b is a positive number and $b \neq 1$, then

1. $\log_b 1 = 0$ **2.** $\log_b b = 1$

3. $\log_b b^x = x$ **4.** $b^{\log_b x} = x$ $(x > 0)$

Properties 1 through 4 follow directly from the definition of logarithm.

1. $\log_b 1 = 0$, because $b^0 = 1$.
2. $\log_b b = 1$, because $b^1 = b$.
3. $\log_b b^x = x$, because $b^x = b^x$.
4. $b^{\log_b x} = x$, because $\log_b x$ is the exponent to which b is raised to get x.

Properties 3 and 4 also indicate that the composition of the exponential and logarithmic functions (in both directions) is the identity function. This is expected, because the exponential and logarithmic functions with the same base are inverse functions.

EXAMPLE 1   **Using Properties of Logarithms to Simplify Expressions**

Simplify each expression: **a.** $\log_3 1$ **b.** $\log_4 4$ **c.** $\log_7 7^3$ **d.** $b^{\log_b 3}$

SOLUTION   We can simplify each expression using the properties of logariathms.

a. By Property 1, $\log_3 1 = \mathbf{0}$, because $3^0 = 1$.

b. By Property 2, $\log_4 4 = \mathbf{1}$, because $4^1 = 4$.

c. By Property 3, $\log_7 7^3 = \mathbf{3}$, because $7^3 = 7^3$.

d. By Property 4, $b^{\log_b 3} = \mathbf{3}$, because $\log_b 3$ is the power to which b is raised to get 3.

Self Check 1   Simplify: **a.** $\log_4 1$ **b.** $\log_3 3$ **c.** $\log_2 2^4$ **d.** $5^{\log_5 2}$

Now Try Exercise 13.

The four properties also hold for natural logarithms.

1. $\ln 1 = 0$, because $e^0 = 1$.
2. $\ln e = 1$, because $e^1 = e$.
3. $\ln e^x = x$, because $e^x = e^x$.
4. $e^{\ln x} = x$, because $\ln x$ is the exponent to which e is raised to get x.

The next two properties state that

The logarithm of a product is the sum of the logarithms.

The logarithm of a quotient is the difference of the logarithms.

The Logarithm of a Product and a Difference

If M, N, and b are positive numbers and $b \neq 1$, then

5. $\log_b MN = \log_b M + \log_b N$ **6.** $\log_b \dfrac{M}{N} = \log_b M - \log_b N$

Property 5 is known as the **Product Rule**.
Property 6 is known as the **Quotient Rule**.

PROOF To prove Property 5, we let $x = \log_b M$ and $y = \log_b N$ and use the definition of logarithm to write each equation in exponential form.

$$M = b^x \qquad \text{and} \qquad N = b^y$$

Then $MN = b^x b^y$ and a property of exponents gives

$$MN = b^{x+y} \qquad b^x b^y = b^{x+y}; \text{ keep the base and add the exponents.}$$

We write this exponential equation in logarithmic form as

$$\log_b MN = x + y$$

Substituting the values of x and y completes the proof.

$$\log_b MN = \log_b M + \log_b N$$

The proof of Property 6 is similar. You will be asked to do it in an Exercise 111. ∎

Caution

By Property 5 of logarithms, the logarithm of a *product* is equal to the *sum* of the logarithms. The logarithm of a product is **not** the product of the logarithms:

$$\log_b MN = (\log_b M)(\log_b N)$$

Caution

The logarithm of a sum or a difference usually does **not** simplify. In general,

$$\log_b(M + N) = \log_b M + \log_b N$$

and

$$\log_b(M - N) = \log_b M - \log_b N$$

Caution

By Property 6, the logarithm of a *quotient* is equal to the *difference* of the logarithms. The logarithm of a quotient is **not** the quotient of the logarithms:

$$\log_b \frac{M}{N} = \frac{\log_b M}{\log_b N}$$

Properties 5 and 6 also hold for natural logarithms.

5. $\ln MN = \ln M + \ln N$ Product Rule

6. $\ln \dfrac{M}{N} = \ln M - \ln N$ Quotient Rule

EXAMPLE 2 **Using the Product and Quotient Rules**

Assume that x, y, z, and b are positive numbers and $b \neq 1$. Write each expression in terms of the logarithms of x, y, and z: **a.** $\log_b xyz$ **b.** $\ln \dfrac{x}{yz}$

SOLUTION   We can use the properties of logarithms to write each logarithm as the sum or difference of several logarithms.

a. $\log_b xyz = \log_b(xy)z$

$$= \log_b(xy) + \log_b z \qquad \text{The log of a product is the sum of the logs.}$$

$$= \log_b x + \log_b y + \log_b z \qquad \text{The log of a product is the sum of the logs.}$$

b. $\ln \dfrac{x}{yz} = \ln x - \ln(yz)$ The ln of a quotient is the difference of the natural logs.

$$= \ln x - (\ln y + \ln z) \qquad \text{The ln of a product is the sum of the natural logs.}$$

$$= \ln x - \ln y - \ln z \qquad \text{Remove parentheses.}$$

Self Check 2   Write the expression in terms of the logarithms of x, y, and z: $\log_b \dfrac{xy}{z}$

Now Try Exercise 27.

ACCENT ON TECHNOLOGY   **Product and Quotient Rules**

A graphing calculator can be used to verify the Product and Quotient Rules.

- Using the Product Rule, we can write the expression

    $$\ln(75 \cdot 20) = \ln 75 + \ln 20$$

    Using a graphing calculator, we can verify that these two are equal as shown in Figure 4-29.

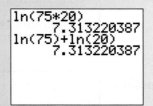

FIGURE 4-29

- Using the Quotient Rule, we can write the expression

    $$\log \frac{240}{13} = \log 240 - \log 13.$$

    Using a graphing calculator, we can verify that these two are equal as shown in Figure 4-30.

FIGURE 4-30

Two more properties state that

The logarithm of a power is the power times the logarithm.

If the logarithms of two numbers are equal, the numbers are equal.

More Properties of Logarithms   If M and b are positive numbers and $b \neq 1$, then

7. $\log_b M^p = p \log_b M$ **8.** If $\log_b x = \log_b y$, then $x = y$.

Property 7 is known as the **Power Rule**.
Property 8 is known as the **One-to-One Property**.

PROOF   To prove Property 7, we let $x = \log_b M$, write the expression in exponential form, and raise both sides to the pth power:

$$M = b^x$$

$$(M)^p = (b^x)^p \qquad \text{Raise both sides to the } p\text{th power.}$$

$$M^p = b^{px} \qquad \text{Keep the base and multiply the exponents.}$$

Using the definition of logarithm gives

$$\log_b M^p = px$$

Substituting the value for x completes the proof.

$$\log_b M^p = p \log_b M$$

Property 8 follows from the fact that the logarithmic function is a one-to-one function. Property 8 will be important in the next section when we solve logarithmic equations.

Properties 7 and 8 also hold for natural logarithms.

7. $\ln M^p = p \ln M$ Power Rule
8. If $\ln x = \ln y$, then $x = y$.

We can use the properties of logarithms to write a logarithm as the sum or difference of several logarithms.

EXAMPLE 3   **Using the Power Rule and Expanding a Logarithmic Expression**

Assume that x, y, z, and b are positive numbers and $b \neq 1$. Write each expression in terms of the logarithms of x, y, and z:

a. $\log_b(x^3 y^2 z)$ **b.** $\ln \dfrac{y^2 \sqrt{z}}{x}$

SOLUTION   We can use the properties of logarithms to write each logarithm as the sum or difference of several logarithms.

a. $\log_b(x^3 y^2 z) = \log_b x^3 + \log_b y^2 + \log_b z$ The log of a product is the sum of the logs.

$\qquad\qquad\qquad = 3 \log_b x + 2 \log_b y + \log_b z$ The log of a power is the power times the log.

b. $\ln \dfrac{y^2 \sqrt{z}}{x} = \ln\left(y^2 \sqrt{z}\right) - \ln x$ The ln of a quotient is the difference of the natural logs.

$\qquad\qquad = \ln y^2 + \ln z^{1/2} - \ln z$ The ln of a product is the sum of the natural logs: $\sqrt{z} = z^{1/2}$.

$\qquad\qquad = 2 \ln y + \dfrac{1}{2} \ln z - \ln z$ The ln of a power is the power times the ln.

Self Check 3   Write the expression in terms of the logarithms of x, y, and z: $\log_b \sqrt[3]{\dfrac{x^2 y}{z}}$.

Now Try Exercise 35.

Applying the Power Rule

The development of calculators has made computations extremely easy and accurate, but hand-held calculators have a limit to the size of numbers they can display and evaluate.

- For example, if we were to try to calculate $\log 12^{1000}$, see Figure 4-31(a), we would most likely get an overflow error message as shown in Figure 4-31(b). However, by applying the Power Rule, the number can be evaluated with our calculator as shown in Figure 4-31(c).

(a)   (b)   (c)

FIGURE 4-31

We can use the properties of logarithms to combine several logarithms into one logarithm.

EXAMPLE 4   **Combining Logarithmic Expressions**

Assume that x, y, z, and b are positive numbers and $b \neq 1$. Write each expression as one logarithm:

a. $2 \log_b x + \dfrac{1}{3} \log_b y$ **b.** $\dfrac{1}{2} \log_b(x - 2) - \log_b y + 3 \log_b z$

SOLUTION   We can use the properties of logarithms to combine several logarithms into one logarithm.

a. $2 \log_b x + \dfrac{1}{3} \log_b y = \log_b x^2 + \log_b y^{1/3}$ A power times a log is the log of the power.

$$= \log_b(x^2 y^{1/3})$$ The sum of two logs is the log of the product.

$$= \log_b\left(x^2 \sqrt[3]{y}\right)$$

b. $\dfrac{1}{2} \log_b(x - 2) - \log_b y + 3 \log_b z$

$$= \log_b(x - 2)^{1/2} - \log_b y + \log_b z^3$$ A power times a log is the log of the power.

$$= \log_b \frac{(x - 2)^{1/2}}{y} + \log_b z^3$$ The difference of two logs is the log of the quotient.

$$= \log_b \frac{z^3 \sqrt{x - 2}}{y}$$ The sum of two logs is the log of the product.

Self Check 4   Write as one logarithm: $2 \ln x + \dfrac{1}{2} \ln y - 3 \ln(x - y)$.

Now Try Exercise 45.

We summarize the eight properties of logarithms and natural logarithms as follows.

Summary of the Properties of Logarithms

If b, M, and N are positive numbers and $b \neq 1$, then

1. $\log_b 1 = 0$
2. $\log_b b = 1$
3. $\log_b b^x = x$
4. $b^{\log_b x} = x$
5. $\log_b MN = \log_b M + \log_b N$
6. $\log_b \dfrac{M}{N} = \log_b M - \log_b N$
7. $\log_b M^p = p \log_b M$
8. If $\log_b x = \log_b y$, then $x = y$.

1. $\ln 1 = 0$
2. $\ln e = 1$
3. $\ln e^x = x$
4. $e^{\ln x} = x$
5. $\ln MN = \ln M + \ln N$
6. $\ln \dfrac{M}{N} = \ln M - \ln N$
7. $\ln M^p = p \ln M$
8. If $\ln x = \ln y$, then $x = y$.

EXAMPLE 5   **Using Properties of Logarithms to Find Approximations**

Given that $\log_{10} 2 \approx 0.3010$ and $\log_{10} 3 \approx 0.4771$, find approximations for

a. $\log_{10} 18$ **b.** $\log_{10} 2.5$

SOLUTION   We will use the properties of logarithms to write each expression in terms of known logarithms. Then, we can substitute the values of the known logarithms and simplify.

a. $\log_{10} 18 = \log_{10}(2 \cdot 3^2)$

$\qquad = \log_{10} 2 + \log_{10} 3^2$ 　　　　The log of a product is the sum of the logs.

$\qquad = \log_{10} 2 + 2\log_{10} 3$ 　　　　The log of a power is the power times the log.

$\qquad \approx 0.3010 + 2(0.4771)$

$\qquad \approx 1.2552$

b. $\log_{10} 2.5 = \log_{10}\left(\dfrac{5}{2}\right)$

$\qquad = \log_{10} 5 - \log_{10} 2$ 　　　　The log of a quotient is the difference of the logs.

$\qquad = \log_{10} \dfrac{10}{2} - \log_{10} 2$ 　　　　Write 5 as $\dfrac{10}{2}$.

$\qquad = \log_{10} 10 - \log_{10} 2 - \log_{10} 2$ 　　　　The log of a quotient is the difference of the logs.

$\qquad = 1 - 2\log_{10} 2$ 　　　　$\log_{10} 10 = 1$

$\qquad \approx 1 - 2(0.3010)$

$\qquad \approx 0.3980$

Self Check 5   Use the information given in Example 5 to find an approximation for $\log_{10} 0.75$.

Now Try Exercise 77.

Comment   In Example 5, it is important to note that we can use logarithmic values that are not stated in the problem. For example, log 10 = 1, log 100 = 2, and log 1000 = 3.

2. Use the Change-of-Base Formula

We have seen how to use a calculator to find base-10 and base-e logarithms. To use a calculator to find logarithms with different bases, such as $\log_7 63$, we can divide the base-10 (or base-e) logarithm of 63 by the base-10 (or base-e) logarithm of 7.

$$\log_7 63 = \frac{\log 63}{\log 7} \qquad\qquad \log_7 63 = \frac{\ln 63}{\ln 7}$$

$$\approx 2.129150068 \qquad\qquad\qquad \approx 2.129150068$$

To check the result, we verify that $7^{2.129150063} \approx 63$. This example suggests that if we know the base-a logarithm of a number, we can find its logarithm to some other base b.

Change-of-Base Formula   If a, b, and x are positive numbers and $a \neq 1$ and $b \neq 1$, then

$$\log_b x = \frac{\log_a x}{\log_a b}$$

To prove this formula, we begin with the equation $\log_b x = y$.

$$y = \log_b x$$

$$x = b^y \qquad\qquad \text{Change the equation from logarithmic to exponential form.}$$

$$\log_a x = \log_a b^y \qquad\qquad \text{Take the base-}a\text{ logarithm of both sides.}$$

$$\log_a x = y \log_a b \qquad\qquad \text{The log of a power is the power times the log.}$$

$$y = \frac{\log_a x}{\log_a b} \qquad\qquad \text{Divide both sides by } \log_a b.$$

$$\log_b x = \frac{\log_a x}{\log_a b} \qquad\qquad \text{Refer to the first equation and substitute } \log_b x \text{ for } y.$$

Caution

$\dfrac{\log_a x}{\log_a b}$ means that one logarithm is to be divided by the other. They are **not** to be subtracted.

If we know logarithms to base a (for example, $a = 10$), we can find the logarithm of x to a new base b by dividing the base-a logarithm of x by the base-a logarithm of b.

EXAMPLE 6   **Using the Change-of-Base Formula**

Use the Change-of-Base Formula to find $\log_3 5$.

SOLUTION   We can substitute 3 for b, 10 for a, and 5 for x into the Change-of-Base Formula and simplify.

$$\log_b x = \frac{\log_a x}{\log_a b}$$

$$\log_3 5 = \frac{\log_{10} 5}{\log_{10} 3} \qquad\qquad \text{Divide the base-10 logarithm of 5 by the base-10 logarithm of 3.}$$

$$\approx 1.464973521$$

To four decimal places, $\log_3 5 = 1.4650$.

Self Check 6   Find $\log_5 3$ to four decimal places.

Now Try Exercise 89.

ACCENT ON TECHNOLOGY   **Using the Change-of-Base Formula to Graph a Logarithmic Function**

We can use the Change-of-Base Formula to graph logarithmic functions that do not have a base of 10. Consider the function

$$f(x) = \log_4(x + 1)$$

- Since the base of the function $f(x) = \log_4(x + 1)$ is 4, to graph this function on a calculator, we use the Change-of-Base Formula and rewrite the function as

$$f(x) = \log_4(x + 1) = \frac{\log(x + 1)}{\log 4}$$

We enter the function as shown in Figure 4-32(a). We can also use the natural logarithm function if we prefer.
- The domain of this function is $(-1, \infty)$, so there will be a vertical asymptote of $x = -1$. This will help us set our [WINDOW]. We then [GRAPH] the function. See Figures 4-32(b) and (c).
- Notice that the graph appears to "stop" at $x = -1$, but that is not the case. We can have the calculator evaluate the function at a value very close to $x = -1$ to verify this. Notice in Figure 4-32(d) that the point on the graph is not shown, but the value is indicated.

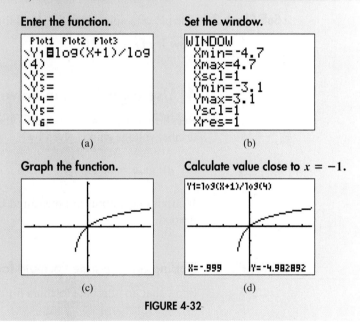

Enter the function.   Set the window.

(a)   (b)

Graph the function.   Calculate value close to $x = -1$.

(c)   (d)

FIGURE 4-32

We can use properties of logarithms to solve many problems.

3. Use Logarithms to Solve pH Problems

The more acidic a chemical solution, the greater the concentration of hydrogen ions. Chemists measure this concentration indirectly by the **pH scale,** or the **hydrogen ion index.**

pH of a Solution   If $[H^+]$ is the hydrogen ion concentration in gram-ions per liter, then

$$pH = -\log[H^+]$$

Since pure water has approximately 10^{-7} gram-ions per liter, its pH is

$$pH = -\log[H^+]$$
$$pH = -\log \mathbf{10^{-7}}$$
$$= -(-7)\log 10 \qquad \text{The log of a power is the power times the log.}$$
$$= -(-7) \cdot 1 \qquad \text{Use Property 2 of logarithms: } \log_b b = 1.$$
$$= 7$$

EXAMPLE 7   Solving an Application Involving pH

Seawater has a pH of approximately 8.5. Find its hydrogen ion concentration.

SOLUTION   We can substitute 8.5 for pH and solve the equation $pH = -\log[H^+]$ for $[H^+]$.

$$8.5 = -\log[H^+]$$
$$-8.5 = \log[H^+]$$
$$[H^+] = 10^{-8.5} \qquad \text{Change the equation from logarithmic form to exponential form.}$$

We can then use a calculator to find that $[H^+] \approx 3.2 \times 10^{-9}$ gram-ions per liter.

Self Check 7   The pH of a solution is 5.7. Find the hydrogen ion concentration.

Now Try Exercise 101.

4. Use Logarithms to Solve Problems in Electronics

Recall that if E_O is the output voltage of a device and E_I is the input voltage, the decibel voltage gain is given by

$$(1) \quad \textbf{dB gain} = \textbf{20 log } \frac{E_O}{E_I}$$

If input and output are measured in watts instead of volts, a different formula is needed.

EXAMPLE 8   Finding an Alternate Formula for Voltage Gain

Show that an alternate formula for dB voltage gain is

$$dB \text{ gain} = 10 \log \frac{P_O}{P_I}$$

where P_I is the power input and P_O is the power output.

SOLUTION   Power is directly proportional to the square of the voltage. So for some constant k,

$$P_I = k(E_I)^2 \qquad \text{and} \qquad P_O = k(E_O)^2$$

and

$$\frac{P_O}{P_I} = \frac{k(E_O)^2}{k(E_I)^2} = \left(\frac{E_O}{E_I}\right)^2$$

We raise both sides to the $\frac{1}{2}$ power to get

$$\frac{E_O}{E_I} = \left(\frac{P_O}{P_I}\right)^{1/2}$$

which we substitute into Equation (1) for dB gain.

$$\text{dB gain} = 20 \log \frac{E_O}{E_I}$$

$$= 20 \log \left(\frac{P_O}{P_I} \right)^{1/2}$$

$$= 20 \cdot \frac{1}{2} \log \frac{P_O}{P_I} \qquad \text{The log of a power is the power times the log.}$$

$$\text{dB gain} = 10 \log \frac{P_O}{P_I} \qquad \text{Simplify.}$$

Self Check 8   Find the dB gain of a device to the nearest hundredth when $P_O = 30$ watts and $P_I = 2$ watts.

Now Try Exercise 103.

5. Use Logarithms to Solve Physiology Problems

In physiology, experiments suggest that the relationship between the loudness and the intensity of sound is a logarithmic one known as the Weber–Fechner Law.

Weber–Fechner Law   If L is the apparent loudness of a sound and I is the intensity, then

$$L = k \ln I$$

EXAMPLE 9   **Using the Weber-Fechner Law**

What increase in the intensity of a sound is necessary to cause a doubling of the apparent loudness?

SOLUTION   We use the formula $L = k \ln I$. To double the apparent loudness, we multiply both sides of the equation by 2 and use Property 7 of logarithms.

$$L = k \ln I$$

$$2L = 2k \ln I$$

$$= k \ln I^2$$

To double the apparent loudness, we must square the intensity. ∎

Self Check 9   What increase is necessary to triple the apparent loudness?

Now Try Exercise 105.

Self Check Answers   1. a. 0   b. 1   c. 4   d. 2   2. $\log_b x + \log_b y - \log_b z$

3. $\frac{1}{3} (2 \log_b x + \log_b y - \log_b z)$ 4. $\ln \frac{x^2 \sqrt{y}}{(x - y)^3}$

5. -0.1249 6. 0.6826 7. 2×10^{-6} 8. 11.76 decibels

9. Cube the intensity.

Exercises **4.5**

Getting Ready

You should be able to complete these vocabulary and concept statements before you proceed to the practice exercises.

Fill in the blanks.

1. $\log_b 1 = $ ___
2. $\log_b b = $ ___
3. $\log_b MN = \log_b$ ___ $+ \log_b$ ___
4. $b^{\log_b x} = $ ___
5. If $\log_b x = \log_b y$, then ___ $= $ ___.
6. $\log_b \dfrac{M}{N} = \log_b M$ ___ $\log_b N$
7. $\log_b x^p = p \cdot \log_b$ ___
8. $\log_b b^x = $ ___
9. $\log_b(A + B)$ ___ $\log_b A + \log_b B$
10. $\log_b A + \log_b B$ ___ $\log_b AB$

Simplify each expression.

11. $\log_4 1 = $ ___
12. $\log_4 4 = $ ___
13. $\log_4 4^7 = $ ___
14. $4^{\log_4 8} = $ ___
15. $5^{\log_5 10} = $ ___
16. $\log_5 5^2 = $ ___
17. $\log_5 5 = $ ___
18. $\log_5 1 = $ ___

Practice

Use a calculator to verify each equation.

19. $\log[(3.7)(2.9)] = \log 3.7 + \log 2.9$
20. $\ln \dfrac{9.3}{2.1} = \ln 9.3 - \ln 2.1$
21. $\ln(3.7)^3 = 3 \ln 3.7$
22. $\log\sqrt{14.1} = \dfrac{1}{2} \log 14.1$
23. $\log 3.2 = \dfrac{\ln 3.2}{\ln 10}$
24. $\ln 9.7 = \dfrac{\log 9.7}{\log e}$

Assume that x, y, z, and b are positive numbers and b ≠ 1. Use the properties of logarithms to write each expression in terms of the logarithms of x, y, and z.

25. $\log_b 2xy$
26. $\log_b 3xz$
27. $\log_b \dfrac{2x}{y}$
28. $\log_b \dfrac{x}{yz}$
29. $\log_b x^2 y^3$
30. $\log_b x^3 y^2 z$
31. $\log_b(xy)^{1/3}$

32. $\log_b x^{1/2} y^3$
33. $\log_b x\sqrt{z}$
34. $\log_b \sqrt{xy}$
35. $\log_b \dfrac{\sqrt[3]{x}}{\sqrt[3]{yz}}$
36. $\log_b \sqrt[4]{\dfrac{x^3 y^2}{z^4}}$
37. $\ln x^7 y^8$
38. $\ln \dfrac{4x}{y}$
39. $\ln \dfrac{x}{y^4 z}$
40. $\ln x\sqrt{y}$

Assume that x, y, z, and b are positive numbers and b ≠ 1. Use the properties of logarithms to write each expression as the logarithm of one quantity.

41. $\log_b(x + 1) - \log_b x$
42. $\log_b x + \log_b(x + 2) - \log_b 8$
43. $2 \log_b x + \dfrac{1}{3} \log_b y$
44. $-2 \log_b x - 3 \log_b y + \log_b z$
45. $-3 \log_b x - 2 \log_b y + \dfrac{1}{2} \log_b z$
46. $3 \log_b(x + 1) - 2 \log_b(x + 2) + \log_b x$
47. $\log_b\left(\dfrac{x}{z} + x\right) - \log_b\left(\dfrac{y}{z} + y\right)$
48. $\log_b(xy + y^2) - \log_b(xz + yz) + \log_b z$
49. $\ln x + \ln(x + 5) - \ln 9$
50. $5 \ln x + \dfrac{1}{5} \ln y$
51. $-6 \ln x - 2 \ln y + \ln z$
52. $-2 \ln x - 3 \ln y + \dfrac{1}{3} \ln z$

Determine whether each statement is true or false.

53. $\log_b ab = \log_b a + 1$
54. $\log_b \dfrac{1}{a} = -\log_b a$
55. $\log_b 0 = 1$
56. $\log_b 2 = \log_2 b$
57. $\log_b(x + y) \neq \log_b x + \log_b y$

58. $\log_b xy = (\log_b x)(\log_b y)$

59. If $\log_a b = c$, then $\log_b a = c$.

60. If $\log_a b = c$, then $\log_b a = \dfrac{1}{c}$.

61. $\log_7 7^7 = 7$

62. $7^{\log_7 7} = 7$

63. $\log_b(-x) = -\log_b x$

64. If $\log_b a = c$, then $\log_b a^p = pc$.

65. $\dfrac{\log_b A}{\log_b B} = \log_b A - \log_b B$

66. $\log_b(A - B) = \dfrac{\log_b A}{\log_b B}$

67. $\log_b \dfrac{1}{5} = -\log_b 5$

68. $3 \log_b \sqrt[3]{a} = \log_b a$

69. $\dfrac{1}{3} \log_b a^3 = \log_b a$

70. $\log_{4/3} y = -\log_{3/4} y$

71. $\log_b y + \log_{1/b} y = 0$

72. $\log_{10} 10^3 = 3(10^{\log_{10} 3})$

73. $\ln xy = (\ln x)(\ln y)$

74. $\dfrac{\ln A}{\ln B} = \ln A - \ln B$

75. $\dfrac{1}{5} \ln a^5 = \ln a$

76. $\ln y = -\ln \dfrac{1}{y}$

Given that $\log_{10} 4 \approx 0.6021$, $\log_{10} 7 \approx 0.8451$, and $\log_{10} 9 \approx 0.9542$, use these values and the properties of logarithms to approximate each value. Do not use a calculator.

77. $\log_{10} 28$

78. $\log_{10} \dfrac{7}{4}$

79. $\log_{10} 2.25$

80. $\log_{10} 36$

81. $\log_{10} \dfrac{63}{4}$

82. $\log_{10} \dfrac{4}{63}$

83. $\log_{10} 252$

84. $\log_{10} 49$

85. $\log_{10} 112$

86. $\log_{10} 324$

87. $\log_{10} \dfrac{144}{49}$

88. $\log_{10} \dfrac{324}{63}$

Use a calculator and the Change-of-Base Formula to find each logarithm. Round to four decimal places.

89. $\log_3 7$

90. $\log_7 3$

91. $\log_\pi 3$

92. $\log_3 \pi$

93. $\log_3 8$

94. $\log_5 10$

95. $\log_{\sqrt{2}} \sqrt{5}$

96. $\log_\pi e$

Applications

97. **pH of water slide** The water in the Abyss, a water slide at the Atlantis Resort in the Bahamas, has a hydrogen ion concentration of 6.3×10^{-8} gram-ions per liter. Find the pH.

98. **pH of swimming pool** The ideal pH for a swimming pool is 7.2, the same pH as our eyes. The swimming pool at the local YMCA has a hydrogen ion concentration of 1.6×10^{-7} gram-ions per liter. Find the pH of the pool. Is this ideal?

99. **pH of a solution** Find the pH of a solution with a hydrogen ion concentration of 1.7×10^{-5} gram-ions per liter.

100. **pH of calcium hydroxide** Find the hydrogen ion concentration of a saturated solution of calcium hydroxide whose pH is 13.2.

101. **pH of apples** The pH of apples can range from 2.9 to 3.3. Find the range in the hydrogen ion concentration.

Mazzur/Shutterstock.com

102. **pH of sour pickles** The hydrogen ion concentration of sour pickles is 6.31×10^{-4}. Find the pH.

© Istockphoto.com/James McQuillan

103. **dB gain** An amplifier produces a 40-watt output with a $\frac{1}{2}$-watt input. Find the dB gain.

104. **dB loss** Losses in a long telephone line reduce a 12-watt input signal to an output of 3 watts. Find the dB gain. (Because it is a loss, the "gain" will be negative.)

105. **Weber–Fechner Law** What increase in intensity is necessary to quadruple the loudness?

106. **Weber–Fechner Law** What decrease in intensity is necessary to make a sound half as loud?

107. Isothermal expansion If a certain amount E of energy is added to one mole of a gas, it expands from an initial volume of 1 liter to a final volume V without changing its temperature according to the formula

$$E = 8,300 \ln V$$

Find the volume if twice that energy is added to the gas.

108. Richter scale By what factor must the amplitude of an earthquake change to increase its severity by 1 point on the Richter scale? Assume that the period remains constant. The Richter scale is given by

$$R = \log \frac{A}{P}$$

where A is the amplitude and P the period of the tremor.

Discovery and Writing

109. Simplify: $3^{4\log_3 2} + 5^{\frac{1}{2}\log_5 25}$

110. Find the value of $a - b$:

$$5 \log x + \frac{1}{3} \log y - \frac{1}{2} \log x - \frac{5}{6} \log y = \log(x^a y^b)$$

111. Prove Property 6 of logarithms:

$$\log_b \frac{M}{N} = \log_b M - \log_b N$$

112. Show that $-\log_b x = \log_{1/b} x$.

113. Show that $e^{x \ln a} = a^x$.

114. Show that $e^{\ln x} = x$.

115. Show that $\ln(e^x) = x$.

116. If $\log_b 3x = 1 + \log_b x$, find b.

117. Explain why $\ln(\log 0.9)$ is undefined.

118. Explain why $\log_b(\ln 1)$ is undefined.

In Exercises 119–120, A and B both are negative. Thus, AB and $\frac{A}{B}$ are positive, and log AB and log $\frac{A}{B}$ are defined.

119. Is it still true that $\log AB = \log A + \log B$? Explain.

120. Is it still true that $\log \frac{A}{B} = \log A - \log B$? Explain.

Review
Determine whether each equation defines a function.

121. $y = 3x - 1$

122. $y = \dfrac{x + 3}{x - 1}$

123. $y^2 = 4x$

124. $y = 4x^2$

Find the domain of each function.

125. $f(x) = x^2 - 4$

126. $f(x) = \dfrac{1}{x^2 - 4}$

127. $f(x) = \sqrt{x^2 + 4}$

128. $f(x) = \sqrt{x^2 - 4}$

4.6 Exponential and Logarithmic Equations

In this section, we will learn to

1. Use like bases to solve exponential equations.
2. Use logarithms to solve exponential equations.
3. Solve logarithmic equations.
4. Solve carbon-14 dating problems.
5. Solve population growth problems.

Coffee is one of the most popular beverages worldwide. Suppose we visit a coffee shop and order a white chocolate mocha. After blending smooth white chocolate with rich espresso and steamed milk and topping it with whipped cream, the coffee is served to us at a temperature of 180°F.

Suppose the exponential function $T = 70 + 110e^{-0.2t}$ models the temperature T of the mocha after t minutes. If we are interested in determining how long it will take for the temperature of the coffee to reach 80°F, we would substitute 80 in for

T and solve the resulting equation $80 = 70 + 110e^{-0.2t}$ for *t*. This equation is called an *exponential equation* because the variable *t* occurs as an exponent.

In this section, we will learn to solve exponential equations. When we complete Exercise 105, we will see that it takes the mocha approximately 12 minutes to reach a temperature of 80°F. We will also learn to solve *logarithmic equations* in this section.

An **exponential equation** is an equation with a variable in one of its exponents. Some examples of exponential equations are

$$3^x = 5 \qquad e^{2x} = 7 \qquad 6^{x-3} = 2^x \qquad 3^{2x+1} - 10(3^x) + 3 = 0$$

A **logarithmic equation** is an equation with logarithmic expressions that contain a variable. Some examples of logarithmic equations are

$$\log 2x = 25 \qquad \ln x - \ln(x - 12) = 24 \qquad \log x = \log \frac{1}{x} + 4$$

1. Use Like Bases to Solve Exponential Equations

Some exponential equations can be solved by using like bases.

One-to-One Property of Exponents	If $b^x = b^y$ and $b \neq 1$, $b \neq 0$, then $x = y$. That is, equal quantities with like bases have equal exponents.

EXAMPLE 1   Using Like Bases to Solve an Exponential Equation

Solve: $4^{x+3} = 8^{2x}$.

SOLUTION   We will use like bases to solve the exponential equation.

$$4^{x+3} = 8^{2x}$$

$$(2^2)^{x+3} = (2^3)^{2x} \qquad \text{Write 4 as } 2^2 \text{ and 8 as } 2^3.$$

$$2^{2(x+3)} = 2^{6x} \qquad \text{Multiply exponents.}$$

$$2(x + 3) = 6x \qquad \text{Equal quantities with like bases have equal exponents.}$$

$$2x + 6 = 6x \qquad \text{Use the Distributive Property.}$$

$$-4x = -6 \qquad \text{Subtract 6 and } 6x \text{ from both sides.}$$

$$x = \frac{3}{2} \qquad \text{Divide both sides by } -4 \text{ and simplify.}$$

∎

Self Check 1   Solve: $3^{3x-5} = 81$.

Now Try Exercise 13.

EXAMPLE 2   Using Like Bases to Solve an Exponential Equation

Solve: $2^{x^2+2x} = \frac{1}{2}$.

SOLUTION   Since $\frac{1}{2} = 2^{-1}$, we can write the equation in the form $2^{x^2+2x} = 2^{-1}$ and use like bases to solve the equation.

Because equal quantities with like bases have equal exponents, we have

$$x^2 + 2x = -1$$

$$x^2 + 2x + 1 = 0 \qquad \text{Add 1 to both sides.}$$

$$(x + 1)(x + 1) = 0 \qquad \text{Factor the trinomial.}$$

$$x + 1 = 0 \quad \text{or} \quad x + 1 = 0 \qquad \text{Set each factor equal to 0.}$$

$$x = -1 \quad \bigg| \quad \qquad x = -1$$

Verify that -1 satisfies the equation. ∎

Self Check 2 Solve: $3^{x^2+2x} = 27$.

Now Try Exercise 19.

ACCENT ON TECHNOLOGY **Verifying Solutions of an Exponential Equation**

We can verify that -1 satisfies the equation shown in Example 2 by using the TABLE feature on a graphing calculator.

- Enter each side of the equation as shown in Figure 4-33.
- Go to TABLE and enter $x = -1$. Notice in Figure 4-33 that the y-values are exactly the same. This confirms that $x = -1$ is a solution.

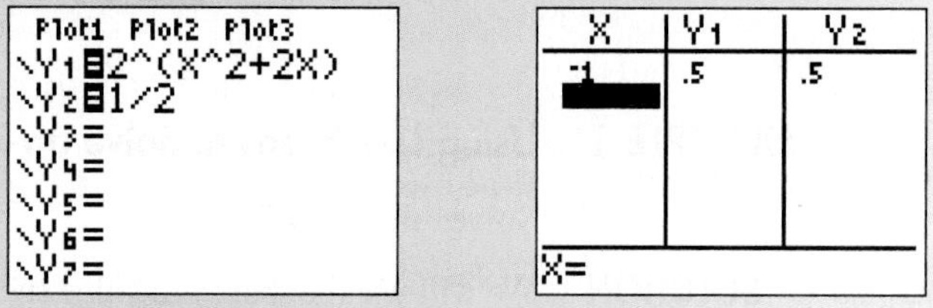

FIGURE 4-33

EXAMPLE 3 **Using Like Bases to Solve an Exponential Equation**

Solve: $e^{6x^2} = e^{-x+1}$.

SOLUTION We can use like bases to solve the exponential equation.
Because equal quantities with like bases have equal exponents, we have

$$6x^2 = -x + 1$$

$$6x^2 + x - 1 = 0 \qquad \text{Add } x - 1 \text{ to both sides.}$$

$$(3x - 1)(2x + 1) = 0 \qquad \text{Factor the trinomial.}$$

$$3x - 1 = 0 \quad \text{or} \quad 2x + 1 = 0 \qquad \text{Set each factor equal to 0.}$$

$$x = \frac{1}{3} \quad \bigg| \quad x = -\frac{1}{2} \qquad \text{Solve each equation.}$$

Verify that $\frac{1}{3}$ and $-\frac{1}{2}$ satisfy the equation. ∎

Self Check 3 Solve: $e^{4x^2} = e^{24}$.

Now Try Exercise 23.

2. Use Logarithms to Solve Exponential Equations

If the bases of the terms in an exponential equation are not equal, as is often the case, we use logarithms to solve the equation.

Since logarithmic functions are one-to-one, we can take the logarithm of both sides of an equation. This allows us to use the properties of logarithms to help solve the equation. A logarithm of any base may be used, but it is most efficient to use natural logarithms or common logarithms because a calculator is often needed to get an approximate answer. A strategy for solving exponential equations with different bases is given next.

Strategy for Solving Exponential Equations with Different Bases

To solve an exponential equation we can follow these steps:

Step 1: Isolate the exponential expression.

Step 2: Take the same logarithm of both sides.

Step 3: Simplify using the rules of logarithms.

Step 4: Solve for the variable.

When using the strategy given, please keep the following things in mind:

- If the equation contains base e, it is most efficient to use natural logarithms because

$$\ln e^x = x$$

- If the equation contains base 10, it is most efficient to use common logarithms because

$$\log 10^x = x$$

- If the equation contains exponential terms with bases other than base e or 10, use either natural logarithms or common logarithms.

EXAMPLE 4   **Using Logarithms to Solve an Exponential Equation with Different Bases**

Solve the exponential equation: $3^x = 5$.

SOLUTION   We will use the strategy outlined above to solve the exponential equation. First, we note that the exponential expression 3^x is isolated. Since logarithms of equal numbers are equal, we can take the common logarithm of each side of the equation. We can then use the Power Rule and move the variable x from its position as an exponent to a position as a coefficient and solve the equation.

$$3^x = 5$$

$$\log 3^x = \log 5 \qquad \text{Take the common logarithm of each side.}$$

$$x \log 3 = \log 5 \qquad \text{The log of a power is the power times the log.}$$

$$(1) \qquad x = \frac{\log 5}{\log 3} \qquad \text{Divide both sides by log 3.}$$

$$\approx 1.464973521 \qquad \text{Use a calculator.}$$

To four decimal places, $x = 1.4650$.

Self Check 4   Solve $5^x = 3$ to four decimal places.

Now Try Exercise 27.

Caution

> A careless reading of Equation 1 leads to a common error. The right side of the equation calls for a division, **not** a subtraction.
>
> $$\frac{\log 5}{\log 3} \quad \text{means} \quad (\log 5) \div (\log 3)$$
>
> It is the expression $\log \frac{5}{3}$ that means $\log 5 - \log 3$.

EXAMPLE 5 **Using Logarithms to Solve an Exponential Equation with Different Bases**

Solve the exponential equation: $6^{x-3} = 2^x$.

SOLUTION We will use logarithms to solve the exponential equation and apply the strategy stated earlier.

$$6^{x-3} = 2^x$$

$\log 6^{x-3} = \log 2^x$ Take the common logarithm of each side.

$(x-3)\log 6 = x \log 2$ The log of a power is the power times the log.

$x \log 6 - 3 \log 6 = x \log 2$ Use the Distributive Property.

$x \log 6 - x \log 2 = 3 \log 6$ Add 3 log 6 and subtract x log 2 from both sides.

$x(\log 6 - \log 2) = 3 \log 6$ Factor out x on the left-hand side.

$$x = \frac{3 \log 6}{\log 6 - \log 2}$$ Divide both sides by log 6 − log 2.

$x \approx 4.892789261$ Use a calculator.

To four decimal places, $x = 4.8928$. ∎

Self Check 5 Solve: $5^{x+3} = 3^x$.

Now Try Exercise 29.

Comment

We can use common logarithms or natural logarithms to solve many exponential equations. In Examples 4 and 5, we took the common logarithm of both sides of an equation to solve the equation. We could just as well have taken the natural logarithm of both sides and obtained the same decimal result.

In the next example, we will take the natural logarithm of both sides of an exponential equation. However, before we do, it is a good idea to review the following properties of natural logarithms.

Natural Logarithm Properties		
$\ln 1 = 0$	$\ln e = 1$	$\ln e^x = x$
Product Rule: $\ln MN = \ln M + \ln N$		
Quotient Rule: $\ln \dfrac{M}{N} = \ln M - \ln N$		
Power Rule: $\ln M^p = p \ln M$		

EXAMPLE 6 **Using Natural Logarithms to Solve Exponential Equations**

Use natural logarithms to solve **a.** $e^x - 7 = 0$ **b.** $4^{x+3} = 8^{2x}$

SOLUTION   For each problem, we will use the strategy stated earlier. We will isolate the exponential expression, take the natural logarithm of both sides, simplify using the natural logarithms properties, and solve for x.

a. $e^x - 7 = 0$

$$e^x - 7 + 7 = 0 + 7 \qquad \text{Isolate the exponential expression. Add 7 to both sides.}$$

$$e^x = 7$$

$$\ln e^x = \ln 7 \qquad \text{Take the natural logarithm of both sides.}$$

$$x = \ln 7 \qquad \text{Substitute } x \text{ for } \ln e^x: \ln e^x = x \, .$$

$$x \approx 1.95 \qquad \text{Use a calculator and round to two decimal places.}$$

b. $4^{x+3} = 8^{2x}$

$$\ln 4^{x+3} = \ln 8^{2x} \qquad \text{Take the natural logarithm of both sides.}$$

$$(x + 3)\ln 4 = (2x)\ln 8 \qquad \text{The log of a power is the power times the log.}$$

$$x \ln 4 + 3 \ln 4 = 2x \ln 8 \qquad \text{Use the Distributive Property on the left side.}$$

$$x \ln 4 - 2x \ln 8 = -3 \ln 4 \qquad \text{Subtract } 2x \ln 8 \text{ and } 3 \ln 4 \text{ from both sides.}$$

$$x(\ln 4 - 2 \ln 8) = -3 \ln 4 \qquad \text{Factor out } x \text{ on the left-hand side.}$$

$$x = \frac{-3 \ln 4}{\ln 4 - 2 \ln 8} \qquad \text{Divide both sides by } \ln 4 - 2 \ln 8.$$

$$x = 1.5 \qquad \text{Use a calculator.}$$

> **Comment**
>
> As we see in Example 6a, if an exponential equation has a term with a base of e, then taking the natural logarithm of both sides is a good choice.

> **Comment**
>
> Note that the exponential equations in Examples 1 and 6b are identical and the answers are the same. As we see, sometimes we can use like bases or logarithms to solve the same equation.

Self Check 6   Use natural logarithms to solve: **a.** $3e^x = 12$ **b.** $8^{x+1} = 4^{2x}$

Now Try Exercise 39.

ACCENT ON TECHNOLOGY   **Finding Approximate Solutions to an Exponential Equation**

If only a decimal approximation of a solution is needed, a graphing calculator can be used to solve exponential equations. Our strategy is to graph each side of the equation and then find the point of intersection.

We will use a graphing calculator to solve the exponential equation $\sqrt{3}^{\,-x} = 2^{\pi x - 3}$.

• Enter each side of the equation in the graph menu and find the point of intersection. We see in the figure below that the solution is $x \approx 1.2770786$.

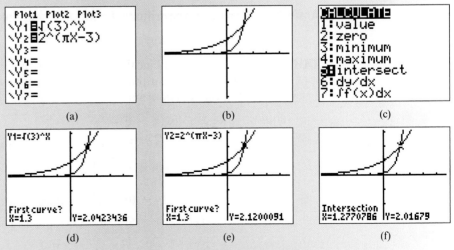

FIGURE 4-34

3. Solve Logarithmic Equations

We can use the One-to-One Property of logarithms to change some logarithmic equations into algebraic equations that we can solve. Here's a reminder of that property.

One-to-One Property of Logarithms
If $\log_b x = \log_b y$, then $x = y$. That is, logarithms of equal numbers are equal.

EXAMPLE 7 **Using the One-to-One Property of Logarithms to Solve a Logarithmic Equation**

Solve: $\log_5(3x + 2) = \log_5(2x - 3)$.

SOLUTION We will use the property of logarithms stated above to solve the equation. Then we will check our solutions.

$$\log_5(3x + 2) = \log_5(2x - 3)$$

$$3x + 2 = 2x - 3 \qquad \text{If the logs of two numbers are equal, the numbers are equal.}$$

$$x = -5 \qquad \text{Subtract } 2x \text{ and 2 from both sides.}$$

Check: $\qquad \log_5(3x + 2) = \log_5(2x - 3)$

$$\log_5[3(-5) + 2] \stackrel{?}{=} \log_5[2(-5) - 3]$$

$$\log_5(-13) \stackrel{?}{=} \log_5(-13)$$

Since the logarithm of a negative number does not exist, -5 is extraneous and must be discarded. The equation has no roots.

Self Check 7 Solve: $\log_3(5x + 2) = \log_3(6x + 1)$.

Now Try Exercise 57.

Comment It is incorrect to state that the logarithms in Example 7 are canceled. The logarithms aren't canceled; the One-to-One Property is used.

Caution Example 7 illustrates that we **must** check the solutions of a logarithmic equation. If we fail to do so, we risk having incorrect answers.

EXAMPLE 8 **Solving a Logarithmic Equation Using the Properties of Logarithms**

Solve: $\log x + \log(x - 3) = 1$.

SOLUTION We will first combine the two logarithms as a single logarithm. Next, we will use the definition of logarithm to write the equation in exponential form. Finally, we will solve the equation and check the solutions.

$$\log x + \log(x - 3) = 1$$

$$\log x(x - 3) = 1 \qquad \text{The sum of two logs is the log of a product.}$$

$$x(x - 3) = 10^1 \qquad \text{Use the definition of logarithms to change the equation to exponential form.}$$

$$x^2 - 3x - 10 = 0 \qquad \text{Remove parentheses and subtract 10 from both sides.}$$

$$(x + 2)(x - 5) = 0 \qquad \text{Factor the trinomial.}$$

$$x + 2 = 0 \quad \text{or} \quad x - 5 = 0$$

$$x = -2 \quad | \quad x = 5$$

Check: The number -2 is not a solution, because it does not satisfy the equation (a negative number does not have a logarithm). We check the remaining number, 5.

$$\log x + \log(x - 3) = 1$$

$$\log 5 + \log(5 - 3) \overset{?}{=} 1 \qquad \text{Substitute 5 for } x.$$

$$\log 5 + \log 2 \overset{?}{=} 1$$

$$\log 10 \overset{?}{=} 1 \qquad \text{The sum of two logs is the log of a product.}$$

$$1 = 1 \qquad \log_b b = 1$$

Since 5 does check, it is a root.

Self Check 8   Solve: $\log x + \log(x - 15) = 2$.

Now Try Exercise 61.

EXAMPLE 9   **Solving Logarithmic Equations**

Solve: $\ln(2x + 5) - \ln(x + 5) = \ln \dfrac{5}{3}$.

SOLUTION   First, we combine the two logarithms on the left. Then we will use the One-to-One Property and solve for x.

$$\ln(2x + 5) - \ln(x + 5) = \ln \frac{5}{3}$$

$$\ln \frac{2x + 5}{x + 5} = \ln \frac{5}{3} \qquad \text{Use the Quotient Rule and combine logs.}$$

$$\frac{2x + 5}{x + 5} = \frac{5}{3} \qquad \text{Use the One-to-One Property of logs.}$$

$$3(2x + 5) = 5(x + 5) \qquad \text{Multiply both sides by } 3(x + 5).$$

$$6x + 15 = 5x + 25$$

$$x = 10$$

Check: The solution $x = 10$ checks as we see here.

$$\ln(2x + 5) - \ln(x + 5) \overset{?}{=} \ln \frac{5}{3}$$

$$\ln[2(10) + 5] - \ln[(10) + 5] \overset{?}{=} \ln \frac{5}{3} \qquad \text{Substitute 10 in for } x.$$

$$\ln 25 - \ln 15 \overset{?}{=} \ln \frac{5}{3}$$

$$\ln \frac{25}{15} \overset{?}{=} \ln \frac{5}{3} \qquad \text{Use the Quotient Rule.}$$

$$\ln \frac{5}{3} = \ln \frac{5}{3} \qquad \text{This is a true statement.}$$

Self Check 9   Solve: $\log_7(2x + 2) - \log_7(x + 6) = \log_7 \dfrac{4}{3}$.

Now Try Exercise 69.

EXAMPLE 10 **Solving Logarithmic Equations**

Solve **a.** $\ln x = 5$ **b.** $\dfrac{\ln(5x - 6)}{\ln x} = 2$

SOLUTION For part (a), we will write the natural logarithm as $\log_e x = 5$ and use the definition of logarithm to solve the equation. For part (b), we will multiply both sides by $\ln x$ and then use properties of logarithms to write an algebraic equation. We will then solve the equation. We must verify the solutions to both parts.

a. $\ln x = 5$

$\log_e x = 5$ Rewrite using the definition of natural logarithm.

$x = e^5$ Write in exponential form.

Verify that e^5 satisfies the solution.

b. We multiply both sides of $\dfrac{\ln(5x - 6)}{\ln x} = 2$ by $\ln x$ to get

$\ln(5x - 6) = 2 \ln x$

and apply the Power Rule of logarithms to get

$\ln(5x - 6) = \ln x^2$

By the property of logarithms stated above $5x - 6 = x^2$, because they have equal logarithms. So

$$x^2 = 5x - 6$$

$$x^2 - 5x + 6 = 0$$

$$(x - 3)(x - 2) = 0$$

$$x - 3 = 0 \quad \text{or} \quad x - 2 = 0$$

$$x = 3 \quad | \quad x = 2$$

Verify that both 2 and 3 satisfy the equation.

Self Check 10 Solve: **a.** $6 \ln x = 12$ **b.** $\dfrac{\ln(8x - 15)}{\ln x} = 2$

Now Try Exercise 79.

The strategy we used to solve a logarithmic equation was dependent upon the nature of the equation. A summary of these strategies appears here.

Summary of Strategies Used to Solve Logarithmic Equations

1. For logarithmic equations containing one logarithm, we can use the definition of the logarithm. We write the logarithm in exponential form and then solve for the variable.

   Note: This strategy can be used to solve the following logarithmic equations.

   $$\log_3(x + 5) = 2, \log(x^2 - 4) = 5, \ln(4x) = 7$$

2. For more complicated logarithmic equations, we can combine and isolate the logarithmic expressions. We can then use the definition of the logarithm and write the logarithm in exponential form to solve for the variable.

Note: This strategy can be used to solve the following logarithmic equations.

$$\log(x) + \log(x - 1) = 3, \log_5(3x + 2) - \log_5(2x - 3) = 2,$$
$$\ln x - \ln(x + 3) = 4$$

3. We can use the One-to-One Property of logarithms for logarithmic equations of the form $\log_b x = \log_b y$.

   Note: This strategy can be used to solve the following logarithmic equations.

   $$\log(5x - 6) = \log x^2, \ln(2x - 7) = \ln(3x + 5)$$

4. For other types of logarithmic equations, we sometimes combine logarithms on one or both sides and then use the One-to-One Property of logarithms to solve the equations.

   Note: This strategy can be used to solve the following logarithmic equations.

   $$\ln(x^2) + \ln(x - 3) = \ln(x + 4), \log_4 x - \log_4(x + 2) = \log_4 5 + \log_4(3x)$$

ACCENT ON TECHNOLOGY　**Finding Approximate Solutions to a Logarithmic Equation**

We can use a graphing calculator to approximate the solutions of a logarithmic equation.

Consider the log equation $\ln x^2 + \ln(x - 3) = \ln(x + 4)$.

- Enter each side of the equation in the graph menu, and then find the point of intersection. We see that in Figure 4-35 that $x \approx 3.5891327$. Use ▸ZOOM and select ZDecimal to graph.

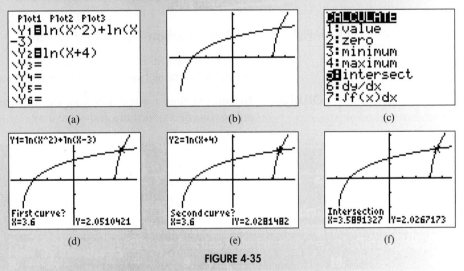

FIGURE 4-35

4. Solve Carbon-14 Dating Problems

When a living organism dies, the oxygen/carbon dioxide cycle common to all living things ceases; then carbon-14, a radioactive isotope with a half-life of 5,700 years, is no longer absorbed. By measuring the amount of carbon-14 present in ancient objects, archaeologists can estimate the object's age.

The Shroud of Turin, the piece of linen long believed to have been wrapped around Jesus's body after the crucifixion, is much older than the date suggested by radiocarbon tests, according to new microchemical research. Published in the current issue of *Thermochimica Acta*, a chemistry peer-reviewed scientific journal, the study dismisses the results of the 1988 carbon-14 dating. At that time, three reputable laboratories in Oxford, Zurich, and Tucson, Ariz., concluded that the cloth on which the smudged outline of the body of a man is indelibly impressed was a medieval fake dating from 1260 to 1390 and not the burial cloth wrapped around the body of Christ.

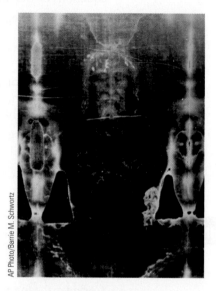

AP Photo/Barrie M. Schwortz

The amount A of radioactive material present at time t is given by the model

$$A = A_0 2^{-t/h}$$

where A_0 is the amount present initially and h is the half-life of the material.

EXAMPLE 11   **Solving a Carbon-14 Dating Problem**

An archeologist finds a wooden statue in the tomb of an ancient Egyptian ruler. If the statue contains two-thirds of its original carbon-14 content, how old is it?

SOLUTION   To find the time t when $A = \frac{2}{3}A_0$, we substitute $\frac{2A_0}{3}$ for A and 5,700 for h in the radioactive decay formula and solve for t:

$$A = A_0 2^{-t/h}$$

$$\frac{2A_0}{3} = A_0 2^{-t/5,700}$$

$$1 = \frac{3}{2}\left(2^{-t/5,700}\right) \qquad \text{Divide both sides by } A_0 \text{ and multiply both sides by } \frac{3}{2}.$$

$$\log 1 = \log \frac{3}{2}\left(2^{-t/5,700}\right) \qquad \text{Take the common logarithm of each side.}$$

$$0 = \log \frac{3}{2} + \log 2^{-t/5,700} \qquad \text{The log of a product is the sum of the logs.}$$

$$-\log \frac{3}{2} = -\frac{t}{5,700}\log 2 \qquad \text{Subtract } \log \frac{3}{2} \text{ from both sides and use the Power Rule of logarithms.}$$

$$5,700\left(\frac{\log \frac{3}{2}}{\log 2}\right) = t \qquad \text{Multiply both sides by } -\frac{5,700}{\log 2}.$$

$$t \approx 3,334.286254 \qquad \text{Use a calculator.}$$

The wooden statue is approximately 3,300 years old.

Self Check 11   How old is an artifact that has 60% of its original carbon-14 content?

Now Try Exercise 95.

5. Solve Population Growth Problems

When there is sufficient food and space, populations of living organisms tend to increase exponentially according to the Malthusian growth model

$$P = P_0 e^{kt}$$

where P_0 is the initial population at $t = 0$ and k depends on the rate of growth.

EXAMPLE 12   **Solving a Population Growth Problem**

Streptococcus bacteria in a laboratory culture increased from an initial population of 500 to 1,500 in 3 hours. Find the time it will take for the population to reach 10,000.

SOLUTION   We will apply the following two steps to solve the problem:

Step 1: Substitute 1,500 for P, 500 for P_0, and 3 for t into the Malthusian growth model and find k.

Step 2: Substitute 10,000 for P, 500 for P_0, and the value of k found in Step 1 into the model and use logarithms to solve for t.

1. Substitute 1,500 for P, 500 for P_0, and 3 for t into the Malthusian growth model and find k:

$$P = P_0 e^{kt}$$

$$1{,}500 = 500 (e^{k3}) \qquad \text{Substitute 1,500 for } P, \text{ 500 for } P_0, \text{ and 3 for } t.$$

$$3 = e^{3k} \qquad \text{Divide both sides by 500.}$$

$$3k = \ln 3 \qquad \text{Take the natural log of both sides.}$$

$$k = \frac{\ln 3}{3} \qquad \text{Divide both sides by 3.}$$

2. To find out when the population will reach 10,000, we substitute 10,000 for P, 500 for P_0, and $\frac{\ln 3}{3}$ for k in the equation $P = P_0 e^{kt}$, and solve for t:

$$P = P_0 e^{kt}$$

$$10{,}000 = 500 e^{\left(\frac{\ln 3}{3}\right)t}$$

$$20 = e^{\left(\frac{\ln 3}{3}\right)t} \qquad \text{Divide both sides by 500.}$$

$$\ln(20) = \ln\left[e^{\left(\frac{\ln 3}{3}\right)t}\right] \qquad \text{Take the natural log of both sides.}$$

$$\ln(20) = \frac{\ln 3}{3} t \qquad \begin{array}{l}\text{Simplify the right side using the natural log property}\\ \ln(e^x) = x.\end{array}$$

$$t = \frac{3 \ln 20}{\ln 3} \qquad \text{Multiply both sides by } \frac{3}{\ln 3}.$$

$$\approx 8.180499084 \qquad \text{Use a calculator.}$$

The culture will reach 10,000 bacteria in a little more than 8 hours. ■

Self Check 12   If the population increases from 1,000 to 3,000 in 3 hours, how long will it take to reach 20,000?

Now Try Exercise 103.

Self Check Answers   **1.** 3   **2.** 1, -3 **3.** $\pm\sqrt{6}$ **4.** $x = 0.6826$ **5.** -9.4520
6. a. $\ln 4$ **b.** 3   **7.** 1   **8.** 20; -5 is extraneous   **9.** 9
10. a. e^2 **b.** 3, 5   **11.** about 4,200 years   **12.** about 8 hr

Exercises **4.6**

Getting Ready
You should be able to complete these vocabulary and concept statements before you proceed to the practice exercises.

Fill in the blanks.

1. An equation with a variable in its exponent is called a(n) _____ equation.
2. An equation with a logarithmic expression that contains a variable is a(n) _____ equation.
3. The formula for carbon dating is $A = $ _____.
4. The formula for population growth is $P = $ _____.

Practice
Solve each exponential equation using like bases.

5. $2^{3x+2} = 16^x$

6. $32^{x+2} = 27^{7x+12}$

7. $27^{x+1} = 32^{2x+1}$

8. $3^{x-1} = 9^{2x}$

9. $5^{4x+1} = 25^{-x-2}$

10. $5^{2x+1} = 125^x$

11. $4^{x-2} = 8^x$

12. $16^{x+1} = 8^{2x+1}$

13. $81^{2x} = 27^{2x-5}$

14. $625^{x-9} = 125^{x-12}$

15. $2^{x^2-2x} = 8$

16. $5^{x^2+3x} = 625$

17. $36^{x^2} = 216^{x^2-3}$

18. $25^{x^2-5x} = 3{,}125^{4x}$

19. $7^{x^2+3x} = \dfrac{1}{49}$

20. $3^{x^2+4x} = \dfrac{1}{81}$

21. $e^{-x+6} = e^x$

22. $e^{2x+1} = e^{3x-11}$

23. $e^{x^2-1} = e^{24}$

24. $e^{x^2+7x} = \dfrac{1}{e^{12}}$

Solve each exponential equation using logarithms. Give the answer in decimal form, rounding to four decimal places.

25. $4^x = 5$

26. $7^x = 12$

27. $13^{x-1} = 2$

28. $5^{x+1} = 3$

29. $2^{x+1} = 3^x$

30. $5^{x-3} = 3^{2x}$

31. $2^x = 3^x$

32. $3^{2x} = 4^x$

33. $7^{x^2} = 10$

34. $8^{x^2} = 11$

35. $8^{x^2} = 9^x$

36. $5^{x^2} = 2^{5x}$

Find the exact solution to each exponential equation.

37. $e^x = 10$

38. $8e^x = 16$

39. $4e^{2x} = 24$

40. $2e^{5x} = 18$

Solve each equation. If an answer is not exact, give the answer in decimal form. Round to four decimal places.

41. $4^{x+2} - 4^x = 15$ (*Hint:* $4^{x+2} = 4^x 4^2$.)

42. $3^{x+3} + 3^x = 84$ (*Hint:* $3^{x+3} = 3^x 3^3$.)

43. $2(3^x) = 6^{2x}$

44. $2(3^{x+1}) = 3(2^{x-1})$

45. $2^{2x} - 10(2^x) + 16 = 0$ (*Hint:* Let $y = 2^x$.)

46. $3^{2x} - 10(3^x) + 9 = 0$ (*Hint:* Let $y = 3^x$.)

47. $2^{2x+1} - 2^x = 1$ (*Hint:* $2^{a+b} = 2^a 2^b$.)

48. $3^{2x+1} - 10(3^x) + 3 = 0$ (*Hint:* $3^{a+b} = 3^a 3^b$.)

Solve each logarithmic equation. Use the definition of logarithm or the definition of natural logarithm.

49. $\log x^2 = 2$

50. $\log x^3 = 3$

51. $\log \dfrac{4x + 1}{2x + 9} = 0$

52. $\log \dfrac{5x + 2}{2(x + 7)} = 0$

53. $\ln x = 6$

54. $\ln x = 3$

55. $\ln(2x - 7) = 4$

56. $\ln(3x - 5) = 7$

Solve each logarithmic equation.

57. $\log(2x - 3) = \log(x + 4)$

58. $\log(3x + 5) - \log(2x + 6) = 0$

59. $\log x + \log(x - 48) = 2$

60. $\log x + \log(x + 9) = 1$

61. $\log x + \log(x - 15) = 2$

62. $\log x + \log(x + 21) = 2$

63. $\log(x + 90) = 3 - \log x$

64. $\log(x - 3) - \log(6) = 2$

65. $\log(5{,}000) - \log(x - 2) = 3$

66. $\log(2x - 3) - \log(x - 1) = 0$

67. $\log_7 x + \log_7(x - 5) = \log_7 6$

68. $\ln x + \ln(x - 2) = \ln 120$

69. $\ln 15 - \ln(x - 2) = \ln x$

70. $\ln 10 - \ln(x - 3) = \ln x$

71. $\log_6 8 - \log_6 x = \log_6(x - 2)$

72. $\log(x - 6) - \log(x - 2) = \log \dfrac{5}{x}$

73. $\log(x - 1) - \log 6 = \log(x - 2) - \log x$

74. $\log x^2 = (\log x)^2$

75. $\log(\log x) = 1$

76. $\log_3(\log_3 x) = 1$

77. $\dfrac{\log(3x - 4)}{\log x} = 2$

78. $\dfrac{\ln(8x - 7)}{\ln x} = 2$

79. $\dfrac{\ln(5x + 6)}{2} = \ln x$

80. $\dfrac{1}{2}\log(4x + 5) = \log x$

81. $\log_3 x = \log_3\left(\dfrac{1}{x}\right) + 4$

82. $\log_5(7 + x) + \log_5(8 - x) - \log_5 2 = 2$

83. $2\log_2 x = 3 + \log_2(x - 2)$

84. $2\log_3 x - \log_3(x - 4) = 2 + \log_3 2$

85. $\ln(7y + 1) = 2\ln(y + 3) - \ln 2$

86. $2\log(y + 2) = \log(y + 2) - \log 12$

Use a graphing calculator to solve each equation. If an answer is not exact, give the result to the nearest hundredth.

87. $\log x + \log(x - 15) = 2$

88. $\log x + \log(x + 3) = 1$

89. $2^{x+1} = 7$

90. $\ln(2x + 5) - \ln 3 = \ln(x - 1)$

Applications

Use a calculator to help solve each problem.

91. **Tritium decay** The half-life of tritium is 12.4 years. How long will it take for 25% of a sample of tritium to decompose?

92. **Radioactive decay** In 2 years, 20% of a radioactive element decays. Find its half-life.

93. **Thorium decay** An isotope of thorium, ^{227}Th, has a half-life of 18.4 days. How long will it take 80% of the sample to decompose?

94. **Lead decay** An isotope of lead, ^{201}Pb, has a half-life of 8.4 hours. How many hours ago was there 30% more of the substance?

95. **Carbon-14 dating** A cloth fragment is found in an ancient tomb. It contains 70% of the carbon-14 that it is assumed to have had initially. How old is the cloth?

96. **Carbon-14 dating** Only 25% of the carbon-14 in a wooden bowl remains. How old is the bowl?

97. **Compound interest** If $500 is deposited in an account paying 8.5% annual interest, compounded semiannually, how long will it take for the account to increase to $800?

98. **Continuous compound interest** In Exercise 97, how long will it take if the interest is compounded continuously?

99. **Compound interest** If $1,300 is deposited in a savings account paying 9% interest, compounded quarterly, how long will it take the account to increase to $2,100?

100. **Compound interest** A sum of $5,000 deposited in an account grows to $7,000 in 5 years. Assuming annual compounding, what interest rate is being paid?

101. **Rule of Seventy** A rule of thumb for finding how long it takes an investment to double is called the **Rule of Seventy.** To apply the rule, divide 70 by the interest rate (expressed as a percent). At 5%, it takes $\frac{70}{5} = 14$ years to double the investment. At 7%, it takes $\frac{70}{7} = 10$ years. Explain why this formula works.

102. **Oceanography** The intensity I of a light a distance x meters beneath the surface of a lake decreases exponentially. If the light intensity at 6 meters is 70% of the intensity at the surface, at what depth will the intensity be 20%?

103. **Bacterial growth** A staphylococcus bacterial culture grows according to the formula $P = P_0 a^t$. If it takes 5 days for the culture to triple in size, how long will it take to double in size?

104. **Rodent control** The rodent population in a city is currently estimated at 30,000. If it is expected to double every 5 years, when will the population reach 1 million?

105. **Temperature of coffee** Refer to the section opener and find the time it takes for the white chocolate mocha to reach a temperature of 80°F.

106. **Time of death** The exponential function $T(t) = 17e^{-0.0626t} + 20$ models the temperature T in °C of a person's body t hours after death. If a dead body is discovered at 8:30 a.m. and the body's temperature is 30°C, what was the person's approximate time of death?

107. **Newton's Law of Cooling** Water whose temperature is at 100°C is left to cool in a room where the temperature is 60°C. After 3 minutes, the water temperature is 90°. If the water temperature T is a function of time t given by $T = 60 + 40e^{kt}$, find k.

108. **Newton's Law of Cooling** Refer to Exercise 107 and find the time for the water temperature to reach 70°C.

109. Newton's Law of Cooling A block of steel, initially at $0°C$, is placed in an oven heated to $300°C$. After 5 minutes, the temperature of the steel is $100°C$. If the steel temperature T is a function of time t given by $T = 300 - 300e^{kt}$, find the value of k.

110. Newton's Law of Cooling Refer to Exercise 109 and find the time for the steel temperature to reach $200°C$.

Discovery and Writing

111. Explain why it is necessary to check the solutions of a logarithmic equation.

112. Use the population growth formula to show that the doubling time for population growth is given by $t = \frac{\ln 2}{r}$.

113. Use the population growth formula to show that the tripling time for population growth is given by $t = \frac{\ln 3}{r}$.

114. Can you solve $x = \log x$ algebraically? Can you find an approximate solution?

Find x.

115. $\log_2(\log_5(\log_7 x)) = 2$

116. $\log_8\left[16\sqrt[3]{4{,}096}\right]^{\frac{1}{6}} = x$

Review

Find the inverse of the function defined by each equation.

117. $y = 3x + 2$

118. $y = \dfrac{1}{x - 3}$

Let $f(x) = 5x - 1$ and $g(x) = x^2$. Find each value.

119. $(f \circ g)(2)$

120. $(g \circ f)(2)$

121. $(f \circ g)(x)$

122. $(g \circ f)(x)$

CHAPTER REVIEW

SECTION **4.1**   Exponential Functions and Their Graphs

Definitions and Concepts	Examples
An **exponential function** with base b is defined by the equation $y = f(x) = b^x \qquad (b > 0, b \neq 1)$	**Graph:** $f(x) = 6^x$. We will find several points (x, y) that satisfy the equation, plot the points, and join them with a smooth curve, as shown in the figure. 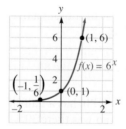

$f(x) = 6^x$		
x	$f(x)$	$(x, f(x))$
-1	$\dfrac{1}{6}$	$\left(-1, \dfrac{1}{6}\right)$
0	1	$(0, 1)$
1	6	$(1, 6)$

Definitions and Concepts	Examples
Compound interest formula: If P dollars are deposited in an account earning interest at an annual rate r, compounded n times each year, the amount A in the account after t years is given by $$A = P\left(1 + \frac{r}{n}\right)^{nt}$$	The grandparents of a newborn child invest \$10,000 in an educational savings plan that earns 8% interest, compounded quarterly. If the money is left untouched, how much will the child have in the account in 18 years? We will substitute 10,000 for P, 0.08 for r, and 18 for t in the formula for compound interest. Because quarterly compounding means four times per year, we will substitute 4 for n. $$A = P\left(1 + \frac{r}{n}\right)^{nt}$$ $$A = 10{,}000\left(1 + \frac{0.08}{4}\right)^{4 \cdot 18}$$ $$= 10{,}000(1.02)^{72}$$ $$\approx 41{,}611.40 \quad \text{Use a calculator and round to two decimals.}$$ In 18 years, the account will be worth \$41,611.40.
The number $e \approx 2.718281828$. The graph of $f(x) = e^x$ is: 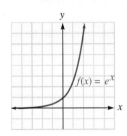 **Continuous compound interest formula:** If P dollars are deposited in an account earning interest at an annual rate r, compounded continuously, the amount A after t years is given by the formula $$A = Pe^{rt}$$	If the grandparents of the newborn child in the previous example had invested \$10,000 at an annual rate of 8%, compounded continuously, how much would the child have in the account in 18 years? We will substitute \$10,000 for P, 0.08 for r, and 18 in for t in the continuous compound interest formula $A = Pe^{rt}$. $$A = Pe^{rt}$$ $$A = 10{,}000e^{(0.08)(18)}$$ $$= 10{,}000e^{1.44}$$ $$\approx 42{,}206.96 \quad \text{Use a calculator and round to two decimal places.}$$ In 18 years, the balance will be \$42,206.96.

EXERCISES

Use properties of exponents to simplify.

1. $5^{\sqrt{2}} \cdot 5^{\sqrt{2}}$

2. $\left(2^{\sqrt{5}}\right)^{\sqrt{2}}$

Graph the function defined by each equation.

3. $f(x) = 3^x$

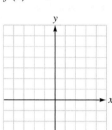

4. $f(x) = \left(\dfrac{1}{3}\right)^x$

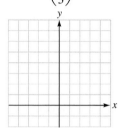

5. The graph of $f(x) = 7^x$ will pass through the points $(0, p)$ and $(1, q)$. Find p and q.

6. Give the domain and range of the function $f(x) = b^x$, with $b > 0$ and $b \neq 1$.

Use translations to help graph each function.

7. $g(x) = \left(\dfrac{1}{2}\right)^x - 2$

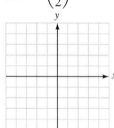

8. $g(x) = \left(\dfrac{1}{2}\right)^{x+2}$

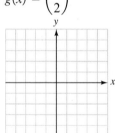

Graph each function.

9. $f(x) = -5^x$

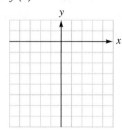

10. $f(x) = -5^x + 4$

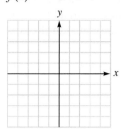

11. $f(x) = e^x + 1$

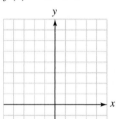

12. $f(x) = e^{x-3}$

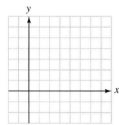

 13. Compound interest How much will \$10,500 become if it earns 9% per year for 60 years, compounded quarterly?

14. Continuous compound interest If \$10,500 accumulates interest at an annual rate of 9%, compounded continuously, how much will be in the account in 60 years?

SECTION **4.2**   Applications of Exponential Functions

Definitions and Concepts	Examples
Radioactive decay formula: The amount A of radioactive material present at time t is given by $$A = A_0 2^{-t/h}$$ where A_0 is the amount that was present initially (at $t = 0$) and h is the material's half-life.	The half-life of radium is approximately 1,600 years. To find how much of a 1-gram sample will remain after 500 years, we will substitute $A_0 = 1$, $h = 1,600$, and $t = 500$ into the formula for radioactive decay and simplify. $$A = A_0 2^{-t/h}$$ $$A = 1 \cdot 2^{-500/1{,}600}$$ $$\approx 0.81 \qquad \text{Use a calculator. Round to two decimal places.}$$ After 500 years, approximately 0.81 gram of radium will remain.

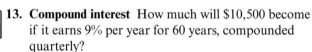

Definitions and Concepts	Examples
Intensity of light formula: The intensity I of light (in lumens) at a distance x meters below the surface of a body of water decreases exponentially according to the formula $I = I_0 k^x$ where I_0 is the intensity of light above the water and k is a constant that depends on the clarity of the water.	At one location in the Atlantic Ocean, the intensity of light above water I_0 is 10 lumens and $k = 0.4$. To find the intensity of light at a depth of 3 meters, we will substitute 10 for I_0, 0.4 for k, and 3 for x into the formula for light intensity and simplify. $I = I_0 k^x$ $I = 10(0.4)^3$ $I = 0.64$ At a depth of 3 meters, the intensity of the light is 0.64 lumen.
Malthusian model of population growth: If b is the annual birth rate, d is the annual death rate, t is the time (in years), P_0 is the initial population at $t = 0$, and P is the current population, then $P = P_0 e^{kt}$ where $k = b - d$ is the **annual growth rate,** the difference between the annual birth rate and death rate.	The population of the United States is approximately 300 million people. Assuming that the annual birth rate is 19 per 1,000 and the annual death rate is 7 per 1,000, what does the Malthusian model predict the U.S. population will be in 30 years? We can use the stated information to write the Malthusian model for U.S. population. We will then substitute into the model to predict the population in 30 years. Since k is the difference between the birth and death rates, we have $k = b - d$ $k = \dfrac{19}{1,000} - \dfrac{7}{1,000}$ Substitute $\dfrac{19}{1,000}$ for b and $\dfrac{7}{1,000}$ for d. $k = 0.019 - 0.007$ $ = 0.012$ We can substitute 300,000,000 for P_0, 30 for t, and 0.012 for k in the formula for the Malthusian model of population growth and simplify. $P = P_0 e^{kt}$ $P = (300,000,000)e^{(0.012)(30)}$ $ = (300,000,000)e^{0.36}$ $ \approx 429,998,824.4$ Use a calculator. After 30 years, the U.S. population will be approximately 430 million people.

EXERCISES

15. The half-life of a radioactive material is about 34.2 years. How much of the material is left after 20 years?

16. Find the intensity of light at a depth of 12 meters if $I_0 = 14$ and $k = 0.7$.

17. The population of the United States is approximately 300,000,000 people. Find the population in 50 years if $k = 0.015$.

18. Spread of hepatitis In a city with a population of 450,000, there are currently 1,000 cases of hepatitis. If the spread of the disease is projected by the following logistic function, how many people will contract the hepatitis virus after 5 years?

$$P = \frac{450,000}{1 + (450 - 1)e^{-0.2t}}$$

SECTION **4.3**   Logarithmic Functions and Their Graphs

Definitions and Concepts	Examples

Logarithmic functions:
If $b > 0$ and $b \neq 1$, the **logarithmic function with base b** is defined by $y = \log_b x$ if and only if $x = b^y$.

The **domain of the logarithmic function** is the interval $(0, \infty)$. The **range** is the interval $(-\infty, \infty)$.

Base-10 logarithms are called **common logarithms**. The notation $\log x$ represents $\log_{10} x$.

Examples

$\log_5 125 = 3$ because $5^3 = 125$
$\log_8 1 = 0$ because $8^0 = 1$

$\log_9 3 = \dfrac{1}{2}$ because $9^{1/2} = \sqrt{9} = 3$

$\log_2 \dfrac{1}{16} = -4$ because $2^{-4} = \dfrac{1}{16}$

$\log 10{,}000 = 4$ because $10^4 = 10{,}000$

To find x in the equation $\log_8 \frac{1}{64} = x$, we note that $\log_8 \frac{1}{64} = x$ is equivalent to $8^x = \frac{1}{64}$. Since $8^{-2} = \frac{1}{64}$, we have $8^x = 8^{-2}$ and $x = -2$.

To graph $f(x) = \log_5 x$. We will find several points (x, y) that satisfy the equation, plot the points, and join them with a smooth curve, as shown in the figure.

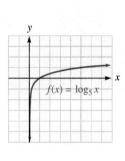

$f(x) = \log_5 x$		
x	$f(x)$	$(x, f(x))$
$\dfrac{1}{25}$	-2	$\left(\dfrac{1}{25}, -2\right)$
$\dfrac{1}{5}$	-1	$\left(\dfrac{1}{5}, -1\right)$
1	0	$(1, 0)$
5	1	$(5, 1)$
25	2	$(25, 2)$

Natural logarithms:
Base-e logarithmms are called **natural logarithms**. The notation $\ln x$ means $\log_e x$.

The graph of $f(x) = \ln x$ is:

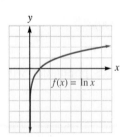

Since the functions $y = \ln x$ and $y = e^x$ are inverses

$\ln e^x = x$ and $e^{\ln x} = x$

To graph $g(x) = \ln(x + 4)$, we will translate the graph of $f(x) = \ln x$ four units to the left as shown in the figure.

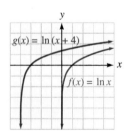

$\ln e^{-5x} = -5x$ $e^{\ln(7x)} = 7x$

EXERCISES

19. Give the domain and range of the logarithmic function $f(x) = \log_3 x$.

20. Give the domain and range of the natural logarithm function, $f(x) = \ln x$.

Find each value.

21. $\log_3 9$

22. $\log_9 \dfrac{1}{3}$

23. $\log_x 1$

24. $\log_5 0.04$

25. $\log_a \sqrt{a}$

26. $\log_a \sqrt[3]{a}$

Find x.

27. $\log_2 x = 5$

28. $\log_{\sqrt{3}} x = 4$

29. $\log_{\sqrt{2}} x = 6$

30. $\log_{0.1} 10 = x$

31. $\log_x 2 = -\dfrac{1}{3}$

32. $\log_x 32 = 5$

33. $\log_{0.25} x = -1$

34. $\log_{0.125} x = -\dfrac{1}{3}$

35. $\log_{\sqrt{2}} 32 = x$

36. $\log_{\sqrt{5}} x = -4$

37. $\log_{\sqrt{3}} 9\sqrt{3} = x$

38. $\log_{\sqrt{5}} 5\sqrt{5} = x$

Graph each function.

39. $f(x) = \log(x - 2)$

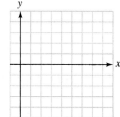

40. $f(x) = 3 + \log x$

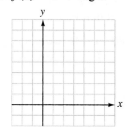

Graph each pair of equations on one set of coordinate axes.

41. $y = 4^x$ and $y = \log_4 x$

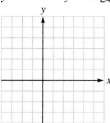

42. $y = \left(\dfrac{1}{3}\right)^x$ and $y = \log_{1/3} x$

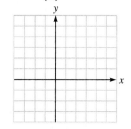

Use a calculator to find each value to four decimal places.

43. $\ln 452$

44. $\ln(\log 7.85)$

Use a calculator to solve each equation. Round each answer to four decimal places.

45. $\ln x = 2.336$

46. $\ln x = \log 8.8$

Graph each function.

47. $f(x) = 1 + \ln x$

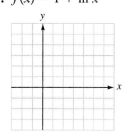

48. $f(x) = \ln(x + 1)$

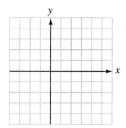

Simplify each expression.

49. $\ln(e^{12})$

50. $e^{\ln 14x}$

SECTION 4.4   Applications of Logarithmic Functions

Definitions and Concepts	Examples
Decibel voltage gain: If E_O is the output voltage of a device and E_I is the input voltage, the **decibel voltage gain** is given by $$\text{dB gain} = 20 \log \frac{E_O}{E_I}$$	To find the dB gain of an amplifier with an input of 0.4 volt and an output of 50 volts, we can substitute the given values into the dB gain formula. $$\text{dB gain} = 20 \log \frac{E_O}{E_I}$$ $$\text{dB gain} = 20 \log \frac{50}{0.4} \quad \text{Substitute 50 for } E_O \text{ and } 0.4 \text{ for } E_I.$$ $$= 20 \log 125$$ $$\approx 41.93820026 \quad \text{Use a calculator.}$$ To the nearest decibel, the dB gain is 42 decibels.
Richter scale: If R is the intensity of an earthquake, A is the amplitude (measured in micrometers), and P is the period (the time of one oscillation of the Earth's surface, measured in seconds), then $$R = \log \frac{A}{P}$$	To find the intensity of an earthquake with amplitude of 5,000 micrometers $\left(\frac{1}{2}\,\text{cm}\right)$ and a period of 0.08 second, we substitute 5,000 for A and 0.08 for P in the Richter scale formula and simplify. $$R = \log \frac{A}{P}$$ $$R = \log \frac{5{,}000}{0.08}$$ $$= \log 62{,}500$$ $$\approx 4.795880017 \quad \text{Use a calculator}$$ To the nearest tenth, the earthquake measures 4.8 on the Richter scale.
Charging batteries: If M is the theoretical maximum charge that a battery can hold and k is a positive constant that depends on the battery and the charger, the length of time (in minutes) required to charge the battery to a given level C is given by $$t = -\frac{1}{k} \ln\left(1 - \frac{C}{M}\right)$$	To find how long will it take to bring a fully discharged battery to 80% of full charge, we will assume that $k = 0.02$ and that time is measured in minutes. Since 80% of full charge means 80% of M, we can substitute $0.80M$ for C and 0.02 for k in the formula for charging batteries and find t. $$t = -\frac{1}{k} \ln\left(1 - \frac{C}{M}\right)$$ $$t = -\frac{1}{0.02} \ln\left(1 - \frac{0.80M}{M}\right)$$ $$= -50 \ln(1 - 0.8)$$ $$= -50 \ln(0.2)$$ $$\approx 80.47189562 \quad \text{Use a calculator.}$$ The battery will reach 80% charge in about 80 minutes.

Definitions and Concepts	Examples
Population doubling time: If r is the annual growth rate and t is the time (in years) required for a population to double, then $$t = \frac{\ln 2}{r}$$	The population of the Earth is growing at the approximate rate of 2.5% per year. If this rate continues, how long will it take the population to double? Because the population is growing at the rate of 2.5% per year, we can substitute 0.025 for r in the formula for doubling time and simplify. $$t = \frac{\ln 2}{r}$$ $$t = \frac{\ln 2}{0.025}$$ $$\approx 27.72588722$$ It will take about 28 years for the Earth's population to double.
Isothermal expansion: If the temperature T is constant, the energy E required to increase the volume of 1 mole of gas from an initial volume V_i to a final volume V_f is given by $$E = RT \ln\left(\frac{V_f}{V_i}\right)$$ E is measured in joules and T in Kelvins. R is the universal gas constant, which is 8.314 joules/mole/K.	To find the amount of energy that must be supplied to double the volume of 1 mole of gas at a constant temperature of 300 K, we substitute 8.314 for R and 300 for T in the formula. Since the final volume is to be two times the initial volume, we also substitute $2V_i$ for V_f. $$E = RT \ln\left(\frac{V_f}{V_i}\right)$$ $$E = (8.314)(300)\ln\left(\frac{2V_i}{V_i}\right)$$ $$= 2{,}494.2 \ln 2$$ $$\approx 1728.847698$$ Approximately 1,729 joules of energy must be added to double the volume.

EXERCISES

51. Decibel gain An amplifier has an output of 18 volts when the input is 0.04 volt. Find the dB gain.

52. Intensity of an earthquake An earthquake had a period of 0.3 second and an amplitude of 7,500 micrometers. Find its measure on the Richter scale.

53. Charging batteries How long will it take a dead battery to reach an 80% charge? (Assume $k = 0.17$.)

54. Doubling time How long will it take the population of the United States to double if the growth rate is 3% per year?

55. Isothermal energy Find the amount of energy that must be supplied to double the volume of 1 mole of gas at a constant temperature of 350K. (*Hint:* $R = 8.314$.)

SECTION **4.5** Properties of Logarithms

Definitions and Concepts	Examples
Properties of logarithms: If b is a positive number and $b \neq 1$,	
1. $\log_b 1 = 0$	By Property 1, $\log_9 1 = 0$, because $9^0 = 1$.
2. $\log_b b = 1$	By Property 2, $\log_{11} 11 = 1$, because $11^1 = 11$.
3. $\log_b b^x = x$	By Property 3, $\log_4 4^3 = 3$, because $4^3 = 4^3$.
4. $b^{\log_b x} = x$	By Property 4, $6^{\log_6 3} = 3$, because $\log_6 3$ is the power to which 6 is raised to get 3.
5. Product Rule:	
$\quad \log_b MN = \log_b M + \log_b N$	By Property 5, $\log(2 \cdot 3) = \log 2 + \log 3$
6. Quotient Rule:	
$\quad \log_b \dfrac{M}{N} = \log_b M - \log_b N$	By Property 6, $\log \dfrac{17}{5} = \log 17 - \log 5$
7. Power Rule:	
$\quad \log_b M^p = p \log_b M$	By Property 7, $\log 3^2 = 2 \log 3$
8. One-to-One Property:	
$\quad$ If $\log_b x = \log_b y$, then $x = y$.	By Property 8, if $\log 8 = \log y$, then $8 = y$.
The properties of logarithms also hold for natural logarithms.	
Properties of natural logarithms:	
1. $\ln 1 = 0$	By Property 1, $\ln 1 = 0$, because $e^0 = 1$.
2. $\ln e = 1$	By Property 2, $\ln e = 1$, because $e^1 = e$.
3. $\ln e^x = x$	By Property 3, $\ln(e^{13}) = 13$.
4. $e^{\ln x} = x$	By Property 4, $e^{\ln 23} = 23$.
5. Product Rule:	
$\quad \ln MN = \ln M + \ln N$	By Property 5, $\ln(2 \cdot 3) = \ln 2 + \ln 3$.
6. Quotient Rule:	
$\quad \ln \dfrac{M}{N} = \ln M - \ln N$	By Property 6, $\ln \dfrac{15}{5} = \ln 15 - \ln 5$.
7. Power Rule:	
$\quad \ln M^p = p \ln M$	By Property 7, $\ln 3^2 = 2 \ln 3$.
8. One-to-One Property:	
$\quad$ If $\ln x = \ln y$, then $x = y$.	By Property 8, if $\ln 8 = \ln y$, then $8 = y$.

Definitions and Concepts	Examples
Expanding or condensing logarithmic expressions: The properties of logarithms can be used to expand or condense logarithmic expressions.	Write the expression $\log_2 \dfrac{x^2}{y}$ in terms of logarithms of x and y. $$\log_2 \frac{x^2}{y} = \log_2 x^2 - \log_2 y \qquad \text{Use the Quotient Rule.}$$ $$= 2 \log_2 x - \log_2 y \qquad \text{Use the Power Rule.}$$ Write $5 \ln x + \dfrac{1}{2} \ln y$ as a single natural logarithm. $$5 \ln x + \frac{1}{2} \ln y = \ln x^5 + \ln y^{1/2} \qquad \text{Use the Power Rule.}$$ $$= \ln(x^5 y^{1/2}) \qquad \text{Use the Product Rule.}$$ $$= \ln(x^5 \sqrt{y})$$
Change-of-Base Formula: $$\log_b y = \frac{\log_a y}{\log_a b}$$	Use the Change-of-Base Formula to find $\log_3 10$. We will substitute 3 for b and 10 for y in the Change-of-Base Formula. $$\log_3 10 = \frac{\log_{10} 10}{\log_{10} 3} = \frac{1}{0.4771212547} = 2.095903274$$
pH of a solution: If $[\text{H}^+]$ is the hydrogen ion concentration in gram-ions per liter, then $$\text{pH} = -\log[\text{H}^+]$$	Lemon juice in a bottle has a pH of approximately 4. To find its hydrogen ion concentration, we can substitute 4 for pH in the pH formula and solve it for $[\text{H}^+]$. $$4 = -\log[\text{H}^+]$$ $$-4 = \log[\text{H}^+]$$ $$[\text{H}^+] = 10^{-4} \qquad \text{Change the equation from logarithmic form to exponential form.}$$ We can then use a calculator to find that $[\text{H}^+] = 1 \times 10^{-4}$ gram-ions per liter.
Weber–Fechner Law: If L is the apparent loudness of a sound and I is the intensity, then $$L = k \ln I$$	To find what increase in the intensity of a sound is necessary to cause a quadrupling of the apparent loudness, we can use the formula $L = k \ln I$. To quadruple the apparent loudness, we multiply both sides of the equation by 4 and use Property 7 of natural logarithms. $$L = k \ln I$$ $$4L = 4k \ln I$$ $$= k \ln I^4$$ To quadruple the apparent loudness, we must raise the intensity to the fourth power.

EXERCISES

Simplify each expression.

56. $\log_7 1$

57. $\log_7 7$

58. $\log_7 7^3$

59. $7^{\log_7 4}$

60. $\ln e^4$

61. $\ln 1$

62. $10^{\log_{10} 7}$

63. $e^{\ln 3}$

64. $\log_b b^4$

65. $\ln e^9$

Write each expression in terms of the logarithms of x, y, and z.

66. $\log_b \dfrac{x^2 y^3}{z^4}$

67. $\log_8 \sqrt{\dfrac{x}{yz^2}}$

68. $\ln \dfrac{x^4}{y^5 z^6}$

69. $\ln \sqrt[3]{xyz}$

Write each expression as the logarithm of one quantity.

70. $3 \log_b x - 5 \log_b y + 7 \log_b z$

71. $\dfrac{1}{2} (\log_b x + 3 \log_b y) - 7 \log_b z$

72. $4 \ln x - 5 \ln y - 6 \ln z$

73. $\dfrac{1}{2} \ln x + 3 \ln y - \dfrac{1}{3} \ln z$

Given that $\log a \approx 0.6$, $\log b \approx 0.36$, *and* $\log c \approx 2.4$, *approximate the value of each expression.*

74. $\log abc$

75. $\log a^2 b$

76. $\log \dfrac{ac}{b}$

77. $\log \dfrac{a^2}{c^3 b^2}$

78. To four decimal places, find $\log_5 17$.

79. pH of grapefruit The pH of grapefruit juice is about 3.1. Find its hydrogen ion concentration.

80. Loudness of sound Find the decrease in loudness if the intensity is cut in half.

SECTION 4.6   Exponential and Logarithmic Equations

Definitions and Concepts	Examples
One-to-One Property of exponents: If $b^x = b^y$ then $x = y$. That is, equal quantities with like bases have equal exponents. The One-to-One Property of exponents can be used to solve exponential equations with like bases.	Solve using like bases: $5^{x^2 - 2x} = \dfrac{1}{5}$. Since $\dfrac{1}{5} = 5^{-1}$, we can write the equation in the form $5^{x^2 - 2x} = 5^{-1}$ and use like bases to solve the equation. Because equal quantities with like bases have equal exponents, we have $x^2 - 2x = -1$ $x^2 - 2x + 1 = 0 \qquad$ Add 1 to both sides. $(x - 1)(x - 1) = 0 \qquad$ Factor the trinomial. $x - 1 = 0 \quad$ or $\quad x - 1 = 0 \qquad$ Set each factor equal to 0. $\qquad x = 1 \quad \mid \quad x = 1$

Definitions and Concepts	Examples
Solving exponential equations with different bases: Exponential equations with different bases can be solve using logarithms.	Solve the exponential equation: $3^x = 15$. Since logarithms of equal numbers are equal, we can take the common logarithm of each side of the equation. We can then use the Power Rule and move the variable x from its position as an exponent to a position as a coefficient and solve the equation. $$3^x = 5$$ $\mathbf{log}\ 3^x = \mathbf{log}\ 15$ Take the common logarithm of each side. $x \log 3 = \log 15$ Use the Power Rule. $x = \dfrac{\log 15}{\log 3}$ Divide both sides by log 3. ≈ 2.464973521 Use a calculator. To four decimal places, $x = 2.4650$.
Solving exponential equations with e: Use natural logarithms to solve exponential equations with a base of e.	Solve $e^x = 19$. We will take the natural logarithm of both sides, simplify using the natural logarithms properties, and solve for x. $$e^x = 19$$ $\mathbf{ln}\ e^x = \mathbf{ln}\ 19$ Take the natural logarithm of both sides. $x = \ln 19$ Use the property $\ln(e^x) = x$. $x \approx 2.94$ Use a calculator and round to two decimal places.
Solving logarithmic equations: The One-to-One Property of logarithms can be used to solve some logarithmic equations. **One-to-One Property:** If $\log_b x = \log_b y$, then $x = y$. The solution(s) to a logarithmic equation must be checked.	Solve: $\log_4(5x + 3) = \log_4(-5x + 23)$. We will use the One-to-One Property to solve the equation. $$\log_4(5x + 3) = \log_4(-5x + 23)$$ $5x + 3 = -5x + 23$ Use the log property stated above. $10x = 20$ Add $5x$ and subtract 3 from both sides. $x = 2$ Divide both sides by 10. **Check:** $\log_4(5x + 3) = \log_4(-5x + 23)$ $\log_4[5(2) + 3] \stackrel{?}{=} \log_4[-5(2) + 23]$ $\log_4(13) = \log_4(13)$ Since $x = 2$ checks, it is the solution.

Definitions and Concepts	Examples
Solving logarithmic equations: Logarithmic properties can be used to solve many logarithmic equations.	Solve: $\log x + \log(x - 9) = 1$. $\log x + \log(x - 9) = 1$ $\log x(x - 9) = 1$ Use the Product Rule. $x(x - 9) = 10^1$ Use the definition of logarithm and write in exponential form. $x^2 - 9x - 10 = 0$ Remove parentheses and subtract 10 from both sides. $(x - 10)(x + 1) = 0$ Factor the trinomial. $x - 10 = 0 \quad \text{or} \quad x + 1 = 0$ $x = 10 \qquad\qquad x = -1$ **Check:** The number -1 is not a solution, because it does not satisfy the equation (a negative number does not have a logarithm). We check the remaining number, 10. $\log x + \log(x - 9) = 1$ $\log 10 + \log(10 - 9) \stackrel{?}{=} 1$ Substitute 10 for x. $\log 10 + \log 1 \stackrel{?}{=} 1$ $1 + 0 \stackrel{?}{=} 1$ $\log_{10} 10 = 1$, $\log_{10} 1 = 0$ $1 = 1$ Since 10 does check, it is a root.
Carbon dating: The amount A of radioactive material present at time t is given by the model $A = A_0 2^{-t/h}$ where A_0 is the amount present initially and h is the half-life of the material.	To find the age of an artifact that contains three-fourths of its original carbon-14 content, we substitute $\frac{3}{4}A_0$ for A and 5,700 for h in the formula for carbon dating and solve for t. $A = A_0 2^{-t/h}$ $\frac{3}{4}A_0 = A_0 2^{-t/5,700}$ $\frac{3}{4} = 2^{-t/5,700}$ Divide both sides by A_0. $\log \frac{3}{4} = \log 2^{-t/5,700}$ Take the base-10 log of both sides. $-0.1249387366 = -\frac{t}{5,700} \log 2$ Use Property 7 of logarithms. $t \approx 2,366$ Solve for t. The artifact is about 2,300 years old.

EXERCISES

Solve each equation for x.

81. $81^{x+2} = 27$

82. $2^{x^2+4x} = \dfrac{1}{8}$

83. $e^x = e^{-6x+14}$

84. $e^{2x^2} = e^{18}$

85. $3^x = 7$

86. $2^x = 3^{x-1}$

87. $2e^x = 16$

88. $-5e^x = -35$

Solve each equation for x.

89. $\log_7(-7x + 2) = \log_7(3x + 32)$

90. $\ln(x + 3) = \ln(-5x + 51)$

91. $\log x + \log(29 - x) = 2$

92. $\log_2 x + \log_2(x - 2) = 3$

93. $\log_2(x + 2) + \log_2(x - 1) = 2$

94. $\dfrac{\log(7x - 12)}{\log x} = 2$

95. $\ln x + \ln(x - 5) = \ln 6$

96. $\log 3 - \log(x - 1) = -1$

97. $e^{x \ln 2} = 9$

98. $\ln x = \ln(x - 1)$

99. $\ln x - 3 = 4$

100. $\ln x = \ln(x - 1) + 1$

101. $\ln x = \log_{10} x$ (*Hint:* Use the Change-of-Base Formula.)

102. **Carbon-14 dating** A wooden statue found in Egypt has a carbon-14 content that is two-thirds of that found in living wood. If the half-life of carbon-14 is 5,700 years, how old is the statue?

CHAPTER TEST

Graph each function.

1. $f(x) = 2^x + 1$

2. $f(x) = e^{x-2}$

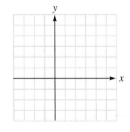

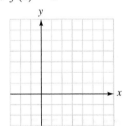

Solve each problem.

3. **Radioactive decay** A radioactive material decays according to the formula $A = A_0(2)^{-t}$. How much of a 3-gram sample will be left in 6 years?

4. **Compound interest** An initial deposit of $1,000 earns 6% interest, compounded twice a year. How much will be in the account in one year?

5. **Continuous compound interest** An account contains $2,000 and has been earning 8% interest, compounded continuously. How much will be in the account in 10 years?

Find each value.

6. $\log_7 343$

7. $\log_3 \dfrac{1}{27}$

8. $\log_{10} 10^{12} + 10^{\log_{10} 5}$

9. $\log_{3/2} \dfrac{9}{4}$

10. $\log_{2/3} \dfrac{27}{8}$

Graph each function.

11. $f(x) = \log(x - 1)$

12. $f(x) = 2 + \ln x$

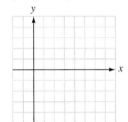

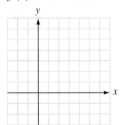

Write each expression in terms of the logarithms of a, b, and c.

13. $\log a^2bc^3$

14. $\ln\sqrt{\dfrac{a}{b^2c}}$

Write each expression as a logarithm of a single quantity.

15. $\dfrac{1}{2} \log(a + 2) + \log b - 2 \log c$

16. $\dfrac{1}{3} (\ln a - 2 \ln b) - \ln c$

Given that $\log 2 \approx 0.3010$ and $\log 3 \approx 0.4771$, approximate each value. Do not use a calculator.

17. $\log 24$

18. $\log \dfrac{8}{3}$

Use the Change-of-Base Formula to find each logarithm.
Do not attempt to simplify the answer.

19. $\log_7 3$

20. $\log_\pi e$

Determine whether each statement is true or false.

21. $\log_a ab = 1 + \log_a b$

22. $\dfrac{\log a}{\log b} = \log a - \log b$

Find the solution.

23. pH of a solution Find the pH of a solution with a hydrogen ion concentration of 3.7×10^{-7}.
(*Hint:* $\text{pH} = -\log[\text{H}^+]$.)

24. Decibel gain Find the dB gain of an amplifier when $E_O = 60$ volts and $E_I = 0.3$ volt.
(*Hint:* dB gain $= 20 \log(E_O/E_I)$.)

Solve each equation.

25. $3^{x^2 - 2x} = 27$

26. $3^{x-1} = 100^x$

27. $5e^x = 45$

28. $\ln(5x + 2) = \ln(2x + 5)$

29. $\log x + \log(x - 9) = 1$

30. $\log_6 18 - \log_6(x - 3) = \log_6 3$

Solving Polynomial Equations

5

CAREERS AND MATHEMATICS: Teacher

AVAVA/Shutterstock.com

Teachers play an important role in fostering the intellectual and social development of students. Whether in elementary school, middle school, or high school, teachers provide the tools and the environment for their students to develop into responsible adults. Teaching is often cited as one of the most challenging, yet fulfilling careers.

Education and Mathematics Required

• Teachers begin their careers by earning a bachelor's degree. Some also earn a master's degree in education. Requirements for regular licenses to teach kindergarten through grade 12 vary by state.

• Elementary and middle school teachers are required to take College Algebra, Survey of the Real Number System, Concepts of Geometry, Measurement and Probability, and Problem Solving. Additional courses such as Trigonometry and Calculus I and II are often required for teachers who plan to concentrate in the mathematics area.

• High school math teachers are required to take Calculus I and II, Linear Algebra, Calculus of Several Variables, Ordinary Differential Equations, Theory of Analysis, Abstract Algebra, and Statistics.

How Teachers Use Math and Who Employs Them

• Teachers teach problem solving skills to students. Middle school math teachers and high school math teachers demonstrate sample math problems in class, create math assignments for students, and write and grade math tests.

• Most teachers work in public or private schools.

Career Outlook and Salary

• Employment of kindergarten, elementary, middle, and secondary school teachers is expected to grow by 13 percent between 2008 and 2018.

• Median annual wages of kindergarten, elementary, middle, and secondary school teachers ranged from \$47,100 to \$51,180 in May 2008.

For more information see: www.bls.gov/oco

A goal throughout all of algebra has been to solve equations. In this chapter, we will develop methods to solve polynomial equations of any degree.

5.1 The Remainder and Factor Theorems; Synthetic Division

In this section, we will learn to

1. Understand the definition of a zero of a polynomial function.
2. Use the Remainder Theorem.
3. Use the Factor Theorem.
4. Use synthetic division to divide polynomials.
5. Use synthetic division to evaluate polynomial functions.
6. Use synthetic division to solve polynomial equations.

The cheetah is the fastest land animal and is known for its speed and stealth. Cheetahs can run at speeds between 70 mph and 75 mph. They have the ability to accelerate from 0 mph to 68 mph in three seconds. This acceleration is greater than that of most sports cars.

Although we will never run as fast as a cheetah, we do live in a society that wants convenience and speed. We demand fast service, fast food, fast communication, fast technology, and fast vehicles.

In this section, we will learn about synthetic division, a fast way to divide certain polynomials. Synthetic division can help us find the roots of many polynomial equations.

1. Understand the Definition of a Zero of a Polynomial Function

We have seen that many polynomial equations can be solved by factoring. For example, to solve the third-degree polynomial equation $x^3 - 3x^2 + 2x = 0$, we proceed as follows:

$$x^3 - 3x^2 + 2x = 0$$

$$x(x^2 - 3x + 2) = 0 \qquad \text{Factor out } x.$$

$$x(x - 1)(x - 2) = 0 \qquad \text{Factor } x^2 - 3x + 2.$$

$$x = 0 \quad \text{or} \quad x - 1 = 0 \quad \text{or} \quad x - 2 = 0 \qquad \text{Set each factor equal to 0.}$$
$$x = 1 \qquad\qquad x = 2$$

The solution set of this equation is $\{0, 1, 2\}$.

In this chapter, we will solve higher-degree polynomial equations. In general, an nth-degree polynomial equation is defined as follows:

Polynomial Equation

A **polynomial equation** is an equation that can be written in the form $P(x) = 0$, where

$$P(x) = a_n x^n + a_{n-1}x^{n-1} + a_{n-2}x^{n-2} + \cdots + a_1 x + a_0$$

and where n is a natural number and the polynomial is of degree n.

Any number that makes $P(x) = 0$ when substituted for x is called a *zero of the polynomial function.*

Zero of a Polynomial Function

A **zero of the polynomial function** $P(x)$ is any number r for which $P(r) = 0$.

The real numbers 0, 1, and 2 are zeros of the polynomial function

$P(x) = x^3 - 3x^2 + 2x$ because

$$P(0) = 0^3 - 3(0)^2 + 2(0) \qquad P(1) = 1^3 - 3(1)^2 + 2(1) \qquad P(2) = 2^3 - 3(2)^2 + 2(2)$$

$$= 0 - 0 + 0 \qquad\qquad\qquad = 1 - 3 + 2 \qquad\qquad\qquad = 8 - 12 + 4$$

$$= 0 \qquad\qquad\qquad\qquad\quad = 0 \qquad\qquad\qquad\qquad\quad = 0$$

As shown, the roots 0, 1, and 2 of the equation $x^3 - 3x^2 + 2x = 0$ are also zeros of the polynomial function $P(x) = x^3 - 3x^2 + 2x$. In general, the roots of the polynomial equation $P(x) = 0$ are the zeros of the polynomial function $P(x)$.

2. Use the Remainder Theorem

There is a relationship between a zero r of a polynomial function $P(x)$ and the results of a long division of $P(x)$ by $x - r$. We will examine this relationship in this section.

EXAMPLE 1 **Evaluating a Polynomial Function and Long Division**

Let $P(x) = 3x^3 - 5x^2 + 3x - 10$.
a. Find $P(1)$ **b.** Divide $P(x)$ by $x - 1$

SOLUTION   **a.** To find $P(1)$, we will substitute 1 for x in the polynomial function.

$$P(1) = 3(1)^3 - 5(1)^2 + 3(1) - 10$$

$$= 3 - 5 + 3 - 10$$

$$= -9$$

b. To divide $P(x)$ by $x - 1$, we proceed as follows:

$$
\require{enclose}
\begin{array}{r}
3x^2 - 2x + 1 \\
x - 1 \enclose{longdiv}{3x^3 - 5x^2 + 3x - 10} \\
\underline{3x^3 - 3x^2} \\
-2x^2 + 3x \\
\underline{-2x^2 + 2x} \\
+\ x - 10 \\
\underline{x - 1} \\
-\ 9
\end{array}
$$

Note that the remainder is equal to $P(1)$.

Self Check 1   Let $P(x) = 2x^2 - 3x + 5$. Find $P(2)$ and divide $P(x)$ by $x - 2$. What do you notice about the results?

Now Try Exercise 11.

EXAMPLE 2 **Evaluating a Polynomial Function and Long Division**

Let $P(x) = 3x^3 - 5x^2 + 3x - 10$.
a. Find $P(-2)$ **b.** Divide $P(x)$ by $x + 2$

SOLUTION **a.** To find $P(-2)$, we will substitute -2 for x in the polynomial.

$$P(-2) = 3(-2)^3 - 5(-2)^2 + 3(-2) - 10$$
$$= 3(-8) - 5(4) + 3(-2) - 10$$
$$= -24 - 20 - 6 - 10$$
$$= -60$$

b. To divide $P(x)$ by $x + 2$, we proceed as follows:

$$
\begin{array}{r}
3x^2 - 11x + 25 \\
x + 2\overline{)3x^3 - 5x^2 + 3x - 10} \\
\underline{3x^3 + 6x^2} \\
-11x^2 + 3x \\
\underline{-11x^2 - 22x} \\
25x - 10 \\
\underline{25x + 50} \\
-60
\end{array}
$$

Note that the remainder is equal to $P(-2)$.

Self Check 2 Let $P(x) = 2x^2 - 3x + 5$. Find $P(-3)$ and divide $P(x)$ by $x + 3$. What do you notice about the results?

Now Try Exercise 13.

When $P(x)$ was divided by $x - 1$ in Example 1, the remainder was $P(1) = -9$. When $P(x)$ was divided by $x + 2$, or $x - (-2)$, in Example 2, the remainder was $P(-2) = -60$. These results are not coincidental. The **Remainder Theorem** states that a division of any polynomial function $P(x)$ by $x - r$ gives $P(r)$ as the remainder.

The Remainder Theorem If $P(x)$ is a polynomial function, r is any number, and $P(x)$ is divided by $x - r$, the remainder is $P(r)$.

PROOF To divide $P(x)$ by $x - r$, we must find a quotient $Q(x)$ and a remainder $R(x)$ such that

Dividend $=$ divisor $\cdot$ quotient $+$ remainder

$$P(x) = (x - r) \cdot Q(x) + R(x)$$

Since the degree of the remainder $R(x)$ must be less than the degree of the divisor $x - r$, and the degree of $x - r$ is 1, $R(x)$ must be a constant R.

In the equation

$$P(x) = (x - r)Q(x) + R$$

the polynomial on the left side is the same as the polynomial on the right side, and the values that they assume for any number x are equal. If we replace x with r, we have

$$P(r) = (r - r)Q(r) + R$$
$$= (0)Q(r) + R$$
$$= R$$

Thus, $P(r) = R$.

EXAMPLE 3   **Using the Remainder Theorem**

Use the Remainder Theorem to find the remainder that will occur when $P(x) = 2x^4 - 10x^3 + 17x^2 - 14x - 3$ is divided by $x - 3$.

SOLUTION   By the Remainder Theorem, the remainder will be $P(3)$.

$$P(x) = 2x^4 - 10x^3 + 17x^2 - 14x - 3$$

$$P(3) = 2(3)^4 - 10(3)^3 + 17(3)^2 - 14(3) - 3 \qquad \text{Substitute 3 for } x.$$

$$= 0$$

The remainder will be 0. Although this calculation is tedious, it is easy to do with a calculator.   ∎

Self Check 3   Find the remainder when $P(x)$ is divided by $x - 2$.

Now Try Exercise 17.

3. Use the Factor Theorem

If $R = P(r) = 0$ in the equation $P(x) = (x - r)Q(x) + R$, then $P(x)$ factors as $(x - r)Q(x)$. This fact can help us factor polynomials.

The Factor Theorem   If $P(x)$ is a polynomial function and r is any number, then

If $P(r) = 0$, then $x - r$ is a factor of $P(x)$.

If $x - r$ is a factor of $P(x)$, then $P(r) = 0$.

PROOF   **Part 1:** First, we assume that $P(r) = 0$ and prove that $x - r$ is a factor of $P(x)$. If $P(r) = 0$, then $R = 0$, and the equation $P(x) = (x - r)Q(x) + R$ becomes

$$P(x) = (x - r)Q(x) + 0$$

$$P(x) = (x - r)Q(x)$$

Therefore, $x - r$ divides $P(x)$ exactly, and $x - r$ is a factor of $P(x)$.

Part 2: Conversely, we assume that $x - r$ is a factor of $P(x)$ and prove that $P(r) = 0$. Because, by assumption, $x - r$ is a factor of $P(x)$, $x - r$ divides $P(x)$ exactly, and the division has a remainder of 0. By the Remainder Theorem, this remainder is $P(r)$. Hence, $P(r) = 0$.   ∎

Below, we state the Factor Theorem in a slightly different way.

Alternate Form of the Factor Theorem   If r is a zero of the polynomial function $P(x)$, then $x - r$ is a factor of $P(x)$.

If $x - r$ is a factor of $P(x)$, then r is a zero of the polynomial.

It is important to note that for each factor of a polynomial function, there corresponds a zero. Several possible factors of a polynomial function are shown next with their corresponding zeros.

Factor	Zero
$x - 5$	5
$x + 2$	-2
$x - \dfrac{1}{2}$	$\dfrac{1}{2}$
$x + \dfrac{2}{3}$	$-\dfrac{2}{3}$

EXAMPLE 4   **Using the Factor Theorem**

Determine whether $x + 2$ is a factor of $P(x) = x^4 - 7x^2 - 6x$.

SOLUTION   By the Factor Theorem, $x + 2$, or $x - (-2)$, is a factor of $P(x)$ if $P(-2) = 0$. So we find the value of $P(-2)$.

$$P(x) = x^4 - 7x^2 - 6x$$

$$P(-2) = (-2)^4 - 7(-2)^2 - 6(-2) \qquad \text{Substitute } -2 \text{ for } x.$$

$$= 16 - 28 + 12$$

$$= 0$$

Comment

In Example 4, $x + 2$ is a factor of the polynomial function $P(x) = x^4 - 7x^2 - 6x$. By the Factor Theorem, we know that -2 is a zero of the polynomial function.

Since $P(-2) = 0$, we know that $x - (-2)$, or $x + 2$, is a factor of $P(x)$. ∎

Self Check 4   Determine whether $x + 3$ is a factor of $P(x)$.

Now Try Exercise 27.

4. Use Synthetic Division to Divide Polynomials

Synthetic division is an easy way to divide higher-degree polynomials by binomials of the form $x - r$, and it is much faster than long division. To see how it works, we consider the following long division. On the left is a complete division. On the right is a modified version in which the variables have been removed.

$$
\begin{array}{r}
2x^2 + 10x + 27 \\
x - 3\overline{)2x^3 + 4x^2 - 3x + 10} \\
\underline{2x^3 - 6x^2} \\
10x^2 - 3x \\
\underline{10x^2 - 30x} \\
27x + 10 \\
\underline{27x - 81} \\
\text{(remainder) } 91
\end{array}
\qquad
\begin{array}{r}
2 + 10 + 27 \\
1 - 3\overline{)2 + 4 - 3 + 10} \\
\underline{2 - 6} \\
10 - 3 \\
\underline{10 - 30} \\
27 + 10 \\
\underline{27 - 81} \\
\text{(remainder) } 91
\end{array}
$$

We can shorten the work even more by omitting the numbers printed in color.

$$
\begin{array}{r}
2 + 10 + 27 \\
-3\overline{)2 + 4 - 3 + 10} \\
\underline{-6} \\
10 - \\
\underline{-30} \\
27 \\
\underline{-81} \\
\text{(remainder) } 91
\end{array}
$$

We can then compress the work vertically to get

$$
\begin{array}{r}
2 + 10 + 27 \\
-3\overline{)2 + 4 - 3 + 10} \\
-6 - 30 - 81 \\
\hline
10 \quad 27 \quad 91
\end{array}
$$

If we write the 2 in the quotient on the bottom line, the bottom line gives both the coefficients of the quotient and the remainder. The top line can now be eliminated, and the division appears as

$$
\begin{array}{r}
\underline{-3}\,|\,2 + 4 - 3 + 10 \\
-6 - 30 - 81 \\
\hline
2 \quad 10 \quad 27 \quad 91
\end{array}
$$

The bottom line was obtained by subtracting the middle line from the top line. If we replace the -3 in the divisor with $+3$, the signs of each number in the middle line will be reversed in the division process. Then the bottom line can be obtained by addition, and we have the final form of the synthetic division.

$$
\begin{array}{r}
\underline{+3}\,|\,2 + 4 - 3 + 10 \\
+6 + 30 + 81 \\
\hline
2 \quad 10 \quad 27 \quad 91
\end{array}
$$

These are the coefficients of the dividend.

These are the coefficients of the quotient and the remainder.

Thus,

$$
\frac{2x^3 + 4x^2 - 3x + 10}{x - 3} = 2x^2 + 10x + 27 + \frac{91}{x - 3}
$$

EXAMPLE 5   **Using Synthetic Division to Divide Polynomials**

Use synthetic division to divide $10x + 3x^4 - 8x^3 + 3$ by $x - 2$.

SOLUTION   We will first write the terms in descending powers of x:

$$3x^4 - 8x^3 + 10x + 3$$

We then write the coefficients of the dividend, with its terms in descending powers of x, and the 2 from the divisor in the following form:

$$\underline{2}\,|\,3 \quad -8 \quad 0 \quad 10 \quad 3$$

Write 0 for the coefficient of the missing x^2 term.

Then we follow these steps:

$$
\begin{array}{r}
\underline{2}\,|\,3 \quad -8 \quad 0 \quad 10 \quad 3 \\
\downarrow \\
\hline
3
\end{array}
$$

Bring down the 3.

$$
\begin{array}{r}
\underline{2}\,|\,3 \quad -8 \quad 0 \quad 10 \quad 3 \\
6 \\
\hline
3 \quad -2
\end{array}
$$

Multiply 2 and 3 together to get 6, and add 6 and -8 to get -2.

$$
\begin{array}{r}
\underline{2}\,|\,3 \quad -8 \quad 0 \quad 10 \quad 3 \\
6 \quad -4 \\
\hline
3 \quad -2 \quad -4
\end{array}
$$

Multiply 2 and -2 together to get -4, and add -4 and 0 to get -4.

$$\begin{array}{r|rrrr} 2 & 3 & -8 & 0 & 10 & 3 \\ & & 6 & -4 & -8 & \\ \hline & 3 & -2 & -4 & 2 \end{array}$$

Multiply 2 and -4 together to get -8, and add -8 and 10 to get 2.

$$\begin{array}{r|rrrrr} 2 & 3 & -8 & 0 & 10 & 3 \\ & & 6 & -4 & -8 & 4 \\ \hline & 3 & -2 & -4 & 2 & 7 \end{array}$$

Multiply 2 and 2 together to get 4, and add 4 and 3 to get 7. These are the coefficients of the quotient and the remainder.

Thus,

$$\frac{10x + 3x^4 - 8x^3 + 3}{x - 2} = 3x^3 - 2x^2 - 4x + 2 + \frac{7}{x - 2}$$

Self Check 5   Divide $2x^3 - 4x^2 + 5x - 7$ by $x - 3$.

Now Try Exercise 43.

5. Use Synthetic Division to Evaluate Polynomial Functions

EXAMPLE 6   **Using Synthetic Division to Evaluate a Polynomial Function**

Use synthetic division to find $P(-2)$ when $P(x) = 5x^3 + 3x^2 - 21x - 1$.

SOLUTION   Because of the Remainder Theorem, $P(-2)$ is the remainder when $P(x)$ is divided by $x - (-2)$. Earlier in the section, we would find the remainder by using long division. We can now simplify the work and find the remainder by using synthetic division.

$$\begin{array}{r|rrrr} -2 & 5 & 3 & -21 & -1 \\ & & -10 & & \\ \hline & 5 & -7 & & \end{array}$$

$-2(5) = -10;\ 3 + (-10) = -7$

$$\begin{array}{r|rrrr} -2 & 5 & 3 & -21 & -1 \\ & & -10 & 14 & \\ \hline & 5 & -7 & -7 & \end{array}$$

$-2(-7) = 14;\ -21 + 14 = -7$

$$\begin{array}{r|rrrr} -2 & 5 & 3 & -21 & -1 \\ & & -10 & 14 & 14 \\ \hline & 5 & -7 & -7 & 13 \end{array}$$

$-2(-7) = 14;\ -1 + 14 = 13$

Because the remainder is 13, $P(-2) = 13$.

Self Check 6   Find $P(3)$.

Now Try Exercise 49.

EXAMPLE 7   **Using Synthetic Division to Evaluate a Polynomial Function at i**

If $P(x) = x^3 - x^2 + x - 1$, find $P(i)$, where $i = \sqrt{-1}$.

SOLUTION We will use synthetic division and find the remainder.

$$
\begin{array}{r|rrrr}
\underline{i} & 1 & -1 & 1 & -1 \\
 & & i & -1-i & 1 \\
\hline
 & 1 & i-1 & -i & 0
\end{array}
$$

Comment

Example 7 illustrates that synthetic division can be used with complex numbers.

Since the remainder is 0, $P(i) = 0$ and i is a zero of $P(x)$. ∎

Self Check 7 Find $P(-i)$.

Now Try Exercise 53.

6. Use Synthetic Division to Solve Polynomial Equations

EXAMPLE 8 **Solving a Polynomial Equation When Given One Solution**

Let $P(x) = 3x^3 - 5x^2 + 3x - 10$. Completely solve the polynomial equation $P(x) = 0$ given that 2 is one solution.

SOLUTION Since 2 is a solution of the equation $P(x) = 0$, we know that 2 is a zero of $P(x)$. We will use synthetic division and divide $P(x)$ by $x - 2$, obtaining a remainder of 0. Then, we will use the result of the synthetic division to help factor the polynomial and solve the equation.

1. We use synthetic division to divide $P(x)$ by $x - 2$.

$$
\begin{array}{r|rrrr}
\underline{2} & 3 & -5 & 3 & -10 \\
 & & 6 & 2 & 10 \\
\hline
 & 3 & 1 & 5 & 0
\end{array}
$$

2. We then write the quotient and factor it.

$$3x^3 - 5x^2 + 3x - 10 = (x - 2)(3x^2 + x + 5)$$

3. Finally, we solve the polynomial equation $P(x) = 0$.

$$3x^3 - 5x^2 + 3x - 10 = 0$$

$$(x - 2)(3x^2 + x + 5) = 0$$

To solve for x, we set each factor equal to 0 and apply the Quadratic Formula to the equation $3x^2 + x + 5 = 0$.

$$x - 2 = 0 \quad \text{or} \quad 3x^2 + x + 5 = 0$$

$$x = 2 \qquad\qquad x = \dfrac{-1 \pm \sqrt{1^2 - 4(3)(5)}}{2(3)}$$

$$x = \dfrac{-1 \pm i\sqrt{59}}{6}$$

The solution set is $\left\{ 2, -\dfrac{1}{6} + \dfrac{\sqrt{59}}{6}i, -\dfrac{1}{6} - \dfrac{\sqrt{59}}{6}i \right\}$. ∎

Self Check 8 Solve: $x^3 - 1 = 0$ given that 1 is one solution.

Now Try Exercise 75.

EXAMPLE 9   **Finding a Polynomial Function When Given the Zeros**

Find a third-degree polynomial function $P(x)$ with three zeros of 3, 3, and -5.

SOLUTION   We will use the Factor Theorem and write the three factors that correspond to the three zeros of 3, 3, and -5. We will then multiply the resulting binomials.

If 3, 3, and -5 are the three zeros of $P(x)$, then $x - 3$, $x - 3$, and $x - (-5)$ are the three factors of $P(x)$.

$$P(x) = (x - 3)(x - 3)(x + 5)$$
$$P(x) = (x^2 - 6x + 9)(x + 5) \qquad \text{Multiply } x - 3 \text{ and } x - 3.$$
$$P(x) = x^3 - x^2 - 21x + 45 \qquad \text{Multiply using the Distributive Property.}$$

The polynomial function $P(x) = x^3 - x^2 - 21x + 45$ has zeros of 3, 3, and -5. Because 3 occurs twice as a zero, we say that 3 is a **zero of multiplicity 2.**

Self Check 9   Find a polynomial function $P(x)$ with zeros of -2, 2, and 3.

Now Try Exercise 85.

Self Check Answers   **1.** $P(2)$ is the remainder.   **2.** $P(-3)$ is the remainder.   **3.** -11 **4.** no

5. $2x^2 + 2x + 11 + \dfrac{26}{x - 3}$ **6.** 98   **7.** 0   **8.** $1, -\dfrac{1}{2} + \dfrac{\sqrt{3}}{2}i, -\dfrac{1}{2} - \dfrac{\sqrt{3}}{2}i$

9. $P(x) = x^3 - 3x^2 - 4x + 12$

Exercises **5.1**

Getting Ready
You should be able to complete these vocabulary and concept statements before you proceed to the practice exercises.

Fill in the blanks.

1. The variables in a polynomial have _____ -number exponents.

2. A zero of $P(x)$ is any number r for which _____.

3. The Remainder Theorem holds when r is _____ number.

4. If $P(x)$ is a polynomial function and $P(x)$ is divided by _____, the remainder will be $P(r)$.

5. If $P(x)$ is a polynomial function, then $P(r) = 0$ if and only if $x - r$ is a _____ of $P(x)$.

6. A shortcut method for dividing a polynomial by a binomial of the form $x - r$ is called _____ division.

Practice
Use long division to perform each division.

7. $\dfrac{4x^3 - 2x^2 - x + 1}{x - 1}$

8. $\dfrac{2x^3 + 3x^2 - 5x + 1}{x + 3}$

9. $\dfrac{2x^4 + x^3 + 2x^2 + 15x - 5}{x + 2}$

10. $\dfrac{x^4 + 6x^3 - 2x^2 + x - 1}{x - 1}$

Find each value by substituting the given value of x into the polynomial and simplifying. Then find the value by performing long division and finding the remainder.

11. $P(x) = 3x^3 - 2x^2 - 5x - 7; \; P(2)$

12. $P(x) = 5x^3 + 4x^2 + x - 1; \; P(-2)$

13. $P(x) = 7x^4 + 2x^3 + 5x^2 - 1; \; P(-1)$

14. $P(x) = 2x^4 - 2x^3 + 5x^2 - 1; \; P(2)$

15. $P(x) = 2x^5 + x^4 - x^3 - 2x + 3; \; P(1)$

16. $P(x) = 3x^5 + x^4 - 3x^2 + 5x + 7; \; P(-2)$

Use the Remainder Theorem to find the remainder that occurs when $P(x) = 3x^4 + 5x^3 - 4x^2 - 2x + 1$ is divided by each binomial. Use a calculator on Exercises 21–24.

17. $x + 2$ 18. $x - 1$

19. $x - 2$ 20. $x + 1$

21. $x - 12$ 22. $x + 15$

23. $x + 3.25$ 24. $x - 7.12$

Use the Factor Theorem to determine whether each statement is true. If the statement is not true, so indicate.

25. $x - 1$ is a factor of $P(x) = x^7 - 1$.

26. $x - 2$ is a factor of $P(x) = x^3 - x^2 + 2x - 8$.

27. $x - 1$ is a factor of $P(x) = 3x^5 + 4x^2 - 7$.

28. $x + 1$ is a factor of $P(x) = 3x^5 + 4x^2 - 7$.

29. $x + 3$ is a factor of $P(x) = 2x^3 - 2x^2 + 1$.

30. $x - 3$ is a factor of
$P(x) = 3x^5 - 3x^4 + 5x^2 - 13x - 6$.

31. $x - 1$ is a factor of
$P(x) = x^{1,984} - x^{1,776} + x^{1,492} - x^{1,066}$.

32. $x + 1$ is a factor of
$P(x) = x^{1,984} + x^{1,776} - x^{1,492} - x^{1,066}$.

Use synthetic division to express the polynomial function $P(x) = 3x^3 - 2x^2 - 6x - 4$ in the form (divisor)(quotient) + remainder for each divisor.

33. $x - 1$

34. $x - 2$

35. $x - 3$

36. $x - 4$

37. $x + 1$

38. $x + 2$

39. $x + 3$

40. $x + 4$

Use synthetic division to perform each division.

41. $\dfrac{x^3 + x^2 + x - 3}{x - 1}$

42. $\dfrac{x^3 - x^2 - 5x + 6}{x - 2}$

43. $\dfrac{7x^3 - 3x^2 - 5x + 1}{x + 1}$

44. $\dfrac{2x^3 + 4x^2 - 3x + 8}{x - 3}$

45. $\dfrac{4x^4 - 3x^3 - x + 5}{x - 3}$

46. $\dfrac{x^4 + 5x^3 - 2x^2 + x - 1}{x + 1}$

47. $\dfrac{3x^5 - 768x}{x - 4}$

48. $\dfrac{x^5 - 4x^2 + 4x + 4}{x + 3}$

Let $P(x) = 5x^3 + 2x^2 - x + 1$. Use synthetic division to find each value.

49. $P(2)$ **50.** $P(-2)$

51. $P(-5)$ **52.** $P(3)$

53. $P(i)$ **54.** $P(-i)$

Let $P(x) = 2x^4 - x^2 + 2$. Use synthetic division to find each value.

55. $P\left(\dfrac{1}{2}\right)$ **56.** $P\left(\dfrac{1}{3}\right)$

57. $P(i)$ **58.** $P(-i)$

Let $P(x) = x^4 - 8x^3 + 8x + 14x^2 - 15$. Write the terms of $P(x)$ in descending powers of x and use synthetic division to find each value.

59. $P(1)$ **60.** $P(0)$

61. $P(-3)$ **62.** $P(-1)$

63. $P(-i)$ **64.** $P(i)$

Let $P(x) = 8 - 8x^2 + x^5 - x^3$. Write the terms of $P(x)$ in descending powers of x and use synthetic division to find each value.

65. $P(i)$ **66.** $P(-i)$

67. $P(-2i)$ **68.** $P(2i)$

A partial solution set is given for each polynomial equation. Find the complete solution set.

69. $x^3 + 3x^2 - 13x - 15 = 0; \{-1\}$

70. $x^3 + 6x^2 + 5x - 12 = 0; \{1\}$

71. $2x^3 + x^2 - 18x - 9 = 0; \left\{-\dfrac{1}{2}\right\}$

72. $2x^3 - 3x^2 - 11x + 6 = 0; \left\{\dfrac{1}{2}\right\}$

73. $x^3 - 6x^2 + 7x + 2; \{2\}$

74. $x^3 + x^2 - 8x - 6; \{-3\}$

75. $x^3 - 3x^2 + x + 57; \{-3\}$

76. $2x^3 - x^2 + x - 2; \{1\}$

77. $x^4 - 2x^3 - 2x^2 + 6x - 3 = 0; \{1, 1\}$

78. $x^5 + 4x^4 + 4x^3 - x^2 - 4x - 4 = 0; \{1, -2, -2\}$

79. $x^4 - 5x^3 + 7x^2 - 5x + 6 = 0; \{2, 3\}$

80. $x^4 + 2x^3 - 3x^2 - 4x + 4 = 0; \{1, -2\}$

Find a polynomial function $P(x)$ with the given zeros.

81. $4, 5$

82. $-3, 5$

83. $1, 1, 1$

84. $1, 0, -1$

85. $2, 4, 5$

86. $7, 6, 3$

87. $1, -1, \sqrt{2}, -\sqrt{2}$

88. $0, 0, 0, \sqrt{3}, -\sqrt{3}$

89. $\sqrt{2}, i, -i$

90. $i, i, 2$

91. $0, 1 + i, 1 - i$

92. $i, 2 + i, 2 - i$

Discovery and Writing

93. If 0 is a zero of
$P(x) = a_n x^n + a_{n-1} x^{n-1} + \cdots + a_1 x + a_0$,
find a_0.

94. If 0 occurs twice as a zero of
$P(x) = a_n x^n + a_{n-1} x^{n-1} + \cdots + a_1 x + a_0$,
find a_1.

95. If $P(2) = 0$ and $P(-2) = 0$, explain why $x^2 - 4$ is a factor of $P(x)$.

96. If $P(x) = x^4 - 3x^3 + kx^2 + 4x - 1$ and $P(2) = 11$, find k.

Review

Find the quadrant in which each point lies.

97. $P(3, -2)$

98. $Q(-2, -5)$

99. $R(8, \pi)$

100. $S(-9, 9)$

Find the distance between each pair of points.

101. $A(3, -3), B(-5, 3)$

102. $C(-8, 2), D(2, -22)$

Find the slope of the line passing through each pair of points.

103. $E(3, 5), F(-5, -3)$

104. $G\left(3, \dfrac{3}{5}\right), H\left(-\dfrac{3}{5}, 1\right)$

5.2 Descartes' Rule of Signs and Bounds on Roots

In this section, we will learn to

1. Understand the Fundamental Theorem of Algebra.

2. Use the Conjugate Pairs Theorem.

3. Use Descartes' Rule of Signs.

4. Find integer bounds on roots.

Englishman David Beckham currently plays major league soccer for the Los Angeles Galaxy. Beckham is one of the highest paid players in the world, and he has twice been runner-up for the *International Federation of Association Football's* player of the year award.

In soccer, it is important for players to know the boundaries of the playing field. They need to know when it is advantageous to kick the ball out of bounds and when it is advantageous to keep the ball in bounds.

Knowing where the boundaries are can also be helpful in mathematics. Establishing integer bounds on roots can be an important aid in locating and finding the roots of polynomial equations. In this section, we will learn how to find integer bounds for the roots.

The Remainder Theorem and synthetic division provide a way of verifying that a particular number is a root of a polynomial equation, but they do not provide the roots. We need some guidelines that indicate how many roots to expect, what kind of roots to expect, and where they are located. This section develops several theorems that provide such guidelines.

1. Understand the Fundamental Theorem of Algebra

Before attempting to find the roots of a polynomial equation, it would be useful to know whether any roots exist. This question was answered by Carl Friedrich Gauss (1777–1855) when he proved the **Fundamental Theorem of Algebra.**

The Fundamental Theorem of Algebra	If $P(x)$ is a polynomial function with positive degree, then $P(x)$ has at least one zero.

The Fundamental Theorem of Algebra guarantees that polynomials such as

$$P(x) = 2x^3 - 2x^2 + 7x - 3 \text{ and } P(x) = 32x^{11} + x^3 - 2x - 5$$

all have zeros. Since all polynomials with positive degree have zeros, their corresponding polynomial equations all have roots. They are the zeros of the polynomial function.

The next theorem will help us show that every nth-degree polynomial equation $P(x) = 0$ has exactly n roots.

The Polynomial Factorization Theorem	If $n > 0$ and $P(x)$ is an nth-degree polynomial function, then $P(x)$ has exactly n linear factors: $$P(x) = a_n(x - r_1)(x - r_2)(x - r_3) \cdot \cdots \cdot (x - r_n)$$ where $r_1, r_2, r_3, \ldots, r_n$ are numbers and a_n is the leading coefficient of $P(x)$.

PROOF   Let $P(x)$ be a polynomial function of degree n $(n > 0)$. Because of the Fundamental Theorem of Algebra, we know that $P(x)$ has a zero r_1 and that the equation $P(x) = 0$ has r_1 for a root. By the Factor Theorem, we know that $x - r_1$ is a factor of $P(x)$. Thus,

$$P(x) = (x - r_1)Q_1(x)$$

If the leading coefficient of the nth-degree polynomial function $P(x)$ is a_n, then $Q_1(x)$ is a polynomial function of degree $n - 1$ whose leading coefficient is also a_n.

By the Fundamental Theorem of Algebra, we know that $Q_1(x)$ also has a zero, r_2. By the Factor Theorem, $x - r_2$ is a factor of $Q_1(x)$, and

$$P(x) = (x - r_1)(x - r_2)Q_2(x)$$

where $Q_2(x)$ is a polynomial function of degree $n - 2$ with leading coefficient a_n.

This process can continue only to n factors of the form $x - r_i$ until the final quotient $Q_n(x)$ is a polynomial function of degree $n - n$, or degree 0. Thus, the polynomial function $P(x)$ factors completely as

$$(1) \qquad P(x) = a_n(x - r_1)(x - r_2)(x - r_3) \cdot \cdots \cdot (x - r_n) \qquad \blacksquare$$

We can use the Polynomial Factorization Theorem and make the following conclusions:

1. If we substitute any one of the numbers $r_1, r_2, r_3, \ldots, r_n$ for x in Equation 1, $P(x)$ will equal 0. Thus, each value of r is a zero of $P(x)$ and a root of the equation $P(x) = 0$.

2. There can be no other roots, because no single factor in Equation 1 is 0 for any value of x not included in the list $r_1, r_2, r_3, \ldots, r_n$.
3. The values of r in the previous list need not be distinct. Any number r_i that occurs k times as a root of a polynomial equation is called a **root of multiplicity k.**

The following theorem summarizes the previous discussion.

Number of Roots Theorem   If multiple roots are counted individually, the polynomial equation $P(x) = 0$ with degree n $(n > 0)$ has exactly n roots among the complex numbers.

The previous theorems are illustrated by the examples shown below.

Polynomial Function	Properties
$P(x) = x - 6$	**Degree:** First **Number of Linear Factors:** 1 **Linear Factor:** $x - 6$ **Number of Zeros:** 1 **Zero:** 6 **Multiplicity of Each Zero:** • 6 has a multiplicity of one.
$P(x) = x^2 - 8x + 16$ $\quad = (x - 4)(x - 4)$	**Degree:** Second **Number of Linear Factors:** 2 **Linear Factors:** $x - 4$, $x - 4$ **Number of Zeros:** 2 **Zeros:** 4 and 4 **Multiplicity of Each Zero:** • 4 has a multiplicity of two.
$P(x) = x^3 + 100x$ $\quad = x(x^2 + 100)$ $\quad = x(x - 10i)(x + 10i)$	**Degree:** Third **Number of Linear Factors:** 3 **Linear Factors:** $x, x - 10i$, and $x + 10i$ **Number of Zeros:** 3 **Zeros:** 0, $10i$, and $-10i$ **Multiplicity of Each Zero:** • 0 has a multiplicity of one. • $10i$ has a multiplicity of one. • $-10i$ has a multiplicity of one.
$P(x) = x^4 - 7x^2$ $\quad = x^2(x^2 - 7)$ $\quad = x \cdot x\left(x + \sqrt{7}\right)\left(x - \sqrt{7}\right)$	**Degree:** Fourth **Number of Linear Factors:** 4 **Linear Factors:** $x, x, x + \sqrt{7}, x - \sqrt{7}$ **Number of Zeros:** 4 **Zeros:** $0, 0, -\sqrt{7}$, and $\sqrt{7}$ **Multiplicity of Each Zero:** • 0 has a multiplicity of two. • $-\sqrt{7}$ has a multiplicity of one. • $\sqrt{7}$ has a multiplicity of one.

Comment   Remember that every nth degree polynomial function has exactly n linear factors and exactly n zeros. The n zeros are also the roots of the polynomial equation $P(x) = 0$.

2. Use the Conjugate Pairs Theorem

Recall that the complex numbers $a + bi$ and $a - bi$ are called **complex conjugates** of each other. The next theorem points out that *complex roots of polynomial equations with real coefficients occur in complex conjugate pairs.*

The Conjugate Pairs Theorem   If a polynomial equation $P(x) = 0$ with real-number coefficients has a complex root $a + bi$ with $b \neq 0$, then its conjugate $a - bi$ is also a root.

A strategy is given next for finding a polynomial equation.

Strategy for Finding a Polynomial Equation with Real Coefficients

To find a polynomial equation with real coefficients, apply the following two steps.

1. Use the Conjugate Pairs Theorem to determine the roots of the polynomial equation.
2. Write the linear factorization of the polynomial and form the equation. Then multiply.

EXAMPLE 1   **Finding a Polynomial Equation**

Find a second-degree polynomial equation with real coefficients that has a root of $2 + i$.

SOLUTION   We can use the Conjugate Pairs Theorem to determine the second root. Then, we will use the two roots to write the linear factorization and equation of the polynomial.

1. **Determine the roots.**
   Because $2 + i$ is a root, its complex conjugate $2 - i$ is also a root. The solution set is $\{2 + i, 2 - i\}$.

2. **Write the linear factorization of the polynomial and form the equation. Then multiply.**
   Because $2 + i$ is a root, $x - (2 + i)$ is a factor. Because $2 - i$ is a root, $x - (2 - i)$ is a factor. The polynomial equation is

   $$[x - (2 + i)][x - (2 - i)] = 0$$

   $$(x - 2 - i)(x - 2 + i) = 0$$

   $$x^2 - 4x + 5 = 0 \qquad \text{Multiply.}$$

   The equation $x^2 - 4x + 5 = 0$ will have roots of $2 + i$ and $2 - i$. ∎

Self Check 1   Find a second-degree polynomial equation with real coefficients that has a root of $1 - i$.

Now Try Exercise 23.

EXAMPLE 2   **Finding a Polynomial Equation**

Find a fourth-degree polynomial equation with real coefficients and i as a root of multiplicity 2.

SOLUTION   We can use the Conjugate Pairs Theorem to determine the other two roots. Then, we will use the four roots to write the linear factorization and equation of the polynomial.

1. Determine the roots.

Because i is a root twice and a fourth-degree polynomial equation has four roots, we must find the other two roots. According to the Conjugate Pairs Theorem, the missing roots are the conjugates of the given roots. Thus, the complete solution set is

$$\{i, i, -i, -i\}$$

2. Write the linear factorization of the polynomial and form the equation. Then multiply.

Because i is a root of multiplicity 2, the factor $x - i$ occurs twice. Because $-i$ is a root of multiplicity 2, the factor $x - (-i)$ occurs twice. The equation is

$$(x - i)(x - i)[x - (-i)][x - (-i)] = 0$$
$$(x - i)(x + i)(x - i)(x + i) = 0$$
$$(x^2 + 1)(x^2 + 1) = 0 \qquad \text{Multiply.}$$
$$x^4 + 2x^2 + 1 = 0 \qquad \text{Multiply.}$$

Comment

The Conjugate Pairs Theorem applies only to polynomials with real-number coefficients. In Examples 1 and 2, the theorem does apply and the results were polynomials with real-number coefficients.

Self Check 2   Find a fourth-degree polynomial equation with real coefficients and $-i$ as a root of multiplicity 2.

Now Try Exercise 29.

EXAMPLE 3   **Finding a Quadratic Equation with a Double Root**

Find a second-degree equation with a double root of i.

SOLUTION   We will use the two given roots of i to write the linear factorization of the polynomial and quadratic equation.

Because i is a root twice, the linear factorization of the polynomial and the quadratic equation are

$$(x - i)(x - i) = 0$$
$$x^2 - 2ix - 1 = 0 \qquad \text{Multiply.}$$

Caution

In Example 3, the Conjugate Pairs Theorem does **not** apply because the resulting polynomial equation does **not** have real-number coefficients.

In this equation, the coefficient of x is $-2i$, which is not a real number. Therefore, it is not surprising that the roots are not in complex conjugate pairs.

Self Check 3   Find a second-degree equation with a double root of $-i$.

Now Try Exercise 33.

2. Use Descartes' Rule of Signs

René Descartes (1596–1650) is credited with a theorem known as **Descartes' Rule of Signs,** which enables us to estimate the number of positive, negative, and nonreal roots of a polynomial equation.

If a polynomial is written in descending powers of x and we scan it from left to right, we say that a variation in sign occurs whenever successive terms have opposite signs. For example, the polynomial function

$$P(x) = \overbrace{3x^5 - 2x^4}^{+ \text{ to } -} \overbrace{- 5x^3 + x^2}^{- \text{ to } +} \overbrace{- x - 9}^{+ \text{ to } -}$$

has three variations in sign, and the polynomial function

$$P(-x) = 3(-x)^5 - 2(-x)^4 - 5(-x)^3 + (-x)^2 - (-x) - 9$$

$$= -3x^5 \overbrace{- 2x^4 + 5x^3}^{- \text{ to } +} + x^2 \overbrace{+ x - 9}^{+ \text{ to } -}$$

has two variations in sign.

Descartes' Rule of Signs	Let $P(x)$ be a polynomial function with real coefficients.

- The number of positive roots of $P(x) = 0$ is either equal to the number of variations in sign of $P(x)$ or less than that by an even number.

- The number of negative roots of $P(x) = 0$ is either equal to the number of variations in sign of $P(-x)$ or less than that by an even number.

EXAMPLE 4   **Using Descartes' Rule of Signs**

Find the number of possible positive, negative, and nonreal roots of $3x^3 - 2x^2 + x - 5 = 0$.

SOLUTION   We can use Descartes' Rule of Signs to determine the number of possible positive, negative, and nonreal roots.

1. **Determine the number of possible positive roots.**
   Let $P(x) = 3x^3 - 2x^2 + x - 5$. Since there are three variations of sign in $P(x) = 3x^3 - 2x^2 + x - 5$, there can be either 3 positive roots or only 1 (1 is less than 3 by the even number 2).

2. **Determine the number of possible negative roots.**
   Because

$$P(-x) = 3(-x)^3 - 2(-x)^2 + (-x) - 5$$

$$= -3x^3 - 2x^2 - x - 5$$

Comment

To obtain $P(-x)$ quickly, we can simply multiply each odd degree term of $P(x)$ by negative one.

has no variations in sign, there are 0 negative roots. Furthermore, 0 is not a root, because the terms of the polynomial do not have a common factor of x.

3. **Determine the number of nonreal roots.**
   If there are 3 positive roots, then all of the roots are accounted for. If there is 1 positive root, the 2 remaining roots must be nonreal complex numbers. The following chart shows these possibilities.

Number of positive roots	Number of negative roots	Number of nonreal roots
3	0	0
1	0	2

The number of nonreal complex roots is the number needed to bring the total number of roots up to 3.

Self Check 4   Discuss the possibilities for the roots of $5x^3 + 2x^2 - x + 3 = 0$.

Now Try Exercise 37.

EXAMPLE 5   **Using Descartes' Rule of Signs**

Find the number of possible positive, negative, and nonreal roots of $5x^5 - 3x^3 - 2x^2 + x - 1 = 0$.

SOLUTION   We can use Descartes' Rule of Signs to determine the number of possible positive, negative, and nonreal roots.

1. **Determine the number of possible positive roots.**
   Let $P(x) = 5x^5 - 3x^3 - 2x^2 + x - 1$. Since there are three variations of sign in $P(x)$, there are either 3 or 1 positive roots.

2. **Determine the number of possible negative roots.**
   Because $P(-x) = -5x^5 + 3x^3 - 2x^2 - x - 1$ has two variations in sign, there are 2 or 0 negative roots.

3. **Determine the number of nonreal roots.**
   The possibilities are shown as follows:

Number of positive roots	Number of negative roots	Number of nonreal roots
1	0	4
3	0	2
1	2	2
3	2	0

In each case, the number of nonreal complex roots is an even number. This is expected, because this polynomial has real coefficients, and its nonreal complex roots will occur in conjugate pairs. ∎

Self Check 5   Discuss the possibilities for the roots of $5x^5 - 2x^2 - x - 1 = 0$.

Now Try Exercise 41.

3. Find Integer Bounds on Roots

A final theorem provides a way to find **bounds** on the roots of a polynomial equation, enabling us to look for roots where they can be found.

Upper and Lower Bounds on Roots   Let $P(x)$ be a polynomial function with real coefficients and a positive leading coefficient.

1. If $P(x)$ is synthetically divided by a positive number c and each term in the last row of the division is nonnegative, then no number greater than c can be a root of $P(x) = 0$. (c is called an **upper bound** of the real roots.)

2. If $P(x)$ is synthetically divided by a negative number d and the signs in the last row alternate,* no value less than d can be a root of $P(x) = 0$. (d is called a **lower bound** of the real roots.)

*If 0 appears in the third row, that 0 can be assigned either a + or a − sign to help the signs alternate.

EXAMPLE 6   **Finding Integer Bounds on Roots**

Establish integer bounds for the roots of $2x^3 + 3x^2 - 5x - 7 = 0$.

SOLUTION   We will use the upper and lower bound rules previously stated to establish the integer bounds for the roots.

1. We will perform several synthetic divisions by positive integers, looking for nonnegative values in the last row.
   Trying 1 first gives

$$\begin{array}{r|rrrr} 1 & 2 & 3 & -5 & -7 \\ & & 2 & +5 & +0 \\ \hline & +2 & +5 & 0 & -7 \end{array}$$

Because one of the signs in the last row is negative, we cannot claim that 1 is an upper bound. We now try 2.

$$\begin{array}{r|rrrr} 2 & 2 & 3 & -5 & -7 \\ & & +4 & 14 & 18 \\ \hline & +2 & +7 & +9 & +11 \end{array}$$

Because the last row is entirely nonnegative, we can claim that 2 is an upper bound. That is, no number greater than 2 can be a root of the equation.

2. We perform several synthetic divisions by negative divisors, looking for alternating signs in the last row.
   We begin with -3.

$$\begin{array}{r|rrrr} -3 & 2 & 3 & -5 & -7 \\ & & -6 & 9 & -12 \\ \hline & +2 & -3 & +4 & -19 \end{array}$$

Since the signs in the last row alternate, -3 is a lower bound. That is, no number less than -3 can be a root. To see whether there is a greater lower bound, we try -2.

$$\begin{array}{r|rrrr} -2 & 2 & 3 & -5 & -7 \\ & & -4 & 2 & 6 \\ \hline & +2 & -1 & -3 & -1 \end{array}$$

Since the signs in the last row do not alternate, we cannot claim that -2 is a lower bound.

Since -3 is a lower bound and 2 is an upper bound, then all of the real roots must be in the interval $(-3, 2)$. ∎

Self Check 6   Establish integer bounds for the roots of $2x^3 + 3x^2 - 11x - 7 = 0$.

Now Try Exercise 53.

It is important to understand what the theorem on the bounds of roots says and what it doesn't say. The following two explanations will help clarify that for us.

1. If we divide synthetically by a positive number c and the last row of the synthetic division is entirely nonnegative, the theorem guarantees that c is an upper bound of the roots. However, if the last row contains some negative values, c could still be an upper bound.

2. If we divide by a negative number d and the signs in the last row alternate, the theorem guarantees that d is a lower bound of the roots. However, if the signs in the last row do not alternate, d could still be a lower bound. This is illustrated in Example 6. It can be shown that the smallest negative root of the equation is

approximately -1.81. Thus, -2 is a lower bound for the roots of the equation. However, when we checked -2, the last row of the synthetic division did not have alternating signs. Unfortunately, the theorem does not always determine the best bounds for the roots of the equation.

Self Check Answers   1. $x^2 - 2x + 2 = 0$ 2. $x^4 + 2x^2 + 1 = 0$ 3. $x^2 + 2ix - 1 = 0$

4.

Num. of pos. roots	Num. of neg. roots	Num. of nonreal roots
2	1	0
0	1	2

5.

Num. of pos. roots	Num. of neg. roots	Num. of nonreal roots
1	2	2
1	0	4

6. Roots are in $(-4, 3)$.

Exercises **5.2**

Getting Ready
You should be able to complete these vocabulary and concept statements before you proceed to the practice exercises.

Fill in the blanks.

1. If $P(x)$ is a polynomial function with positive degree, then $P(x)$ has at least one _____.

2. The statement in Exercise 1 is called the

   _____.

3. The _____ of $a + bi$ is $a - bi$.

4. The polynomial $6x^4 + 5x^3 - 2x^2 + 3$ has ___ variations in sign.

5. The polynomial $(-x)^3 - (-x)^2 - 4$ has ___ variations in sign.

6. The equation $7x^4 + 5x^3 - 2x + 1 = 0$ can have at most ___ positive roots.

7. The equation $7x^4 + 5x^3 - 2x + 1 = 0$ can have at most ___ negative roots.

8. Complex roots occur in complex _____ pairs. (Assume that the equation has real coefficients.)

9. If no number less than d can be a root of $P(x) = 0$, then d is called a(n) _____.

10. If no number greater than c can be a root of $P(x) = 0$, then c is called a(n) _____.

Practice
Determine how many roots each equation has.

11. $x^{10} = 1$

12. $x^{40} = 1$

13. $3x^4 - 4x^2 - 2x = -7$

14. $-32x^{111} - x^5 = 1$

15. One root of $x(3x^4 - 2) = 12x$ is 0. How many other roots are there?

16. Two roots of $3x^2(x^7 - 14x + 3) = 0$ are 0. How many other roots are there?

Determine how many linear factors and zeros each polynomial function has.

17. $P(x) = x^4 - 81$

18. $P(x) = x^{40} + x^{39}$

19. $P(x) = 4x^5 + 8x^3$

20. $P(x) = x^3 + 144x$

Write a second-degree polynomial equation with real coefficients and the given root.

21. $2i$

22. $-3i$

23. $3 - i$

24. $4 + 2i$

Write a third-degree polynomial equation with real coefficients and the given roots.

25. $3, -i$

26. $1, i$

27. $2, 2 + i$

28. $-2, 3 - i$

Write a fourth-degree polynomial equation with real coefficients and the given roots.

29. $3, 2, i$

30. $1, 2, 1 + i$

31. $i, 1 - i$

32. $i, 2 - i$

Write a second-degree equation with the given double root.

33. $2i$

34. $-2i$

Use Descartes' Rule of Signs to find the number of possible positive, negative, and nonreal roots of each equation.

35. $3x^3 + 5x^2 - 4x + 3 = 0$

36. $3x^3 - 5x^2 - 4x - 3 = 0$

37. $2x^3 + 7x^2 + 5x + 5 = 0$

38. $-2x^3 - 7x^2 - 5x - 4 = 0$

39. $8x^4 = -5$

40. $-3x^3 = -5$

41. $x^4 + 8x^2 - 5x - 10 = 0$

42. $5x^7 + 3x^6 - 2x^5 + 3x^4 + 9x^3 + x^2 + 1 = 0$

43. $-x^{10} - x^8 - x^6 - x^4 - x^2 - 1 = 0$

44. $x^{10} + x^8 + x^6 + x^4 + x^2 + 1 = 0$

45. $x^9 + x^7 + x^5 + x^3 + x = 0$ (Is 0 a root?)

46. $-x^9 - x^7 - x^5 - x^3 - x = 0$ (Is 0 a root?)

47. $-2x^4 - 3x^2 + 2x + 3 = 0$

48. $-7x^5 - 6x^4 + 3x^3 - 2x^2 + 7x - 4 = 0$

Find integer bounds for the roots of each equation. Answers can vary.

49. $x^2 - 2x - 4 = 0$ **50.** $9x^2 - 6x - 1 = 0$

51. $18x^2 - 6x - 1 = 0$ **52.** $2x^2 - 10x - 9 = 0$

53. $6x^3 - 13x^2 - 110x = 0$

54. $12x^3 + 20x^2 - x - 6 = 0$

55. $x^5 + x^4 - 8x^3 - 8x^2 + 15x + 15 = 0$

56. $3x^4 - 5x^3 - 9x^2 + 15x = 0$

57. $3x^5 - 11x^4 - 2x^3 + 38x^2 - 21x - 15 = 0$

58. $3x^6 - 4x^5 - 21x^4 + 4x^3 + 8x^2 + 8x + 32 = 0$

Discovery and Writing

59. Explain why the Fundamental Theorem of Algebra guarantees that every polynomial equation of positive degree has at least one root.

60. Explain why the Fundamental Theorem of Algebra and the Factor Theorem guarantee that an *n*th-degree polynomial equation has *n* roots.

61. Prove that any odd-degree polynomial equation with real coefficients must have at least one real root.

62. If *a, b, c,* and *d* are positive numbers, prove that $ax^4 + bx^2 + cx - d = 0$ has exactly two nonreal roots.

Review

Assume that k represents a positive real number, and complete each sentence.

63. The graph of $y = f(x - k)$ looks like the graph of $y = f(x)$, except that it has been translated

_____.

64. The graph of $y = f(x) - k$ looks like the graph of $y = f(x)$, except that it has been translated

_____.

65. The graph of $y = f(-x)$ looks like the graph of $y = f(x)$, except that it has been

_____.

66. The graph of $y = -f(x)$ looks like the graph of $y = f(x)$, except that it has been

_____.

67. If $k > 1$, the graph of $y = kf(x)$ looks like the graph of $y = f(x)$, except that it has been

_____.

68. If $0 < k < 1$, the graph of $y = kf(x)$ looks like the graph of $y = f(x)$, except that it has been

_____

5.3 Roots of Polynomial Equations

In this section, we will learn to

1. Find possible rational roots of polynomial equations.

2. Find rational roots of polynomial equations.

3. Find real and nonreal roots of polynomial equations.

4. Solve applications.

FedEx is a company that offers reliable shipping services across town and across the globe. It has a wide range of envelopes and boxes available to accommodate the shipping needs of most individuals and companies.

Suppose a FedEx box has the following characteristics:

- The length of the box is 6 inches more than its height.
- The width of the box is 3 inches more than its height.
- The volume of the box is 2,080 cubic inches.

To find the dimensions of the box, we can let h represent the height (in inches). Then $h + 6$ will represent the length and $h + 3$ will represent the width. Since the volume of the box is given to be 2,080, we have

$$V = l \cdot w \cdot h$$

$$2,080 = (h + 6)(h + 3)h \qquad \text{Substitute.}$$

$$2,080 = h^3 + 9h^2 + 18h \qquad \text{Multiply.}$$

$$0 = h^3 + 9h^2 + 18h - 2,080 \qquad \text{Subtract 2,080 from both sides.}$$

To find the height h, we must find the roots of the polynomial equation. One root of the equation is $x = 10$, because 10 satisfies the equation:

$$(10)^3 + 9(10)^2 + 18(10) - 2,080 = 0$$

Therefore, the height of the box is $h = 10$ inches, the length is $h + 6 = 16$ inches, and the width is $h + 3 = 13$ inches.

In this section, we will develop a strategy for finding roots of higher-degree polynomial equations.

1. Find Possible Rational Roots of Polynomial Equations

Recall that a rational number is any number that can be written in the form $\frac{p}{q}$, where p and q are integers and $q \neq 0$. The following theorem enables us to list the possible rational roots of such polynomial equations.

Rational Root Theorem   Let the polynomial function

$$P(x) = a_n x^n + a_{n-1}x^{n-1} + a_{n-2}x^{n-2} + \cdots + a_1 x + a_0$$

have integer coefficients. If the rational number $\frac{p}{q}$ (written in lowest terms) is a root of $P(x) = 0$, then p is a factor of the constant a_0, and q is a factor of the leading coefficient a_n.

PROOF   Let $\frac{p}{q}$ (written in lowest terms) be a rational root of $P(x) = 0$. Then the equation is satisfied by $\frac{p}{q}$:

(1) $a_n\left(\dfrac{p}{q}\right)^n + a_{n-1}\left(\dfrac{p}{q}\right)^{n-1} + a_{n-2}\left(\dfrac{p}{q}\right)^{n-2} + \cdots + a_1\left(\dfrac{p}{q}\right) + a_0 = 0$

We can clear Equation 1 of fractions by multiplying both sides by q^n.

(2) $a_n p^n + a_{n-1}p^{n-1}q + a_{n-2}p^{n-2}q^2 + \cdots + a_1 pq^{n-1} + a_0 q^n = 0$

We can factor p from all but the last term and subtract $a_0 q^n$ from both sides to get

$$p(a_n p^{n-1} + a_{n-1}p^{n-2}q + a_{n-2}p^{n-3}q^2 + \cdots + a_1 q^{n-1}) = -a_0 q^n$$

Since p is a factor of the left side, it is also a factor of the right side. So p is a factor of $-a_0 q^n$, but because $\dfrac{p}{q}$ is written in lowest terms, p cannot be a factor of q^n. Therefore, p is a factor of a_0.

We can factor q from all but the first term of Equation 2 and subtract $a_n p^n$ from both sides to get

$$q(a_{n-1}p^{n-1} + a_{n-2}p^{n-2}q + a_{n-3}p^{n-3}q^2 + \cdots + a_0 q^{n-1}) = -a_n p^n$$

Since q is a factor of the left side, it is also a factor of the right side. Because q is not a factor of p^n, it must be a factor of a_n. ∎

To illustrate the Rational Root Theorem, we consider the equation

$$\frac{1}{2}x^4 + \frac{2}{3}x^3 + 3x^2 - \frac{3}{2}x + 3 = 0$$

Because the theorem requires integer coefficients, we multiply both sides of the equation by 6 to clear it of fractions.

$$3x^4 + 4x^3 + 18x^2 - 9x + 18 = 0$$

By the previous theorem, the only possible numerators for the rational roots of the equation are the factors of the constant term 18:

$$\pm 1, \ \pm 2, \ \pm 3, \ \pm 6, \ \pm 9, \text{ and } \pm 18$$

The only possible denominators are the factors of the leading coefficient 3:

$$\pm 1 \quad \text{and} \quad \pm 3$$

We can form a list of all possible rational solutions by listing the combinations of possible numerators and denominators:

$$\pm\frac{1}{1}, \ \pm\frac{2}{1}, \ \pm\frac{3}{1}, \ \pm\frac{6}{1}, \ \pm\frac{9}{1}, \ \pm\frac{18}{1}, \ \pm\frac{1}{3}, \ \pm\frac{2}{3}, \ \pm\frac{3}{3}, \ \pm\frac{6}{3}, \ \pm\frac{9}{3}, \ \pm\frac{18}{3}$$

Since several of these possibilities are duplicates, we can condense the list to get

Comment

When we list possible rational roots for a polynomial equation, our list will not contain radicals or complex numbers.

Possible rational roots

$$\pm 1, \quad \pm 2, \quad \pm 3, \quad \pm 6, \quad \pm 9, \quad \pm 18, \quad \pm\frac{1}{3}, \quad \pm\frac{2}{3}$$

2. Find Rational Roots of Polynomial Equations

To find the rational roots of a polynomial equation, we will use the following steps.

Strategy for Finding Rational Roots

1. Use Descartes' Rule of Signs to determine the number of possible positive, negative, and nonreal roots.
2. Use the Rational Root Theorem to list possible rational roots.
3. Use synthetic division to find a root.
4. If there are more roots, repeat the previous steps.

EXAMPLE 1 **Finding the Rational Roots of a Polynomial Equation**

Find the solution set of $2x^3 + 3x^2 - 8x + 3 = 0$.

SOLUTION We will use the steps previously outlined to solve the polynomial equation. Since the equation is of third degree, it has 3 roots.

1. To determine the number of possible positive, negative, and nonreal roots, we will use Descartes' Rule of Signs. We find that there are two possible combinations of positive, negative, and nonreal roots. They are as follows:

Number of positive roots	Number of negative roots	Number of nonreal roots
2	1	0
0	1	2

2. We then find the possible rational roots that have the form $\frac{\text{factor of the constant 3}}{\text{factor of the leading coefficient 2}}$. They are

$$\pm\frac{3}{1}, \quad \pm\frac{1}{1}, \quad \pm\frac{3}{2}, \quad \pm\frac{1}{2}$$

or, written in order of increasing size,

$$-3, \quad -\frac{3}{2}, \quad -1, \quad -\frac{1}{2}, \quad \frac{1}{2}, \quad 1, \quad \frac{3}{2}, \quad 3$$

3. We then use synthetic division and check each possibility to see whether it is a root. We can start with $\frac{3}{2}$.

$$\begin{array}{r|rrrr} \frac{3}{2} & 2 & 3 & -8 & 3 \\ & & 3 & 9 & \frac{3}{2} \\ \hline & 2 & 6 & 1 & \frac{9}{2} \end{array}$$

Since the remainder is not 0, $\frac{3}{2}$ is not a root and we can cross it off the list. Since every number in the last row of the synthetic division is positive, $\frac{3}{2}$ is an upper bound. So 3 cannot be a root either and we can cross it off the list.

$$-3, \quad -\frac{3}{2}, \quad -1, \quad -\frac{1}{2}, \quad \frac{1}{2}, \quad 1, \quad \cancel{\frac{3}{2}}, \quad \cancel{3}$$

We now try $\frac{1}{2}$:

$$\begin{array}{r|rrrr} \frac{1}{2} & 2 & 3 & -8 & 3 \\ & & 1 & 2 & -3 \\ \hline & 2 & 4 & -6 & 0 \end{array}$$ This row represents the quotient $2x^2 + 4x - 6$.

Since the remainder is 0, $\frac{1}{2}$ is a root and the binomial $x - \frac{1}{2}$ is a factor of $P(x)$.

4. The remaining roots must be supplied by the remaining factor, which is the quotient $2x^2 + 4x - 6$. We can find the other roots by solving the equation $2x^2 + 4x - 6 = 0$, called the **depressed equation**.

$$2x^2 + 4x - 6 = 0$$

$$x^2 + 2x - 3 = 0 \qquad \text{Divide both sides by 2.}$$

$$(x - 1)(x + 3) = 0 \qquad \text{Factor } x^2 + 2x - 3.$$

$$x - 1 = 0 \quad \text{or} \quad x + 3 = 0$$

$$x = 1 \qquad \qquad x = -3$$

Comment

Example 1 illustrates that the upper and lower bound theorem is helpful when the list of possible rational roots is long.

The solution set of the equation is $\left\{\frac{1}{2}, 1, -3\right\}$. Note that two positive roots and one negative root is a predicted possibility. ∎

Self Check 1   Find the solution set of $3x^3 - 10x^2 + 9x - 2 = 0$.

Now Try Exercise 19.

ACCENT ON TECHNOLOGY   **Confirming Roots of a Polynomial Equation**

We can confirm that the roots found in Example 1 are correct by graphing the function $P(x) = 2x^3 + 3x^2 - 8x + 3$ and locating the resulting x-intercepts of the graph. If we use a graphing WINDOW as shown in Figure 5-1(a) and GRAPH the function, we will obtain the graph shown in Figure 5-1(b). From the graph, we can see that the graph crosses the x-axis at $x = -3$, $x = \frac{1}{2}$, and $x = 1$. These are the roots of the polynomial equation.

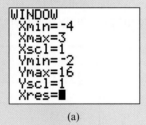

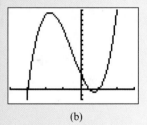

(a)   (b)

FIGURE 5-1

3. Find Real and Nonreal Roots of Polynomial Equations

EXAMPLE 2   **Finding All the Solutions of a Polynomial Equation**

Find the solution set of $x^7 - 2x^6 - 5x^5 + 6x^4 - x^3 + 2x^2 + 5x - 6 = 0$.

SOLUTION   We will use the steps outlined above to find the roots of the polynomial equation. Because the equation is of seventh degree, it has 7 roots.

1. To determine the number of possible positive, negative, and nonreal roots, we will use Descartes' Rule of Signs. We find that there are several possible combinations of positive, negative, and nonreal roots. They are as follows:

Number of positive roots	Number of negative roots	Number of nonreal roots
5	2	0
3	2	2
1	2	4
5	0	2
3	0	4
1	0	6

2. We then find the possible rational roots that have the form $\frac{\text{factor of the constant} -6}{\text{factor of the leading coefficient } 1}$. They are as follows:

$$-6, \quad -3, \quad -2, \quad -1, \quad 1, \quad 2, \quad 3, \quad 6$$

3. We then use synthetic division and check each possibility to see whether it is a root. We can start with -3.

$$
\begin{array}{r|rrrrrrrr}
-3 & 1 & -2 & -5 & 6 & -1 & 2 & 5 & -6 \\
   &   & -3 & 15 & -30 & 72 & -213 & 633 & -1{,}914 \\
\hline
   & 1 & -5 & 10 & -24 & 71 & -211 & 638 & -1{,}920
\end{array}
$$

Since the last number in the synthetic division is not 0, -3 is not a root and can be crossed off the list. Since the signs in the last row alternate, -3 is a lower bound, and we can cross off -6 as well.

$$-\cancel{6}, \quad -\cancel{3}, \quad -2, \quad -1, \quad 1, \quad 2, \quad 3, \quad 6$$

We now try -2:

$$
\begin{array}{r|rrrrrrr}
-2 & 1 & -2 & -5 & 6 & -1 & 2 & 5 & -6 \\
   &   & -2 & 8 & -6 & 0 & 2 & -8 & 6 \\
\hline
   & 1 & -4 & 3 & 0 & -1 & 4 & -3 & 0
\end{array}
$$

Since the remainder is 0, -2 is a root.

4. Because the root -2 is negative, we can revise the chart of possibilities to eliminate the possibility that there are 0 negative roots.

Number of positive roots	Number of negative roots	Number of nonreal roots
5	2	0
3	2	2
1	2	4

The remaining roots must be supplied by the remaining factor, which is the quotient in the synthetic division shown above: $x^6 - 4x^5 + 3x^4 - x^2 + 4x - 3 = 0$. We can find the other roots by repeating steps 1–3 and solving this equation, called the **depressed equation.**

Because the constant term of this depressed equation is different from the constant term of the original equation, we can cross off other possible rational roots. For example, the number -2 cannot be a root a second time, because it is not a factor of -3. The numbers 2 and 6 are no longer possible roots, because neither is a factor of -3.

The list of possible roots is now

$$-\cancel{6}, \quad -\cancel{3}, \quad -2, \quad -1, \quad 1, \quad 2, \quad 3, \quad \cancel{6}$$

Since we know that there is one more negative root, we will synthetically divide the coefficients of the depressed equation by -1.

$$
\begin{array}{r|rrrrrr}
-1 & 1 & -4 & 3 & 0 & -1 & 4 & -3 \\
   &   & -1 & 5 & -8 & 8 & -7 & 3 \\
\hline
   & 1 & -5 & 8 & -8 & 7 & -3 & 0
\end{array}
$$

Since the remainder is 0, -1 is a root, and the solution set so far is $\{-2, -1, \ldots\}$. The number -1 cannot be a root again, because we have found both negative roots. When we cross off -1, we have only two possibilities left.

$$-\cancel{6}, \quad -\cancel{3}, \quad -2, \quad -\cancel{1}, \quad 1, \quad 2, \quad 3, \quad \cancel{6}$$

The depressed equation is now $x^5 - 5x^4 + 8x^3 - 8x^2 + 7x - 3 = 0$.

We can synthetically divide the coefficients of this equation by 1 to get

$$\underline{1|} \quad 1 \quad -5 \quad 8 \quad -8 \quad 7 \quad -3$$
$$\phantom{\underline{1|} \quad 1} \quad 1 \quad -4 \quad 4 \quad -4 \quad 3$$
$$\overline{\phantom{\underline{1|} \quad} 1 \quad -4 \quad 4 \quad -4 \quad 3 \quad 0}$$

The depressed equation is now $x^4 - 4x^3 + 4x^2 - 4x + 3 = 0$.

Since the remainder is 0, 1 joins the solution set $\{-2, -1, 1, \ldots\}$. To see whether 1 is a root a second time, we synthetically divide the coefficients of the new depressed equation by 1.

$$\underline{1|} \quad 1 \quad -4 \quad 4 \quad -4 \quad 3$$
$$\phantom{\underline{1|} \quad 1} \quad 1 \quad -3 \quad 1 \quad -3$$
$$\overline{\phantom{\underline{1|} \quad} 1 \quad -3 \quad 1 \quad -3 \quad 0}$$

The depressed equation is now $x^3 - 3x^2 + x - 3 = 0$.

Again, 1 is a root, and the solution set is now $\{-2, -1, 1, 1, \ldots\}$.

To see whether 1 is a root a third time, we synthetically divide the coefficients of the new depressed equation by 1.

$$\underline{1|} \quad 1 \quad -3 \quad 1 \quad -3$$
$$\phantom{\underline{1|} \quad 1} \quad 1 \quad -2 \quad -1$$
$$\overline{\phantom{\underline{1|} \quad} 1 \quad -2 \quad -1 \quad -4}$$

Since the remainder is not 0, the number 1 is not a root for a third time, and we can cross 1 off the list of possibilities, leaving only 3.

$$-\cancel{6}, \quad -\cancel{3}, \quad -2, \quad -\cancel{1}, \quad \cancel{1}, \quad 2, \quad 3, \quad \cancel{6}$$

To see whether 3 is a root, we synthetically divide the coefficients of $x^3 - 3x^2 + x - 3 = 0$ by 3.

$$\underline{3|} \quad 1 \quad -3 \quad 1 \quad -3$$
$$\phantom{\underline{3|} \quad 1} \quad 3 \quad 0 \quad 3$$
$$\overline{\phantom{\underline{3|} \quad} 1 \quad 0 \quad 1 \quad 0}$$

Since the remainder is 0, 3 joins the solution set, which is now $\{-2, -1, 1, 1, 3 \ldots\}$.

The depressed equation is now $x^2 + 1 = 0$, which can be solved as a quadratic equation.

$$x^2 + 1 = 0$$
$$x^2 = -1$$
$$x = i \quad \text{or} \quad x = -i$$

The complete solution set is $\{-2, -1, 1, 1, 3, i, -i\}$. The solution set contains 3 positive roots, 2 negative roots, and 2 nonreal roots that are complex conjugates. This combination was one of the predicted possibilities. ∎

Comment

Example 2 illustrates that finding the seven roots of a seventh-degree polynomial equation can be a lengthy process. In such cases, the four-step process for finding rational roots must be repeated.

Self Check 2   Find the solution set of $x^5 - x^4 + 3x^3 - 3x^2 - 4x + 4 = 0$.

Now Try Exercise 51.

ACCENT ON TECHNOLOGY   **Confirming Roots of a Polynomial Equation**

We can confirm that the roots found in Example 2 are correct by graphing the function

$$P(x) = x^7 - 2x^6 - 5x^5 + 6x^4 - x^3 + 2x^2 + 5x - 6$$

and locating the resulting x-intercepts of the graph. If we use the graphing 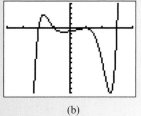 shown in Figure 5-2(a) and 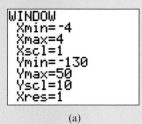 the function, we will obtain the graph shown in Figure 5-2(b). From the graph, we can see that the graph crosses or touches the x-axis at $x = -2$, $x = -1$, $x = 1$, and $x = 3$. We cannot detect the complex roots from the graph.

```
WINDOW
 Xmin=-4
 Xmax=4
 Xscl=1
 Ymin=-130
 Ymax=50
 Yscl=10
 Xres=1
```

(a)   (b)

FIGURE 5-2

4. Solve Applications

EXAMPLE 3   **Solving an Application**

To protect cranberry crops from the damage of early freezes, growers flood the cranberry bogs. Three irrigation sources, used together, can flood a cranberry bog in one day. If the sources are used one at a time, the second source requires one day longer to flood the bog than the first, and the third requires four days longer than the first. If the bog must be flooded before a freeze that is predicted in three days, can the water in the last two sources be diverted to other bogs?

SOLUTION   We can use the given information to write an equation that models the situation. We will then solve the equation.

We can let x represent the number of days it would take the first irrigation source to flood the bog. Then $x + 1$ and $x + 4$ represent the number of days it would take the second and third sources, respectively, to flood the bog.

Because the first source, alone, requires x days to flood the bog, that source could fill $\frac{1}{x}$ of the bog in one day. In one day's time, the remaining sources could flood $\frac{1}{x + 1}$ and $\frac{1}{x + 4}$ of the bog. This gives the equation

The part of the bog the first source can flood in one day	plus	the part the second source can flood in one day	plus	the part the third source can flood in one day	equals	one bog.
$\dfrac{1}{x}$	$+$	$\dfrac{1}{x + 1}$	$+$	$\dfrac{1}{x + 4}$	$=$	1

We multiply both sides of the equation by $x(x + 1)(x + 4)$ to clear it of fractions and then simplify to get

$$x(x + 1)(x + 4)\left(\frac{1}{x} + \frac{1}{x + 1} + \frac{1}{x + 4}\right) = 1 \cdot x(x + 1)(x + 4)$$

$$(x + 1)(x + 4) + x(x + 4) + x(x + 1) = x(x + 1)(x + 4)$$

$$x^2 + 5x + 4 + x^2 + 4x + x^2 + x = x^3 + 5x^2 + 4x$$

$$0 = x^3 + 2x^2 - 6x - 4$$

To solve the equation $x^3 + 2x^2 - 6x - 4 = 0$, we first list its possible rational roots, which are the factors of the constant term, -4.

$$-4, -2, -1, 1, 2, \text{ and } 4$$

One solution of this equation is $x = 2$, because when we synthetically divide by 2, the remainder is 0:

$$
\begin{array}{r|rrrr}
2 & 1 & 2 & -6 & -4 \\
  &   & 2 & 8 & 4 \\
\hline
  & 1 & 4 & 2 & 0
\end{array}
$$
   The depressed equation is $x^2 + 4x + 2 = 0$.

We can find the remaining solutions by using the Quadratic Formula to solve the depressed equation. The two solutions are $-2 + \sqrt{2}$ and $-2 - \sqrt{2}$. Since both of these numbers are negative and the time it takes to flood the bog cannot be negative, these roots must be discarded. The only meaningful solution is 2.

Since the first source, alone, can flood the bog in two days, and it is three days until the freeze, the other two water sources can be diverted to flood other bogs. ■

Self Check 3   If a freeze is predicted in one day, can the water in the last two sources be diverted?

Now Try Exercise 65.

ACCENT ON TECHNOLOGY   **Using the Table Feature on a Graphing Calculator to Find Roots of a Polynomial Equation**

A graphing calculator can be used to reduce the amount of work involved in finding the roots of a polynomial equation. From a list of possible rational zeros, we can utilize the TABLE function and identify the rational roots. Consider the following example.

- We will use the Rational Root Theorem, a calculator, and synthetic division, if necessary, to find all the rational zeros of the polynomial equation $2x^4 + x^3 - 9x^2 - 4x + 4 = 0$.

- First, we list all the possible rational roots. For the polynomial, we see that

$$p = 4, q = 2 \rightarrow \frac{p}{q} = \frac{\text{all integer factors of 4}}{\text{all interger factors of 2}} = \frac{\pm 1, \pm 2, \pm 4}{\pm 1, \pm 2} = \pm 1, \pm 2, \pm 4, \pm \frac{1}{2}$$

- Next, we evaluate the function at these values. The work is shown in Figure 5-3.

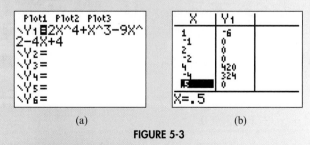

(a)   (b)

FIGURE 5-3

- From the table, we see that the rational roots are $x = -1, 2, -2,$ and $\frac{1}{2}$. Because the polynomial is of fourth degree and we have identified four roots, our work is finished.

It wasn't necessary for us to use synthetic division to solve this equation. If, for example, we had only found two rational roots using the table feature, we could be able to finish the problem using synthetic division. The table feature simply saves us time.

Self Check Answers   **1.** $\left\{1, 2, \dfrac{1}{3}\right\}$ **2.** $\{-1, 1, 1, 2i, -2i\}$ **3.** no

Exercises **5.3**

Getting Ready

You should be able to complete these vocabulary and concept statements before you proceed to the practice exercises.

Fill in the blanks.

1. The rational roots of the equation
   $3x^3 + 4x - 7 = 0$ will have the form $\frac{p}{q}$, where p is a
   factor of _____ and q is a factor of 3.

2. The rational roots of the equation
   $5x^3 + 3x^2 - 4 = 0$ will have the form $\frac{p}{q}$, where p is a
   factor of -4 and q is a factor of ___.

3. Consider the synthetic division of
   $5x^3 - 7x^2 - 3x - 63 = 0$.

   $$\underline{3}\ |\ \begin{array}{cccc} 5 & -7 & -3 & -63 \\ & 15 & 24 & 63 \\ \hline 5 & 8 & 21 & 0 \end{array}$$

   Since the remainder is 0, 3 is a _____ of the equation.

4. In Exercise 3, the depressed equation is

   _____.

Practice

Use the Rational Root Theorem to list all possible rational roots of the polynomial equation.

5. $x^3 + 10x^2 + 5x - 12 = 0$
6. $-x^3 + 3x^2 - 4x - 8 = 0$

7. $2x^4 - x^3 + 10x^2 + 5x - 6 = 0$

8. $3x^4 - x^3 + 7x^2 - 5x - 8 = 0$

9. $4x^5 - x^4 - x^3 + x^2 + 5x - 10 = 0$

10. $6x^4 - 2x^3 + x^2 - x + 3 = 0$

Find all rational roots of each equation.

11. $x^3 - 5x^2 - x + 5 = 0$
12. $x^3 + 7x^2 - x - 7 = 0$
13. $x^3 - 2x^2 - x + 2 = 0$

14. $x^3 + x^2 - 4x - 4 = 0$

15. $x^3 - x^2 - 4x + 4 = 0$

16. $x^3 + 2x^2 - x - 2 = 0$

17. $x^3 - 2x^2 - 9x + 18 = 0$

18. $x^3 + 3x^2 - 4x - 12 = 0$

19. $2x^3 - x^2 - 2x + 1 = 0$

20. $3x^3 + x^2 - 3x - 1 = 0$

21. $3x^3 + 5x^2 + x - 1 = 0$

22. $2x^3 - 3x^2 + 1 = 0$

23. $30x^3 - 47x^2 - 9x + 18 = 0$

24. $20x^3 - 53x^2 - 27x + 18 = 0$

25. $15x^3 - 61x^2 - 2x + 24 = 0$

26. $20x^3 - 44x^2 + 9x + 18 = 0$

27. $24x^3 - 82x^2 + 89x - 30 = 0$

28. $3x^3 - 2x^2 + 12x - 8 = 0$

29. $x^4 - 10x^3 + 35x^2 - 50x + 24 = 0$
30. $x^4 + 4x^3 + 6x^2 + 4x + 1 = 0$

31. $x^4 + 3x^3 - 13x^2 - 9x + 30 = 0$
32. $x^4 - 8x^3 + 14x^2 + 8x - 15 = 0$

33. $4x^4 - 8x^3 - x^2 + 8x - 3 = 0$

34. $3x^4 - 14x^3 + 11x^2 + 16x - 12 = 0$

35. $2x^4 - x^3 - 2x^2 - 4x - 40 = 0$

36. $12x^4 + 20x^3 - 41x^2 + 20x - 3 = 0$

37. $36x^4 - x^2 + 2x - 1 = 0$

38. $12x^4 + x^3 + 42x^2 + 4x - 24 = 0$

39. $x^5 + 3x^4 - 5x^3 - 15x^2 + 4x + 12 = 0$
40. $x^5 - 3x^4 - 5x^3 + 15x^2 + 4x - 12 = 0$

41. $4x^5 - 12x^4 + 15x^3 - 45x^2 - 4x + 12 = 0$

42. $6x^5 - 7x^4 - 48x^3 + 81x^2 - 4x - 12 = 0$

43. $x^7 - 12x^5 + 48x^3 - 64x = 0$

44. $x^7 + 7x^6 + 21x^5 + 35x^4 + 35x^3 + 21x^2 + 7x + 1 = 0$

Find all roots of each polynomial equation.

45. $x^3 - 3x^2 - 2x + 6 = 0$

46. $x^3 + 3x^2 - 3x - 9 = 0$

47. $2x^3 - x^2 + 2x - 1 = 0$

48. $3x^3 + x^2 + 3x + 1 = 0$

49. $x^4 - 2x^3 - 8x^2 + 8x + 16 = 0$

50. $x^4 - 2x^3 - 2x^2 + 2x + 1 = 0$

51. $2x^4 + x^3 + 17x^2 + 9x - 9 = 0$

52. $2x^4 - 4x^3 + 2x^2 + 4x - 4 = 0$

53. $x^5 - 3x^4 + 28x^3 - 76x^2 + 75x - 25 = 0$

54. $x^5 + 3x^4 - 2x^3 - 14x^2 - 15x - 5 = 0$

55. $2x^5 - 3x^4 + 6x^3 - 9x^2 - 8x + 12 = 0$

56. $3x^5 - x^4 + 36x^3 - 12x^2 - 192x + 64 = 0$

In Exercises 57–60, $1 + i$ is a root of each polynomial equation. Find the other roots.

57. $x^3 - 5x^2 + 8x - 6 = 0$

58. $x^3 - 2x + 4 = 0$

59. $x^4 - 2x^3 - 7x^2 + 18x - 18 = 0$

60. $x^4 - 2x^3 - 2x^2 + 8x - 8 = 0$

Solve each equation.

61. $x^3 - \dfrac{4}{3}x^2 - \dfrac{13}{3}x - 2 = 0$

62. $x^3 - \dfrac{19}{6}x^2 + \dfrac{1}{6}x + 1 = 0$

63. $x^{-5} - 8x^{-4} + 25x^{-3} - 38x^{-2} + 28x^{-1} - 8 = 0$

64. $1 - x^{-1} - x^{-2} - 2x^{-3} = 0$

Applications

65. Parallel resistance If three resistors with resistances of R_1, R_2, and R_3 are wired in parallel, their combined resistance R is given by the following formula. The design of a voltmeter requires that the resistance R_2 be 10 ohms greater than the resistance R_1, that the resistance R_3 be 50 ohms greater than R_1, and that their combined resistance be 6 ohms. Find the value of each resistance.

$$\frac{1}{R} = \frac{1}{R_1} + \frac{1}{R_2} + \frac{1}{R_3}$$

66. Fabricating sheet metal The open tray shown in the illustration is to be manufactured from a 12-by-14-inch rectangular sheet of metal by cutting squares from each corner and folding up the sides. If the volume of the tray is to be 160 cubic inches and x is to be an integer, what size squares should be cut from each corner?

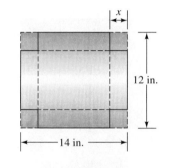

67. FedEx box The length of a FedEx 25-kg box is 7 inches more than its height. The width of the box is 4 inches more than its height. If the volume of the box is 4,420 cubic inches, find the height of the box.

68. Mountain Dew can A Mountain Dew aluminum can is approximately the shape of a cylinder. If the height of the can is 9 centimeters more than its radius and the volume of the can is approximately 108π cubic centimeters, find the radius of the can. The formula for the volume of a cylinder is $V = \pi r^2 h$

69. Hilly terrain We are interested in the nature of some hilly terrain. Computer simulation has told us that for a cross-section from west to east, the height $h(x)$, in feet above sea level is related to the horizontal distance x (in miles) from a fixed point by the function,

$$h(x) = -x^4 + 5x^3 + 91x^2 - 545x + 550, \; x \in [0,9].$$

At what distances from the fixed point is the height 100 feet above sea level?

70. Velocity of a hot-air balloon A hot-air balloon is tethered to the ground and only moves up and down. You and a friend take a ride on the ballon for approximately 25 minutes. On this particular ride the velocity of the balloon, $v(t)$ in feet per minute, as a function of time, t in minutes, is represented by the function

$$v(t) = -t^3 + 34t^2 - 320t + 850$$

At what times is the velocity of the balloon 50 feet per minute?

Steve Bower/Shutterstock.com

Discovery and Writing

71. If n is an even integer and c is a positive constant, show that $x^n + c = 0$ has no real roots.

72. If n is an even positive integer and c is a positive constant, show that $x^n - c = 0$ has two real roots.

73. Precalculus A rectangle is inscribed in the parabola $y = 16 - x^2$, as shown in the illustration. Find the point (x, y) if the area of the rectangle is 42 square units.

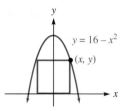

74. Precalculus One corner of the rectangle shown is at the origin, and the opposite corner (x, y) lies in the first quadrant on the curve $y = x^3 - 2x^2$. Find the point (x,y) if the area of the rectangle is 27 square units.

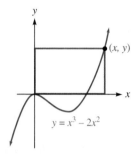

Review

Simplify each radical expression. Assume that all variables represent positive numbers.

75. $\sqrt{72a^3b^5c}$

76. $\dfrac{5a}{\sqrt{5a}}$

77. $\sqrt{18a^2b} + a\sqrt{50b}$

78. $(\sqrt{3} + \sqrt{5b})(\sqrt{12} - \sqrt{45b})$

79. $\dfrac{2}{\sqrt{3} - 1}$

80. $\dfrac{\sqrt{11} + \sqrt{x}}{\sqrt{11} - \sqrt{x}}$

5.4 Approximating Irrational Roots of Polynomial Equations

In this section, we will learn to

1. Use the Intermediate Value Theorem.

2. Use the Bisection Method.

3. Approximate real roots with a graphing calculator.

Sometimes we find ourselves in a car wondering how to get somewhere. Fortunately, many cars are now equipped with a navigation system that can give us directions to a specific address or point of interest.

kaczor58/Shutterstock.com

A navigation system uses satellite signals to determine the exact location of our car. Once the system knows the location of the car, it can guide us to our destination.

When solving more complicated equations, there are several theorems that act like a navigation system and can tell us the location of the roots of the equation. Once we know their location, it is much easier to find them.

Before developing these theorems, we will review some facts about solving equations.

- All first-degree equations are easy to solve by using simple algebraic methods.
- All quadratic equations can be solved by using the Quadratic Formula.
- There are formulas for solving third- and fourth-degree polynomial equations. However, they are complicated.
- No formulas exist for solving polynomial equations of degree 5 or greater. This fact was proved by the Norwegian mathematician Niels Henrik Abel (1802–1829) and (for equations of degree greater than 5) by the French mathematician Evariste Galois (1811–1832).

To solve a higher-degree polynomial equation with integer coefficients, we can use the methods of the previous section to find its *rational roots*. To find any irrational or complex roots, the final depressed equation would have to be a first- or second-degree equation.

In this section, we will discuss ways of approximating *irrational roots* of higher-degree polynomial equations that may not have any rational roots.

EXAMPLE 1   **Showing That an Equation Has Irrational Roots**

Show that $\sqrt{3}$ is a root of $x^4 - 9 = 0$ and that it is irrational.

SOLUTION   We know that $\sqrt{3}$ is a root of the equation because

$$x^4 - 9 = 0$$
$$\left(\sqrt{3}\right)^4 - 9 \stackrel{?}{=} 0 \qquad \text{Substitute } \sqrt{3} \text{ for } x.$$
$$9 - 9 \stackrel{?}{=} 0 \qquad \left(\sqrt{3}\right)^4 = 9$$
$$0 = 0$$

If there were any rational roots of this equation, they must have the form $\frac{\text{factor of the constant } -9}{\text{factor of the leading coefficient } 1}$. Therefore, the only possible rational roots of the equation would be

$$\pm\frac{1}{1}, \qquad \pm\frac{3}{1}, \qquad \text{and} \qquad \pm\frac{9}{1}$$

Since none of the numbers $1, -1, 3, -3, 9,$ or -9 satisfies the equation, no rational number can be a root. Thus, the root of $\sqrt{3}$ must be irrational. ■

Self Check 1   Show that $-\sqrt{5}$ is a root of $x^4 - 25 = 0$ and that it is irrational.

Comment

Roots of polynomial equations can be rational numbers, irrational numbers, or complex numbers. In this section, we will restrict ourselves to roots that are real, that is, rational or irrational numbers.

Now Try Exercise 5.

1. Use the Intermediate Value Theorem

The following theorem, called the **Intermediate Value Theorem**, leads to a way of locating an interval that contains a root.

The Intermediate Value Theorem	Let $P(x)$ be a polynomial function with real coefficients. If $P(a) \neq P(b)$ for $a < b$, then $P(x)$ takes on all values between $P(a)$ and $P(b)$ on the closed interval $[a, b]$.

JUSTIFICATION   This theorem becomes clear when we consider the graph of the polynomial function $y = P(x)$, shown in Figure 5-4. We have seen that graphs of polynomials are *continuous* curves, a technical term that means, roughly, that they can be drawn without lifting the pencil from the paper. If $P(a) \neq P(b)$, the continuous curve joining the points $A(a, P(a))$ and $B(b, P(b))$ must take on all values between $P(a)$ and $P(b)$ in the interval $[a, b]$, because the curve has no gaps in it.

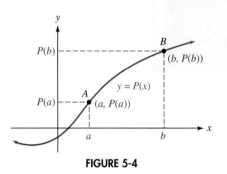

FIGURE 5-4

The next theorem follows from the Intermediate Value Theorem.

The Location Theorem	Let $P(x)$ be a polynomial function with real coefficients. If $P(a)$ and $P(b)$ have opposite signs, there is at least one number r in the interval (a, b) for which $P(r) = 0$.

PROOF   See Figure 5-5. By the Intermediate Value Theorem, $P(x)$ takes on all values between $P(a)$ and $P(b)$. Since $P(a)$ and $P(b)$ have opposite signs, the number 0 lies between them. Thus, there is a number r between a and b for which $P(r) = 0$. This number r is a zero of $P(x)$ and a root of the equation $P(x) = 0$.

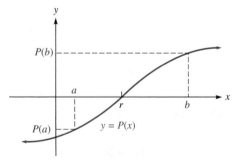

FIGURE 5-5

EXAMPLE 2   Using the Intermediate Value Theorem and Location Theorem

Show that $2x^3 - x^2 - 8x + 4 = 0$ has at least one real root between 0 and 1.

SOLUTION Let $P(x) = 2x^3 - x^2 - 8x + 4$. We will evaluate the polynomial function at $x = 0$ and $x = 1$. If the resulting values have opposite signs, we know that the equation has a root that lies between 0 and 1.

Evaluate $P(0)$ and $P(1)$ as follows:

$$P(0) = 2(0)^3 - (0)^2 - 8(0) + 4 = 4$$

$$P(1) = 2(1)^3 - (1)^2 - 8(1) + 4 = -3$$

Because $P(0)$ and $P(1)$ have opposite signs, we know by the Location Theorem that there is at least one real root between 0 and 1, as shown in Figure 5-6. The root shown is $\frac{1}{2}$.

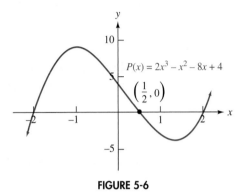

FIGURE 5-6

Self Check 2 Show that $2x^3 - 9x^2 + 7x + 6 = 0$ has at least one real root between -1 and 0.

Now Try Exercise 13.

2. Use the Bisection Method

The Location Theorem provides a method (called the **Bisection Method**) for finding the roots of $P(x) = 0$ to any desired degree of accuracy.

Suppose we find, by trial and error, that the numbers x_l and x_r (for left and right) straddle a root—that is, $x_l < x_r$ and $P(x_l)$ and $P(x_r)$ have opposite signs. (See Figure 5-7.) Further suppose that $P(x_l) < 0$ and $P(x_r) > 0$. We can compute the number c that is halfway between x_l and x_r and then compute $P(c)$. If $P(c) = 0$, we've found a root. However, if $P(c)$ is not 0, we proceed in one of two ways:

1. If $P(c) < 0$, the root r lies between c and x_r, as shown in Figure 5-7(a). In this case, we let c become a new x_l and repeat the procedure.
2. If $P(c) > 0$, the root r lies between x_l and c, as shown in Figure 5-7(b). In this case, we let c become a new x_r and repeat the procedure

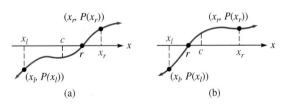

FIGURE 5-7

At any stage in this procedure, the root is contained between the current values of x_l and x_r. If the original bounds were 1 unit apart, after 10 repetitions of this procedure, the root would be between bounds that were 2^{-10} units apart. This means that the zero of $P(x)$ would be within 0.0001 units of either x_l or x_r.

After 20 repetitions, the root would be between bounds that were 2^{-20} units apart. This means that the zero of $P(x)$ would be within 0.000001 of either x_l or x_r.

EXAMPLE 3 **Using the Bisection Method**

Use the Bisection Method to find the positive root of $x^2 - 2 = 0$ to the nearest tenth.

SOLUTION We will use the Location Theorem and the Bisection Method to find the positive root of the equation to the nearest tenth.

Since $P(1) = -1$ and $P(2) = 2$ have opposite signs, there is a root between 1 and 2. We can let $x_l = 1$ and $x_r = 2$ and compute the midpoint c:

$$c = \frac{x_l + x_r}{2} = \frac{1 + 2}{2} = 1.5$$

Because $P(c) = P(1.5) = 0.25$ is a positive number, we let c become a new x_r and calculate a new midpoint, which we will call c_1.

$$c_1 = \frac{x_l + x_r}{2} = \frac{1 + \mathbf{1.5}}{2} = 1.25$$

Because $P(c_1) = P(1.25) = -0.4375$ is a negative number, we let c_1 become a new x_l and calculate a new midpoint, which we will call c_2.

$$c_2 = \frac{x_l + x_r}{2} = \frac{\mathbf{1.25} + 1.5}{2} = 1.375$$

Because $P(c_2) = P(1.375) = -0.109375$ is a negative number, we let c_2 become a new x_l and calculate a new midpoint, which we will call c_3.

$$c_3 = \frac{x_l + x_r}{2} = \frac{\mathbf{1.375} + 1.5}{2} = 1.4375$$

Because $P(c_3) = P(1.4375) = 0.066406$ is a positive number, we let c_3 become a new x_r and calculate a new midpoint, which we will call c_4.

$$c_4 = \frac{x_l + x_r}{2} = \frac{1.375 + \mathbf{1.4375}}{2} = 1.40625$$

From here on, the first two digits of the midpoints will remain 1.4. The root r of the equation (to the nearest tenth) is 1.4. ∎

Self Check 3 Find the negative root of $x^2 - 2 = 0$ to the nearest tenth.

Now Try Exercise 19.

3. Approximate Real Roots with a Graphing Calculator

We can approximate the real roots of an equation either by using the TRACE and ZOOM capabilities of a graphing calculator or by using the ZERO feature.

ACCENT ON TECHNOLOGY **Using a Graphing Calculator to Approximate Real Roots**

We will use the ZERO feature on a graphing calculator to approximate the real roots of the equation $x^4 - 6x^2 + 9 = 0$.

• We first enter the equation, using the WINDOW shown, and GRAPH the equation to obtain the graph shown in Figure 5-8.

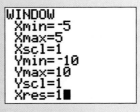
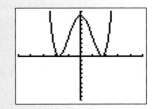

FIGURE 5-8

- We then select ZERO under the CALC menu. To find the positive root, we guess a left bound and press ENTER, guess a right bound and press ENTER, and then press ENTER again. The root of the equation is approximately 1.7320511. See Figure 5-9.

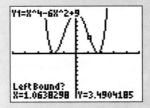

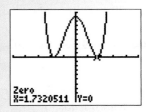

FIGURE 5-9

Because of symmetry, we know that the negative solution is -1.7320511. The real roots are $x \approx \pm 1.7320511$.

The graph shown in Figure 5-9 illustrates a situation in which the Bisection Method does not work. Since the graph doesn't cross the x-axis, the values of the function on opposite sides of the intercepts do not have opposite signs. In this case, the Bisection Method would be useless.

Self Check Answers **1.** $\pm\dfrac{1}{1}$, $\pm\dfrac{5}{1}$, or $\pm\dfrac{25}{1}$ do not satisfy $x^4 - 25 = 0$.   **2.** $P(-1) = -12$; $P(0) = 6$
They have opposite signs.   **3.** -1.4

Exercises **5.4**

Getting Ready
You should be able to complete these vocabulary and concept statements before you proceed to the practice exercises.

Fill in the blanks.

1. If $P(x)$ is a polynomial function with real coefficients and $P(a) \neq P(b)$ for $a < b$, then $P(x)$ takes on all values between _____ in the interval $[a, b]$.

2. If $P(x)$ has real coefficients and $P(a)$ and $P(b)$ have opposite signs, there is at least one number r in (a, b) for which _____.

3. In the Bisection Method, we find two numbers x_l and x_r that straddle a root. We then compute a new guess for the root (a number c) by finding the average of _____.

4. A _____ curve has no gaps in it.

Practice
Show that the given number is a root of the equation and that it is irrational.

5. $x^4 - 4 = 0$; $\sqrt{2}$

6. $x^4 - 36 = 0$; $\sqrt{6}$

7. $x^4 - 49 = 0$; $-\sqrt{7}$

8. $x^4 - 64 = 0$; $-2\sqrt{2}$

Show that each equation has at least one real root between the specified numbers.

9. $2x^2 + x - 3 = 0$; -2 and -1

10. $2x^3 + 17x^2 + 31x - 20 = 0$; -1 and 2

11. $3x^3 - 11x^2 - 14x = 0$; 4 and 5

12. $2x^3 - 3x^2 + 2x - 3 = 0$; 1 and 2

13. $x^4 - 8x^2 + 15 = 0$; 1 and 2

14. $x^4 - 8x^2 + 15 = 0$; 2 and 3

15. $30x^3 + 10 = 61x^2 + 39x$; 2 and 3

16. $30x^3 + 10 = 61x^2 + 39x$; -1 and 0

17. $30x^3 + 10 = 61x^2 + 39x$; 0 and 1

18. $5x^3 - 9x^2 - 4x + 9 = 0$; -1 and 0

Use the Bisection Method to find the following values to the nearest tenth.

19. The positive root of $x^2 - 3 = 0$.

20. The negative root of $x^2 - 3 = 0$.

21. The negative root of $x^2 - 5 = 0$.

22. The positive root of $x^2 - 5 = 0$.

23. The positive root of $x^3 - x^2 - 2 = 0$.

24. The negative root of $x^3 - x + 2 = 0$.

25. The negative root of $3x^4 + 3x^3 - x^2 - 4x - 4 = 0$.

26. The positive root of
$x^5 + x^4 - 4x^3 - 4x^2 - 5x - 5 = 0$.

Use a graphing calculator to find the distinct real solutions of each equation to the nearest tenth. Which roots, if any, would the Bisection Method fail to find?

27. $x^2 - 5 = 0$

28. $x^2 - 10x + 25 = 0$

29. $x^3 - 5x^2 + 8x - 4 = 0$

30. $x^3 - 5x^2 - 2x + 10 = 0$

Applications

31. **Containers** A box has a length of 16 inches, a width of 10 inches, and a height of between 4 inches and 8 inches. Can it have a volume of 1,000 in.³? *Hint:* Model the volume of the box with a function of h.

32. **Lollipops** If a candy company makes spherical-shaped lollipops with radii between 1 and 4 cm, use the Intermediate Value Theorem to determine whether a lollipop can be made with a volume of 200 cubic centimeters. *Hint:* The volume formula for a sphere is $V = \frac{4}{3}\pi r^3$.

33. **Precalculus** Use the Bisection Method or a graphing calculator to find the coordinates of the two points on the graph of $y = x^3$ that lie 1 unit from the origin. (See the illustration.) Give the result to the nearest hundredth.

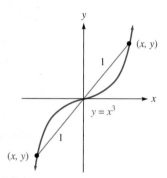

34. **Building crates** The width of the shipping crate shown is to be 2 feet greater than its height, the length is to be 5 feet greater than its width, and its volume is to be 170 cubic feet. Use the Bisection Method or a graphing calculator to find the height of the crate to the nearest tenth of a foot.

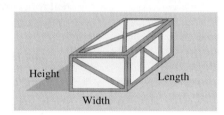

Discovery and Writing

35. Explain when the Bisection Method will not work.

36. Describe how you could tell from a graph that a polynomial equation has no real roots.

Review

Give the equations of all of the asymptotes of each rational function. Don't graph the functions.

37. $y = \dfrac{5x - 3}{x^2 - 4}$

38. $y = \dfrac{5x^2 - 3}{x^2 - 25}$

39. $y = \dfrac{x^2}{x - 2}$

40. $y = \dfrac{x + 1}{x^2 + 1}$

CHAPTER REVIEW

SECTION **5.1**   The Remainder and Factor Theorems; Synthetic Division

Definitions and Concepts	Examples
A **polynomial equation** is an equation that can be written in the form $P(x) = 0$, where $P(x) = a_n x^n + a_{n-1} x^{n-1} + a_{n-2} x^{n-2} + \cdots + a_1 x + a_0$ where n is a natural number and the polynomial is of degree n.	$3x - 10 = 0$; degree: 1 $7x^2 + 3x - 1 = 0$; degree: 2 $4x^3 + 7x^2 + 3x - 1 = 0$; degree: 3 $x^4 - 4x^3 + 7x^2 + 3x - 1 = 0$; degree: 4
Zero of a polynomial function: A **zero of the polynomial** $P(x)$ is any number r for which $P(r) = 0$.	The real number 1 is a zero of the polynomial function $P(x) = x^3 - 4x^2 + 3x$ because $$P(1) = 1^3 - 4(1)^2 + 3(1)$$ $$= 1 - 4 + 3$$ $$= 0$$
The Remainder Theorem: If $P(x)$ is a polynomial function, r is any number, and $P(x)$ is divided by $x - r$, the remainder is $P(r)$.	Let $P(x) = 3x^3 - 5x^2 + 3x - 10$ and use the Remainder Theorem and long division to find $P(2)$. We will divide $P(x)$ by $x - 2$ as follows: $$\begin{array}{r} 3x^2 + x + 5 \\ x - 2 \overline{)3x^3 - 5x^2 + 3x - 10} \\ \underline{3x^3 - 6x^2} \\ x^2 + 3x \\ \underline{x^2 - 2x} \\ 5x - 10 \\ \underline{5x - 10} \\ 0 \end{array}$$ The remainder is 0. $P(2) = 0$.
The Factor Theorem: If $P(x)$ is a polynomial function and r is any number, then if $P(r) = 0$, then $x - r$ is a factor of $P(x)$. If $x - r$ is a factor of $P(x)$, then $P(r) = 0$. **Alternate Form of the Factor Theorem:** If r is a zero of the polynomial function $P(x)$, then $x - r$ is a factor of $P(x)$. If $x - r$ is a factor of $P(x)$, then r is a zero of the polynomial function.	We see in the previous example that $P(2) = 0$. By the Factor Theorem, we know that $x - 2$ is a factor of the polynomial function $P(x) = 3x^3 - 5x^2 + 3x - 10$. We also know that 2 is a zero of the polynomial function.

Definitions and Concepts	Examples
Synthetic division is a fast way to divide higher-degree polynomials by binomials of the form $x - r$.	Divide $P(x) = 5x^3 + 3x^2 - 21x - 1$ by $x - 2$ using synthetic division.

We first write the coefficients of the dividend, with its terms in descending powers of x, and the 2 from the divisor in the following form:

$$\underline{2|}\ \ 5\quad 3\quad -21\quad -1$$

$$\underline{2|}\ \ 5\quad 3\quad -21\quad -1$$
$$\qquad\downarrow$$
$$\qquad 5$$

Bring down the 5.

$$\underline{2|}\ \ 5\quad 3\quad -21\quad -1$$
$$\qquad\ \ 10$$
$$\qquad 5\ \ 13\qquad\qquad 9$$

$2(5) = 10;\ 3 + 10 = 13$

$$\underline{2|}\ \ 5\quad 3\quad -21\quad -1$$
$$\qquad\ \ 10\quad 26$$
$$\qquad 5\ \ 13\quad 5\quad 9$$

$2(13) = 26; -21 + 26 = 5$

$$\underline{2|}\ \ 5\quad 3\quad -21\quad -1$$
$$\qquad\ \ 10\quad 26\ \ 10$$
$$\qquad 5\ \ 13\quad 5\quad 9$$

$2(5) = 10;\ -1 + 10 = 9$

The quotient is $5x^2 + 13x + 5$. The remainder is 9. Thus,

$$\frac{5x^3 - 3x^2 - 21x - 1}{x - 2} = 5x^2 + 13x + 5 + \frac{9}{x - 2}$$

Because the remainder is 9, we know by the Remainder Theorem that $P(2) = 9$. Because $P(2) \neq 0$, we know by the Factor Theorem that $x - 2$ is not a factor of $5x^3 + 3x^2 - 21x - 1$.

Synthetic division can be used to solve polynomial equations.

Let $P(x) = x^3 + x^2 - 3x - 3$. Completely solve the polynomial equation $P(x) = 0$ given that -1 is a root.

1. We use synthetic division to divide $P(x)$ by $x + 1$.

$$\underline{-1|}\ \ 1\quad 1\quad -3\quad -3$$
$$\qquad\quad -1\quad 0\quad 3$$
$$\qquad\ \ 1\quad 0\quad -3\quad 0$$

2. We then write the quotient and factor it.

$$x^3 + x^2 - 3x - 3 = (x + 1)(x^2 - 3).$$

Definitions and Concepts	Examples
	3. Finally, we solve the polynomial equation $P(x) = 0$. $$x^3 + x^2 - 3x - 3 = 0$$ $$(x + 1)(x^2 - 3) = 0$$ We then set each factor equal to 0 and solve for x. $$x + 1 = 0 \quad \text{or} \quad x^2 - 3 = 0$$ $$x = -1 \qquad\qquad x^2 = 3$$ $$x = \pm\sqrt{3}$$ The solution set is $\left\{-1, \sqrt{3}, -\sqrt{3}\right\}$.

EXERCISES

Let $P(x) = 4x^4 + 2x^3 - 3x^2 - 2$. Find the remainder when $P(x)$ is divided by each binomial.

1. $x - 1$

2. $x - 2$

3. $x + 3$

4. $x + 2$

Use the Factor Theorem to determine whether each statement is true.

5. $x - 2$ is a factor of $x^3 + 4x^2 - 2x + 4$.

6. $x + 3$ is a factor of $2x^4 + 10x^3 + 4x^2 + 7x + 21$.

7. $x - 5$ is a factor of $x^5 - 3{,}125$.

8. $x - 6$ is a factor of $x^5 - 6x^4 - 4x + 24$.

Use synthetic division to perform each division.

9. $3x^4 + 2x^2 + 3x + 7; \; x - 3$

10. $2x^4 - 3x^2 + 3x - 1; \; x - 2$

11. $5x^5 - 4x^4 + 3x^3 - 2x^2 + x - 1; \; x + 2$

12. $4x^5 + 2x^4 - x^3 + 3x^2 + 2x + 1; \; x + 1$

Let $P(x) = 5x^3 + 2x^2 - x + 1$. Use synthetic division to find each value.

13. $P(3)$

14. $P(-3)$

15. $P\left(\dfrac{1}{2}\right)$

16. $P(i)$

A partial solution set is given for each polynomial equation. Find the complete solution set.

17. $2x^3 - 3x^2 - 11x + 6 = 0; \; \{3\}$

18. $x^4 + 4x^3 - x^2 - 20x - 20 = 0; \; \{-2, -2\}$

Find the polynomial function of lowest degree with integer coefficients and the given zeros.

19. $-1, 2,$ and $\dfrac{3}{2}$

20. $1, -3,$ and $\dfrac{1}{2}$

21. $2, -5, i,$ and $-i$

22. $-3, 2, i,$ and $-i$

SECTION 5.2   Descartes' Rule of Signs and Bounds on Roots

Definitions and Concepts	Examples
The Fundamental Theorem of Algebra: If $P(x)$ is a polynomial function with positive degree, then $P(x)$ has at least one zero.	The Fundamental Theorem of Algebra guarantees that polynomial functions like $P(x) = 3x^3 - 5x^2 + 7x - 2$ and $P(x) = 4x^5 + x^3 - 6x - 15$ have zeros.

Definitions and Concepts	**Examples**
The Polynomial Factorization Theorem: If $n > 0$ and $P(x)$ is an nth-degree polynomial function, then $P(x)$ has exactly n linear factors: $$P(x) = a_n(x - r_1)(x - r_2)(x - r_3) \cdot \,\cdots\, \cdot (x - r_n)$$ where $r_1, r_2, r_3, \ldots, r_n$ are numbers and a_n is the leading coefficient of $P(x)$.	Consider the following polynomial function of degree $n = 3$. $$P(x) = x^3 + 144x$$ $$= x(x^2 + 144)$$ $$= x(x - 12i)(x + 12i)$$ The Polynomial Factorization Theorem guarantees that we have exactly 3 linear factors. They are $$x, \ x - 12i, \text{ and } x + 12i$$
If multiple roots are counted individually, the polynomial equation $P(x) = 0$ with degree n $(n > 0)$ has exactly n roots among the complex numbers.	The polynomial equation $P(x) = x^3 + 144x = 0$ has exactly 3 roots. They are $$0, \ 12i, \text{ and } -12i$$ Each root occurs once and has a multiplicity of one.
The Conjugate Pairs Theorem: If a polynomial equation $P(x) = 0$ with real-number coefficients has a complex root $a + bi$ with $b \neq 0$, then its conjugate $a - bi$ is also a root.	The Conjugate Pairs Theorem applies to $P(x) = x^3 + 144x = 0$ because the polynomial has real coefficients. The complex roots $12i$ and $-12i$ that occur are conjugate pairs.
Descartes' Rule of Signs: Let $P(x)$ be a polynomial function with real coefficients. • The number of positive roots of $P(x) = 0$ is either equal to the number of variations in sign of $P(x)$ or less than that by an even number. • The number of negative roots of $P(x) = 0$ is either equal to the number of variations in sign of $P(-x)$ or less than that by an even number.	Use Descartes' Rule of Signs to find the number of possible positive, negative, and nonreal roots of $P(x) = 5x^3 - 7x^2 + x - 6 = 0$. **1. Find the number of possible positive roots.** Since there are three variations of sign in $P(x) = 5x^3 - 7x^2 + x - 6 = 0$, there can be either 3 positive roots or only 1 (1 is less than 3 by the even number 2). **2. Find the number of possible negative roots.** Because $$P(-x) = 5(-x)^3 - 7(-x)^2 + (-x) - 6$$ $$= -5x^3 - 7x^2 - x - 6$$ has no variations in sign, there are 0 negative roots. Furthermore, 0 is not a root, because the terms of the polynomial do not have a common factor of x. **3. Find the number of nonreal roots.** If there are 3 positive roots, then all of the roots are accounted for. If there is 1 positive root, the 2 remaining roots must be nonreal complex numbers. The following chart shows these possibilities.

Number of positive roots	Number of negative roots	Number of nonreal roots
3	0	0
1	0	2

The number of nonreal complex roots is the number needed to bring the total number of roots up to 3.

Definitions and Concepts	Examples

Definitions and Concepts

Upper and lower bounds on roots:
Let $P(x)$ be a polynomial function with real coefficients and a positive leading coefficient.

1. If $P(x)$ is synthetically divided by a positive number c and each term in the last row of the division is nonnegative, then no number greater than c can be a root of $P(x) = 0$. (c is called an **upper bound** of the real roots.)

2. If $P(x)$ is synthetically divided by a negative number d and the signs in the last row alternate,* no value less than d can be a root of $P(x) = 0$. (d is called a **lower bound** of the real roots.)

*If 0 appears in the third row, that 0 can be assigned either a + or a − sign to help the signs alternate.

Examples

Use the upper and lower bound rules to establish bounds for the roots of $2x^3 + 2x^2 - 8x - 8 = 0$.

1. We will perform several synthetic divisions by positive integers, looking for nonnegative values in the last row.
   Trying 1 first gives

$$\underline{1|}\quad 2\quad 2\quad -8\quad -8$$
$$\qquad\qquad 2\quad 4\quad -4$$
$$\overline{\quad +2\ +4\ -4\ -12}$$

Because at least one of the signs in the last row is negative, we cannot claim that 1 is an upper bound. We now try 2.

$$\underline{2|}\quad 2\quad 2\quad -8\quad -8$$
$$\qquad\qquad 4\quad 12\quad 8$$
$$\overline{\quad +2\ +6\ +4\quad 0}$$

Because the last row is entirely nonnegative, we can claim that 2 is an upper bound. That is, no number greater than 2 can be a root of the equation.

2. We perform several synthetic divisions by negative divisors, looking for alternating signs in the last row. We begin with −3.

$$\underline{-3|}\quad 2\quad 2\quad -8\quad -8$$
$$\qquad\qquad -6\quad 12\quad -12$$
$$\overline{\quad +2\ -4\ +4\ -20}$$

Since the signs in the last row alternate, −3 is a lower bound. That is, no number less than −3 can be a root. All of the real roots must be in the interval $(-3, 2)$.

EXERCISES

How many roots does each equation have?

23. $3x^6 - 4x^5 + 3x + 2 = 0$

24. $2x^6 - 5x^4 + 5x^3 - 4x^2 + x - 12 = 0$

25. $3x^{65} - 4x^{50} + 3x^{17} + 2x = 0$

26. $x^{1,984} - 12 = 0$

Determine how many linear factors and zeros each polynomial function has.

27. $P(x) = x^4 - 16$

28. $P(x) = x^{40} + x^{30}$

29. $P(x) = 4x^5 + 2x^3$

30. $P(x) = x^3 - 64x$

Find another root of a polynomial equation with real coefficients if the given quantity is one root.

31. $2 + i$

32. $-i$

Write a third-degree polynomial equation with real coefficients and the given roots.

33. $4, -i$

34. $-5, i$

Find the number of possible positive, negative, and nonreal roots for each equation. Do not attempt to solve the equation.

35. $3x^4 + 2x^3 - 4x + 2 = 0$

36. $2x^4 - 3x^3 + 5x^2 + x - 5 = 0$

37. $4x^5 + 3x^4 + 2x^3 + x^2 + x = 7$

38. $3x^7 - 4x^5 + 3x^3 + x - 4 = 0$

39. $x^4 + x^2 + 24{,}567 = 0$

40. $-x^7 - 5 = 0$

Find integer bounds for the roots of each equation. Answers can vary.

41. $5x^3 - 4x^2 - 2x + 4 = 0$

42. $x^4 + 3x^3 - 5x^2 - 9x + 1 = 0$

SECTION **5.3**   Roots of Polynomial Equations

Definitions and Concepts	Examples
Rational Root Theorem: Let the polynomial function $P(x) = a_n x^n + a_{n-1} x^{n-1} + a_{n-2} x^{n-2} + \cdots + a_1 x + a_0$ have integer coefficients. If the rational number $\frac{p}{q}$ (written in lowest terms) is a root of $P(x) = 0$, then p is a factor of the constant a_0, and q is a factor of the leading coefficient a_n.	Consider $2x^3 - 7x^2 - 17x + 10 = 0$ By the Rational Root Theorem, the only possible numerators for the rational roots of the equation are the factors of the constant term 10: $\pm 1, \pm 2, \pm 5,$ and ± 10 The only possible denominators are the factors of the leading coefficient 2: $\pm 1,$ and ± 2 We can form a list of all possible rational solutions by listing the combinations of possible numerators and denominators: $$\pm\frac{1}{1}, \pm\frac{2}{1}, \pm\frac{5}{1}, \pm\frac{10}{1}, \pm\frac{1}{2}, \pm\frac{2}{2}, \pm\frac{5}{2}, \pm\frac{10}{2}$$ Since several of these possibilities are duplicates, we can condense the list to get ***Possible Rational Roots*** $$\pm 1, \pm 2, \pm 5, \pm 10, \pm\frac{1}{2}, \text{ and } \pm\frac{5}{2}$$

Finding rational roots:
1. Use Descartes' Rule of Signs to determine the number of possible positive, negative, and nonreal roots.
2. Use the Rational Root Theorem to list possible rational roots.
3. Use synthetic division to find a root.
4. If there are more roots, repeat the previous steps.
 	
(Once the depressed equation is quadratic, synthetic division isn't required. It can be solved as a quadratic equation.)
	Find the solution set of the polynomial equation $P(x) = 2x^3 - 7x^2 - 17x + 10 = 0$.
	
	1. To determine the number of possible positive, negative, and nonreal roots, we will use Descartes' Rule of Signs.
	    Since there are two variations in sign of $P(x)$, the number of positive roots is either 2 or 0.
	    Next, we consider $P(-x)$.
	
	$P(-x) = -2x^3 - 7x^2 + 17x + 10.$
	
	Since there is one variation in sign of $P(-x)$, the number of negative roots is 1.
	    The following chart shows these possibilities:

Number of positive roots	Number of negative roots	Number of nonreal roots
2	1	0
0	1	2

Definitions and Concepts	Examples

	2. List the possible rational roots.
	The possible rational roots were found in the previous example. They are:
	
	    Possible Rational Roots
	
	    $\pm 1, \ \pm 2, \ \pm 5, \ \pm 10, \ \pm\dfrac{1}{2}, \text{ and } \pm\dfrac{5}{2}$
	
	3. Use synthetic division to find a root. We can start with -2.
	
	$$\begin{array}{r|rrrr} -2 & 2 & -7 & -17 & 10 \\ & & -4 & 22 & -10 \\ \hline & 2 & -11 & 5 & 0 \end{array}$$
	
	Since the remainder is 0, -2 is a root and the binomial $x + 2$ is a factor of $P(x)$.
	
	4. The remaining roots must be supplied by the remaining factor, which is the quotient $2x^2 - 11x + 5$. We can find the other roots by solving the depressed equation $2x^2 - 11x + 5 = 0$.
	
	$2x^2 - 11x + 5 = 0$
	
	$(2x - 1)(x - 5) = 0$ Factor.
	
	$2x - 1 = 0 \quad \text{or} \quad x - 5 = 0$
	
	$2x = 1 \qquad\qquad\quad x = 5$
	
	$x = \dfrac{1}{2}$
	
	The solution set of the equation is $\left\{-2, \frac{1}{2}, 5\right\}$.
	Note that two positive roots and one negative root is a predicted possibility.

EXERCISES

Use the Rational Root Theorem to list all possible rational roots of the polynomial equation.

43. $2x^4 + x^3 - 3x^2 - 5x - 6 = 0$

44. $4x^5 - 2x^4 + 3x^3 - 5x - 10 = 0$

Find all rational roots of each equation.

45. $x^3 - 10x^2 + 29x - 20 = 0$

46. $x^3 - 8x^2 - x + 8 = 0$

47. $2x^3 + 17x^2 + 41x + 30 = 0$

48. $3x^3 + 2x^2 + 2x - 1 = 0$

49. $4x^4 - 25x^2 + 36 = 0$

50. $2x^4 - 11x^3 - 6x^2 + 64x + 32 = 0$

Find all roots of each equation.

51. $3x^3 - x^2 + 48x - 16 = 0$

52. $x^4 - 2x^3 - 9x^2 + 8x + 20 = 0$

SECTION **5.4** Approximating Irrational Roots of Polynomial Equations

Definitions and Concepts	Examples
The Intermediate Value Theorem: Let $P(x)$ be a polynomial function with real coefficients. If $P(a) \neq P(b)$ for $a < b$, then $P(x)$ takes on all values between $P(a)$ and $P(b)$ on the closed interval $[a, b]$.	Consider the polynomial function $P(x) = 4x^3 + 2x^2 - 12x + 3$ and note that $$P(0) = 4(0)^3 + 2(0)^2 - 12(0) + 3 = 3$$ $$P(1) = 4(1)^3 + 2(1)^2 - 12(1) + 3 = -3$$ Because $P(0) \neq P(1)$ and $0 < 1$, the Intermediate Value Theorem guarantees that $P(x)$ takes on all values between -3 and 3 on the closed interval $[0,1]$.
The Location Theorem: Let $P(x)$ be a polynomial function with real coefficients. If $P(a)$ and $P(b)$ have opposite signs, there is at least one number r in the interval (a, b) for which $P(r) = 0$.	To show that the polynomial equation, $4x^3 + 2x^2 - 12x + 3 = 0$ has at least one real root between 0 and 1, we see from above that $P(0) = 3$ and $P(1) = -3$ have opposite signs. Therefore, we know by the Location Theorem that there is at least one real root between 0 and 1, as shown in the figure. 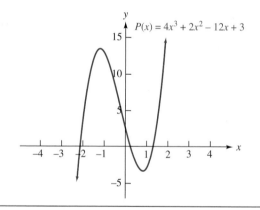
Bisection Method: The Bisection Method is a method for finding the roots of $P(x) = 0$ to any desired degree of accuracy. Suppose we find, by trial and error, that the numbers x_l and x_r (for left and right) straddle a root—that is, $x_l < x_r$ and $P(x_l)$ and $P(x_r)$ have opposite signs. Further suppose that $P(x_l) < 0$ and $P(x_r) > 0$. We can compute the number c that is halfway between x_l and x_r and then compute $P(c)$. If $P(c) = 0$, we've found a root. However, if $P(c)$ is not 0, we proceed in one of two ways: **1.** If $P(c) < 0$, the root r lies between c and x_r. In this case, we let c become a new x_l and repeat the procedure. **2.** If $P(c) > 0$, the root r lies between x_l and c. In this case, we let c become a new x_r and repeat the procedure.	Use the Bisection Method to find the positive root of $x^2 - 5 = 0$ to the nearest tenth. We let $P(x) = x^2 - 5$. Since $P(2) = -1$ and $P(3) = 4$ have opposite signs, there is a root between 2 and 3. We can let $x_l = 2$ and $x_r = 3$ and compute the midpoint c: $$c = \frac{x_l + x_r}{2} = \frac{2 + 3}{2} = 2.5$$ Because $P(c) = P(2.5) = 1.25$ is a positive number, we let c become a new x_r and calculate a new midpoint, which we will call c_1. $$c_1 = \frac{x_l + x_r}{2} = \frac{2 + 2.5}{2} = 2.25$$ Because $P(c_1) = P(2.25) = 0.0625$ is a positive number, we let c_1 become a new x_r and calculate a new midpoint, which we will call c_2. $$c_2 = \frac{x_l + x_r}{2} = \frac{2 + 2.25}{2} = 2.125$$

Definitions and Concepts	Examples
	Because $P(c_2) = P(2.125) = -0.484375$ is a negative number, we let c_2 become a new x_l and calculate a new midpoint, which we will call c_3. $$c_3 = \frac{x_l + x_r}{2} = \frac{\mathbf{2.125} + 2.25}{2} = 2.1875$$ Because $P(c_3) = P(2.1875) = -0.21484375$ is a negative number, we let c_3 become a new x_l and calculate a new midpoint, which we will call c_4. $$c_4 = \frac{x_l + x_r}{2} = \frac{\mathbf{2.1875} + 2.25}{2} = 2.21875$$ Because $P(c_4) = P(2.21875) = -0.0771484375$ is a negative number, we calculate a new midpoint. From here on, the first two digits of the midpoints will remain 2.2. The root r of the equation (to the nearest tenth) is 2.2.
Graphing calculator usage: To approximate the roots of a polynomial equation using a graphing calculator, we can use the ZOOM and TRACE capabilities of the calculator or the ZERO feature of the calculator.	See Accent on Technology in Section 5.4.

EXERCISES

Show that each polynomial has a zero between the two given numbers.

53. $5x^3 + 37x^2 + 59x + 18 = 0$; -1 and 0

54. $6x^3 - x^2 - 10x - 3 = 0$; 1 and 2

Use the Bisection Method to find the positive root of each equation to the nearest tenth.

55. $x^3 - 2x^2 - 9x - 2 = 0$

56. $6x^2 - 13x - 5 = 0$

Use a graphing calculator to find the positive root of each equation to the nearest hundredth.

57. $6x^2 - 7x - 5 = 0$

58. $3x^2 + x - 2 = 0$

59. Designing solar collectors The space available for the installation of three solar collecting panels requires that their lengths differ by the amounts shown in the illustration, and that the total of their widths be 15 meters. To be equally effective, each panel must measure exactly 60 square meters. Find the dimensions of each panel.

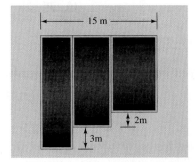

60. Designing a storage tank The design specifications for the cylindrical storage tank shown require that its height be 3 feet greater than the radius of its circular base and that the volume of the tank be 19,000 cubic feet. Use the Bisection Method or a graphing calculator to find the radius of the tank to the nearest hundredth of a foot.

CHAPTER TEST

1. Is $x = -2$ a zero of $P(x) = x^2 + 5x + 6$?

Use long division and the Remainder Theorem to find each value.

2. $P(x) = 3x^3 - 9x - 5$; $P(2)$

3. $P(x) = x^5 + 2$; $P(-2)$

4. Use the Factor Theorem to determine whether $x - 3$ is a factor of $2x^4 - 10x^3 + 4x^2 + 7x + 21$.

Use synthetic division to express
$P(x) = 2x^3 - 3x^2 - 4x - 1$ *in the form*
(divisor)(quotient) + remainder *for each divisor.*

5. $x - 2$

6. $x + 1$

Use synthetic division to perform each division.

7. $\dfrac{2x^2 - 7x - 15}{x - 5}$

8. $\dfrac{3x^3 + 7x^2 + 2x}{x + 2}$

Let $P(x) = 3x^3 - 2x^2 + 4$. Use synthetic division to find each value.

9. $P(1)$

10. $P(-2)$

11. $P\left(-\dfrac{1}{3}\right)$

12. $P(i)$

Find a polynomial function with the given zeros.

13. $5, -1, 0$

14. $i, -i, \sqrt{3}, -\sqrt{3}$

Write a third-degree polynomial equation with real coefficients and the given roots.

15. $2, i$

16. $1, 2 + i$

17. How many linear factors and zeros does $P(x) = 3x^3 + 2x^2 - 4x + 1$ have?

18. If $3 - 2i$ is a root of $P(x) = 0$ where $P(x)$ has real-number coefficients, find another root.

Use Descartes' Rule of Signs to find the number of possible positive, negative, and nonreal roots of each equation.

19. $3x^5 - 2x^4 + 2x^2 - x - 3 = 0$

20. $2x^3 - 5x^2 - 2x - 1 = 0$

Find integer bounds for the roots of each equation. Answers can vary.

21. $x^5 - x^4 - 5x^3 + 5x^2 + 4x - 5 = 0$

22. $2x^3 - 11x^2 + 10x + 3 = 0$

23. Use the Rational Root Theorem to list all possible rational roots of $x^3 + 4x^2 + 3x + 8 = 0$.

24. Use the Rational Root Theorem to list all possible rational roots of $5x^3 + 4x^2 + 3x + 2 = 0$.

25. Find all roots of the equation
$$2x^3 + 3x^2 - 11x - 6 = 0.$$

26. Find all roots of the equation $x^3 - 2x^2 + x - 2 = 0$.

27. Find all roots of the equation
$x^4 + x^3 + 3x^2 + 9x - 54 = 0$.

28. Find all roots of the equation
$x^5 + 2x^4 + 3x^3 + 6x^2 - 4x - 8 = 0$.

29. Does the polynomial $P(x) = 3x^3 + 2x^2 - 4x + 4$ have a zero between the values $x = 1$ and $x = 2$?

Use the Bisection Method to find the positive root of the equation to the nearest tenth.

30. $x^2 - 11 = 0$

CUMULATIVE REVIEW EXERCISES

Graph the function defined by each equation.

1. $f(x) = 3^x - 2$

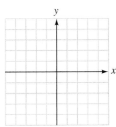

2. $f(x) = 2e^x$

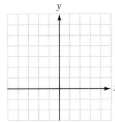

3. $f(x) = \log_3 x$

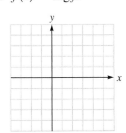

4. $f(x) = \ln(x - 2)$

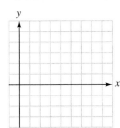

Find each value.

5. $\log_2 64$

6. $\log_{1/2} 8$

7. $\ln e^3$

8. $2^{\log_2 2}$

Write each expression in terms of the logarithms of a, b, and c.

9. $\log abc$

10. $\log \dfrac{a^2 b}{c}$

11. $\log \sqrt{\dfrac{ab}{c^3}}$

12. $\ln \dfrac{\sqrt{ab^2}}{c}$

Write each expression as the logarithm of a single quantity.

13. $3 \ln a - 3 \ln b$

14. $\dfrac{1}{2}\log a + 3 \log b - \dfrac{2}{3}\log c$

Solve each equation.

15. $3^{x+1} = 8$

16. $3^{x-1} = 3^{2x}$

17. $\log x + \log 2 = 3$

18. $\log (x + 1) + \log (x - 1) = 1$

Let $P(x) = 4x^3 + 3x + 2$. Use synthetic division to find each value.

19. $P(1)$

20. $P(-2)$

21. $P\left(\dfrac{1}{2}\right)$

22. $P(i)$

Determine whether each binomial is a factor of $P(x) = x^3 + 2x^2 - x - 2$. Use synthetic division.

23. $x + 1$

24. $x - 2$

25. $x - 1$

26. $x + 2$

Determine how many roots each equation has.

27. $x^{12} - 4x^8 + 2x^4 + 12 = 0$

28. $x^{2,000} - 1 = 0$

Determine the number of possible positive, negative, and nonreal roots of each equation.

29. $x^4 + 2x^3 - 3x^2 + x + 2 = 0$

30. $x^4 - 3x^3 - 2x^2 - 3x - 5 = 0$

Solve each equation.

31. $x^3 + x^2 - 9x - 9 = 0$

32. $x^3 - 2x^2 - x + 2 = 0$

Appendix I: A Proof of the Binomial Theorem

The Binomial Theorem can be proved for positive-integer exponents using mathematical induction.

The Binomial Theorem   If n is a positive integer, then

$$(a + b)^n = a^n + \frac{n!}{1!(n-1)!}a^{n-1}b^1 + \frac{n!}{2!(n-2)!}a^{n-2}b^2 + \cdots$$

$$+ \frac{n!}{r!(n-r)!}a^{n-r}b^r + \cdots + b^n$$

PROOF   As in all induction proofs, there are two parts.

Part 1: Substituting the number 1 for n on both sides of the equation, we have

$$(a + b)^1 = a^1 + \frac{1!}{1!(1-1)!}a^{1-1}b^1$$

$$a + b = a + a^0 b$$

$$a + b = a + b$$

and the theorem is true when $n = 1$. Part 1 is complete.

Part 2: We write expressions for two general terms in the statement of the induction hypothesis. We assume that the theorem is true for $n = k$:

$$(a + b)^k = a^k + \frac{k!}{1!(k-1)!}a^{k-1}b + \frac{k!}{2!(k-2)!}a^{k-2}b^2 + \cdots$$

$$+ \frac{k!}{(r-1)!(k-r+1)!}a^{k-r+1}b^{r-1}$$

$$+ \frac{k!}{r!(k-r)!}a^{k-r}b^r + \cdots + b^k$$

We multiply both sides of this equation by $a + b$ and hope to obtain a similar equation in which the quantity $k + 1$ replaces all of the n values in the Binomial Theorem:

$$(a + b)^k(a + b)$$

$$= (a + b)\left[a^k + \frac{k!}{1!(k-1)!}a^{k-1}b + \frac{k!}{2!(k-2)!}a^{k-2}b^2 + \cdots \right.$$

$$\left. + \frac{k!}{(r-1)!(k-r+1)!}a^{k-r+1}b^{r-1} + \frac{k!}{r!(k-r)!}a^{k-r}b^r + \cdots + b^k \right]$$

We distribute the multiplication first by a and then by b:

$$(a + b)^{k+1} = \left[a^{k+1} + \frac{k!}{1!(k-1)!}a^k b + \frac{k!}{2!(k-2)!}a^{k-1}b^2 + \cdots \right.$$

$$\left. + \frac{k!}{(r-1)!(k-r+1)!}a^{k-r+2}b^{r-1} + \frac{k!}{r!(k-r)!}a^{k-r+1}b^r + \cdots + ab^k \right]$$

$$+ \left[a^k b + \frac{k!}{1!(k-1)!}a^{k-1}b^2 + \frac{k!}{2!(k-2)!}a^{k-2}b^3 + \cdots \right.$$

$$\left. + \frac{k!}{(r-1)!(k-r+1)!}a^{k-r+1}b^r + \frac{k!}{r!(k-r)!}a^{k-r}b^{r+1} + \cdots + b^{k+1} \right]$$

Combining like terms, we have

$$(a + b)^{k+1} = a^{k+1} + \left[\frac{k!}{1!(k - 1)!} + 1\right]a^k b$$

$$+ \left[\frac{k!}{2!(k - 2)!} + \frac{k!}{1!(k - 1)!}\right]a^{k-1}b^2 + \cdots$$

$$+ \left[\frac{k!}{r!(k - r)!} + \frac{k!}{(r - 1)!(k - r + 1)!}\right]a^{k-r+1}b^r + \cdots + b^{k+1}$$

These results may be written as

$$(a + b)^{k+1} = a^{k+1} + \frac{(k + 1)!}{1!(k + 1 - 1)!}a^{(k+1)-1}b + \frac{(k + 1)!}{2!(k + 1 - 2)!}a^{(k+1)-2}b^2$$

$$+ \cdots + \frac{(k + 1)!}{r!(k + 1 - r)!}a^{(k+1)-r}b^r + \cdots + b^{k+1}$$

This formula has precisely the same form as the Binomial Theorem, with the quantity $k + 1$ replacing all of the original n values. Therefore, the truth of the theorem for $n = k$ implies the truth of the theorem for $n = k + 1$. Because both parts of the axiom of mathematical induction are verified, the theorem is proved. ∎

Appendix II: Tables

Table A Powers and Roots

n	n^2	$\sqrt{n}$	n^3	$\sqrt[3]{n}$	n	n^2	$\sqrt{n}$	n^3	$\sqrt[3]{n}$
1	1	1.000	1	1.000	51	2,601	7.141	132,651	3.708
2	4	1.414	8	1.260	52	2,704	7.211	140,608	3.733
3	9	1.732	27	1.442	53	2,809	7.280	148,877	3.756
4	16	2.000	64	1.587	54	2,916	7.348	157,464	3.780
5	25	2.236	125	1.710	55	3,025	7.416	166,375	3.803
6	36	2.449	216	1.817	56	3,136	7.483	175,616	3.826
7	49	2.646	343	1.913	57	3,249	7.550	185,193	3.849
8	64	2.828	512	2.000	58	3,364	7.616	195,112	3.871
9	81	3.000	729	2.080	59	3,481	7.681	205,379	3.893
10	100	3.162	1,000	2.154	60	3,600	7.746	216,000	3.915
11	121	3.317	1,331	2.224	61	3,721	7.810	226,981	3.936
12	144	3.464	1,728	2.289	62	3,844	7.874	238,328	3.958
13	169	3.606	2,197	2.351	63	3,969	7.937	250,047	3.979
14	196	3.742	2,744	2.410	64	4,096	8.000	262,144	4.000
15	225	3.873	3,375	2.466	65	4,225	8.062	274,625	4.021
16	256	4.000	4,096	2.520	66	4,356	8.124	287,496	4.041
17	289	4.123	4,913	2.571	67	4,489	8.185	300,763	4.062
18	324	4.243	5,832	2.621	68	4,624	8.246	314,432	4.082
19	361	4.359	6,859	2.668	69	4,761	8.307	328,509	4.102
20	400	4.472	8,000	2.714	70	4,900	8.367	343,000	4.121
21	441	4.583	9,261	2.759	71	5,041	8.426	357,911	4.141
22	484	4.690	10,648	2.802	72	5,184	8.485	373,248	4.160
23	529	4.796	12,167	2.844	73	5,329	8.544	389,017	4.179
24	576	4.899	13,824	2.884	74	5,476	8.602	405,224	4.198
25	625	5.000	15,625	2.924	75	5,625	8.660	421,875	4.217
26	676	5.099	17,576	2.962	76	5,776	8.718	438,976	4.236
27	729	5.196	19,683	3.000	77	5,929	8.775	456,533	4.254
28	784	5.292	21,952	3.037	78	6,084	8.832	474,552	4.273
29	841	5.385	24,389	3.072	79	6,241	8.888	493,039	4.291
30	900	5.477	27,000	3.107	80	6,400	8.944	512,000	4.309
31	961	5.568	29,791	3.141	81	6,561	9.000	531,441	4.327
32	1,024	5.657	32,768	3.175	82	6,724	9.055	551,368	4.344
33	1,089	5.745	35,937	3.208	83	6,889	9.110	571,787	4.362
34	1,156	5.831	39,304	3.240	84	7,056	9.165	592,704	4.380
35	1,225	5.916	42,875	3.271	85	7,225	9.220	614,125	4.397
36	1,296	6.000	46,656	3.302	86	7,396	9.274	636,056	4.414
37	1,369	6.083	50,653	3.332	87	7,569	9.327	658,503	4.431
38	1,444	6.164	54,872	3.362	88	7,744	9.381	681,472	4.448
39	1,521	6.245	59,319	3.391	89	7,921	9.434	704,969	4.465
40	1,600	6.325	64,000	3.420	90	8,100	9.487	729,000	4.481
41	1,681	6.403	68,921	3.448	91	8,281	9.539	753,571	4.498
42	1,764	6.481	74,088	3.476	92	8,464	9.592	778,688	4.514
43	1,849	6.557	79,507	3.503	93	8,649	9.644	804,357	4.531
44	1,936	6.633	85,184	3.530	94	8,836	9.695	830,584	4.547
45	2,025	6.708	91,125	3.557	95	9,025	9.747	857,375	4.563
46	2,116	6.782	97,336	3.583	96	9,216	9.798	884,736	4.579
47	2,209	6.856	103,823	3.609	97	9,409	9.849	912,673	4.595
48	2,304	6.928	110,592	3.634	98	9,604	9.899	941,192	4.610
49	2,401	7.000	117,649	3.659	99	9,801	9.950	970,299	4.626
50	2,500	7.071	125,000	3.684	100	10,000	10.000	1,000,000	4.642

Table B   Base-10 Logarithms

N	0	1	2	3	4	5	6	7	8	9
1.0	.0000	.0043	.0086	.0128	.0170	.0212	.0253	.0294	.0334	.0374
1.1	.0414	.0453	.0492	.0531	.0569	.0607	.0645	.0682	.0719	.0755
1.2	.0792	.0828	.0864	.0899	.0934	.0969	.1004	.1038	.1072	.1106
1.3	.1139	.1173	.1206	.1239	.1271	.1303	.1335	.1367	.1399	.1430
1.4	.1461	.1492	.1523	.1553	.1584	.1614	.1644	.1673	.1703	.1732
1.5	.1761	.1790	.1818	.1847	.1875	.1903	.1931	.1959	.1987	.2014
1.6	.2041	.2068	.2095	.2122	.2148	.2175	.2201	.2227	.2253	.2279
1.7	.2304	.2330	.2355	.2380	.2405	.2430	.2455	.2480	.2504	.2529
1.8	.2553	.2577	.2601	.2625	.2648	.2672	.2695	.2718	.2742	.2765
1.9	.2788	.2810	.2833	.2856	.2878	.2900	.2923	.2945	.2967	.2989
2.0	.3010	.3032	.3054	.3075	.3096	.3118	.3139	.3160	.3181	.3201
2.1	.3222	.3243	.3263	.3284	.3304	.3324	.3345	.3365	.3385	.3404
2.2	.3424	.3444	.3464	.3483	.3502	.3522	.3541	.3560	.3579	.3598
2.3	.3617	.3636	.3655	.3674	.3692	.3711	.3729	.3747	.3766	.3784
2.4	.3802	.3820	.3838	.3856	.3874	.3892	.3909	.3927	.3945	.3962
2.5	.3979	.3997	.4014	.4031	.4048	.4065	.4082	.4099	.4116	.4133
2.6	.4150	.4166	.4183	.4200	.4216	.4232	.4249	.4265	.4281	.4298
2.7	.4314	.4330	.4346	.4362	.4378	.4393	.4409	.4425	.4440	.4456
2.8	.4472	.4487	.4502	.4518	.4533	.4548	.4564	.4579	.4594	.4609
2.9	.4624	.4639	.4654	.4669	.4683	.4698	.4713	.4728	.4742	.4757
3.0	.4771	.4786	.4800	.4814	.4829	.4843	.4857	.4871	.4886	.4900
3.1	.4914	.4928	.4942	.4955	.4969	.4983	.4997	.5011	.5024	.5038
3.2	.5051	.5065	.5079	.5092	.5105	.5119	.5132	.5145	.5159	.5172
3.3	.5185	.5198	.5211	.5224	.5237	.5250	.5263	.5276	.5289	.5302
3.4	.5315	.5328	.5340	.5353	.5366	.5378	.5391	.5403	.5416	.5428
3.5	.5441	.5453	.5465	.5478	.5490	.5502	.5514	.5527	.5539	.5551
3.6	.5563	.5575	.5587	.5599	.5611	.5623	.5635	.5647	.5658	.5670
3.7	.5682	.5694	.5705	.5717	.5729	.5740	.5752	.5763	.5775	.5786
3.8	.5798	.5809	.5821	.5832	.5843	.5855	.5866	.5877	.5888	.5899
3.9	.5911	.5922	.5933	.5944	.5955	.5966	.5977	.5988	.5999	.6010
4.0	.6021	.6031	.6042	.6053	.6064	.6075	.6085	.6096	.6107	.6117
4.1	.6128	.6138	.6149	.6160	.6170	.6180	.6191	.6201	.6212	.6222
4.2	.6232	.6243	.6253	.6263	.6274	.6284	.6294	.6304	.6314	.6325
4.3	.6335	.6345	.6355	.6365	.6375	.6385	.6395	.6405	.6415	.6425
4.4	.6435	.6444	.6454	.6464	.6474	.6484	.6493	.6503	.6513	.6522
4.5	.6532	.6542	.6551	.6561	.6571	.6580	.6590	.6599	.6609	.6618
4.6	.6628	.6637	.6646	.6656	.6665	.6675	.6684	.6693	.6702	.6712
4.7	.6721	.6730	.6739	.6749	.6758	.6767	.6776	.6785	.6794	.6803
4.8	.6812	.6821	.6830	.6839	.6848	.6857	.6866	.6875	.6884	.6893
4.9	.6902	.6911	.6920	.6928	.6937	.6946	.6955	.6964	.6972	.6981
5.0	.6990	.6998	.7007	.7016	.7024	.7033	.7042	.7050	.7059	.7067
5.1	.7076	.7084	.7093	.7101	.7110	.7118	.7126	.7135	.7143	.7152
5.2	.7160	.7168	.7177	.7185	.7193	.7202	.7210	.7218	.7226	.7235
5.3	.7243	.7251	.7259	.7267	.7275	.7284	.7292	.7300	.7308	.7316
5.4	.7324	.7332	.7340	.7348	.7356	.7364	.7372	.7380	.7388	.7396

Table B (continued)

N	0	1	2	3	4	5	6	7	8	9
5.5	.7404	.7412	.7419	.7427	.7435	.7443	.7451	.7459	.7466	.7474
5.6	.7482	.7490	.7497	.7505	.7513	.7520	.7528	.7536	.7543	.7551
5.7	.7559	.7566	.7574	.7582	.7589	.7597	.7604	.7612	.7619	.7627
5.8	.7634	.7642	.7649	.7657	.7664	.7672	.7679	.7686	.7694	.7701
5.9	.7709	.7716	.7723	.7731	.7738	.7745	.7752	.7760	.7767	.7774
6.0	.7782	.7789	.7796	.7803	.7810	.7818	.7825	.7832	.7839	.7846
6.1	.7853	.7860	.7868	.7875	.7882	.7889	.7896	.7903	.7910	.7917
6.2	.7924	.7931	.7938	.7945	.7952	.7959	.7966	.7973	.7980	.7987
6.3	.7993	.8000	.8007	.8014	.8021	.8028	.8035	.8041	.8048	.8055
6.4	.8062	.8069	.8075	.8082	.8089	.8096	.8102	.8109	.8116	.8122
6.5	.8129	.8136	.8142	.8149	.8156	.8162	.8169	.8176	.8182	.8189
6.6	.8195	.8202	.8209	.8215	.8222	.8228	.8235	.8241	.8248	.8254
6.7	.8261	.8267	.8274	.8280	.8287	.8293	.8299	.8306	.8312	.8319
6.8	.8325	.8331	.8338	.8344	.8351	.8357	.8363	.8370	.8376	.8382
6.9	.8388	.8395	.8401	.8407	.8414	.8420	.8426	.8432	.8439	.8445
7.0	.8451	.8457	.8463	.8470	.8476	.8482	.8488	.8494	.8500	.8506
7.1	.8513	.8519	.8525	.8531	.8537	.8543	.8549	.8555	.8561	.8567
7.2	.8573	.8579	.8585	.8591	.8597	.8603	.8609	.8615	.8621	.8627
7.3	.8633	.8639	.8645	.8651	.8657	.8663	.8669	.8675	.8681	.8686
7.4	.8692	.8698	.8704	.8710	.8716	.8722	.8727	.8733	.8739	.8745
7.5	.8751	.8756	.8762	.8768	.8774	.8779	.8785	.8791	.8797	.8802
7.6	.8808	.8814	.8820	.8825	.8831	.8837	.8842	.8848	.8854	.8859
7.7	.8865	.8871	.8876	.8882	.8887	.8893	.8899	.8904	.8910	.8915
7.8	.8921	.8927	.8932	.8938	.8943	.8949	.8954	.8960	.8965	.8971
7.9	.8976	.8982	.8987	.8993	.8998	.9004	.9009	.9015	.9020	.9025
8.0	.9031	.9036	.9042	.9047	.9053	.9058	.9063	.9069	.9074	.9079
8.1	.9085	.9090	.9096	.9101	.9106	.9112	.9117	.9122	.9128	.9133
8.2	.9138	.9143	.9149	.9154	.9159	.9165	.9170	.9175	.9180	.9186
8.3	.9191	.9196	.9201	.9206	.9212	.9217	.9222	.9227	.9232	.9238
8.4	.9243	.9248	.9253	.9258	.9263	.9269	.9274	.9279	.9284	.9289
8.5	.9294	.9299	.9304	.9309	.9315	.9320	.9325	.9330	.9335	.9340
8.6	.9345	.9350	.9355	.9360	.9365	.9370	.9375	.9380	.9385	.9390
8.7	.9395	.9400	.9405	.9410	.9415	.9420	.9425	.9430	.9435	.9440
8.8	.9445	.9450	.9455	.9460	.9465	.9469	.9474	.9479	.9484	.9489
8.9	.9494	.9499	.9504	.9509	.9513	.9518	.9523	.9528	.9533	.9538
9.0	.9542	.9547	.9552	.9557	.9562	.9566	.9571	.9576	.9581	.9586
9.1	.9590	.9595	.9600	.9605	.9609	.9614	.9619	.9624	.9628	.9633
9.2	.9638	.9643	.9647	.9652	.9657	.9661	.9666	.9671	.9675	.9680
9.3	.9685	.9689	.9694	.9699	.9703	.9708	.9713	.9717	.9722	.9727
9.4	.9731	.9736	.9741	.9745	.9750	.9754	.9759	.9763	.9768	.9773
9.5	.9777	.9782	.9786	.9791	.9795	.9800	.9805	.9809	.9814	.9818
9.6	.9823	.9827	.9832	.9836	.9841	.9845	.9850	.9854	.9859	.9863
9.7	.9868	.9872	.9877	.9881	.9886	.9890	.9894	.9899	.9903	.9908
9.8	.9912	.9917	.9921	.9926	.9930	.9934	.9939	.9943	.9948	.9952
9.9	.9956	.9961	.9965	.9969	.9974	.9978	.9983	.9987	.9991	.9996

Table C (continued)

N	0	1	2	3	4	5	6	7	8	9
5.5	1.7047	7066	7084	7102	7120	7138	7156	7174	7192	7210
5.6	.7228	7246	7263	7281	7299	7317	7334	7352	7370	7387
5.7	.7405	7422	7440	7457	7475	7492	7509	7527	7544	7561
5.8	.7579	7596	7613	7630	7647	7664	7681	7699	7716	7733
5.9	.7750	7766	7783	7800	7817	7834	7851	7867	7884	7901
6.0	1.7918	7934	7951	7967	7984	8001	8017	8034	8050	8066
6.1	.8083	8099	8116	8132	8148	8165	8181	8197	8213	8229
6.2	.8245	8262	8278	8294	8310	8326	8342	8358	8374	8390
6.3	.8405	8421	8437	8453	8469	8485	8500	8516	8532	8547
6.4	.8563	8579	8594	8610	8625	8641	8656	8672	8687	8703
6.5	1.8718	8733	8749	8764	8779	8795	8810	8825	8840	8856
6.6	.8871	8886	8901	8916	8931	8946	8961	8976	8991	9006
6.7	.9021	9036	9051	9066	9081	9095	9110	9125	9140	9155
6.8	.9169	9184	9199	9213	9228	9242	9257	9272	9286	9301
6.9	.9315	9330	9344	9359	9373	9387	9402	9416	9430	9445
7.0	1.9459	9473	9488	9502	9516	9530	9544	9559	9573	9587
7.1	.9601	9615	9629	9643	9657	9671	9685	9699	9713	9727
7.2	.9741	9755	9769	9782	9796	9810	9824	9838	9851	9865
7.3	.9879	9892	9906	9920	9933	9947	9961	9974	9988	2.0001
7.4	2.0015	0028	0042	0055	0069	0082	0096	0109	0122	0136
7.5	2.0149	0162	0176	0189	0202	0215	0229	0242	0255	0268
7.6	.0281	0295	0308	0321	0334	0347	0360	0373	0386	0399
7.7	.0412	0425	0438	0451	0464	0477	0490	0503	0516	0528
7.8	.0541	0554	0567	0580	0592	0605	0618	0631	0643	0656
7.9	.0669	0681	0694	0707	0719	0732	0744	0757	0769	0782
8.0	2.0794	0807	0819	0832	0844	0857	0869	0882	0894	0906
8.1	.0919	0931	0943	0956	0968	0980	0992	1005	1017	1029
8.2	.1041	1054	1066	1078	1090	1102	1114	1126	1138	1150
8.3	.1163	1175	1187	1199	1211	1223	1235	1247	1258	1270
8.4	.1282	1294	1306	1318	1330	1342	1353	1365	1377	1389
8.5	2.1401	1412	1424	1436	1448	1459	1471	1483	1494	1506
8.6	.1518	1529	1541	1552	1564	1576	1587	1599	1610	1622
8.7	.1633	1645	1656	1668	1679	1691	1702	1713	1725	1736
8.8	.1748	1759	1770	1782	1793	1804	1815	1827	1838	1849
8.9	.1861	1872	1883	1894	1905	1917	1928	1939	1950	1961
9.0	2.1972	1983	1994	2006	2017	2028	2039	2050	2061	2072
9.1	.2083	2094	2105	2116	2127	2138	2148	2159	2170	2181
9.2	.2192	2203	2214	2225	2235	2246	2257	2268	2279	2289
9.3	.2300	2311	2322	2332	2343	2354	2364	2375	2386	2396
9.4	.2407	2418	2428	2439	2450	2460	2471	2481	2492	2502
9.5	2.2513	2523	2534	2544	2555	2565	2576	2586	2597	2607
9.6	.2618	2628	2638	2649	2659	2670	2680	2690	2701	2711
9.7	.2721	2732	2742	2752	2762	2773	2783	2793	2803	2814
9.8	.2824	2834	2844	2854	2865	2875	2885	2895	2905	2915
9.9	.2925	2935	2946	2956	2966	2976	2986	2996	3006	3016

Table C   Base-e Logarithms

N	0	1	2	3	4	5	6	7	8	9
1.0	.0000	.0100	.0198	.0296	.0392	.0488	.0583	.0677	.0770	.0862
1.1	.0953	.1044	.1133	.1222	.1310	.1398	.1484	.1570	.1655	.1740
1.2	.1823	.1906	.1989	.2070	.2151	.2231	.2311	.2390	.2469	.2546
1.3	.2624	.2700	.2776	.2852	.2927	.3001	.3075	.3148	.3221	.3293
1.4	.3365	.3436	.3507	.3577	.3646	.3716	.3784	.3853	.3920	.3988
1.5	.4055	.4121	.4187	.4253	.4318	.4383	.4447	.4511	.4574	.4637
1.6	.4700	.4762	.4824	.4886	.4947	.5008	.5068	.5128	.5188	.5247
1.7	.5306	.5365	.5423	.5481	.5539	.5596	.5653	.5710	.5766	.5822
1.8	.5878	.5933	.5988	.6043	.6098	.6152	.6206	.6259	.6313	.6366
1.9	.6419	.6471	.6523	.6575	.6627	.6678	.6729	.6780	.6831	.6881
2.0	.6931	.6981	.7031	.7080	.7129	.7178	.7227	.7275	.7324	.7372
2.1	.7419	.7467	.7514	.7561	.7608	.7655	.7701	.7747	.7793	.7839
2.2	.7885	.7930	.7975	.8020	.8065	.8109	.8154	.8198	.8242	.8286
2.3	.8329	.8372	.8416	.8459	.8502	.8544	.8587	.8629	.8671	.8713
2.4	.8755	.8796	.8838	.8879	.8920	.8961	.9002	.9042	.9083	.9123
2.5	.9163	.9203	.9243	.9282	.9322	.9361	.9400	.9439	.9478	.9517
2.6	.9555	.9594	.9632	.9670	.9708	.9746	.9783	.9821	.9858	.9895
2.7	.9933	.9969	1.0006	.0043	.0080	.0116	.0152	.0188	.0225	.0260
2.8	1.0296	.0332	.0367	.0403	.0438	.0473	.0508	.0543	.0578	.0613
2.9	.0647	.0682	.0716	.0750	.0784	.0818	.0852	.0886	.0919	.0953
3.0	1.0986	.1019	.1053	.1086	.1119	.1151	.1184	.1217	.1249	.1282
3.1	.1314	.1346	.1378	.1410	.1442	.1474	.1506	.1537	.1569	.1600
3.2	.1632	.1663	.1694	.1725	.1756	.1787	.1817	.1848	.1878	.1909
3.3	.1939	.1969	.2000	.2030	.2060	.2090	.2119	.2149	.2179	.2208
3.4	.2238	.2267	.2296	.2326	.2355	.2384	.2413	.2442	.2470	.2499
3.5	1.2528	.2556	.2585	.2613	.2641	.2669	.2698	.2726	.2754	.2782
3.6	.2809	.2837	.2865	.2892	.2920	.2947	.2975	.3002	.3029	.3056
3.7	.3083	.3110	.3137	.3164	.3191	.3218	.3244	.3271	.3297	.3324
3.8	.3350	.3376	.3403	.3429	.3455	.3481	.3507	.3533	.3558	.3584
3.9	.3610	.3635	.3661	.3686	.3712	.3737	.3762	.3788	.3813	.3838
4.0	1.3863	.3888	.3913	.3938	.3962	.3987	.4012	.4036	.4061	.4085
4.1	.4110	.4134	.4159	.4183	.4207	.4231	.4255	.4279	.4303	.4327
4.2	.4351	.4375	.4398	.4422	.4446	.4469	.4493	.4516	.4540	.4563
4.3	.4586	.4609	.4633	.4656	.4679	.4702	.4725	.4748	.4770	.4793
4.4	.4816	.4839	.4861	.4884	.4907	.4929	.4951	.4974	.4996	.5019
4.5	1.5041	.5063	.5085	.5107	.5129	.5151	.5173	.5195	.5217	.5239
4.6	.5261	.5282	.5304	.5326	.5347	.5369	.5390	.5412	.5433	.5454
4.7	.5476	.5497	.5518	.5539	.5560	.5581	.5602	.5623	.5644	.5665
4.8	.5686	.5707	.5728	.5748	.5769	.5790	.5810	.5831	.5851	.5872
4.9	.5892	.5913	.5933	.5953	.5974	.5994	.6014	.6034	.6054	.6074
5.0	1.6094	.6114	.6134	.6154	.6174	.6194	.6214	.6233	.6253	.6273
5.1	.6292	.6312	.6332	.6351	.6371	.6390	.6409	.6429	.6448	.6467
5.2	.6487	.6506	.6525	.6544	.6563	.6582	.6601	.6620	.6639	.6658
5.3	.6677	.6696	.6715	.6734	.6752	.6771	.6790	.6808	.6827	.6845
5.4	.6864	.6882	.6901	.6919	.6938	.6956	.6974	.6993	.7011	.7029

Answers to Selected Exercises

Section 0.1
1. set   **3.** union   **5.** decimal   **7.** 2   **9.** composite
11. decimals   **13.** negative   **5.** $x + (y + z)$
17. $5m + 5 \cdot 2$ **19.** interval   **21.** two   **23.** positive
25. true   **27.** false   **29.** true   **31.** $\{a, b, c, d, e, f, g\}$
33. $\{a, c, e\}$ **35.** terminates   **37.** repeats
39. 1, 2, 6, 7   **41.** $-5, -4, 0, 1, 2, 6, 7$ **43.** $\sqrt{2}$
45. 6   **47.** $-5, 1, 7$
49.
51.
53.
55.
57. $(2, \infty)$
59. $(0, 5)$
61. $(-4, \infty)$
63. $[-2, 2)$
65. $(-\infty, 5]$
67. $(-5, 0]$
69. $[-2, 3]$
71. $[2, 6]$
73. $(-5, \infty) \cap (-\infty, 4)$
75. $[-8, \infty) \cap (-\infty, -3]$
77. $(-\infty, -2) \cup (2, \infty)$
79. $(-\infty, -1] \cup [3, \infty)$
81. 13   **83.** 0   **85.** -8 **87.** -32 **89.** $5 - \pi$
91. 0   **93.** $x + 1$ **95.** $-(x - 4)$ **97.** 5   **99.** 5
101. natural numbers   **103.** integers

Section 0.2
1. factor   **3.** 3, $2x$ **5.** scientific, integer   **7.** x^{m+n}
9. $x^n y^n$ **11.** 1   **13.** 169   **15.** -25
17. $4 \cdot x \cdot x \cdot x$ **19.** $(-5x)(-5x)(-5x)(-5x)$
21. $-8 \cdot x \cdot x \cdot x \cdot x$ **23.** $7x^3$ **25.** x^2 **27.** $-27t^3$
29. $x^3 y^2$ **31.** 10.648   **33.** -0.0625 **35.** x^5 **37.** z^6
39. y^{21} **41.** z^{26} **43.** a^{14} **45.** $27x^3$ **47.** $x^6 y^3$
49. $\dfrac{a^6}{b^3}$ **51.** 1   **53.** 1   **55.** $\dfrac{1}{z^4}$ **57.** $\dfrac{1}{y^5}$ **59.** x^2

61. x^4 **63.** a^4 **65.** x **67.** $\dfrac{m^9}{n^6}$ **69.** $\dfrac{1}{a^9}$ **71.** $\dfrac{a^{12}}{b^4}$
73. $\dfrac{1}{r^4}$ **75.** $\dfrac{x^{32}}{y^{16}}$ **77.** $\dfrac{9x^{10}}{25y^2}$ **79.** $\dfrac{4y^{12}}{x^{20}}$ **81.** $\dfrac{64z^7}{25y^6}$
83. -50 **85.** 4   **87.** -8 **89.** 216   **91.** -12
93. 20   **95.** $\dfrac{3}{64}$ **97.** 3.72×10^5 **99.** -1.77×10^8
101. 7×10^{-3} **103.** -6.93×10^{-7} **105.** 1×10^{12}
107. 937,000   **109.** 0.0000221   **111.** 3.2
113. -0.0032 **115.** 1.17×10^4 **117.** 7×10^4
119. 5.3×10^{19} **121.** 1.986×10^4 meters per min
123. 1.67248×10^{-15} g   **125.** 9.3×10^7 mi, 1.48×10^8 mi
127. polar radius: 6.356750×10^3 km;
equatorial radius: 6.378135×10^3 km
129. x^{n+2} **131.** x^{m-1} **133.** x^{m+4}
137.   **139.** $5 - \pi$

Section 0.3
1. 0   **3.** not   **5.** $a^{1/n}$ **7.** $\sqrt[n]{ab}$ **9.** $\neq$ **11.** 3
13. $\dfrac{1}{5}$ **15.** -3 **17.** 10   **19.** $-\dfrac{3}{2}$
21. not a real number   **23.** $4|a|$ **25.** $2|a|$
27. $-2a$ **29.** $-6b^2$ **31.** $\dfrac{4a^2}{5|b|}$ **33.** $-\dfrac{10x^2}{3y}$
35. 8   **37.** -64 **39.** -100 **41.** $\dfrac{1}{8}$ **43.** $\dfrac{1}{512}$
45. $-\dfrac{1}{27}$ **47.** $\dfrac{32}{243}$ **49.** $\dfrac{16}{9}$ **51.** $10s^2$ **53.** $\dfrac{1}{2y^2 z}$
55. $x^6 y^3$ **57.** $\dfrac{1}{r^6 s^{12}}$ **59.** $\dfrac{4a^4}{25b^6}$ **61.** $\dfrac{100s^8}{9r^4}$ **63.** a
65. 7   **67.** 5   **69.** -5 **71.** $-\dfrac{1}{5}$ **73.** $6|x|$
75. $3y^2$ **77.** $2y$ **79.** $\dfrac{|x|y^2}{|z^3|}$ **81.** $\sqrt{2}$ **83.** $17x\sqrt{2}$
85. $2y^2\sqrt{3y}$ **87.** $12\sqrt[3]{3}$ **89.** $6z\sqrt[4]{3z}$ **91.** $6x\sqrt{2y}$
93. 0   **95.** $\sqrt{3}$ **97.** $\dfrac{2\sqrt{x}}{x}$ **99.** $\sqrt[3]{4}$ **101.** $\sqrt[3]{5a^2}$
103. $\dfrac{2b\sqrt[4]{27a^2}}{3a}$ **105.** $\dfrac{u\sqrt[3]{6uv^2}}{3v}$ **107.** $\dfrac{1}{2\sqrt{5}}$ **109.** $\dfrac{1}{\sqrt[3]{3}}$
111. $\dfrac{b}{32a\sqrt[5]{2b^2}}$ **113.** $\dfrac{2\sqrt{3}}{9}$ **115.** $-\dfrac{\sqrt{2x}}{8}$ **117.** $\sqrt{3}$
119. $\sqrt[5]{4x^3}$ **121.** $\sqrt[6]{32}$ **123.** $\dfrac{\sqrt[4]{12}}{2}$ **129.** $(-2, 5]$
131. -5 **133.** 6.17×10^8

Section 0.4
1. monomial, variables   **3.** trinomial   **5.** one   **7.** like
9. coefficients, variables   **11.** yes, trinomial, 2nd degree

A1

13. no   **15.** yes, binomial, 3rd degree   **17.** yes, monomial, 0th degree   **19.** yes, monomial, no defined degree

21. $6x^3 - 3x^2 - 8x$ **23.** $4y^3 + 14$ **25.** $-x^2 + 14$

27. $-28t + 96$ **29.** $-4y^2 + y$ **31.** $2x^2y - 4xy^2$

33. $8x^3y^7$ **35.** $\dfrac{m^4n^4}{2}$ **37.** $-4r^3s - 4rs^3$

39. $12a^2b^2c^2 + 18ab^3c^3 - 24a^2b^4c^2$ **41.** $a^2 + 4a + 4$

43. $a^2 - 12a + 36$ **45.** $x^2 - 16$ **47.** $x^2 + 2x - 15$

49. $3u^2 + 4u - 4$ **51.** $10x^2 + 13x - 3$

53. $9a^2 - 12ab + 4b^2$ **55.** $9m^2 - 16n^2$

57. $6y^2 - 16xy + 8x^2$ **59.** $9x^3 - x^2y - 27xy + 3y^2$

61. $5z^3 - 5tz + 2tz^2 - 2t^2$ **63.** $-3\sqrt{5}x^2 + x + 2\sqrt{5}$

65. $27x^3 - 27x^2 + 9x - 1$ **67.** $6x^3 + 14x^2 - 5x - 3$

69. $6x^3 - 5x^2y + 6xy^2 + 8y^3$ **71.** $6y^{2n} + 2$

73. $-10x^{4n} - 15y^{2n}$ **75.** $x^{2n} - x^n - 12$

77. $6r^{2n} - 25r^n + 14$ **79.** $xy + x^{3/2}y^{1/2}$

81. $a - b$ **83.** $\sqrt{3} + 1$ **85.** $x(\sqrt{7} - 2)$

87. $\dfrac{x(x + \sqrt{3})}{x^2 - 3}$ **89.** $\dfrac{(y + \sqrt{2})^2}{y^2 - 2}$

91. $\dfrac{\sqrt{3} + 3 - \sqrt{2} - \sqrt{6}}{2}$ **93.** $\dfrac{x - 2\sqrt{xy} + y}{x - y}$

95. $\dfrac{1}{2(\sqrt{2} - 1)}$ **97.** $\dfrac{y^2 - 3}{y^2 + 2y\sqrt{3} + 3}$

99. $\dfrac{1}{\sqrt{x + 3} + \sqrt{x}}$ **101.** $\dfrac{2a}{b^3}$ **103.** $-\dfrac{2z^9}{3x^3y^2}$

105. $\dfrac{x}{2y} + \dfrac{3xy}{2}$ **107.** $\dfrac{2y^3}{5} - \dfrac{3y}{5x^3} + \dfrac{1}{5x^4y^3}$ **109.** $3x + 2$

111. $x - 7 + \dfrac{2}{2x - 5}$ **113.** $2x^2 + 2x + 2 + \dfrac{3}{x - 1}$

115. $x - 3$ **117.** $x^2 - 2 + \dfrac{-x^2 + 5}{x^3 - 2}$

119. $x^4 + 2x^3 + 4x^2 + 8x + 16$ **121.** $6x^2 + x - 12$

123. $(x^2 + 3x - 10)$ ft²   **125.** $(4x^3 - 48x^2 + 144x)$ in.³

133. 27   **135.** $\dfrac{125x^3}{8y^6}$ **137.** $-b\sqrt[3]{2ab}$

Section 0.5

1. factor   **3.** $x(a + b)$ **5.** $(x + y)^2$

7. $(x + y)(x^2 - xy + y^2)$ **9.** $3(x - 2)$ **11.** $4x^2(2 + x)$

13. $7x^2y^2(1 + 2x)$ **15.** $(x + y)(a + b)$

17. $(4a + b)(1 - 3a)$ **19.** $(2x + 3)(2x - 3)$

21. $(2 + 3r)(2 - 3r)$ **23.** $(9x^2 + 1)(3x + 1)(3x - 1)$

25. $(x + z + 5)(x + z - 5)$ **27.** $(x + 4)^2$

29. $(b - 5)^2$ **31.** $(m + 2n)^2$

33. $(4x - 3y)(3x + 2y)$ **35.** $(x + 7)(x + 3)$

37. $(x - 6)(x + 2)$ **39.** $(2p + 3)(3p - 1)$

41. $(t + 7)(t^2 - 7t + 49)$ **43.** $(2z - 3)(4z^2 + 6z + 9)$

45. $3abc(a + 2b + 3c)$ **47.** $(x + 1)(3x^2 - 1)$

49. $t(y + c)(2x - 3)$ **51.** $(a + b)(x + y + z)$

53. $(x + y - z)(x - y + z)$ **55.** $-4xy$

57. $(x^2 + y^2)(x + y)(x - y)$ **59.** $3(x + 2)(x - 2)$

61. $2x(3y + 2)(3y - 2)$ **63.** prime

65. $(6a + 5)(4a - 3)$ **67.** $(3x + 7y)(2x + 5y)$

69. $2(6p - 35q)(p + q)$ **71.** $-(6m - 5n)(m - 7n)$

73. $-x(6x + 7)(x - 5)$ **75.** $x^2(2x - 7)(3x + 5)$

77. $(x^2 + 5)(x^2 - 3)$ **79.** $(a^n - 3)(a^n + 1)$

81. $(3x^n - 2)(2x^n - 1)$ **83.** $(2x^n + 3y^n)(2x^n - 3y^n)$

85. $(5y^n + 2)(2y^n - 3)$ **87.** $2(x + 10)(x^2 - 10x + 100)$

89. $(x + y - 4)(x^2 + 2xy + y^2 + 4x + 4y + 16)$

91. $(2a + y)(2a - y)(4a^2 - 2ay + y^2)(4a^2 + 2ay + y^2)$

93. $(a - b)(a^2 + ab + b^2 + 1)$

95. $(4x^2 + y^2)(16x^4 - 4x^2y^2 + y^4)$

97. $(x - 3 + 12y)(x - 3 - 12y)$

99. $(a + b - 5)(a + b + 2)$

101. $(x + 2)(x^2 - 2x + 4)(x - 1)(x^2 + x + 1)$

103. $(x^2 + 1 + x)(x^2 + 1 - x)$

105. $(x^2 + x + 4)(x^2 - x + 4)$

107. $(2a^2 + a + 1)(2a^2 - a + 1)$

109. $\dfrac{4}{3}\pi(r_1 - r_2)(r_1^2 + r_1r_2 + r_2^2)$ **115.** $2\left(\dfrac{3}{2}x + 1\right)$

117. $2\left(\dfrac{1}{2}x^2 + x + 2\right)$ **119.** $a\left(1 + \dfrac{b}{a}\right)$ **121.** $x^{1/2}(x^{1/2} + 1)$

123. $\sqrt{2}\left(\sqrt{2}x + y\right)$ **125.** $ab(b^{1/2} - a^{1/2})$

127. $(x - 2)(x + 3 + y)$ **129.** $(a + 1)(a^3 + a^2 + 1)$

131. 1   **133.** x^{20} **135.** 1   **137.** $-3\sqrt{5}x$

Section 0.6

1. numerator   **3.** $ad = bc$ **5.** $\dfrac{ac}{bd}$ **7.** $\dfrac{a + c}{b}$

9. equal   **11.** not equal   **13.** $\dfrac{a}{3b}$ **15.** $\dfrac{8x}{35a}$

17. $\dfrac{16}{3}$ **19.** $\dfrac{z}{c}$ **21.** $\dfrac{2x^2y}{a^2b^3}$ **23.** $\dfrac{2}{x + 2}$

25. $-\dfrac{x + 2}{x - 3}$ **27.** $\dfrac{3x - 4}{2x - 1}$ **29.** $\dfrac{x^2 + 2x + 4}{x + a}$

31. $\dfrac{x(x - 1)}{x + 1}$ **33.** $\dfrac{x - 1}{x}$ **35.** $\dfrac{x(x + 1)^2}{x + 2}$

37. $\dfrac{1}{2}$ **39.** $\dfrac{z - 4}{(z + 2)(z - 2)}$ **41.** 1   **43.** 1

45. $\dfrac{x - 5}{x + 5}$ **47.** $\dfrac{x + 5}{x + 3}$ **49.** 4   **51.** $\dfrac{-1}{x - 5}$

53. $\dfrac{5x - 1}{(x + 1)(x - 1)}$ **55.** $\dfrac{2(a - 2)}{(a + 4)(a - 4)}$

57. $\dfrac{2}{(x + 2)(x - 2)}$ **59.** $\dfrac{2(x^2 - 3x + 1)}{(x + 1)^2(x - 1)}$

61. $\dfrac{3y - 2}{y - 1}$ **63.** $\dfrac{1}{x + 2}$ **65.** $\dfrac{2x - 5}{2x(x - 2)}$

67. $\dfrac{2x^2 + 19x + 1}{(x + 4)(x - 4)}$ **69.** 0

71. $\dfrac{-x^4 + 3x^3 - 43x^2 - 58x + 697}{(x + 5)(x - 5)(x + 4)(x - 4)}$ **73.** $\dfrac{b}{2c}$

75. $81a$ **77.** -1 **79.** $\dfrac{y + x}{x^2y^2}$ **81.** $\dfrac{y + x}{y - x}$

83. $\dfrac{a^2(3x - 4ab)}{ax + b}$ **85.** $\dfrac{x - 2}{x + 2}$ **87.** $\dfrac{3x^2y^2}{xy - 1}$

89. $\dfrac{3x^2}{x^2 + 1}$ **91.** $\dfrac{x^2 - 3x - 4}{x^2 + 5x - 3}$ **93.** $\dfrac{x}{x + 1}$

95. $\dfrac{5x + 1}{x - 1}$ **97.** $\dfrac{k_1k_2}{k_2 + k_1}$ **99.** $\dfrac{3x}{3 + x}$

101. $\dfrac{x + 1}{2x + 1}$ **107.** 6   **109.** $\dfrac{y^9}{x^{12}}$ **111.** $-\sqrt{5}$

Chapter Review

1. 3, 6, 8   **2.** 0, 3, 6, 8   **3.** $-6, -3, 0, 3, 6, 8$
4. $-6, -3, 0, \frac{1}{2}, 3, 6, 8$ **5.** $\pi, \sqrt{5}$
6. $-6, -3, 0, \frac{1}{2}, 3, \pi, \sqrt{5}, 6, 8$ **7.** 3   **8.** 6, 8
9. $-6, 0, 6, 8$ **10.** $-3, 3$
11. Associative Property of Addition
12. Commutative Property of Addition
13. Associative Property of Multiplication
14. Distributive Property
15. Commutative Property of Multiplication
16. Commutative Property of Addition
17. Double Negative Rule
18. ← • • • • • → (10 11 12 13 14 15 16 17 18 19 20)
19. ← • • • • • • → (6 7 8 9 10 11 12 13 14)
20. (-3 5]   **21.** (-1 0)
22. (-2 4)   **23.** (-5 2)
24. -4 6]

25. 6   **26.** 25   **27.** $\sqrt{2} - 1$ **28.** $\sqrt{3} - 1$
29. 12   **30.** $-5aaa$ **31.** $(-5a)(-5a)$ **32.** $3t^3$
33. $-6b^2$ **34.** n^6 **35.** p^6 **36.** $x^{12}y^8$
37. $\dfrac{a^{12}}{b^6}$ **38.** $\dfrac{1}{m^6}$ **39.** $\dfrac{q^6}{8p^6}$ **40.** $\dfrac{1}{a^3}$ **41.** $\dfrac{b^6}{a^4}$
42. $\dfrac{y^8}{9}$ **43.** $\dfrac{a^8}{b^{10}}$ **44.** $\dfrac{y^4}{9x^4}$ **45.** $-\dfrac{8m^{12}}{n^3}$
46. 18   **47.** 6.75×10^3 **48.** 2.3×10^{-4}
49. 480   **50.** 0.00025   **51.** 1.5×10^{14} **52.** 11
53. $\dfrac{3}{5}$ **54.** $2x$ **55.** $3|a|$ **56.** $-10x^2$
57. not a real number   **58.** $x^6|y|$ **59.** $\dfrac{y^2}{x^6}$
60. $-c$ **61.** a^2 **62.** 16   **63.** $\dfrac{1}{8}$ **64.** $\dfrac{8}{27}$ **65.** $\dfrac{4}{9}$
66. $\dfrac{9}{4}$ **67.** $\dfrac{125}{8}$ **68.** $36x^2$ **69.** $p^{2a/3}$ **70.** 6
71. -7 **72.** $\dfrac{3}{5}$ **73.** $\dfrac{3}{5}$ **74.** $|x|y^2$ **75.** x
76. $\dfrac{m^2|n|}{p^4}$ **77.** $\dfrac{a^3b^2}{c}$ **78.** $7\sqrt{2}$ **79.** 0   **80.** $x\sqrt[3]{3x}$
81. $\dfrac{\sqrt{35}}{5}$ **82.** $2\sqrt{2}$ **83.** $\dfrac{\sqrt[3]{4}}{2}$ **84.** $\dfrac{2\sqrt[3]{5}}{5}$
85. $\dfrac{2}{5\sqrt{2}}$ **86.** $\dfrac{1}{\sqrt{5}}$ **87.** $\dfrac{2x}{3\sqrt{2x}}$ **88.** $\dfrac{21x}{2\sqrt[3]{49x^2}}$
89. 3rd degree, binomial   **90.** 2nd degree, trinomial
91. 2nd degree, monomial   **92.** 4th degree, trinomial
93. $5x - 6$ **94.** $2x^3 - 7x^2 - 6x$ **95.** $9x^2 + 12x + 4$
96. $6x^2 - 7xy - 3y^2$ **97.** $8a^2 - 8ab - 6b^2$
98. $3z^3 + 10z^2 + 2z - 3$ **99.** $a^{2n} + a^n - 2$
100. $2 + 2x\sqrt{2} + x^2$ **101.** $\sqrt{6} + \sqrt{2} + \sqrt{3} + 1$
102. -5 **103.** $\sqrt{3} + 1$ **104.** $-2\left(\sqrt{3} + \sqrt{2}\right)$

105. $\dfrac{2x\left(\sqrt{x} + 2\right)}{x - 4}$ **106.** $\dfrac{x - 2\sqrt{xy} + y}{x - y}$
107. $\dfrac{x - 4}{5\left(\sqrt{x} - 2\right)}$ **108.** $\dfrac{1 - a}{a\left(1 + \sqrt{a}\right)}$ **109.** $\dfrac{y}{2x}$
110. $2a^2b + 3ab^2$ **111.** $x^2 + 2x + 1$
112. $x^3 + 2x - 3 - \dfrac{6}{x^2 - 1}$ **113.** $3t(t + 1)(t - 1)$
114. $5(r - 1)(r^2 + r + 1)$ **115.** $(3x + 8)(2x - 3)$
116. $(3a + x)(a - 1)$ **117.** $(2x - 5)(4x^2 + 10x + 25)$
118. $2(3x + 2)(x - 4)$ **119.** $(x + 3 + t)(x + 3 - t)$
120. prime   **121.** $(2z + 7)(4z^2 - 14z + 49)$
122. $(7b + 1)^2$ **123.** $(11z - 2)^2$
124. $8(2y - 5)(4y^2 + 10y + 25)$ **125.** $(y - 2z)(2x - w)$
126. $(x^2 + 1 + x)(x^2 + 1 - x)(x^4 + 1 - x^2)$
127. $\dfrac{-1}{x - 2}$ **128.** $\dfrac{a + 3}{a - 3}$ **129.** $(x - 2)(x + 3)$
130. $\dfrac{2y - 5}{y - 2}$ **131.** $\dfrac{t + 1}{5}$ **132.** $\dfrac{p(p + 4)}{(p^2 + 8p + 4)(p - 3)}$
133. $\dfrac{(x - 2)(x + 3)(x - 3)}{(x - 1)(x + 2)^2}$ **134.** 1
135. $\dfrac{3x^2 - 10x + 10}{(x - 4)(x + 5)}$ **136.** $\dfrac{2(2x^2 + 3x + 6)}{(x + 2)(x - 2)}$
137. $\dfrac{3x^3 - 12x^2 + 11x}{(x - 1)(x - 2)(x - 3)}$ **138.** $\dfrac{-5x - 6}{(x + 1)(x + 2)}$
139. $\dfrac{-x^3 + x^2 - 12x - 15}{x^2(x + 1)}$ **140.** $\dfrac{3x}{x + 1}$
141. $\dfrac{20}{3x}$ **142.** $\dfrac{y}{2}$ **143.** $\dfrac{y + x}{xy(x - y)}$ **144.** $\dfrac{y + x}{x - y}$

Chapter Test

1. $-7, 1, 3$ **2.** 3   **3.** Commutative Property of
Addition   **4.** Distributive Property
5. (-4 2]   **6.** [-3 6]
7. 17   **8.** $-(x - 7)$ **9.** 16   **10.** 8   **11.** x^{11}
12. $\dfrac{r}{s}$ **13.** a **14.** x^{24} **15.** 4.5×10^5
16. 3.45×10^{-4} **17.** 3,700   **18.** 0.0012
19. $5a^2$ **20.** $\dfrac{216}{729}$ **21.** $\dfrac{9s^6}{4t^4}$ **22.** $3a^2$
23. $5\sqrt{3}$ **24.** $-4x\sqrt[3]{3x}$ **25.** $\dfrac{x\left(\sqrt{x} + 2\right)}{x - 4}$
26. $\dfrac{x - y}{x + 2\sqrt{xy} + y}$ **27.** $-a^2 + 7$ **28.** $-6a^6b^6$
29. $6x^2 + 13x - 28$ **30.** $a^{2n} - a^n - 6$ **31.** $x^4 - 16$
32. $2x^3 - 5x^2 + 7x - 6$ **33.** $6x + 19 + \dfrac{34}{x - 3}$
34. $x^2 + 2x + 1$ **35.** $3(x + 2y)$ **36.** $(x + 10)(x - 10)$
37. $(5t - 2w)(2t - 3w)$ **38.** $3(a - 6)(a^2 + 6a + 36)$
39. $(x + 2)(x - 2)(x^2 + 3)$ **40.** $(3x^2 - 2)(2x^2 + 5)$
41. 1   **42.** $\dfrac{-2x}{(x + 1)(x - 1)}$ **43.** $\dfrac{(x + 5)^2}{x + 4}$
44. $\dfrac{1}{(x + 1)(x - 2)}$ **45.** $\dfrac{b + a}{a}$ **46.** $\dfrac{y}{y + x}$

Section 1.1

1. root; solution   **3.** no   **5.** linear   **7.** one   **9.** no restrictions   **11.** $x \neq 0$ **13.** $x \neq 3$ or -2
15. $x \neq 3, 4,$ or -4 **17.** 5; conditional equation
19. no solution; contradiction   **21.** 7; conditional equation
23. no solution; contradiction   **25.** all real numbers; identity
27. 6; conditional equation   **29.** all real numbers; identity
31. 1   **33.** 6   **35.** $\dfrac{5}{2}$ **37.** 9   **39.** 10   **41.** -3
43. 3   **45.** $\dfrac{21}{19}$ **47.** $-\dfrac{14}{11}$ **49.** 2   **51.** all real numbers
53. 4   **55.** -4 **57.** no solution   **59.** 17
61. $-\dfrac{2}{5}$ **63.** $\dfrac{2}{3}$ **65.** 3   **67.** 5   **69.** no solution
71. 2   **73.** $p = \dfrac{k}{2.2}$ **75.** $w = \dfrac{P - 2l}{2}$ **77.** $r^2 = \dfrac{3V}{\pi h}$
79. $s = \dfrac{f(P_n - L)}{i}$ **81.** $m = \dfrac{r^2 F}{Mg}$ **83.** $y = b\left(1 - \dfrac{x}{a}\right)$
85. $r = \dfrac{r_1 r_2}{r_1 + r_2}$ **87.** $n = \dfrac{l - a + d}{d}$ **89.** $n = \dfrac{360}{180 - a}$
91. $r_1 = \dfrac{-Rr_2 r_3}{Rr_3 + Rr_2 - r_2 r_3}$ **95.** $5|x|$ **97.** $\dfrac{4y^4}{25x^2}$
99. $5|y|$ **101.** $\dfrac{|ab^3|}{z^2}$

Section 1.2

1. add   **3.** amount   **5.** rate; time   **7.** 84   **9.** 94
11. 7   **13.** 10 ft by 36 ft   **15.** $2\dfrac{1}{2}$ ft   **17.** 20 ft by 8 ft
19. 20 ft   **21.** \$10,000 at 7%, \$12,000 at 6%
23. \$29,100   **25.** 327   **27.** \$79.95   **29.** 200 units
31. 21   **33.** $2\dfrac{6}{11}$ days   **35.** $1\dfrac{1}{3}$ hr   **37.** $21\dfrac{1}{9}$ hr
39. 10 oz   **41.** 1 liter   **43.** $\dfrac{1}{15}$ liter   **45.** about 4.5 gal
47. 4 liters   **49.** 50 lb   **51.** 30 gal   **53.** 39 mph going; 65 mph returning   **55.** $2\dfrac{1}{2}$ hr   **57.** 50 sec   **59.** 12 mph
61. 600 lb barley; 1,637 lb oats; 163 lb soybean meal
63. about 11.2 mm   **67.** $(x + 7)(x - 9)$
69. $(3x - 5)(3x + 1)$ **71.** $(x + 3)^2$
73. $(x + 2)(x^2 - 2x + 4)$

Section 1.3

1. $ax^2 + bx + c = 0$ **3.** $\sqrt{c}; -\sqrt{c}$
5. rational numbers   **7.** $3, -2$ **9.** $12, -12$
11. $2, -\dfrac{5}{2}$ **13.** $2, \dfrac{3}{5}$ **15.** $\dfrac{3}{5}, -\dfrac{5}{3}$ **17.** $\dfrac{3}{2}, \dfrac{1}{2}$
19. $3, -3$ **21.** $5\sqrt{2}, -5\sqrt{2}$ **23.** $2\sqrt{5}, -2\sqrt{5}$
25. $\dfrac{\sqrt{7}}{2}, -\dfrac{\sqrt{7}}{2}$ **27.** $\dfrac{\sqrt{26}}{2}, -\dfrac{\sqrt{26}}{2}$ **29.** $3, -1$
31. $-1 + 2\sqrt{2}, -1 - 2\sqrt{2}$ **33.** $\dfrac{-1 + 3\sqrt{3}}{2}, \dfrac{-1 - 3\sqrt{3}}{2}$
35. $x^2 + 6x + 9$ **37.** $x^2 - 4x + 4$ **39.** $a^2 + 5a + \dfrac{25}{4}$
41. $r^2 - 11r + \dfrac{121}{4}$ **43.** $y^2 + \dfrac{3}{4}y + \dfrac{9}{64}$ **45.** $q^2 - \dfrac{1}{5}q + \dfrac{1}{100}$

47. $5, 3$ **49.** $-6 \pm 2\sqrt{7}$ **51.** $\dfrac{-5 \pm \sqrt{5}}{2}$
53. $\dfrac{10 \pm \sqrt{2}}{2}$ **55.** $\dfrac{-2 \pm \sqrt{7}}{3}$ **57.** $\dfrac{3 \pm \sqrt{17}}{4}$
59. $\pm 2\sqrt{3}$ **61.** $0, 25$ **63.** $3, -\dfrac{5}{2}$ **65.** $\dfrac{-5 \pm \sqrt{13}}{6}$
67. $\dfrac{-7 \pm 3\sqrt{5}}{2}$ **69.** $\dfrac{-3 \pm \sqrt{6}}{3}$ **71.** $\dfrac{-1 \pm \sqrt{61}}{10}$
73. $t = \pm\dfrac{\sqrt{2hg}}{g}$ **75.** $t = \dfrac{8 \pm \sqrt{64 - h}}{4}$
77. $y = \pm\dfrac{b\sqrt{a^2 - x^2}}{a}$ **79.** $a = \pm\dfrac{bx\sqrt{b^2 + y^2}}{b^2 + y^2}$
81. $x = \dfrac{-y \pm y\sqrt{5}}{2}$ **83.** one repeated rational root
85. no real roots   **87.** two different rational roots
89. two different irrational roots   **91.** yes   **93.** 2, 10
95. $3, -4$ **97.** $\dfrac{3}{2}, -\dfrac{1}{4}$ **99.** $\dfrac{1}{2}, -\dfrac{4}{3}$ **101.** $\dfrac{5}{6}, -\dfrac{2}{5}$
103. $1, -1$ **105.** $-\dfrac{1}{2}, 5$ **107.** -2 **109.** $3, -\dfrac{8}{11}$
111. $\dfrac{7 \pm \sqrt{145}}{4}$ **121.** $2x^2 - 8x$ **123.** 12m
125. $3x\sqrt{2x}$

Section 1.4

1. $A = lw$ **3.** 4 ft by 8 ft   **5.** 72 ft by 160 ft   **7.** 9 cm
9. width: $7\dfrac{1}{4}$ ft; length: $13\dfrac{3}{4}$ ft   **11.** 2 in.   **13.** 4 m
15. 10 m and 24 m   **17.** 40.1 in. by 22.6 in.
19. 20 mph going and 10 mph returning   **21.** 7 hr
23. 25 sec   **25.** about 9.5 sec   **27.** 1.6 sec   **29.** 10¢
31. 1,440   **33.** Morgan at 7%; Chloe at 8%   **35.** 10
37. 4 hr   **39.** about 9.5 hr   **41.** 221   **43.** 24.3 ft
45. 1.70 in.   **47.** No   **49.** $\dfrac{x - 6}{x(x - 3)}$ **51.** $\dfrac{x + 3}{x(x + 2)}$
53. $\dfrac{y + x}{y - x}$

Section 1.5

1. imaginary   **3.** imaginary   **5.** $2 - 5i$ **7.** real
9. $12i$ **11.** $-4\sqrt{6}i$ **13.** $\dfrac{5\sqrt{2}}{3}i$ **15.** $-\dfrac{7\sqrt{6}}{4}i$
17. $x = 3; y = 5$ **19.** $x = \dfrac{2}{3}; y = -\dfrac{2}{9}$ **21.** $5 - 6i$
23. $-2 - 10i$ **25.** $4 + 10i$ **27.** $1 - i$
29. $-15 - 25i$ **31.** $56 + 28i$ **33.** $-9 + 19i$
35. $-5 + 12i$ **37.** $52 - 56i$ **39.** $-6 + 17i$
41. $0 + i$ **43.** $0 + \dfrac{4}{3}i$ **45.** $\dfrac{2}{5} - \dfrac{1}{5}i$ **47.** $\dfrac{1}{25} + \dfrac{7}{25}i$
49. $\dfrac{1}{2} + \dfrac{1}{2}i$ **51.** $-\dfrac{7}{13} - \dfrac{22}{13}i$ **53.** $-\dfrac{12}{17} + \dfrac{11}{34}i$
55. $\dfrac{6 + \sqrt{3}}{10} + \dfrac{3\sqrt{3} - 2}{10}i$ **57.** i **59.** -1
61. $-i$ **63.** 1   **65.** -1 **67.** -1 **69.** i **71.** 4
73. 5   **75.** $\sqrt{13}$ **77.** $7\sqrt{2}$ **79.** $\dfrac{\sqrt{2}}{2}$ **81.** 6

83. $\sqrt{2}$　**85.** $\dfrac{3\sqrt{5}}{5}$　**87.** 1　**89.** $\pm 13i$

91. $\pm 3\sqrt{6}i$　**93.** $\pm 3\sqrt{5}i$　**95.** $\pm\dfrac{\sqrt{7}}{3}i$

97. $\pm\dfrac{\sqrt{30}}{2}i$　**99.** $-1 \pm 2\sqrt{3}i$　**101.** $-\dfrac{1}{5} \pm \dfrac{2\sqrt{2}}{5}i$

103. $-5 \pm 2\sqrt{3}i$　**105.** $-\dfrac{11}{2} \pm \dfrac{5\sqrt{3}}{2}i$

107. $1 \pm \dfrac{\sqrt{5}}{3}i$　**109.** $\dfrac{7}{2} \pm \dfrac{\sqrt{11}}{2}i$　**111.** $-1 \pm i$

113. $-2 \pm i$　**115.** $1 \pm 2i$　**117.** $\dfrac{1}{3} \pm \dfrac{1}{3}i$

119. $(x + 2i)(x - 2i)$　**121.** $(5p + 6qi)(5p - 6qi)$
123. $2(y + 2zi)(y - 2zi)$　**125.** $2(5m + ni)(5m - ni)$
127. $21 + 12i$　**129.** $6.2 - 0.7i$　**137.** $4x^2\sqrt{2}$
139. $x - 4\sqrt{x + 1} + 5$　**141.** $\sqrt{5} + 1$

Section 1.6

1. equal　**3.** extraneous　**5.** $0, -5, -4$　**7.** $0, \dfrac{4}{3}, -\dfrac{1}{2}$

9. $5, -5, 1, -1$　**11.** $6, -6, 1, -1$

13. $3\sqrt{2}, -3\sqrt{2}, \sqrt{5}, -\sqrt{5}$　**15.** $0, 1$　**17.** $\dfrac{1}{8}, -8$

19. $1, 144$　**21.** $\dfrac{1}{64}$　**23.** $\dfrac{1}{9}$　**25.** 27　**27.** $-\dfrac{1}{3}$

29. 2　**31.** $-\dfrac{3}{2}, -\dfrac{5}{2}$　**33.** 9　**35.** 20　**37.** 2

39. 2　**41.** $3, 4$　**43.** $\dfrac{1}{5}, -1$　**45.** $3, 5$　**47.** $-2, 1$

49. $2, -\dfrac{5}{2}$　**51.** $0, 4$　**53.** -2　**55.** no solution

57. 3　**59.** 1　**61.** 400 ft　**63.** 8 ft
65. about \$3,109
69.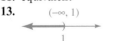

71.
[3, ∞)

73. [−2, 1)

75.
(−∞, 1) ∪ [2,∞)

Section 1.7

1. right　**3.** $a < c$　**5.** $b - c$　**7.** $>$　**9.** linear
11. equivalent
13. (−∞, 1)

15. [1, ∞)

17. (−∞, 1)

19. [1, ∞)

21. (−∞, 3]

23. (−10/3, ∞)

25. (5, ∞)

27. [14, ∞)

29. (−∞, 15/4]

31. (−44/41, ∞)

33. (6, 9]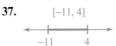

35. (8, 22]

37. [−11, 4]

39. [5, 21]

41. (−∞, 0)

43. (0, ∞)

45. (3, 14/3)

47. (2, ∞)

49. (−4, 5/6)

51. [−2, ∞)

53. (−4, −3)

55. (−∞, 2] ∪ [3, ∞)

57. (−3, −2)

59. (−∞, −1/2] ∪ [−1/3, ∞)

61. (1/3, 1/2)

63. (−∞, −3/2] ∪ [1, ∞)

65. $(-\infty, -\sqrt{3}] \cup [\sqrt{3}, \infty)$

67. $(-\sqrt{11}, \sqrt{11})$

69. (−3, 2)

71. (−∞, −1) ∪ (−1, 0) ∪ (1, ∞)

73. $(-\infty, -3) \cup (-3, -2] \cup (2, \infty)$

75. $(-\infty, -2) \cup (-2, -1/3) \cup (1/2, \infty)$　**77.** (0, 3/2)

79. (−∞, 0) ∪ (3/2, ∞)

81. (−∞, 2) ∪ [13/5, ∞)

83. $(-\infty, -\sqrt{7}) \cup (-1, 1) \cup (\sqrt{7}, \infty)$ 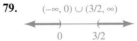　**85.** 19 min　**87.** 12

89. $p \le \$1,124.12$　**91.** $a > \$50,000$
93. anything over \$1,800　**95.** $16\dfrac{2}{3}$ cm $< s <$ 20 cm
97. $40 + 2w < P < 60 + 2w$　**101.** $0, -2, 2$　**103.** 2
105. All are real.

Section 1.8

1. x　**3.** $x = k$ or $x = -k$　**5.** $-k < x < k$
7. $x \le -k$ or $x \ge k$　**9.** 7　**11.** 0　**13.** 2
15. $\pi - 2$　**17.** $x - 5$　**19.** x^3 if $x \ge 0$; $-x^3$ if $x < 0$
21. $0, -4$　**23.** $2, -\dfrac{4}{3}$　**25.** $\dfrac{14}{3}, -2$　**27.** $7, -3$

29. no solution　**31.** 5　**33.** $\dfrac{2}{7}, 2$　**35.** $x \ge 0$
37. $-\dfrac{3}{2}$　**39.** $0, -6$　**41.** 0　**43.** $\dfrac{3}{5}, 3$

45. $-\dfrac{3}{13}, \dfrac{9}{5}$

47. $(-3, 9)$

$$\xleftarrow{\quad} \overset{\displaystyle(\quad\quad)}{\underset{-3 \qquad 9}{\quad}} \xrightarrow{\quad}$$

49. $(-\infty, -9) \cup (3, \infty)$

$$\xleftarrow{\quad} \underset{-9 \qquad 3}{)\qquad(} \xrightarrow{\quad}$$

51. $(-\infty, -7] \cup [3, \infty)$

$$\xleftarrow{\quad} \underset{-7 \qquad 3}{]\qquad[} \xrightarrow{\quad}$$

53. $[-13/3, 1]$

$$\xleftarrow{\quad} \underset{-13/3 \qquad 1}{[\qquad]} \xrightarrow{\quad}$$

55. $(-\infty, -3) \cup (-3, \infty)$

$$\xleftarrow{\quad} \underset{-3}{\times} \xrightarrow{\quad}$$

57. $(-1, 1/5)$

$$\xleftarrow{\quad} \underset{-1 \qquad 1/5}{(\qquad)} \xrightarrow{\quad}$$

59. $(-\infty, -7/9) \cup (13/9, \infty)$

$$\xleftarrow{\quad} \underset{-7/9 \qquad 13/9}{)\qquad(} \xrightarrow{\quad}$$

61. $(-5, 7)$

$$\xleftarrow{\quad} \underset{-5 \qquad 7}{(\qquad)} \xrightarrow{\quad}$$

63. $(-2, -1/2) \cup (-1/2, 1)$

$$\xleftarrow{\quad} \underset{-2 \quad -1/2 \quad 1}{(\quad\times\quad)} \xrightarrow{\quad}$$

65. $(-7/3, -2/3) \cup (4/3, 3)$

$$\xleftarrow{\quad} \underset{-7/3 \;\; -2/3 \;\; 4/3 \;\;\; 3}{(\quad)\;(\quad)} \xrightarrow{\quad}$$

67. $(-7, -1) \cup (11, 17)$

$$\xleftarrow{\quad} \underset{-7 \;\; -1 \;\; 11 \;\; 17}{(\quad)\;(\quad)} \xrightarrow{\quad}$$

69. $(-18, -6) \cup (10, 22)$

$$\xleftarrow{\quad} \underset{-18 \;\; -6 \;\;\; 10 \;\; 22}{(\quad)\;(\quad)} \xrightarrow{\quad}$$

71. $(-10, -7] \cup [5, 8)$

$$\xleftarrow{\quad} \underset{-10 \;\; -7 \qquad 5 \;\;\; 8}{(\quad]\;[\quad)} \xrightarrow{\quad}$$

73. $\left[-\dfrac{1}{2}, \infty\right)$ **75.** $(-\infty, 0)$ **77.** $\left(-\infty, -\dfrac{1}{2}\right)$

79. $[0, \infty)$ **81.** $70° \le t \le 86°$ **83.** $|c - 0.6°| \le 0.5°$

85. $|h - 55| < 17$ **87. a.** 26.45%, 24.76%

b. It is less than or equal to 1%.   **95.** 3.725×10^4

97. 523,000   **99.** $-4xy$

Chapter Review

1. no restrictions   **2.** $x \ne 0$ **3.** $x \ne 1$

4. $x \ne 2, x \ne 3$ **5.** $\dfrac{16}{27}$; conditional equation

6. -14; conditional equation   **7.** $\dfrac{16}{5}$; conditional equation

8. no solution; contradiction   **9.** 7; conditional equation

10. all real numbers except -9; identity

11. 7; conditional equation   **12.** $\dfrac{1}{3}$; conditional equation

13. -2; conditional equation   **14.** 0; conditional equation

15. $F = \dfrac{9}{5}C + 32$ **16.** $f = \dfrac{is}{P_n - l}$ **17.** $f_1 = \dfrac{ff_2}{f_2 - f}$

18. $l = \dfrac{a - S + Sr}{r}$ **19.** 60%   **20.** 22.5 ft by 27.5 ft

21. 3 hr   **22.** 0.5 hr   **23.** 1.5 liters   **24.** about 3.9 hr

25. $5\dfrac{1}{7}$ hr   **26.** $3\dfrac{1}{3}$ oz   **27.** \$4,500 at 11%; \$5,500 at 14%

28. 10   **29.** $2, -\dfrac{3}{2}$ **30.** $\dfrac{1}{4}, -\dfrac{4}{3}$ **31.** $0, \dfrac{8}{5}$

32. $\dfrac{2}{3}, \dfrac{4}{9}$ **33.** $\pm 2\sqrt{2}$ **34.** $\pm\sqrt{5}$

35. $\dfrac{5 \pm 4\sqrt{2}}{4}$ **36.** $\dfrac{7 \pm 3\sqrt{5}}{5}$ **37.** 3, 5

38. $-4, -2$ **39.** $\dfrac{1 \pm \sqrt{21}}{10}$ **40.** $0, \dfrac{1}{5}$ **41.** $2, -7$

42. $9, -\dfrac{2}{3}$ **43.** $\dfrac{-1 \pm \sqrt{21}}{10}$ **44.** $-1 \pm \sqrt{6}$ **45.** 1

46. two different rational numbers   **47.** $\dfrac{1}{3}$ **48.** 10, 2

49. $-\dfrac{5}{2}$ **50.** $\dfrac{8}{5}, 5$ **51.** either 95 by 110 yd or 55 by 190 yd   **52.** 320 mph for prop plane; 440 mph for jet plane

53. 1 sec   **54.** $1\dfrac{1}{2}$ ft   **55.** $-2 - i$ **56.** $-2 - 5i$

57. $5 - 2i$ **58.** $21 - 9i$ **59.** $0 - 3i$ **60.** $0 - 2i$

61. $\dfrac{3}{2} - \dfrac{3}{2}i$ **62.** $-\dfrac{2}{5} + \dfrac{4}{5}i$ **63.** $\dfrac{4}{5} + \dfrac{3}{5}i$ **64.** $\dfrac{1}{2} - \dfrac{5}{2}i$

65. $0 + i$ **66.** $0 - i$ **67.** $\sqrt{10} + 0i$ **68.** $1 + 0i$

69. $\dfrac{1}{3} \pm \dfrac{\sqrt{2}}{3}i$ **70.** $\dfrac{1}{3} \pm \dfrac{\sqrt{11}}{3}i$ **71.** $2, -3$

72. $4, -2$ **73.** $1, 1, -1, -1$ **74.** $1, -1, 6, -6$ **75.** 9

76. $8, -27$ **77.** 5   **78.** 0   **79.** $4, -4$

80. no solution   **81.** $(-\infty, 7)$

$$\xleftarrow{\quad} \overset{)}{\underset{7}{\quad}} \xrightarrow{\quad}$$

82. $[-1/5, \infty)$

$$\xleftarrow{\quad} \underset{-1/5}{[} \xrightarrow{\quad}$$

83. $(-\infty, 5/3)$

$$\xleftarrow{\quad} \underset{5/3}{)} \xrightarrow{\quad}$$

84. $(-\infty, -12/7)$

$$\xleftarrow{\quad} \underset{-12/7}{} \xrightarrow{\quad}$$

85. $[-3, 5)$

$$\xleftarrow{\quad} \underset{-3 \qquad 5}{[\qquad)} \xrightarrow{\quad}$$

86. $(2, \infty)$

$$\xleftarrow{\quad} \underset{2}{(} \xrightarrow{\quad}$$

87. $(-\infty, -2) \cup (4, \infty)$

$$\xleftarrow{\quad} \underset{-2 \qquad 4}{)\qquad(} \xrightarrow{\quad}$$

88. $(-4, 1)$

$$\xleftarrow{\quad} \underset{-4 \qquad 1}{(\qquad)} \xrightarrow{\quad}$$

89. $(-1, 3)$

$$\xleftarrow{\quad} \underset{-1 \qquad 3}{(\qquad)} \xrightarrow{\quad}$$

90. $(-\infty, -3/2) \cup (1, \infty)$

$$\xleftarrow{\quad} \underset{-3/2 \qquad 1}{)\qquad(} \xrightarrow{\quad}$$

91. $(-\infty, -2] \cup (3, \infty)$

$$\xleftarrow{\quad} \underset{-2 \qquad 3}{]\qquad(} \xrightarrow{\quad}$$

92. $(-4, 1]$

$$\xleftarrow{\quad} \underset{-4 \qquad 1}{(\qquad]} \xrightarrow{\quad}$$

93. $[-2, 1] \cup (3, \infty)$

$$\xleftarrow{\quad} \underset{-2 \quad 1 \quad 3}{[\quad]\;(} \xrightarrow{\quad}$$

94. $(-\infty, 0) \cup (5/2, \infty)$

$$\xleftarrow{\quad} \underset{0 \qquad 5/2}{)\qquad(} \xrightarrow{\quad}$$

95. $5, -7$ **96.** 0   **97.** $-\dfrac{4}{3}, -6$ **98.** $-\dfrac{3}{8}, \dfrac{3}{10}$

99. $(-6, 0)$

$$\xleftarrow{\quad} \underset{-6 \qquad 0}{(\qquad)} \xrightarrow{\quad}$$

100. $(-\infty, 2] \cup [8/3, \infty)$

$$\xleftarrow{\quad} \underset{2 \qquad 8/3}{]\qquad[} \xrightarrow{\quad}$$

101. $(-5, 1)$

$$\xleftarrow{\quad} \underset{-5 \qquad 1}{(\qquad)} \xrightarrow{\quad}$$

102. $(-\infty, -29) \cup (35, \infty)$

$$\xleftarrow{\quad} \underset{-29 \qquad 35}{)\qquad(} \xrightarrow{\quad}$$

103. $(-7/2, -2) \cup (-1, 1/2)$

$$\xleftarrow{\quad} \underset{-7/2 \;\; -2 \quad -1 \;\; 1/2}{(\quad)\;(\quad)} \xrightarrow{\quad}$$

104. $(-1, 4/3) \cup (4/3, 11/3)$

$$\xleftarrow{\quad} \underset{-1 \qquad 4/3 \;\; 11/3}{(\quad\times\quad)} \xrightarrow{\quad}$$

Chapter Test

1. $x \ne 0, x \ne 1$ **2.** $x \ne \dfrac{2}{3}$ **3.** $\dfrac{5}{2}$ **4.** 37

5. $x = \mu + z\sigma$ **6.** $a = \dfrac{bc}{c + b}$ **7.** 87.5   **8.** \$14,000

9. $\dfrac{1}{2}, \dfrac{3}{2}$ **10.** $\dfrac{3}{2}, -4$ **11.** $x = \dfrac{-b \pm \sqrt{b^2 - 4ac}}{2a}$

12. $\dfrac{5 \pm \sqrt{133}}{6}$ **13.** $5, -3$ **14.** 8 sec   **15.** $7 - 12i$

16. $-23 - 43i$ **17.** $\dfrac{4}{5} + \dfrac{2}{5}i$ **18.** $0 + i$ **19.** i

20. 1   **21.** 13   **22.** $\dfrac{\sqrt{10}}{10}$ **23.** $2, -2, 3, -3$

24. $1, -\dfrac{1}{32}$ **25.** 139   **26.** -1

27. $(-\infty, 2]$

28. $(-\infty, 5)$

29. $[3, 4)$

30. $(2, \infty)$

31. $(-\infty, -1] \cup [8, \infty)$

32. $[-2, 1)$ **33.** $2, -\dfrac{10}{3}$ **34.** 0

35. $(-\infty, 3/2) \cup (7/2, \infty)$ **36.** $[-9, 6]$

Cumulative Review

1. $-2, 0, 2, 6$ **2.** $2, 5, 11$

3. $[-4, 7)$ **4.** $(-\infty, 0) \cup [2, \infty)$

5. Comm. Prop. of Addition   **6.** Transitive Prop.   **7.** $9a^2$

8. $81a^2$ **9.** a^6b^4 **10.** $\dfrac{9y^2}{x^4}$ **11.** $\dfrac{x^4}{16y^2}$ **12.** $\dfrac{4y^{10}}{9x^6}$

13. a^3b^3 **14.** abc^2 **15.** $\sqrt{3}$ **16.** $\dfrac{\sqrt[3]{2x^2}}{x}$

17. $\dfrac{3(y + \sqrt{3})}{y^2 - 3}$ **18.** $\dfrac{3x(\sqrt{x} + 1)}{x - 1}$ **19.** $5\sqrt{3} - 3\sqrt{5}$

20. $3\sqrt{2}$ **21.** $5 - 2\sqrt{6}$ **22.** 4   **23.** $-8x + 8$

24. $9x^4 - 4x^3$ **25.** $6x^2 + 11x - 35$ **26.** $z^3 + z^2 + 4$

27. $2x^2 - x + 1$ **28.** $3x^2 + 1 - \dfrac{x}{x^2 + 2}$ **29.** $3t(t - 2)$

30. $(3x + 2)(x - 4)$ **31.** $(x + 1)^2(x - 1)^2(x^2 + 1)^2$

32. $(x + 1)(x^2 - x + 1)(x - 1)(x^2 + x + 1)$ **33.** $\dfrac{x - 5}{x + 5}$

34. $(2x + 1)(x + 2)$ **35.** $\dfrac{5x^2 + 17x - 6}{(x + 3)(x - 3)}$

36. $\dfrac{-x^2 + x + 7}{(x + 2)(x + 3)}$ **37.** $b + a$ **38.** $-\dfrac{1}{xy}$

39. $0, 10$ **40.** 34   **41.** $R = \dfrac{R_1 R_2}{R_1 + R_2}$

42. $r = \dfrac{a - S}{1 - S}$ or $r = \dfrac{S - a}{S - 1}$

43. either 8 ft by 24 ft or 12 ft by 16 ft   **44.** \$8,000

45. $\dfrac{3}{5} + \dfrac{4}{5}i$ **46.** $\dfrac{3}{2} - \dfrac{1}{2}i$ **47.** 5   **48.** $0 + 10i$

49. 3   **50.** $2, -2, 3, -3$ **51.** $2, 7$ **52.** $64, 729$

53. $\left(-\infty, \dfrac{11}{5}\right]$ **54.** $(-\infty, 3) \cup (5, \infty)$

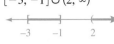

55. $[-3, -1] \cup (2, \infty)$ **56.** $(-\infty, -3) \cup (0, 3)$

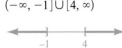

57. $(-\infty, -1] \cup [4, \infty)$ **58.** $\left(\dfrac{1}{3}, 3\right)$

Section 2.1

1. quadrants   **3.** to the right   **5.** first   **7.** linear
9. x-intercept   **11.** horizontal   **13.** midpoint
15. $A(2, 3)$ **17.** $C(-2, -3)$ **19.** $E(0, 0)$
21. $G(-5, -5)$ **23.** QI   **25.** QIII   **27.** QI
29. x-axis

31.

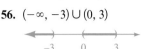

33.

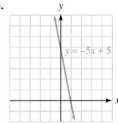

35.

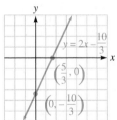

37.

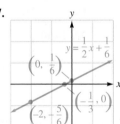

39.

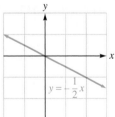

41.

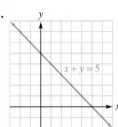

43.

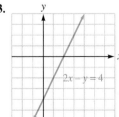

45.

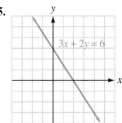

47.

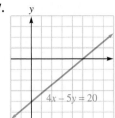

$4x - 5y = 20$

49.

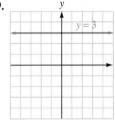

$y = 3$

51.

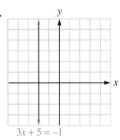

$3x + 5 = -1$

53.

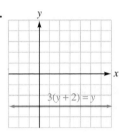

$3(y + 2) = y$

55.

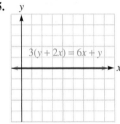

$3(y + 2x) = 6x + y$

57. 1.22   **59.** 4.67   **61.** 5   **63.** $\sqrt{13}$ **65.** $\sqrt{2}$

67. 2   **69.** 5   **71.** 13   **73.** $2\sqrt{13}$ **75.** $8\sqrt{2}$

77. 7   **79.** $(4, 6)$ **81.** $(0, 1)$ **83.** $\left(-1, \frac{1}{2}\right)$

85. $\left(\frac{\sqrt{5}}{2}, \frac{\sqrt{5}}{2}\right)$ **87.** $(5, 6)$ **89.** $(5, 15)$

93. $\sqrt{2}$ units   **97.** \$312,500   **99.** 200   **101.** 100 rpm

103. approx. 171 mi   **109.**

$\xleftarrow{\quad\;|\quad\quad\quad\quad|\;\quad}\atop{-3\quad\quad\quad3}$

111. no graph; the intersection is the empty set   **113.** -12

Section 2.2

1. divided   **3.** run   **5.** the change in   **7.** vertical

9. perpendicular   **11.** 1   **13.** $-\dfrac{5}{12}$ **15.** $-\dfrac{7}{4}$

17. -2 **19.** undefined   **21.** $\dfrac{5}{3}$ **23.** -1 **25.** 3

27. $\dfrac{1}{2}$ **29.** $\dfrac{2}{3}$ **31.** 0   **33.** 0   **35.** undefined

37. negative   **39.** positive   **41.** undefined
43. perpendicular   **45.** parallel   **47.** perpendicular
49. perpendicular   **51.** perpendicular   **53.** parallel
55. perpendicular   **57.** neither   **59.** 5   **61.** 6
63. not on same line   **65.** on same line
67. No two are perpendicular.   **69.** PQ and PR are
perpendicular.   **71.** PQ and PR are perpendicular.
81. 3.5 students per yr   **83.** \$642.86 per year

85. $\frac{\Delta T}{\Delta t}$ is the hourly rate of change in temperature.

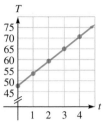

87. The slope is the speed of the plane.

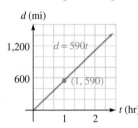

d (mi)

$d = 590t$

(1, 590)

t (hr)

91. $y = -\dfrac{3}{7}x + 3$ **93.** $y = -\dfrac{2}{5}x + 2$

95. $(2p + 3)(3p - 4)$ **97.** $(m + n)(p + q)$

Section 2.3

1. $y - y_1 = m(x - x_1)$ **3.** slope-intercept   **5.** $-\dfrac{A}{B}$

7. $2x - y = 0$ **9.** $4x - 2y = -7$ **11.** $2x - 5y = -7$
13. $y = -3$ **15.** $x = -6$ **17.** $\pi x - y = \pi^2$

19. $2x - 3y = -11$ **21.** $y = x$ **23.** $y = \dfrac{7}{3}x - 3$

25. $y = -\dfrac{9}{5}x + \dfrac{2}{5}$ **27.** $y = 3x - 2$ **29.** $y = 5x - \dfrac{1}{5}$

31. $y = ax + \dfrac{1}{a}$ **33.** $y = ax + a$ **35.** $3x - 2y = 0$

37. $3x + y = -4$ **39.** $\sqrt{2}x - y = -\sqrt{2}$

41. 1, $(0, -1)$ **43.** $\dfrac{2}{3}$, $(0, 2)$

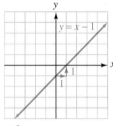

$y = x - 1$

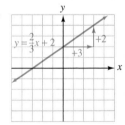

$y = \frac{2}{3}x + 2$

45. $-\dfrac{2}{3}$, $(0, 6)$ **47.** $\dfrac{3}{2}$, $(0, -4)$

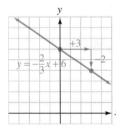

$y = -\frac{2}{3}x + 6$

49. $-\dfrac{1}{3}$, $\left(0, -\dfrac{5}{6}\right)$ **51.** $\dfrac{7}{2}$, $(0, 2)$ **53.** parallel

55. perpendicular   **57.** parallel   **59.** perpendicular
61. perpendicular   **63.** perpendicular   **65.** $y = 4x$

67. $y = 4x - 3$ **69.** $y = \dfrac{4}{5}x - \dfrac{26}{5}$ **71.** $y = -\dfrac{1}{4}x$

73. $y = -\dfrac{1}{4}x + \dfrac{11}{2}$ **75.** $y = -\dfrac{5}{4}x + 3$

77. $m = -\dfrac{4}{5}; (0, 4)$ **79.** $m = -\dfrac{2}{3}; (0, 4)$ **81.** $x = -2$

83. $x = 5$ **85.** $y = -3{,}200x + 24{,}300$

87. $y = 47{,}500x + 475{,}000$ **89.** $y = -\dfrac{710}{3}x + 1{,}900$

91. \$90   **93.** \$890   **95.** \$37,200   **97.** \$230

99. about 838   **101.** $C = \dfrac{5}{9}(F - 32)$

103. $y = -\dfrac{9}{10}x + 47; 11\%$ **105.** $y = 3.75x + 37.5; \$52\dfrac{1}{2}$

107. 1,655 barrels per day

109. a. Chirps/min

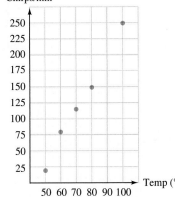

b. $y = \dfrac{23}{5}x - 210$ (answers may vary)

c. 204 (answers may vary)   **111.** $y = 4.44x - 196.62$

123. x^5 **125.** $\dfrac{125}{729}$ **127.** $-\sqrt{3}$ **129.** $\dfrac{5\left(\sqrt{x} - 2\right)}{x - 4}$

Section 2.4

1. x-intercept   **3.** axis of symmetry   **5.** x-axis
7. circle; center   **9.** $x^2 + y^2 = r^2$

11. $(-2, 0), (2, 0); (0, -4)$ **13.** $(0, 0), \left(\dfrac{1}{2}, 0\right); (0, 0)$

15. $(-1, 0), (5, 0); (0, -5)$ **17.** $(1, 0), (-2, 0); (0, -2)$
19. $(-3, 0), (0, 0), (3, 0); (0, 0)$ **21.** $(-1, 0), (1, 0); (0, -1)$
23.   **25.**

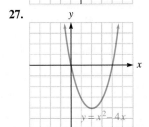

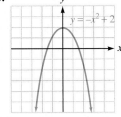

27.   **29.**

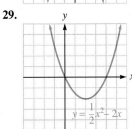

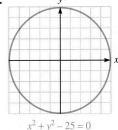

31. about the y-axis   **33.** about the x-axis
35. about the x-axis, the y-axis, and the origin
37. about the y-axis   **39.** none   **41.** about the x-axis
43. about the y-axis   **45.** about the x-axis
47.   **49.**

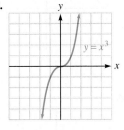

51.   **53.**

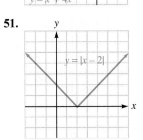

55.   **57.**

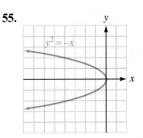

59.   **61.**

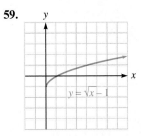

63. $(0, 0); 10$ **65.** $(0, 5); 7$ **67.** $(-6, 0); \dfrac{1}{2}$

69. $(4, 1); 3$ **71.** $\left(\dfrac{1}{4}, -2\right); 3\sqrt{5}$ **73.** $x^2 + y^2 = 25$

75. $x^2 + (y + 6)^2 = 36$ **77.** $(x - 8)^2 + y^2 = \dfrac{1}{25}$

79. $(x + 2)^2 + (y - 12)^2 = 169$ **81.** $x^2 + y^2 - 1 = 0$
83. $x^2 + y^2 - 12x - 16y + 84 = 0$
85. $x^2 + y^2 - 6x + 8y + 23 = 0$
87. $x^2 + y^2 - 6x - 6y - 7 = 0$
89. $x^2 + y^2 + 6x - 8y = 0$ **91.** $(x - 3)^2 + (y + 2)^2 = 9$
93. $(x - 5)^2 + (y - 6)^2 = 4$ **95.** $(x - 2)^2 + (y - 4)^2 = 9$
97.   **99.**

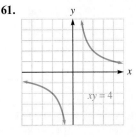

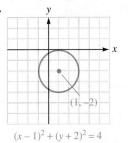

101.

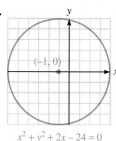

$x^2 + y^2 + 2x - 24 = 0$

103.

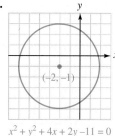

$x^2 + y^2 + 4x + 2y - 11 = 0$

5–8.

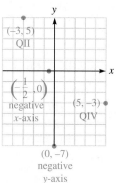

105.

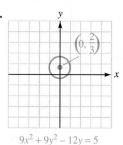

$9x^2 + 9y^2 - 12y = 5$

107.

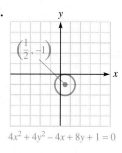

$4x^2 + 4y^2 - 4x + 8y + 1 = 0$

9.

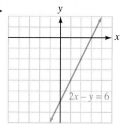

10.

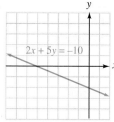

109. $(0.25, 0.88)$

111. $(0.50, 7.25)$

113. ± 2.65

115. 1.44

11.

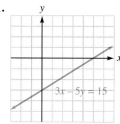

12.

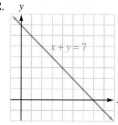

117. 4 sec   **119.**

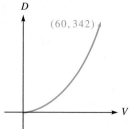

13.

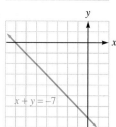

14.

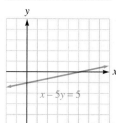

121. $x^2 + y^2 = 36$ **123.** $x^2 + (y - 35)^2 = 900$
125. $x^2 + y^2 - 14x - 8y + 40 = 0$ **127.** $(-3, 2)$
129. a single point   **131.** 6   **133.** -1 **135.** 20 oz

15.

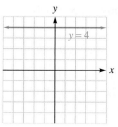

16.

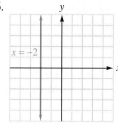

Section 2.5
1. quotient   **3.** means   **5.** extremes; means
7. inverse   **9.** joint   **11.** 14   **13.** 2, -3 **15.** 18
17. $\dfrac{1}{2}$ **19.** 1,000   **21.** 1   **23.** $\dfrac{21}{4}$ **25.** 6   **27.** -8
29. direct variation   **31.** neither   **33.** 202, 172, 136
35. about 2 gal   **37.** $14\dfrac{6}{11}$ ft^3 **39.** 3 sec   **41.** $\sqrt{2}$ m
43. 15 lumens   **45.** It is multiplied by 18.
47. The force is multiplied by $\dfrac{9}{4}$.   **49.** $\dfrac{\sqrt{3}}{4}$
57. $\dfrac{3x + 5}{(x + 2)(x + 1)}$ **59.** $\dfrac{(x + 4)(x + 1)}{x - 1}$ **61.** $-\dfrac{1}{x}$

Chapter Review
1. $(2, 0)$ **2.** $(-2, 1)$ **3.** $(0, -1)$ **4.** $(3, -1)$

17. $12,150   **18.** $332,500   **19.** 10   **20.** $4\sqrt{2}$
21. 2   **22.** $2\sqrt{2}|a|$ **23.** $(0, 3)$ **24.** $\left(-6, \dfrac{15}{2}\right)$
25. $\left(\sqrt{3}, 8\right)$ **26.** $(0, 0)$ **27.** -6 **28.** 2   **29.** -1
30. 1   **31.** 3   **32.** 5   **33.** 0   **34.** undefined
35. negative   **36.** positive   **37.** perpendicular
38. neither   **39.** $y = 7$ **40.** $x = 3$ **41.** 200 ft/min
42. $48,750 per year   **43.** $7x + 5y = 0$
44. $4x + y = -7$ **45.** $x + 5y = -3$ **46.** $2x + y = 9$
47. $y = \dfrac{2}{3}x + 3$ **48.** $y = -\dfrac{3}{2}x - 5$

49.

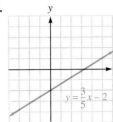

50.

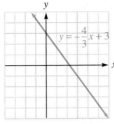

51. $\frac{3}{2}$, $(0, -5)$ **52.** $-\frac{1}{2}$, $(0, -2)$ **53.** $\frac{3}{2}$, $(0, -5)$

54. $-\frac{1}{2}$, $(0, -2)$ **55.** $-\frac{5}{2}$, $\left(0, \frac{7}{2}\right)$ **56.** $\frac{3}{4}$, $\left(0, -\frac{7}{2}\right)$

57. $y = 17$ **58.** $x = -5$ **59.** $y = \frac{3}{4}x - \frac{3}{2}$

60. $y = -7x + 47$ **61.** $y = 3x + 5$

62. $y = \frac{1}{7}x - 3$ **63.** parallel   **64.** perpendicular

65. $(0, 0)$, $\left(\frac{1}{2}, 0\right)$; $(0, 0)$ **66.** $(12, 0)$, $(-2, 0)$; $(0, -24)$

67. about the x-axis   **68.** about the y-axis
69. about the y-axis   **70.** none
71.

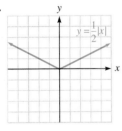

72. $y = x^3 - 2$

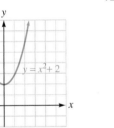

73.

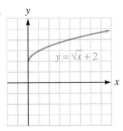

74.

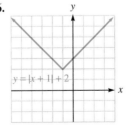

75.

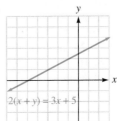

76.

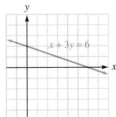

77.   **78.**

79.   **80.**

81. $(0, 0)$; 8   **82.** $(0, 6)$; 10   **83.** $(-7, 0)$; $\frac{1}{2}$

84. $(5, -1)$; 3   **85.** $x^2 + y^2 = 49$ **86.** $(x - 3)^2 + y^2 = \frac{1}{25}$

87. $(x + 2)^2 + (y - 12)^2 = 25$ **88.** $\left(x - \frac{2}{7}\right)^2 + (y - 5)^2 = 81$

89. $(x + 3)^2 + (y - 4)^2 = 144$ or $x^2 + y^2 + 6x - 8y - 119 = 0$

90. $\left(x + \frac{1}{2}\right)^2 + \left(y - \frac{5}{2}\right)^2 = \frac{121}{2}$ or $x^2 + y^2 + x - 5y - 54 = 0$

91. $(x + 3)^2 + (y - 2)^2 = 9$
92. $(x - 2)^2 + (y - 4)^2 = 25$
93.

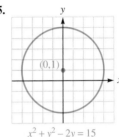

94.

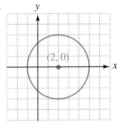

95.

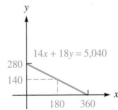

96.

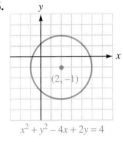

97. 3.32, −3.32   **98.** −1, 0, 1   **99.** 1.73, −1.73, 1, −1
100. 4.19, −1.19   **101.** $x = 2, x = 5$ **102.** $x = -5, x = 5$

103. $\frac{9}{5}$ lb   **104.** $\frac{25}{9}$ **105.** $333\frac{1}{3}$ cc   **106.** 1

107. about 117 ohms   **108.** $385
109. 140 hr

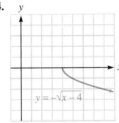

Chapter Test
1. QII   **2.** y-axis
3.

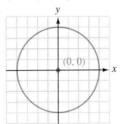

4.

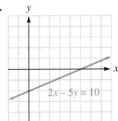

5.

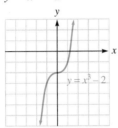

6.

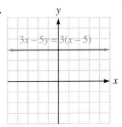

7.

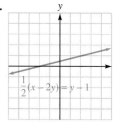

8.

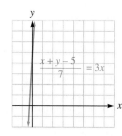

9. $\sqrt{41}$ **10.** approximately 4.44   **11.** $(0, 0)$

12. $\left(\sqrt{2}, 2\sqrt{2}\right)$ **13.** $-\dfrac{5}{4}$ **14.** $\dfrac{\sqrt{3}}{3}$ **15.** neither

16. perpendicular   **17.** $y = 2x - 11$ **18.** $y = 3x + \dfrac{1}{2}$

19. $y = 2x + 5$ **20.** $y = -\dfrac{1}{2}x + 5$ **21.** $y = 2x - \dfrac{11}{2}$

22. $x = 3$ **23.** $(-4, 0), (0, 0), (4, 0); (0, 0)$
24. $(4, 0); (0, 4)$ **25.** about the x-axis
26. about the y-axis

27.

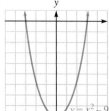

28.

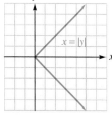

29.

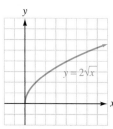

30.

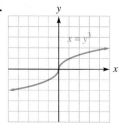

31. $(x - 5)^2 + (y - 7)^2 = 64$ **32.** $(x - 2)^2 + (y - 4)^2 = 32$
33.

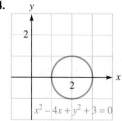

34.

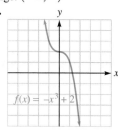

35. $y = kz^2$ **36.** $w = krs^2$ **37.** $P = \dfrac{35}{2}$ **38.** $x = \dfrac{27}{32}$

39. $x = 2.65$ **40.** $x = 5.85$

Section 3.1
1. function   **3.** domain   **5.** $y = f(x)$ **7.** x
9. vertical; once   **11.** function   **13.** not a function
15. function   **17.** not a function   **19.** not a function
21. function   **23.** domain: $(-\infty, \infty)$ **25.** domain: $(-\infty, \infty)$
27. domain: $[2, \infty)$ **29.** domain: $(-\infty, 4]$
31. domain: $(-\infty, -1] \cup [1, \infty)$ **33.** domain: $(-\infty, \infty)$
35. domain: $(-\infty, -1) \cup (-1, \infty)$

37. domain: $(-\infty, 3) \cup (3, \infty)$
39. domain: $(-\infty, -2) \cup (-2, 2) \cup (2, \infty)$
41. domain: $(-\infty, -1) \cup (-1, 5) \cup (5, \infty)$

43. $4; -11; 3k - 2; 3k^2 - 5$ **45.** $4; \dfrac{3}{2}; \dfrac{1}{2}k + 3; \dfrac{1}{2}k^2 + \dfrac{5}{2}$

47. $4; 9; k^2; k^4 - 2k^2 + 1$ **49.** $9; -1; k^2 + 3k - 1;$
$k^4 + k^2 - 3$ **51.** $5; 10; k^2 + 1; k^4 - 2k^2 + 2$

53. $\dfrac{1}{3}; 2; \dfrac{2}{k + 4}; \dfrac{2}{k^2 + 3}$ **55.** $\dfrac{1}{3}; \dfrac{1}{8}; \dfrac{1}{k^2 - 1}; \dfrac{1}{k^4 - 2k^2}$

57. $\sqrt{5}; \sqrt{10}; \sqrt{k^2 + 1}; \sqrt{k^4 - 2k^2 + 2}$ **59.** 3
61. $2x + h$ **63.** $8x + 4h$ **65.** $2x + h + 3$
67. $4x + 2h - 4$ **69.** $3x^2 + 3xh + h^2$
71.

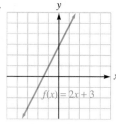

73.

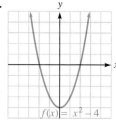

domain: $(-\infty, \infty)$ domain: $(-\infty, \infty)$
range: $(-\infty, \infty)$ range: $(-\infty, \infty)$
75.

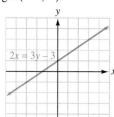

77.

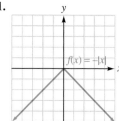

domain: $(-\infty, \infty)$ domain: $(-\infty, \infty)$
range: $(-\infty, \infty)$ range: $[-4, \infty]$
79.

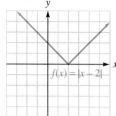

81.

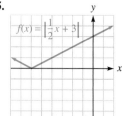

domain: $(-\infty, \infty)$ domain: $(-\infty, \infty)$
range: $(-\infty, \infty)$ range: $(-\infty, 0]$
83.

domain: $(-\infty, \infty)$ domain: $(-\infty, \infty)$
range: $[0, \infty)$ range: $[0, \infty)$

85.

87.
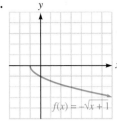
domain: $[-1, \infty)$
range: $(-\infty, 0]$

89.
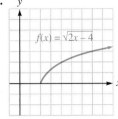
domain: $[2, \infty)$
range: $[0, \infty)$

91.
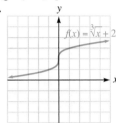
domain: $(-\infty, \infty)$
range: $(-\infty, \infty)$

93.

95. function   **97.** function   **99.** function

101.

domain: $(-\infty, \infty)$;
range: $[0, \infty)$

103.

domain: $(-\infty, \infty)$;
range: $(-\infty, \infty)$

105. a. $C(x) = 8x + 75$ **b.** $\$755$
107. a. $C(t) = 0.07t + 9.99$ **b.** $\$11.39$
109. a. $C(f) = 95f + 14{,}000$ **b.** $\$261{,}000$
111. $C(x) = 0.10x + 7$ **113.** $n(t) = 1{,}692t + 6{,}400$

115. $E(D) = 1.3911D$ **117.** $-\dfrac{2}{3}$ **123.** 1, 7, 8   **125.** 7

127. $(-4, 7]$ **129.**

Section 3.2

1. $f(x) = ax^2 + bx + c$ **3.** $(3, 5)$ **5.** upward
7. $-\dfrac{b}{2a}$ **9.** upward; minimum
11. downward; maximum   **13.** downward; maximum
15. $(0, -1)$ **17.** $(3, 5)$ **19.** $(-6, -4)$ **21.** $(3, 0)$
23. $(2, 0)$ **25.** $(-3, -12)$ **27.** $(3, 1)$

29. $\left(\dfrac{2}{3}, \dfrac{11}{3}\right)$ **31.** $(-4, -11)$

33.
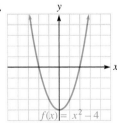
$f(x) = x^2 - 4$

35.
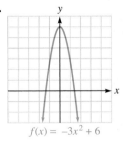
$f(x) = -3x^2 + 6$

37.
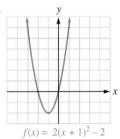
$f(x) = -\dfrac{1}{2}x^2 + 8$

39.
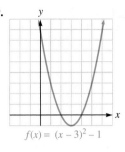
$f(x) = (x - 3)^2 - 1$

41.
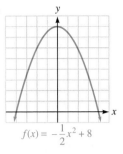
$f(x) = 2(x + 1)^2 - 2$

43.
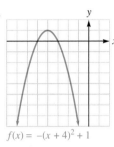
$f(x) = -(x + 4)^2 + 1$

45.
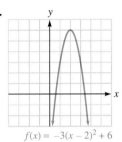
$f(x) = -3(x - 2)^2 + 6$

47.
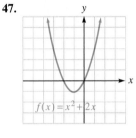
$f(x) = x^2 + 2x$

49.
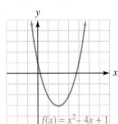
$f(x) = x^2 - 4x + 1$

51.
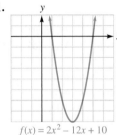
$f(x) = 2x^2 - 12x + 10$

53.
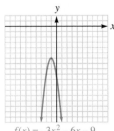
$f(x) = -3x^2 - 6x - 9$

55. 75 ft by 75 ft, $5{,}625$ ft^2

57. 25 ft by 25 ft   **59.** $w = 12$ in.; $d = 6$ in.   **61.** 20 ft
63. 15.4 ft   **65.** 208 ft
67. 48 digital cameras; minimum cost $\$2{,}400$ **69.** $\$95$
71. $\dfrac{5}{2}$ sec   **73.** 100 ft   **75.** $(-2.25, -66.13)$

77. $(3.3, -68.5)$ **79.** 2, 3   **81.** 6 by $4\dfrac{1}{2}$ units

83. Both numbers are 3.   **87.** $a^2 - 3a$; $a^2 + 3a$
89. $(5 - a)^2$; $(5 + a)^2$ **91.** 7; 7

Section 3.3

1. 4   **3.** $n - 1$ **5.** odd   **7.** piecewise-defined   **9.** 3

11.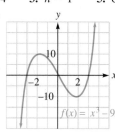
$f(x) = x^3 - 9x$

13.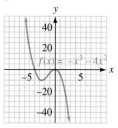
$f(x) = -x^3 - 4x^2$

15.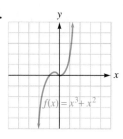
$f(x) = x^3 + x^2$

17.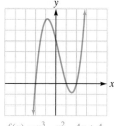
$f(x) = x^3 - x^2 - 4x + 4$

19.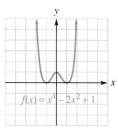
$f(x) = x^4 - 2x^2 + 1$

21.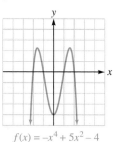
$f(x) = -x^4 + 5x^2 - 4$

23. even   **25.** neither   **27.** odd   **29.** odd
31. even   **33.** odd   **35.** neither
37. decreasing on $(-\infty, 0)$; increasing on $(0, \infty)$
39. increasing on $(-\infty, 0)$; decreasing for $(0, \infty)$
41. decreasing on $(-\infty, -2)$; constant on $(-2, 2)$; increasing on $(2, \infty)$ **43.** decreasing on $(-\infty, 2)$; increasing on $(2, \infty)$
45. a. -2 **b.** 3   **47. a.** 2   **b.** 1   **c.** 3

49.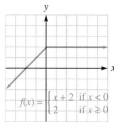
$f(x) = \begin{cases} x + 2 & \text{if } x < 0 \\ 2 & \text{if } x \geq 0 \end{cases}$

51.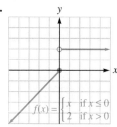
$f(x) = \begin{cases} x & \text{if } x \leq 0 \\ 2 & \text{if } x > 0 \end{cases}$

53.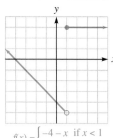
$f(x) = \begin{cases} -4 - x & \text{if } x < 1 \\ 3 & \text{if } x \geq 1 \end{cases}$

55.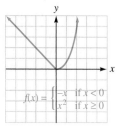
$f(x) = \begin{cases} -x & \text{if } x < 0 \\ x^2 & \text{if } x \geq 0 \end{cases}$

57.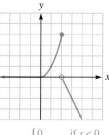
$f(x) = \begin{cases} 0 & \text{if } x < 0 \\ x^2 & \text{if } 0 \leq x \leq 2 \\ 4 - 2x & \text{if } x > 2 \end{cases}$

59. a. 3   **b.** -4
c. -3 **61. a.** 2   **b.** 3
c. 4

63.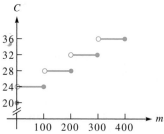
$y = [\![2x]\!]$

65.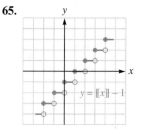
$y = [\![x]\!] - 1$

67. B

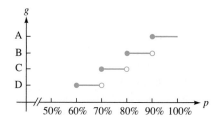

69. \$32

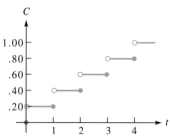

71. \$1.60

73.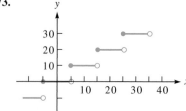

75. no; this is not defined at $x = 0$

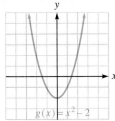

81. $3x + 5$; $3x + 3$ **83.** $\dfrac{3x - 8}{5}$; $\dfrac{3x + 1}{5} - 3$ or $\dfrac{3x - 14}{5}$

85. $-1, \dfrac{3}{2}$

Section 3.4

1. upward **3.** to the right **5.** 2; downward **7.** y-axis
9. horizontally

11.

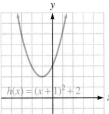

$g(x) = x^2 - 2$

13.

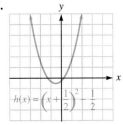

$g(x) = (x + 3)^2$

15.

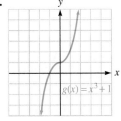

$h(x) = (x + 1)^2 + 2$

17.

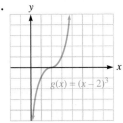

$h(x) = \left(x + \dfrac{1}{2}\right)^2 - \dfrac{1}{2}$

19.

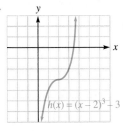

$g(x) = x^3 + 1$

21.

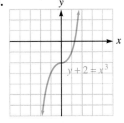

$g(x) = (x - 2)^3$

23.

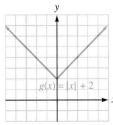

$h(x) = (x - 2)^3 - 3$

25.

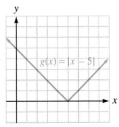

$y + 2 = x^3$

27.

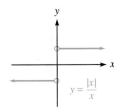

$g(x) = |x| + 2$

29.

$g(x) = |x - 5|$

31.

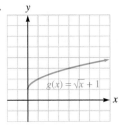

$f(x) = |x + 2| - 1$

33.

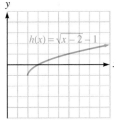

$g(x) = \sqrt{x} + 1$

35.

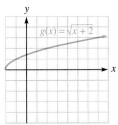

$g(x) = \sqrt{x} + 2$

37.

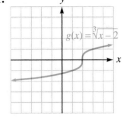

$h(x) = \sqrt{x - 2} - 1$

39.

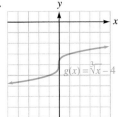

$g(x) = \sqrt[3]{x} - 4$

41.

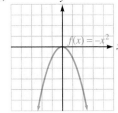

$g(x) = \sqrt[3]{x} - 2$

43.

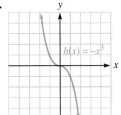

$h(x) = \sqrt[3]{x + 1} - 1$

45.

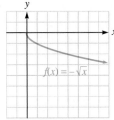

$f(x) = -x^2$

47.

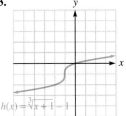

$h(x) = -x^3$

49.

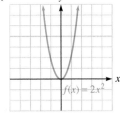

$f(x) = -\sqrt{x}$

51.

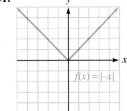

$f(x) = |-x|$

53.

$f(x) = 2x^2$

55.
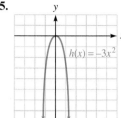

57.

83.
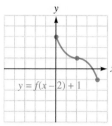

85.
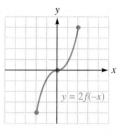

93. $\dfrac{x-2}{x+2}$ **95.** $(-\infty, 3) \cup (3, \infty)$ **97.** $x + 2 + \dfrac{-2}{x+1}$

Section 3.5
1. asymptote   **3.** vertical   **5.** x-intercept   **7.** same
9. horizontal; vertical   **11.** vertical asymptote: $x = 2$;
horizontal asymptote: $y = 1$; domain: $(-\infty, 2) \cup (2, \infty)$; range:
$(-\infty, 1) \cup (1, \infty)$ **13.** 20 hr   **15.** 12 hr   **17.** $5,555.56
19. $50,000   **21.** $(-\infty, 2) \cup (2, \infty)$
23. $(-\infty, -5) \cup (-5, 5) \cup (5, \infty)$
25. $(-\infty, -1) \cup (-1, 0) \cup (0, 1) \cup (1, \infty)$ **27.** $(-\infty, \infty)$
29. $x = 3$ **31.** $x = 1, x = -1$ **33.** $x = -2, x = 3$
35. none   **37.** $y = 2$ **39.** $y = \dfrac{1}{2}$ **41.** $y = 0$
43. none   **45.** $y = x - 3$ **47.** $y = 2x + 3$
49. $y = x + 2$

59.
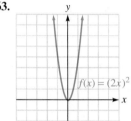

61.
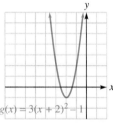

51.
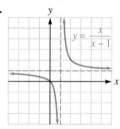

53.
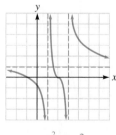

63.
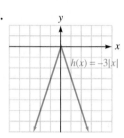

65.
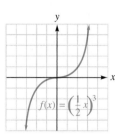

55.
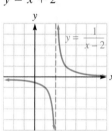

57.
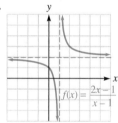

67.
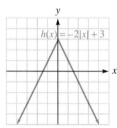

69.
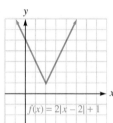

59.
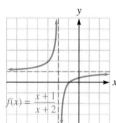

61.
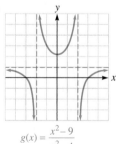

71.
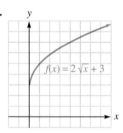

73.
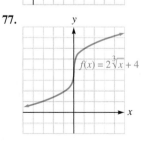

75.
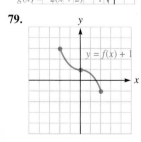

77.
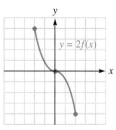

79.

81.

63.
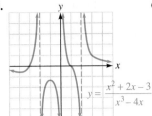
$$y = \frac{x^2 + 2x - 3}{x^3 - 4x}$$

65.
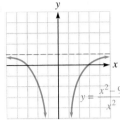
$$y = \frac{x^2 - 9}{x^2}$$

67.
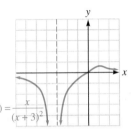
$$f(x) = \frac{x}{(x+3)^2}$$

69.
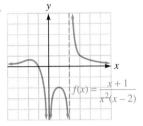
$$f(x) = \frac{x+1}{x^2(x-2)}$$

71.
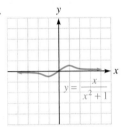
$$y = \frac{x}{x^2+1}$$

73.
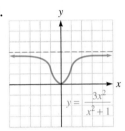
$$y = \frac{3x^2}{x^2+1}$$

75.
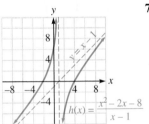
$$y = x - 1$$
$$y = x + 1$$
$$h(x) = \frac{x^2 - 2x - 8}{x-1}$$

77.
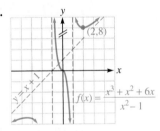
$$(2,8)$$
$$y = x + 1$$
$$f(x) = \frac{x^3 + x^2 + 6x}{x^2 - 1}$$

79.
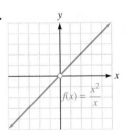
$$f(x) = \frac{x^2}{x}$$

81.
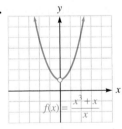
$$f(x) = \frac{x^3 + x}{x}$$

83.
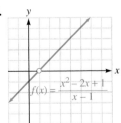
$$f(x) = \frac{x^2 - 2x + 1}{x - 1}$$

85.
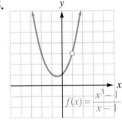
$$f(x) = \frac{x^3 - 1}{x - 1}$$

87. a. $C(x) = 3.25x + 700$ **b.** \$2,325

c. $\overline{C}(x) = \dfrac{3.25x + 700}{x}$ **d.** \$4.65   **e.** \$3.95   **f.** \$3.60

89. a. $C(n) = 0.095n + 8.50$ **b.** $\overline{C}(n) = \dfrac{0.095n + 8.50}{n}$

c. 10.5¢   **101.** $3x^2 + x$ **103.** $10x^2 + 29x + 10$
105. $3x + 5$

Section 3.6

1. $f(x) + g(x)$ **3.** $f(x)g(x)$ **5.** intersection
7. $g(f(x))$ **9.** commutative
11. $(f + g)(x) = 5x - 1$; $(-\infty, \infty)$
13. $(f \cdot g)(x) = 6x^2 - x - 2$; $(-\infty, \infty)$
15. $(f - g)(x) = x + 1$; $(-\infty, \infty)$
17. $(f/g)(x) = \dfrac{x^2 + x}{x^2 - 1} = \dfrac{x}{x - 1}$; $(-\infty, -1) \cup (-1, 1) \cup (1, \infty)$
19. $(f + g)(x) = x^2 + \sqrt{x} - 7$; $[0, \infty)$
21. $(f/g)(x) = \dfrac{x^2 - 7}{\sqrt{x}}$; $(0, \infty)$ **23.** 7   **25.** 1   **27.** 12
29. undefined   **31.** $f(x) = 3x^2$; $g(x) = 2x$
33. $f(x) = 3x^2$; $g(x) = x^2 - 1$ **35.** $f(x) = 3x^3$; $g(x) = -x$
37. $f(x) = x + 9$; $g(x) = x - 2$ **39.** 11   **41.** -17
43. 190   **45.** 145   **47.** $(-\infty, \infty)$; $(f \circ g)(x) = 3x + 3$
49. $(-\infty, \infty)$; $(f \circ f)(x) = 9x$ **51.** $(-\infty, \infty)$; $(g \circ f)(x) = 2x^2$
53. $(-\infty, \infty)$; $(g \circ g)(x) = 4x$
55. $[-1, \infty)$; $(f \circ g)(x) = \sqrt{x + 1}$
57. $[0, \infty)$; $(f \circ f)(x) = \sqrt[4]{x}$ **59.** $[-1, \infty)$; $(g \circ f)(x) = x$
61. $(-\infty, \infty)$; $(g \circ g)(x) = x^4 - 2x^2$
63. $(-\infty, 2) \cup (2, 3) \cup (3, \infty)$; $(f \circ g)(x) = \dfrac{x - 2}{3 - x}$
65. $(-\infty, 1) \cup (1, 2) \cup (2, \infty)$; $(f \circ f)(x) = \dfrac{x - 1}{2 - x}$
67. $f(x) = x - 2$; $g(x) = 3x$
69. $f(x) = x - 2$; $g(x) = x^2$
71. $f(x) = x^2$; $g(x) = x - 2$
73. $f(x) = \sqrt{x}$; $g(x) = x + 2$
75. $f(x) = x + 2$; $g(x) = \sqrt{x}$
77. $f(x) = x$; $g(x) = x$ **79.** 0   **81.** 0   **83.** 1
85. 1   **87.** 8   **89.** 9
91. a. $(R - C)(x) = 260x - 60,000$ **b.** \$70,000
93. $A(t) = \dfrac{9}{4}\pi t^2$; 101,787.6 square inches   **95.** $P = 4\sqrt{A}$

105. $y = \dfrac{x + 7}{3}$ **107.** $y = \dfrac{3x}{1 - x}$

Section 3.7

1. one-to-one   **3.** identity   **5.** one-to-one
7. not one-to-one   **9.** not one-to-one
11. not one-to-one   **13.** not one-to-one   **15.** one-to-one
17. one-to-one   **19.** not a function
25. $f^{-1}(x) = \dfrac{1}{3}x$ **27.** $f^{-1}(x) = \dfrac{x - 2}{3}$
29. $f^{-1}(x) = \sqrt[3]{x - 2}$ **31.** $f^{-1}(x) = x^5$
33. $f^{-1}(x) = \dfrac{1}{x} - 3$ **35.** $f^{-1}(x) = \dfrac{1}{2x}$

37.
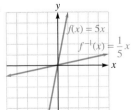
$$f(x) = 5x$$
$$f^{-1}(x) = \frac{1}{5}x$$

39.
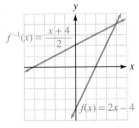
$$f^{-1}(x) = \frac{x + 4}{2}$$
$$f(x) = 2x - 4$$

41.

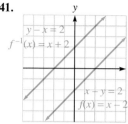

$y - x = 2$
$f^{-1}(x) = x + 2$
$x - y = 2$
$f(x) = x - 2$

43.

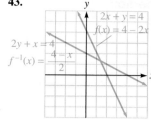

$2x + y = 4$
$f(x) = 4 - 2x$
$2y + x = 4$
$f^{-1}(x) = \dfrac{4 - x}{2}$

17. function   **18.** not a function   **19. a.** $I(h) = 3.5h - 50$
b. $650   **20.** $C(x) = 0.15x + 5$ **21.** upward; minimum
22. downward; maximum   **23.** $(1, 6)$ **24.** $(-4, -5)$
25. $(-3, -13)$ **26.** $\left(\dfrac{1}{2}, -8\right)$

45.

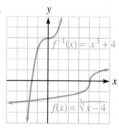

$f^{-1}(x) = x^3 + 4$
$f(x) = \sqrt[3]{x - 4}$

47.

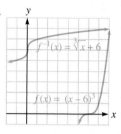

$f^{-1}(x) = \sqrt[3]{x} + 6$
$f(x) = (x - 6)^3$

27.

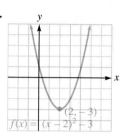

$(2, -3)$
$f(x) = (x - 2)^2 - 3$

28.

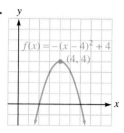

$f(x) = -(x - 4)^2 + 4$
$(4, 4)$

49.

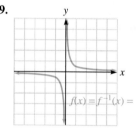

$f(x) = f^{-1}(x) = \dfrac{1}{2x}$

51.

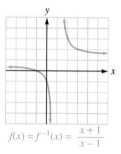

$f(x) = f^{-1}(x) = \dfrac{x + 1}{x - 1}$

29.

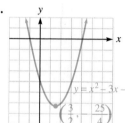

$y = x^2 - x$
$\left(\dfrac{1}{2}, -\dfrac{1}{4}\right)$

30.

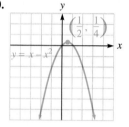

$\left(\dfrac{1}{2}, \dfrac{1}{4}\right)$
$y = x - x^2$

31.

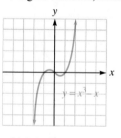

$y = x^2 - 3x - 4$
$\left(\dfrac{3}{2}, -\dfrac{25}{4}\right)$

32.

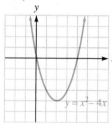

$y = 3x^2 - 8x - 3$
$\left(\dfrac{4}{3}, -\dfrac{25}{3}\right)$

53. $f^{-1}(x) = -\sqrt{x + 3}$ $(x \geq -3)$
55. $f^{-1}(x) = \sqrt[4]{x + 8}$ $(x \geq -8)$
57. $f^{-1}(x) = \sqrt{4 - x^2}$ $(0 \leq x \leq 2)$
59. domain: $(-\infty, 2) \cup (2, \infty)$; range: $(-\infty, 1) \cup (1, \infty)$
61. domain: $(-\infty, 0) \cup (0, \infty)$; range: $(-\infty, -2) \cup (-2, \infty)$
63. a. $f(x) = 0.75x + 8.50$ **b.** $11.50

c. $f^{-1}(x) = \dfrac{x - 8.50}{0.75}$ **d.** 2   **67.** 0   **69.** $a \geq 0$

71. 8   **73.** 4   **75.** $\dfrac{5}{4}$ **77.** $\dfrac{1}{7}$

33. 300 units   **34.** Both numbers are $\dfrac{1}{2}$.
35. 350 ft by 350 ft; 122,500 ft^2
36. 50 digital cameras; minimum cost $1,100

Chapter Review
1. function   **2.** function   **3.** not a function   **4.** function
5. domain: $(-\infty, \infty)$ **6.** domain: $(-\infty, 5] \cup [5, \infty)$
7. domain: $[1, \infty)$ **8.** domain: $(-\infty, \infty)$ **9.** $8; -17; 5k - 2$
10. $-2; -\dfrac{3}{4}; \dfrac{6}{k - 5}$ **11.** $0; 5; |k - 2|$ **12.** $\dfrac{1}{7}; \dfrac{1}{2}; \dfrac{k^2 - 3}{k^2 + 3}$
13. 5   **14.** $4x + 2h - 7$

15.

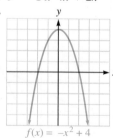

$f(x) = -x^2 + 4$

domain: $(-\infty, \infty)$
range: $(-\infty, 4]$

16.

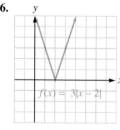

$f(x) = 3|x - 2|$

domain: $(-\infty, \infty)$
range: $[0, \infty)$

37.

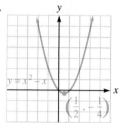

$y = x^3 - x$

odd function

38.

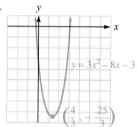

$y = x^2 - 4x$

neither even nor odd

39.

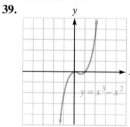

$y = x^3 - x^2$

neither even nor odd

40.

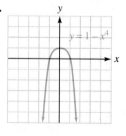

$y = 1 - x^4$

even function

41. a. 2   **b.** 9   **42. a.** $\dfrac{1}{2}$ **b.** 3

43.

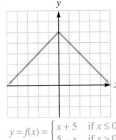

$$y = f(x) = \begin{cases} x+5 & \text{if } x \le 0 \\ 5-x & \text{if } x > 0 \end{cases}$$

increasing on $(-\infty, 0)$; decreasing on $(0, \infty)$

44.

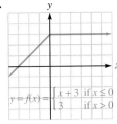

$$y = f(x) = \begin{cases} x+3 & \text{if } x \le 0 \\ 3 & \text{if } x > 0 \end{cases}$$

increasing on $(-\infty, 0)$; constant on $(0, \infty)$

45. 3   **46.** -1

47.

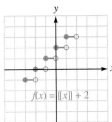

$f(x) = [\![x]\!] + 2$

48.

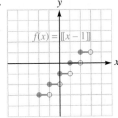

$f(x) = [\![x-1]\!]$

49. \$44   **50.** \$26

51.

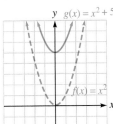

$g(x) = x^2 + 5$; $f(x) = x^2$

52.

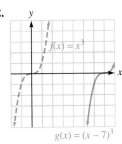

$f(x) = x^3$; $g(x) = (x-7)^3$

53.

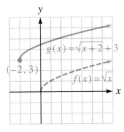

$g(x) = \sqrt{x+2} + 3$; $(-2, 3)$; $f(x) = \sqrt{x}$

54.

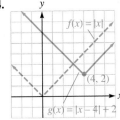

$f(x) = |x|$; $g(x) = |x-4| + 2$; $(4, 2)$

55.

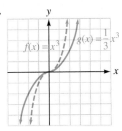

$f(x) = x^3$; $g(x) = \frac{1}{3}x^3$

56.

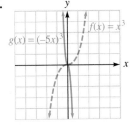

$g(x) = (-5x)^3$; $f(x) = x^3$

57.

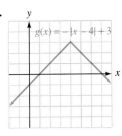

$g(x) = -|x-4| + 3$

58.

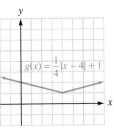

$g(x) = \frac{1}{4}|x-4| + 1$

59.

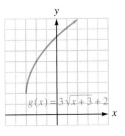

$g(x) = 3\sqrt{x+3} + 2$

60.

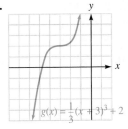

$g(x) = \frac{1}{3}(x+3)^3 + 2$

61.

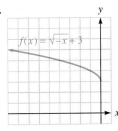

$f(x) = \sqrt{-x} + 3$

62.

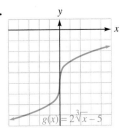

$g(x) = 2\sqrt[3]{x} - 5$

63. $(-\infty, -5) \cup (-5, 5) \cup (5, \infty)$ **64.** $(-\infty, \infty)$

65. $x = 1, x = -1$ **66.** $x = -7$ **67.** $x = 2, x = -3$

68. $x = 4, x = -1$ **69.** $y = \frac{1}{2}$ **70.** $y = -5$

71. $y = 0$ **72.** none   **73.** $y = 2x + 3$ **74.** none

75.

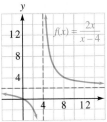

$f(x) = \frac{2x}{x-4}$

76.

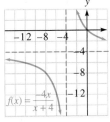

$f(x) = \frac{-4x}{x+4}$

77.

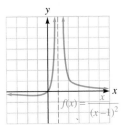

$f(x) = \frac{x}{(x-1)^2}$

78.

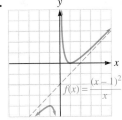

$f(x) = \frac{(x-1)^2}{x}$

79.

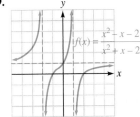

$f(x) = \frac{x^2 - x - 2}{x^2 + x - 2}$

80.

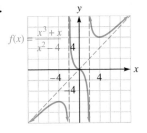

$f(x) = \frac{x^3 + x}{x^2 - 4}$

81. $(f + g)(x) = x^2 + 2x$; domain: $(-\infty, \infty)$

82. $(f \cdot g)(x) = 2x^3 + x^2 - 2x - 1$; domain: $(-\infty, \infty)$

83. $(f - g)(x) = x^2 - 2x - 2$; domain: $(-\infty, \infty)$

84. $(f/g)(x) = \frac{f(x)}{g(x)} = \frac{x^2 - 1}{2x + 1}$; domain: $\left(-\infty, -\frac{1}{2}\right) \cup \left(-\frac{1}{2}, \infty\right)$

85. 10   **86.** 60   **87.** 21   **88.** undefined

89. $(f \circ g)(x) = f(g(x)) = 4x^2 + 4x$; domain: $(-\infty, \infty)$

90. $(g \circ f)(x) = g(f(x)) = 2x^2 - 1$; domain: $(-\infty, \infty)$

91. 20   **92.** -2 **93.** $f(x) = x^2$; $g(x) = x - 5$
94. $f(x) = x^3$; $g(x) = x + 6$ **95.** not one-to-one
96. one-to-one   **97.** one-to-one   **98.** not one-to-one

101. $f^{-1}(x) = \dfrac{x + 1}{7}$ **102.** $f^{-1}(x) = 2 - \dfrac{1}{x}$

103. $f^{-1}(x) = \dfrac{x}{x + 1}$ **104.** $f^{-1}(x) = \sqrt[3]{\dfrac{3}{x}} = \dfrac{\sqrt[3]{3x^2}}{x}$

105. $f^{-1}(x) = \dfrac{x + 5}{2}$ **106.** $\left(-\infty, \dfrac{2}{5}\right) \cup \left(\dfrac{2}{5}, \infty\right)$

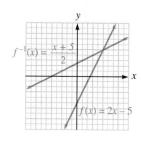

21. 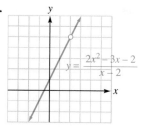   **22.**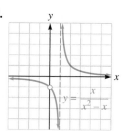

23. $(f + g)(x) = f(x) + g(x) = x^2 + 3x + 2$
24. $(g \circ f)(x) = g(f(x)) = 9x^2 + 2$
25. $(f/g)(x) = \dfrac{f(x)}{g(x)} = \dfrac{3x}{x^2 + 2}$
26. $(f \circ g)(x) = f(g(x)) = 3x^2 + 6$
27. $f^{-1}(x) = \dfrac{x + 1}{x - 1}$ **28.** $f^{-1}(x) = \sqrt[3]{x + 3}$
29. range: $(-\infty, -2) \cup (-2, \infty)$
30. range: $(-\infty, 3) \cup (3, \infty)$

Chapter Test

1. domain: $(-\infty, 5) \cup (5, \infty)$ **2.** domain: $[-3, \infty)$
3. $\dfrac{1}{2}, 2$ **4.** $\sqrt{6}, 3$ **5.** $(7, -3)$ **6.** $(1, -4)$
7. $(4, -10)$ **8.** $(-2, 9)$
9. 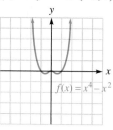   **10.**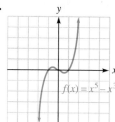

11. $\dfrac{25}{8}$ sec   **12.** $\dfrac{625}{4}$ ft   **13.** 10 ft   **14.** 110 ft
15. 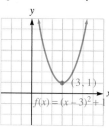   **16.**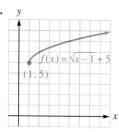

17. vertical asymptotes: $x = -3$ and $x = 3$; horizontal
asymptote: $y = 0$
18. vertical asymptote: $x = 3$; horizontal asymptote: none;
slant asymptote: $y = x - 2$
19. 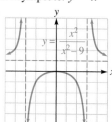   **20.**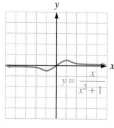

Cumulative Review

1. 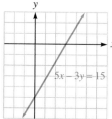   **2.**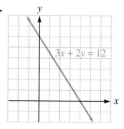

3. $\sqrt{41}$; $\left(\dfrac{1}{2}, \dfrac{3}{2}\right)$; $-\dfrac{4}{5}$ **4.** $2\sqrt{29}$; $(-2, 5)$; $\dfrac{2}{5}$
5. $y = -2x - 1$ **6.** $y = \dfrac{7}{2}x - \dfrac{11}{4}$ **7.** $y = \dfrac{3}{5}x + 6$
8. $y = -4x$
9. 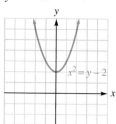   **10.**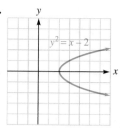

11. 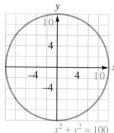   **12.**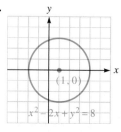

13. $x = 1$, $x = 10$ **14.** $x = 2$, $x = 8$ **15.** \$62.50
16. $\dfrac{25}{4}$ **17.** function   **18.** function   **19.** function
20. not a function   **21.** domain: $(-\infty, \infty)$
22. domain: $(-\infty, -2) \cup (-2, \infty)$ **23.** domain: $[2, \infty)$

24. domain: $[-4, \infty)$ **25.** $\left(-\dfrac{5}{2}, -\dfrac{49}{4}\right)$ **26.** $\left(\dfrac{5}{2}, \dfrac{49}{4}\right)$

27.
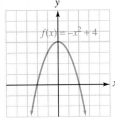

28.
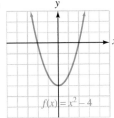

29.
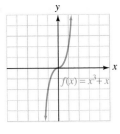

30.
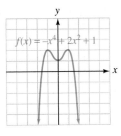

31.

32. $f(x) = \dfrac{x^2 - 1}{x^2 - 9}$

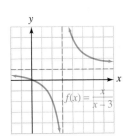
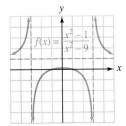

33. $(f + g)(x) = f(x) + g(x) = x^2 + 3x - 3$
domain: $(-\infty, \infty)$
34. $(f - g)(x) = f(x) - g(x) = -x^2 + 3x - 5$
domain: $(-\infty, \infty)$
35. $(f \cdot g)(x) = f(x) \cdot g(x) = 3x^3 - 4x^2 + 3x - 4$
domain: $(-\infty, \infty)$
36. $(f/g)(x) = \dfrac{f(x)}{g(x)} = \dfrac{3x - 4}{x^2 + 1}$ domain: $(-\infty, \infty)$
37. $(f \circ g)(2) = 11$ **38.** $(g \circ f)(2) = 5$
39. $(f \circ g)(x) = 3x^2 - 1$ **40.** $(g \circ f)(x) = 9x^2 - 24x + 17$
41. $f^{-1}(x) = \dfrac{x - 2}{3}$ **42.** $f^{-1}(x) = \dfrac{1}{x} + 3$
43. $f^{-1}(x) = \sqrt{x - 5}$ **44.** $f^{-1}(x) = \dfrac{x + 1}{3}$
45. $y = kwz$ **46.** $y = \dfrac{kx}{t^2}$

Section 4.1
1. exponential   **3.** $(-\infty, \infty)$ **5.** $(0, \infty)$ **7.** asymptote
9. increasing   **11.** 2.72   **13.** increasing   **15.** 11.0357
17. 451.8079   **19.** $5^{2\sqrt{2}} = 25^{\sqrt{2}}$ **21.** a^4 **23.** 1, 25
25. 1, 9

27.
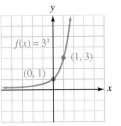

29.
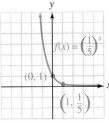

31.
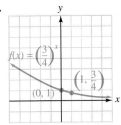

33.
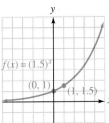

35.
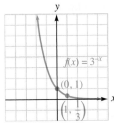

37.
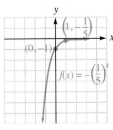

39. yes   **41.** no   **43.** $b = \dfrac{1}{2}$ **45.** no value of b
47. $b = 2$ **49.** $b = e$

51.
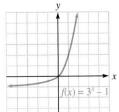

53.
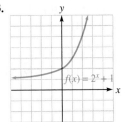

55.
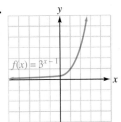

57.
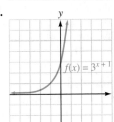

59.
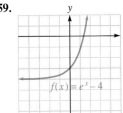

61.
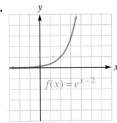

63.
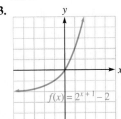

65.
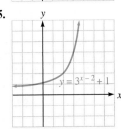

67.

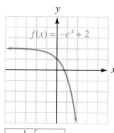

$f(x) = -3^x + 1$

69.

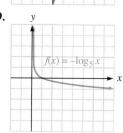

$f(x) = 2^{-x} - 3$

103. $b = 2$

105.

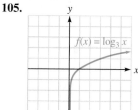

$f(x) = \log_3 x$

107.

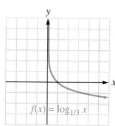

$f(x) = \log_{1/3} x$

71.

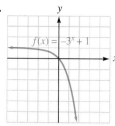

$f(x) = -e^x + 2$

73.

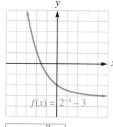

109.

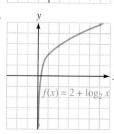

$f(x) = -\log_5 x$

111.

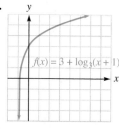

$f(x) = 2 + \log_2 x$

75.

77.

113.

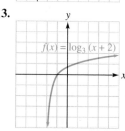

$f(x) = \log_3 (x + 2)$

115.

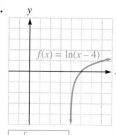

$f(x) = 3 + \log_3(x + 1)$

79.

81. $22,080.40 **83.** $15.79 **85.** $2,273,996.13
87. $1,263.77 **89.** $13,375.68 **91.** $7,647.95
from continuous compounding, $7,518.28 from annual
compounding **93.** $291.27 **95.** $12,155.61
97. 13,228 **101.** 125 **103.** $x^2(1 + 9x^2)$
105. $(x + 4)(x - 3)$

117.

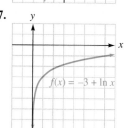

$f(x) = -3 + \ln x$

119.

$f(x) = \ln(x - 4)$

Section 4.2
1. birth; death **3.** 0.1868 g **5.** about 10 kg
7. about 35.4% **9.** 0.1575 unit **11.** 2 lumens
13. 8 lumens **15.** about 56,570 **17.** $61.9°C$
19. 315 **21.** 13 **23.** 10.6 billion **25.** 2.6
27. about 0.07% **29.** 0 **31.** 18,394
33. about 24,060 **35.** about 492 **37.** 19.0 mm
39. 49 mps **41.** about 7 million **43.** about 72.2 years
49. 8 **51.** 3 **53.** 2 **55.** 3

121.

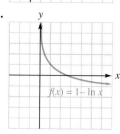

$f(x) = 1 - \ln x$

123.

Section 4.3
1. $x = b^y$ **3.** range **5.** inverse **7.** exponent
9. $(b, 1)$; $(1, 0)$ **11.** $\log_e x$ **13.** $(-\infty, \infty)$ **15.** 10
17. $\log_8 64 = 2$ **19.** $\log_4 \dfrac{1}{16} = -2$ **21.** $\log_{1/2} 32 = -5$
23. $\log_x z = y$ **25.** $3^4 = 81$ **27.** $\left(\dfrac{1}{2}\right)^3 = \dfrac{1}{8}$

125.

127.

129.

29. $4^{-3} = \dfrac{1}{64}$ **31.** $\pi^1 = \pi$ **33.** 3 **35.** −3 **37.** 3

39. $\dfrac{1}{2}$ **41.** −3 **43.** 64 **45.** 7 **47.** 5 **49.** $\dfrac{1}{25}$

131. b is larger. **135.** $\left(-\dfrac{7}{2}, -\dfrac{37}{4}\right)$
137. 425 ft by 850 ft **139.** $y = 5x$

51. $\dfrac{1}{6}$ **53.** 5 **55.** $\dfrac{3}{2}$ **57.** 4 **59.** $\dfrac{2}{3}$ **61.** 5
63. 4 **65.** 0.5119 **67.** −2.3307 **69.** 3.8221
71. −0.4055 **73.** 3.5596 **75.** 2.0664 **77.** −0.2752
79. undefined **81.** 25.2522 **83.** 1.9498×10^{-4}
85. 4.0645 **87.** 69.4079 **89.** 0.0245 **91.** 120.0719
93. 4 **95.** −3 **97.** 7 **99.** 4 **101.** $b = 2$

Section 4.4
1. $20 \log \dfrac{E_0}{E_I}$ **3.** $t = -\dfrac{1}{k} \ln\left(1 - \dfrac{C}{M}\right)$

5. $E = RT \ln\left(\dfrac{V_f}{V_i}\right)$ **7.** 55 dB **9.** 29 dB **11.** 49.5 dB

13. 4.4   **15.** 4   **17.** no   **19.** 19.8 min
21. about 5.8 yr   **23.** about 9.2 yr
25. about 3,654 joules   **27.** about 99% per year
29. 3 yr old   **31.** about 10.8 yr   **33.** about 208,000 V
35. 1 mi   **39.** $y = 7x + 3$ **41.** $x = 2$
43. $\dfrac{1}{2x - 3}$ **45.** $\dfrac{x + 1}{3(x - 2)}$

Section 4.5

1. 0   **3.** $M; N$ **5.** $x; y$ **7.** x **9.** $\neq$ **11.** 0
13. 7   **15.** 10   **17.** 1   **25.** $\log_b 2 + \log_b x + \log_b y$
27. $\log_b 2 + \log_b x - \log_b y$ **29.** $2 \log_b x + 3 \log_b y$
31. $\dfrac{1}{3}(\log_b x + \log_b y)$ **33.** $\log_b x + \dfrac{1}{2} \log_b z$
35. $\dfrac{1}{3} \log_b x - \dfrac{1}{3} \log_b y - \dfrac{1}{3} \log_b z$ **37.** $7 \ln x + 8 \ln y$
39. $\ln x - 4 \ln y - \ln z$ **41.** $\log_b \dfrac{x + 1}{x}$
43. $\log_b x^2 \sqrt[3]{y}$ **45.** $\log_b \dfrac{\sqrt{z}}{x^3 y^2}$ **47.** $\log_b \dfrac{\dfrac{x}{z} + x}{\dfrac{y}{z} + y} = \log_b \dfrac{x}{y}$
49. $\ln \dfrac{x(x + 5)}{9}$ **51.** $\ln \dfrac{z}{x^6 y^2}$ **53.** true   **55.** false
57. true   **59.** false   **61.** true   **63.** false   **65.** false
67. true   **69.** true   **71.** true   **73.** false   **75.** true
77. 1.4472   **79.** 0.3521   **81.** 1.1972   **83.** 2.4014
85. 2.0493   **87.** 0.4682   **89.** 1.7712   **91.** 0.9597
93. 1.8928   **95.** 2.3219   **97.** 7.20   **99.** 4.77
101. from 5.01×10^{-4} to 1.26×10^{-3} **103.** 19 dB
105. The original intensity must be raised to the 4th power.
107. The volume V is squared.   **121.** yes   **123.** no
125. $(-\infty, \infty)$ **127.** $(-\infty, \infty)$

Section 4.6

1. exponential   **3.** $A_0 2^{-t/h}$ **5.** 2   **7.** -2 **9.** $-\dfrac{5}{6}$
11. -4 **13.** $-\dfrac{15}{2}$ **15.** $3, -1$ **17.** ± 3
19. $-2, -1$ **21.** 3   **23.** ± 5 **25.** 1.1610
27. 1.2702   **29.** 1.7095   **31.** 0   **33.** ± 1.0878
35. $0, 1.0566$ **37.** $\ln 10$ **39.** $\dfrac{1}{2} \ln 6$ **41.** 0
43. 0.2789   **45.** $1, 3$ **47.** 0   **49.** $10, -10$ **51.** 4
53. e^6 **55.** $\dfrac{1}{2}(e^4 + 7)$ **57.** 7   **59.** 50   **61.** 20
63. 10   **65.** 7   **67.** 6   **69.** 5   **71.** 4   **73.** $3, 4$
75. 10^{10} **77.** no solution   **79.** 6   **81.** 9   **83.** 4
85. $7, 1$ **87.** 20   **89.** 1.81   **91.** about 5.1 yr
93. about 42.7 days   **95.** about 2,900 yr
97. about 5.6 yr   **99.** about 5.4 yr
101. because $\ln 2 \approx 0.70$ **103.** about 3.2 days
105. about 12 min   **107.** $k = \dfrac{\ln 0.75}{3}$ **109.** $\dfrac{\ln\left(\dfrac{2}{3}\right)}{5}$
117. $f^{-1}(x) = \dfrac{x - 2}{3}$ **119.** 19   **121.** $5x^2 - 1$

Chapter Review

1. $5^{2\sqrt{2}}$ **2.** $2^{\sqrt{10}}$

3. 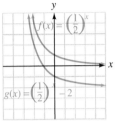   **4.**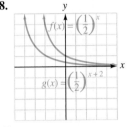

5. $p = 1, q = 7$ **6.** domain: $(-\infty, \infty)$; range: $(0, \infty)$

7. 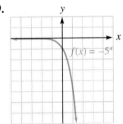   **8.**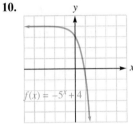

9. 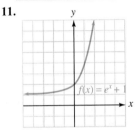   **10.**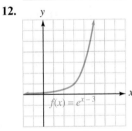

11. 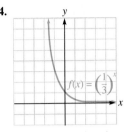   **12.**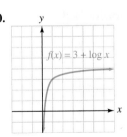

13. $2,189,703.45   **14.** $2,324,767.37   **15.** $\dfrac{2}{3}$
16. 0.19 lumen   **17.** about 635,000,000   **18.** about 2,708
19. $(0, \infty); (-\infty, \infty)$ **20.** $(0, \infty); (-\infty, \infty)$ **21.** 2
22. $-\dfrac{1}{2}$ **23.** 0   **24.** -2 **25.** $\dfrac{1}{2}$ **26.** $\dfrac{1}{3}$ **27.** 32
28. 9   **29.** 8   **30.** -1 **31.** $\dfrac{1}{8}$ **32.** 2   **33.** 4
34. 2   **35.** 10   **36.** $\dfrac{1}{25}$ **37.** 5   **38.** 3

39. 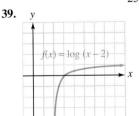   **40.**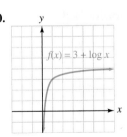

41.
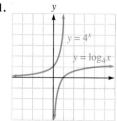

42.
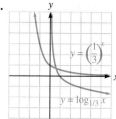

11.
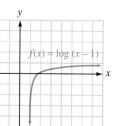

12.
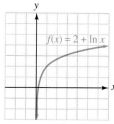

43. 6.1137　　**44.** -0.1111　　**45.** 10.3398　　**46.** 2.5715

47.
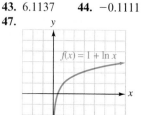

48.
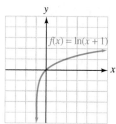

13. $2 \log a + \log b + 3 \log c$　　**14.** $\frac{1}{2}(\ln a - 2 \ln b - \ln c)$

15. $\log \dfrac{b\sqrt{a + 2}}{c^2}$　　**16.** $\ln \dfrac{\sqrt[3]{\dfrac{a}{b^2}}}{c}$　　**17.** 1.3801

18. 0.4259　　**19.** $\dfrac{\log 3}{\log 7}$ or $\dfrac{\ln 3}{\ln 7}$　　**20.** $\dfrac{\log e}{\log \pi}$ or $\dfrac{1}{\ln \pi}$

21. true　　**22.** false　　**23.** 6.4　　**24.** 46 dB　　**25.** $-1, 3$

26. $\dfrac{\log 3}{\log 3 - 2} \approx -0.3133$　　**27.** $\ln 9 \approx 2.1972$　　**28.** 1

29. 10　　**30.** 9

49. 12　　**50.** $14x$　　**51.** 53 dB　　**52.** 4.4　　**53.** $9\frac{1}{2}$ min

54. 23 yr　　**55.** 2,017 joules　　**56.** 0　　**57.** 1　　**58.** 3

59. 4　　**60.** 4　　**61.** 0　　**62.** 7　　**63.** 3　　**64.** 4

65. 9　　**66.** $2 \log_b x + 3 \log_b y - 4 \log_b z$

67. $\frac{1}{2}(\log_8 x - \log_8 y - 2 \log_8 z)$

68. $4 \ln x - 5 \ln y - 6 \ln z$　　**69.** $\frac{1}{3}(\ln x + \ln y + \ln z)$

70. $\log_b \dfrac{x^3 z^7}{y^5}$　　**71.** $\log_b \dfrac{\sqrt{xy^3}}{z^7}$　　**72.** $\ln \dfrac{x^4}{y^5 z^6}$

73. $\ln \dfrac{y^3 \sqrt{x}}{\sqrt[3]{z}}$　　**74.** 3.36　　**75.** 1.56　　**76.** 2.64

77. -6.72　　**78.** 1.7604

79. about 7.94×10^{-4} gram-ions per liter　　**80.** $k \ln 2$ less

81. $-\dfrac{5}{4}$　　**82.** $-1, -3$　　**83.** 2　　**84.** $3, -3$

85. $\dfrac{\log 7}{\log 3} \approx 1.7712$　　**86.** $\dfrac{\log 3}{\log 3 - \log 2} \approx 2.7095$

87. $\ln 8 \approx 2.0794$　　**88.** $\ln 7 \approx 1.9459$　　**89.** -3

90. 8　　**91.** 25, 4　　**92.** 4　　**93.** 2　　**94.** 4, 3　　**95.** 6

96. 31　　**97.** $\dfrac{\ln 9}{\ln 2} \approx 3.1699$　　**98.** no solution

99. $e^7 \approx 1,096.6332$　　**100.** $\dfrac{e}{e - 1} \approx 1.5820$　　**101.** 1

102. about 3,300 yr

Chapter Test

1.
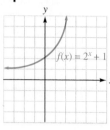

2.
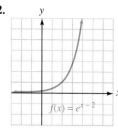

3. $\dfrac{3}{64}$ g　　**4.** \$1,060.90　　**5.** \$4,451.08　　**6.** 3

7. -3　　**8.** 17　　**9.** 2　　**10.** -3

Section 5.1

1. whole　　**3.** any　　**5.** factor

7. $4x^2 + 2x + 1 + \dfrac{2}{x - 1}$

9. $2x^3 - 3x^2 + 8x - 1 + \dfrac{-3}{x + 2}$

11. -1　　**13.** 9　　**15.** 3　　**17.** -3　　**19.** 69

21. 70,249　　**23.** 128.3085938　　**25.** true　　**27.** true

29. false　　**31.** true　　**33.** $(x - 1)(3x^2 + x - 5) - 9$

35. $(x - 3)(3x^2 + 7x + 15) + 41$

37. $(x + 1)(3x^2 - 5x - 1) - 3$

39. $(x + 3)(3x^2 - 11x + 27) - 85$

41. $x^2 + 2x + 3$　　**43.** $7x^2 - 10x + 5 + \dfrac{-4}{x + 1}$

45. $4x^3 + 9x^2 + 27x + 80 + \dfrac{245}{x - 3}$

47. $3x^4 + 12x^3 + 48x^2 + 192x$　　**49.** 47　　**51.** -569

53. $-1 - 6i$　　**55.** $\dfrac{15}{8}$　　**57.** 5　　**59.** 0　　**61.** 384

63. $-28 - 16i$　　**65.** $16 + 2i$　　**67.** $40 - 40i$

69. $\{-1, -5, 3\}$　　**71.** $\left\{-\dfrac{1}{2}, 3, -3\right\}$

73. $\left\{2, 2 + \sqrt{5}, 2 - \sqrt{5}\right\}$

75. $\left\{-3, 3 + \sqrt{10}i, 3 - \sqrt{10}i\right\}$

77. $\left\{1, 1, \sqrt{3}, -\sqrt{3}\right\}$　　**79.** $\{2, 3, i, -i\}$

81. $P(x) = x^2 - 9x + 20$　　**83.** $P(x) = x^3 - 3x^2 + 3x - 1$

85. $P(x) = x^3 - 11x^2 + 38x - 40$

87. $P(x) = x^4 - 3x^2 + 2$

89. $P(x) = x^3 - \sqrt{2}x^2 + x - \sqrt{2}$

91. $P(x) = x^3 - 2x^2 + 2x$　　**93.** 0　　**97.** QIV　　**99.** QI

101. 10　　**103.** 1

Section 5.2

1. zero   **3.** conjugate   **5.** 0   **7.** 2
9. lower bound   **11.** 10   **13.** 4   **15.** 4   **17.** 4, 4
19. 5, 5   **21.** $x^2 + 4 = 0$ **23.** $x^2 - 6x + 10 = 0$
25. $x^3 - 3x^2 + x - 3 = 0$
27. $x^3 - 6x^2 + 13x - 10 = 0$
29. $x^4 - 5x^3 + 7x^2 - 5x + 6 = 0$
31. $x^4 - 2x^3 + 3x^2 - 2x + 2 = 0$
33. $x^2 - 4xi - 4 = 0$
35. 0 or 2 positive; 1 negative; 0 or 2 nonreal
37. 0 positive; 1 or 3 negative; 0 or 2 nonreal
39. 0 positive; 0 negative; 4 nonreal
41. 1 positive; 1 negative; 2 nonreal
43. 0 positive; 0 negative; 10 nonreal
45. 0 positive; 0 negative; 8 nonreal; yes
47. 1 positive; 1 negative; 2 nonreal   **49.** $(-2, 4)$
51. $(-1, 1)$ **53.** $(-4, 6)$ **55.** $(-4, 3)$ **57.** $(-2, 4)$
63. k units to the right   **65.** reflected about the y-axis
67. stretched vertically by a factor of k

Section 5.3

1. -7 **3.** root   **5.** $\pm 1, \pm 2, \pm 3, \pm 4, \pm 6, \pm 12$
7. $\pm 1, \pm 2, \pm 3, \pm 6, \pm \dfrac{1}{2}, \pm \dfrac{3}{2}$
9. $\pm 1, \pm 2, \pm 5, \pm 10, \pm \dfrac{1}{2}, \pm \dfrac{5}{2}, \pm \dfrac{1}{4}, \pm \dfrac{5}{4}$
11. 1, -1, 5   **13.** 1, 2, -1 **15.** 1, 2, -2 **17.** 3, -3, 2
19. 1, -1, $\dfrac{1}{2}$ **21.** -1, -1, $\dfrac{1}{3}$ **23.** $\dfrac{3}{2}, \dfrac{2}{3}, -\dfrac{3}{5}$
25. $\dfrac{2}{3}, -\dfrac{3}{5}, 4$ **27.** $\dfrac{3}{2}, \dfrac{2}{3}, \dfrac{5}{4}$ **29.** 1, 2, 3, 4   **31.** 2, -5
33. 1, -1, $\dfrac{1}{2}, \dfrac{3}{2}$ **35.** $-2, \dfrac{5}{2}$ **37.** $\dfrac{1}{3}, -\dfrac{1}{2}$
39. 1, -1, 2, -2, -3 **41.** 3, $\dfrac{1}{2}, -\dfrac{1}{2}$
43. 0, 2, 2, 2, -2, -2, -2 **45.** 3, $\sqrt{2}, -\sqrt{2}$
47. $\dfrac{1}{2}, i, -i$ **49.** 2, -2, $1 + \sqrt{5}, 1 - \sqrt{5}$
51. $\dfrac{1}{2}, -1, 3i, -3i$ **53.** 1, 1, 1, 5i, $-5i$
55. $\dfrac{3}{2}, 1, -1, 2i, -2i$ **57.** 3, $1 - i$ **59.** 3, -3, $1 - i$
61. $-\dfrac{2}{3}, 3, -1$ **63.** $\dfrac{1}{2}, \dfrac{1}{2}, \dfrac{1}{2}, 1, 1$ **65.** 10, 20, 60 ohms
67. 13 in.   **69.** 1, 5, and 9 miles
73. (3, 7) or approx. (1.54, 13.63)   **75.** $6ab^2\sqrt{2abc}$
77. $8a\sqrt{2b}$ **79.** $\sqrt{3} + 1$

Section 5.4

1. $P(a)$ and $P(b)$ **3.** x_l and x_r
9. $P(-2) = 3$; $P(-1) = -2$ The signs of the results are opposites.
11. $P(4) = -40$; $P(5) = 30$ The signs of the results are opposites.
13. $P(1) = 8$; $P(2) = -1$ The signs of the results are opposites.

15. $P(2) = -72$; $P(3) = 154$ The signs of the results are opposites.
17. $P(0) = 10$; $P(1) = -60$ The signs of the results are opposites.
19. 1.7   **21.** -2.2 **23.** 1.7   **25.** -1.2 **27.** $-2.2, 2.2$
29. 1, 2; The Bisection Method fails to find the solution 2.
31. yes   **33.** (0.83, 0.56), $(-0.83, -0.56)$
37. vertical: $x = 2$, $x = -2$; horizontal: $y = 0$; slant: none
39. vertical: $x = 2$; horizontal: none; slant: $y = x + 2$

Chapter Review

1. 1   **2.** 66   **3.** 241   **4.** 34   **5.** false   **6.** false
7. true   **8.** true   **9.** $3x^3 + 9x^2 + 29x + 90 + \dfrac{277}{x - 3}$
10. $2x^3 + 4x^2 + 5x + 13 + \dfrac{25}{x - 2}$
11. $5x^4 - 14x^3 + 31x^2 - 64x + 129 + \dfrac{-259}{x + 2}$
12. $4x^4 - 2x^3 + x^2 + 2x + \dfrac{1}{x + 1}$ **13.** 151   **14.** -113
15. $\dfrac{13}{8}$ **16.** $-1 - 6i$ **17.** $\left\{ 3, \dfrac{1}{2}, -2 \right\}$
18. $\left\{ -2, -2, \sqrt{5}, -\sqrt{5} \right\}$
19. $P(x) = 2x^3 - 5x^2 - x + 6$
20. $P(x) = 2x^3 + 3x^2 - 8x + 3$
21. $P(x) = x^4 + 3x^3 - 9x^2 + 3x - 10$
22. $P(x) = x^4 + x^3 - 5x^2 + x - 6$ **23.** 6   **24.** 6
25. 65   **26.** 1,984   **27.** 4, 4   **28.** 40, 40   **29.** 5, 5
30. 3, 3   **31.** $2 - i$ **32.** i
33. $x^3 - 4x^2 + x - 4 = 0$ **34.** $x^3 + 5x^2 + x + 5 = 0$
35. 0 or 2 positive; 0 or 2 negative; 0, 2, or 4 nonreal
36. 1 or 3 positive; 1 negative; 0 or 2 nonreal
37. 1 positive; 0, 2, or 4 negative; 0, 2, or 4 nonreal
38. 1 or 3 positive; 0 or 2 negative; 2, 4, or 6 nonreal
39. 0 positive; 0 negative; 4 nonreal
40. 0 positive; 1 negative; 6 nonreal
41. $(-1, 2)$ **42.** $(-5, 2)$ **43.** $\pm 1, \pm 2, \pm 3, \pm 6, \pm \dfrac{1}{2}, \pm \dfrac{3}{2}$
44. $\pm 1, \pm 2, \pm 5, \pm 10, \pm \dfrac{1}{2}, \pm \dfrac{5}{2}, \pm \dfrac{1}{4}, \pm \dfrac{5}{4}$ **45.** 1, 4, 5
46. 1, -1, 8   **47.** $-5, -\dfrac{3}{2}, -2$ **48.** $\dfrac{1}{3}$
49. 2, -2, $\dfrac{3}{2}, -\dfrac{3}{2}$ **50.** 4, 4, -2, $-\dfrac{1}{2}$ **51.** $\dfrac{1}{3}, 4i, -4i$
52. 2, -2, $1 + \sqrt{6}, 1 - \sqrt{6}$ **53.** $P(-1) = -9$; $P(0) = 18$
54. $P(1) = -8$; $P(2) = 21$ **55.** 4.2   **56.** 2.5
57. 1.67   **58.** 0.67
59. 10 m by 6 m; 12 m by 5 m; 15 m by 4 m   **60.** 17.27 ft

Chapter Test

1. yes   **2.** 1   **3.** -30 **4.** no
5. $(x - 2)(2x^2 + x - 2) - 5$
6. $(x + 1)(2x^2 - 5x + 1) - 2$ **7.** $2x + 3$ **8.** $3x^2 + x$
9. 5   **10.** -28 **11.** $\dfrac{11}{3}$ **12.** $6 - 3i$
13. $P(x) = x^3 - 4x^2 - 5x$ **14.** $P(x) = x^4 - 2x^2 - 3$
15. $x^3 - 2x^2 + x - 2 = 0$ **16.** $x^3 - 5x^2 + 9x - 5 = 0$
17. 3, 3   **18.** $3 + 2i$

19. 1 or 3 positive; 0 or 2 negative; 0, 2, or 4 nonreal
20. 1 positive; 0 or 2 negative; 0 or 2 nonreal   **21.** $(-3, 3)$
22. $(-1, 6)$

23. $\pm 1, \pm 2, \pm 4, \pm 8$ **24.** $\pm 1, \pm 2, \pm\dfrac{1}{5}, \pm\dfrac{2}{5}$
25. $2, -3, -\dfrac{1}{2}$ **26.** $2, i, -i$ **27.** $2, -3, 3i, -3i$
28. $1, -1, -2, 2i, -2i$ **29.** no   **30.** 3.3

Cumulative Review

1.
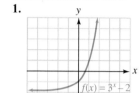

2.
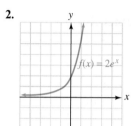
$f(x) = 2e^x$

3.
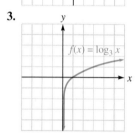
$f(x) = \log_3 x$

4.
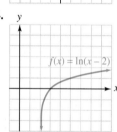
$f(x) = \ln(x-2)$

5. 6   **6.** -3 **7.** 3   **8.** 2
9. $\log a + \log b + \log c$ **10.** $2\log a + \log b - \log c$
11. $\dfrac{1}{2}(\log a + \log b - 3\log c)$ **12.** $\dfrac{1}{2}\ln a + \ln b - \ln c$
13. $\ln\dfrac{a^3}{b^3}$ **14.** $\log\dfrac{\sqrt{ab^3}}{\sqrt[3]{c^2}}$ **15.** $x = \dfrac{\log 8}{\log 3} - 1$ **19.** 9
16. $x = -1$ **17.** $x = 500$ **18.** $x = \sqrt{11}$
20. -36 **21.** 4   **22.** $2 - i$ **23.** a factor
24. not a factor   **25.** a factor   **26.** a factor   **27.** 12
28. 2,000   **29.** 2 or 0 positive; 2 or 0 negative; 4, 2, or 0
nonreal   **30.** 1 positive; 3 or 1 negative; 2 or 0 nonreal
31. $-1, -3, 3$ **32.** $-1, 1, 2$

Section 6.1

1. system   **3.** consistent   **5.** independent
7. consistent   **9.** dependent   **11.** is
13.
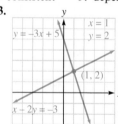

15.
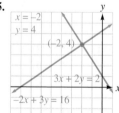

17.
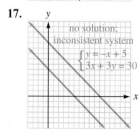

19.
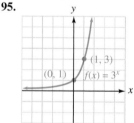

21. $(2.2, -4.7)$ **23.** $(1.7, 0.3)$ **25.** $(-1, -2)$
27. $(3, -2)$ **29.** $\left(\dfrac{1}{2}, \dfrac{1}{3}\right)$ **31.** no solution; inconsistent
system   **33.** dependent equations; a general solution is
$(x, 3x - 6)$ **35.** $(3, 1)$ **37.** $(3, 2)$ **39.** $(-3, 0)$
41. $\left(1, -\dfrac{1}{2}\right)$ **43.** dependent equations; a general solution
is $(x, 5 - 2x)$ **45.** no solution; inconsistent system
47. $(4, -7)$ **49.** $(2, -3)$ **51.** $(9, -1)$ **53.** $(1, 2, 0)$
55. $\left(0, -\dfrac{1}{3}, -\dfrac{1}{3}\right)$ **57.** $(1, 2, -1)$ **59.** $(1, 0, 5)$
61. no solution; inconsistent system   **63.** $(0, 1, 0)$
65. $\left(\dfrac{2}{3}, \dfrac{1}{4}, \dfrac{1}{2}\right)$ **67.** $\left(\dfrac{1}{2}, \dfrac{1}{2}, \dfrac{1}{2}\right)$
69. dependent equations; a general solution is $(x, 2 - x, 1)$
71. hamburger $2; fries $1
73. 225 acres of corn; 125 acres of soybeans   **75.** 8 kph
77. 40 g and 20 g   **79.** 10 ft
81. $E(x) = 43.53x + 742.72$, $R(x) = 89.95x$; 16 pairs/day
83. 15 hr cooking hamburgers, 10 hr pumping gas, 5 hr
janitorial   **85.** 1.05 million in $0-14$ group, 1.56 million in
$15-49$ group, 0.39 million in 50-and-older group
87. $30°, 50°, 100°$
95.

97. 8   **99.** $\log x - 2\log y - \log z$ **101.** $\log\dfrac{xy^3}{\sqrt{z}}$

Section 6.2

1. matrix   **3.** coefficient   **5.** equation
7. row equivalent   **9.** interchanged
11. adding, multiple   **13.** $(2, 3)$ **15.** $(7, 6)$
17. $(1, 2, 3)$ **19.** $(1, 1, 3)$ **21.** row echelon form
23. reduced row echelon form   **25.** $(2, -1)$
27. $(-2, 0)$ **29.** $(3, 1)$
31. no solution; inconsistent system   **33.** $(0, -7)$
35. $(1, 0, 2)$ **37.** $(2, -2, 1)$ **39.** $(1, 1, 2)$

41. dependent equations; a general solution is
$\left(x, -x + \dfrac{7}{2}, -\dfrac{3}{2}\right)$ **43.** no solution; inconsistent system
45. $(13, 3)$ **47.** $(-13, 7, -2)$ **49.** $(10, 3)$
51. $(0, 0)$ **53.** $(3, 1, -2)$ **55.** $\left(\dfrac{1}{4}, 1, \dfrac{1}{4}\right)$

57. $(1, 2, 1, 1)$ **59.** $(1, 2, 0, 1)$ **61.** $\left(\dfrac{9}{4}, -3, \dfrac{3}{4}\right)$

63. $\left(0, \dfrac{20}{3}, -\dfrac{1}{3}\right)$ **65.** dependent equations; a general

solution is $\left(x, \dfrac{1}{2}x - 3, 0\right)$ **67.** $(1, -3)$

69. dependent equations; a general solution is

$\left(\dfrac{8}{7} + \dfrac{1}{7}z, \dfrac{10}{7} - \dfrac{4}{7}z, z\right)$ **71.** dependent equations; a general

solution is $(1 + z, -z, -1 - z, z)$

73. no solution; inconsistent system   **75.** 1,300 mi

77. Dictionaries 4.5 in; atlases 3.5 in; thesauruses 4 in

79. 2 niacin, 4 zinc, 6 vitamin C

85. $x = \pm 2; y = \pm 1; z = \pm 3$ **87.** $y = mx + b$

89. equal   **91.** $y = 2x + 7$ **93.** $x = 2$

Section 6.3

1. i, j **3.** corresponding   **5.** columns, rows

7. additive identity   **9.** $x = 2, y = 5$ **11.** $x = 1, y = 2$

13. $\begin{bmatrix} -1 & 2 & 1 \\ -6 & 0 & 0 \end{bmatrix}$ **15.** $\begin{bmatrix} -5 & 2 & -7 \\ 5 & 0 & -3 \\ 2 & -3 & 5 \end{bmatrix}$

17. $\begin{bmatrix} -6 & 5 & 0 \\ 1 & -1 & 0 \end{bmatrix}$ **19.** $\begin{bmatrix} 15 & -15 \\ 0 & -10 \end{bmatrix}$

21. $\begin{bmatrix} 25 & 75 & -10 \\ -10 & -25 & 5 \end{bmatrix}$ **23.** $\begin{bmatrix} 18 & -1 & -4 \\ -35 & 0 & -1 \end{bmatrix}$

25. $\begin{bmatrix} 2 & -2 \\ 3 & 10 \end{bmatrix}$ **27.** $\begin{bmatrix} -22 & -22 \\ -105 & 126 \end{bmatrix}$

29. $\begin{bmatrix} 4 & 2 & 10 \\ 5 & -2 & 4 \\ 2 & -2 & 1 \end{bmatrix}$ **31.** $\begin{bmatrix} 4 & -5 & -6 \\ -8 & 10 & 12 \\ -12 & 15 & 18 \end{bmatrix}$

33. not possible   **35.** $\begin{bmatrix} 16 \\ 12 \\ 12 \end{bmatrix}$ **37.** not possible

39. $\begin{bmatrix} -36.29 \\ 16.2 \\ -19.26 \end{bmatrix}$ **41.** $\begin{bmatrix} -16.11 & 4.71 & 33.64 \\ -19.6 & 20.35 & 6.4 \\ -100.6 & 72.82 & 62.71 \end{bmatrix}$

47. $\begin{bmatrix} 4 & 5 \\ -7 & -1 \end{bmatrix}$ **49.** not possible

51. $\begin{bmatrix} 24 & 16 \\ 39 & 26 \end{bmatrix}$ **53.** $\begin{bmatrix} 47 \\ 81 \end{bmatrix}$

55. $QC = \begin{bmatrix} 2,000 \\ 1,700 \end{bmatrix}$ The cost to Supplier 1 is \$2,000. The cost to Supplier 2 is \$1,700.

57. $QP = \begin{bmatrix} 584.50 \\ 709.25 \\ 1,036.75 \end{bmatrix}$ Adult males spent \$584.50. Adult females spent \$709.25. Children spent \$1,036.75.

59. $\begin{bmatrix} 1 & 1 & 0 \\ 0 & 1 & 1 \\ 1 & 0 & 0 \end{bmatrix}$

61. $A^2 = \begin{bmatrix} 5 & 1 & 2 & 2 \\ 1 & 5 & 2 & 2 \\ 2 & 2 & 6 & 0 \\ 2 & 2 & 0 & 4 \end{bmatrix}$ indicates the number of ways two cities can be linked with exactly one intermediate city to relay messages.

63. No; if $A = \begin{bmatrix} 1 & 1 \\ 1 & 1 \end{bmatrix}$ and $B = \begin{bmatrix} 1 & 0 \\ 0 & 0 \end{bmatrix}$, then $(AB)^2 \neq A^2 B^2$.

65. Let $A = \begin{bmatrix} 1 & 2 \\ 1 & 2 \end{bmatrix}$ and $B = \begin{bmatrix} 2 & 2 \\ -1 & -1 \end{bmatrix}$. Neither is the zero matrix, yet $AB = 0$.

67. $6x^2 - 4x - 8$ **69.** $\dfrac{x+1}{x-1}$ **71.** $a = \dfrac{2s}{n} - l$

Section 6.4

1. $AB = BA = I$ **3.** $[I \mid A^{-1}]$ **5.** $\begin{bmatrix} 3 & 4 \\ 2 & 3 \end{bmatrix}$

7. $\begin{bmatrix} 5 & -7 \\ -2 & 3 \end{bmatrix}$ **9.** $\begin{bmatrix} -2 & 3 & -3 \\ -5 & 7 & -6 \\ 1 & -1 & 1 \end{bmatrix}$

11. $\begin{bmatrix} 4 & 1 & -3 \\ -5 & -1 & 4 \\ -1 & -1 & 1 \end{bmatrix}$ **13.** no inverse

15. $\begin{bmatrix} 1 & -2 & 1 \\ 0 & 1 & -2 \\ 0 & 0 & 1 \end{bmatrix}$ **17.** no inverse

19. $\begin{bmatrix} 1 & -2 & 1 & 0 \\ 0 & 1 & -2 & 1 \\ 0 & 0 & 1 & -2 \\ 0 & 0 & 0 & 1 \end{bmatrix}$ **21.** $\begin{bmatrix} 8 & -2 & -6 \\ -5 & 2 & 4 \\ 2 & 0 & -2 \end{bmatrix}$

23. $\begin{bmatrix} -2.5 & 5 & 3 & 5.5 \\ 5.5 & -8 & -6 & -9.5 \\ -1 & 3 & 1 & 3 \\ -5.5 & 9 & 6 & 10.5 \end{bmatrix}$

25. $x = 23, y = 17$ **27.** $x = 0, y = 0$

29. $x = 1, y = 2, z = 2$ **31.** $x = 54, y = -37, z = -49$

33. $x = 2, y = 1$ **35.** $x = 1, y = 2, z = 1$

37. 2 of model A, 3 of model B   **39.** hi   **41.** no

45. $X = \begin{bmatrix} 0 \\ 0 \\ 0 \end{bmatrix}$ **51.** $(-\infty, -2) \cup (-2, 2) \cup (2, \infty)$

53. $(-\infty, \infty)$ **55.** $[0, \infty)$ **57.** $(-\infty, \infty)$

Section 6.5

1. $|A|$, $\det(A)$ **3.** 0   **5.** 0   **7.** 8   **9.** 1

11. -42 **13.** 13   **15.** 42   **17.** 13   **19.** -54

21. -7 **23.** 86   **25.** -2 **27.** 2   **29.** 120

31. true   **33.** false   **35.** 3   **37.** 3   **39.** $(1, 2)$

41. $(3, 0)$ **43.** $(1, 0, 1)$ **45.** $(1, -1, 2)$ **47.** $(6, 6, 12)$

49. $\left(\dfrac{5}{6}, \dfrac{2}{3}, \dfrac{1}{2}, \dfrac{5}{2}\right)$ **51.** $3x - 2y = 0$ **53.** $6x + 7y = 9$

55. 30 sq. units   **57.** 73 sq. units   **63.** 8   **65.** -1

67. \$5,000 in HiTech, \$8,000 in SaveTel, \$7,000 in OilCo

69. 10   **71.** 24   **73.** 8

77. domain: $n \times n$ matrices; range: all real numbers

79. yes   **81.** 21.468   **83.** $(x - 1)(x + 4)$

85. $x(3x + 1)(3x - 1)$ **87.** $\dfrac{4x - 5}{(x - 2)(2x - 1)}$

89. $\dfrac{x^2 + 2x + 1}{x(x^2 + 1)}$

Section 6.6

1. first-degree; second-degree   **3.** $\dfrac{1}{x} + \dfrac{2}{x - 1}$

5. $\dfrac{5}{x} - \dfrac{3}{x - 3}$ **7.** $\dfrac{1}{x + 1} + \dfrac{2}{x - 1}$ **9.** $\dfrac{2}{x} - \dfrac{2}{x - 2}$

11. $\dfrac{1}{x - 3} - \dfrac{3}{x + 2}$ **13.** $\dfrac{8}{x + 3} - \dfrac{5}{x - 1}$

15. $\dfrac{5}{2x - 3} + \dfrac{2}{x - 5}$ **17.** $\dfrac{2}{x} + \dfrac{3}{x - 1} - \dfrac{1}{x + 1}$

19. $\dfrac{1}{x} + \dfrac{1}{x^2 + 3}$ **21.** $\dfrac{3}{x + 1} + \dfrac{2}{x^2 + 2x + 3}$

23. $\dfrac{3}{x} + \dfrac{2}{x + 1} + \dfrac{1}{(x + 1)^2}$ **25.** $\dfrac{1}{x} + \dfrac{2}{x^2} - \dfrac{3}{x - 1}$

27. $\dfrac{2}{x} + \dfrac{1}{x - 3} + \dfrac{2}{(x - 3)^2}$ **29.** $\dfrac{1}{x - 1} - \dfrac{4}{(x - 1)^3}$

31. $\dfrac{1}{x} + \dfrac{1}{x^2} + \dfrac{2}{x^2 + x + 1}$ **33.** $\dfrac{3}{x} + \dfrac{4}{x^2} + \dfrac{x + 1}{x^2 + 1}$

35. $-\dfrac{1}{x + 1} - \dfrac{3}{x^2 + 2}$ **37.** $\dfrac{x + 1}{x^2 + 2} + \dfrac{2}{x^2 + x + 2}$

39. $\dfrac{1}{x} + \dfrac{x}{x^2 + 2x + 5} + \dfrac{x + 2}{(x^2 + 2x + 5)^2}$

41. $x - 3 - \dfrac{1}{x + 1} + \dfrac{8}{x + 2}$ **43.** $1 + \dfrac{2}{3x + 1} + \dfrac{1}{x^2 + 1}$

45. $1 + \dfrac{1}{x} + \dfrac{x}{x^2 + x + 1}$ **47.** $2 + \dfrac{1}{x} + \dfrac{3}{x - 1} + \dfrac{2}{x^2 + 1}$

49. No, it's the sum of two cubes.   **51.** $2|a|\sqrt{2ab}$

53. $3x^2\sqrt{2x}$ **55.** $x = 9$

Section 6.7

1. half-plane; boundary   **3.** is not

5.
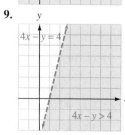

7.
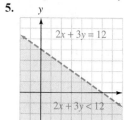

9.
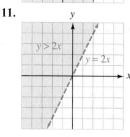

11.
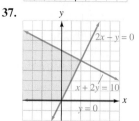

13.
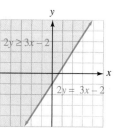

15.
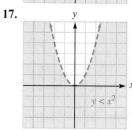

17.
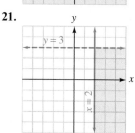

19.
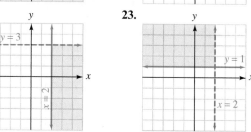

21.
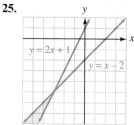

23.
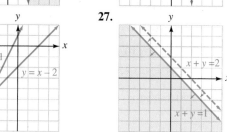

25.
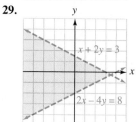

27.
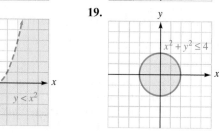

29.
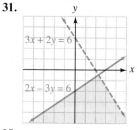

31.
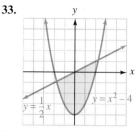

33.
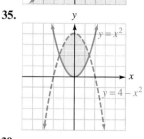

35.

37.
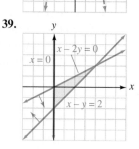

39.

41.

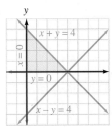

43.

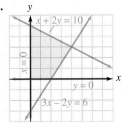

45. $\begin{cases} 6s + 4l \le 60 \\ s \ge 0 \\ l \ge 0 \end{cases}$ **53.** one; one   **55.** 0

Section 6.8

1. constraints   **3.** objective   **5.** $P = 12$ at $(0, 4)$
7. $P = \dfrac{13}{6}$ at $\left(\dfrac{5}{3}, \dfrac{4}{3}\right)$ **9.** $P = \dfrac{18}{7}$ at $\left(\dfrac{3}{7}, \dfrac{12}{7}\right)$
11. $P = 3$ at $(1, 0)$ **13.** $P = 0$ at $(0, 0)$
15. $P = 0$ at $(0, 0)$ **17.** $P = -12$ at $(-2, 0)$
19. $P = -2$ at the edge joining $(1, 2)$ and $(-1, 0)$
21. 3 tables, 12 chairs; $1,260
23. 30 IBMs, 30 Apple; $2,700
25. 15 DVRs, 30 TVs; $1,560
27. $150,000 in stocks, $50,000 in bonds; $17,000
29. 2 buses, 2 trucks; $1,100

33. $\begin{bmatrix} 1 & 0 & 0 \\ 0 & 1 & 0 \\ 0 & 0 & 1 \\ 0 & 0 & 0 \end{bmatrix}$ **35.** $\left(\dfrac{1}{3} - 3y, y, -\dfrac{1}{3}\right)$

Chapter Review

1.

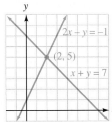

2.

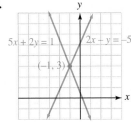

3.

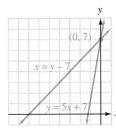

4.

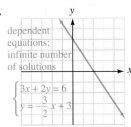

a general solution is
$\left(x, -\dfrac{3}{2}x + 3\right)$

5.

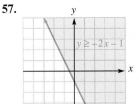

6. $x = 2, y = -1$ **7.** $x = 0, y = -3$ **8.** $x = 1, y = 1$
9. dependent equations; a general solution is $(x, 3x - 4)$
10. no solution; inconsistent system   **11.** $x = -3, y = 2$
12. $x = -2, y = 5$ **13.** $x = 2, y = -1$
14. no solution; inconsistent system   **15.** dependent
equations; a general solution is $(x, 3x - 4)$
16. $x = 1, y = 0, z = 1$ **17.** $x = 1, y = 1, z = -1$
18. $x = 0, y = 1, z = 2$ **19.** $10,400
20. 900 adult tickets, 450 senior tickets, 450 children's tickets
21. $x = 1, y = 1$ **22.** dependent equations; a general
solution is $(x, 3x + 4)$ **23.** $x = 3, y = 1, z = -2$
24. $x = -10, y = 1, z = 10$
25. no solution; inconsistent system
26. $x = -4, y = 3$ **27.** $\begin{bmatrix} 1 & 3 & 4 \\ 4 & 0 & 2 \end{bmatrix}$

28. $\begin{bmatrix} 2 & 5 & 4 \\ -2 & -6 & 6 \\ -4 & 5 & -3 \end{bmatrix}$ **29.** $\begin{bmatrix} 4 & -1 \\ -7 & -7 \end{bmatrix}$

30. $\begin{bmatrix} -17 & 19 \\ 10 & -12 \end{bmatrix}$ **31.** $[5]$ **32.** $\begin{bmatrix} 2 & -1 & 1 & 3 \\ 4 & -2 & 2 & 6 \\ 2 & -1 & 1 & 3 \\ 10 & -5 & 5 & 15 \end{bmatrix}$

33. not possible   **34.** $[-24]$ **35.** $\begin{bmatrix} 0 \\ -6 \end{bmatrix}$

36. $\begin{bmatrix} 5 & -3 \\ -3 & 2 \end{bmatrix}$ **37.** $\begin{bmatrix} 1 & 0 & 0 \\ -\dfrac{3}{2} & \dfrac{1}{2} & \dfrac{1}{2} \\ 1 & -\dfrac{1}{2} & 0 \end{bmatrix}$

38. $\begin{bmatrix} 9 & 16 & -56 \\ -3 & -5 & 18 \\ -1 & -2 & 7 \end{bmatrix}$ **39.** No inverse exists.
40. No inverse exists.   **41.** $x = 1, y = 2, z = -1$
42. $w = 1, x = 1, y = 0, z = -1$ **43.** -7 **44.** -6
45. 3   **46.** -25 **47.** $x = 1, y = -2$
48. $x = 1, y = 0, z = -2$ **49.** $x = 1, y = -1, z = 3$
50. $w = 1, x = 0, y = -1, z = 2$ **51.** 21   **52.** 7
53. $\dfrac{3}{x} + \dfrac{4}{x + 1}$ **54.** $\dfrac{3}{x} + \dfrac{2}{x^2} + \dfrac{x - 1}{x^2 + 1}$
55. $\dfrac{1}{x} - \dfrac{1}{x^2 + x + 5}$ **56.** $\dfrac{1}{x + 1} - \dfrac{2}{(x + 1)^2} + \dfrac{2}{(x + 1)^3}$

57.

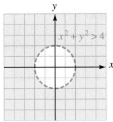

58.

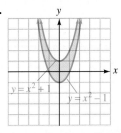

59.

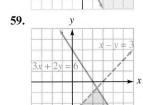

60.

61. $P = 6$ at $(3, 0)$ **62.** $P = 12$ at $(0, -4)$

63. $P = 2$ at $(1, 1)$ **64.** $P = 3$ at $\left(-\dfrac{2}{3}, \dfrac{5}{3}\right)$

65. 1,000 bags of x, 1,400 bags of y

Chapter Test

1.

 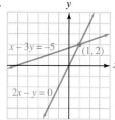

2.

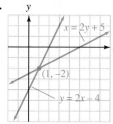

3. $x = 1, y = -3$ **4.** $x = 3, y = 5$

5. 6 liters of 20% solution, 4 liters of 45% solution

6. CD World 100 units, Ace 25 units, Hi-Fi 50 units

7. $x = 2, y = 1$ **8.** $x = 1, y = 2, z = 1$

9. $x = 1, y = 0, z = -2$

10. $x = -\dfrac{2}{5}y + \dfrac{7}{5}, z = -\dfrac{8}{5}y - \dfrac{7}{5}, y = $ any number

11. $\begin{bmatrix} 16 & -14 & 20 \\ 0 & -6 & -13 \end{bmatrix}$ **12.** $[-1]$ **13.** $\begin{bmatrix} -\dfrac{7}{3} & \dfrac{19}{3} \\ \dfrac{2}{3} & -\dfrac{5}{3} \end{bmatrix}$

14. $\begin{bmatrix} -13 & -3 & 14 \\ 4 & 1 & -4 \\ 12 & 3 & -13 \end{bmatrix}$ **15.** $x = \dfrac{17}{3}, y = -\dfrac{4}{3}$

16. $x = -36, y = 11, z = 34$ **17.** -12 **18.** -24

19. $-\dfrac{5}{4}$ **20.** 1   **21.** $\dfrac{3}{2x - 3} + \dfrac{1}{x + 1}$

22. $\dfrac{1}{x} + \dfrac{2x + 1}{x^2 + 2}$

23.

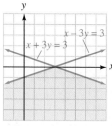

24.

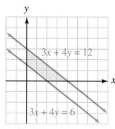

25. $P = 7$ at $(1, 2)$ **26.** $P = -8$ at $(8, 0)$

Section 7.1

1. $2, -5, 3$ **3.** $0, 0, \sqrt{5}$ **5.** to the left

7. downward   **9.** directrix, focus   **11.** circle

13. parabola   **15.** $x^2 + y^2 = 49, x^2 + y^2 - 49 = 0$

17. $(x - 2)^2 + (y + 2)^2 = 17, x^2 + y^2 - 4x + 4y - 9 = 0$

19. $(x - 1)^2 + (y + 2)^2 = 36, x^2 + y^2 - 2x + 4y - 31 = 0$

21.

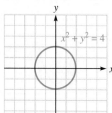

23.

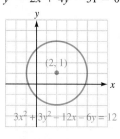

25. $(0, 0), (0, 3), y = -3$ **27.** $(0, 3), (5, 3), x = -5$

29. $(-2, 1), (-2, -5), y = 7$ **31.** $x^2 = 12y$

33. $y^2 = -12x$ **35.** $(x - 3)^2 = -12(y - 5)$

37. $(x - 3)^2 = -28(y - 5)$ **39.** $x^2 = -4(y - 2)$

41. $(y + 5)^2 = 8(x - 1)$

43. $(y - 2)^2 = -2(x - 2)$ or $(x - 2)^2 = -2(y - 2)$

45. $(x + 4)^2 = -\dfrac{16}{3}(y - 6)$ or $(y - 6)^2 = \dfrac{9}{4}(x + 4)$

47. $(y - 8)^2 = -4(x - 6)$ **49.** $(x - 3)^2 = \dfrac{1}{2}(y - 1)$

51.

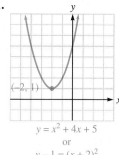

$y = x^2 + 4x + 5$
or
$y - 1 = (x + 2)^2$

53.

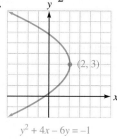

$y^2 + 4x - 6y = -1$
or
$(y - 3)^2 = -4(x - 2)$

55.

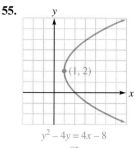

$y^2 - 4y = 4x - 8$
or
$(y - 2)^2 = 4(x - 1)$

57.

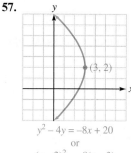

$y^2 - 4y = -8x + 20$
or
$(y - 2)^2 = -8(x - 3)$

59.

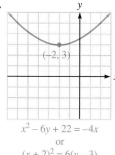

$x^2 - 6y + 22 = -4x$
or
$(x + 2)^2 = 6(y - 3)$

61.

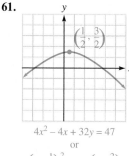

$4x^2 - 4x + 32y = 47$
or
$\left(x - \dfrac{1}{2}\right)^2 = -8\left(y - \dfrac{3}{2}\right)$

63.

65. yes   **67.** 60 mi   **69.** $(x - 4)^2 + y^2 = 16$

71. $(x - 7)^2 + y^2 = 9$ **73.** 2 ft   **75.** $x^2 = -\dfrac{45}{2}y$

77. 1 ft   **79.** 300 ft   **81.** about 12.6 cm

83. about 520 ft   **85.** $0x^2 + 0xy + y^2 - 8x - 4y + 12 = 0$

87. $x^2 + (y - 3)^2 = 25$ **89.** $y = x^2 + 4x + 3$ **93.** 4

95. $\dfrac{49}{4}$ **97.** $x = 1, -5$ **99.** $x = -2, 9$

Section 7.2

1. sum, constant   **3.** vertices   **5.** $a, 0; -a, 0$

7. 26-in string; thumbtacks 24 in. apart   **9.** circle

11. parabola   **13.** ellipse   **15.** $\dfrac{x^2}{16} + \dfrac{y^2}{9} = 1$

17. $\dfrac{x^2}{25} + \dfrac{y^2}{16} = 1$ **19.** $\dfrac{9x^2}{16} + \dfrac{9y^2}{25} = 1$

21. $\dfrac{x^2}{7} + \dfrac{y^2}{16} = 1$ **23.** $\dfrac{(x-3)^2}{4} + \dfrac{(y-4)^2}{9} = 1$

25. $\dfrac{(x-3)^2}{9} + \dfrac{(y-4)^2}{4} = 1$

27. $\dfrac{(x-3)^2}{41} + \dfrac{(y-4)^2}{16} = 1$ **29.** $\dfrac{x^2}{36} + \dfrac{(y-4)^2}{20} = 1$

31. $\dfrac{x^2}{100} + \dfrac{y^2}{64} = 1$

33.
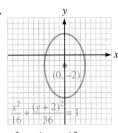

35.
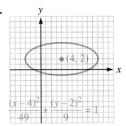

37.
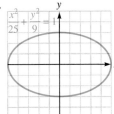

39.
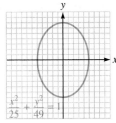

41. $\dfrac{x^2}{4} + \dfrac{(y-1)^2}{16} = 1$ **43.** $\dfrac{(x+1)^2}{4} + \dfrac{(y+2)^2}{9} = 1$

45.
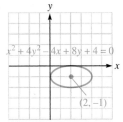

47.
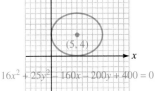

49.

51. $\dfrac{x^2}{900} + \dfrac{y^2}{400} = 1$

53. $\dfrac{x^2}{2,500} + \dfrac{y^2}{900} = 1$; 36 m   **55.** about 20.8 in.

57. 199,395 mi   **63.** $\dfrac{x^2}{9} + \dfrac{y^2}{8} = 1$

67. 4 of the 75 planets shown have eccentricity 0.

69. $\begin{bmatrix} 1 & 3 \\ 1 & 1 \\ -3 & 1 \end{bmatrix}$ **71.** $\begin{bmatrix} -2 & 4 \\ 2 & 2 \\ -4 & 5 \end{bmatrix}$

Section 7.3

1. difference, constant   **3.** $a, 0; -a, 0$ **5.** transverse axis

7. circle   **9.** parabola   **11.** ellipse   **13.** hyperbola

15. $\dfrac{x^2}{25} - \dfrac{y^2}{24} = 1$ **17.** $\dfrac{(x-2)^2}{4} - \dfrac{(y-4)^2}{9} = 1$

19. $\dfrac{(y-3)^2}{9} - \dfrac{(x-5)^2}{9} = 1$ **21.** $\dfrac{y^2}{9} - \dfrac{x^2}{16} = 1$

23. $\dfrac{(x-1)^2}{4} - \dfrac{(y-4)^2}{32} = 1$ **25.** $\dfrac{x^2}{10} - \dfrac{3y^2}{20} = 1$

27. 24 sq. units   **29.** 12 sq. units

31. $\dfrac{(x+2)^2}{4} - \dfrac{4(y+4)^2}{81} = 1$ or $\dfrac{(y+4)^2}{4} - \dfrac{4(x+2)^2}{81} = 1$

33. $\dfrac{x^2}{36} - \dfrac{16y^2}{25} = 1$

35.
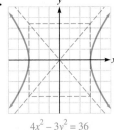

37.
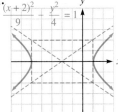

$4x^2 - 3y^2 = 36$

39.
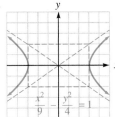

$y^2 - x^2 = 1$

41.
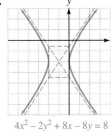

$\dfrac{(x+2)^2}{9} - \dfrac{y^2}{4} = 1$

43.
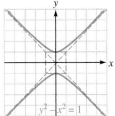

$4(y-2)^2 - 9(x+1)^2 = 36$

45.
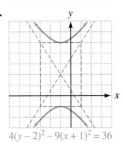

$4x^2 - 2y^2 + 8x - 8y = 8$

47.
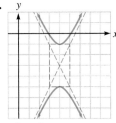

$y^2 - 4x^2 + 6y + 32x = 59$

49.
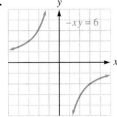

$-xy = 6$

51. $\dfrac{(x-3)^2}{9} - \dfrac{(y-1)^2}{16} = 1$ **53.** $4x^2 - 5y^2 - 60y = 0$

55.   **57.** 24   **59.** 3 units

61. hyperbola; $\dfrac{x^2}{144} - \dfrac{y^2}{25} = 1$

63. hyperbola; $\dfrac{x^2}{36} - \dfrac{y^2}{64} = 1$

69. $f^{-1}(x) = \dfrac{x+2}{3}$ **71.** $f^{-1}(x) = \dfrac{2x}{5-x}$

73. $f(g(x)) = (x+1)^4 + 1$
75. $f(f(x)) = (x^2 + 1)^2 + 1$

Section 7.4
1. graphs
3.
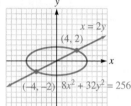
5.
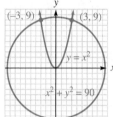

7.
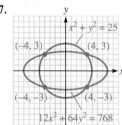
9.
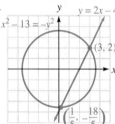

11.
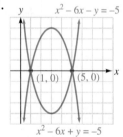

13. $(1, 2), (-1, 0)$ **15.** $(1, 0.67), (-1, -0.67)$
17. $(3, 0), (0, 5)$ **19.** $(1, 1)$ **21.** $(1, 2), (2, 1)$
23. $(-2, 3), (2, 3)$ **25.** $\left(\sqrt{5}, 5\right), \left(-\sqrt{5}, 5\right)$
27. $\left(\sqrt{3}, 0\right), \left(-\sqrt{3}, 0\right)$
29. $(2, 4), (2, -4), (-2, 4), (-2, -4)$
31. $\left(-\sqrt{15}, 5\right), \left(\sqrt{15}, 5\right), (-2, -6), (2, -6)$
33. $(0, -4), (-3, 5), (3, 5)$
35. $(-2, 3), (2, 3), (-2, -3), (2, -3)$
37. $(3, 3)$ **39.** $(6, 2), (-6, -2), \left(\sqrt{42}, 0\right), \left(-\sqrt{42}, 0\right)$
41. $\left(\dfrac{1}{2}, \dfrac{1}{3}\right), \left(\dfrac{1}{3}, \dfrac{1}{2}\right)$ **43.** 7 cm by 9 cm
45. 80 ft by 100 ft or 50 ft by 160 ft
47. either $750 at 9% or $900 at 7.5%
49. $(30, 3)$ **51.** yes at $(-2, 4)$ and $(1, 1)$
53. about 23 mi
61. vertical: $x = 1$; horizontal: $y = 3$
63. vertical: $x = 1, x = -1$; horizontal: $y = 0$
65. y-axis   **67.** origin

Chapter Review
1. $x^2 + y^2 = 16$ **2.** $x^2 + y^2 = 100$
3. $(x - 3)^2 + (y + 2)^2 = 25$ **4.** $(x + 2)^2 + (y - 4)^2 = 25$
5. $(x - 5)^2 + (y - 10)^2 = 85$
6. $(x - 2)^2 + (y - 2)^2 = 89$ **7.** $(x - 3)^2 + (y + 2)^2 = 16$

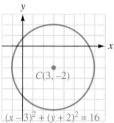

8. $(x + 2)^2 + (y - 5)^2 = 16$ **9.** $y^2 = -2x$

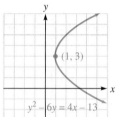

10. $x^2 = 16y$ **11.** $(x + 2)^2 = -\dfrac{4}{11}(y - 3)$

12.
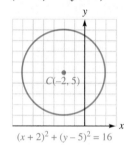
13.
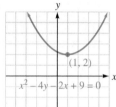

14. $\dfrac{x^2}{36} + \dfrac{y^2}{16} = 1$ **15.** $\dfrac{x^2}{4} + \dfrac{y^2}{25} = 1$
16. $\dfrac{(x + 2)^2}{16} + \dfrac{(y - 3)^2}{9} = 1$

17. $(x - 2)^2 + \dfrac{(y + 1)^2}{4} = 1$

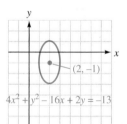

18. $\dfrac{x^2}{4} - \dfrac{y^2}{12} = 1$ **19.** $\dfrac{y^2}{9} - \dfrac{x^2}{16} = 1$ **20.** $\dfrac{x^2}{9} - \dfrac{(y - 3)^2}{16} = 1$
21. $\dfrac{y^2}{9} - \dfrac{(x - 3)^2}{16} = 1$ **22.** $y = \pm\dfrac{4}{5}x$

23. $\dfrac{(x-1)^2}{4} - \dfrac{(y+2)^2}{9} = 1$

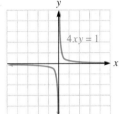

$9x^2 - 4y^2 - 16y - 18x = 43$

24.

25.

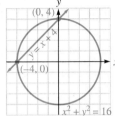

26.

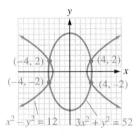

$x^2 - y^2 = 12 \qquad 3x^2 + y^2 = 52$

27.

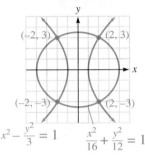

$x^2 - \dfrac{y^2}{3} = 1 \qquad \dfrac{x^2}{16} + \dfrac{y^2}{12} = 1$

28. $(-4, 2), (-4, -2), (4, 2), (4, -2)$

29. $(0, 4), \left(2\sqrt{3}, -2\right)$ **30.** $(-2, 3), (-2, -3), (2, 3), (2, -3)$

Chapter Test

1. $(x-2)^2 + (y-3)^2 = 9$ **2.** $(x-2)^2 + (y-3)^2 = 41$

3. $(x-2)^2 + (y+5)^2 = 169$

4.

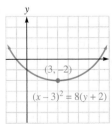

5. $(x-3)^2 = 16(y-2)$ **6.** $(y+6)^2 = -4(x-4)$

7. $(x-2)^2 = \dfrac{4}{3}(y+3)$ or $(y+3)^2 = -\dfrac{9}{2}(x-2)$

8.

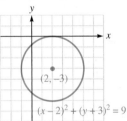

9. $\dfrac{x^2}{100} + \dfrac{y^2}{64} = 1$ **10.** $\dfrac{x^2}{169} + \dfrac{y^2}{144} = 1$

11. $\dfrac{(x-2)^2}{4} + \dfrac{(y-3)^2}{36} = 1$ **12.**

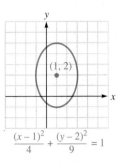

$\dfrac{(x-1)^2}{4} + \dfrac{(y-2)^2}{9} = 1$

13. $\dfrac{x^2}{25} - \dfrac{y^2}{144} = 1$ **14.** $\dfrac{x^2}{36} - \dfrac{4y^2}{25} = 1$

15. $\dfrac{(x-2)^2}{64} - \dfrac{(y+1)^2}{36} = 1$

16.

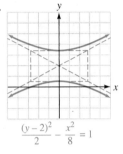

$\dfrac{(y-2)^2}{2} - \dfrac{x^2}{8} = 1$

17. $\left(\sqrt{7}, 4\right), \left(-\sqrt{7}, 4\right)$

18. $\left(3\sqrt{2}, 3\right), \left(-3\sqrt{2}, 3\right), \left(3\sqrt{2}, -3\right), \left(-3\sqrt{2}, -3\right)$

19. $(y-2)^2 = 6(x+3)$; parabola

20. $\dfrac{(x-1)^2}{3} + \dfrac{(y+2)^2}{2} = 1$; ellipse

Cumulative Review

1. 16 **2.** $\dfrac{1}{2}$ **3.** y^2 **4.** $x^{17/12}$ **5.** $x^{4/3} - x^{2/3}$

6. $\dfrac{1}{x} + 2 + x$ **7.** $-3x$ **8.** $4t\sqrt{3t}$ **9.** $4x$

10. $x + 3$ **11.** $7\sqrt{2}$ **12.** $-12\sqrt[4]{2} + 10\sqrt[4]{3}$

13. $-18\sqrt{6}$ **14.** $\dfrac{5\sqrt[3]{x^2}}{x}$ **15.** $\dfrac{x + 3\sqrt{x} + 2}{x - 1}$

16. $\sqrt{xy}$ **17.** $2, 7$ **18.** $\dfrac{1}{4}$ **19.** $1, -\dfrac{3}{2}$

20. $\dfrac{-2 \pm \sqrt{7}}{3}$ **21.** $7 + 2i$ **22.** $-5 - 7i$

23. $13 + 0i$ **24.** $12 - 6i$ **25.** $-12 - 10i$

26. $\dfrac{3}{2} + \dfrac{1}{2}i$ **27.** $\sqrt{13}$ **28.** $\sqrt{61}$ **29.** -2

30. $\left(1, \dfrac{1}{2}\right)$ **31.** $(-\infty, -2) \cup (3, \infty)$ **32.** $[-2, 3]$

33. 5 **34.** 27 **35.** $12x^2 - 12x + 5$ **36.** $6x^2 + 3$

37. $2^y = x$ **38.** $\log_3 a = b$ **39.** 5 **40.** 3

41. $\dfrac{1}{27}$ **42.** 1 **43.** $y = 2^x$ **44.** x

45. 1.9912 **46.** 0.301 **47.** 1.6902 **48.** 0.1461

49. $\dfrac{2\log 2}{\log 3 - \log 2}$ **50.** 16 **51.** $2{,}848.31$

52. 1.16056 **53.** (2, 1) **54.** (1, 1) **55.** (2, −2)
56. (3, 1) **57.** −1 **58.** −1 **59.** (−1, −1, 3)
60. (1, 1, 1)

Section 8.1

1. power **3.** first **5.** $7 \cdot 6 \cdot 5 \cdot 4 \cdot 3 \cdot 2 \cdot 1$
7. $(n - 1)!$ **9.** 120 **11.** 4,320 **13.** 1,440
15. $\dfrac{1}{1,320}$ **17.** $\dfrac{5}{3}$ **19.** 18,564
21. $a^5 + 5a^4b + 10a^3b^2 + 10a^2b^3 + 5ab^4 + b^5$
23. $x^3 - 3x^2y + 3xy^2 - y^3$ **25.** $a^3 + 3a^2b + 3ab^2 + b^3$
27. $a^5 - 5a^4b + 10a^3b^2 - 10a^2b^3 + 5ab^4 - b^5$
29. $8x^3 + 12x^2y + 6xy^2 + y^3$
31. $x^3 - 6x^2y + 12xy^2 - 8y^3$
33. $16x^4 + 96x^3y + 216x^2y^2 + 216xy^3 + 81y^4$
35. $x^4 - 8x^3y + 24x^2y^2 - 32xy^3 + 16y^4$
37. $x^5 - 15x^4y + 90x^3y^2 - 270x^2y^3 + 405xy^4 - 243y^5$
39. $\dfrac{x^4}{16} + \dfrac{x^3y}{2} + \dfrac{3x^2y^2}{2} + 2xy^3 + y^4$ **41.** $6a^2b^2$
43. $35a^3b^4$ **45.** $-b^5$ **47.** $2,380a^{13}b^4$ **49.** $-4\sqrt{2}a^3$
51. $1,134a^5b^4$ **53.** $\dfrac{3x^2y^2}{2}$ **55.** $-\dfrac{55r^2s^9}{2,048}$
57. $\dfrac{n!}{3!(n-3)!}a^{n-3}b^3$ **59.** $\dfrac{n!}{(r-1)!(n-r+1)!}a^{n-r+1}b^{r-1}$
63. -252 **71.** $3xyz^2(x^2yz^2 - 2z^3 + 5x)$
73. $(a^2 + b^2)(a + b)(a - b)$ **75.** $\dfrac{3 + x}{3 - x}$

Section 8.2

1. domain **3.** series **5.** infinite
7. Summation notation **9.** 6 **11.** $5c$
13. 0, 10, 30, 60, 100, 150 **15.** 21 **17.** $a + 4d$
19. 15 **21.** 15 **23.** 15 **25.** $\dfrac{242}{243}$ **27.** 35
29. 3, 7, 15, 31 **31.** $-4, -2, -1, -\dfrac{1}{2}$ **33.** k, k^2, k^4, k^8
35. $8, \dfrac{16}{k}, \dfrac{32}{k^2}, \dfrac{64}{k^3}$ **37.** an alternating infinite series
39. not an alternating infinite series **41.** 30 **43.** −50
45. 40 **47.** 500 **49.** $\dfrac{7}{12}$ **51.** 160 **53.** 3,725
59. 6 cm **61.** 26 ft

Section 8.3

1. $(n - 1)$ **3.** infinite **5.** $a_n = a + (n - 1)d$
7. Arithmetic means **9.** 1, 3, 5, 7, 9, 11
11. $5, \dfrac{7}{2}, 2, \dfrac{1}{2}, -1, -\dfrac{5}{2}$ **13.** $9, \dfrac{23}{2}, 14, \dfrac{33}{2}, 19, \dfrac{43}{2}$
15. 318 **17.** 6 **19.** 370 **21.** 44 **23.** $\dfrac{25}{2}, 15, \dfrac{35}{2}$
25. $-\dfrac{82}{15}, -\dfrac{59}{15}, -\dfrac{12}{5}, -\dfrac{13}{15}$ **27.** 285 **29.** 555
31. $157\dfrac{1}{2}$ **33.** 20,100 **35.** 1080°; 1800° **37.** $460
39. $1,587,500 **41.** 80 ft **43.** 210 **49.** 6 **51.** −1
53. $1, -1, i, -i$

Section 8.4

1. r^{n-1} **3.** ar^{n-1} **5.** infinite **7.** Geometric means
9. 10, 20, 40, 80 **11.** $-2, -6, -18, -54$
13. $3, 3\sqrt{2}, 6, 6\sqrt{2}$ **15.** 2, 6, 18, 54 **17.** 256
19. -162 **21.** $10\sqrt[4]{2}, 10\sqrt[4]{4}$ or $10\sqrt{2}, 10\sqrt[4]{8}$
23. 8, 32, 128, 512 **25.** 124 **27.** $-29,524$
29. $\dfrac{1,995}{32}$ **31.** 18 **33.** 8 **35.** $\dfrac{5}{9}$ **37.** $\dfrac{25}{99}$
39. 23 **41.** 12.96 ft, 400 ft **43.** 5.13 m
45. $69.82 **47.** about $\dfrac{1}{3}C$ **49.** about 1.03×10^{20}
51. $180,176.87 **53.** $2,001.60 **55.** $2,013.62
57. $264,094.58 **59.** 5,000 **61.** 1.8447×10^{19} grains
63. no **65.** $5 - 3i$ **67.** $10 + 0i$ **69.** $\dfrac{1}{5} + \dfrac{8}{5}i$
71. $0 - i$

Section 8.5

1. two **3.** $n = k + 1$
5. $5 = 5; 15 = 15; 30 = 30; 50 = 50$
7. $7 = 7; 17 = 17; 30 = 30; 46 = 46$
29. no
37. a. 1 **b.** 3 **c.** 7 **d.** 15
39. 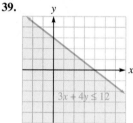 **41.**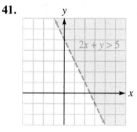

Section 8.6

1. 24 **3.** $\dfrac{n!}{(n-r)!}$ **5.** 1 **7.** $\dfrac{n!}{r!(n-r)!}$ **9.** 1
11. $\dfrac{n!}{a!b!\cdots}$ **13.** 840 **15.** 35 **17.** 120 **19.** 5
21. 1 **23.** 1,200 **25.** 40 **27.** 2,278 **29.** 144
31. 8,000,000 **33.** 240 **35.** 6 **37.** 40,320
39. 14,400 **41.** 24,360 **43.** 5,040 **45.** 48
47. 96 **49.** 210 **51.** 24 **53.** 5,040 **55.** 60
57. 2,721,600 **59.** 1,120 **61.** 59,400 **63.** 272
65. 28 **67.** 252 **69.** 142,506 **71.** 66 **73.** 256
75. 56 **83.** 2 **85.** 4 **87.** true **89.** true

Section 8.7

1. experiment **3.** $\dfrac{n(E)}{n(S)}$
5. $\{(1, H), (2, H), (3, H), (4, H), (5, H), (6, H), (1, T), (2, T),$
$(3, T), (4, T), (5, T), (6, T)\}$ **7.** $\{a, b, c, d, e, f, g, h, i, j, k, l,$
$m, n, o, p, q, r, s, t, u, v, w, x, y, z\}$ **9.** $\dfrac{1}{6}$ **11.** $\dfrac{2}{3}$
13. $\dfrac{19}{42}$ **15.** $\dfrac{13}{42}$ **17.** $\dfrac{3}{8}$ **19.** 0 **21.** $\dfrac{1}{12}$
23. $\dfrac{1}{169}$ **25.** $\dfrac{5}{12}$ **27.** about 6.3×10^{-12} **29.** 0
31. $\dfrac{3}{13}$ **33.** $\dfrac{1}{6}$ **35.** $\dfrac{1}{8}$ **37.** $\dfrac{5}{16}$

39. {SSSS, SSSF, SSFS, SFSS, FSSS, SSFF, SFSF, FSSF, SFFS, FSFS, FFSS, SFFF, FSFF, FFSF, FFFS, FFFF}

41. $\frac{1}{4}$ **43.** $\frac{1}{4}$ **45.** 1 **47.** $\frac{32}{119}$ **49.** $\frac{1}{3}$

51. 0.18 **53.** 0.14 **55.** about 33% **57.** no

59. $-10, 4$ **61.** $(-4, 10)$

Chapter Review

1. 720 **2.** 30,240 **3.** 8 **4.** $\frac{280}{3}$

5. $x^3 + 3x^2y + 3xy^2 + y^3$

6. $p^4 + 4p^3q + 6p^2q^2 + 4pq^3 + q^4$

7. $a^5 - 5a^4b + 10a^3b^2 - 10a^2b^3 + 5ab^4 - b^5$

8. $8a^3 - 12a^2b + 6ab^2 - b^3$ **9.** $56a^5b^3$ **10.** $80x^3y^2$

11. $84x^3y^6$ **12.** $439,040x^3$ **13.** 63 **14.** 9

15. 5, 17, 53, 161 **16.** $-2, 8, 128, 32,768$ **17.** 90

18. 60 **19.** 1,718 **20.** -360 **21.** 117 **22.** 281

23. -92 **24.** $-\frac{135}{2}$ **25.** $\frac{7}{2}, 5, \frac{13}{2}$

26. 25, 40, 55, 70, 85 **27.** 3,320 **28.** 5,780

29. $-5,220$ **30.** $-1,540$ **31.** $\frac{1}{729}$ **32.** 13,122

33. $\frac{9}{16,384}$ **34.** $\frac{8}{15,625}$ **35.** $2\sqrt{2}, 4, 4\sqrt{2}$

36. $4, -8, 16, -32$ **37.** 16 **38.** $\frac{3,280}{27}$ **39.** 6,560

40. $\frac{2,295}{128}$ **41.** $\frac{520,832}{78,125}$ **42.** $\frac{3,280}{3}$ **43.** $16\sqrt{2}$

44. $\frac{2}{3}$ **45.** $\frac{3}{25}$ **46.** no sum **47.** 1 **48.** $\frac{1}{3}$

49. 1 **50.** $\frac{17}{99}$ **51.** $\frac{5}{11}$ **52.** $4,775.81

53. 6,516; 3,134 **54.** $3,486.78

55. $1 = 1; 9 = 9; 36 = 36; 100 = 100$

57. 6,720 **58.** 35 **59.** 1 **60.** 4,050 **61.** 564,480

62. 840 **63.** 21 **64.** 66 **65.** 120 **66.** $\frac{56}{1,287}$

67. $\frac{1}{6}$ **68.** $\frac{33}{66,640}$ **69.** 20,160 **70.** 90,720

72. 24 **73.** $\frac{1}{108,290}$ **74.** $\frac{108,289}{108,290}$

75. about 6.3×10^{-12} **76.** $\frac{60}{143}$ **77.** $\frac{1}{2}$ **78.** $\frac{7}{13}$

79. $\frac{1}{2,598,960}$ **80.** $\frac{33}{16,660}$ **81.** $\frac{15}{16}$

Chapter Test

1. 144 **2.** 384 **3.** $10x^4y$ **4.** $112a^2b^6$ **5.** 27

6. -36 **7.** 155 **8.** -130 **9.** 9, 14, 19

10. $-6, -18$ **11.** 255.75 **12.** about 9

13. about $0.42C$ **14.** about $1.46C$ **16.** 800,000

17. 42 **18.** 24 **19.** 28 **20.** 1 **21.** 576

22. 120 **23.** 60 **24.** {(H, H, H), (H, H, T), (H, T, H), (H, T, T), (T, H, H), (T, H, T), (T, T, H), (T, T, T)}

25. $\frac{1}{6}$ **26.** $\frac{2}{13}$ **27.** $\frac{33}{66,640}$ **28.** $\frac{1}{9}$ **29.** $\frac{87}{245}$

30. $\frac{12}{19}$

Index

Answers for Exam 1 Additional Practice Problems:

1. $A = \dfrac{BC + B}{C - 1}$ **2.** $b = \dfrac{ac}{a + 2c}$ **4.** No Solution   **5.** 15 l of 70% solution; 5 l of 30% solution

6. .86 liters   **7.** 6 miles   **8.** 700 \$30 tickets; 200 \$20 tickets   **9.** 1.08 hrs (65 minutes)

10. $\dfrac{4 \pm \sqrt{100 - h}}{4}$ **11.** $\dfrac{43 \pm \sqrt{4297}}{34}$ **12.** 7.90 seconds   **13.** 12 and 13; -16 and -15

14. 2 hours 13 minutes   **15.** \$260 (or \$100)   **16.** (a) $-2 + 2i$; (b) $x^2 - 6x + 13$; (c) $-i$

17. (a) $\dfrac{1}{10} + \dfrac{3}{10}i$; (b) $-i$; (c) $0 - \dfrac{\sqrt{5}}{5}i$ or $0 - \dfrac{1}{5}i\sqrt{5}$ **19.** $\{-2i, 5i\}$

20. (a) $c > 4$; (b) $c = 0$; (c) $c < 4$ **21.** $\{1, -2\}$ **22.** $\{6\}$ **23.** $\left\{-\dfrac{5}{3}, \dfrac{1}{2}\right\}$ **24.** (a) $\{-9, 4\}$;

(b) $\{\pm 3i, \pm 2\}$;   (c) $\{16\}$;   (d) $\left\{-\dfrac{1}{9}, \dfrac{1}{4}\right\}$;   (e) $\{-729, 64\}$;   (f) $\left\{-\dfrac{1}{9}, \dfrac{1}{4}\right\}$;   (g) $\{-6, 7\}$

25. 96% or higher   **26.** Omit Problem (Refer to Problem #1 in Exam 2 Practice Problems.)

Answers for Exam 2 Additional Practice Problems:

1. 90.5% or higher   **2.** $76\% \le x \le 96\%$ **3.** $0 \text{ ft} < w < 7.5 \text{ ft}$ **4.** $|c - 726| \le 235$; $491 \le c \le 961$

5. (a) $(-\infty, -2] \cup (-1, 3]$;   (b) $[-4, -2)$;   (c) $(-\infty, -3) \cup (-1, 1)$;   (d) $(-\infty, -3] \cup (0, 3]$;   (e) ϕ

6. (a) $(-\infty, -2] \cup [5, \infty)$;   (b) $\left(\dfrac{1}{2}, 1\right]$ **7.** (a) Refer to graph:

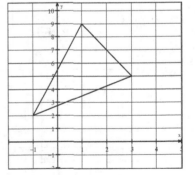

(b) – (d)

Side	Length	Slope	Equation
AB	5	$\dfrac{3}{4}$	$y = \dfrac{3}{4}x + \dfrac{11}{4}$
BC	$2\sqrt{5}$	-2	$y = -2x + 11$
AC	$\sqrt{53}$	$\dfrac{7}{2}$	$y = \dfrac{7}{2}x = \dfrac{11}{2}$

(e) $y = -2x$;   (f) $y = \dfrac{1}{2}x + \dfrac{5}{2}$ **8.** 2.9 minutes/day   **9.** (a) $(3, 8)$; (b) $r = \sqrt{61}$;

(c) $(x - 3)^2 + (y - 8)^2 = 61$ **10.** $y = -\dfrac{1}{5}x - \dfrac{13}{5}$

Answers for Exam 3 Additional Practice Problems:

1. (a) $2x$; (b) $-4x - 2h - 1$, $h \neq 0$ **2.** $\dfrac{-1}{3x(x+h)}$, $h \neq 0$

3.

Equation	$f(x) = -(x-4)^2 - 5$	$f(x) = 3x^2 + 8x - 3$
Vertex	(4, -5)	$\left(-\dfrac{4}{3}, -\dfrac{25}{3}\right)$
Line of Symmetry	$x = 4$	$x = -\dfrac{4}{3}$
x-intercept(s)	none	$\dfrac{1}{3}, -3$
y-intercept	-21	-3
Domain	$(-\infty, \infty)$	$(-\infty, \infty)$
Range	$(-\infty, -5)$	$\left[-\dfrac{25}{3}, \infty\right)$

4. (a) 50; initial height of rocket (height of tower); (b) $h(3) = 266$ ft; (c) 3.75 seconds; (d) 275 ft;

(e) 7.06 seconds   **5.** 150 ft by 300 ft; 45,000 ft^2 **6.** 187.5 ft by 250 ft; 46,875 ft^2 **7.** -26

8.

	(a)	(b)	(c)	(d)	(e)	(f)	(g)
$f(x)$	rises left rises right	0 5 -2	1 1 4	C C T	0	6	neither
$g(x)$	falls left rises right	3 -3 5	1 1 1	C C C	-45	3	neither
$h(x)$	rises left falls right	0 1 -1	3 1 1	C C C	0	5	odd

9. 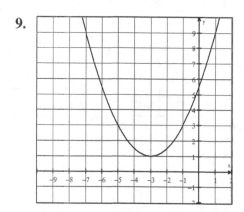   **10.** 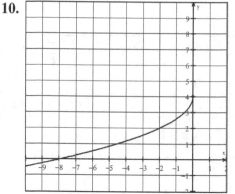   **11.** (1, 6)

12. $(2a, -b)$

13. (a) $\left(-\infty, \dfrac{5}{2}\right]$ (b) $(-\infty, -5) \cup (-5, 4) \cup (4, \infty)$ (e) $(-2, \infty)$ (d) $[-2, 3) \cup (3, \infty)$

14. (a) Domain: $(-\infty, 2)\cup(2, \infty)$; Range: $\left(-\infty, -\dfrac{2}{3}\right)\cup\left(-\dfrac{2}{3}, \infty\right)$; (b) $x = 2$; (c) $y = -\dfrac{2}{3}$;

(d) $-\dfrac{2}{3}$; 2^+; 2^- **15.** (a) $f\circ g = \dfrac{1}{1-2x}$; Domain: $(-\infty, 0)\cup\left[0, \dfrac{1}{2}\right)\cup\left(\dfrac{1}{2}, \infty\right)$;

(b) $g\circ f = \dfrac{x-2}{x}$; Domain: $(-\infty, 0)\cup(0, 2)\cup(2, \infty)$ **16.** (a) $f^{-1}(x) = \dfrac{-2x}{x-1}$;

(b) $f^{-1}(x) = -\dfrac{4x-7}{x-5}$; (c) $f^{-1}(x) = x^5 + 3$ **17.** (a) 2; (b) 10; (c) 4; (d) 3; (e) 5; (f) 1;

(g) -3; (h) 16; (i) [-2, 4]; (j) [-2, 2)$\cup$(2, 4]; (k) (-5, -3)$\cup$(3, 5); (l) 3; $x = -3$; (m) $y = f(x)$;

(n) odd

Answers for Exam 4 Additional Practice Problems:

1. (a) $f^{-1}(x) = e^{3x}$; (b) $f^{-1}(x) = 10^{4-x} + 3$; (c) $f^{-1}(x) = \dfrac{1}{4}\ln x$ **2.** (a) Domain: $(-\infty, \infty)$;

Range: $(-27, \infty)$; (b) $y = -27$; (c) -18; (d) $x = -2$; $y = 216$; (e) $f^{-1}(x) = \log_3(x + 27) - 5$;

(f) Domain: $(-27, \infty)$; Range: $(-\infty, \infty)$ **3.** (a) Domain: $(3, \infty)$; Range: $(-\infty, \infty)$; (b) $x = 3$;

(c) 6; (d) $x = \dfrac{49}{16}$; no y-intercept; (e) $f^{-1}(x) = 2^{x-4} + 3$; (f) Domain: $(-\infty, \infty)$; Range: $(3, \infty)$

4. (a) $2A + B$; (b) $3B - A - \dfrac{1}{4}C$ **5.** (a) Domain: $\left(-\dfrac{3}{2}, \infty\right)$; Range: $(-\infty, \infty)$;

(b) Domain: $(-\infty, -3)\cup(3, \infty)$; Range: $(-\infty, \infty)$; (c) Domain: $(-\infty, 2)$; Range: $(-\infty, \infty)$;

(d) Domain: $(-\infty, 3)\cup(3, \infty)$; Range: $(-\infty, \infty)$ **6.** 0 **7.** (a) $x^2 y^5$; (b) $\dfrac{x^7 y^3 z^3}{(x-3)^2}$

8. (a) $\left\{-\dfrac{9}{7}\right\}$; (b) $\left\{\dfrac{1}{2}\right\}$; (c) $\{0\}$ **9.** $4 + \dfrac{1}{2}\ln(e^2 + 1) - 2\ln(e + 1)$ **10.** $\log\left[\dfrac{x(x-2)}{5}\right]$

11. (a) \$143,146.99; (b) \$143,733.65; (c) \$141,909.86; (d) \$141,449.77 (with n = 360); \$141,450.37

(with n = 365); (e) \$140,053.29 **12.** 13.5% **13.** (a) 29.44%; (b) 27.47% **14.** 3.65%

15. quarterly: \$13,263.31; continuously: \$13,250.70 **16.** (a) 50.68 million people;

(b) 6.02 years after 1987 (in 1993) **17.** (a) $\left\{\dfrac{e^2}{5}\right\}$; (b) $\{2\}$; (c) $\left\{-4 + 2\sqrt{7}\right\}$; (d) $\{9, -1\}$;

(e) $\{e^6\}$; (f) $\{3+\log_5 12\}$; (g) $\left\{\dfrac{2\ln 10}{3-\ln 10}\right\}$ or $\left\{\dfrac{2}{3\log e-1}\right\}$; (h) $\{\ln 10\}$; (i) $\{\log_3 6\}$; (j) $\{2\}$; (k) ϕ;

(l) $\left\{25,\dfrac{1}{400}\right\}$ **18.** (a) $A=A_0 e^{-0.086643t}$; (b) 27 days   **19.** (a) $A=A_0 e^{-0.000121t}$;

(b) No, the painting is 252 years old (painted ≈ 1759); DaVinci died in 1519.   **20.** (a) $A=50e^{0.057762t}$;

(b) ≈ 283 (282.84) flies; (c) ≈ 64 (63.86) days   **21.** (a) 20 deer; (b) 22 (22.33) deer;

(c) 50 (50.34) deer; (d) 7 (7.01) years; (e) 100 deer

Answers for Chapter 5 Additional Practice Problems:

1. $(x-2)(x^2+2x+4)$ **2.** 6   **3.** $\left\{\dfrac{1}{2},-\dfrac{1}{3},-\dfrac{3}{2}\right\}$ **4.** $k=16$ **5.** $h=2$ in; $w=4$ in; $l=9$ in

6. $\{1,2\pm 3i\}$ **7.** $\{-1,4\}$; $f(x)=(x+1)^2(x-4)^2$ **8.** $f(x)=x^4-6x^3+34x^2-150x+225$

9. $f(x)=-3x^4+15x^3-15x^2-75x+78$

Answers for Exam 1 Sample Exam

1. $C = \dfrac{B}{AB-1}$

2. No Solution ($x = -7$ is a restriction)

3. (a) $-\dfrac{7}{13} + \dfrac{22}{13}i$

   (b) $12 - 3i$

4. $x = -\dfrac{2}{3} \pm \dfrac{\sqrt{13}}{3}$ or $x = \dfrac{-2 \pm \sqrt{13}}{3}$

5. $\left\{ -\dfrac{1}{3}, \dfrac{1}{4} \right\}$

6. $\left\{ \pm 2, -\dfrac{3}{2} \right\}$

7. 50 mph and 60 mph

8. $\{4\}$

9. $\dfrac{1}{3}x + x + x + 48 = 692$; Lana: 276 miles; Tony: 92 miles; Mark: 324 miles

10. T

11. F

12. F

13. T

14. T

Name:_____ **Section:** _____ **Date:** _____

Signature: _____

Do not start this test until your instructor says so. You will have exactly **50 minutes** from this time to finish the test. Please show all work in the space provided.

There are 13 questions on 6 pages. Without fully opening the exam. please check to make sure your exam is complete before you begin. **Please read instructions carefully**.

1. You will need a pencil or pen, one calculator and this exam paper. Please clear everything else from your desk. (Please turn off or switch cell phones to manner mode and place in backpack – NOT on your person.)

2. Calculators such as the TI89, TI-NSPIRE, Voyage 200, TI92, TI92+, HP49G, HP49G+, HP50G, Casio algefx2.0, algefx2.0pls or laptop, handheld, desktop and palmtop computers are NOT to be used. The use of cell phones, PDAs, 'Blackberries', 'head sets', or any other wireless device in the classroom is specifically prohibited. **Please do not ask your instructor/proctors about calculator use or mode.** You must clear all programs from your calculator memory. **Calculators cannot be shared.**

3. Please look at overhead/board for possible corrections to this exam.

4. Do not spend too much time on a particular problem. Work the problems that are simplest for you first.

5. There is no partial credit on the multiple choice questions. On the open-ended questions, the Grading will be based on your method. **SHOW ALL OF YOUR WORK.**
   NO WORK = NO CREDIT! DO NOT SKIP STEPS! You must show all work that supports your answer in order to get credit. Be neat and orderly. If you need additional space, use the back of the exam pages.

6. Circle the letter of the correct answer for multiple choice or true false questions. For open-ended questions, place your work and answers in the space provided. Answers can be in any form unless otherwise specified. For problems asking for EXACT answers, leave in terms of fractions, π and $\sqrt{number}$. No credit will be given for decimal approximations when an **EXACT** answer is required. When solving equations, please put your answer in solution set form. For decimal answers, **round to 4 decimal places** unless otherwise specified.

7. Give measurement units (feet, degrees, etc.) when appropriate.

Answers for Exam 4 Sample Exam

1. 7.6% ~~~~ ~hly is the better investment since it yields $36,402.90. (7.5% continuously only yi~~~~ `35,853.51.)

2. $3x^4y^6$

3. $2A - B - \dfrac{1}{2}C$

4. $\{3\}$

5. $\left\{\dfrac{\log 5}{\log 3}\right\}$ or $\left\{\dfrac{\ln 5}{\ln 3}\right\}$

6. (a) Domain: (-9, inf); Range: (-in.

   (b) x = -9

   (c) Increasing

   (d) -5

   (e) x-intercept: 234; y-intercept: -3

7. 7.70%

8. a

9. d

10. c

11. a

12. (a) T
    (b) F
    (c) T
    (d) F

Please note that this is only a sample exam. The format of your exam may be slightly different and may have different problems.

Michigan State University
Department of Mathematics

Name : _____   PID: _____

Signature: _____   Section #: _____

Page #	Possible Points	Actual Points	Page #	Possible Points	Actual Points
2	24		7	20	
3	22		8	25	
4	18		9	25	
5	18		10	30	
6	18				
Total Exam Points				**200**	

INSTRUCTIONS: ALL CELL PHONES MUST BE TURNED OFF AND PUT AWAY. Failure to do so will result in a score of "0" on your exam.

1. DO NOT OPEN THIS EXAM UNTIL YOU ARE INSTRUCTED TO DO SO.
2. *Without fully opening the exam,* check that you have pages 1-10 and that none is blank.
3. Fill in the information at the top of the page.
4. You will need a pen or pencil, one **approved** calculator and this test booklet for the exam. Please clear everything else from your desk.
5. Calculators are NOT to be shared. Please do not ask your instructor any questions about the use of your calculator.
6. **Please look to the board or overhead for possible corrections to this exam.**
7. Do not spend too much time on a particular problem. Work the easier problems first.
8. Show your work in the space provided. If you need additional space, use the backs of the exam pages. **YOU MUST SHOW ALL OF YOUR WORK!** Points may be withdrawn for answers given without substantiation. (There are problems, however, that will be graded on a right-wrong basis.)
9. Place your answers in the boxes, where provided. **Answers must be simplified** and can be in any form unless specified otherwise. Use standard notation for imaginary number solutions.
10. You will be given **exactly** 120 minutes for this exam.

1. Solve the equation $A = \dfrac{B+2}{B-3}$ for B.

8

2. Find the solution set for the equation $2x - \sqrt{x} - 1 = 0$.

8

3. Solve the inequality $\left| 5 - \dfrac{x}{2} \right| \geq 10$. Write your answer using interval notation.

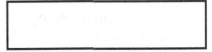

8

24

4. Find the equation of the line whose x-intercept is 8 and y-intercept is -2. Write the equation in slope-intercept form.

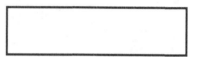

| 6 | |

5. The endpoints of the diameter of a circle are (1, -5) and (7, 3). Find the coordinates of the center, radius, domain and range of this circle.

Center:

Radius:

Domain:

Range:

| 8 | |

6. Refer to the graph of the polynomial function $y = f(x)$ below to determine whether each of the statements below is true or false.

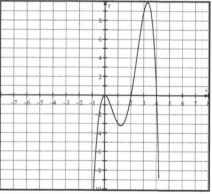

   T F   The sign of the leading coefficient
           is positive.

   T F   The degree of the polynomial is odd.

   T F   At $x = 2$, f has a zero of odd
           multiplicity.

   T F f has a relative maximum at $x = 0$.

| 8 | |

| 22 | |

3

7. A rectangular field is to be fenced off and divided into three sections as shown below. 800 feet of fencing is to be used.

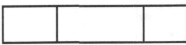

   (a) Find the dimensions of the field that maximize the total enclosed area.

   (b) What is the maximum area?

10

---

8. Given $f(x) = 14x^3 - 23x^2 + 10x - 1$.

   (a) List all possible rational zeros of f.

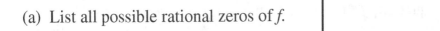

   (b) Find all of the zeros of f.

8

18

9. Find the domain of f given $f(x) = \dfrac{\sqrt{x}}{x^2 - 25}$ using interval notation.

6

10. If $f(x) = \log_5(x+8) - 3$, find the x-intercept of the graph of f.

6

11. Find the inverse equation, $f^{-1}(x)$, of the function $f(x) = \log_3(x-5) + 3$.

$f^{-1}(x) =$

6

18

5

12. Solve $\log(x+3) + \log(x-2) = \log 14$.

| 8 | |

---

13. The federal minimum wage was \$3.10 in 1980. It was \$7.25 in 2009.

(a) Find an exponential growth model, $A = A_0 e^{kt}$, in which t is the number of years after 1980. Round the k value in your model to 6 decimal places.

$$A =$$

(b) Based on your model, estimate the minimum wage in 2012.

| 10 | |

| 18 | |

14. Refer to the graphs of *f* and *g* below to answer questions (a) – (j). Use **interval notation** where applicable.

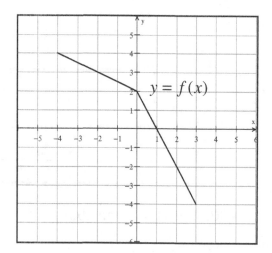
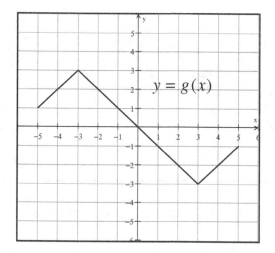

(a) Find $f(-2) - g(2)$

(f) On which interval(s) is *g* decreasing?

(b) Evaluate $f^{-1}(-2)$.

(g) State the domain of $\dfrac{g}{f}$.

(c) Evaluate $\dfrac{f}{g}(3)$.

(h) Find the slope of the steepest line segment on the graph of *f*.

(d) Evaluate $(g \circ f)(-4)$.

(i) Which function is a one-to-one function?

(e) State the range of *f*.

(j) Is *g* even, odd or neither?

20

Problems 15 – 30 are **MULTIPLE CHOICE** problems.   Each problem is worth 5 points. **Circle one answer for each question.**

15. Find the solution set: $\dfrac{1}{x+7}+\dfrac{3}{x+4}=\dfrac{-3}{x^2+11x+28}$.

    (a) $\{-4, -7\}$ (b) $\{-4\}$ (c) $\{-7\}$ (d) $\{7\}$ (e) ϕ

---

16. Which of the following are solutions of $x^2+10=6x$?

    (a) $3\pm 2i$ (b) $6\pm i$ (c) $3\pm i$ (d) $-3\pm i$ (e) none of these

---

17. $(5+6i)^2$ is equal to

    (a) 41 (b) -11 (c) $61+60i$ (d) $-11+60i$ (e) none of these

---

18. Which of the following ordered pairs satisfies the equation $3y-2x=-4$?

    (a) $(5, 2)$ (b) $(5, -2)$ (c) $(-5, 2)$ (d) $(-5, -2)$ (e) none of these

---

19. If $f(x)=\dfrac{2}{x-3}$ and $g(x)=\dfrac{1}{x}$, then $(f\circ g)(x)$ is equal to

    (a) $\dfrac{2x}{1-3x}$ (b) $\dfrac{2}{x-3}$ (c) $\dfrac{2x}{x-3}$ (d) $\dfrac{1-3x}{2-x}$ (e) none of these

---

25	

20. If $(5, 2)$ is a point on the graph of $y = f(x)$, what are the coordinates of the corresponding point on the graph of $y = 2f(-x-3)$?

(a) $(2, -4)$ (b) $(-2, 4)$ (c) $(8, -4)$ (d) $(-8, 4)$ (e) none of these

---

21. The slope of all lines perpendicular to the line $2x - 3y - 7 = 0$ is

(a) $\dfrac{2}{3}$ (b) $-\dfrac{2}{3}$ (c) $\dfrac{3}{2}$ (d) $-\dfrac{3}{2}$ (e) none of these

---

22. The remainder when $f(x) = x^4 - 2x^2 + 3x + 5$ is divided by $x + 2$ is

(a) -17 (b) 7 (c) 11 (d) 19 (e) none of these

---

23. Find the horizontal asymptote(s) of g, given $g(x) = \dfrac{2x^2}{x^2 - 7x + 6}$.

(a) $y = 0$ (b) $y = 6$ and $y = 1$ (c) $y = 2$ (d) $x = 2$

(e) g has no horizontal asymptotes

---

24. The solution for the rational inequality $\dfrac{x+4}{x-8} \le 0$ is

(a) $[-4, 8)$ (b) $(-4, 8)$ (c) $[-4, 8]$ (d) $(-4, 8]$ (e) none of these

---

25

25. Which of the following equations is equivalent to the equation $y = 8(2)^x$?

   (a) $y = 8x \ln 2$ (b) $y = 8e^{\ln(2x)}$ (c) $y = 8e^{2 \ln x}$ (d) $y = 8e^{(\ln 2)x}$

   (e) none of these

---

26. $e^{5 \ln xy - 2 \ln y + \ln 2}$ is equal to

   (a) $\dfrac{x^5 y^3}{2}$ (b) $2xy^3$ (c) $2x^5 y^3$ (d) $\dfrac{x^3 y^3}{2}$ (e) none of these

---

27. \$2400 is invested for 5 years in an account that earns 8.5% annual interest compounded quarterly. The accumulated value (to the nearest cent) is

   (a) \$3520.00   (b) \$3608.76   (c) \$3671.00   (d) \$3654.71   (e) none of these

---

28. How long (to the nearest tenth of a year) will it take an investment to double if compounded continuously with an interest rate of 4.3%?

   (a) 1.6 years (b) 7.0 years   (c) 16.1 years   (d) 46.5 years   (e) none of these

---

29. The asymptote of the graph of $h(x) = 2^{x-5} + 3$ is

   (a) $x = 3$ (b) $x = -3$ (c) $y = -3$ (d) $y = 3$ (e) none of these

---

30. Which of the following is the best estimate of $\log_5 (\ln \pi)$?

   (a) .084   (b) .289   (c) .800   (d) .758   (e) .998

30	

Answers for Sample Final Exam

1. $B = \dfrac{3A + 2}{A - 1}$

2. $\{1\}$

3. $(-\infty, -10] \cup [30, \infty)$

4. $y = \dfrac{1}{4}x - 2$

5. Center: $(4, -1)$; Radius: 5; Domain: $[-1, 9]$; Range: $[-6, 4]$

6. F, F, T, T

7. (a) 100 feet by 200 feet; (b) 20,000 square feet

8. (a) $\pm 1, \pm\dfrac{1}{2}, \pm\dfrac{1}{7}, \pm\dfrac{1}{14}$ (b) $\left\{1, \dfrac{1}{2}, \dfrac{1}{7}\right\}$

9. $[0, 5) \cup (5, \infty)$

10. 117

11. $f^{-1}(x) = 3^{x-3} + 5$

12. $\{4\}$

13. (a) $A = 3.10e^{.029297t}$ (b) \$7.92

14. (a) 5   (b) 2   (c) $\dfrac{4}{3}$ (d) -2   (e) $[-4, 4]$ (f) $(-3, 3)$ (g) $[-4, 1) \cup (1, 3]$
    (h) -2 (i) f (j) odd

15. e   16. c   17. d   18. a   19. a   20. d   21. d   22. b

23. c   24. a   25. d   26. c   27. d   28. c   29. d   30. a

Properties of Logarithms

If b is a positive number and $b \neq 1$,

1. $\log_b 1 = 0$
2. $\log_b b = 1$
3. $\log_b b^x = x$
4. $b^{\log_b x} = x$
5. **Product Rule:**
   $\log_b MN = \log_b M + \log_b N$
6. **Quotient Rule:**
   $\log_b \dfrac{M}{N} = \log_b M - \log_b N$
7. **Power Rule:**
   $\log_b M^p = p \log_b M$
8. **One-to-One Property:**
   If $\log_b x = \log_b y$, then $x = y$.

Graphs of $f(x) = e^x$ and $f(x) = \ln x$

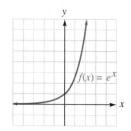
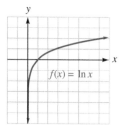

Natural Logarithm Properties

1. $\ln 1 = 0$
2. $\ln e = 1$
3. $\ln e^x = x$
4. $e^{\ln x} = x$

Theorems

Remainder Theorem:
If $P(x)$ is a polynomial, r is any number, and $P(x)$ is divided by $x - r$, the remainder is $P(r)$.

Factor Theorem:
If $P(x)$ is a polynomial and r is any number, then

If $P(r) = 0$, then $x - r$ is a factor of $P(x)$.

If $x - r$ is a factor of $P(x)$, then $P(r) = 0$.

Binomial Theorem:
If n is any positive integer, then

$$(a + b)^n = a^n + \frac{n!}{1!(n-1)!}a^{n-1}b$$

$$+ \frac{n!}{2!(n-2)!}a^{n-2}b^2 + \frac{n!}{3!(n-3)!}a^{n-3}b^3$$

$$+ \frac{n!}{r!(n-r)!}a^{n-r}b^r + \cdots + b^n$$

Parabolas

A **parabola** is the set of all points in a plane equidistant from a line l (called the **directrix**) and fixed point F (called the **focus**) that is not on line l.

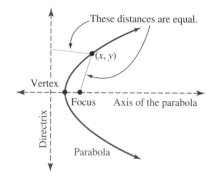

Parabola opening	Vertex at origin
Right	$y^2 = 4px$ $(p > 0)$
Left	$y^2 = 4px$ $(p < 0)$
Upward	$x^2 = 4py$ $(p > 0)$
Downward	$x^2 = 4py$ $(p < 0)$

- For a parabola that opens right or left with vertex at the origin, the directrix is $x = -p$ and the focus is $(p, 0)$.
- For a parabola that opens upward or downward with vertex at the origin, the directrix is $y = -p$ and the focus is $(0, p)$.

Parabola opening	Vertex at $V(h, k)$
Right	$(y - k)^2 = 4p(x - h)$ $(p > 0)$
Left	$(y - k)^2 = 4p(x - h)$ $(p < 0)$
Upward	$(x - h)^2 = 4p(y - k)$ $(p > 0)$
Downward	$(x - h)^2 = 4p(y - k)$ $(p < 0)$

- For a parabola that opens right or left with vertex at (h, k), the directrix is $x = -p + h$ and the focus is $(h + p, k)$.
- For a parabola that opens upward or downward with vertex at (h, k), the directrix is $y = -p + k$ and the focus is $(h, k + p)$.

Ellipses

An **ellipse** is the set of all points P in a plane such that the sum of the distances from P to two other fixed points F and F' is a positive constant.

The standard equations of an ellipse with center (h, k) and major axis horizontal is

$$\frac{(x - h)^2}{a^2} + \frac{(y - k)^2}{b^2} = 1,$$

where $a > b > 0$.

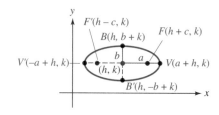